Palaeozoic Amalgamation of Central Europe

Special Publications reviewing procedures

The Society makes every effort to ensure that the scientific and production quality of its books matches that of its journals. Since 1997, all book proposals have been refereed by specialist reviewers as well as by the Society's Books Editorial Committee. If the referees identify weaknesses in the proposal, these must be addressed before the proposal is accepted.

Once the book is accepted, the Society has a team of Book Editors (listed above) who ensure that the volume editors follow strict guidelines on refereeing and quality control. We insist that individual papers can only be accepted after satisfactory review by two independent referees. The questions on the review forms are similar to those for *Journal of the Geological Society.* The referees' forms and comments must be available to the Society's Book Editors on request.

Although many of the books result from meetings, the editors are expected to commission papers that were not presented at the meeting to ensure that the book provides a balanced coverage of the subject. Being accepted for presentation at the meeting does not guarantee inclusion in the book.

Geological Society Special Publications are included in the ISI Science Citation Index, but they do not have an impact factor, the latter being applicable only to journals.

More information about submitting a proposal and producing a Special Publication can be found on the Society's web site: www.geolsoc.org.uk.

It is recommended that reference to all or part of this book should be made in one of the following ways:

WINCHESTER, J. A., PHARAOH, T. C. & VERNIERS, J. (eds) 2002. *Palaeozoic Amalgamation of Central Europe.* Geological Society, London, Special Publications, **201**.

COCKS, L. R. M. 2002. Key Lower Palaeozoic faunas from near the Trans-European Suture Zone. *In*: WINCHESTER, J. A., PHARAOH, T. C. & VERNIERS, J. (eds) 2002. *Palaeozoic Amalgamation of Central Europe.* Geological Society, London, Special Publications, **201**, 37–46.

GEOLOGICAL SOCIETY SPECIAL PUBLICATION NO. 201

Palaeozoic Amalgamation of Central Europe

EDITED BY

J. A. WINCHESTER
Keele University, UK

T. C. PHARAOH
British Geological Survey, Notts, UK

and

J. VERNIERS
University of Ghent, Belgium

2002
Published by
The Geological Society
London

THE GEOLOGICAL SOCIETY

The Geological Society of London (GSL) was founded in 1807. It is the oldest national geological society in the world and the largest in Europe. It was incorporated under Royal Charter in 1825 and is Registered Charity 210161.

The Society is the UK national learned and professional society for geology with a worldwide Fellowship (FGS) of 9000. The Society has the power to confer Chartered status on suitably qualified Fellows, and about 2000 of the Fellowship carry the title (CGeol). Chartered Geologists may also obtain the equivalent European title, European Geologist (EurGeol). One fifth of the Society's fellowship resides outside the UK. To find out more about the Society, log on to www.geolsoc.org.uk.

The Geological Society Publishing House (Bath, UK) produces the Society's international journals and books, and acts as European distributor for selected publications of the American Association of Petroleum Geologists (AAPG), the American Geological Institute (AGI), the Indonesian Petroleum Association (IPA), the Geological Society of America (GSA), the Society for Sedimentary Geology (SEPM) and the Geologists' Association (GA). Joint marketing agreements ensure that GSL Fellows may purchase these societies' publications at a discount. The Society's online bookshop (accessible from www.geolsoc.org.uk) offers secure book purchasing with your credit or debit card.

To find out about joining the Society and benefiting from substantial discounts on publications of GSL and other societies worldwide, consult www.geolsoc.org.uk, or contact the Fellowship Department at: The Geological Society, Burlington House, Piccadilly, London W1J 0BG: Tel. +44 (0)20 7434 9944; Fax +44 (0)20 7439 8975; Email: enquiries@geolsoc.org.uk.

For information about the Society's meetings, consult *Events* on www.geolsoc.org.uk. To find out more about the Society's Corporate Affiliates Scheme, write to enquiries@geolsoc.org.uk.

Published by The Geological Society from:
The Geological Society Publishing House
Unit 7, Brassmill Enterprise Centre
Brassmill Lane
Bath BA1 3JN,
UK

(*Orders*: Tel. +44 (0)1225 445046
Fax +44 (0)1225 442836)
Online bookshop: http://bookshop.geolsoc.org.uk

The publishers make no representation, express or implied, with regard to the accuracy of the information contained in this book and cannot accept any legal responsibility for any errors or omissions that may be made.

British Library Cataloguing in Publication Data
A catalogue record for this book is available from the British Library.

ISBN 1–86239–118–1

Typeset by Type Study, Scarborough, UK
Printed by the Alden Press, Oxford, UK.

Distributors
USA
AAPG Bookstore
PO Box 979
Tulsa
OK 74101–0979
USA
Orders: Tel. +1 918 584-2555
Fax +1 918 560-2652
E-mail *bookstore@aapg.org*

India
Affiliated East–West Press PVT Ltd
G-1/16 Ansari Road, Daryaganj,
New Delhi 110 002
India
Orders: Tel. +91 11 327-9113
Fax +91 11 326-0538
E-mail *affiliat@nda.vsnl.net.in*

Japan
Kanda Book Trading Co.
Cityhouse Tama 204
Tsurumaki 1-3-10
Tama-shi
Tokyo 206–0034
Japan
Orders: Tel. +81 (0)423 57-7650
Fax +81 (0)423 57-7651

Contents

Seismic traverses and deep crustal structure

Palaeozoic amalgamation of Central Europe: an introduction and synthesis of new results from recent geological and geophysical investigations

J. A. WINCHESTER[1], T. C. PHARAOH[2] & J. VERNIERS[3]

[1]*School of Earth Sciences and Geography, Keele University, Staffs ST5 5BG, UK; j.a.winchester@esci.keele.ac.uk*

[2]*British Geological Survey, Kingsley Dunham Centre, Keyworth, Notts NG12 5GG, UK*

[3]*Laboratorium voor Palaontologie, Krijgslaan 281/S8, B 9000, Gent, Belgium*

Abstract: Multidisciplinary studies undertaken within the EU-funded PACE Network have permitted a new 3-D reassessment of the relationships between the principal crustal blocks abutting Baltica along the Trans-European Suture Zone (**TESZ**). The simplest model indicates that accretion was in three stages: end-Cambrian accretion of the Bruno-Silesian, Łysogóry and Małopolska terranes; late Ordovician accretion of Avalonia, and early Carboniferous accretion of the Armorican Terrane Assemblage (**ATA**), which had coalesced during Late Devonian – Early Carboniferous time. All these accreted blocks contain similar Neoproterozoic basement indicating a peri-Gondwanan origin: Palaeozoic plume-influenced metabasite geochemistry in the Bohemian Massif in turn may explain their progressive separation from Gondwana before their accretion to Baltica, although separation of the Bruno-Silesian and related blocks from Baltica during the Cambrian is contentious.

Inherited ages from both the Bruno-Silesian crustal block and Avalonia contain a 1.5 Ga 'Rondonian' component arguing for proximity to the Amazonian craton at the end of the Neoproterozoic: such a component is absent from Armorican terranes, which suggests that they have closer affinities with the West African craton.

Models showing the former locations of these terranes and the larger continents from which they rifted, or to which they became attached, must conform to the above constraints, as well as those provided by palaeomagnetic data. Hence, at the end of the Proterozoic and in the early Palaeozoic, these smaller terranes, some of which contain Neoproterozoic ophiolitic marginal basin and magmatic arc remnants, probably occurred within the end-Proterozoic supercontinent as part of a 'Pacific-type' margin, which became dismembered and relocated as the supercontinent fragmented.

The SW margin of the East European Craton, the Trans-European Suture Zone (TESZ) is traceable from the Black Sea coast of Romania to the mouth of the River Oder on the Baltic Sea, despite being everywhere concealed beneath thick sedimentary cover. Further to the NW the continuation of this suture bends westwards, passes south of Denmark, and, traversing the SE North Sea (here known as the Thor-Tornquist Suture: Berthelsen 1998; Pharaoh 1999) curves NW to meet the Iapetus Suture at a triple point junction 300 km east of Dundee (Pharaoh 1999). It is therefore arguably one of the most prominent lithospheric features of Europe. Originally defined by Berthelsen (1993), as a collage of crustal blocks that separates the more than 850 Ma old Precambrian crust of the East European Craton (**EEC**) from the Variscan and Alpine mobile belts of western Europe, the term TESZ is now understood to be a broad zone incorporating the major shear zones forming the margin of the EEC, including the Teisseyre–Tornquist Zone in Poland, the Sorgenfrei Thrust Zone in Sweden and the Thor Suture west of Denmark (Gee & Zeyen 1996). It is marked by a major geophysical anomaly, separating the strongly magnetized East European Craton from the contrasting weakly magnetized crustal blocks to the SW (Banka *et al.* 2002; Williamson *et al.* 2002).

The EU-funded Training and Mobility of Researchers (TMR) Network 'Palaeozoic Amalgamation of Central Europe' No. ERBFMRXCT97–0136 (PACE) was set up to improve understanding of how central Europe was assembled. Despite the difficulties caused by the extensive post-accretion Mesozoic sedimentary cover the main objective of the study was achieved by collating the geological and geophysical evidence for the sequence of

From: WINCHESTER, J. A., PHARAOH, T. C. & VERNIERS, J. 2002. *Palaeozoic Amalgamation of Central Europe.* Geological Society, London, Special Publications, **201**, 1–18. 0305-8719/02/$15.00

collisions which produced the present configuration of crustal blocks accreted to central Europe. The sixteen contributions to this volume record aspects of the multidisciplinary work done, and are listed under five separate subject-related headings: (1) biostratigraphy and provenance evidence; (2) isotopic constraints; (3) petrology and geochemistry; (4) structural evolution; and (5) seismic traverses and deep crustal structure.

Despite several co-ordination meetings of the Network, different shades of opinion remain. It is not the purpose of this volume to minimize debates: indeed it is partly intended to emphasize and focus on the main discussion points so that discussion can continue to be made as broad as possible.

One important debate concerns the continental affinities of the Bruno-Silesian Block together with the possibly associated Łysogóry and Małopolska blocks of the Holy Cross Mountains in Poland. On one hand Cocks (2002) claims that there is no faunal evidence to suggest that these blocks were ever separated from Baltica, and that, since the discovery of late Neoproterozoic ('Cadomian' or 'Panafrican') deformed basement to the Uralides in the east of Baltica (Glasmacher *et al.* 1999), the presence of Panafrican-age detrital muscovites is not proof of Gondwanan affinities. By contrast Belka *et al.* (2000) combined the presence of detrital muscovites with a claim that some Cambrian faunas have an affinity with Gondwana. Both agree that since the end of the Cambrian these blocks were attached to the Baltica margin. This precludes any possibility of them being part of Avalonia, which was still attached to Gondwana in the early Ordovician. Equally clearly any 'Central European Caledonides' should not include the Late Cambrian Sandomierz Deformation (Samsonowicz 1926).

A second debate highlighted is the affinity of basement blocks accreted to the East European Craton, and how they may be distinguished. Many papers continue to be published suggesting that, for example, Avalonian basement underlies parts of the Bohemian Massif (e.g. Finger *et al.* 2000). In this debate establishing the late Ordovician timing of the accretion of Avalonia to Baltica (Vecoli & Samuelsson 2001; Samuelsson *et al.* 2002*a*) is a crucial piece of evidence that the basement of 'Far Eastern Avalonia' (Fig. 1) has Avalonian affinities, although rendered possibly suspect with respect to the main part of Avalonia by the Anglo-Brabant Deformation Belt, characterized by calc-alkaline magmatism (Pharaoh *et al.* 1993). Further evidence concerning the Anglo-Brabant Deformation Belt and the likely basement of Far Eastern Avalonia is provided in this volume by Verniers *et al.* (2002) and Samuelsson *et al.* (2002*b*).

Isotopic evidence from the Bohemian Massif for the timing of ophiolite generation and deformation (Marheine *et al.* 2002; Crowley *et al.* 2002*a*) clearly shows that accretion dates for these crustal blocks is much later than that of Avalonia, and that Devonian and Carboniferous subduction and collision of constituent blocks of the Armorican Terrane Assemblage predated accretion to the Laurussian supercontinent, comprising Laurentia, Baltica and Avalonia. However, as observed by Aleksandrowski and Mazur (2002) individual crustal blocks within the Armorican Terrane Assemblage appear to be continuous for long distances to the west, negating suggestions that separate 'Armorican' and 'Perunican' blocks existed. More focused studies of a single meta-ophiolitic body, the Marianske Lazne Complex, have produced differing conclusions. A dominantly petrological study (Štědrá *et al.* 2002) has produced a different assessment of the margins and affinities of gabbros at the southern margin of the complex than that reached, by means of a mainly geochemical study (Crowley *et al.* 2002*b*), even though authors are common to both papers. Clearly there is scope for more detailed studies of these rocks, as also indicated by a study of the Ślęża Ophiolite (Floyd *et al.* 2002), which reports for the first time on pillow lavas in its discussion of an otherwise well-studied ophiolite.

On the large scale structural interpretations vary widely, usually reflecting the part of central Europe with which the authors are most familiar. Thus, based on considerable knowledge, structural reconstructions provided in this volume by both Aleksandrowski and Mazur, and by Franke and Żelaźniewicz, present widely differing models.

Assistance is also provided by the abundance of seismic traverses. These reveal that, whether below the thick late Palaeozoic–Mesozoic sedimentary cover in the Polish Trough (Grad *et al.* 2002) or further to the NW beneath the southeastern North Sea (Scheck *et al.* 2002) an important feature of deep Central European geology is the shallow-dipping wedge of Baltican basement which, attenuating steadily, projects far to the SW of its sub-Permian position. This evidence shows that the major suture lines in Central Europe are shallow-dipping. A final survey, further to the west (Sintubin & Everaerts 2002) provides further evidence for the Lower Palaeozoic Anglo-Brabant Deformation Belt in Belgium.

Faced with these debates and the mass of

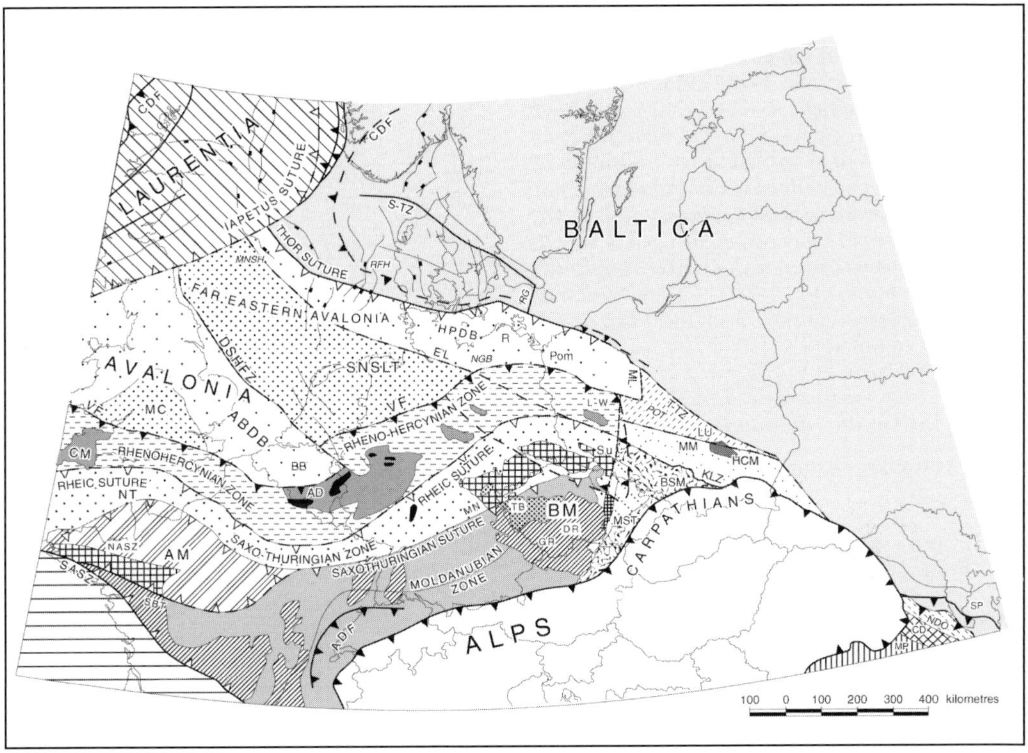

Fig. 1. A map showing the distribution of crustal blocks and Palaeozoic deformation belts in Central Europe. Key to abbreviations: ABDB, Anglo-Brabant Deformation Belt; AD, Ardennes; ADF, Alpine Deformation Front; AM, Armorican Massif; BB, Brabant; BM, Bohemian Massif; BSM, Bruno-Silesian Massif; CD, Central Dobrogea; CDF, Caledonian Deformation Front; CM, Cornubian Massif; DR, Dronsendorf Unit; EA, Ebbe Anticline; EFZ, Elbe Fault Zone; EL, Elbe Lineament; GF, Gföhl Unit; HCM, Holy Cross Mountains; HM, Harz Mountains; HPDB, Heligoland-Pomerania Deformation Belt; KLZ, Krakow-Lubliniec Zone; LU, Łysogóry Unit; L-W, Leszno-Wolsztyn High; MC, Midlands Microcraton; MM, Małopolska Massif; MN, Münchberg Nappe; MNSH, Mid-North Sea High; MP, Moesian Platform; MST, Moravo-Silesian Terrane; NASZ, North Armorican Shear Zone; NBT, North Brittany Terrane; NDO, North Dobrogea; NGB, North German Basin; Pom, Pomerania; POT, Polish Trough; R, Rügen Island; RFH, Rynkøbing-Fyn High; RG, Rønne Graben; RM, Rhenish Massif; SASZ, South Armorican Shear Zone; SBT, South Brittany Terrane; SH, South Hunsruck Massif; SNF, Sveconorwegian Front; SNSLT, South North Sea – Luneberg Terrane; SP, Scythian Platform; S-TZ, Sorgenfrei-Tornquist Zone; Su, Sudetes; TB, Teplá-Barrandia; T-TZ, Teisseyre–Tornquist Line; VF, Variscan Front.

supporting data, much of it new, a co-ordinated summary of the Palaeozoic Amalgamation of Central Europe, reconciling the differences, is needed. In this introduction the lines of evidence cited in the remainder of the volume are brought together in an attempt to explain these processes as part of a more global framework. Compromises have been sought where disagreements appear to be fundamental; hence the suggestion that during the Cambrian the Bruno-Silesian Block may have acted as a 'bridge' between Baltica and the Amazonian part of Gondwana – the Amazonian link indicated by the inherited Proterozoic dates obtained by Friedl *et al.* (2000). It was also necessary to account for the series of oceanic openings and closures which produced the crustal blocks, and to explain the mechanisms controlling their sequential rifting from the Palaeozoic Gondwana margin. This in turn required establishment of a series of global models consistent with the geological histories of these microcontinental blocks, as well as those of the principal continents, and these are explained next.

Major accreted crustal block assemblages

The outline structure of central Europe north of the Alpine-Carpathian Front and west of the approximate course of the TESZ has long been known. Using evidence from geophysical compilations, geological information provided by deep boreholes and outcrops of Palaeozoic and older rocks across central Germany and in the Bohemian Massif, the principal crustal blocks have long been distinguished. Recently summarized (Pharaoh 1999), they include Bruno-Silesia, Avalonia and the Armorican Terrane Assemblage.

Bruno-Silesia and associated blocks

These comprise Bruno-Silesia itself and, partly exposed in the Holy Cross Mountains, the Łysogóry and Małopolska terranes. Further to the SE and apparently sharing a similar Palaeozoic geological history are the Central and Southern Dobrogea terranes and the Moesian Platform of southern Romania, the latter only known from boreholes. Because it contains late Proterozoic magmatic rocks in the sub-Phanerozoic basement, Bruno-Silesia was considered by Moczydlowska (1997) to be a possible eastward extension of Avalonia. At the time, the presence of such rocks, sometimes termed 'Cadomian' in western Europe, or more generally, 'Panafrican', was accepted as an indication of attachment to the southern continents, collectively known as Gondwana. By contrast the Łysogóry and Małopolska terranes were interpreted as fragments of Baltica (Dadlez 1996; Pharaoh 1996), because their Ordovician faunas have Baltican affinities. However, a link between the Bruno-Silesian Block and the Małopolska Block is inferred because the Cambrian sequence on the latter has been interpreted as an accretionary wedge to the Bruno-Silesian Block. If true, the two blocks have always been closely linked. Initially the presence of a Panafrican-type, late Neoproterozoic deformed basement, with evidence of end-Proterozoic deformation, was taken as evidence of a Gondwanan origin, as opposed to Baltican, where, at the time, it was supposed that no Panafrican deformation had occurred. However, the discovery of widespread end-Proterozoic deformation along the Uralide margin of Baltica showed that the presence of Cadomian deformation was therefore not continent-specific. Hence the presence of Cadomian-type basement in the Bruno-Silesian block and the derivation of sediment in the Małopolska Block from a 'Cadomian' source (Belka *et al.* 2000), does not now prove a specifically Gondwanan origin for this crustal block. Furthermore, the existence of a late Proterozoic orogenic belt along this margin of Baltica may indicate that Baltica was still (albeit fleetingly) attached to Gondwana at the end of the Proterozoic, although strong faunal differences between Gondwana and Baltica show it was surely detached by the early Cambrian. However, Belka *et al.* (2000) show that the early Cambrian brachiopod faunal assemblages of the Małopolska block mostly have Gondwanan affinities with only a single Baltican species, *Westonia bottnica*, present, a conclusion disputed by Cocks (2002). With progressive introduction of Baltican brachiopod species and the ingress of sediment derived from Baltican sources during the middle Cambrian (Jendryka-Fuglewicz 1998), the Małopolska Block must at that time have been adjacent to Baltica. Actual docking is recorded by the Sandomierz Phase of deformation in the late Cambrian (Belka *et al.* 2000). However, the absence of calc-alkaline volcanic rocks in the Cambrian succession of both the Bruno-Silesian and Małopolska blocks argues against them having had a tectonically independent existence: it seems likely that displacement relative to adjacent continents may have involved major strike-slip movement, and the conflicting faunal evidence suggests that they acted as a link between Baltica and Gondwana during the Cambrian. In the Łysogóry Block, Middle to Upper Cambrian rocks contain fossils which do not occur in Baltica. Inarticulate brachiopods include forms with Gondwanan affinity (Belka pers. comm.) and trilobite trace fossils are identical to those from Gondwanan and peri-Gondwanan microplates (Seilacher 1983). Ordovician faunas, well documented in the southern part of the Holy Cross Mountains (Dzik *et al.* 1994) show essentially Baltican affinities, confirming that a connection of the Małopolska and Bruno-Silesian blocks with Baltica was established by the end of the Cambrian.

Palaeomagnetic and structural data (Lewandowski 1993; Mizerski 1995) suggest dextral strike-slip displacement of the Małopolska Block along the SW margin of the EEC. Provenance of clastic material, sedimentary history and palaeomagnetic data (Nawrocki 1999; Belka *et al.* 2000) show that amalgamation of the Małopolska and Łysogóry blocks was attained during the late Silurian. However, the presence of Devonian arc-related magmatism in the Jesiniky Mountains suggests that with SE-directed subduction on the NW margin of the Bruno-Silesian Block, its displacement along the

TESZ margin of Baltica may have continued into late Palaeozoic time.

Where exposed, structures along the western margin of the Bruno-Silesian Block are tectonic. They show highly oblique (dextral sense of shear) complex overthrusting to the east (Moldanubian Thrust) in the early Carboniferous between 350–330 Ma (Schulmann & Gayer 2000).

Attempts have been made to trace this junction northwards beneath the thick sedimentary cover of the Polish Trough. Because of the thickness of Mesozoic and Cenozoic sedimentary cover rocks, this has proved difficult and controversial, and depends largely on the results of seismic profiling. Both the Polonaise P1 and TTZ profiles (Jensen *et al.* 1999; Grad *et al.* 1999) show a clear change of mid-crustal structure north of the Moldanubian Thrust, suggesting that it continues northward as a major feature termed the Moravian Line by Winchester *et al.* (2002). In the TTZ profile the mid-crustal break illustrated is displaced eastwards compared to Polonaise P1: this may suggest dextral displacement of the Moravian Line by strike-slip faulting between the two profiles, perhaps along the Dolsk Line (Grad *et al.* 2002).

To the SE a possible link between the Moesian Platform and the Bruno-Silesian blocks has been suggested. According to Dudek (1980), the Bruno-Silesian Block continues under the Carpathians to the SE, presumably as far as the Peri-Pieniny lineament (Carpathian suture). Its southwestern extent is also not reliably constrained, but Dudek (1980) supposed that it extends to the Danube, approximately as far as the Krems-Vienna Line in Austria. Further work is therefore needed to establish the relationship with the Moesian Platform and other crustal blocks in SE Europe.

Avalonia

Precambrian and early Palaeozoic basement exposed in central England, Belgium and western Germany is widely accepted as part of Avalonia, the Palaeozoic microcontinent extending west as far as New England, and best exposed in the Avalon Peninsula of Newfoundland, after which it is named. Avalonian basement in central England, which typically consists of late Proterozoic intrusive, volcanic and sedimentary rocks (e.g. Thorpe *et al.* 1984; Pharaoh & Gibbons 1994; Strachan *et al.* 1996) was, like the Bruno-Silesian Block, affected by end-Proterozoic/pre-Lower Cambrian deformation. Because this area has been affected so little by later movements, and is overlain by a thin early Palaeozoic shallow marine sedimentary sequence succeeded conformably by Devonian terrestrial deposits: the ‘Old Red Sandstone’, it has sometimes been called the ‘Midlands Microcraton’ (e.g. Turner 1949; Pharaoh *et al.* 1987).

Boreholes in eastern England reveal that the Midlands Microcraton is bounded to the NE by a Caledonian deformation belt (Pharaoh *et al.* 1987; Noble *et al.* 1993). Late Ordovician calc-alkaline volcanic rocks are present within this belt and extend from eastern England to Belgium (André *et al.* 1986; Pharaoh *et al.* 1991). The southern end of this belt is exposed in the Brabant Massif of Belgium, and hence it has been termed (Winchester *et al.* 2002) the Anglo-Brabant Deformation Belt (ABDB). The deformation belt is inferred to have developed in early Devonian (Acadian) time above a zone of crustal suturing inherited from the late Ordovician soft collision of Avalonia and Baltica.

The presence of the ABDB questions whether the basement further east, NE of the Dowsing – South Hewett Fault Zone – Lower Rhine Lineament (Pharaoh 1999), is also part of Avalonia. Pharaoh *et al.* (1993) suggested that this lineament may separate crusts with differing structures, juxtaposed by late Ordovician subduction, the inferred cause of the calc-alkaline volcanism identified above. In this area the crystalline basement is generally not exposed. Far to the south, the 574 ± 3 Ma Wartenstein Gneiss (Molzahn *et al.* 1998), cropping out in the south Hunsrück at the SE margin of the Rhenish Massif and the 560 Ma Ecker Gneiss in the Harz Mountains (Baumann *et al.* 1991), both lying south of the Variscan Front, may be the only exposures of crystalline basement in this crustal block. The typically calc-alkaline composition and late Neoproterozoic age of these gneisses is broadly comparable to Avalonian basement exposed in central England and hence, despite the presence of the intervening ABDB, the basement of this area is generally linked with that of Avalonia. However, as so many pieces of crustal basement in both Avalonia and the Variscides of Central Europe appear to record late Proterozoic Cadomian deformation, it is the timing of the docking of these individual crustal blocks with Baltica which is most likely to decide their affinities.

Fossil evidence and sediment provenance data obtained from the G14 borehole, north of the Caledonian Deformation Front close to Rügen, NE Germany show that sediments with clear Gondwanan fossil associations and Cadomian mineral ages are first encountered in the Ashgill. The presence of reworked acritarchs of Llanvirn age and peri-Gondwanan affinity in the

Ashgill stratal sequences on the SW margin of the EEC (Samuelsson *et al.* 2002*b*) proves that an elevated area was being eroded in latest Ordovician time. Sediment provenance studies (Vecoli *et al.* 1999) show that the uplifted area was part of the Danish–North German–Polish 'Caledonides' which formed at the NE margin of Avalonia on its collision with Baltica (Berthelsen 1992; Dallmeyer *et al.* 1999). Hence, the timing of closure of the Tornquist Ocean and Avalonia–Baltica collision must have taken place between the middle Caradoc and the Rawtheyan. This interval of approximately 10 Ma was apparently sufficient for the development of the deformation belt separating the North Sea basement from Baltica and its partial erosion. The timing of this collision only slightly predates Avalonian convergence with Laurentia, based on evidence from Atlantic Canada (e.g. Cawood *et al.* 1994), and the onset of Windermere Supergroup sedimentation in the English Lake District (Cooper *et al.* 1993). The basement to the southern North Sea (the Southern North Sea–Luneberg Terrane (SNSLT) of Pharaoh *et al.* (1995) was probably an extension of Avalonia, possibly separated from Avalonia proper by a small, perhaps marginal, oceanic basin. If so, the ABDB was an intra-Avalonian mobile belt, perhaps developed in the Acadian orogenic phase, when Avalonia was moulding itself on to the margins of Baltica and Laurentia.

The lack of significant volcanism in the Heligoland-Pomerania Deformation Belts (HPDB) in either the passive margin sediments on the Baltican side or those on the Avalonian side (with the exception of volcanogenic clasts in sediment which could have originated from ashfall from distant volcanism) suggests that, in view of the earlier rapid northward motion of Avalonia, continental convergence was probably very oblique.

Finally, at the time of Avalonian convergence with Baltica, the Bruno-Silesian and related blocks must have formed a promontory. At the time of convergence, more easterly portions of Avalonia may have been detached and displaced eastwards, lending credence to the accounts that 'Celtic' (e.g. Avalonian) faunas in the Zonguldak Terrane of Turkey (Dean *et al.* 2000; Kozur & Göncüoğlu 1998).

Armorican Terrane Assemblage

The Armorican Terrane Assemblage (*sensu* Franke 2000; Tait *et al.* 2000) is exposed in a series of massifs across much of middle Europe from Spain to Poland. The largest and most significant areas of critical exposure in central Europe are in the Bohemian Massif, west of the Moldanubian Thrust. Here several different crustal blocks have been recognized, though their relations to each other have been far from clear. Of these, three have become widely recognized as distinctive: Saxothuringia, Teplá-Barrandia, and Moldanubia. A fourth crustal block in the Bohemian Massif, Bruno-Silesia, is recognized as having a completely separate geological history and believed to have formed part of a separate microcontinent (see above), but distinctions between the histories of the other terranes have not been fully explored because for a long time it was thought that the palaeomagnetic data from Teplá-Barrandia was typical of the entire massif. Recent work (e.g. Franke 2000; Franke *et al.* 1995) showing division of the Bohemian Massif into independently moving blocks suggests that this is not valid.

Numerous papers provide evidence of the complexity of relationships between independent terranes of the Bohemian Massif (e.g. in the summary provided by Aleksandrowski and Mazur, this volume), but they generally lack evidence of end-Ordovician/early Silurian collision seen in eastern Avalonia, and the Rheic Suture, interpreted to mark the southern margin of Avalonia, is shown on most reconstructions to pass north of the exposed Palaeozoic rocks in the Bohemian Massif (e.g. Franke 1995). Though Early Devonian ('Caledonian', but historically and collectively termed Eo-Variscan elsewhere in Hercynian Europe, e.g. Faure *et al.* 1997; Shelley & Bossière 2000)) metamorphism and magmatism has been recorded locally in the northern Bohemian Massif, it is mostly confined to high-grade metamorphic rocks in the Góry Sowie Block (**GSB**) (Brueckner *et al.* 1996; O'Brien *et al.* 1997) and the Münchberg klippe (395–390 Ma: Kreuzer *et al.* 1989; Stosch & Lugmair 1990) and may record some local tectonothermal and hence collisional activity between migrating platelets of the ATA, with subsequent exhumation. Although often portrayed as an exotic fault-bounded block, recent results from the GSB are not inconsistent with other parts of the West Sudetes. Although high pressure metamorphism was initiated somewhat earlier than further west, as indicated by growth of metamorphic (granulite facies) zircon at 402+0.8 Ma (O'Brien *et al.* 1997), other ages obtained indicate that further high temperature/medium pressure metamorphism occurred around *c.* 380 Ma, with later minor stages around 370 Ma consistent with a more widespread event in the Sudetes (Timmermann *et al.* 2000). Pre-400 Ma metamorphic events outside NW Europe otherwise seem to be almost entirely limited to the Anglo-Brabant and Heligoland-Pomerania Deformation Belts (Winchester *et al.* 2002), where Baltican Lower Palaeozoic passive margin shelf sediments have been folded, thrust and eventually overridden by high-density crust interpreted as Avalonian basement.

Subsequent late Devonian high temperature/medium pressure metamorphism in the GSB is well-constrained by U-Pb monazite ages (van Breemen *et al.* 1988; Bröcker *et al.* 1998; Timmermann *et al.* 2000) and appears to be contemporary with high pressure/low temperature metamorphism along the contact zone of the Saxothuringian and Teplá-Barrandian blocks between 380–365 Ma. In this event the orogenic wedge in the West Sudetes generally propagated from east to west. In the Karkonosze-Izera complex (central West Sudetes) this is shown by: a) early kinematic indicators in mylonitic ductile shear zones (Mazur 1995; Seston *et al.* 2000); b) the decrease in metamorphic grade from garnet zone in the east to chlorite zone in the NW (Baranowski *et al.* 1990; Kachlík & Patočka 1998; Collins *et al.* 2000); c) the decrease of $^{40}Ar–^{39}Ar$ cooling ages towards the west (Marheine *et al.* 1999); d) diminishing ages of flysch sedimentation onsets towards the west showing that tectonic exhumation was much earlier in the east. In addition, pre-late Devonian unconformities occur in the central West Sudetes between the Kłodzko metamorphic complex and the Bardo Unit (Hladil *et al.* 1998; Kryza *et al.* 2000), while late Devonian coarse-grained clastic sedimentary fills derived from exhumed metamorphic complexes to the east were deposited in syntectonic basins (Aleksandrowski & Mazur 2002). These processes, which started in pre-late Devonian times in the central West Sudetes (e.g. Hladil *et al.* 1998) continued until the Tournaisian in both the northwesternmost frontal parts of the West Sudetic orogenic wedge, where mélanges formed in the Kaczawa Complex (Collins *et al.* 2000), and in the metamorphic core of the complex such as the Orlica-Snieznik area where high pressure metamorphism produced eclogites. This range of dates suggests that a plethora of small-scale collisional events occurred, consistent with jostling of the small platelets of the ATA.

The term ‘Variscan Orogeny’ has been used to describe the deformation associated with the closure of the Rheic Ocean. However, this closure was complex, and although only younger Early–Middle Carboniferous dates (350–330 Ma) prevailing along the Rheic and Moravian suture lines may relate to final closure, both earlier (from mid-Devonian onwards) and later dates, up to post-Stephanian age, are regularly described as Variscan. In the West Sudetes Carboniferous metamorphism is recorded as well as an earlier Devonian event, and the latter was followed by tectonic exhumation of deeply-buried crustal slices (353–350 Ma) and the superimposition of a greenschist to lower amphibolite facies overprint dated at 345–340 Ma). $^{40}Ar–^{39}Ar$ dating (325–320 Ma) suggests that metamorphism was complete by the middle to late Carboniferous (Marheine *et al.* 2000), a timing supported by the age of deposition in adjacent intramontane basins. It is these Carboniferous events which are generally considered to reflect the docking of the amalgamated ATA with the Avalonian and Bruno-Silesian margin of the growing Laurussian supercontinent. The range of dates suggests that collision was not a simple process: it probably began earlier where the accreting ATA first impinged on promontories, such as that of the Bruno-Silesian Block, and occurred later further west.

Deformation of the Laurussian margin as a result of this collision produced the only significant late Palaeozoic deformation to affect both Avalonia and Bruno-Silesia: the continuity of this event has led some workers to equate the Rhenohercynian deformation zone with that in Bruno-Silesia. The northern junction of the ATA is generally marked by the Northern Phyllite Zone in Germany. However, ophiolitic fragments assigned to the Giessen-Werra-Südharz Unit (e.g. Franke 2000), which are spatially related to this suture, appear to mark the closure of an early Devonian successor basin, the Lizard-Giessen-Harz ‘ocean’, which apparently developed on the south side of the Rheic Ocean, and was, on collision, overthrust to the north, so that the ophiolitic fragments resulting from the obduction of this successor basin are now situated within the Giessen-Werra-Südharz/Selke Nappe, north of the Rheic Suture. The Mid-German Crystalline High (**MGCH**) marks the position of both, below the Rheic Suture late Silurian–Devonian arc magmatism on the Avalonian margin, and, now spatially superimposed upon it, but above the south-dipping Rheic Suture, Carboniferous age volcanism (Oncken 1997). Small magnetic highs seem to indicate a continuation of the volcanic centres within the MGCH eastwards into Poland as far as a point just NE of the Leszno-Wolsztyn High, corresponding to the location of the Moravian Line. The metamorphism which followed the closure of the Rheic Suture is Visean (350–330 Ma). As it approached Laurussia, subduction was south-dipping beneath the leading edge of the ATA, leading to the formation of an arc edifice (volcanic rocks of the MGCH?) with its associated oceanic back-arc basin – the Lizard-Giessen-Harz 'ocean'. Subduction of this successor back-arc basin occurred in Devono-Carboniferous time, with obduction of small remaining fragments, now thrust on to the northern side of the Rheic Suture.

Still unanswered is the question whether the ATA included crustal blocks which converged with Laurussia further east, and were thus accreted to the southern margin of Bruno-Silesian Block. Because the latter area is overprinted by the Carpathian/Alpine movements, and basement inliers are well-scattered within the Carpathian arc, further work is needed before this question can be answered. However, rocks apparently subjected to Variscan-age metamorphism, often intruded by mid-Carboniferous post-orogenic granitoids, do occur further east, south of the Bruno-Silesian Block, and are found both in inliers of basement in the Carpathians, such as the Tatra Mountains, and further east: there are for example reports of 'Celtic' (e.g. Avalonian) faunas in the Zonguldak Terrane of northwestern Turkey (Dean *et al.* 2000; Kozur & Göncüoğlu 1998). In the Tatra Mountains, metamorphic rocks containing amphibolites with similar chemistry to those in the West Sudetes (Gawęda *et al.* 2000) are cut by post-metamorphic Variscan granitoid rocks, dated by both ^{40}Ar–^{39}Ar and Rb–Sr methods at 300–330 Ma (Burchart 1968; Janak 1994). If these rocks form part of the European Variscides, the distance of its eastward continuation is uncertain.

In the northern Bohemian Massif extensive bimodal magmatism occurred in the early Ordovician, with bursts of magmatism continuing until the Devonian. Early, mainly acidic magmatism of Cambro-Ordovician age (e.g. Philippe *et al.* 1995; Hammer *et al.* 1997; Korytowski *et al.* 1993; Kröner *et al.* 1994) shows calc-alkaline chemistry, which was interpreted by some as evidence for an arc or active continent margin tectonic setting (e.g. Oliver *et al.* 1993; Kröner & Hegner 1998). Others suggested that the absence of supporting geological evidence for an arc edifice at the time makes it more likely that chemical characteristics of the intrusions were inherited from extensive melting of the calc-alkaline Cadomian basement (Kryza & Pin 1997; Aleksandrowski *et al.* 2000; Floyd *et al.* 2000). Subsequent dominantly basic magmatism was associated with clastic basin-fill metasedimentary rocks, typical of magmatism associated with an extensional tectonic setting. Minor associated felsic volcanic rocks are shown by Sm–Nd systematics and their REE distribution to result from continued melting of continental crust (Furnes *et al.* 1994; Patočka *et al.* 1997, 2002; Dostal *et al.* 2000). Analytical results from the basic rocks, using a database of over 600 full analyses (e.g. Floyd *et al.* 1996, 2000; Winchester *et al.* 1995, 1998), argue that the magmatic range is more likely to result from the interaction of an enriched plume with both asthenospheric and sediment-contaminated lithospheric mantle sources (Floyd *et al.* 2000). Although the volume of magmatism preserved is smaller than younger plume-influenced magmatic provinces, it has widespread correlatives in many parts of Western Europe including the Massif Central (Briand *et al.* 1991, 1995) and Massif des Maures (Briand pers. comm.) in France and from NW Spain (e.g. Peucat *et al.* 1990). Floyd *et al.* (2000) also suggested that plume-induced magmatism can also explain the amount of heat needed to melt substantial volumes of lower crust to produce the major granitoid bodies, and provides one possible mechanism for the fragmentation of the Armorican Terrane Assemblage (ATA) as it separated from Gondwana, and the repeated rifting of crustal fragments from the Gondwana margin, including Avalonia and the ATA.

On the basis of faunal distinctions and putative timing of rifting from Gondwana it has been argued that the Bohemian and Armorican Massifs were on separate microcontinents in the middle of the Palaeozoic: to the former the term 'Perunica' was given. However, the absence of any definitive collision zone between these crustal blocks, and the continuing uncertainty in defining which blocks were rifting apart renders such distinctions putative at best. It is likely instead that the ATA comprised several related crustal blocks (though not as many as indicated in the so-called Hun Superterrane; Stampfli 1996), which migrated towards Baltica *en bloc* after rifting from their former peri-Gondwanan position.

Global Models: the affinities and likely wander paths of the accreted mid-European crustal blocks

Such models are necessarily speculative, as much work remains to be done. In this section a series of sketch models are presented in order to explain how the main crustal blocks became accreted to the TESZ margin of Baltica, and their likely origins. These models do not draw upon the wealth of palaeomagnetic data used by

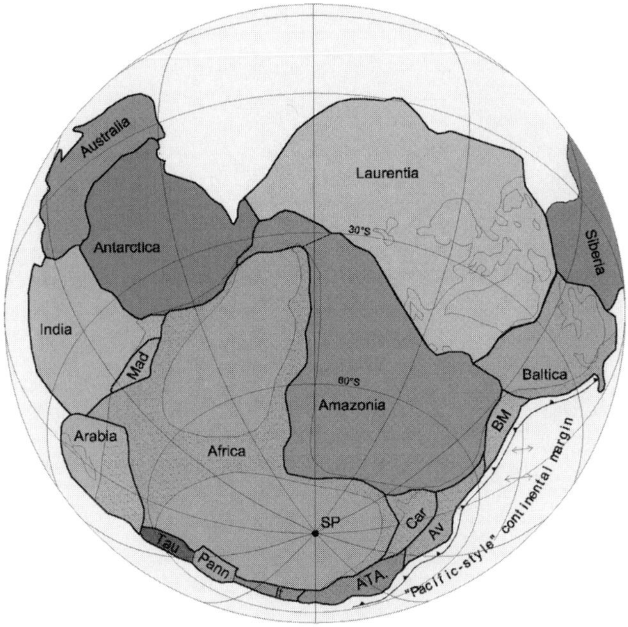

Fig. 2. Schematic reconstruction of the 'Pannotia' supercontinent in the late Proterozoic 600 Ma ago. Stippled areas are those deformed during the Panafrican event. Toothed lines represent areas of arc development or active continental margins Abbreviated microcontinent names are: ATA, Armorican Terrane Assemblage; Av, Avalonia; BM, Bruno-Silesia-Moesia; Car, Carolina; It, Italy; Pann, Pannonia; Tau, Taurus.

others, notably Dalziel (1997) and Torsvik *et al.* (1996). The models presented here are primarily designed to be 'correct' in terms of the likely location of microcontinent derivation and timing of accretion to Baltica.

An initial model (600 Ma) predates the opening of the Iapetus and related oceans and represents the fleeting development of the 'Pannotian' supercontinent (Dalziel 1997) resulting from the continental collisions recorded by the Panafrican orogenic events (Fig. 2). This model shows the main pre-Alpine Central European microcontinents forming an active continental margin (ACM) to the supercontinent, with the Bruno-Vistulian basement and Avalonia both adjacent to the Amazonian craton, based on the presence of inherited 1.5 Ga 'Rondonian' ages obtained from rocks in NE Austria (Friedl *et al.* 2000), Nova Scotia (Nance & Murphy 1994) and central England (Tucker & Pharaoh 1991). Baltica is shown adjacent to Bruno-Vistulia, and the end-Proterozoic magmatic belt extending the length of the Urals, and into the Timanides is shown as an extension of the ACM. However, if the orientation of Baltica at this time was as is claimed by Torsvik & Rehnström (2001), it is possible that Baltica was situated on the opposite side of the Panafrican mobile belt from Amazonia. In such a scenario the belt would represent a collisional zone rather than an ACM. In the opposite direction the ACM extends through the ATA, shown adjacent to the north African craton as it lacks inherited 'Rondonian' ages, and other blocks thought to have separated from their peri-Gondwanan positions later, notably the basements of Italy, the Pannonian blocks, and the Tauride basement of southern Turkey. The presence of late Neoproterozoic minor ophiolitic fragments within this ACM (e.g. Scarrow *et al.* 2001) attests to the obduction of successor basins and suggests that it originally formed a West Pacific-type rather than Andean-type continental margin.

A second model (Fig. 3) represents changes taking place at the end of the Proterozoic. It shows a narrow, but widening Iapetus Ocean, formed by the rifting of Laurentia in the early break-up of the end-Proterozoic supercontinent. Similar rifting of Baltica has occurred, with the Brunosilesia-Moesia crustal block occupying a position between it and peri-Gondwanan terranes. At this stage the Pacific-type margin of the supercontinent remains active, as recorded by voluminous calc-alkaline volcanism.

The third model (Fig. 4) shows an Early Cambrian setting with the Iapetus Ocean now wide,

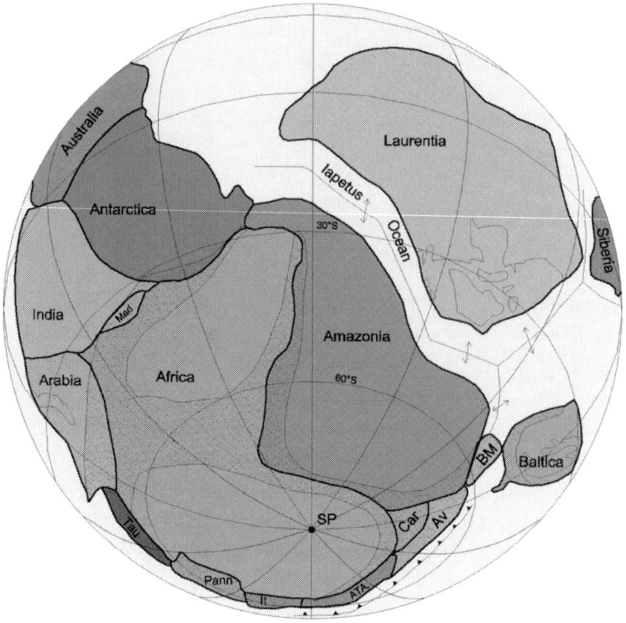

Fig. 3. Sketch reconstruction of continental distribution showing the opening of Iapetus at the end of the Proterozoic at 550 Ma ago. Ornament and abbreviations are as in Figure 2.

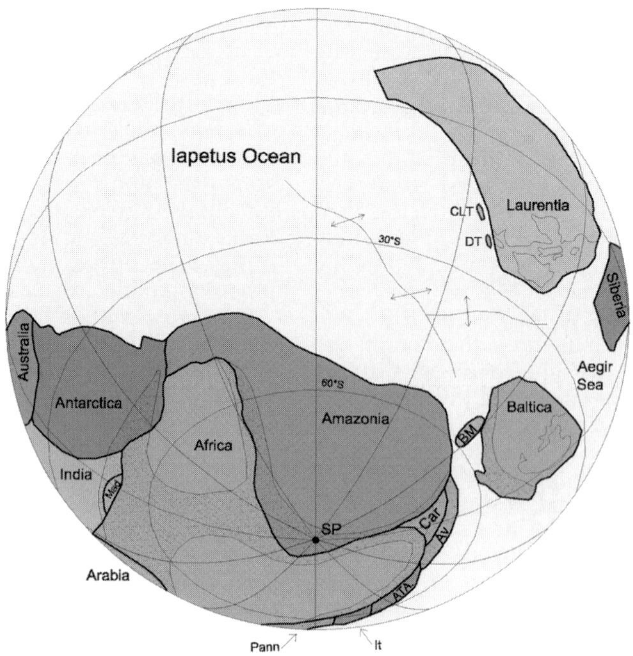

Fig. 4. Sketch reconstruction of continental distribution in the Early Cambrian at 520 Ma ago showing the Iapetus Ocean at its maximum width. Ornament and abbreviations are as in Figure 2. In addition CLT, Chain Lakes Terrane; DT, Dashwood Terrane.

and a cessation of magmatic activity along the Gondwana West Pacific-type margin. Brunosilesia-Moesia continues to act as a bridge between Gondwana and Baltica, and the latter continent remains in middle to high latitudes, as indicated by Torsvik & Rehnström (2001).

The start of the Ordovician Period (Fig. 5) reveals several changes. Rapid closure of the Iapetus Ocean has begun with subduction on both the Laurentian (Taconic Arc) and Gondwana (Gander Arc, off Avalonia) margins. At the same time Brunosilesia-Moesia is now detached from Gondwana and attached to Baltica at a location still far SE of its present position. Avalonia and the ATA remain attached to the Gondwana margin, with shelf sedimentation.

During the Llanvirn Stage (Fig. 6), renewed arc magmatism marks the detachment of Avalonia from the Gondwana margin and its rapid northward migration, narrowing the Iapetus Ocean. At the same time a widening Rheic Ocean is developing between Avalonia and the Gondwana margin, from which the ATA is already separating as a series of linked blocks. Avalonia is also depicted as migrating as more than one block, to allow for the possible presence of ophiolitic material in the Anglo-Brabant Deformation Belt, which would indicate that the evidence for the attachment of the easternmost part of the microcontinent is not secure.

By the early Silurian (Fig. 7), Avalonian docking with the TESZ margin of Baltica is shown, with imminent closure of the relic of the Iapetus Ocean between these continents and Laurentia. Speculatively, the collision of Avalonia with the Bruno-Silesian promontory has detached its easternmost portion, which is now displaced sinistrally: it could potentially form part of the western Pontides, if 'Celtic' faunas do indeed occur there. The ATA is now shown separated from Gondwana, while the Rheic Ocean is already starting to close, with subduction along the southern margin of Avalonia, marking the earlier stage of volcanism in the Mid-German Crystalline High. The new ocean separating the ATA from the Gondwana margin is now the Proto-Tethys Ocean.

By the early Carboniferous (Fig. 8), later, southward subduction, marked by renewed volcanism in the Mid-German Crystalline High, illustrates the final stage in the approach of the ATA to Baltica, also impelled by Gondwanan convergence. Contact has already been made with the Bruno-Silesian promontory, with dextral strike-slip faulting along its western margin, and detachment of the easternmost Variscides, which are displaced eastwards by

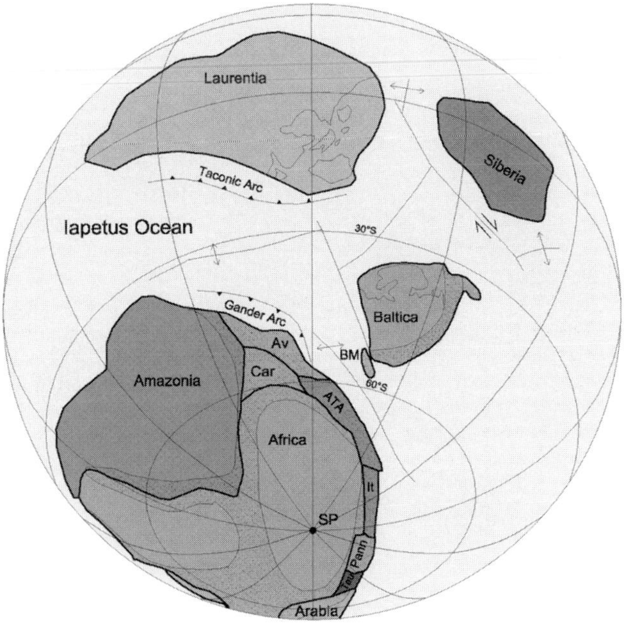

Fig. 5. Sketch reconstruction of continental distribution in the Cambro-Ordovician at 490 Ma ago as the Iapetus Ocean began to close. Ornament and abbreviations are as in Figure 2.

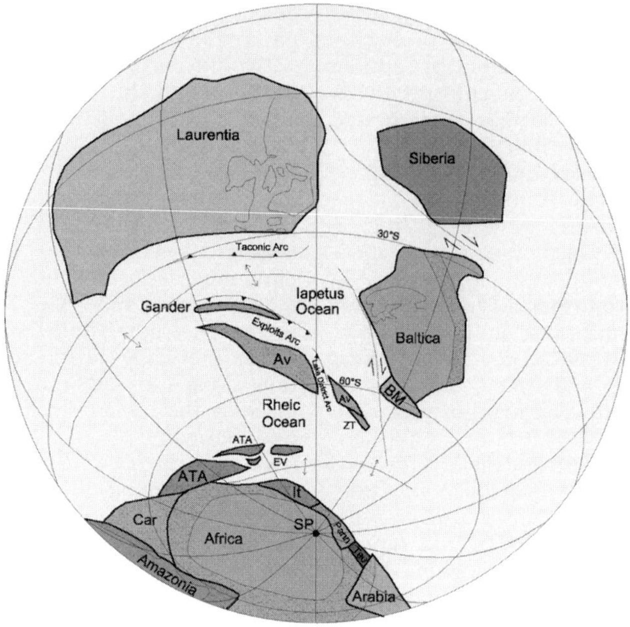

Fig. 6. Sketch reconstruction of continental distribution in the Llanvirn, 465 Ma ago as Avalonia migrated northwards. Ornament and abbreviations are as in Figure 2. In addition EV, Eastern Variscides; ZT, Zonguldak Terrane.

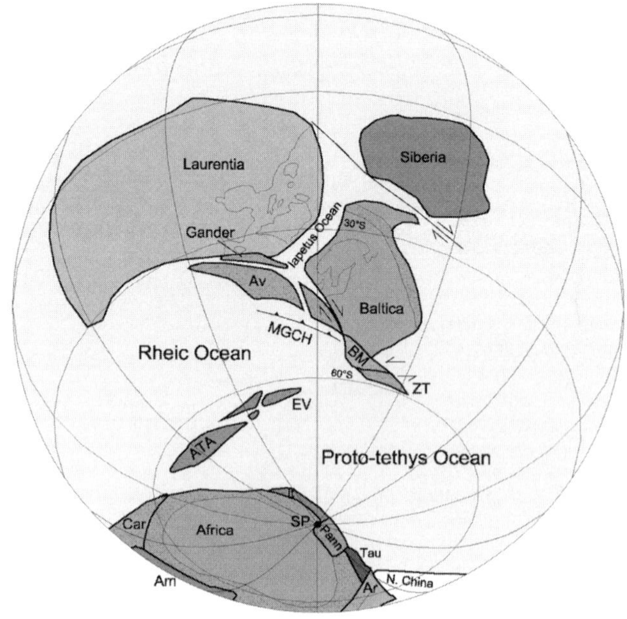

Fig. 7. Sketch reconstruction of continental distribution in the Early Silurian 440 Ma ago showing the accretion of Avalonia. Ornament and abbreviations are as in Figure 6. In addition Ar, Arabia; MGCH, Mid-German Crystalline High.

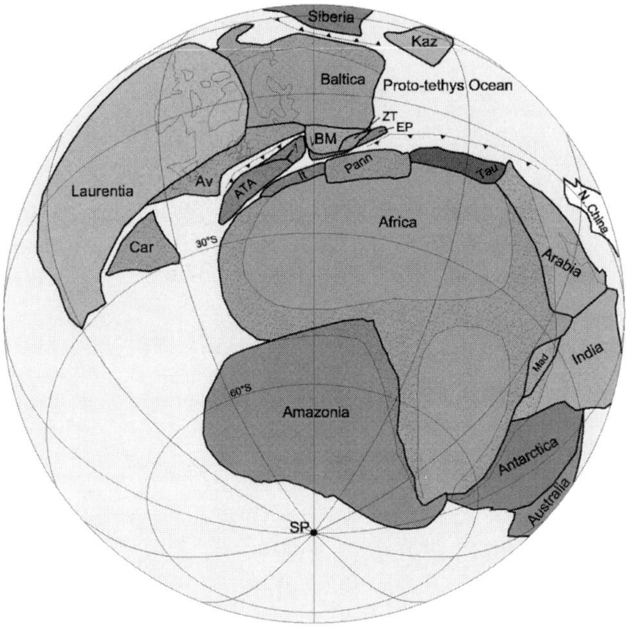

Fig. 8. Sketch reconstruction of continental distribution in the Early Carboniferous 350 Ma ago, as Gondwana and Laurasia converged. Ornament and abbreviations are as in Figure 6. Kaz, Kazakhstan.

sinistral faulting to form the Variscide basement seen in Carpathian inliers and in the Zonguldak and Istanbul Terranes of NW Turkey.

These investigations, and the collation of information was supported by the EU-funded PACE TMR Network, no. ERBFMRXCT97–0136. The contribution of T.C. Pharaoh appears with permission of the Executive Director, British Geological Survey (NERC).

The editors would also like to give especial thanks to N. Bakun-Czubarow, Z. Belka, A. Berthelsen, B. Briand, R. Dadlez, W. Dörr, J. A. Evans, F. Finger, A. Galdeano, V. Kachlík, G. Katzung, P. Krzywiec, D. Laduron, P. D. Lane, A. Lassen, B. Leveridge, J. Maletz, H. Maluski, P. Matte, R. Meissner, J. Menuge, S. G. Molyneux, F. Neubauer, M. Okrusch, S. R. Noble, F. Patočka, C. Pin, L. Popov, R. A. Strachan, N. H. Woodcock, R. Wrona, J. Zalasiewicz, A. Żelaźniewicz and three anonymous reviewers.

References

ALEKSANDROWSKI, P. & MAZUR, S. 2002. Collage tectonics in the northeasternmost part of the Variscan Belt (Sudetes, Bohemian Massif): facts, interpretations and open questions. *In:* WINCHESTER, J. A., PHARAOH, T. C. & VERNIERS, J. (eds) *Palaeozoic Amalgamation of Central Europe,* Geological Society London, Special Publication, **201**, 237–277.

ALEKSANDROWSKI, P., KRYZA, R., MAZUR, S. PIN, C. & ZALASIEWICZ, J. 2000. The Polish Sudetes: Caledonian or Variscan? *Transactions of the Royal Society of Edinburgh,* **90**, 127–146.

ANDRÉ, L, HERTOGEN, J. & DEUTSCH, S. 1986. Ordovician-Silurian magmatic provinces in Belgium and the Caledonian Orogeny in Middle Europe. *Geology*, **14**, 879–882.

BANKA, D., PHARAOH, T. C., WILLIAMSON, J. P. and TESZ Project Potential Field Core Group. 2002. Potential field imaging of Palaeozoic orogenic structure in northern and central Europe. *Tectonophysics*, in press.

BARANOWSKI, Z., HAYDUKIEWICZ, A., KRYZA, R., LORENC, S., MUSZYŃSKI, A., SOLECKI, A. & URBANEK, Z. 1990. Outline of the geology of the Góry Kaczawskie (Sudetes, Poland). *Neues Jahrbuch für Geologie und Paläontologie, Abhandlungen,* **179**, 223–257.

BAUMANN, A., GRAUERT, B., MECKLENBERG, S. & VINX, R. 1991. Isotopic age determinations of crystalline rocks of the Upper Harz Mountains, Germany. *Geologische Rundschau*, **80**, 669–690.

BELKA, Z., AHRENDT, H., FRANKE, W., SCHÄFER, J. & WEMMER, K. 2000. The Baltica–Gondwana suture in Central Europe: evidence from K/Ar ages of detrital muscovites and biogeographical data. *In:* FRANKE, W., ALTHERR, R., HAAK, V., ONCKEN, O. & TANNER, D. (eds) Geological Society London, Special Publications, **179**, 87–102.

BELKA, Z., VALVERDE-VAQUERO, P., AHRENDT, H., WEMMER, K., FRANKE, W. & SCHÄFER, J. 2002. Accretion of first Gondwana-derived terranes at the margin of Baltica. *In:* WINCHESTER, J. A., PHARAOH, T. C. & VERNIERS, J. (eds) *Palaeozoic Amalgamation of Central Europe,* Geological Society London, Special Publication, **201**, 19–36.

BERTHELSEN, A. 1992. Mobile Europe. In: BLUNDELL,

D. J., FREEMAN, R., MUELLER, St. (eds.) *A continent revealed – the European Geotraverse.* Cambridge University Press, 11–32.

BERTHELSEN, A. 1993. Where different geological philosophies meet: the Trans-European Suture Zone. *Publications of the Institute of Geophysics, Polish Academy of Sciences,* **A20 255**, 19-31.

BERTHELSEN, A. 1998. The Tornquist Zone northwest of the Carpathians: an intraplate pseudosuture. *Geologiske Föreningen Förhandlinger*, **120**, 223–230.

BRIAND, B., PIBOULE, M., SANTALLIER, D. & BOUCHARDON, J.L. 1991. Geochemistry and tectonic implications of two Ordovician bimodal igneous complexes, southern French Massif Central. *Journal of the Geological Society, London,* **148**, 959–971.

BRIAND, B., BOUCHARDON, J. L., OUALI, H., PIBOULE, M. & CAPIEZ, P. 1995. Geochemistry of bimodal amphibolitic-felsic gneiss complexes from eastern Massif Central, France. *Geological Magazine,* **132**, 321–337.

BRÖCKER, H. K., ŻELAŹNIEWICZ, A. & ENDERS, M. 1998. Rb-Sr and U-Pb geochronology of migmatitic gneisses from the Góry Sowie, (West Sudetes, Poland): the importance of mid-Late Devonian metamorphism. *Journal of the Geological Society, London*, **155**, 1025–1035.

BRUECKNER, H. K., BLUSZTAJN, J. & BAKUN-CZUBAROW, N. 1996. Trace element and Sm-Nd 'age' zoning in garnets from peridotites of the Caledonian and Variscan Mountains and tectonic implications. *Journal of Metamorphic Geology,* **14**, 61–73.

BURCHART, J. 1968. Rb-Sr isochron ages of the crystalline core of the Tatra Mountains, Poland. *American Journal of Science,* **266**, 895–907.

CAWOOD, P. A., DUNNING, G. R., LUX, D. & VAN GOOL, J. A. M. 1994. Timing of peak metamorphism and deformation along the Appalachian margin of Laurentia in Newfoundland: Silurian not Ordovician. *Geology,* **22**, 399–402.

COCKS, L. R. M. 2002. Key Lower Palaeozoic faunas from near the Trans-European Suture Zone. *In:* WINCHESTER, J. A., PHARAOH, T. C. & VERNIERS, J. (eds) *Palaeozoic Amalgamation of Central Europe,* Geological Society London, Special Publication, **201**, 37–46.

COLLINS, A. S., KRYZA, R. & ZALASIEWICZ, J. A., 2000. Macrofabric fingerprints of Late Devonian-Early Carboniferous subduction in the Polish Variscides, the Kaczawa Complex, Sudetes. *Journal of the Geological Society, London*, **157**, 283–288.

COOPER, A. H., MILLWARD, D., JOHNSON, E. W. & SOPER, N. J. 1993. The early Palaeozoic evolution of northwest England. *Geological Magazine,* **130**, 711–724.

CROWLEY, Q. G., ŠTĚDRÁ, V., FLOYD, P. A., WINCHESTER, J. A., KACHLIK, V. & HOLLAND, J. G. 2002a. Mariánské-Lázně Complex, NW Bohemian Massif: development and destruction of an early Palaeozoic seaway. *In:* WINCHESTER, J. A., PHARAOH, T. C. & VERNIERS, J. (eds) *Palaeozoic Amalgamation of Central Europe,* Geological Society London, Special Publication, **201**, 177–195.

CROWLEY, Q. G., TIMMERMANN, H., NOBLE, S. R. & HOLLAND, J. G. 2002b. Palaeozoic terrane amalgamation in Central Europe: a REE and Sm-Nd isotopic study of the pre-Variscan basement, NE Bohemian Massif. *In:* WINCHESTER, J. A., PHARAOH, T. C. & VERNIERS, J. (eds) *Palaeozoic Amalgamation of Central Europe,* Geological Society London, Special Publication, **201**, 157–176.

DADLEZ, R. 1996. Seismic profile of LT-7 (NW Poland): geological implications. *Geological Magazine,* **134**, 635–659.

DALLMEYER, R. D., GIESE, U., GLASMACHER, U. & PICKEL, W. 1999. First $^{40}Ar/^{39}Ar$ age constraints for the Caledonian evolution of the Trans-European Suture Zone in NE Germany. *Journal of the Geological Society, London*, **156**, 279–290.

DALZIEL, I. W. D. 1997. Neoproterozoic-Palaeozoic geography and tectonics: review, hypothesis, environmental speculation. *Geological Society of America Bulletin,* **109**, 16–42.

DEAN, W. T., MONOD. O., RICKARDS, B., DEMIR, O. & BULTYNCK, P. 2000. Lower Palaeozoic stratigraphy and palaeontology, Karadere-Zirze area, Pontus Mountains, northern Turkey. *Geological Magazine,* **137**, 555–582.

DOSTAL, J., PATOČKA, F. & PIN, C. 2000. Early Palaeozoic intracontinental rifting and early sea-floor spreading in the central West Sudetes (Bohemian Massif): geochemistry and Sr-Nd isotopic study on metavolcanic rocks of the East Krkonoše Complex. *Geolines,* **10**, 19–20.

DUDEK, A. 1980. The crystalline basement block of the Outer Carpathians in Moravia: Bruno-Vistulicum. *Rozprawy Ceskoslovenský Akademie Ved,* **90**, 1-85.

DZIK, J., OLEMPSKA, E., & PISERA, A. 1994. Ordovician carbonate platform ecosystem of the Holy Cross Mountains. *Palaeontologia Polonica* **53**, 317.

FAURE, M., LELOIX, Ch. & ROIG, J-Y. 1997. L'évolution polycyclique de la chaîne hercynienne. *Bulletin de la Société géologique de la France*, **168**, 695–705.

FINGER, F., HANZL, P., PIN, C., VON QUADT, A. & STEYRER, H.P. 2000. The Brunovistulian: Avalonian Precambrian at the eastern end of the Central European Variscides? . *In:* FRANKE, W., HAAK, V., ONCKEN, O. & TANNER, D. (eds) *Orogenic processes: Quantification and Modelling in the Variscan Belt.* Geological Society, London, Special Publications, **179**, 103–112.

FLOYD, P. A., WINCHESTER, J. A., CIESIELCZUK, J., LEWANDOWSKA, A., SZCZEPAŃSKI, J. & TURNIAK, K. 1996. Geochemistry of early Palaeozoic amphibolites from the Orlica-Śnieżnik dome, Bohemian massif: petrogenesis and palaeotectonic aspects. *Geologische Rundschau*, **85**, 225–238.

FLOYD, P. A., WINCHESTER, J. A., SESTON, R., KRYZA, R. & CROWLEY, Q. G. 2000. Review of geochemical variation in Lower Palaeozoic metabasites from the NE Bohemian Massif: intracratonic rifting and plume-ridge interaction. *In:* FRANKE, W., HAAK, V., ONCKEN, O. & TANNER, D. (eds) *Orogenic processes: Quantification and modelling in the Variscan Belt.* Geological Society, London, Special Publications, **179**, 155–174.

FLOYD, P. A., CROWLEY, Q.G., WINCHESTER, J.A., KOZDROJ, W. & CYMERMAN, Z. 2002. Pillow lavas in the Ślęża Ophiolite and their geochemical relationship to other Sudetic mafic components. *In:* WINCHESTER, J. A., PHARAOH, T. C. & VERNIERS, J. (eds) *Palaeozoic Amalgamation of Central Europe,* Geological Society London, Special Publications, **201**, 197–215.

FRANKE, W. 1995. Rhenohercynian foldbelt: autochthon and nonmetamorphic nappe units-stratigraphy. *In:* DALLMEYER, D., FRANKE, W. & WEBER, K. (eds) *Pre-Permian geology of Central and Western Europe*. Springer, Berlin, Heidelberg, New York, 33–49.

FRANKE, W. 2000. The mid-European segment of the Variscides: tectono-stratigraphic units, terrane boundaries and plate tectonic evolution. *In:* FRANKE, W., HAAK, V., ONCKEN, O. & TANNER, D. (eds) *Orogenic processes: Quantification and modelling in the Variscan Belt.* Geological Society, London, Special Publication, **179**, 35–62.

FRANKE, W. & ŻELAŹNIEWICZ, A. 2002. Structure and evolution of the Bohemian Arc. *In:* WINCHESTER, J. A., PHARAOH, T. C. & VERNIERS, J. (eds) *Palaeozoic Amalgamation of Central Europe,* Geological Society London, Special Publications, **201**, 279–293.

FRANKE, W., KREUZER, H., OKRUSCH, M., SCHÜSSLER, U. & SEIDEL, E. 1995. Saxothuringian Basin: exotic metamorphic nappes: stratigraphy, structure and igneous activity. *In:* DALLMEYER, D., FRANKE, W. & WEBER, K. (eds) *Pre-Permian geology of Central and Western Europe*. Springer, Berlin, Heidelberg, New York, 275–294.

FRIEDL, G., FINGER, F., MCNAUGHTON, N.J. & FLETCHER, I.R. 2000. deducing the ancestry of terranes: SHRIMP evidence for South America-derived Gondwana fragments in Central Europe. *Geology,* **28**, 1035–1038.

FURNES, H., KRYZA, R., MUSZYŃSKI, A., PIN, C. & GARMANN, L. B. 1994. Geochemical evidence for progressive rift-related volcanism in the eastern Variscides. *Journal of the Geological Society, London*, **151**, 91–109.

GAWĘDA, A., WINCHESTER, J. A., KOZŁOWSKI, K., NARĘBSKI, W. & HOLLAND, J.G. 2000. Geochemistry and palaeotectonic setting of amphibolites from the Western Tatra Mountains, southern Poland. *Geological Journal* **35**, 69–85.

GEE, D. G. & ZEYEN, H. 1996. *EUROPROBE 1996 – Lithosphere dynamics: origin and evolution of continents*. EUROPROBE Secretariat and Strasbourg European Science Foundation, Uppsala.

GLASMACHER, U. A., GIESE, U., STRONIK, L., REYNOLDS, P., ALEKSEYEV, A., PUCHKOV, V. & BAUER, W. 1999. Neoproterozoic terrane at the Eastern margin of Baltica – implications for Late Proterozoic palaeogeography andstructural evolution of SW Urals, Russia. *Journal of Conference Abstracts* EUG10, Strasbourg, 108.

GRAD, M., JANIK, T., YLINIEMI, J., GUTERCH, A., LUOSTO, U., TIIRA, T., KOMMINAHO, K., SRODA, P., HÖING, K., MAKRIS, J. & LUND, C-E. 1999. Crustal structure of the Mid-Polish Trough beneath the Teisseyre-Tornquist Zone seismic profile, *Tectonophysics* **314**, 145–160.

GRAD, M., GUTERCH, A. & MAZUR, S. 2002. Seismic refraction evidence for the crustal structure in the central part of the Trans-European Suture Zone in Poland. *In:* WINCHESTER, J. A., PHARAOH, T. C. & VERNIERS, J. (eds) *Palaeozoic Amalgamation of Central Europe,* Geological Society London, Special Publication, **201**, 295–309.

HAMMER, J., BRÖCKER, M. & KRAUSS, M. 1997. Alter und geologische Signifikanz von Deformationszonen im östlichen Teil des Lausitzer Granitoidkomplexes. *Terra Nostra*, **97/5**, 62–63.

HLADIL, J., MAZUR, S., GALLE, A. & EBERT, S. R. 1998. Revised age of the Mały Bołków limestone in the Kłodzko metamorphic belt (early Givetian; late Middle Devonian): implications for the geology of the Sudetes. *Geolines* **6**, 22–24.

JANAK, M. 1994. Variscan uplift of the crystalline basement of the Tatra Mts, central western Carpathians: evidence from $^{40}Ar/^{39}Ar$ laser probe dating of biotite and P-T-t paths. *Geologica Carpathica,* **45**, 239–300.

JENDRYKA-FUGLEWICZ, B. 1998. Kambryjska eksplozja zycia. Najstarsze zespoly brachiopodow w profilach geologicznich Polski. *Abstracts, XVI Palaeontological Meeting, Wiktorowo*, 18–19.

JENSEN, S. L., JANIK, T., THYBO, H., and the POLONAISE profile P1 Working Group. 1999. Seismic structure of the Palaeozoic Platform along POLONAISE '97 profile P1 in northwestern Poland. *Tectonophysics,* **314**, 123–143.

KACHLÍK, V. & PATOČKA, F. 1998. Lithostratigraphy and tectonomagmatic evolution of the Železný Brod Crystalline Unit: Some constraints for the palaeotectonic development of the W Sudetes (NE Bohemian Massif). *Geolines*, **6**, 34–35.

KORYTOWSKI, A., DÖRR, W. & ŻELAŹNIEWICZ, A. 1993. U-Pb dating of (meta)granitoids in the NW Sudetes (Poland) and their bearing on tectonostratigraphic correlation. *Terra Nova*, **5**, 331–332.

KOZUR, H. & GÖNCUÖĞLU, M. C. 1998. Facies development and thermal alteration of Silurian rocks in Turkey. *In:* GUITERREZ-MARCO, J. C. & RABANO, I (eds) *Proceedings, 1998 field meeting,* IUGS Subcommission on Silurian stratigraphy, Temas Geologico-Mineros ITGE 23, 87–90.

KREUZER, H., SEIDEL, E., SCHŰSSLER, U., OKRUSCH, M., LENZ, K-L. & RASCHKA, H. 1989. K-Ar geochronology of different tectonic units at the northwestern margin of the Bohemian Massif. *Tectonophysics,* **157**, 149–178.

KRÖNER, A. & HEGNER, E. 1998. Geochemistry, single zircon ages and Sm-Nd systematics of granitoid rocks from the Góry Sowie (Owl Mts), Polish West Sudetes: evidence for early Palaeozoic arc-related plutonism. *Journal of the Geological Society, London*, **155**, 711–724.

KRÖNER, A., JAECKEL, P. & OPLETAL, M. 1994. Pb-Pb and U-Pb zircon ages for orthogneisses from eastern Bohemia: further evidence for a major Cambro-Ordovician magmatic event. *Journal of the Czech Geological Society,* **39**, 61.

KRYZA, R. & PIN, C. 1997. Cambrian/Ordovician magmatism in the Polish Sudetes: no evidence for subduction-related setting. *Terra Nova*, **7**, 144.

KRYZA, R., MAZUR, S. & ALEKSANDROWSKI, P. 2000. Pre-Late Devonian unconformity in the Kłodzko area excavated: a record of Eo-Variscan metamorphism and exhumation in the Sudetes. *Geologica Sudetica,* **32**, 127–137.

LEWANDOWSKI, M. 1993. Paleomagnetism of the Paleozoic rocks of the Holy Cross Mts (Central Poland) and the origin of the Variscan orogen. *Publications of the Institute of Geophysics, Polish Academy of Sciences* **A23**, **265**, 3-85.

MARHEINE, D., KACHLÍK, V., PATOČKA, F. & MALUSKI, H. 1999. The Palaeozoic polyphase tectonothermal record in the Krkonoše-Jizera Crystalline Unit (West Sudetes, Czech Republic). *Geolines* **9**, 133–135.

MARHEINE, D., KACHLÍK, V., PATOČKA, F., MALUSKI, H. & ŻELAŹNIEWICZ, A. 2000. Nouveaux ages hercyniens Ar-Ar dans les Sudètes occidentales (Massif de Bohème). RST, *18ème Réunion des Sciences de la Terre, abstract,* 189.

MARHEINE, D., KACHLÍK, V., MALUSKI, H., PATOČKA, F. & ŻELAŹNIEWICZ, A. 2002. New Ar-Ar ages in the West Sudetes (Bohemian Massif) – constraints on the Variscan polyphase tectonothermal development. *In:* WINCHESTER, J. A., PHARAOH, T. C. & VERNIERS, J. (eds) *Palaeozoic Amalgamation of Central Europe,* Geological Society London, Special Publications, **201**, 133–155.

MAZUR, S. 1995. Structural and metamorphic evolution of the country rocks at the eastern contact of the Karkonosze Granite in the southern Rudawy Janowickie Mts and Lasocki Range. *Geologia Sudetica,* **29**, 31–98.

MIZERSKI, W. 1995. Geotectonic evolution of the Holy Cross Mts in central Europe. *Biuletyn Panstwowego Instytutu Geologicznego* **372**, 5-47.

MOCZYDLOWSKA, M. 1997. Proterozoic and Cambrian successions in Upper Silesia: an Avalonian terrane in southern Poland. *Geological Magazine,* **134**, 679-689.

MOLZAHN, M., ANTHES, G. & REISCHMANN, T. 1998. Single zircon Pb/Pb age geochronology and isotope systematics of the Rhenohercynian basement. *Terra Nostra,* **98/1**, 67–68.

NANCE, R. D. & MURPHY, J. B. 1994. Contrasting basement isotopicsignatures and the palinspastic restoration of peripheral orogens: example from the Neoproterozoic Avalonian-Cadomian belt. *Geology,* **22**, 617–620.

NAWROCKI, J. 1999. Prefolding remanent magnetization of diabase intrusion from the Bardo syncline in the Holy Cross Mts (Central Poland). *Przeglad Geologiczny* **47**, 1001-1104.

NOBLE, S.R., TUCKER, R. D. & PHARAOH, T. C. 1993. Lower Palaeozoic and Precambrian igneous rocks from eastern England, and their bearing on Late Ordovician closure of the Tornquist Sea: constraints from U-Pb and Nd isotopes. *Geological Magazine,* **130**, 835–846.

O'BRIEN, P. J., KRÖNER, A., JAECKEL, P., HEGNER, E., ŻELAŹNIEWICZ, A. & KRYZA, R. 1997. Petrological and isotopic studies on Paleozoic high pressure granulites, Góry Sowie Mts, Polish Sudetes. *Journal of Petrology*, **38**, 433–456.

OLIVER, G. J. H., CORFU, F. & KROGH, T. E. 1993. U-Pb ages from SW Poland: evidence for a Caledonian suture zone between Baltica and Gondwana. *Journal of the Geological Society, London*, **150**, 355–369.

ONCKEN, O. 1997. Transformation of a magmatic arc and orogenic root during the oblique collision and its consequences for the evolution of the European Variscides (Mid-German Crystalline Rise). *Geologische Rundschau,* **86**, 2–21.

PATOČKA, F. DOSTAL, J. & PIN, C. 1997. Early Palaeozoic intracontinental rifting in the central West Sudetes, Bohemian Massif: geochemical and Sr-Nd isotope study on felsic-mafic metavolcanics of the Rýchory Mts. Crystalline complex (W Sudetes, Bohemian Massif*). Neues Jahrbuch Mineral Abhandlungen,* **170**, 313–330.

PATOČKA, F., FAJST, M. & KACHLÍK, V. 2000. Mafic-felsic to mafic-ultramafic Early Palaeozoic magmatism of the West Sudetes (NE Bohemian Massif): the South Krkonoše Complex. *Zeitschrift für geologische Wissenschaften*, **28**, 177–210.

PEUCAT, J. J., BERNARD-GRIFFITHS, J., GIL IBARGUCHI, J. I., DALLMEYER, R.D., MENOT, R. P., CORNICHET, J. & IGLESIAS PONCE DE LEON, M. 1990. Geochemical and geochronological cross section of the deep Variscan crust: The Cabo Ortegal high-pressure nappe (north western Spain). *Tectonophysics*, **177**, 263–292.

PHARAOH, T. C. 1996. The Trans-European Suture Zone: Phanerozoic accretion and the evolution of contrasting continental lithospheres. *In:* GEE, D. G. & ZEYEN, H. (eds) EUROPROBE 1996 – *Lithosphere dynamics: origin and evolution of continents.* Uppsala, 40–55.

PHARAOH, T. C. 1999. Palaeozoic terranes and their lithospheric boundaries within the Trans-European Suture Zone (TESZ): a review. *Tectonophysics*, **314**, 17–41.

PHARAOH, T. C. & GIBBONS, W. 1994. Precambrian rocks in England and Wales south of the Menai Strait Fault System. In: GIBBONS, W. & HARRIS, A. L. (eds) A revised correlation of Precambrian rocks in the British Isles. *Special Report of the Geological Society, London*, **22**, 85–97.

PHARAOH, T. C., MERRIMAN, R. J., WEBB, P. C. & BECKINSALE, R. D. 1987. The concealed Caledonides of eastern England: preliminary results of a multidisciplinary study. *Proceedings of the Yorkshire Geological Society,* **46**, 355–369.

PHARAOH, T. C., MERRIMAN, R. J., EVANS, J. A., BREWER, T. S., WEBB, P. C. & SMITH, N. J. P. 1991. Early Palaeozoic arc-related volcanism in the concealed Caledonides of southern Britain. *Annales de la Société Géologique de Belgique,* **114**, 63–91.

PHARAOH, T. C., BREWER, T. S. & WEBB, P. C. 1993. Subduction-related magmatism of Late Ordovician age in eastern England. *Geological Magazine,* **130**, 647–656.

PHARAOH, T. C., ENGLAND, R. & LEE, M. K. 1995. The concealed Caledonide basement of eastern England and the southern North Sea – a review. In: GEE, D. G. & BECKHOLMEN, M. (eds) The Trans-European Suture Zone. Europrobe in Liblice 1993. *Studia Geophysica et Geodaetica* **39**, 330–346.

PHILIPPE, S., HAACK, U., ŻELAŹNIEWICZ, A., DÖRR, W. & FRANKE, W. 1995. Preliminary geochemical and geochronological results on shear zones in the Izera-Karkonosze Block (Sudetes, Poland). *Terra Nostra*, **95/8**, 122.

SAMSONOWICZ, J. 1926. Remarques sur la tectonique et la paleogeographie du massif paleozoïque de Swiety Krzyz. *Posiedzenia Naukowe Panstwowego Instytutu Geologicznego*, **15**, 44–46.

SAMUELSSON, J., VECOLI, M. & BEIER, H. 2001. Ordovician-Silurian palynostratigraphy (chitinozoa and acritarchs) of the G 14–1/86 borehole, southern Baltic Sea. *Neues Jahrbuch für Geologie und Paläontologie, Abhandlungen*, **222**, 259–290.

SAMUELSSON, J., GERDES, A., KOCH, L., SERVAIS, T & VERNIERS, J. 2002*a*. Chitinozoa and Nd isotope stratigraphy of the Ordovician rocks in the Ebbe Anticline, NW Germany. *In:* WINCHESTER, J. A., PHARAOH, T. C. & VERNIERS, J. (eds) *Palaeozoic Amalgamation of Central Europe,* Geological Society London, Special Publication, **201**, 115–131.

SAMUELSSON, J., VECOLI, M., BEDNARCZYK, W. S. & VERNIERS, J. 2002*b*. Timing of the Avalonia – Baltica plate convergence as inferred from palaeogeographic and stratigraphic data of chitinozoa assemblages in Pomerania, N. Poland. *In:* WINCHESTER, J. A., PHARAOH, T. C. & VERNIERS, J. (eds) *Palaeozoic Amalgamation of Central Europe,* Geological Society London, Special Publication, **201**, 95–113.

SCARROW, J. H., PEASE, V., FLEUTELOT, C. & DUSHIN, V. 2001. The late Neoproterozoic Enganepe ophiolite, Polar Urals, Russia: an extension of the Cadomian arc? *Precambrian Research*, **110**, 255–275.

SCHECK, M., THYBO, H., LASSEN, A., ABRAMOVITZ, T. & LAIGLE, M. 2002. Basement structures in the southern North Sea, offshore Denmark, based on seismic interpretation. *In:* WINCHESTER, J. A., PHARAOH, T. C. & VERNIERS, J. (eds) *Palaeozoic Amalgamation of Central Europe,* Geological Society London, Special Publication, **201**, 311–326.

SCHULMANN, K. & GAYER, R. 2000. A model for a continental accretionary wedge developed by oblique collision: the NE Bohemian Massif. *Journal of the Geological Society, London*, **157**, 401–416.

SEILACHER, A. 1983. Upper Palaeozoic trace fossils from the Gilf Kebir-Abu Ras area in southwestern Egypt. *Journal of African Earth Sciences* **1**, 21–34.

SESTON, R., WINCHESTER, J. A., PIASECKI, M. A. J., CROWLEY, Q. G. & FLOYD, P. A. 2000. A structural model for the Western-Central Sudetes: a deformed stack of Variscan thrust sheets. *Journal of the Geological Society, London*, **157**, 1155–1167.

SHELLEY, D. & BOSSIÈRE, G. 2000. A new model for the Hercynian orogen of Gondwanan France and *Iberian Journal of Structural Geology*, **22**, 757–776.

SINTUBIN, M. & EVERAERTS, M. 2002. A compressional wedge model for the Lower Palaeozoic Anglo-Brabant Belt, Belgium, based on potential field data. *In:* WINCHESTER, J. A., PHARAOH, T. C. & VERNIERS, J. (eds) *Palaeozoic Amalgamation of Central Europe,* Geological Society London, Special Publication, **201**, 327–343.

STAMPFLI, G. 1996. The Intra-Alpine terrain: a Palaeotethyan remnant in the Alpine Variscides. *Eclogae geologicae Helvetiae,* **86**, 13–42.

ŠTĚDRÁ, V., KACHLÍK, V. & KRYZA, R. 2002. Coronitic metagabbros from the Mariánské-Lázně Complex, Teplá Crustal Unit and Góry Sowie Block, northern part of the Bohemian Massif: protoliths, metamorphism and tectonic position. *In:* WINCHESTER, J. A., PHARAOH, T. C. & VERNIERS, J. (eds) *Palaeozoic Amalgamation of Central Europe,* Geological Society London, Special Publication, **201**, 217–236.

STOSCH, H. G. & LUGMAIR, G. W. 1990. Geochemistry and evolution of MORB-type eclogites from the Münchberg Massif, southern Germany. *Earth and Planetery Science Letters*, **99**, 230–249.

STRACHAN, R. A., NANCE, R. D., DALLMEYER, R. D., D'LEMOS, R. S., MURPHY, J. B. & WATT, G. R. 1996. Late Precambrian tectonothermal evolution of the Malverns Complex. *Journal of the Geological Society, London*, **153**, 589–600.

TAIT, J., SCHÄTZ, M., BACHTADSE, V., & SOFFEL, H. C. 2000. Palaeomagnetism and Palaeozoic palaeogeography of Gondwana and European terranes. *In:* FRANKE, W., HAAK, V., ONCKEN, O. & TANNER, D. (eds) *Orogenic processes: Quantification and modelling in the Variscan Belt.* Geological Society, London, Special Publications, **179**, 21–34.

THORPE, R. S., BECKINSALE, R. D., PATCHETT, P. J., PIPER, J. D. A., DAVIES, G. R. & EVANS, J. A. 1984. Crustal Growth and Late Precambrian- Early Palaeozoic plate tectonic evolution of England and Wales. *Journal of the Geological Society, London*, **141**, 521–536.

TIMMERMANN, H., PARRISH, R. R., NOBLE, S. R. & KRYZA, R. 2000. New U-Pb monazite and zircon data from the Sudetes Mountains in SW Poland: evidence for a single-cycle Variscan orogeny. *Journal of the Geological Society, London*, **157**, 265–268.

TORSVIK, T. H. & REHNSTRÖM, E. F. 2001. Cambrian palaeomagnetic data from Baltica: implications for true polar wander and Cambrian palaeogeography. *Journal of the Geological Society, London*, **158**, 321–329.

TORSVIK, T. H., SMETHURST, M. A., MEERT, J. G., VAN DER VOO, R., MCKERROW, W. S., BRASIER, M. D., STURT, B. A. & WALDERHAUG, H. J. 1996. Continental break-up and collision in the Neoproterozoic and Palaeozoic: a tale of Baltica and Laurentia. Earth Science Reviews, **40**, 229–258.

TUCKER, R. D. & PHARAOH, T. C. 1991. U-Pb zircon ages for Late Precambrian igneous rocks in southern Britain. *Journal of the Geological Society, London,* **148**, 435–443.

TURNER, J. S. 1949. The deeper structure of central and northern England. *Proceedings of the Yorkshire Geological Society,* **27**, 280–297.

VAN BREEMEN, O., BOWES, D. R., AFTALION, M. & ŻELAŹNIEWICZ, A. 1988. Devonian tectonothermal activity in the Góry Sowie gneissic block, Sudetes, southwestern Poland: evidence from Rb-Sr and U-Pb isotopic studies. *Annales Societatis Geologorum Poloniae*, **58**, 3–19.

VECOLI, M. & SAMUELSSON, J. 2001. Quantitative evaluation of microplankton palaeobiogeography in the Ordovician – Early Silurian of the northern TESZ (Trans-European Suture Zone): implications for the timing of the Avalonia-Baltica collision. *Review of Palaeobotany and Palynology,* **115**, 43–68.

VECOLI, M., SAMUELSSON, J. & GIESE, U. 1999.

Provenance analysis of Ordovician-Silurian clastic sediments at the southwestern margin of theBaltic Platform: implications for the timing of closure of the Torquist Ocean. *PACE Network mid-term review meeting, Copenhagen, Programme with Abstracts* 13.

VERNIERS, J., PHARAOH, T. C., ANDRÉ, L., DEBACKER, T., DE VOS, M., EVERAERTS, M., HERBOSCH, A., SAMUELSSON, J., SINTUBIN, M. & VECOLI, M. 2002. Lower Palaeozoic basin development and collision history of Eastern Avalonia. *In:* WINCHESTER, J. A., PHARAOH, T. C. & VERNIERS, J. (eds) *Palaeozoic Amalgamation of Central Europe,* Geological Society London, Special Publication, **201**, 47–93.

WILLIAMSON, J. P., PHARAOH, T. C., BANKA, D., THYBO, H., LAIGLE, M. & LEE, M. K. 2002. Potential field modelling of the Baltica-Avalonia (Thor-Tornquist) suture beneath the southern North Sea. *Tectonophysics,* in press.

WINCHESTER, J. A., FLOYD, P. A., CHOCYK, M., HORBOWY, K. & KOZDROJ, W. 1995. Geochemistry and tectonic environment of Ordovician meta-igneous rocks in the Rudawy Janowickie Complex, SW Poland. *Journal of the Geological Society, London*, **152**, 105–115.

WINCHESTER, J. A., FLOYD, P. A., AWDANKIEWICZ, M., PIASECKI, M. A. J., AWDANKIEWICZ, H., GUNIA, P. & GLIWICZ, T. 1998. Geochemistry and tectonic significance of metabasic suites in the Góry Sowie Block, SW Poland. *Journal of the Geological Society, London*, **155**, 155–164.

WINCHESTER, J. A. & PACE TMR NETWORK. 2002. Palaeozoic Amalgamation of Central Europe: new results from recent geological and geophysical investigations. *Tectonophysics,* in press.

List of participants in the PACE (Palaeozoic Amalgamation of Central Europe) TMR Network

Keele University, UK	J.A. Winchester (co-ordinator)
	P.A. Floyd
	M.A.J. Piasecki (decd 1999)
	Q.G. Crowley (visiting researcher)
British Geological Survey, UK	M.K. Lee
	T.C. Pharaoh
	J.P. Williamson
	D. Banka (visiting researcher)
Ghent University, Belgium	J. Verniers
	J. Samuelsson (visiting researcher)
GeoForschungsZentrum, Potsdam, Germany	U. Bayer
	C. Krawczyk
	A.-M. Marotta (visiting researcher)
	J. Lamarche (visiting researcher)
Justus Liebig University, Giessen, Germany	W. Franke
	W. Dörr
	P. Valverde-Vaquero (visiting researcher)
Martin Luther University, Halle, Germany	U. Giese
	M. Vecoli (visiting researcher)
	R. Handler (visiting researcher)
Copenhagen University, Denmark	H. Thybo
	A. Lassen
	M. Laigle (visiting researcher)
	M. Scheck (visiting researcher)
CNRS-Montpellier, France	H. Maluski
	D. Marheine (visiting researcher)
NERC Isotope Geoscience Laboratory, UK	R.R. Parrish
	S. Noble
	J.A. Evans
	H. Timmermann (visiting researcher)
	A. Gerdes (visiting researcher)
Polish Academy of Sciences, Warsaw, Poland	A. Guterch
	M. Grad
PGI, Wrocław, Poland	S. Cwojdzínski
	Z. Cymerman
	W. Kozdroj
Wrocław University, Poland	R. Kryza
	P. Aleksandrowski
	S. Mazur
CGU, Prague, Czech Republic	V. Štědrá
	J. Kotková

Accretion of first Gondwana-derived terranes at the margin of Baltica

Z. BELKA[1], P. VALVERDE-VAQUERO[2,3], W. DÖRR[3], H. AHRENDT[4†], K. WEMMER[4], W. FRANKE[3] & J. SCHÄFER[3]

[1]*Institut für Geologische Wissenschaften und Geiseltalmuseum, Martin-Luther-Universität Halle-Wittenberg, Domstr. 5, D-06108 Halle, Germany (e-mail: belka@geologie.uni-halle.de)*

[2]*Continental Geoscience Division, Geological Survey of Canada, 615 Booth St, Ottawa, K1A 0E8, Canada*

[3]*Institut für Geowissenschaften und Lithosphärenforschung, Justus-Liebig-Universität-Giessen, Senckenbergstrasse 3, D-35390 Giessen, Germany*

[4]*Geowissenschaftliches Zentrum (GZG), Universität Göttingen, Goldschmidtstr. 1–3, D-37077 Göttingen, Germany*

Abstract: In central Europe, three crustal units, i.e. the Małopolska, the Łysogóry and the Bruno-Silesia, can be recognized by basement data, faunas and provenance of clastic material in the Cambrian clastic rocks. They are now situated within the Trans-European Suture Zone, a tectonic collage of continental terranes bordering the Tornquist margin of the palaeocontinent of Baltica, but during the Cambrian their position in relation to each other and to Baltica was different from today. These units are exotic terranes in respect to Baltica and are interpreted as having been derived from the Cadomian margin of Gondwana. Their detachment is probably related to the final break-up of the supercontinent Rodinia at *c.* 550–590 Ma.

New detrital zircon and muscovite age data provide evidence that Małopolska was derived from the segment of the Cadomian orogen that bordered the Amazonian Craton. It must have already separated from Gondwana in Early Cambrian time (some 40–50 Ma before Avalonia became detached and began its rapid drift). The accretion of Małopolska to Baltica occurred between late mid-Cambrian and Tremadocian times. Both palaeontological and provenance evidence demonstrate that Małopolska and not Avalonia was the first terrane to join the Baltica palaeocontinent. This event initiated the progressive crustal growth of the European lithosphere, which continued during Phanerozoic times and led to the formation of modern Europe.

A complex suture zone, the Trans-European Suture Zone (TESZ), stretching from Denmark through Poland and Ukraine to Romania, separates the ancient East European Craton (EEC) from the young mobile belts (Variscan and Alpine) of western Europe. It is a tectonic collage of continental terranes which were accreted to the margin of the palaeocontinent of Baltica (that is, the EEC) in Early Palaeozoic time. There is still a debate about the number of crustal units involved and their provenance (e.g. Pożaryski *et al.* 1992; Dadlez 1995; Franke 1995; Pharaoh 1999). According to Berthelsen (1993), who created the concept of this domain, the TESZ comprises terranes from the western edge of the EEC (that is, the Teisseyre–Tornquist Line in central Europe) to basement blocks underlying the Rheno-Hercynian and Moravo-Silesian belts. Pharaoh (1999), however, extended the TESZ to include also Armorican terranes of the Variscan orogen. In this paper, we follow the original Berthelsen concept for the TESZ, because this is clearly justified by the general subdivision of the crystalline basement (see Dadlez 1997).

In its northern segment the TESZ generally has a very thick Upper Palaeozoic, Mesozoic and Cenozoic cover, thus direct geological information (isotopic, palaeomagnetic, biogeographical, and structural data) from the Precambrian basement and the Palaeozoic is confined to a small number of borehole sites (see Frost *et al.* 1982; Giese *et al.* 1997). This is why the identity and affinity of individual terranes incorporated within the TESZ is only roughly known there. However, in southern

From: WINCHESTER, J. A., PHARAOH, T. C. & VERNIERS, J. 2002. *Palaeozoic Amalgamation of Central Europe.* Geological Society, London, Special Publications, **201**, 19–36. 0305-8719/02/$15.00

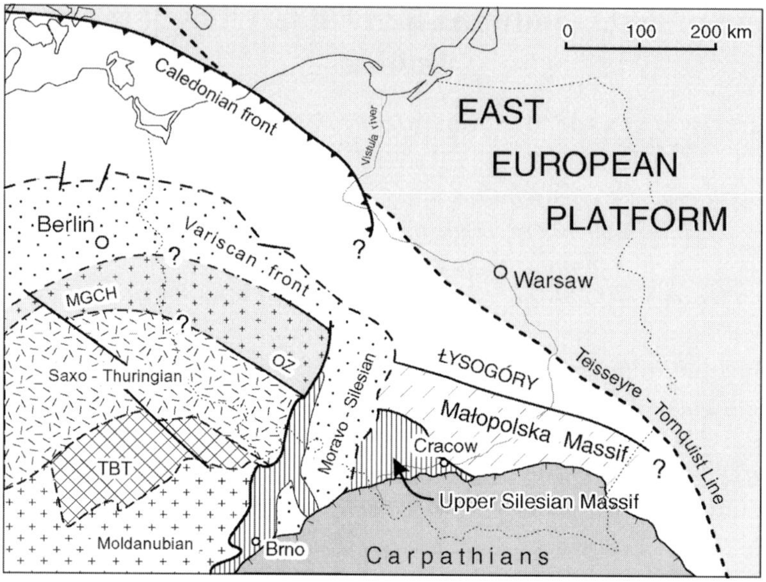

Fig. 1. Simplified structural map of central Europe showing the crustal units in the Variscan foreland of southern Poland and in the eastern part of the Variscan Belt. The Upper Silesian Massif constitutes the NE part of the Bruno-Silesia block. Dotted line indicates the Polish border. TBT, Teplá-Barrandian Terrane; OZ, Odra Zone; MGCH, Mid German Crystalline High.

Poland the structure and evolution of the TESZ are relatively well constrained. Extensive data from outcrops and hundreds of boreholes provide evidence for three fault-bounded crustal units: the Łysogóry Unit, the Małopolska Massif and the Upper Silesian Massif (Fig. 1). They are characterized by an individual tectonic and sedimentary evolution during the Early Palaeozoic (for summary, see Belka *et al.* 2000). Although there is currently general agreement that these units display the character of terranes, their geotectonic affinities and the accretionary history are still a matter of controversy. This is because records of faunal elements in the Cambrian suggested linkages both to the Baltic palaeocontinent and to the Peri-Gondwanan plates (e.g. Orłowski 1975; Seilacher 1983; Jendryka-Fuglewicz 1992; Żylińska 2001, 2002). Unfortunately, the results of palaeomagnetic studies are still unsatisfactory, as they allow fundamentally different palaeogeographical interpretation for the Early Palaeozoic (for discussion, see Lewandowski (1995) and Nawrocki (1995)). Recently, first provenance studies (dating of detrital zircon and muscovites in the Cambrian) provided additional strong arguments that the Łysogóry, Małopolska and Upper Silesia blocks are exotic terranes derived from the Gondwana margin (Belka *et al.* 2000; Valverde-Vaquero *et al.* 2000). Due to a combination of available faunal and provenance data, it is now evident that the position of these terranes in relation to each other during the Cambrian was fundamentally different from that today. However, there is, still not enough data on the age of detritus to reconstruct the accretionary history of these terranes in detail.

In this paper we provide the first detrital zircon ages from the Cambrian of the Małopolska Massif and new K–Ar cooling ages of detrital muscovites. The latter ages supplement the mica provenance data set from the Early Palaeozoic of Małopolska (Belka *et al.* 2000), which is still very limited (see Table 1 and Fig. 2). This is because the Cambrian clastic rocks here are characterized by a very low content of detrital muscovites and there are only a few horizons with enough material for K–Ar dating. The new data help to evaluate the potential source regions for the sediment and, combined with the biogeographical information, they provide a basis to reconstruct the palaeogeographical evolution of Małopolska during Early Palaeozoic time. Comparison with characteristics of other terranes within the TESZ allows us to propose an accretionary setting for central Europe in which Małopolska constituted the first Gondwana-derived microplate that docked at the margin of Baltica.

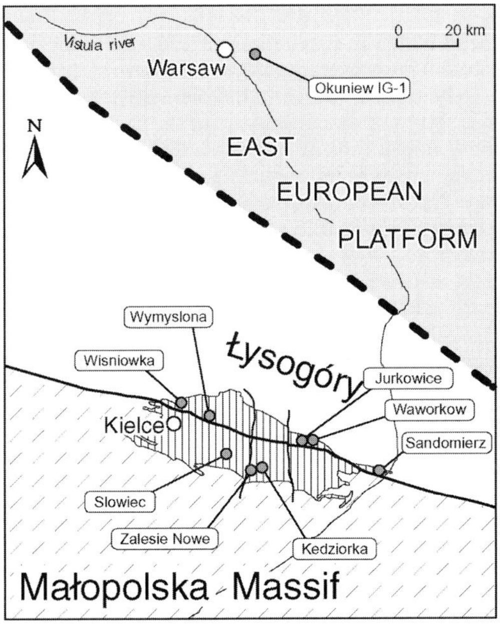

Fig. 2. Location map of samples presented or discussed in the paper. The Palaeozoic rocks of the Holy Cross Mountains are hatched.

Geological background

There is general agreement that at least three crustal units, namely the Łysogóry Unit, the Małopolska Massif, and the Upper Silesian Massif can be distinguished within the TESZ in southern Poland (Fig. 1). The geophysical data suggest that the subdivision is probably more complex (for summary see Dadlez 2001). Recently, Unrug *et al.* (1999) offered a different subdivision of the TESZ in southern Poland. However, this concept includes geological information which is inconsistent with detailed data known from the literature. The authors present a model but the evidence on which it is based is lacking or is not substantiated by proof. Therefore, we do not intend to discuss this concept here and give only one example of inaccuracy. Unrug *et al.* (1999, p. 140) described the Silurian succession in the western margin of Małopolska as composed of coarsening upwards turbiditic greywackes up to 12 km thick. In an earlier paper, a thickness of more than 6 km was suggested (Harańczyk 1994). However, until now, no evidence for such a thick Silurian sequence has been presented. The succession is not exposed and is known only from boreholes. Belka & Siewniak (1996) have evaluated the cores and found no greywackes or coarse-grained clastic rocks in the predominantly shaly succession. Moreover, they noted that the Silurian is less than 900 m thick. During the present study we examined the relevant boreholes and found only very small amounts of detrital mica in the siltstone interbeds, not enough for K–Ar dating.

Many boreholes and outcrops offer insight into the Palaeozoic sequence of southern Poland. Thus, there is plenty of literature on various aspects of geology and palaeontology. The evaluation and understanding of published data, however, are sometimes difficult for foreign geologists, because various local geographical

Table 1. *Summary of K–Ar cooling age data for detrital muscovites from Cambrian and Ordovician clastic rocks of the Holy Cross Mountains*

No.	Sample locality	Stratigraphy	Age (Ma)
	Łysogóry		
1	Wisniowka Quarry	Upper Cambrian	777.1 ± 22.9
2	Wisniowka Quarry	Upper Cambrian	613.7 ± 12.6
3	Wisniowka Quarry	Upper Cambrian	1319.1 ± 52.1
4	Wisniowka Quarry	Upper Cambrian	929.4 ± 23.6
5	Wymyślona	Upper Cambrian	539.1 ± 14.9
6	Waworkow Quarry	Upper Cambrian	848.4 ± 19.4
7	Jurkowice	Upper Cambrian	1745.3 ± 35.3
8	Sandomierz	Upper Cambrian	1721.1 ± 37.9
	Małopolska		
9	Kędziorka	Lower Cambrian	534.1 ± 18.6
10	Kędziorka	Lower Cambrian	547.1 ± 14.9
11	Słowiec	Middle Cambrian	557.0 ± 11.8
12	Słowiec	Middle Cambrian	618.0 ± 13.4
13	Zalesie Nowe	Ashgill	650.3 ± 18.5

Sample numbers 1–10 are taken from Belka *et al.* (2000); the stratigraphic ages of samples are updated.

names and terms are often used to describe the same tectonic unit in the area. The Lower Palaeozoic facies development, stratigraphy, fauna, and tectonics have been summarized in numerous papers (e.g. Mizerski 1995; Buła *et al.* 1997; Żaba 1999; Belka *et al.* 2000; Dadlez 2001). The reader is referred to these for details; general features only are presented here. Moreover, we focus attention upon the new biostratigraphic data as well as on features and points that are uncertain or controversial.

The Łysogóry Unit

The unit is composed of thick continental crust; the Moho depth ranges from 44 km in the west to more than 50 km in the east. Nothing is known about the age of the crystalline basement, which appears to occur at depths below 6–7 km (Semenov *et al.* 1998). The Palaeozoic succession of the Łysogóry Unit is exposed in the northern part of the Holy Cross Mountains (Fig. 2), where it is bounded to the south by the Holy Cross Fault (HCF). This is a brittle fault zone, recognizable in the Moho topography, which separates the Łysogóry Unit from the Małopolska Massif. It can be traced in the field throughout the Holy Cross Mountains but its position east of Opatow is unclear. Traditionally, the HCF is positioned north of the Pieprzowe Mountains, a small area where strongly deformed Cambrian shales are exposed in the gorge of the Vistula river at Sandomierz (Fig. 2). New acritarch and thermal maturation data suggest that this shaly succession has much more in common with the Cambrian of the Łysogóry Unit than with that of the Małopolska Massif (Szczepanik 2001). Thus, it is very likely that either the HCF runs south of the Pieprzowe Mountains, or the Cambrian sequence at Sandomierz represent an overthrust fragment of Łysogóry rocks resting on the margin of the Małopolska Massif. Incorporation of the Pieprzowe Mountains into the Łysogóry Unit also seems to be justified in terms of lithostratigraphy: in the existing lithostratigraphic scheme (Orłowski 1992) the Gory Pieprzowe Shale Formation was the only unit distinguished on both sides of the HCF. In the revised concept, the entire Gory Pieprzowe Shale Formation occurs within the Łysogóry Unit.

The extent of the Łysogóry Unit to the west, north and east is obscure. Recent geophysical soundings seem to indicate that it constitutes a very narrow zone, only about 30 km wide, which is delimited to the NE by a prominent, almost vertical deep fault (Semenov *et al.* 1998). It is therefore not clear whether the Łysogóry Unit is directly in contact with the crust of the East European Platform or whether the block adjacent to it possibly represents another terrane.

The most significant and distinctive feature of the sedimentary succession of the Łysogóry Unit with respect to the adjacent Małopolska Massif is the absence of angular unconformities in the pre-Carboniferous stratigraphy (Fig. 3). The oldest rocks in the Łysogóry Unit are strongly deformed shales that occur along the HCF; a recent study of acritarchs has provided evidence for their Late Cambrian age (Szczepanik 2001). However, this is at variance with the previously suggested Mid or Early Cambrian age based on trilobites and trace fossils occurring in the overlying sandstone sequence (Orłowski 1992; Kowalczewski 1995). This sequence, up to 1400 m thick, is famous for a rich assemblage of trilobite trace fossils (e.g. Radwański & Roniewicz 1963). It is dominated by sandstones with very mature compositions deposited in a shallow-water offshore environment (Radwański & Roniewicz 1960). More information on the Cambrian succession of the Łysogóry Unit is provided by recent publications of Żylińska (2001, 2002).

In understanding the amalgamation history of the TESZ in southern Poland the Silurian sequence of the Łysogóry Unit is of great importance. It includes a prominent clastic unit consisting of up to 1500 m of fine-grained greywackes with numerous volcaniclastic interbeds. Fauna and sedimentary structures indicate an upward evolution from deep-water to predominantly very shallow-water deposition (Tomczyk 1970). A similar complex of greywackes is also present on the opposite side of the HCF, in the marginal part of the Małopolska Massif (Fig. 3).

The Małopolska Massif

This unit most probably represents a composite terrane, composed of the Małopolska and the San Blocks, that exhibit individual tectonic and sedimentary development during Early Palaeozoic time (Belka *et al.* 2000). Although the boundary between these blocks cannot be defined precisely, differences in the crustal thickness suggest an NE–SW boundary line very close and parallel to the Vistula River (see Dadlez 2001). The western part, which is the Małopolska Massif *sensu stricto*, is characterized by the pre-Ordovician deformation of Cambrian and Precambrian rocks (Fig. 3). Moreover, the crustal thickness here of 35 to 38 km is about 10 km thinner than that of the San Block. In this paper we provide new provenance data from the

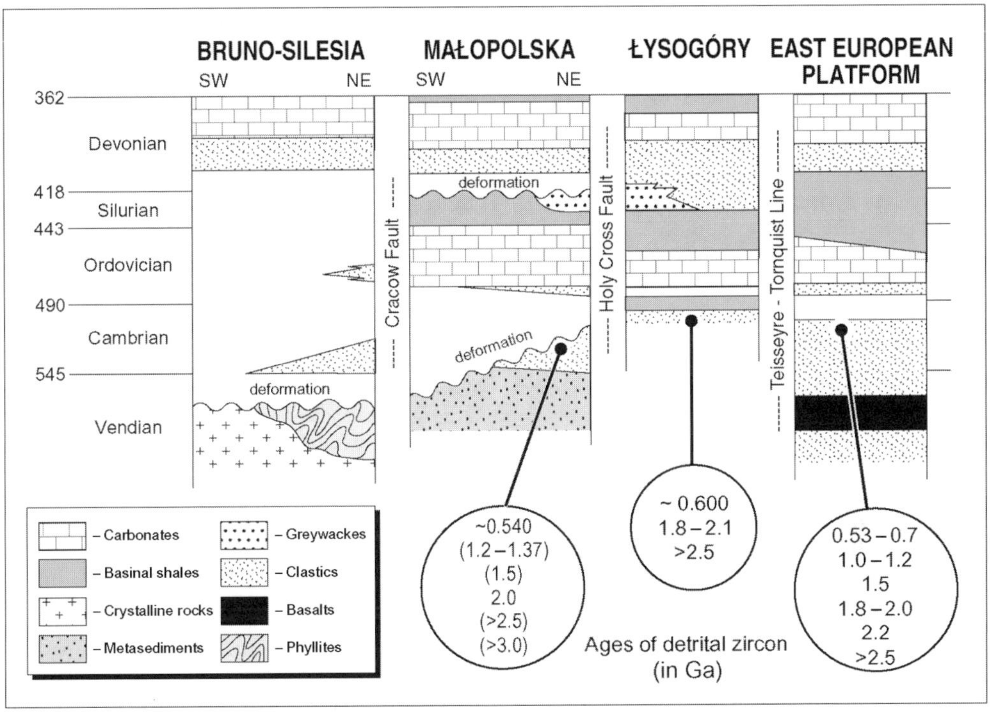

Fig. 3. Stratigraphic columns of the crustal units forming the Variscan foreland in southern Poland and the stratigraphic succession for the marginal part of the East European Platform in the Warsaw region. Note the differences in occurrence of stratigraphic gaps (white spaces) and deformation phenomena. Circles are to show the age spectra of detrital zircon in the clastic material of Cambrian rocks.

Małopolska Massif (*sensu stricto*) only and discuss its accretionary history. Belka *et al.* (2000) provide a recent review of the geological features of the San Block.

The Małopolska Massif is clearly delimited to the north and to the SW by the HCF and the Cracow Fault, respectively (Figs 1 and 3). The crystalline basement is unknown. In several places deep boreholes pierced a succession of weakly metamorphosed polymictic clastic rocks more than 3000 m thick below the Palaeozoic (Kowalczewski 1990). A thermal overprint seems to result from deep burial and was promoted by the immature composition of clastic material. The highly immature detritus from mafic and felsic igneous rocks, the large amount of arkosic material, sedimentary features related to turbidites and debris flows, and the extreme thickness led Belka *et al.* (2000) to interpret the basement of Małopolska as a late Precambrian forearc-trench system linked to the Avalonian/Cadomian active continental margin. The age of this succession is poorly constrained. Rare and badly preserved acritarchs predominantly indicate a Vendian age. Compston *et al.* (1995) dated a volcanic tuff at 549 ± 3 Ma from the uppermost part of the succession in the central part of the Małopolska Massif. Further NE, where parts of the succession are exposed in the southern Holy Cross Mountains, it ranges up to the Middle Cambrian in age. This youngest portion is unmetamorphosed and exhibits a shallowing-upward trend associated with a much maturer clastic lithology (Fig. 3).

The entire Precambrian to Middle Cambrian succession of the Małopolska Massif is folded, and discordantly overlain by the Lower Ordovician (Fig. 3). Moreover, the rocks in Małopolska underlying the Ordovician tend to be progressively older and more highly metamorphosed from NE to SW. This demonstrates that the whole Małopolska block was tilted prior to the Ordovician and erosion removed much of the succession in the western part of the massif, close to the Cracow Fault.

The Ordovician represents a transgressive sequence, with offshore sandstones, open-marine carbonates and shales. The Silurian is developed in a monotonous facies of graptolites shales followed by up to 1200 m thick greywackes. Thus, it resembles the Silurian sequence of the Łysogóry Unit but is generally thinner. Sedimentary structures in the greywackes suggest a turbidite transport of clastic material that derived almost exclusively from local volcanic sources (Przybylowicz & Stupnicka 1991). Some small dolerite dykes also occur within the Silurian of the southern Holy Cross Mountains (Nawrocki 1999).

Unlike in the Łysogóry Unit and Upper Silesia, the Lower Devonian of the Małopolska Massif rests discordantly on deformed Lower Palaeozoic strata (Fig. 3). Locally its basal surface rests directly on rocks as old as the Lower Cambrian (Szulczewski 1995). Moreover, the marine Lower Devonian succession oversteps the Holy Cross Fault.

The Upper Silesian Massif

The Upper Silesian Massif (USM) constitutes the northern part of a larger block called the Brunovistulian (BV) (*sensu* Dudek 1980) or Bruno-Silesia as a palaeogeographical unit (Belka *et al.* 2000). However, there is some uncertainty concerning the inferred outline and the position of terrane boundaries because large parts of the Brunovistulian (BV) are hidden under the Variscan overthrust (the Moravo-Silesian belt) and under the Carpathians in the SE (Fig. 1). To the NE, the Upper Silesian Massif is separated from the Małopolska Massif by the Cracow Fault (Belka & Siewniak 1996; Źaba 1999), which is sometimes incorrectly considered as a prolongation of the Odra Fault (e.g. Finger *et al.* 2000). The latter is an intra-Variscan structure, a fault between the Franconian and the Saxothuringian terranes (Franke 2000; Franke & Żelaźniewicz 2000). The western limit of the Brunovistulian unit coincides with the eastern boundaries of the Moldanubian and the Saxothuringian terranes. According to Dudek (1980), the BV may extend to the SE under the Carpathians as far as the Peri-Pieniny lineament and to the SW approximately to the Krems-Vienna Line in Austria.

The Brunovistulian unit possesses an extremely thick autochthonous Palaeozoic sedimentary cover which rests on a crystalline basement of Cadomian age. The basement includes both intrusive rocks and their metamorphic cover (Dudek 1995). Isotopic data give evidence for plutonic activity at *c.* 580–590 Ma and metamorphism between 580 and 610 Ma (Finger *et al.* 2000, and references therein). The lithology and crustal evolution of these Proterozoic rocks closely resembles the Avalon terrane of Newfoundland. The Cadomian basement of the BV is discordantly overlain by a thick sequence of Lower/Middle Cambrian clastic rocks, primarily preserved in Upper Silesia (Fig. 3), which are undeformed and display only a moderate thermal overprint (Belka 1993). Sedimentary structures and trace fossils indicate deposition in a tide-dominated shallow offshore environment that grades upwards into a moderately deep offshore setting. The stratigraphic gap, comprising at least the Ordovician and the Silurian, is a regional feature observed in Upper Silesia (Fig. 3). The only exception is the marine Ordovician sequence present at the northern margin of the USM. The Lower Devonian succession is very thin and composed of both alluvial and shallow-water marine clastic rocks that truncate the Lower Palaeozoic rocks. There is, however, no angular unconformity between the Devonian and the Lower Palaeozoic. In the southwestern part of the BV, however, the Lower Devonian clastic rocks directly overlie the crystalline basement (Matte *et al.* 1990).

Sample preparation and analytical procedures

Detrital zircons

To obtain the age spectrum of detrital zircons in the Cambrian of the Małopolska Massif a 40 kg sample was collected from the Ocieseki Sandstone Formation exposed at Kędziorka (Fig. 2). In this locality, also known as Chojnów Dół, fine- to medium-grained sandstones are intercalated with thin shale and siltstone layers. The quartz sandstones are immature with abundant lithic fragments, feldspars, and various heavy minerals (Michniak 1969). This locality was selected because detrital muscovites had already been dated there (Belka *et al.* 2000) and the sequence is well constrained biostratigraphically. Macrofauna represented by inarticulate brachiopods and trilobites is indicative of the *Protolenus* Zone of the Lower Cambrian.

The rock samples were processed and analysed at the University of Giessen, Germany. Following crushing and pulverization, the sample was separated by flotation with water, bromoform, and diiodomethane into light and heavy mineral fractions. Magnetic separation using a Frantz magnetic separator obtained heavy mineral fractions of different para-magnetic quality. The non-magnetic, zircon-bearing, fraction was sieved to concentrate the larger crystals. The selected zircons were hand picked under a microscope for analysis. These grains were selected on the basis of colour and crystal quality. Most were single-grain analyses, but given the small size of most zircons, some analyses consisted of fractions of 2 to 4 grains. To avoid losing material, the air abrasion technique of Krogh (1982) was not used. After the zircon crystals were cleaned with 4 N HNO_3, double distilled H_2O, and ultrapure acetone, they were weighed and dissolved in Krogh-type Teflon® dissolution bombs with HF.

The samples were spiked with a mixed ^{205}Pb–^{235}U spike, which was added directly to the sample prior to dissolution. The Pb and U separation was carried out using a scaled-down version of the ion exchange chemistry of Krogh (1973). The purified Pb and U were collected, separately, with H_3PO_4 and loaded on separate single Re filaments using a mixture of silica gel and H_3PO_4. The isotopic ratios were measured in static and peak-jumping mode using a Finnigan MAT 261 mass spectrometer equipped with a Spectromat ion counting system. The silica gel–H_3PO_4 mixture provided an ionization efficiency of approximately 1mv signal per 3.5 pg of ^{205}Pb. In all cases this level of sample ionization allowed static mode measurements; the ^{204}Pb was measured with the calibrated ion counter. All isotopic ratios were corrected for mass fractionation (1.12 ± 0.18‰ /a.m.u.), blank (3 to 10 pg Pb, 1 pg U) and initial common Pb estimated after the model of Stacey & Kramers (1975). The isotopic ages were calculated using the decay constants of Jaffey *et al.* (1971). The atomic ratios were calculated using PBDAT (Ludwig 1991). In all cases uncertainties are reported at the 2σ level. Total uncertainties on individual points are represented by 2σ uncertainty ellipses.

Detrital muscovites

Samples for K–Ar dating of detrital mica were taken from two localities in the southern Holy Cross Mountains (= Małopolska). On the slope of Słowiec Hill (Fig. 2) two samples were collected from the Middle Cambrian. Sample *Słowiec 1* is from fine- to medium-grained immature sandstones from the lower part of the succession. These rocks most probably represent the upper part of the Ociesęki Sandstone Formation and resemble sandstones exposed at Kędziorka. Sample *Słowiec 2* is from the top of Słowiec Hill, where coarse-grained, very mature sandstones with thin conglomerate intercalations occur. In addition, a single sample was collected from the Ordovician sequence at Zalesie Nowe (Fig. 2). This sequence is well constrained biostratigraphically (Dzik 1999) and composed predominantly of shales and dolomites. Detrital mica is very rare; only one horizon within the Ashgill portion of the section had enough detrital grains for K–Ar dating. Sample information is presented in Table 1. Although there is no direct information on the thermal overprint of the sampled rocks, the acritarch record in the Cambrian of the southern Holy Cross Mountains (Szczepanik 1997) suggests burial temperatures below 100°C for the studied area. Therefore, all the ages obtained were interpreted as cooling ages of the source areas.

The rock samples were processed and analysed at the University of Göttingen, Germany. After cleaning, the rock samples were passed through a jaw crusher twice. The loosened grain fabric was fractionated by dry sieving. The grain size with the highest amount of macroscopically visible mica was treated further in order to obtain at least 120 mg of pure muscovite for analysis. The micas were then enriched by applying the 'mica jet' method (Horstmann 1987). Subsequently, they were treated by different separation procedures, e.g. separation after grain shape on a self-constructed dry shaking device and/or separation using magnetic susceptibility with a Frantz magnetic separator. The muscovite concentrate was cleaned under a binocular microscope by hand-picking in order to prevent contamination by organic particles or unwanted minerals. Following hand selection the muscovite concentrate was rubbed with alcohol in a porcelain mortar to remove weathered grain margins and then sieved using an 80 μm disposable sieve. A final inspection followed under a binocular microscope so that the material for analysis comprised only the fresh cores of the muscovite grains.

The argon isotopic composition was measured in a Pyrex glass extraction and purification line coupled to a VG 1200 C noble gas mass spectrometer operating in static mode. The amount of radiogenic ^{40}Ar was determined by isotope dilution method using a highly enriched ^{38}Ar spike from Schumacher, Bern (Schumacher 1975). The spike is calibrated against the biotite standard HD-B1 (Fuhrmann *et al.* 1987). The age calculations are based on the constants recommended by the IUGS quoted in Steiger & Jäger (1977). Potassium was determined in duplicate by flame photometry using an Eppendorf Elex 63/61. The samples were dissolved in a mixture of HF and HNO_3 according to the technique of Heinrichs & Herrmann (1990). CsCl and LiCl were added as ionization buffer and internal standard respectively. The analytical error for the K–Ar age calculations is given at a 95% confidence level (2σ).

Provenance data

The Małopolska Massif

A total of 15 analyses of detrital zircon is reported. All analyses are single zircon grains,

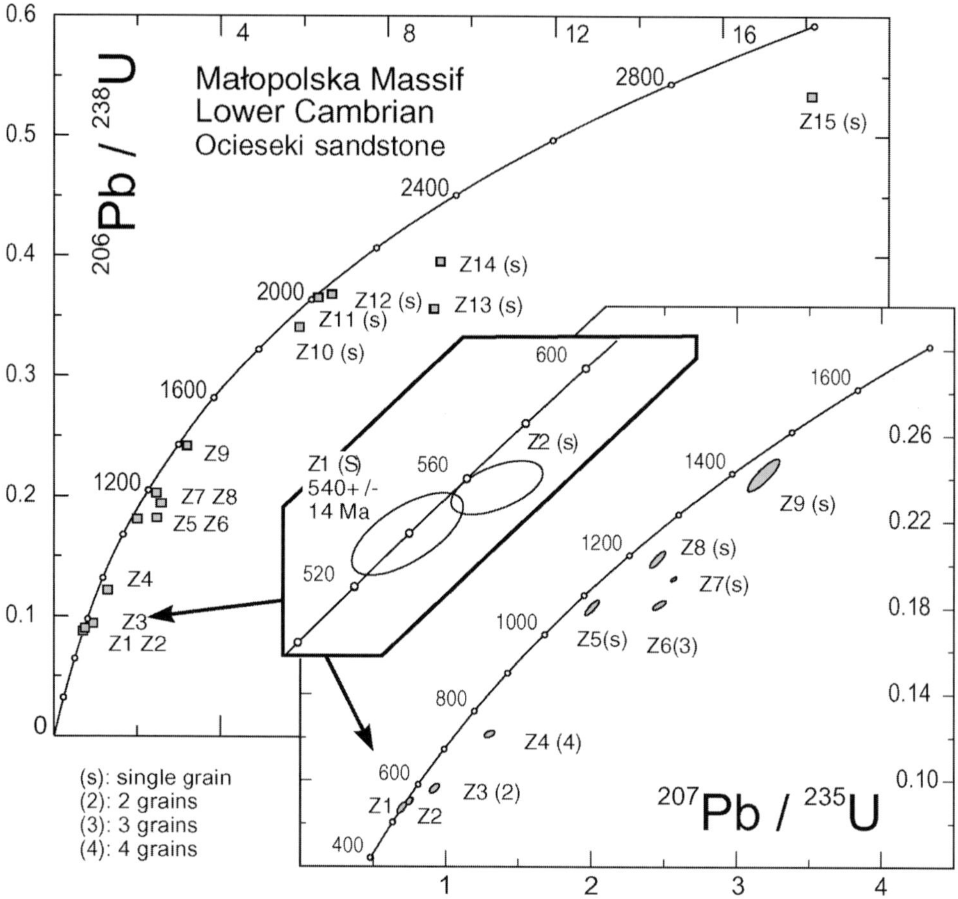

Fig. 4. U–Pb concordia diagram for the detrital zircon from the Ocieseki Sandstone Formation of Kędziorka (Lower Cambrian, Małopolska). Insets to view the relevant segments of the diagram.

except analysis Z3, Z4 and Z6. Zircon Z1 is concordant, and its U–Pb ages give an age of 540 ± 14 Ma (Fig. 4 and Table 2). This indicates, together with the analysis Z2, the presence of Late Proterozoic zircon component with an age close to the Precambrian–Cambrian boundary. Additionally, the concordant analysis Z1 provides a maximum age for the deposition of the Lower Cambrian Ocieseki Sandstone. The ^{207}Pb/^{206}Pb ages of analyses Z5 and Z8, which are 13 and 15% discordant respectively (Fig. 4), suggest the presence of a 1.2–1.37 Ga component, assuming that the discordance is exclusively due to recent Pb loss. Under the same assumption the ^{207}Pb/^{206}Pb analysis Z9 (13% discordant) suggests the presence of an approximately 1.56 Ga 'Gothian' zircon. To use the ^{207}Pb/^{206}Pb ages in this case might appear simplistic, but more complex interpretations should be considered. These three analyses, Z5, Z8, and Z9, plot left of the mixing line between 540 Ma and 2.0 Ga (not shown) and therefore cannot be mixtures of these two components. If these grains were partial mixtures of 540 Ma zircon overgrowth around older components, these components would have to be between 1.3 Ga and 1.6 Ga, namely the upper intercepts of the mixing lines between these points and point Z1. Also, even if they were mixtures of Early Proterozoic–Archean cores and Mid Proterozoic overgrowths, the lower intercepts of such hypothetical mixing lines would reinforce the presence of a 1.0–1.3 Ga 'Sveconorwegian-Grenvillian' zircon component. Therefore, in the case of these analyses, which are 15% discordant or less, we consider that the ^{207}Pb/^{206}Pb ages provide a reasonable estimate of the zircon age. Analyses Z3, Z4, Z6, and Z7 are between 37 and 29% discordant; the ^{207}Pb/^{206}Pb ages of these analyses suggest the presence of Middle

Table 2. *U–Pb data for detrital zircon from the Ocieseki Sandstone Formation (Lower Cambrian; Kędziorka, Małopolska Massif)*

Sample	Concentration				Measured*		Corrected atomic ratios†						Ages				(Ma)
	weight (mg)	U (ppm)	Pb_r (ppm)	Pb_c^t pgr	$^{206}Pb/^{204}Pb$	$^{208}Pb/^{206}Pb$	$^{206}Pb/^{238}U$	±2s	$^{207}Pb/^{235}U$	±2s	$^{207}Pb/^{206}Pb$	±2s	$^{206}Pb/^{238}U$	$^{207}Pb/^{235}U$	$^{207}Pb/^{206}Pb$		Discordia (%)
Z1 clr turbid rd 125	0.005	75	7.1	8	715	0.2366	0.0873	*21*	0.701	*27*	0.0582	*17*	540	539	537	±63	*−0.5*
Z2 clr rd w/ incl. 125	0.009	131	13.5	11	5499	0.2723	0.0902	*13*	0.753	*22*	0.0606	*14*	557	570	624	±52	*11.2*
Z3 2 clr rd frag. 80–60	0.008	55	6.3	19	192	0.4765	0.0962	*17*	0.924	*28*	0.0697	*16*	592	665	918	±49	*37.1*
Z4 4 prisms 60	0.007	94	12.8	20	341	0.3169	0.1218	*13*	1.302	*29*	0.0775	*15*	741	846	1135	±38	*36.7*
Z5 pink-grey rd 125	0.01	59	10.9	11	4852	0.1281	0.1803	*27*	2.005	*36*	0.08066	*75*	1068	1117	1213	±18	*12.9*
Z6 3 clr rd stub. pr. 80–60	0.007	69	13.8	24	294	0.304	0.1814	*18*	2.464	*35*	0.0985	*95*	1074	1261	1596	±18	*35.4*
Z7 pink rd 80–60	0.005	232	48.1	27	630	0.1968	0.19363	*81*	2.566	*17*	0.0961	*46*	1141	1291	1550	±9	*28.8*
Z8 pink rd 80–60	0.003	183	45.6	12	3223	0.3524	0.2024	*29*	2.451	*43*	0.08781	*88*	1188	1258	1378	±19	*15.0*
Z9 clr pink rd 125	0.018	16	4.7	15	891	0.3699	0.2418	*60*	3.181	*89*	0.0954	*12*	1396	1453	1563	±23	*13.1*
Z10 pink rd 80	0.005	109	41.6	18	1374	0.1999	0.3407	*34*	5.885	*64*	0.1253	*51*	1890	1959	2033	±7	*8.1*
Z11 pink rd 80–60	0.002	410	156	12	2535	0.1006	0.3654	*13*	6.334	*29*	0.1257	*35*	2008	2023	2039	±5	*1.8*
Z12 dark pink 80–60	0.003	180	76.8	14	2843	0.2249	0.3678	*30*	6.652	*64*	0.13116	*67*	2019	2066	2114	±9	*5.2*
Z13 clr subrd w/ incl. 80	0.002	192	79.8	19	1459	0.1901	0.3563	*12*	9.127	33	0.18579	*24*	1965	2351	2705	±2	*31.6*
Z14 pink 80	0.005	233	103	33	1203	0.1486	0.3956	*14*	9.265	35	0.16984	*20*	2149	2365	2556	±2	*18.7*
Z15 pink rd 80	0.008	75	48.8	14	4641	0.1785	0.5334	*28*	18.162	96	0.24695	*31*	2755	2998	3165	±2	*15.8*

All crystals are 5°, 1.6Å non-magnetic (Frantz) fraction. clr: clear; rd: rounded; subrd: subrounded; pr.: prism (1: 3 to 1: 5 width–length ratio); stub.: stubby (<1: 3 width–length ratio); w/: with; incl.: inclusions; 125, 125 μm; 80, 80 μm; 60, 60 μm. Error on the sample weight ± 0.003 mg. Estimated error on the concentrations given in ppm: 10–25%. Pb_r, radiogenic Pb; Pb_c^t, total common Pb. * Measured ratio corrected for blank and fractionation (1.12 ± 0.18 ‰ / a.m.u.). † Atomic ratios corrected for fractionation, spike (^{205}Pb–^{235}U spike), laboratory blanks (5–10 pg Pb and 1 pg U) and initial common Pb (Stacey and Kramers, 1975).

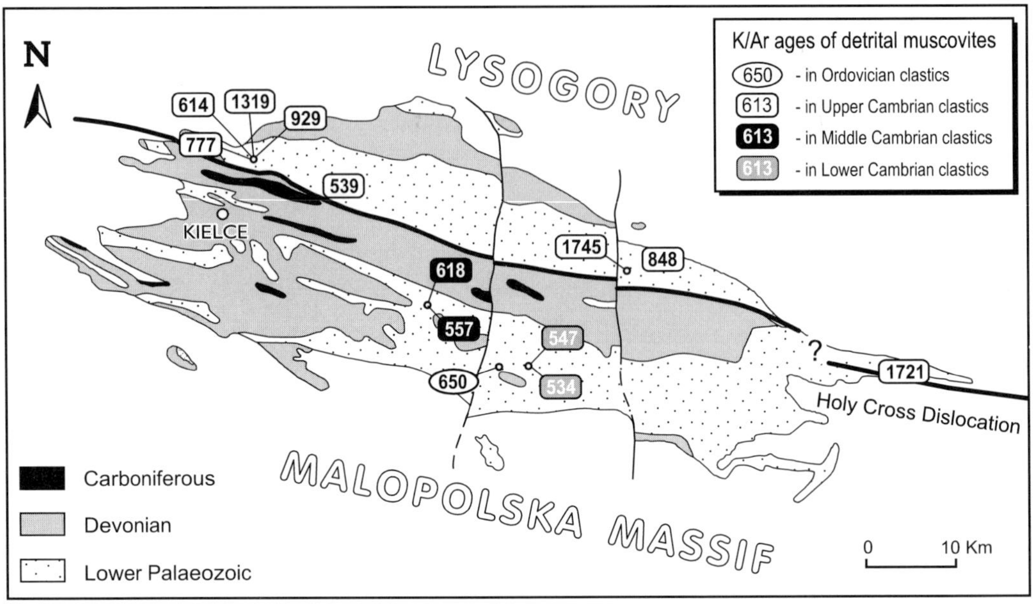

Fig. 5. K–Ar cooling ages of detrital muscovites in the Cambrian and Ordovician rocks of the Holy Cross Mountains (for more details on samples and localities, see Table 1 and Fig. 2).

Proterozoic (0.9–1.6 Ga) zircon, despite the high degree of uncertainty. But due to their position in the concordia plot (Fig. 4), analyses Z3, Z4, and Z6, which contain 2 to 4 zircon grains, could be mixtures of about 540 Ma and Early Proterozoic components.

The presence of Early Proterozoic zircon in the detritus is indicated by analyses Z10, Z11, and Z12. Analysis Z11 is subconcordant (1.8% discordant) and has a $^{207}Pb/^{206}Pb$ age of 2039 ± 5 Ma. This age coincides within error with the 2033 ± 7 Ma $^{207}Pb/^{206}Pb$ age of analysis Z10 (8% discordant). Analysis Z12 (5% discordant) has a $^{207}Pb/^{206}Pb$ age of 2114 ± 9 Ma. Analyses Z13 and Z14, despite their degree of discordance, indicate the presence of zircon with an Archaean age older than 2.5 Ga. Analysis Z15 has a $^{207}Pb/^{235}U$ age of 2998 Ma, which demonstrates that Archaean zircon older than 3.0 Ga was incorporated to the clastic component of the Lower Cambrian sandstones of the Małopolska Massif (Fig. 4).

In summary, the data set from the Lower Cambrian of Kędziorka provides evidence for detrital zircon with ages of about 540 Ma and 2.0 Ga, and suggests the presence of *c.* 1.2–1.37 Ga, 1.5 Ga, >2.5 Ga and > 3.0 Ga zircon (Fig. 3). This wide age spectrum correlates well with detrital zircon ages known from Neoproterozoic rocks of West Avalonia and with basement isotopic signatures of the Amazonian Craton (see for review Nance & Murphy 1996). A very similar geochronological fingerprint has also been recognized in the basement of the Brunovistulian (Friedl *et al.* 2000). Detrital zircon alone, however, cannot be an unequivocal argument for palaeogeographical correlations and distinction between Gondwanan and Baltic sources (Valverde-Vaquero *et al.* 2000). Nevertheless, the provenance of clastic material of the Lower Cambrian at Kędziorka from Cadomian sources is apparent. This is because the presence of detrital mica grains showing K–Ar cooling ages of about 535–545 Ma (Fig. 5 and Table 1) coincide well with the youngest population of detrital zircon and is only about 20–25 Ma older than the sedimentary age of the host rock. Therefore, muscovite grains were interpreted as representing a uniform population supplied from a single source region with a Cadomian imprint. The apparent absence of older detrital muscovite grains, equivalent to Mid Proterozoic to Archaean ages in the zircon population, suggests that an igneous–metamorphic overprint associated with the *c.* 540 Ma age has presumably obliterated any older muscovite ages. The age of 557 ± 12 Ma, recognized in the lower portion of the Middle Cambrian sequence of the Słowiec Hill (Fig. 5), suggests the same source. Sandstones at the top of the sequence, however, contain slightly older material, with a K–Ar cooling age of 618 ± 13 Ma. Using the multigrain analysis method, it is not possible to ascertain the integrity of this mica population; but its main component (if not all grains) constitutes material from a Cadomian source.

Belka *et al.* (2000) suggested a change in provenance from Cadomian to Baltic sources in the Małopolska Massif during the mid Cambrian, based on the assumption that the Pieprzowe Mountains belongs to the Małopolska Massif. Now, however, with the present interpretation in which the Pieprzowe Mountains are included in the Łysogóry Unit, there is no evidence for any other clastic material in the Cambrian of Małopolska except that from Cadomian sources. An input of older detritus appears to happen during the Ordovician. The age of 650 ± 18 Ma recognized in the Ashgill at Zalesie Nowe (Fig. 5) is presumably due to mixing of different muscovite populations. The dominant Cadomian component could be derived by recycling from the local Cambrian rocks. This scenario offers a convincing explanation why detrital mica grains are far scarcer and smaller in the Ordovician, rather than in the Cambrian clastic rocks of the Małopolska Massif.

The Łysogóry Unit

During the past few years, the first provenance studies have been carried out in the Cambrian of the Łysogóry Unit (Belka *et al.* 2000; Valverde-Vaquero *et al.* 2000). Detrital zircon and muscovite data revealed a complex provenance pattern and derivation of clastic material both from Cadomian and Baltic sources. The poor biostratigraphic resolution of the Cambrian succession only permitted a formulation of the accretionary history of the Łysogóry Unit in rough outline. However, Źylińska (2001, 2002) reinvestigated the Cambrian trilobite record in the Łysogóry Unit and Szczepanik (2001) used acritarchs to date Cambrian rocks precisely in several localities. A new interpretation has emerged from the re-evaluated stratigraphic framework and the available provenance data.

The oldest rocks in the Łysogóry Unit, the Upper Cambrian shales of the Gory Pieprzowe Formation, contain detrital muscovites with K–Ar cooling ages of about 1720–1740 Ma (Fig. 5). Their geochronological signatures and chemical composition, characterized by low Si, closely resemble those of mica grains recovered from the Middle Cambrian sandstones of the marginal part of the East European Platform (Belka *et al.* 2000). This detritus is therefore interpreted to have been derived from the Svecofennian basement of Baltica. Higher in the sequence, in the shallow-water Wisniowka Sandstone Formation, a change in provenance of clastic material can be observed. A strong variation in the ages of detrital muscovites suggests a mixing of different mica populations (Fig. 5). Belka *et al.* (2000) postulated a bimodal composition of detritus due to contribution from both Baltic and Cadomian sources. The input from a Cadomian source is clearly documented by a cooling age of 539 ± 15 Ma. The Baltic provenance, however, is still uncertain and needs to be confirmed by Ar/Ar single-grain dating. Valverde-Vaquero *et al.* (2000) investigated detrital zircon population associated with Cadomian mica grains at Wymyslona (Figs 2 and 3), and obtained geochronological signatures (*c.* 600 Ma, 1.8–2.1 Ma and >2.5 Ga) known from both Gondwanan and Baltic sources (e.g. Nance & Murphy 1996). However, the apparent absence of Sveconorwegian detritus (*c.* 1.0–1.2 Ga), which is a significant component of the Middle Cambrian clastics in certain areas of the EEC (Fig. 3), questions the derivation of the zircon detritus from Baltic sources (Valverde-Vaquero *et al.* 2000).

The Upper Silesian Massif

Abundant detrital muscovites in the Lower Cambrian sandstones of Upper Silesia show a tight range of K–Ar cooling ages from 542 Ma to 566 Ma, with approximately ± 12 Ma uncertainties (Belka *et al.* 2000). Although based only on five samples, the pattern seems to be regionally stable. The Cadomian detritus may have been derived from the crystalline basement of the southern part of the Brunovistulian unit, where cooling ages of 540–590 Ma are known (Dudek & Melkova 1975). In addition to concurrent ages, sedimentary trends and the distribution of the Cambrian rocks in the Brunovistulian unit provide supporting evidence for this derivation. However, microprobe analysis reveals that detrital mica grains were supplied from two sources with different cooling histories and petrological characteristics. Four samples, which gave ages from 542 Ma to 555 Ma, include detritus characterized by a low Si content; this population is derived from magmatic or low-pressure metamorphic rocks. In the sample with a slightly higher age of 566 ± 13 Ma a bimodal composition has been detected. The detrital mica characterized by a low Si content is associated with detritus from older high-pressure metamorphic rocks, thereby explaining the higher age obtained from the multigrain sample.

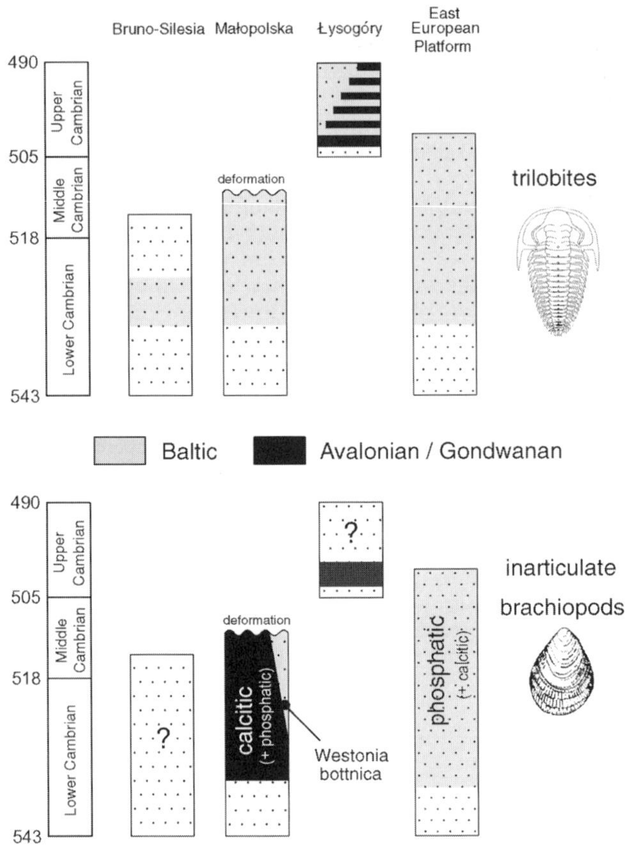

Fig. 6. Stratigraphic distribution and biogeographic affinity of the Cambrian trilobite and brachiopod faunas of Poland. Columns indicate the stratigraphic ranges of the Cambrian sequences in the different crustal units. Note the predominant phosphatic shell mineralogy of inarticulate brachiopods on the shelf of Baltica (EEP) and the calcitic one in the Cambrian of Małopolska. Brachiopod data are after Jendryka-Fugelwicz (1992, 1998); data on biogeographical affinity of trilobites in Łysogóry are after Żylińska (2001).

Biogeographical data

The Cambrian trilobites and brachiopod fauna display a peculiar composition in southern Poland. This is because taxa typical of both a Baltic zoogeographical province and a Peri-Gondwanan affinity are present. In the past, the records of trilobites diagnostic of the Baltic realm in Małopolska and Upper Silesia (Orłowski 1975, 1985) were considered to be conclusive proof that these crustal blocks together with Łysogóry were integral parts of the palaeocontinent of Baltica (e.g. Bergström 1984; Vidal & Moczydlowska 1995). A short summary of the Cambrian faunal data and their biogeographical significance has been given by Belka *et al.* (2000). They showed that the seemingly incompatible faunal records are results of very complex, individual evolution of the particular terranes during their amalgamation. Figure 6 summarizes the occurrence and biogeographical affinities of trilobites and inarticulate brachiopods in the Cambrian of Poland.

In terms of biogeography, the most enigmatic block is certainly the Łysogóry Unit. Although now located close to, or even in contact with the East European Craton, it yields Cambrian fossils, most of which are unknown from Baltica. The famous assemblage of trilobite trace fossils described in several papers (e.g. Radwański & Roniewicz 1963) is identical to those distributed throughout Gondwana and the Peri-Gondwanan microplates (Seilacher 1983, and pers. comm.). Żylińska (2001, 2002) reinvestigated the Cambrian trilobite fauna of the Łysogóry Unit and revealed a specific faunal succession. She also confirmed the Late Cambrian age of the entire succession and found no evidence for the previously suggested presence of the Middle Cambrian in the Łysogóry Unit (e.g. Orłowski 1992). The oldest trilobites from the *Olenus* and *Agnostus* zones are very rare and represented by forms showing an Avalonian affinity (Fig. 6). The inarticulate brachiopods also show a strong link to Avalonian faunas (Jendryka-Fuglewicz pers. comm.). Higher in the section, the trilobite fauna is predominantly composed of immigrant species both from Avalonia and Baltica, which are associated with a small number of endemic taxa. Of great importance for the palaeogeography is the presence of a few 'exotic' forms that are identical to those described from the Cambrian of South America (Żylińska 2001, 2002). The Ordovician fauna of the Łysogóry Unit generally has a Baltic character, but in the Caradoc sediments some chitinozoan taxa typical of Gondwana are still present (Wrona pers. comm.).

The Cambrian sediments of the Małopolska Massif contain a specific faunal succession. Baltic olenellid trilobites occur in the Lower Cambrian (Orłowski 1985). The fauna is distinct and endemic at species level. In contrast, the Early Cambrian inarticulate brachiopods show Avalonian affinities. There is only one Baltic species, *Westonia bottnica,* present in this Avalonian assemblage (Fig. 6), a fauna dominated by forms with calcitic shells, in contrast to the phosphatic mineralogy of most Baltic brachiopods. During the mid-Cambrian, a progressive migration of Baltic brachiopods to the Małopolska area has been recorded (Jendryka-Fuglewicz 1998). The Ordovician faunas, which are perfectly documented in the southern part of the Holy Cross Mountains (Dzik *et al.* 1994), belong essentially to the Baltic province (Cocks & Fortey 1998), despite enlarged endemicity of ostracods (Olempska 1994) and some links to other continents amongst the conodonts (Dzik 1989).

There are only a few records of Early Palaeozoic benthic fossils in Upper Silesia. Orłowski (1975) reported the occurrence of the Early Cambrian trilobites typical of the Baltic realm. Unfortunately, the associated brachiopod fauna has not been described until now. The conodont fauna recovered from the Middle Ordovician clastic rocks suggest, as do Cambrian trilobites, that the area was positioned within the Baltic zoogeographic province.

Accretionary history

Modern Europe is a complex mosaic of crustal fragments that once constituted parts of three palaeocontinents Laurentia, Baltica, and Gondwana. These continents originated as result of the break-up of the Neoproterozoic supercontinent Rodinia between about 725–750 Ma and 550–590 Ma (Bond *et al.* 1984; Dalziel 1992; Powell *et al.* 1993; Storey 1993; Soper 1994). Their Vendian and Cambrian drift history is poorly documented, but it is widely accepted that Laurentia, Baltica and Gondwana remained separated by oceans at least until late Ordovician time (e.g. Torsvik *et al.* 1996). The existing data prove that Gondwana underwent extensive rifting along its Cadomian margin during the Ordovician, and several terranes became detached and drifted away. In this concept Avalonia is traditionally treated as the first microplate that rifted away from Gondwana, crossed the Iapetus Ocean and amalgamated with Laurentia and Baltica (e.g. Torsvik & Trench 1991; Tait *et al.* 2000). However, recently presented data from southern Poland (Belka *et al.* 2000; Valverde-Vaquero *et al.* 2000) revealed the presence of terranes with Cadomian/Gondwanan linkages already close to the Baltica margin during the Cambrian. Understaning the complex history of these blocks, i.e. the Łysogóry, the Małopolska and the Bruno-Silesia, is difficult because they are not entirely suspect in regard to their palaeocontinent of Baltica. Moreover, their position in relation to each other and to Baltica during the Cambrian was fundamentally different from that of today. Although neither the time nor the place of the dispersion can be defined with great precision at present, their detachment seems to be related to the final break-up of Rodinia, when Laurentia and Baltica were separated from Amazonia and Africa.

A major conclusion of the recent provenance studies (Belka *et al.* 2000; Friedl *et al.* 2000; Valverde-Vaquero *et al.* 2000) and of the work reported here is that Małopolska, Łysogóry and Bruno-Silesia, although yielding faunal linkages to Baltica, are exotic terranes derived from the Gondwana margin and not displaced fragments of the Baltic crust. Their apparent faunal linkages to Baltica during the Cambrian are because they were situated closer to Baltica than any other part of Gondwana at that time (Fig. 7). Moreover, these blocks have individual and disparate drift-histories.

Zircon provenance data presented here suggest that Małopolska most probably derived from the segment of the Cadomian orogen that bordered the Amazonian Craton, i.e. it was originally positioned in the immediate vicinity of the western termination of Avalonia. Both palaeontological and provenance evidence show that Małopolska was already separated from the Gondwana margin in Early Cambrian time,

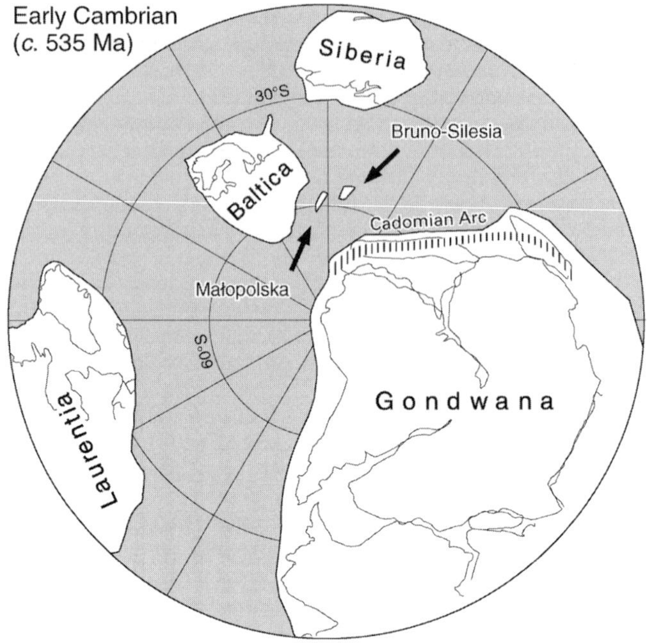

Fig. 7. Early Cambrian (*c.* 535 Ma) palaeogeographic reconstruction to show the position of Małopolska and Bruno-Silesia in relation to Baltica and Gondwana. The general positions of Laurentia, Siberia, and Gondwana are taken from the reconstruction presented by Torsvik & Rehnström (2001) with exception of Baltica, which is not faced with the Teisseyre–Tornquist margin to the north but to the west. This position of Baltica is favoured because it is much more compatible with facies and faunal trends.

some 40–50 Ma before Avalonia became detached and started its rapid northward drift. The coexistence of Baltic trilobites and inarticulate brachiopods with Avalonian affinity emphasizes the 'midway' position of Małopolska, between Baltica and the Cadomian margin of Gondwana (Fig. 7). The calcitic mineralogy of brachiopods points additionally to ecological conditions (temperature?) similar to those of the Avalonian segment of Gondwana but different from those on the Tornquist margin of Baltica. The fact that Małopolska shared the Baltic province trilobites at the same time allows to draw the following inferences: (1) The distance of Małopolska to Baltica was presumably smaller than to Gondwana in the Early Cambrian; (2) Cambrian trilobites were less sensitive to ecological factors than inarticulate brachiopods; and (3) Ecological gradients or unfavourable sea-water circulation hindered the dispersion of brachiopod larvae in the strait between Małopolska and the Baltica margin.

As Małopolska progressively approached the Baltica margin during the mid-Cambrian, increasing numbers of Baltic brachiopod taxa were able to reach the shallow-water realm of Małopolska. Finally, continuing convergence caused the closure of the strait and Małopolska collided with the Baltica margin. This tectonic event, which took place between late mid-Cambrian time and late Tremadocian time, was presumably an oblique collision which resulted in deformation of Cambrian and Precambrian rocks ('Sandomierz Phase') and in tilting of the entire Małopolska crustal block. There is no evidence for any magmatic activity or other significant thermal events at that time.

Unfortunately, the provenance data obtained from sediments deposited before and after the accretion provide no specific information to constrain the place where Małopolska joined the Baltica margin. Structural and some palaeomagnetic data (Lewandowski 1993; Mizerski 1995) point to strike-slip displacement of Małopolska along the Tornquist margin during the Early Palaeozoic. This implies a more southeasterly accretion of Małopolska than its present position in relation to the EEC. Recent palaeomagnetic data (Nawrocki 1999) and the facies pattern of the Silurian rocks in the Holy Cross Mountains testify to amalgamation of Małopolska with Łysogóry during late Silurian.

The ancestry of Bruno-Silesia is well constrained by geochronological and provenance data (e.g. Finger *et al.* 2000 and references therein; Friedl *et al.* 2000), which point to its original location within the Cadomian orogen of Gondwana, close to the position of West Avalonia. However, similarities in Precambrian crustal evolution and also the analogous position within the Variscan Belt (Avalonia and Bruno-Silesia, both formed the southern passive continental margin of Euramerica), do not mean that Bruno-Silesia is a part of Avalonia, as has been considered in the past (e.g. Moczydlowska 1997). Although Bruno-Silesia and Avalonia share the same geotectonic derivation, they have disparate drift-histories. During the Early Cambrian the former was already separated from the Gondwanan margin and close to Baltica. Except for the single record of Baltic conodonts in the Ordovician of Upper Silesia there is no other evidence for the palaeogeographical association of Bruno-Silesia after post-middle Cambrian and pre-early Devonian times. Because of contrast with the adjacent Małopolska Massif in pre-Devonian stratigraphy and facies development, Belka *et al.* (2000) suggests an amalgamation of Bruno-Silesia with Małopolska during the Early Devonian.

In contrast with Małopolska and Bruno-Silesia, the origin of the Łysogóry block is enigmatic. Clear faunal and provenance linkages to Avalonian and Armorican terranes during the Late Cambrian provide strong arguments in favour of a Peri-Gondwanan derivation. Łysogóry does not constitute the easternmost prolongation of East Avalonia in central Europe; it was located close to Baltica during the Late Cambrian, i.e. for a long time before Avalonia became detached from thc Gondwana margin and began its rapid northward drift. Provenance information from the Ordovician and the Silurian successions of Łysogóry are still needed to constrain the place and timing of its accretion.

In summary, the available stratigraphic, palaeomagnetic, provenance, and biogeographical data from southern and central Poland lead to the conclusion that Małopolska was the first terrane to join the Baltica palaeocontinent. Its accretion commenced the progressive crustal growth continued by subsequent events, chiefly by the Caledonian, Variscan and Alpine orogenies, which have led to formation of the modern European lithosphere.

This research has been carried out within the PACE TMR Network of the European Union. Financial support for this study was also provided by the German Research Council (DFG), grant Be 1296/5–3 and is greatly appreciated. This is a contribution to the Special Research Programme 'Orogenic Processes' funded by the DFG. We thank R. Dadlez, Ph. Matte and J. Verniers for their comments that improved the manuscript.

References

BELKA, Z. 1993. Remarks on thermal maturity level in the subsurface of the Upper Silesian Coal Basin. *Acta Geologica Polonica*, **43**, 95–101.

BELKA, Z. & SIEWNIAK, A. 1996. Thermal maturation of the Lower Palaeozoic strata in the southwestern margin of the Małopolska Massif, southern Poland: no evidence for Caledonian regional metamorphism. *Geologische Rundschau*, **85**, 775–781.

BELKA, Z., AHRENDT, H., FRANKE, W. & WEMMER, K. 2000. The Baltica-Gondwana suture in central Europe: evidence from K/Ar ages of detrital muscovites. *In*: FRANKE, W., ALTHERR, R., HAAK, V. & ONCKEN, O. (eds) *Orogenic processes: Quantification and modelling in the Variscan Belt of central Europe*, Geological Society, London, Special Publications, **179**, 87–102.

BERGSTRÖM, J. 1984. Strike-slip faulting and Cambrian biogeography around the Tornquist Zone. *Geologiska Föreningens i Stockholm Förhandlingar*, **106**, 382–283.

BERTHELSEN, A. 1993. Where different geological philosophies meet: the Trans-European Suture Zone. *Publications of the Institute of Geophysics, Polish Academy of Sciences*, **A20** (255), 19–31.

BOND, G. C., NICKESON, P. A. & KOMINZ, M. A. 1984. Breakup of a supercontinent between 625 Ma and 555 Ma: New evidence and implications for continental histories. *Earth and Planetary Science Letters*, **70**, 325–345.

BUŁA, Z., JACHOWICZ, M. & ŹABA, J. 1997. Principal characteristic of the Upper Silesian Block and Małopolska Block border zone (southern Poland). *Geological Magazine*, **134**, 669–677.

COCKS, L. R. M. & FORTEY, R. A. 1998. The Lower Palaeozoic margins of Baltica. *Geologiska Föreningens Forhandligar*, **120**, 173–179.

COMPSTON, W, SAMBRIDGE, M. S., REINFRANK, R. F., MOCZYDLOWSKA, M., VIDAL, G. & CLAESSON, S. 1995. Numerical ages of volcanic rocks and the earliest faunal zone within the Late Precambrian of east Poland. *Journal of the Geological Society, London*, **152**, 599–611.

DADLEZ, R. 1995. Debates about the pre-Variscan tectonics of Poland. *Studia geophysica et geodaetica*, **39**, 227–234.

DADLEZ, R. 1997. Seismic profile LT-7 (northwest Poland): geological implications. *Geological Magazine*, **134**, 653–659.

DADLEZ, R. 2001. Holy Cross Mts area – crustal structure, geophysical data and general geology. *Geological Quarterly*, **45**, 99–106.

DALZIEL, I. W. D. 1992. On the organization of American plates in the Neoproterozoic and the breakout of Laurentia. *Geological Society of America Today*, **2**, 237–241.

DUDEK, A. 1980. The crystalline basement block of

the Outer Carpathians in Moravia: Bruno-Vistulicum. *Rozprawy Ceskoslovenské Akademie Ved*, **90**, 1–85.

DUDEK, A. 1995. Moravo-Silesian Zone – metamorphic evolution. In: DALLMEYER, R. D., FRANKE, W. & WEBER, K. (eds) *Pre-Permian geology of Central and Eastern Europe*. Springer-Verlag, Berlin, Heidelberg. 508–511.

DUDEK, A. & MELKOVA, J. 1975. Radiometric age determination in the crystalline basement of the Carpathian Foredeep and of the Moravian Flysch. *Vestnik ustredniho ustavu geologickeho*, **50,** 257–264.

DZIK, J. 1989. Conodont evolution in high latitudes of the Ordovician. *Courier Forschungsinstitut Senckenberg*, **117**, 1–28.

DZIK, J. 1999. The Ordovician in the Holy Cross Mountains. *In*: *International Symposium on the Ordovician System, Excursion guide Poland and Germany*, 3–7.

DZIK, J., OLEMPSKA, E. & PISERA, A. 1994. Ordovician carbonate platform ecosystem of the Holy Cross Mountains. *Palaeontologia Polonica*, **53**, 1–315.

FINGER, F., HANZL, P., PIN, C., VON QUADT, A. & STEYRER, H. P. 2000. The Brunovistulian: Precambrian sequence at the eastern end of the Central European Variscides? *In*: FRANKE, W., ALTHERR, R., HAAK, V. & ONCKEN, O. (eds) *Orogenic processes – Quantification and modelling in the Variscan Belt of central Europe*, Geological Society, London, Special Publications, **179**, 103–112.

FRANKE, D. 1995. Caledonian terranes along the southwestern border of the East European Platform – evidence, speculation and open questions. *Studia geophysica et geodaetica*, **39**, 241–256.

FRANKE, W. 2000. The mid-European segment of the Variscides: tectonostratigraphic units, terrane boundaries and plate tectonic evolution. *In*: FRANKE, W., ALTHERR, R., HAAK, V. & ONCKEN, O. (eds) *Orogenic processes; Quantification and modelling in the Variscan Belt of central Europe*, Geological Society, London, Special Publications, **179**, 35–61.

FRANKE, W. & ŻELAŹNIEWICZ, A. 2000. The eastern termination of the Variscides: terrane correlation and kinematic evolution. *In*: FRANKE, W., ALTHERR, R., HAAK, V. & ONCKEN, O. (eds) *Orogenic processes: Quantification and modelling in the Variscan Belt of central Europe*, Geological Society, London, Special Publications, **179**, 63–86.

FRIEDL, G., FINGER, F., MCNAUGHTON, N. J. & FLETCHER, I. R. 2000. Deducing the ancestry of terranes: SHRIMP evidence for South America-derived Gondwana fragments in central Europe. *Geology*, **28**, 1035–1038.

FROST, R. T. C., FITCH, F. J. & MILLER, J. A. 1982. The age and nature of the crystalline basement of the North Sea Basin. *In*: ILLING, L. V. & HOBSON, G. D. (eds) *Petroleum Geology of the Continental Shelf of North-West Europe*, 43–57.

FUHRMANN, U., LIPPOLT, H. J. & HESS, J. C. 1987. Examination of some proposed K-Ar standards: $^{40}Ar/^{39}Ar$ analyses and conventional K-Ar-data. *Chemical Geology*, **66,** 41–51.

GIESE, U., KATZUNG, G., WALTER, R. & WEBER, J. 1997. The Caledonian deformation of the Brabant Massif and the Early Palaeozoic in northeast Germany: compared. *Geological Magazine*, **134**, 637–652.

HARAŃCZYK, C. 1994. Significance of Zawiercie-Rzeszotary terrane suture for recognition of Caledonian transpressive orogen of the Krakowidy. *Field Guide, 65th Annual Meeting, Geological Society of Poland*, 69–79.

HEINRICHS, H. & HERRMANN, A.G. 1990. *Praktikum der Analytischen Geochemie,* Springer, Berlin.

HORSTMANN, U. E. 1987. Die metamorphe Entwicklung im Damara Orogen, Südwestafrika/Namibia, abgeleitet aus K/Ar-Datierungen an detritischen Hellglimmern aus Molassesedimenten der Nama Group. *Göttinger Arbeiten zur Geologie und Paläontologie*, **32**, 1–95.

JAFFEY, A. H., FLYNN, K. F., GLENDENIN, L. E., BENTLEY, W. C. & ESSLING, A. M. 1971. Precision measurement of half-lives and specific activities of ^{235}U and ^{238}U. *Physics Reviews Section C, Nuclear Physics*, **4**, 1889–1906.

JENDRYKA-FUGLEWICZ, B. 1992. Analiza porownawcza ramienionogow z utworow kambru Gor Swietokrzyskich i platformy prekambryjskiej w Polsce. *Przeglad Geologiczny*, **467**, 150–155.

JENDRYKA-FUGLEWICZ, B. 1998. Kambryjska eksplozja zycia. Najstarsze zespoly brachiopodow w profilach geologicznych Polski. *Abstracts of the XVI Palaeontological Meeting*, Wiktorowo, 18–19.

KOWALCZEWSKI, Z. 1990. Grubookruchowe skaly kambru na srodkowym poludniu Polski. *Prace Panstwowego Instytutu Geologicznego*, **131**, 1–82.

KOWALCZEWSKI, Z. 1995. Fundamental stratigraphic problems of the Cambrian in the Holy Cross Mts. *Geological Quarterly*, **39**, 449–470.

KROGH, T. E. 1973. A low-contamination method for hydrothermal decomposition of zircon and extraction of U and Pb for isotopic age determinations. *Geochimica Cosmochimica Acta*, **37**, 485–494.

KROGH, T. E. 1982. Improved accuracy of U-Pb zircon ages by the creation of more concordant systems using an air abrasion technique. *Geochimica Cosmochimica Acta*, **46**, 637–649.

LEWANDOWSKI, M. 1993. Paleomagnetism of the Paleozoic rocks of the Holy Cross Mts (Central Poland) and the origin of the Variscan orogen. *Publications of the Institute of Geophysics, Polish Academy of Sciences*, **A23** (265), 3–85.

LEWANDOWSKI, M. 1995. Palaeomagnetic constraints for Variscan mobilism of the Upper Silesian and Małopolska Massifs, southern Poland – reply. *Geological Quarterly*, **39**, 283–292.

LUDWIG, K.R. 1991. PBDAT version 1.23.

MATTE, P. H., MALUSKI, H., RAJLICH, P. & FRANKE, W. 1990. Terrane boundaries in the Bohemian Massif: result of large scale Variscan shearing. *Tectonophysics*, **177**, 151-170.

MICHNIAK, R. 1969. Petrography of the Late Precambrian (Riphean) and Cambrian of the eastern part of the Holy Cross Mountains. *Studia Geologica Polonica*, **30,** 1–106.

MIZERSKI, W. 1995. Geotectonic evolution of the Holy Cross Mts in central Europe. *Biuletyn Panstwowego Instytutu Geologicznego*, **372**, 5–47.

MOCZYDLOWSKA, M. 1997. Proterozoic and Cambrian successions in Upper Silesia: an Avalonian terrane in southern Poland. *Geological Magazine*, **134**, 679–689.

NANCE, R. D. & MURPHY, J. B. 1996. Basement isotopic signatures and the Neoproterozoic palaeogeography of Avalonian-Cadomian and related terranes in the circum- North Atlantic. *In*: NANCE, R. D. and THOMPSON, M. D. (eds) Avalonian and Related Peri-Gondwanan Terranes of the Circum-North Atlantic. *Geological Society of America, Special Paper*, **304**, 333–346.

NAWROCKI, J. 1995. Palaeomagnetic constraints for Variscan mobilism of the Upper Silesian and Małopolska Massifs, southern Poland – discussion. *Geological Quarterly*, **39**, 271–282.

NAWROCKI, J. 1999. Prefolding remanent magnetization of diabase intrusion of the Bardo syncline in the Holy Cross Mountains (central Poland). *Przeglad Geologiczny*, **47**, 1101–1104.

OLEMPSKA, E. 1994. Ostracods of the Mojcza Limestone. *Palaeontologia Polonica*, **53**, 129–212.

ORŁOWSKI, S. 1975. Lower Cambrian trilobites from Upper Silesia (Goczalkowice borehole). *Acta Geologica Polonica*, **25**, 377–383.

ORŁOWSKI, S. 1985. Lower Cambrian and its trilobites in the Holy Cross Mts. *Acta Geologica Polonica*, **35**, 231–250.

ORŁOWSKI, S. 1992. Cambrian stratigraphy and stage subdivision in the Holy Cross Mountains, Poland. *Geological Magazine,* **129**, 471–474.

PHARAOH, T.C. 1999. Palaeozoic terranes and their lithospheric boundaries within the Trans-European Suture Zone (TESZ): a review. *Tectonophysics*, **314**, 17–41.

POWELL, C. MCA., LI, Z. X., MCELHINNY, M. W., MEERT, J. G. & PARK, J. K. 1993. Palaeomagnetic constraints on timing of the Neoproterozoic breakup of Rodinia and the Cambrian formation of Gondwana. *Geology*, **21**, 889–892.

POŻARYSKI, W., GROCHOLSKI, A., TOMCZYK, H., KARNKOWSKI, P. & MORYC, W. 1992. The tectonic map of Poland in the Variscan epoch. *Przegląd Geologiczny*, **40**, 643–651.

PRZYBYLOWICZ, T. & STUPNICKA, E. 1991. Manifestation of volcanism in Ordovician and Silurian of the southern part of Swietokrzyskie Mountains. *Archiwum Mineralogiczne*, **47**, 137–154.

RADWAŃSKI, A. & RONIEWICZ, P. 1960. Ripple marks and other sedimentary structures of the Upper Cambrian at Wielka Wisniowka (Holy Cross Mountains). *Acta Geologica Polonica*, **10**, 371–399.

RADWAŃSKI, A. & RONIEWICZ, P. 1963. Upper Cambrian trilobite ichnocoenosis from Wielka Wisniowka (Holy Cross Mountains, Poland). *Acta Palaeontologica Polonica*, **8,** 259-280.

SCHUMACHER, E. 1975. Herstellung von 99.9997% ^{38}Ar für die ^{40}K/^{40}Ar Geochronologie. *Geochronologia Chimia*, **24,** 441–442.

SEILACHER, A. 1983. Upper Paleozoic trace fossils from the Gilf Kebir-Abu Ras area in southwestern Egypt. *Journal of African Earth Sciences*, **1**, 21–34.

SEMENOV, V. Y., JANKOWSKI, J., ERNST, T., JOZWIAK, W., PAWLISZYN, J. & LEWANDOWSKI, M. 1998. Electromagnetic soundings across the Holy Cross Mountains, Poland. *Acta Geophysica Polonica*, **46**, 171–185.

SOPER, N. J. 1994. Neoproterozoic sedimentation on the northeast margin of Laurentia and the opening of Iapetus. *Geological Magazine*, **131**, 291–299.

STACEY, J. S. & KRAMERS, J. D. 1975. Approximation of terrestrial lead isotope evolution by a two stage model. *Earth and Planetary Science Letters*, **6**, 15–25.

STEIGER, R. H. & JÄGER, E. 1977. Subcommission on Geochronology: Convention on the use of decay constants in geo- and cosmochronology. *Earth and Planetary Science Letters*, **36**, 359–362.

STOREY, B. B. 1993. The changing face of the Precambrian and early Palaeozoic reconstructions. *Journal of the Geological Society, London, Special Paper*, 665–668.

SZCZEPANIK, Z. 1997. Preliminary results of thermal alteration investigations of the Cambrian acritarchs in the Holy Cross Mts. *Geological Quarterly*, **41**, 257–264.

SZCZEPANIK, Z. 2001. Acritarchs from Cambrian deposits of the southern part of the Łysogóry unit in the Holy Cross Mountains, Poland. *Geological Quarterly*, **45**, 117-130.

SZULCZEWSKI 1995. Depositional evolution of the Holy Cross Mts. (Poland) in the Devonian and Carboniferous – a review. *Geological Quarterly*, **39**, 471-488.

TAIT, J., SCHAETZ, M., BACHTADSE, V. & SOFFEL, H. 2000. Palaeomagnetism and Palaeozoic palaeogeography of Gondwana and European terranes. *In*: FRANKE, W., ALTHERR, R., HAAK, V. & ONCKEN, O. (eds) *Orogenic processes: Quantification and modelling in the Variscan Belt of central Europe*, Geological Society, London, Special Publications, **179**, 21–34.

TOMCZYK, H. 1970. Silurian. *In:* Geology of Poland, vol. 1, Stratigraphy: part 1: Precambrian and Palaeozoic, 237–319.

TORSVIK, T. H. & REHNSTRÖM, E.F. 2001. Cambrian palaeomagnetic data from Baltica: implications for true polar wander and Cambrian palaeogeography. *Journal of the Geological Society, London*, **158**, 321–329.

TORSVIK, T. H. & TRENCH, A. 1991. The Ordovician history of the Iapetus Ocean in Britain: New palaeomagnetic constraints. *Journal of the Geological Society, London*, **148**, 423–425.

TORSVIK, T. H., SMETHURST, M. A., MEERT, J. G., VAN DER VOO, R., MCKERROW, W. S., BRASIER, M. D., STURT, B. A. & WALDERHAUG, H. J. 1996. Continental break-up and collision in the Neoproterozoic and Palaeozoic – a tale of Baltica and Laurentia. *Earth Science Reviews*, **40**, 229–258.

UNRUG, R., HARAŃCZYK, C. & CHOCYK-JAMINSKA, M.

1999. Easternmost Avalonian and Armorican-Cadomian terranes of central Europe and Caledonian-Variscan evolution of the polydeformed Krakow mobile belt: geological constraints. *Tectonophysics*, **302**, 133–157.

VALVERDE-VAQUERO, P., DÖRR, W., BELKA, Z., FRANKE, W., WISZNIEWSKA, J. & SCHASTOK, J. 2000. U-Pb single-grain dating of detrital zircon in the Cambrian of central Poland: implications for Gondwana versus Baltica provenance studies. *Earth and Planetary Science Letters*, **184**, 225–240.

VIDAL, G. & MOCZYDLOWSKA, M. 1995. The Neoproterozoic of Baltica – stratigraphy, palaeobiology and general geological evolution. *Precambrian Research*, **73**, 197–216.

ŻABA, J. 1999. Structural evolution of Lower Palaeozoic deposits in the boundary zone of the Upper Silesia and Małopolska Blocks. *Prace Panstwowego Instytutu Geologicznego*, **166**, 1–162.

ŻYLIŃSKA, A. 2001. Late Cambrian trilobites from the Holy Cross Mountains, Central Poland. *Acta Geologica Polonica*, **51**, 333–383.

ŻYLIŃSKA, A. 2002. Stratigraphic and biogeographic significance of Late Cambrian trilobites from the Holy Cross Mountains. *Acta Geologica Polonica*, **52**, 217–238.

Key Lower Palaeozoic faunas from near the Trans-European Suture Zone

L. ROBIN M. COCKS

Department of Palaeontology, The Natural History Museum, Cromwell Road, London SW7 5BD, UK

(e-mail: r.cocks@nhm.ac.uk)

Abstract: Following recognition of the Vendian to mid-Ordovician rotation of Baltica, with more than 55° of that rotation occurring in the Upper Cambrian and Lower Ordovician, the Tornquist Margin of Baltica must have faced northwards towards Laurentia and the Panthalassic Ocean, rather than, as now, southwestwards towards Gondwana (including Avalonia). Unequivocally Baltic endemic trilobite, brachiopod and other faunas are known from both the Cambrian and the Ordovician of the Holy Cross Mountains, Poland, and from both parts of them, i.e. the Małopolska Block and the Łysogóry Block. Whether or not these two blocks were united into a single terrane or were separate as two terranes is equivocal from the faunal evidence, and there is no faunal evidence of substantial strike-slip faulting of the blocks in relation to the main Baltic craton: they are perceived as having made up part of the margin of Baltica itself. However, both Holy Cross Mountain blocks were different and palaeogeographically separate from the Bruno-Silesian Block, whose continental origins are yet to be finally determined. The Ordovician clastic sediments at both Rügen, north Germany, and Pomerania, NW Poland, have yielded no macrofossils other than graptolites, but microfossils (acritarchs and chitinozoa) are interpreted as having been deposited at relatively high palaeolatitudes, i.e. at a higher palaeolatitude than Baltica, and may have been deposited in an ocean basin within the Tornquist Ocean between Baltica and Avalonia.

Today the Trans-European Suture Zone (TESZ) is the most fundamental structural line in Europe, stretching as it does from a triple junction in the North Sea 300 km east of Aberdeen, through southern Denmark, northern Germany, across Poland and on past the Carpathian Front to the Black Sea. In general the TESZ marks the sutures between the substantial Lower Palaeozoic continent of Baltica (sometimes termed the East European Craton) to the NE and the various fragments which once formed part of Gondwana to the SW (Cocks 2000). However, due to various post-Lower Palaeozoic tectonics, there are fragments of Baltica to the SW of the TESZ. Chief among them are the two or more areas, which today make up the core of the Holy Cross Mountains of Poland and the adjacent Bruno-Silesian Block (Fig. 1). There is also an equivocal area which underlies Rügen, a German island in the southern Baltic Sea, and some new evidence from adjacent Pomerania in NW Poland. The terminology of the various blocks bordering on the TESZ largely follows Winchester *et al.* (2002). Figure 2 shows the outline of Lower Palaeozoic Baltica on modern geography: it differs from that originally postulated by Cocks & Fortey (1998) in two ways. Firstly, the Taimyr peninsula of northern Siberia (the unnumbered star on Fig. 2) was previously shown as forming part of Baltica; it is now known, both from the faunas (Rushton *et al.* 2002; Fortey & Cocks 2002) and from the tectonics and palaeomagnetism (Torsvik & Rehnström 2001), that the northern part of Taimyr, together with Severnaya Zemlya, formed a separate Kara Terrane, whereas the central and southern areas of Taimyr formed an integral part of the main palaeocontinent of Siberia. Secondly, the present SE part of Baltica, near the Caspian Sea, has a revised margin which follows the terrane disposition shown in Cocks (2000, fig. 6).

To understand the geography of the TESZ margin (termed the Tornquist Margin by Torsvik & Rehnström 2001) of Baltica in the Lower Palaeozoic, it is necessary to appreciate that the whole palaeocontinent rotated by over 100 degrees between the Vendian and the mid-Ordovician (Torsvik & Rehnström 2001 and references therein). In particular, 55 degrees of that rotation occurred in the late Cambrian and early Ordovician. Thus in Cambrian times the Tornquist Margin did not face Gondwana but northwards towards the vast Panthalassic Ocean. Therefore any peri-Baltic terranes which existed on the Tornquist Margin, if any, would

From: WINCHESTER, J. A., PHARAOH, T. C. & VERNIERS, J. 2002. *Palaeozoic Amalgamation of Central Europe.* Geological Society, London, Special Publications, **201**, 37–46. 0305-8719/02/$15.00

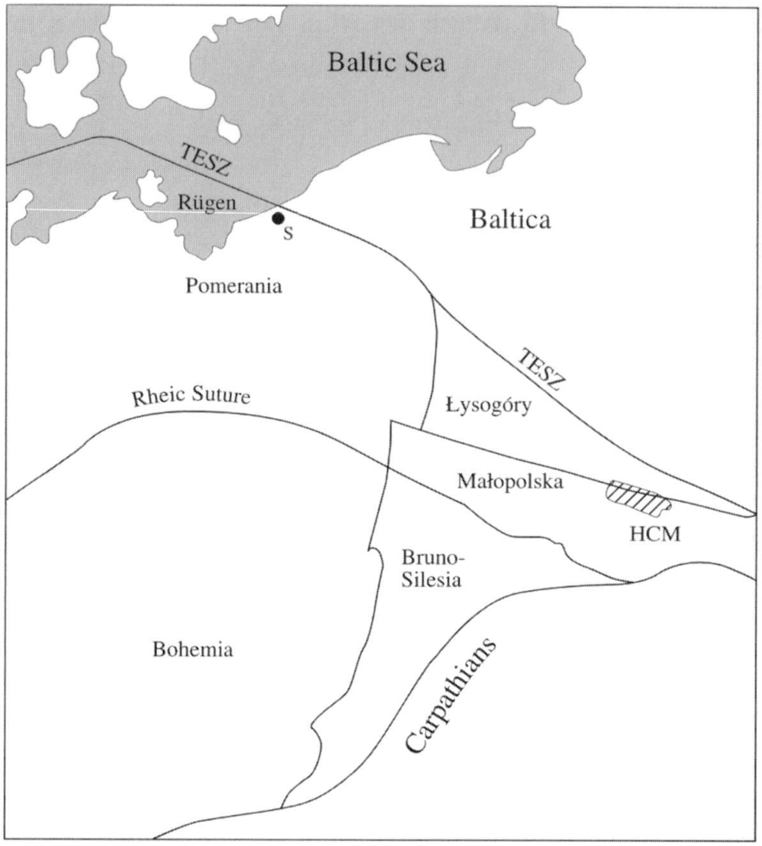

Fig. 1. Central Europe, with the modern locations of the Trans-European Suture Zone (TESZ), the island of Rügen, Pomerania, the Holy Cross Mountains (HCM), the Małopolska Block, the Łysogóry Block, and the Bruno-Silesian Block, as well as Bohemia, the Rheic Suture and the Carpathians, largely following Winchester *et al.* (2002). S, Skibno borehole, Pomerania.

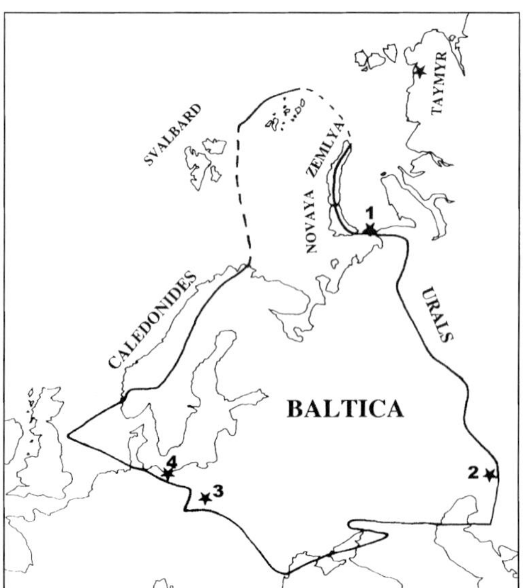

Fig. 2. Location of the present-day boundaries of Lower Palaeozoic Baltica (thick line), together with some of its neighbouring areas. The modern boundaries are indicative of post-Silurian tectonics and are not the continental edges during the Lower Palaeozoic. 1, Arenig of Pai-Khoe; 2, Tremadoc-Arenig of Aktyubinsk, Kazakhstan; 3, the Holy Cross Mountains of Poland; 4, Rügen island, north Germany. The unnumbered star is the position of the late Ashgill fauna from Taimyr now no longer considered to be part of Baltica.

not have displayed any indication of Gondwanan affinities in the Cambrian or early Ordovician. If they had migrated from another terrane the nearest palaeocontinent would have been Laurentia, not Gondwana. Winchester *et al.* (2002) suggested that this rotation is in question because of palaeomagnetic data by Nawrocki (1999) which indicates that Baltica (or more precisely the Małopolska terrane) has not rotated much from its present position. The fact that Nawrocki (1999) did not detect rotation is irrelevant, since his data came from a late Silurian (probably late Ludlow) diabase in the Bardo-Pragowiec section of the Holy Cross Mountains, and Baltica's rotation had finished by that time. The faunal evidence from these contentious areas will be considered in turn.

The Holy Cross Mountains

Within the Holy Cross Mountains of south-central Poland different authors have followed different terrane interpretations. Orłowski (1975*b*) published a map of the Cambrian of the whole Holy Cross area (see stratigraphic details below) and, whilst recognizing the substantial Holy Cross Thrust, which runs from east to west and divides the Małopolska and Łysogóry blocks, nevertheless treated the whole area as belonging to a single terrane. Dzik *et al.* (1994) in a substantial and impressive monograph, postulated only one terrane, which they termed the Małopolska microcontinent, which they divided into the Upper Silesian Massif (termed the Bruno-Silesian Block in this paper) and the Małopolska (*sensu stricto*) Massif. Within the latter they recognized three 'facies', the Kielce facies in the centre and east (and forming the southern part of the Holy Cross Mountains), the Łysogóry facies to the north (including the northern Holy Cross Mountains) and the Lagów facies to the west. Each of these 'facies' had different lithological successions. The formations of most interest here are the early and middle Ordovician Bukówka Sandstone and the Mójcza Limestone, both of which have yielded brachiopods and trilobites of unequivocal Baltic affinity. Various other workers (refs in Winchester *et al.* 2002, including Belka *et al.* 2000, 2002) have postulated that the Małopolska and Łysogóry blocks were separate and independent terranes in Lower Palaeozoic time. All research workers are agreed that the Małopolska and Łysogóry blocks had joined before the unconformable Lower Devonian (Emsian) marine rocks were deposited – these overlie both blocks without lateral discontinuity.

Baltican rotation, as discussed above, falsifies the conclusions of Belka *et al.*, who postulated (2000, pp. 98–9 – their sequence numbers follow here) that the Małopolska Block: 1, separated from Gondwana before the early Cambrian, but got close enough to Baltica to accommodate Baltic trilobites, but retaining 'Avalonian' inarticulated brachiopods; 2, became closer to Baltica so that progressively more Baltic brachiopods could settle there, and at the same time the 'supply of Cadomian clastic material' ceased and was replaced by Baltic sources; 3, collided with Baltica between mid-Cambrian and late Tremadoc time; 4, became covered by a shallow sea in the early Ordovician resulting in the more endemic ostracode faunas; and 5, underwent amalgamation with the Łysogóry terrane in the late Silurian. Belka *et al.* (2000, p. 99) state that the Małopolska terrane was 'in the immediate proximity of Baltica already during Early Cambrian time'. They also state that although the Łysogóry Block was 'probably also situated very close (or was even attached) to Baltica from late Cambrian, there is a paradox that the Cambrian fauna shows Gondwanan rather than Baltic affinities'.

I believe that the conclusions of Belka and colleagues (2000, 2002) are wrong and that both the Małopolska and the Łysogóry blocks of the Holy Cross Mountains formed integral parts of Baltica during the whole of the Lower Palaeozoic. The faunal and sedimentary evidence, including that adduced by Belka (2000) and Belka *et al.* (2000, 2002) to support their scenario, will now be reviewed in turn.

Cambrian trilobites and stratigraphy

Orłowski (1975*b*) established (in ascending order) the Lower Cambrian Czarna Shale, Ociesêki Sandstone and Kamieniec Shale formations, all occurring only to the south of the Holy Cross Thrust (the Małopolska block); the Middle Cambrian Slowiec Sandstone and Usarzow Sandstone as only occurring south of the thrust (with an unconformity between them), and the Góry Pieprzowe Shale Formation as occurring both north and south of the thrust. From north of the thrust (in the Łysogóry block), in addition to the Middle Cambrian Pieprzowe Shale, he identified the Upper Cambrian Wisniówka Sandstone Formation and the Klonowka Shale Formation. However, later work by Szczepanik (2001) on acritarchs and thermal maturation suggests that the Pieprzowe Shale is entirely confined to the Łysogóry Block north of the thrust (and may even include some Lower Cambrian), but no macrofossils have been recorded from the lower levels of the formation. Thus Lower Cambrian macrofauna do not occur in the Łysogóry Block and the Upper Cambrian does not occur in the Małopolska Block. Following descriptions of the Lower Cambrian trilobites (Orłowski 1985a), the Middle Cambrian trilobites (Orłowski 1985b) and the Upper Cambrian trilobite and other faunas (Orłowski 1968), Orłowski (1992) summarized the biostratigraphy of the Holy Cross

Mountains and determined that the Ociesêki Formation in fact extended upwards into the early Middle Cambrian and the Wisniówka Formation downwards into the late Middle Cambrian. The Lower and Middle Cambrian trilobites are of undoubted Baltic affinity 'similar or identical to those of Scandinavia' (Orłowski 1992, p.473).

The Upper Cambrian trilobites are worth detailed consideration here since their identifications and palaeobiogeographical interpretations have been controversial. Orłowski (1985*b*) determined them as *Olenus rarus* sp. nov., *Protopeltura olenusorum* sp. nov., *Protopeltura* sp., *Sphaerophthalmus alatus* (Boeck), *Peltura scarabeoides scarabeoides* (Wahlenberg), *Peltura? protopeltorum* sp. nov., *Agnostus (Homagnostus) pseudobesus* sp. nov., *Beltella irae* sp. nov, *Beltella* sp. and *Acerocare? klonowkae* sp. nov. These trilobites have now been revised by Zylínska (2001), and Dr A.W.A. Rushton also kindly comments on each as follows. *Olenus rarus* has now been transferred to *Aphelaspis*, a genus which is common in the shallow-water cratons of Laurentia and Kazakhstan and a single specimen is known from North Wales: it is thus of widespread palaeogeographic signal. The species *olenusorum* is endemic, but *Protopeltura* is known from Severnaya Zemlya (Rushton *et al.* 2002), Baltica (Norway), Wales, New Brunswick, Nova Scotia and Kazakhstan; it too was essentially cosmopolitan. Likewise the species *Sphaerophthalmus alatus* is widespread, being dominant in the olenid facies of the Baltic and elsewhere. *Peltura scarabeoides* is also widespread in the olenid facies, although the species *protopeltorum*, whilst confirmed as truly attributable to *Peltura* rather than the queried identification of Orłowski, is endemic to the Holy Cross Mountains. *Agnostus (Homagnostus) pseudobesus* is now seen to be a junior synonym of *Trilobagnostus rudis* (Salter), which, since it could swim, traversed oceans and is no palaeogeographical indicator. *Beltella irae* is now recognized as an endemic species of *Leptoplastides*. Similarly, *Acerocare? klonowkae* is now recognized as an endemic species of *Acerocarina*. The species *bella*, hitherto considered an endemic species of *Parabolina*, is now within the synonymy of *Parabolina (Neoparabolina) lapponica* Westergård, which is known only from Sweden and the Holy Cross Mountains. *Leptoplastides, Acerocarina* and *Parabolina (Neoparabolina)* are further members of the olenid facies. Thus, overall the Late Cambrian trilobites are strongly identifiable with the olenid facies. This facies covered most of Baltica in Late Cambrian time, but, although many of the included species and genera were originally described from the Baltic, the distribution of the olenid facies was probably chiefly defined and constrained by poorly-ventilated ocean bottom waters, and the facies is much more widely distributed than in Baltica alone (Shergold 1988). However, bearing in mind the Baltic affinities of the Lower and Middle Cambrian trilobites and the overlying Arenig brachiopods, the total faunas conclusively endorse the Baltic affinities of the Łysogóry Block – they certainly cannot be termed 'Avalonian'.

Cambrian trace fossils

These trace fossils are particularly well exposed in Wiélka Wiœniówka Quarry in the Łysogóry Block, in which shallow-water sandstones more than 350 m thick may be seen (Belka 2000). The ichnofacies have been well described by Radwanski & Roniewicz (1960, 1963). However, Belka *et al.* (2000, p. 94) have interpreted these as 'identical to those distributed throughout Gondwana and the Peri-Gondwanan microplates (Seilacher 1983)'. This is mysterious since Seilacher's excellent work (1983) describes some Upper Palaeozoic, probably Carboniferous, trace fossils from Egypt (then part of Gondwana); but it does not call them of Gondwanan or peri-Gondwanan affinity, and in fact makes close comparison chiefly with Devonian and Carboniferous ichnofacies of North America. Comparable, often trilobite-generated, trace fossils are well known from many parts of the world, including Lower Palaeozoic Baltica, and are of no use in assessing palaeogeographical affinities in this case.

Cambrian inarticulated brachiopods

The only brachiopod named by Belka *et al.* (2000) (as a 'Baltic' species) is *Westonia bottnica*. *Westonia* is a genus recorded from the Middle and Upper Cambrian of Canada, USA, Russia, Spain, China and Australia according to Holmer & Popov (2000). In other words it was essentially cosmopolitan, and cannot be called 'Baltic', particularly since the form from the Holy Cross Mountains has yet to be properly monographed. Belka *et al.* (2000, 2002) and Winchester *et al.* (2002) term these brachiopods 'Avalonian'. To support this they lean heavily on a short paper by Jendryka-Fuglewicz (1992), who lists the following brachiopods, but does not figure or systematically describe them:

Lower Cambrian: *Westonia bottnica* (Wiman), *Trematobolus* sp., *Acrothele* cf.

granulata Linnarsson, *Obolella rotundata* Kiaer, *Mickwitzia* sp., *Lingulella sp.*
Middle Cambrian: *Lingulella vistulae* (Gurich), *Westonia* sp. nov., *Acrothele granulata* Linnarsson, *Acrotreta* sp., *Mickwitzia* sp., *Trematobolus pristinus* (Matthew), *Lingulella* sp.
Upper Cambrian: *Lingulella davisii* (M'Coy), *Lingulella lepis* (Salter), *Lingulella ferruginea* Salter, *Lingulella* sp., *Orusia* cf. *lenticularis* (Wahlenberg), *Eoorthis* sp., *Acrotreta multa* Orlowski, *Acrotreta* sp., *Acrothele* sp.

Without seeing the material, it is not easy to know what these specimens or their affinities truly are. It is certain that *Lingulella* only occurs in the Upper Cambrian and Ordovician and so the Lower and Middle Cambrian records are misidentifications (the ranges quoted here are those of Holmer & Popov 2000); similarly *Acrotreta* is known only from the Ordovician, not the Cambrian, and *Mickwitzia* only from the Lower and not the Middle Cambrian. Thus it is only *Acrothele*, which is cosmopolitan, and *Trematobolus*, which is recorded from Lower and Middle Cambrian Laurentia, Siberia and the Altai Mountains of Russia as well as the Gondwanan localities of Spain and Morocco, which remain as possibly correctly identified in the Middle Cambrian records of Jendryka-Fuglewicz (1992).

Comparably, of the four possible remaining Lower Cambrian names, *Acrothele* and *Mickwitzia* are cosmopolitan (with the latter abundant in Estonia and other parts of Baltica), *Obolella* occurs in many areas, including Norway, and only *Westonia* and *Trematobolus* are as yet unknown from Baltica, but could just as well be termed 'Laurentian' or 'Kazakhstanian' as 'Gondwanan'. Not many Lower Cambrian faunas have yet been identified and described from Baltica. In the Upper Cambrian *Obolus* is an exclusively Baltic endemic, *Orusia* occurs in Baltica as well as Laurentia, China and the Welsh and Argentinian parts of Gondwana; *Lingulella*, *Eoorthis* and *Acrothele* are all cosmopolitan forms. Dr L.E. Popov informs me that *Obolus rotundata* Kiaer is more likely to belong to the mainly Baltic obolellide *Magnicaulis* rather than to *Obolella*, but its distribution requires revision and he doubts whether it truly occurs in the Lower Cambrian.

In contrast to the above unverifiable listings, Orłowski (1968) figured *Obolus* sp., *Orusia* cf. *lenticularis* (Wahlenberg) and *'Acrotreta' multa* sp. nov. from the Upper Cambrian of Chabowe Doly and Waworków in the Łysogóry block, all known from elsewhere in Baltica. Thus, contrary to the assertions of Belka *et al.* (2000, 2002) and Winchester *et al.* (2002), the Cambrian inarticulated brachiopods of the Łysogóry and Małopolska blocks do not show 'Avalonian affinities' (the term 'Avalonian' is again incorrect here since Avalonia did not exist until the early Ordovician – during the Cambrian the Avalonian area was simply an integral part of Gondwana). The Holy Cross Mountain brachiopods, particularly when interpreted with the trilobites found in the same sediments, are perfectly in accord with the interpretation of a Baltic origin.

Ordovician faunas and stratigraphy

The stratigraphy of the Małopolska Block (termed the Kielce facies by Dzik *et al.* 1994) consists of late Cambrian shales followed by Tremadoc Miedzygórz Beds (a marly limestone), the Arenig Bukówka Sandstone, the middle Ordovician Mójcza Limestone, the Ashgill Zalesie Formation, above which are Silurian graptolitic shales, including the basal Llandovery *acuminatus* graptolite Biozone. The faunas from the 'Tremadoc chalcedonites' of Biernat (1973) have many widespread genera, but several species have many similarities with the inarticulated brachiopods decribed by Popov & Holmer (1994) from the southern Ural part of Baltica. The Holy Cross Ordovician brachiopods are best found in the Mójcza locality of the Małopolska block, and in the Arenig Bukówka Sandstone are dominated by *Antigonambonites planus* (Pander) and *Lycophoria nucella* (Dalman), mentioned by Cocks & Fortey (1982) and figured by Cocks & Fortey (1998), Cocks (2000) and again here (Fig. 3), together with *Paurorthis* sp., *Plectella* sp, and *Syntrophina?* sp. In addition Dzik *et al.* (1994, pl. 7) illustrated *Productorthis obtusa* (Pander), *Orthis kielcensis* Roemer and *Orthambonites calligramma* (Dalman) from the higher beds in the Bukówka Sandstone. This all provides conclusive evidence that the Małopolska Block formed part of the craton of Baltica during Ordovician time. In particular *Lycophoria* occurs only in Baltica, and often in rock-forming abundance. It is also the only genus within the family Lycophoriidae, whose place within the Order Pentamerida is uncertain since it is so far removed from other families in that order: in other words, as certain a conclusive endemic palaeobiogeographical signal as is possible. The overlying middle Ordovician Mójcza Limestone faunas also 'show close Baltic similarities which allows a quite precise correlation with the Baltic region' (Dzik *et al.* 1994, p. 34). Above these the late

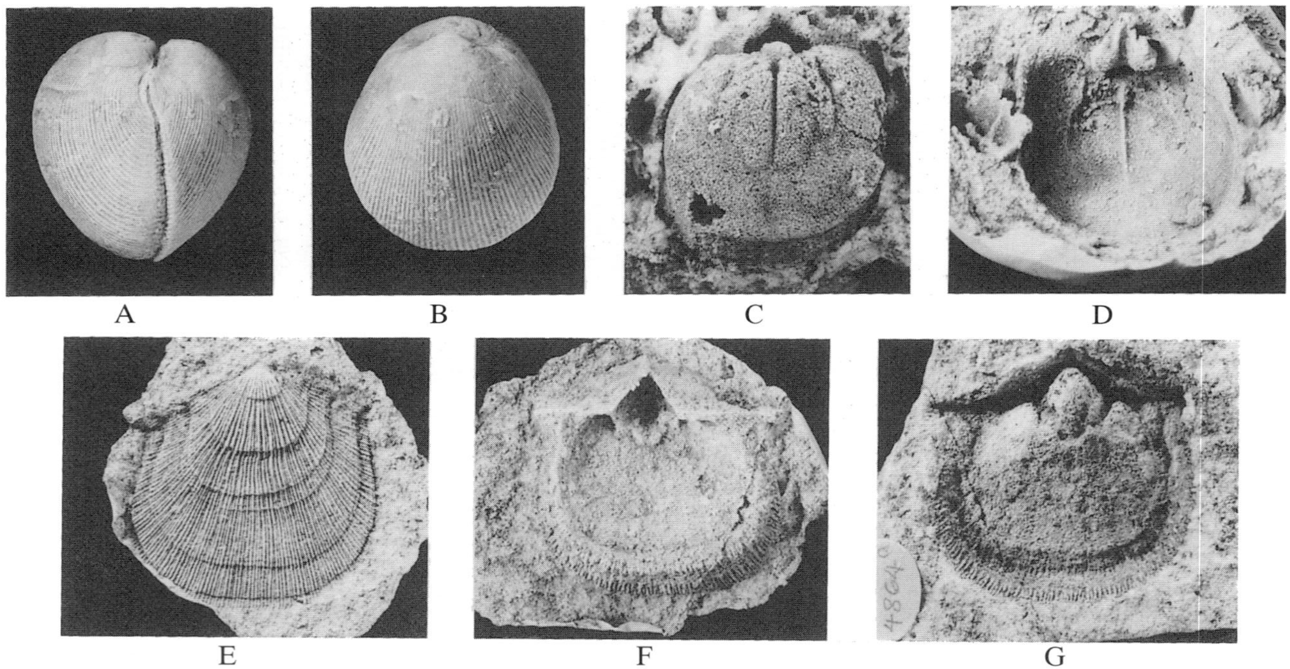

Fig. 3. Lower Ordovician (Arenig) brachiopods of Baltic affinity from Russia and the Bukówka Sandstone Formation, Mójcza Quarry, Małopolska Block, Holy Cross Mountains, Poland. A–D, *Lycophoria nucella* (Dalman), A,B, conjoined valves, Pavlovsk, St Petersburg, B 987, × 1.5: C,D, internal mould and latex cast of dorsal valve, Mójcza, BC 4883, × 3.0. E–G, *Antigonambonites planus* (Pander), latex of external mould, internal mould and latex cast of ventral interior, Mójcza, BC 4864, × 1.5. All specimens in The Natural History Museum, London.

Ordovician Zalesie Formation yields both a deeper-water *Foliomena* Fauna (Cocks & Rong 1988) and above that a shallower-water *Hirnantia* fauna (Temple 1965), the latter the first to be identified as such since the early work of Frederick M'Coy in North Wales in the mid-nineteenth century. Both these Ashgill age brachiopod-dominated faunas are geographically widespread and indicate little affinity to any particular terrane. However, in summary, the Ordovician brachiopods of the Holy Cross Mountains together constitute a fauna typical of the Baltica terrane.

Sediments

The sediments of the Holy Cross Mountains are equally ambiguous in palaeocontinental affinity. Since the recognition of late Proterozoic 'Cadomian' basement in the Uralides of Baltica (Glasmacher *et al.* 1999; Winchester *et al.* 2002), no weight can be placed on the allegedly 'Cadomian' micas identified as 'Avalonian' by Belka *et al.* (2000) from both parts of the Holy Cross Mountains. The mixture and variability of ages and cooling patterns revealed by their provenance studies are surely what are to be expected on the margins of a large and variable palaeocontinent. However, all authors (including Belka 2000; Belka *et al.* 2000, 2002) are agreed on the high degree of maturity of the sediments, particularly in the Upper Cambrian Wiœniówka Sandstone Formation, which favours origin from a relatively large palaeocontinent (e.g. Baltica) rather than a small terrane such as the Łysogóry Block today, supporting the thesis that the Łysogóry Block formed an integral part of Baltica itself. The Holy Cross Mountains must have been near the margin of the old palaeocontinent, as may be deduced from the substantial (1500 m) late Silurian turbidites present (Belka *et al.* 2002) in the Łysogóry Block and the similar (1200 m) contemporary sediments on the Małopolska Block.

Thus, after analysis of all the varied faunas and other data available, the conclusion can be reached that both blocks of the Holy Cross Mountains formed part of the margins of Baltica itself in Lower Palaeozoic times. Whether or not the Małopolska Block and the Łysogóry Block were adjacent to each other within Baltica in the Lower Palaeozoic is equivocal from the stratigraphical and faunal evidence. Bergström (1984), following an analysis of the distribution of Cambrian trilobites, deduced that any strike-slip faulting of the Holy Cross Mountains in relation to the rest of Baltica is unlikely to have been very substantial, and not the thousands of kilometres suggested by some authors.

The Bruno-Silesian Block

Despite linkage of the Bruno-Silesian (often termed the Upper Silesian) Block and the Małopolska Block by some authors (who have sometimes termed the two together as the Brunovistulicum or Brunovistulian block; although, confusingly, other authors have used Brunovistulicum as a synonym for the Bruno-Silesian Block alone), Buła *et al.* (1997) have convincingly demonstrated that the two blocks have separate stratigraphical and tectonic developments. Even taking into account the records of Cambrian acritarchs (Buła *et al.* 1997) and a few Cambrian trilobites (Orłowski 1975a), the evidence for any particular faunal geographic affinity for the Bruno-Silesian Block seems inconclusive and will not be considered further here; it could originally have been either peri-Gondwanan or peri-Baltic. Finger *et al.* (2000) treat the 'Brunovistulian' as one block, with interpretation of it as a former part of 'Avalonia' (the latter term is again used in error since in the Precambrian Avalonia did not exist except as part of Gondwana), although the rocks which they considered 'Cadomian' (i.e. Gondwanan) in origin are only exposed in the Bruno-Silesian part. To interpret the Małopolska Block as an accretionary wedge to the Bruno-Silesian Block is certainly mistaken, since the Małopolska Block formed part of Baltica (see above).

Rügen and Pomerania

Cocks *et al.* (1997) in their review of the boundaries of the terrane of Avalonia, concluded that its eastern margin with Baltica was at the Elbe Line. However, Franke & Żelaźniewicz (1997) have traced facies and structural styles across the Elbe Line, and it now seems probable that the Avalonia–Baltica suture today corresponds to the TESZ in southern Denmark and northern Germany. A soft docking of Baltica with Avalonia occurred at about 443 Ma in the latest Ordovician (Ashgill). Yet there is equivocal palaeogeographical evidence from the Ordovician deposits of the north German island of Rügen in the southern Baltic Sea. There relatively undeformed Mesozoic and later rocks overlie much-folded Ordovician strata which consist of clastic deposits. No shallow-water benthic faunas are known, and the only macrofossils collected are graptolites of the Llanvirn *artus*, *murchisoni* and *teretiusculus* Biozones and the early Caradoc *gracilis* Biozone (Jaeger

1967). Beneath the graptolite-bearing rocks, acritarchs date the beds as from latest Tremadoc and early Arenig age (Servais & Molyneux 1997) as well as the mid-Ordovician. Servais & Molyneux (1997) have also demonstrated that the early Ordovician acritarchs present at Rügen are the same as from the English Lake District of Avalonia, particularly the early Arenig *messaoudensis-trifidum* Assemblage. There are also many species in common with those from Spain, the Taurides of Turkey and Bohemia, which were all Gondwanan or peri-Gondwanan in the early Ordovician. A few elements of the assemblage are also to be found in Baltica and South China, but most of the Rügen species indicate fairly high latitude, rather than the intermediate palaeolatitudes of much of Baltica. Katzung *et al.* (1995) found structural and sedimentological similarities between Rügen and the Condroz Inlier of Belgium and the Rhenish Massif of NW Germany, both within Avalonia. However, as stated by Cocks & Verniers (2000), acritarchs are planktonic and therefore their distribution was linked to palaeotemperature-controlled water masses and consequently but approximately to palaeolatitude, and thus cannot properly be said to be of 'affinity' to any particular palaeocontinent.

Giese *et al.* (1994) undertook detailed petrographic and provenance studies for the Rügen rocks and record a substantial proportion of carbonate grains. This is puzzling, since no carbonate deposits are known from high-latitude Gondwanan or peri-Gondwanan terranes in the early Ordovician. Dallmeyer *et al.* (1999), on the basis of Ar–Ar dates, considered that Rügen was chiefly 'Cadomian' in origin and thrust over Baltic rocks between 450 and 425 Ma during the Avalonia–Baltic docking period. However, as stated above under the Holy Cross Mountains sediments, since 'Cadomian' rocks are now known from the Baltic Urals, 'Cadomian' does not necessarily imply affiliation to Gondwana/Avalonia. It therefore seems most probable, as suggested briefly by Cocks & Verniers (2000), that the Rügen sediments were deposited in a medium to high-latitude sedimentary basin within the Tornquist Ocean, which lay between Avalonia and Baltica (Cocks & Fortey 1982).

Graptolites of the late Llanvirn *teretiusculus* and early Caradoc *gracilis* Biozones have also been recovered from the Skibno 1 borehole in Pomerania, NW Poland (Fig. 1). That borehole has also yielded chitinozoans and acritarchs of dominantly 'Avalonian' (i.e. higher-latitude) affinity (Wrona *et al.* 2001; Samuelsson *et al.* 2002) from clastic rocks including turbidites. Thus it is possible that the sedimentary basin in which these rocks were deposited may have been the same as that at Rügen.

Conclusions

1. Because of the pre mid-Ordovician rotation of Baltica, with more than 55 degrees of it occurring in the late Cambrian and early Ordovician, today's Tornquist Margin would have faced northwards towards the Panthalassic Ocean and Laurentia, and not southwards across the Tornquist Ocean to face Gondwana/Avalonia.

2. Although more than one Lower Palaeozoic terrane has been postulated to exist in the Holy Cross Mountains area of Poland, this has not yet been proven beyond doubt. However, whether there were separate terranes (the Małopolska and Łysogóry blocks), or a single united area: there is no doubt from their contained fossils that both blocks formed part of the old continent of Baltica in Lower Palaeozoic times. There is no faunal evidence to support the theory that either block was part of or connected with Gondwana (Avalonia) until after the Avalonia–Baltica docking in the latest Ordovician. In any case the blocks were certainly united by early Devonian time since identical Emsian sediments and faunas cover both.

3. There seems little doubt that neither the Małopolska nor the Łysogóry blocks formed part of the same terrane in the Lower Palaeozoic as the Bruno-Silesian Block. However, the faunal evidence from the latter is currently inconclusive as to whether the latter formed part of Baltica or Gondwana or was a separate and independent terrane.

4. Whether or not any substantial strike-slip occurred between the Holy Cross Mountains and the main East European Craton of Baltica along the Tornquist Margin of Baltica remains uncertain: no such movement is required to explain the evidence from palaeontology, but faunal constraints in the Lower Cambrian indicate that any movement would have only been of up to a few hundred kilometres, rather than the thousands of kilometres postulated by some authors.

5. Acritarch and chitinozoan evidence from the Rügen and Pomerania areas suggest relatively high-latitude deposition of the sediments there, but the carbonate grains present at Rügen and absent from Avalonia suggest that the Rügen and Pomeranian sediments were deposited in an ocean basin within the Tornquist Ocean, rather than on either Baltica or Avalonia themselves.

I am most grateful to Adrian Rushton (The Natural History Museum) and Trond Torsvik (Trondheim, Norway) for comments on Cambrian trilobites and palaeomagnetics respectively, and to John Winchester (Keele, England), Zdzislaw Belka (Halle, Germany) and Elena Sokiran (Sosnowiec, Poland) for access to published and unpublished papers. Thanks also to European Science Foundation (Europrobe) for funding travel to conferences.

References

BELKA, Z. 2000. *Excursion guidebook The Holy Cross Mountains, joint meeting of Europrobe and PACE projects.* Warsaw, 38 pp.

BELKA, Z., AHRENDT, H., FRANKE, W. & WEMMER, K. 2000. The Baltica-Gondwana suture in central Europe: evidence from K-Ar ages of detrital muscovites and biogeographical data. *Geological Society, London, Special Publications*, **179**, 87–102.

BELKA, Z., VALVERDE-VAQUERO, P., DÖRR, W., AHRENDT, H., WEMMER, K. & FRANKE, W. 2002. Accretion of first Gondwana-derived terranes at the margin of Baltica. *Geological Society, London, Special Publications,* **201**, 00–01.

BERGSTRÖM, J. 1984. Strike-slip faulting and Cambrian biogeography around the Tornquist Zone. *Geologiska Föreningens Förhandligar*, **106**, 382–383.

BIERNAT, G. 1973. Ordovician inarticulate brachiopods from Poland and Estonia. *Palaeontologia Polonica*, **28**, 1–120, pls 1–40.

BUŁA, Z, JACHOWICZ, M. & ŻABA, J. 1997. Principal characteristics of the Upper Silesian Block and Małopolska Block border zone (southern Poland). *Geological Magazine*, **134**, 669–677.

COCKS, L. R. M. 2000. The early Palaeozoic geography of Europe. *Journal of the Geological Society, London,* **157**, 1–10.

COCKS, L. R. M. & FORTEY, R. A. 1982. Faunal evidence for oceanic separations in the Palaeozoic of Britain. *Journal of the Geological Society, London*, **139**, 465–478.

COCKS, L. R. M. & FORTEY, R. A. 1998. The Lower Palaeozoic margins of Baltica. *Geologiska Föreningens Förhandligar*, **120**, 173–179.

COCKS, L. R. M. & RONG J.-Y. 1988. A review of the late Ordovician *Foliomena* brachiopod fauna with new data from China, Wales and Poland. *Palaeontology*, **31**, 53–67.

COCKS, L. R. M. & VERNIERS, J. 2000. Applicability of planktic and nektic fossils to palaeogeographic reconstructions. *Acta Universitatis Carolinae-Geologica*, **42**, 399–400.

COCKS, L. R. M., MCKERROW, W. S. & VAN STAAL, C. R. 1997. The margins of Avalonia. *Geological Magazine*, **134**, 627–636.

DALLMEYER, R. D., GIESE, U., GLASSMACHER, U. & PICKEL, W. 1999. First $^{40}Ar/^{39}Ar$ age constraints for the Caledonian evolution of the Trans-European Suture Zone in NE Germany. *Journal of the Geological Society, London*, **156**, 279–290.

DZIK, J., OLEMPSKA, E. & PISERA, A. 1994. Ordovician carbonate platform ecosystem of the Holy Cross Mountains. *Palaeontologica Polonica*, **53**, 1–317.

FINGER, F. HANZL, P., PIN, C., VON QUADT, A. & STEYRER, H.P. 2000. The Brunovistulian: Avalonian Precambrian sequence at the eastern end of the Central European Variscides. *Geological Society, London, Special Publications,* **179**, 103–112.

FORTEY, R. A. & COCKS, L. R. M. 2002. Palaeontological evidence bearing on Ordovician-Silurian continental reconstructions. *Earth Science Reviews*, in press.

FRANKE, W. & ŻELAŹNIEWICZ, A. 1997. The Sudetes seen from the west: terrane correlation across the Elbe Zone. *Terra Nostra,* **97**(11), 46–50.

GIESE, U., KATZUNG, G. & WALTER, R. 1994. Detrital composition of Ordovician sandstones from the Rügen boreholes: implications for the evolution of the Tornquist Ocean. *Geologisches Rundschau*, **83**, 293–308.

GLASMACHER, U. A., GIESE, U. STROINK, L., REYNOLDS, P., ALEKSEYEV, A., PUCHKOV, V. & BAUER, W. 1999. Neoproterozoic terrane at the eastern margin of Baltica – implications for Late Proterozoic palaeogeography and structural evolution of SW Urals, Russia. *Journal of Conference Abstracts EUG,* **10**, 108.

HOLMER, L. E. & POPOV, L. E. 2000. Lingulata. In KAESLER, R. L. (ed.) *Treatise on Invertebrate Paleontology, H Brachiopoda Revised*, Vol. 2. Geological Society of America and University of Kansas, Boulder, 423 pp.

JAEGER, H. 1967. Ordoviz auf Rügen. Datierung und Vergleich mit anderen Gebieten (Vorläufige Mitteilung). *Berichte der deutschen Gesellschaft für Geologische Wissenschaften. A, Geologie und Paläontologie*, **12**, 156–176.

JENDRYKA-FUGLEWICZ, B. 1992. Analisa porównawca ramienionogów z utworów Kambru Gór Swietokrzynkich i Platformy Prekambryjskiej w Polsce. *Przeglad Geologiczny*, **467**, 150–155.

KATZUNG, G., GIESE, U., MALETZ, J., SERVAIS, T. & VAN GROOTEL, G. 1995. The eastern end of Avalonia: continuation into northern central Europe. In COOPER, J. D., DROSER, M. L. & FINNEY, S. C. (eds) *Ordovician Odyssey.* SEPM, Fullerton, California, pp 233–236.

NAWROCKI, J. 1999. Prefolding remanent magnetisation of diabase intrusion from the Bardo syncline in the Holy Cross Mts (central Poland). *Przeglad Geologiczny*, **47**, 1001–1004.

ORŁOWSKI, S. 1968. Upper Cambrian fauna of the Holy Cross Mts. *Acta Geologica Polonica*, **18**, 258–292.

ORŁOWSKI, S. 1975*a*. Lower Cambrian trilobites from Upper Silesia (Gocza kowice borehole). *Acta Geologica Polonica*, **25**, 377–383.

ORŁOWSKI, S. 1975*b*. Cambrian and upper Precambrian lithostratigraphic units in the Holy Cross Mts. *Acta Geologica Polonica,* **25**, 431–448.

ORŁOWSKI, S. 1985*a*. Lower Cambrian and its trilobites in the Holy Cross Mts. *Acta Geologica Polonica,* **35**, 231–250.

ORŁOWSKI, S. 1985*b*. New data on the Middle Cambrian trilobites and stratigraphy in the Holy Cross Mts. *Acta Geologica Polonica*, **35**, 251–263.

ORŁOWSKI, S. 1992. Cambrian stratigraphy and stage subdivision in the Holy Cross Mountains, Poland. *Geological Magazine*, **129**, 471–474.

POPOV, L.E. & HOLMER, L.E. 1994. Cambrian-Ordovician lingulate brachiopods from Scandinavia, Kazakhstan, and South Ural Mountains. *Fossils and Strata*, **35**, 1–156.

RADWANSKI, A. & RONIEWICZ, P. 1960. Ripple marks and other sedimentary structures of the Upper Cambrian at Wielka Wiœniówka (Holy Cross Mts). *Acta Geologica Polonica*, **10**, 371–400.

RADWANSKI, A. & RONIEWICZ, P. 1963. Upper Cambrian trilobite ichnocoenosis from Wielka Wiœniówka (Holy Cross Mountains, Poland). *Acta Palaeontologica Polonica*, **8**, 259–280.

RUSHTON, A. W. A., COCKS, L. R. M. & FORTEY, R. A. 2002. Upper Cambrian trilobites and brachiopods from Severnaya Zemlya, Arctic Russia, and their implications for correlation and biogeography. *Geological Magazine* (in press).

SAMUELSSON, J., VECOLI, M., BEDNARCZYK, W. S. & VERNIERS, J. 2002. Timing of the Avalonia-Baltica plate convergence as inferred from palaeogeographic and stratigraphic data of chitinozoan assemblages in West Pomerania, northern Poland. *Geological Society, London, Special Publication* **201**, 95–114.

SEILACHER, A. 1983. Upper Paleozoic trace fossils from the Gilf Kebir – Abu Ras area in southwestern Egypt. *Journal of African Earth Sciences*, **1**, 21–34.

SERVAIS, T. & MOLYNEUX, S. G. 1997. The *messaoudensis-trifidum* acritarch assemblage (Ordovician: late Tremadoc – early Arenig) from the subsurface of Rügen (Baltic Sea, NE Germany). *Palaeontographica Italica*, **84**, 113–161.

SHERGOLD, J. H. 1988. Review of trilobite biofacies distributions at the Cambrian-Ordovician boundary. *Geological Magazine*, **125**, 363–380.

SZCZEPANIK, Z. 2001. Acritarchs from Cambrian deposits of the southern part of the Łysogóry Unit in the Holy Cross Mountains, Poland. *Geological Quarterly*, **45**, 117–130.

TEMPLE, J. T. 1965. Upper Ordovician brachiopods from Poland and Britain. *Acta Palaeontologia Polonica*, **10**, 379–427.

TORSVIK, T. H. & REHNSTRÖM, E. F. 2001. Cambrian palaeomagnetic data from Baltica: implications for true polar wander and Cambrian palaeogeography. *Journal of the Geological Society, London,* **158**, 321–329.

WINCHESTER, J. A. & THE PACE TMR NETWORK TEAM, 2002. Palaeozoic amalgamation of central Europe: new results from recent geological and geophysical investigations. *Tectonophysics*, in press.

WRONA, R., BEDNARCZYK, W. S. & STEMPIEN-SALEK, M. 2001. Chitinozoans and acritarchs from the Ordovician of the Skibno 1 borehole, Pomerania, Poland: implications for stratigraphy and palaeontology. *Acta Geologica Polonica*, **51**, 317–331.

ŻYLÍNSKA, A. 2001. Late Cambrian trilobites from the Holy Cross Mountains, central Poland. *Acta Geologica Polonica*, **51**, 333–383.

The Cambrian to mid Devonian basin development and deformation history of Eastern Avalonia, east of the Midlands Microcraton: new data and a review

J. VERNIERS[1], T. PHARAOH[2], L. ANDRÉ[3], T. N. DEBACKER[1,7], W. DE VOS[4], M. EVERAERTS[5], A. HERBOSCH[6], J. SAMUELSSON[1,9], M. SINTUBIN[7] & M. VECOLI[8]

[1]*Laboratory of Palaeontology, Ghent University, Krijgslaan 281 building S8, B-9000 Ghent, Belgium (e-mail: Jacques. Verniers@rug.ac.be)*

[2]*British Geological Survey, Nottingham NG12 5GG Keyworth, UK*

[3]*Royal Museum for Central Africa, Steenweg op Leuven, 13, B-3080 Tervuren, Belgium*

[4]*Geological Survey of Belgium, Jennerstraat 13, B-1000 Brussels, Belgium*

[5]*Royal Observatory of Belgium, Avenue Circulaire 3, B-1180 Brussels, Belgium*

[6]*Département des Sciences de la Terre et de l'Environnement, Université Libre de Bruxelles, Avenue F. Roosevelt 50 CP160/02, B-1050 Brussels, Belgium*

[7]*Structural Geology & Tectonics Group, Katholieke Universiteit Leuven, Redingenstraat 16, B-3000 Leuven, Belgium*

[8]*Institut für Geologische Wissenschaften, Martin-Luther Universität, Domstrasse 5, Halle (Saale) D-06108, Germany; present address: Lab. Paléontologie, UMR 6538 Domaines Océaniques, Université de Bretagne Occidentale, 6 av. Le Gorgeu, BP809 F-29285 Brest cedex, France*

[9]*Present address: Uppsala University, Institute of Earth Sciences, Historical Geology and Palaeontology, Norbyvägen 22, S-752 36, Uppsala, Sweden*

Abstract: A review is given of recently published and new data on Avalonia east of the Midlands Microcraton. The three megasequences from Cambrian to mid Devonian described in Wales and Welsh Borderland are also present east of the Midlands Microcraton (Brabant Massif, Condroz, Ardennes, Remscheid and Ebbe inliers, Krefeld high). The three megasequences are caused by a tectonic driving mechanism and are explained by three different geodynamic contexts: an earlier phase with extensional basins or rifting and rather thick sequences, when Avalonia was still attached to Gondwana; a second phase with a shelf basin with moderately thin sequences when Avalonia was a separate continent and a later phase with a shelf or foreland basin development and thick sequences. Deformation of the megasequences 1 and 2 or 1 to 3 varies between areas. In Wales and the Lake District the Acadian phase is long-lived and active from early to mid Devonian. In the Ardennes inliers a deformation is active between the late Ordovician and the Silurian (Ardennian Phase), with a similar intensity as the core of the Brabant Massif, when present erosion levels are compared. The Brabant Massif is partly deformed by the long-lived Brabantian Phase from late Silurian till early mid Devonian. Both the Ardennes inliers and the Brabant Massif are not classic orogenic belts, only slate belts where no more than the epizone is reached at present erosion levels. Areas supposedly close to the microcraton or basement are nearly undeformed (SW Brabant Massif and central Condroz). A model of anticlockwise rotation of Avalonia of about 55° from Caradoc to Emsian is proposed to explain the deposition setting of megasequence 3 and the subsequent Acadian and Brabantian deformation. Immediately after the Avalonian microcontinent touched Baltica in Caradoc times it created a short-lived subduction magmatic event from The Wash to the Brabant Massif and soon after the magmatism ended a foreland basin developed. Possibly during and after that development a long-lived and slow compressional event occurred, leading to the deformation of the Anglo-Brabant Deformation Belt. In the early Devonian, contemporaneous with the shortening of the Anglo-Brabant Deformation Belt, extension occurred in the Rheno-Hercynian Zone, possibly caused by the same slow rotation of Avalonia. More evidence emerges that Avalonia east of the Midlands Microcraton comprises not one but probably two terranes: the remnant of the palaeocontinent Avalonia, and what is called the palaeocontinent Far Eastern Avalonia; the latter is only occasionally observed in the few deep boreholes into the

From: WINCHESTER, J. A., PHARAOH, T. C. & VERNIERS, J. 2002. *Palaeozoic Amalgamation of Central Europe*. Geological Society, London, Special Publications, **201**, 47–93. 0305-8719/02/$15.00

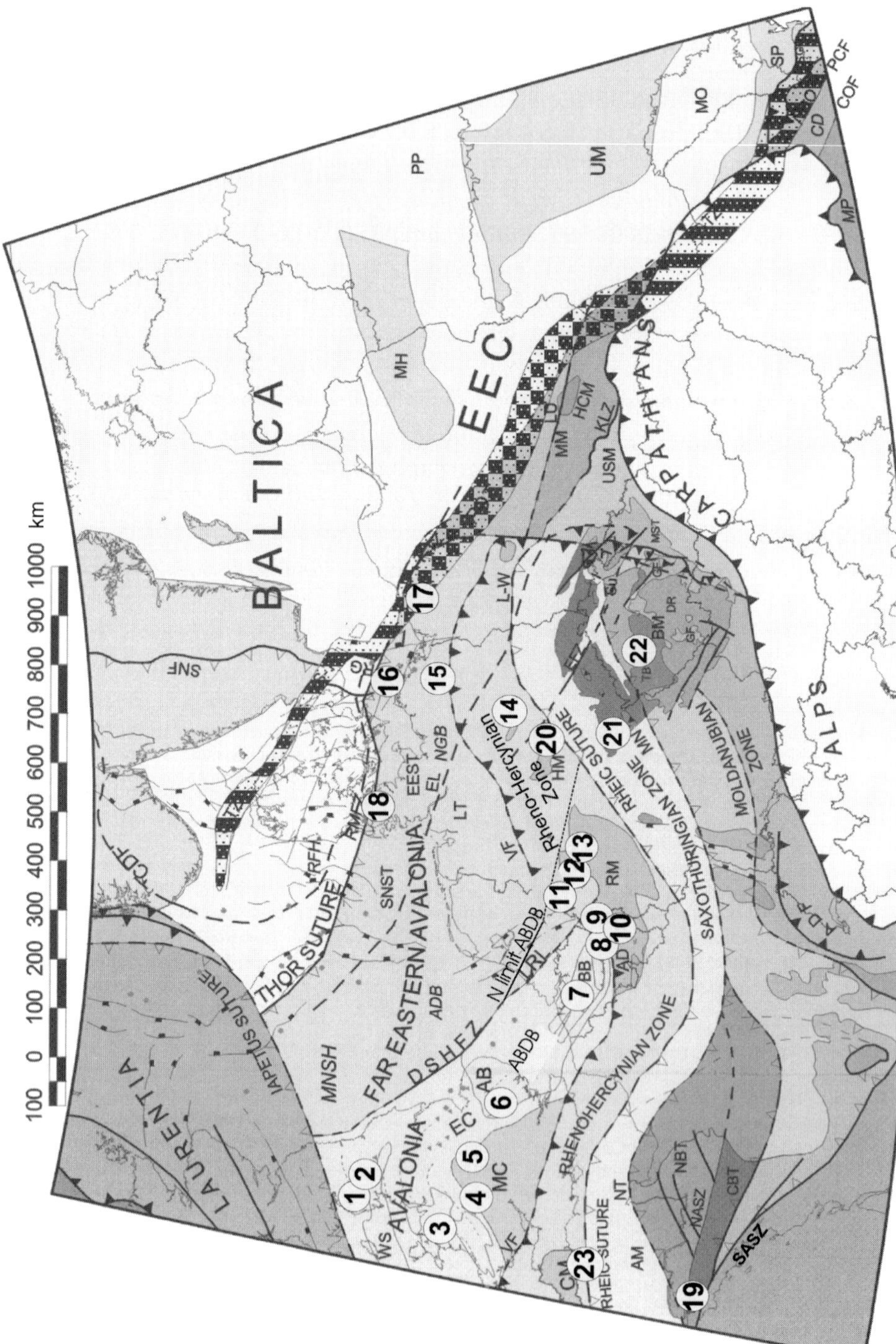
100 0 100 200 300 400 500 600 700 800 900 1000 km
BALTICA
EEC
LAURENTIA
IAPETUS SUTURE
THOR SUTURE
AVALONIA
FAR EASTERN AVALONIA
DSHFZ
N limit ABDB
Rheno-Hercynian Zone
RHEIC SUTURE
RHENOHERCYNIAN ZONE
RHEIC ZONE
SAXOTHURINGIAN ZONE
MOLDANUBIAN ZONE
CARPATHIANS
ALPS
SNF
CDF
MNSH
ADB
ABDB
SNST
EEST
EL
NGB
LT
VF
L-W
HM
RM
MN
BB
AD
AB
EC
MC
Ws
CM
AM
NT
NBT
NASZ
CBT
SASZ
MH
PP
UM
MO
SP
CD
PCF
COF
MP
LU
MM
HCM
KLZ
USM
BM
DR
TB
MST
RFH
RMF

Heligoland–Pomerania Deformation Belt, in southern Denmark, NE Germany and NW Poland, with scant available indirect data in between indicating only Proterozoic basement and no Caledonian deformation. For Far Eastern Avalonia a similar palaeogeographical history is postulated as Avalonia, with rifting from Gondwana in Arenig or earlier times, collision with Baltica before the mid-Ashgill and deformation between the late Ordovician and latest Silurian. The Avalonia concept might need to be expanded to an 'Avalonian Terrane Assemblage' with cratonic cores and small short-lived oceans as in the Armorican Terrane Assemblage.

The Cambrian to Devonian palaeogeographical history of the Avalonia microcontinent and its relation to the palaeocontinents Laurentia, Baltica and Gondwana has been established and widely accepted in recent years (Cocks 2000 and see below). The margins of Avalonia have been discussed by Cocks *et al.* (1997) and the composition and history of the Avalonia microcontinent and related microplates are reviewed in Pharaoh (1999). The presence of Neoproterozoic basement has been proven in Eastern Avalonia (see Pharaoh 1999), beneath the North German Basin and in the Harz Mountains by Breitkreuz and Kennedy (1999) and in the south Hunsrück by Molzahn *et al.* (1998). Pharaoh (1999) defined the Southern North Sea–Lüneburg Terrane (SNSLT), and Winchester *et al.* (in press), the microplate, Far Eastern Avalonia (Fig. 1). The latter authors established the northern boundary west of Denmark, and renamed the collision zone situated north and NE of the SNSLT, as the Heligoland–Pomerania Deformation Belt (HPDB), formerly called the North German–Polish Caledonides (see Ziegler 1982), the English North German–Polish Caledonides (Berthelsen 1992) or the Danish–North German–Polish Caledonides (see Katzung 2001). The 'Caledonian' orogenic belt in northern Central Europe was divided into the Schleswig, the Rügen and the Pomeranian Caledonides by Katzung (2001). Winchester *et al.* (in press) located the southern boundary at the Anglo-Brabant Deformation Belt (ABDB), corroborating the hypothesis of Van Grootel *et al.* (1997) of two separate orogenic belts, ABDB and HPDB.

The more detailed history of Eastern Avalonia, from the margins of the Midlands Microcraton up to the Heligoland–Pomerania Deformation Belt, has not yet been reviewed. The study area comprises the Welsh Basin, the Lake District and Pennine inliers, the Midlands Platform, the Eastern England Caledonides and the Anglian Basin, the Brabant Massif, the

Fig. 1. Study areas, numbers and names superimposed on a basement tectonic sketch map of NW Europe modified after Pharaoh (1999) and Winchester *et al.* (in press). Numbers give locations of the stratigraphical columns in Fig. 2. 1: Lake District; 2: Pennine inliers; 3: Welsh Basin; 4: Midlands Platform West; 5: Midlands Platform East; 6: Eastern England Caledonides; 7: Brabant Massif; 8: central part of the Condroz Inlier; 9: Bolland borehole; 10: Stavelot-Venn Inlier; 11: Krefeld high in subsurface; 12: Remscheid Inlier; 13: Ebbe anticline Inlier; 14: Flechtingen high; 15: Penkun borehole; 16: Rügen boreholes; 17: Pomerania boreholes; 18: Danish boreholes; 19: Crozon peninsula, Armorican Massif (part of the Armorican Terrane Assemblage, A.T.A.); 20: NW Harz mountains; 21: Thüringenwald; 22: Prague Basin; 23: Lizard Point.
Key: Oceanic sutures, line with open triangles; orogenic frontal zones, line with filled triangles; key boreholes, solid dots; Ordovician arc volcanic rocks in Avalonia, triangles. **Post-Palaeozoic**: ADB, Anglo-Dutch Basin; ADF, Alpine Deformation Front; CD, Central Dobrogea; MNSH, Mid-North Sea High; MP, Moesian Platform; NDO, North Dobrogea Orogen; NGB, North German Basin; POT, Polish Trough; RFH, Rynkøbing-Fyn High; RG, Rønne Graben; RMFZ, Rømø-Møn Fracture Zone; SP, Scythian Platform. **Postulated Palaeozoic terranes and possible terrane/sub-terrane boundaries**: DSHFZ, Dowsing-South Hewett Fault Zone; EEST, East Elbian Suspect Terranes; EL, Elbe Lineament; KLZ, Kraków-Lubliniec Zone; LRL, Lower Rhine Lineament; LT, Lüneburg Terrane; LU, Lysogory Block; MM, Malopolska Block; MST, Moravo-Silesian Terrane; NT, Norannian Terrane; PCF, Peceneaga-Camena Fault; SNST, Southern North Sea Terrane; SGF, Sfantu Gheorghe Fault. **Proterozoic-Palaeozoic tectonic elements**: ABDB: Anglo-Brabant Deformation Belt; AB, Anglian Basin; AD, Ardennes Massifs; AM, Armorican Massif; BB, Brabant Massif; BM, Bohemian Massif; CBT, Central Brittany Terrane; CDF, front of Caledonian deformation; CM, Cornubian Massif; COF, Capidava-Ovidiu Fault; DR, Drosendorf Unit (of BM); EC, Eastern England Caledonides; EEC, East European Craton; EFZ, Elbe Fault Zone; GF, Gföhl Unit (of BM); HM, Harz Mountains; HCM, Holy Cross Mountains; L-W, Leszno-Wolsztyn Basement High; MC, Midlands Microcraton; MH, Mazurska High; MN, Münchberg Nappe (of BM); MO, Moldavian Platform; NASZ, North Armorican Shear Zone; NBT, North Brittany Terrane; PP, Pripyat Trough; RM, Rhenish Massif; USM, Upper Silesian Massif (= MST); SNF, Sveconorwegian Front; SASZ, South Armorican Shear Zone; S-TZ, Sorgenfrei-Tornquist Zone; Su, Sudetes Mountains; TB, Tepla-Barrandian Basin (of BM); T-TZ, Teisseyre-Tornquist Zone; UM, Ukrainian Massif; VF, Variscan Front; WS, Windermere Supergroup.

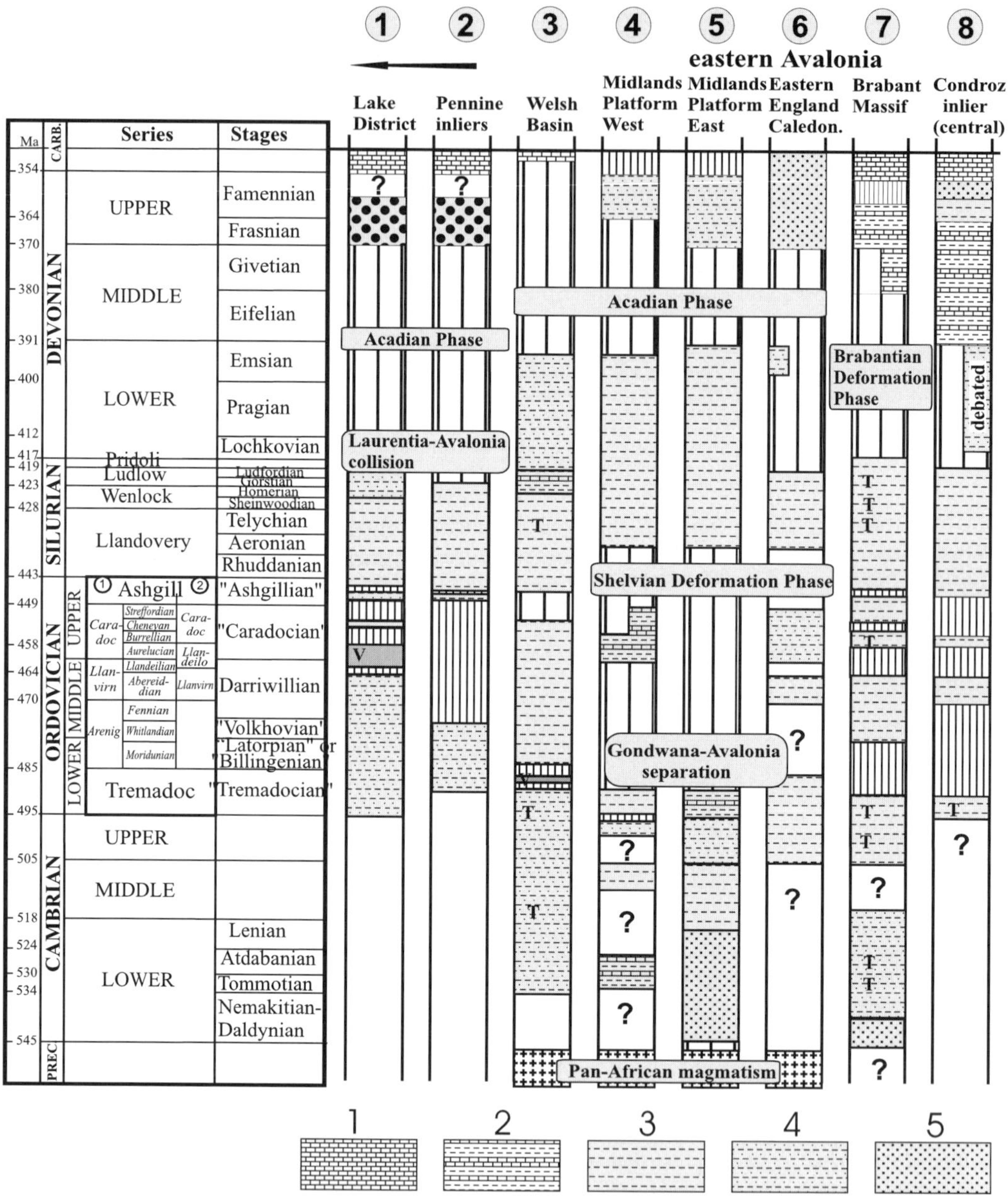

Fig. 2. Stratigraphical columns of the Cambrian–Devonian succession in different parts of Eastern Avalonia and surrounding areas, with indication of times of deposition, sedimentation gaps, unconformities and time range of deformation. The following publications are used: Lake District (Cooper *et al.* 1993); Pennine inliers (Kirby *et al.* 2000); Welsh Basin (Cocks *et al.* 1992; Fortey *et al.* 2000; Woodcock & Strachan 2000); Midlands platform East (Woodcock 1991); Midlands platform West (Bridge *et al.* 1998); Eastern England Caledonides (Molyneux 1991; Woodcock 1991; Woodcock & Pharaoh 1993); Belgium (Verniers *et al*: in press); Krefeld High (Ahrendt *et al.* 2001*a*,*b*); Ebbe and Remscheid (Samuelsson *et al.* 2002*b*); Danish and north German subsurface (this study); Rügen (Samuelsson *et al.* 2001); Pomerania (Samuelsson *et al.* 2002*a*); Armorican Massif (Paris & Le Hérissé 1992); Saxothuringia (Linneman *et al.* 1998); Prague Basin (Chlupac *et al.* 1998). 1: carbonate dominated; 2: carbonate and mudstone; 3: mudstone dominated; 4: mudstone and sandstone; 5: sandstone dominated; 6: conglomerate; 7: Proterozoic basement; 8: sedimentation hiatus; 9: gap of sedimentation caused by tectonic deformation; 10: period of intrusion; T: turbiditic facies; V: volcanic rocks; ?: unknown; time scale according to Tucker & McKerrow (1995); ① after Fortey *et al.* (1995); ② older division.

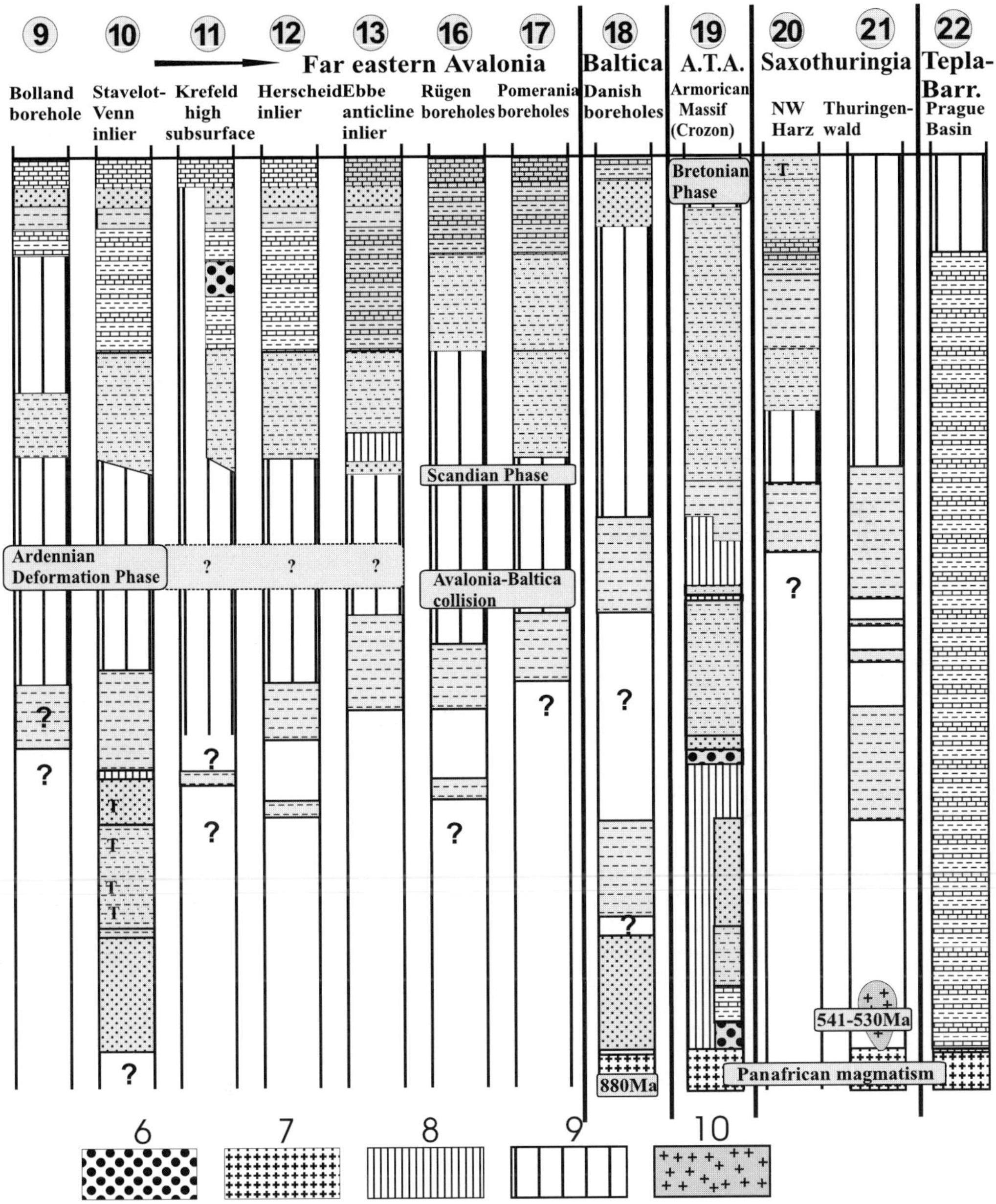

Condroz Inlier, the Ardennes inliers, the Krefeld High, the Ebbe and Remscheid inliers, the North Sea deep boreholes, and the southern Danish and northern Germany subcrop, including Rügen and Pomerania (Fig. 1).

Recently, new data and reviews were published on the study area following new studies of the litho- and biostratigraphy in the British Isles (Fortey *et al.* 2000; Cocks *et al.* 1992), Brabant, Condroz, Stavelot-Venn (Verniers *et al.* 2001), Ebbe (Samuelsson *et al.* 2002*b*), Rügen (Vecoli & Samuelsson 2001*b*; Servais *et al.* 2001) and Pomerania (Samuelsson *et al.* 2002*a*). This allows better dating of deposition, sedimentation gaps and breaks caused by deformation phases (Fig. 2) and a first analysis of the basin evolution. Nd isotope studies allow the provenance of the sediments in the different areas to be determined (Gerdes *et al.* 2000, 2001*a*,*b*). New, detailed structural, geochemical and metamorphic studies in the Brabant Massif (Debacker 2001*a*; Piessens 2001) and Ardennes inliers (Belanger 1998; Schroyen 2000) constrain the tectonometamorphic history more accurately. New studies on potential field data in SE Avalonia provide physical parameters and limits for the different parts of the crust (Everaerts 2000).

The purpose of this paper is to provide detailed new data on Belgium, southern Denmark and northern Germany and relate them within an overview of the recent data and relevant references to other parts of Eastern Avalonia. Observations are reviewed in the early sections, followed by the interpretations in the discussion. The constraints to which a model for the basin and deformation history of Eastern Avalonia should comply are summarized, followed by a tentative model.

Nomenclature

For some geological or tectonic units it has been necessary to make a distinction between the environment of sedimentation (shelf, slope or basin), the areas that underwent deformation (orogenic or deformation belts), and the present day outcrop or subcrop areas (as inliers or massifs), which reflect the results of a Variscan or later history of deformation, uplift and erosion.

Throughout this paper the geological/tectonic units are defined as follows: the Midlands Microcraton (MC, Fig. 1) (Turner 1949; Soper *et al.* 1987) comprises the shallow Proterozoic basement beneath central England, unconformably covered by weakly deformed Palaeozoic strata of the Midlands Platform, here divided into eastern and western parts (see Figs 1 & 2). The Anglo-Brabant Deformation Belt (ABDB, Fig. 1) (Winchester *et al.* in press), previously referred to as the Caledonides of the Midlands and Brabant Massif (Verniers *et al.* 1991), the Caledonides of the Anglo-Brabant Massif (Pharaoh *et al.* 1993*b*) and the Anglo-Brabant fold belt (Van Grootel *et al.* 1997), forms a folded, faulted and weakly metamorphosed belt in the subcrop of East Anglia. According to interpreted potential field images and a few deep boreholes, this continues under the North Sea (Lee *et al.* 1993; Pharaoh *et al.* 1993*b*) into the Brabant Massif. The English part of the fold belt is called the concealed Caledonides of eastern England (Pharaoh *et al.* 1987). The Anglian Basin (AB, Fig. 1) (Woodcock & Pharaoh 1993) is defined as the predominantly Silurian sedimentary basin in the subsurface of East Anglia, which extends offshore to the Dowsing-South Hewett Fault Zone (DSHFZ, Fig. 1) (Pharaoh 1999). It was deformed by the Acadian Phase and became part of the Anglo-Brabant Deformation Belt (Van Grootel *et al.* 1997).

The Brabant Massif (BB, Fig. 1) (Dumont 1847) is defined by the present day outcrop and (sub-Mesozoic or Cenozoic) subcrop of Lower Palaeozoic rocks in central and west Belgium, northern France and southwestern Netherlands, which are unconformably covered by Middle Devonian strata. The depositional basin is here called the Brabant Basin, with, in late Ordovician and Silurian time, a shelf area in the southwestern part, south of the Ronse – Veurne line (Fig. 6), the southwestern Brabant Shelf; and a basinal area north of it, the central and northern Brabant Basin. Most of the Brabant Massif, except for its SW part, was folded, faulted and weakly metamorphosed during the Brabantian Deformation Phase, coeval with the Acadian Phase and hence belongs to the Anglo-Brabant Deformation Belt (Van Grootel *et al.* 1997). Within the Brabant Massif three tectonic domains are recognized: (1) a southwestern undeformed domain, south of the Brabantian deformation front (Sintubin 1999); (2) a southern domain (Sintubin 1997*a*), or Ordovician–Silurian domain (Sintubin 1999), and (3) a northern domain (Sintubin 1997*a*), or Cambrian core domain (Sintubin 1999) or steep belt (Sintubin & Everaerts 2002). Sintubin (1997*b*), using the aeromagnetic data of the Belgian Geological Survey (1994), distinguished three domains: a southeastern, a central and a southwestern domain. Mansy *et al.* (1999), also using potential field data, distinguished only the northern and southern Brabant subdomains, which reflect differences in the composition and structure of the crust.

The Condroz Inlier (Dumont 1847), previously also called Condroz ridge or strip, *Bande d'entre Sambre et Meuse* (d'Omalius d'Halloy 1842 p. 27), or Sambre and Meuse strip (Verniers & Van Grootel 1991), is a long and narrow Ordovician–Silurian inlier, south of the Sambre and Meuse rivers and north of the Condroz Plateau. The small Oxhe Inlier is also included. It contains different sedimentation areas, mostly of the Condroz Shelf, some with a different tectonic history, and juxtaposed by Variscan thrust faults (Verniers *et al.* 2001).

A distinction is made between (micro-)plates, which are still moving, from terranes which are bordered by fault contacts and represent the collided or docked parts of previous (micro-)plates. Hence the term Southern North Sea–Lüneburg Terrane (SNSLT, Fig. 1) (Pharaoh *et al.* 1995; Franke 1995*a*) represents the present-day expression of the Far Eastern Avalonian microplate (Winchester *et al.* in press). For the definition of other terms on Figure 1 we refer to Pharaoh (1999) and for the definition of Heligoland–Pomerania Deformation Belt we refer to Winchester *et al.* (in press). Recently the Polish part of this orogen was called Pomeranian Caledonides by Dadlez (2000).

In their redefinition of the Caledonian Orogeny, McKerrow *et al.* (2000) stated that it includes 'all Cambrian to Devonian tectonic events associated with the development and the closure of the Iapetus Ocean, situated between Laurentia, Baltica and Avalonia'. The deformation phase present in the Brabant Massif was included by the latter authors in the new definition and, with more reservation, was also linked to the events associated with the closure of the Tornquist Sea. In this study local geographic names are used to indicate previously defined, local orogenic belts and their deformations at the different margins of Avalonia: e.g. Shelveian Phase, Brabantian Phase, Condrozian Phase, Ardennian Phase, Scandian Phase (Fig. 2). The term Acadian Phase is used to group all broadly early Devonian deformation phases on the east coast of North America, in Ireland and UK during the final collision of Avalonia, Baltica and Laurentia (McKerrow *et al.* 2000). Throughout the paper the chronostratigraphy by Gradstein & Ogg (1996) is followed, and the time scale of Tucker & McKerrow (1995), except where indicated.

Faunal and floral provinciality from Cambrian to Devonian

Cocks and Fortey (1982) were the first to demonstrate by means of benthic fossils that in Cambrian to early Ordovician times, southern Britain, Ireland and eastern Newfoundland, later called Avalonia (see below) were situated at high latitudes attached to Gondwana, whereas northern Britain and western Newfoundland were at equatorial latitudes attached to the North American continent. Both areas were separated by the Iapetus Ocean. At temperate latitudes the Baltica palaeocontinent was situated in between both other continents. It implies that 'Avalonia' moved in the Ordovician from Gondwana in the direction of Baltica with an ocean in between, called the Tornquist Sea (Cocks & Fortey 1982). They showed that certainly from the late Caradoc, faunas from southern Britain are very similar to those from Baltica. In Soper and Hutton (1984) the New England part and eastern Newfoundland are named for the first time Avalonia, and the southern Britain part Cadomia; whereas in Soper (1986) the former is named 'Western Avalonia' and the latter 'Eastern Avalonia'. Cocks *et al.* (1997) review the margins of Avalonia and argue that no faunal evidence exists for a separation in the Ordovician between the eastern and western parts, terms that can only be used in the present day geographical sense, following Mesozoic–Cenozoic opening of the Atlantic Ocean.

The faunal studies proving the provinciality of Eastern Avalonia in the mid to late Ordovician are situated mostly in Wales, the Welsh Borderland and the Lake District (Cocks & Fortey 1982, 1990). Benthic faunas further east are only found in the Condroz Inlier (Verniers & De Vos 1995) and the Ebbe Inlier. In the former area Llanvirn trilobites in the Oxhe Inlier show an affinity with Bohemia, northern England and Wales (Dean 1991) and middle Caradoc trilobites in the Oxhe Inlier indicate a 'NW Europe Terrane' (Dean 1991), distinct from Baltica. In the Ebbe Inlier, Llanvirn to Caradoc trilobite faunas indicate a southern (Mediterranean) faunal province (Koch 1999; Samuelsson *et al.* 2002*b*). Evidence from benthic faunas recording the first contact of Avalonia with Baltica appears in the Ashgill of Fosses (Condroz Inlier). Brachiopods, trilobites and cystoids show a North European affinity closest to Scandinavia and the Baltic area with other brachiopods well represented in Ireland. The faunal relationship with Bohemia and the Armorican Massif is limited (Regnell 1951; Lespérance & Sheehan 1988; Sheehan 1987). The Fosses bryozoans are most similar to those from Wales with some similarity with Baltica. Rugose and tabulate corals show affinities with the Baltic area and less obviously with Wales and northern England. Algae and corals indicate a tropical position (Tourneur *et al.* 1993).

On the Southern North Sea–Lüneburg Terrane no benthic fossils have been described that could prove the palaeoprovinciality of that supposed part of Avalonia. Planktonic fossils cannot be used to indicate palaeoprovinciality but they can indicate palaeolatitudes (Cocks & Verniers 2000). The planktonic acritarchs and chitinozoans in the Middle Ordovician of the Rügen and Pomerania subsurface are clearly of a high latitude and not of a low latitude Baltic affinity (Servais 1994; Servais & Katzung 1993; Servais & Fatka 1997; Samuelsson *et al.* 2000, 2001; Vecoli & Samuelsson 2001*a contra* Cocks *et al.* 1997). They indicate a southern provenance for this microplate and hence it is thought to be Gondwana-derived (Servais 1994; Vecoli & Samuelsson 2001*b*; Samuelsson *et al.* 2002*a*).

From late Llandovery to Ludlow time, planktonic chitinozoans indicate a close relationship between the Brabant Massif, Condroz Inlier, Welsh Borderland, Baltica and Bohemia and a much smaller similarity to the Armorican Terrane Assembly, northern Gondwana or Laurentia (Verniers 1983).

Borehole G14, north of Rügen in the Baltic Sea, contains sediments deposited on Baltica, as proven by its Middle Ordovician low latitude fossils or fauna endemic to Baltica (Berthelsen 1992). In the mid Ashgill strata of the borehole, Samuelsson *et al.* (2001) found reworked Middle Ordovician acritarchs, with a high-latitude provenance. The sediments were derived from the erosion of an accretionary wedge south of the Thor-Tornquist Suture, called the Heligoland–Pomerania Deformation Belt (Winchester *et al.* in press). Samuelsson *et al.* (2001) concluded that the Tornquist Sea had closed and by mid Ashgill time and possibly earlier, Gondwana-derived shelf sediments could overstep the former oceanic suture. From faunal and floral evidence (see above) a mid Caradoc to mid Ashgill closure of the ocean was proposed (Vecoli and Samuelsson 2001*b*), compatible with the evidence of metamorphic ages from North Sea deep boreholes (Frost *et al.*1981; Pharaoh *et al.* 1995).

Palaeogeographical reconstructions of Avalonia based on palaeomagnetism

Using palaeomagnetic evidence, Trench and Torsvik (1991) were the first to prove that southern Britain had moved northwards away from Gondwana. Torsvik *et al.* (1993) reconstructed the palaeogeographical position of Avalonia based on palaeomagnetic data from Wales, England and Western Avalonia and the neighbouring palaeocontinents Baltica, Gondwana and Laurentia. From Cambrian to Tremadocian time Avalonia was attached or close to northern Gondwana, with, in the Cambrian, a very high latitude position and in the Tremadocian, an approximately 60° south position for Wales. By Llanvirn time Avalonia was moving away from Gondwana, leading to the opening of the Rheic Ocean between Avalonia and Gondwana. A 45° south position is estimated for Avalonia at the end of the mid Ordovician. By the late Ordovician a position of Avalonia between 30 and 40° south was deduced and close to Baltica, with an inferred closure of the Tornquist Sea. By the Wenlock, Avalonia was located at 13° south. Baltica rotated substantially anticlockwise from the late Cambrian to the early Ordovician (about 100°) and from late Ordovician to mid Silurian (about 50°) (Trench & Torsvik 1992; Trench *et al.* 1992; Torsvik *et al.* 1992, 1993, 1996; Tait *et al.* 1997, 2000; Torsvik 1998).

Piper (1997) compared palaeomagnetic signals from the Lake District with those from Wales and deduced that the Lake District rotated 55° anticlockwise with respect to Wales from the Caradoc–early Ashgill until the mid Devonian. As the Anglo-Brabant Deformation Belt was situated at the eastern side of the Midlands Microcraton, a similar rotation is expected to have occurred during the same period. The geological consequences of this supposed rotation need to be accounted for in any model.

Stratigraphy, sedimentology, sediment provenance and subsidence and basin evolution

Welsh Basin (Figs 1 & 2, column 3)

Three major unconformities are found, separating three sedimentation megasequences that correspond roughly to the Dyfed, Gwynedd and Powys supergroups of the Welsh Basin (Woodcock 1990*a*, 1991). The term megasequence is used as defined by Woodcock (1991). Megasequence 1 (early Cambrian–Tremadocian), the Dyfed Supergroup, comprises marine strata deposited on the episodically-rifted, southern margin of the Iapetus Ocean, while Eastern Avalonia was still attached (or proximal) to Gondwana. An early phase of shallow marine deposition, associated with a pulsed transgression (Woodcock & Strachan 2000), recognized throughout the basin, was followed by turbidite deposition in localized areas with rapid subsidence, most notably in the Harlech Dome. More than 3600 m of volcanic and related strata were deposited at Rhobell Fawr, in the Welsh Basin (Kokelaar 1986), recording the onset of subduction of Iapetus Ocean crust beneath Avalonia. Megasequence 2 (Arenig–Ashgill), the Gwynedd Supergroup, comprises a mainly muddy offshore marine sequence with a variable contribution of volcanogenic detritus from a developing arc system, associated with subduction of the Iapetus Ocean beneath Avalonia. Initially (Tremadocian–Arenig), calc-alkaline in aspect, the volcanism had by Llanvirn and Caradoc times become dominantly bimodal and submarine in character (Howells *et al.* 1990), and the Welsh Basin had transformed into a back-arc marginal basin (Kokelaar *et al.* 1984). Megasequence 2 is terminated by the cessation of volcanism related to the subduction of the Iapetus Ocean and Tornquist Sea in Ashgill time, and by a regional unconformity most strongly developed adjacent to ancient basement lineaments such as the Welsh Borderland Fault System and the Tywi Anticline, which may have been reactivated by Shelveian (Toghill 1992) dextral strike-slip deformation (Lynas 1988). The cause of this deformation was probably the docking/collision of Avalonia with Baltica (Pharaoh *et al.* 1995) and will be described in more detail below. Megasequence 3, the Powys Supergroup (Ashgill–Emsian), comprises mainly turbiditic strata deposited in rapidly subsiding deepwater successor basins, e.g. in central Wales. Volcanogenic rocks are absent, with the exception of rather thin bentonites representing air fall ashes. The later strata record a shallowing (Ludlow) followed by emergence (Přídolí–Lochkovian), with deposition of the 'Lower Old Red Sandstone' alluvial facies.

Lake District and Pennine inliers (Figs 1 & 2, columns 1 & 2)

The sequence stratigraphic scheme developed by Woodcock (1990*b*) can also be applied in northern England, making allowance for the closer proximity of the latter to the trench associated with subduction of the Iapetus Ocean. The earliest exposed strata here are the Skiddaw and Ingleton Groups of Tremadocian–Llanvirn age, comprising turbiditic sandstones and olistostromes (Cooper *et al.* 1993) deposited in deep water in a fore-arc basin at the northern margin of Eastern Avalonia, and incorporated into the subduction–accretion complex there. In Llanvirn to early Caradoc time eruption of the thick (up to 6000 m) Borrowdale Volcanic Group, a calc-alkaline suite comprising andesite lavas, tuffs and voluminous felsic ignimbrites, reflects a further phase in subduction of the Iapetus Ocean. A regional unconformity separates megasequence 2 from the overlying Windermere Supergroup (Ashgill–Ludlow) of megasequence 3, particularly well demonstrated in the Pennine inliers. Fold structures developed in the underlying Borrowdale Group are attributed to volcano-tectonism (Branney & Soper 1988), rather than Shelveian deformation, which is apparently less intense in northern England than in Wales. The Windermere Supergroup comprises mainly muddy strata deposited in a rapidly deepening foreland basin during and following closure of the Iapetus Ocean. It contains post-collision Laurentia-derived overstep sequences of mid Wenlock age (King 1994). As in Wales, Ludlow strata record the shallowing-up of this basin towards the end of Silurian time, but Lower Devonian strata are not preserved.

Subsidence curves were calculated for the Lower Palaeozoic in the Avalonian part of the UK by King (1994) for the Upper Ordovician, Silurian and Lower Devonian, and by Prigmore *et al.* (1997) for the Cambrian and Ordovician. They recognized a transtension in the early Cambrian (545–518 Ma) after the Neoproterozoic Orogeny in Avalonia, followed in mid Cambrian to early Tremadocian (495–485 Ma) by a transtension associated with the opening of the Iapetus Ocean. During the late Ordovician (Caradoc 461–449 Ma) pronounced subsidence occurred associated with back-arc rifting. During the Silurian (443–417 Ma) development of a foreland basin is deduced from the convex upwards curves in the Lake District and also in the Craven Inlier. At the same time extensional basins with concave upwards subsidence curves developed in Wales and the Welsh Borderland (UK). Non conclusive straight subsidence curves were recorded on the Midlands Microcraton (King 1994).

Midlands Microcraton and Platform (Figs 1 & 2, columns 4 & 5)

A weakly metamorphosed Neoproterozoic basement comprises volcanic arc and marginal basin sequences (e.g. Charnian Supergroup at Nuneaton) accreted to the Rodinia-Pannotia Supercontinent (subsequently, Gondwana) between 680 and 545 Ma (Pharaoh & Gibbons 1994). Isotopic studies suggest derivation from near the Amazonian Craton (Nance & Murphy 1994; Murphy *et al.* 2000) and hint that a more ancient crystalline crust may be present, up to about 1450 Ma old (Tucker & Pharaoh 1991; Noble *et al.* 1993). Major crustal lineaments form the boundary to the microcraton in the Welsh Borderland (Woodcock & Gibbons 1988; Gibbons 1990). The Cambrian overstep sequence (megasequence 1) is directly correlated with Avalonian strata in SE Newfoundland (Brasier *et al.* 1992). A phase of rapid subsidence in Tremadocian time, notably in the West Midlands (Smith & Rushton 1993), may presage the outbreak of arc-related volcanism in the Welsh Basin. Megasequence 2 is largely absent from the microcraton, which at this time was an emergent region lying behind the volcanic arc established in the Welsh Basin. Strata of megasequence 3 (Silurian) are shelf mudstone and limestone which form a strong contrast with the contemporaneous deep water, largely turbiditic strata, deposited in Wales, northern and eastern England (Molyneux 1991; Woodcock & Pharaoh 1993).

Eastern England Caledonides and Anglian Basin (Figs 1 & 2, column 6)

Neoproterozoic basement here is petrographically and geochemically distinct from that of the Midlands Microcraton, and more akin to the Arvonian basement of North Wales. These characteristics suggest that the boundary between the MMC and EEC may date from latest Neoproterozoic time. Numerous boreholes prove quartzitic metasediments, so far undated, but of possible Cambrian age in view of lithological similarity to the Tubize Formation in Belgium. A late Ordovician calc-alkaline arc is traced from north England to Belgium and is inferred to result from subduction of Iapetus/Tornquist oceanic lithosphere beneath Eastern Avalonia (Pharaoh *et al.* 1993*a*; Noble *et al.* 1993; see below). In the Anglian Basin the strata of megasequence 3 are of a deep water facies (Woodcock & Pharaoh 1993) and much more strongly deformed than the contemporaneous strata of the Midlands Microcraton. These deep-water basins inverted during the Acadian Phase, and now form the Acadian slate belts of Wales, north and east England (Turner 1949; Soper *et al.* 1987). In these regions, large granitic intrusions were emplaced in two phases, first associated with Ordovician subduction-related magmatism (Wales, Lake District) and second, following crustal-thickening during Acadian deformation, in early Devonian time (north England). The belt of strong Acadian deformation, the Anglo-Brabant Deformation Belt, extends into Belgium (Lee *et al.* 1993).

Brabant Massif (Figs 1 & 2, column 7)

The three megasequences can also be distinguished in Belgium. Vanguestaine (1992) subdivided the Lower Palaeozoic into what he called megacycle I with Lower Cambrian to Tremadocian units, megacycle II with Arenig to Caradoc units and megacycle III with Ashgill to Silurian units. He concluded that they are comparable to the three megasequences described by Woodcock (1990*b*) in the Welsh Basin (see above). Renewed litho- and biostratigraphical studies with acritarchs, chitinozoans and graptolites in Belgium (Servais *et al.* 1993; Van Grootel *et al.* 1997; Maletz & Servais 1998; Samuelsson & Verniers 2000; Verniers *et al.* 2001) allow a more precise description and detailed dating of the megasequences. Verniers & De Vos (1995) argued, using faunal data, that the base of megasequence 1 is basal Cambrian and not uppermost Neoproterozoic. As elsewhere in Belgium, the hiatus between megasequences 1 and 2 is long, from early Tremadocian to late mid Arenig. The next significant hiatus and/or condensed section is situated between the Llanvirn and early-mid Caradoc, as in the Welsh Basin. However, in the central and northern Brabant Basin sedimentation continued from mid Caradoc to mid Ashgill, unlike the shelf area of the Welsh Borderland. The central and northern Brabant Basin records sedimentation through this transition, as in the deeper parts of the Welsh Basin (Fortey *et al.* 2000). The time interval corresponds to the proposed docking of Eastern Avalonia with Baltica, closing the Tornquist Sea (Woodcock 1990*a*; Pharaoh *et al.* 1995; Vecoli & Samuelsson 2001*b*).

The base of megasequence 1 on a supposed Neoproterozoic craton has not been observed. The presence of a Proterozoic crystalline basement beneath or near the Brabant Massif is inferred from inclusions and xenoliths in the Upper Ordovician magmatic rocks, from lithic fragments with a predominance of metavolcanic rocks in the Lower Cambrian sediments (Vander Auwera & André 1985; André 1991) and also from the presence in the Blanmont sandstone Formation of mainly rounded to euhedral colourless zircons which were formed during a magmatotectonic event dated as latest Neoproterozoic (U–Pb data from 530 to 600 Ma; Von Hoegen *et al.* 1990). ε^t Nd(t) studies by André (1991) and Gerdes *et al.* (2000, 2001*a*,*b*) indicated initial erosion of a Neoproterozoic tholeiitic metavolcanic crust. From mid Cambrian onwards, an older Proterozoic crust was the source for the sediments in the Brabant Massif. Megasequence 1 contains earliest Cambrian (Blanmont, Tubize, Oisquercq, Mousty Formations) to early Tremadocian (Chevlipont Formation), often thick terrigenous and deep sea sediments with greywacke, sandstone and thick sequences of pelagic to hemipelagic mudstone. Coarse-grained sediments only occur in the basal part of the megasequence (Blanmont and parts of the Tubize Formations). The latter formation is interpreted as turbiditic (Vander Auwera & André 1985) for the coarser-grained middle member, and pelagic to hemipelagic for the lower and upper member (Herbosch *et al.* 2001). Complete sections and contacts between these formations or members are nowhere observed. There might be a hiatus in the mid Cambrian between the green to purple

Oisquercq mudstone and the dark grey and black Mousty shale Formation. Higher in the Mousty and Chevlipont Formations the sedimentation is fine-grained in a deep anoxic environment and seems to be continuous (Herbosch *et al.* 2001). The often turbiditic nature of the megasequence points to an environment deeper than the shelf. The thickness can be estimated at minimum 3700 m but is probably much thicker (Verniers *et al.* 2001). At the top of megasequence 1 a substantial hiatus occurs from the Lower Tremadocian to the upper part of the Middle Arenig (estimated hiatus of about 12 Ma, Verniers *et al.* 2001).

The terrigenous shallow shelf sediments of megasequence 2 (Abbaye de Villers and lower half of the Tribotte Formations) with an intertidal period (upper half of the Tribotte Formation, middle Arenig to middle Caradoc, Cheneyan) are much thinner than those of megasequence 1. The succeeding dark grey mudstone (Rigenée Formation) records a relatively rapidly subsiding shelf. Only at the top of the megasequence 2 does a drastic change in environment occur. After a long interval (hiatus or condensed sedimentation over about 7 Ma), a deeper environment was initiated rapidly with sedimentation of distinct turbidites of the Ittre Formation passing into less energetic turbidites of the Bornival Formation (Burrellian, early mid Caradoc; Servais 1991; Herbosch *et al.* 2001). A Bouma-type turbiditic sedimentation in Caradoc time is only known around the Midlands Microcraton in the Brabant Massif and with minor importance in West Wales (Poppit Sands Formation). The thickness of megasequence 2 is estimated at minimum 850 m (Verniers *et al.* 2001). The stratigraphic contact with the overlying megasequence 3 is unknown because of the presence of faults.

The transition from megasequence 2 to 3 witnesses a drastic change from deep water to shelf with only a short hiatus in time, less than 1 Ma, as estimated from chitinozoan biostratigraphy (Verniers *et al.* 2001). Megasequence 3 records shelf deposition in the upper Caradoc to middle Llandovery which evolves into a thick foreland basin deposit in the upper Llandovery (upper Telychian) to Přídolí. At the bottom muddy sandstone with bioclasts deposited on a shelf with detritus derived from a quite distant carbonate platform (Huet Formation) evolved into an anoxic dark grey graptolitic mudstone (Fauquez Formation, transition Caradoc–Ashgill; Herbosch *et al.* 1991), which is followed by many volcanic and volcanosedimentary rocks deposited on a shallow shelf (Madot and Brutia Formations, Ashgill to Lower Llandovery; Mortelmans 1952; Van Grootel *et al.* 1997). From mid Llandovery time two distinct basinal areas developed. In the southwestern Brabant Massif (south of the Ronse – Veurne Line) a deep shelf environment persisted, the southwestern Brabant Shelf, while in the outcrop area, the north and central Brabant Basin, a turbiditic regime was present on a slope or deep basin. Thick turbidite sequences are well developed from the Upper Telychian (Latinne, Hosdin and Fallais Formations). They were distal at first and, from the base of the middle Wenlock more proximal. This type of sedimentation lasted until the early Ludlow in the basinal parts and until the Přídolí in the southwestern Brabant Shelf. The thickness of the megasequence is estimated at more than 3200 m in the outcrop area of the Brabant Massif and much more than 470 m in the southwestern Brabant Shelf (Verniers 1983; Verniers & Van Grootel 1991; Verniers *et al.* 2001). Debacker (2001*a*) constructed a cumulative thickness curve for the central and north Brabant Basin, using the new stratigraphical thickness estimates of Verniers *et al.* (2001) (Fig. 3). Although the curve is not corrected for compaction or tectonic thickening, it shows a clear concave-up form in megasequence 1, indicating an extensional or rift basin in the early Cambrian to Tremadocian of the Brabant Massif. The curve is nearly straight and not conclusive in megasequence 2 and is convex-up in megasequence 3, indicating a foreland basin development, as already proposed in Van Grootel *et al.* (1997) with corrected subsidence curves.

Condroz Inlier (Figs 1 & 2, column 8)

Only the upper part of megasequence 1, i.e. the Chevlipont Formation, is found in the Condroz Inlier (Wépion borehole, Graulich 1961) with distinct low-density distal turbidites, identical in facies to the same formation in the Brabant Massif (Herbosch *et al.* 1991; Verniers *et al.* 2001; Herbosch *et al.* 2001). The unconformity at the top of megasequence 1 is well constrained, with a hiatus from the early Tremadocian to Llanvirn (Graulich 1961; Servais & Maletz 1992).

Megasequence 2 begins with a basal conglomerate, followed by Llanvirn graptolitic mudstone with benthic macrofauna (Huy and Sart-Bernard Formations; Graulich 1961; Servais & Maletz 1992). After a hiatus or condensed section of about 6 Ma it is covered by mid Caradoc (Burrellian) micaceous siltstones with occasional quartzitic beds and graptolitic levels (Vitrival-Bruyère Formation; Michot 1954; Herbosch *et al.* 2001). In a lateral facies, in the small

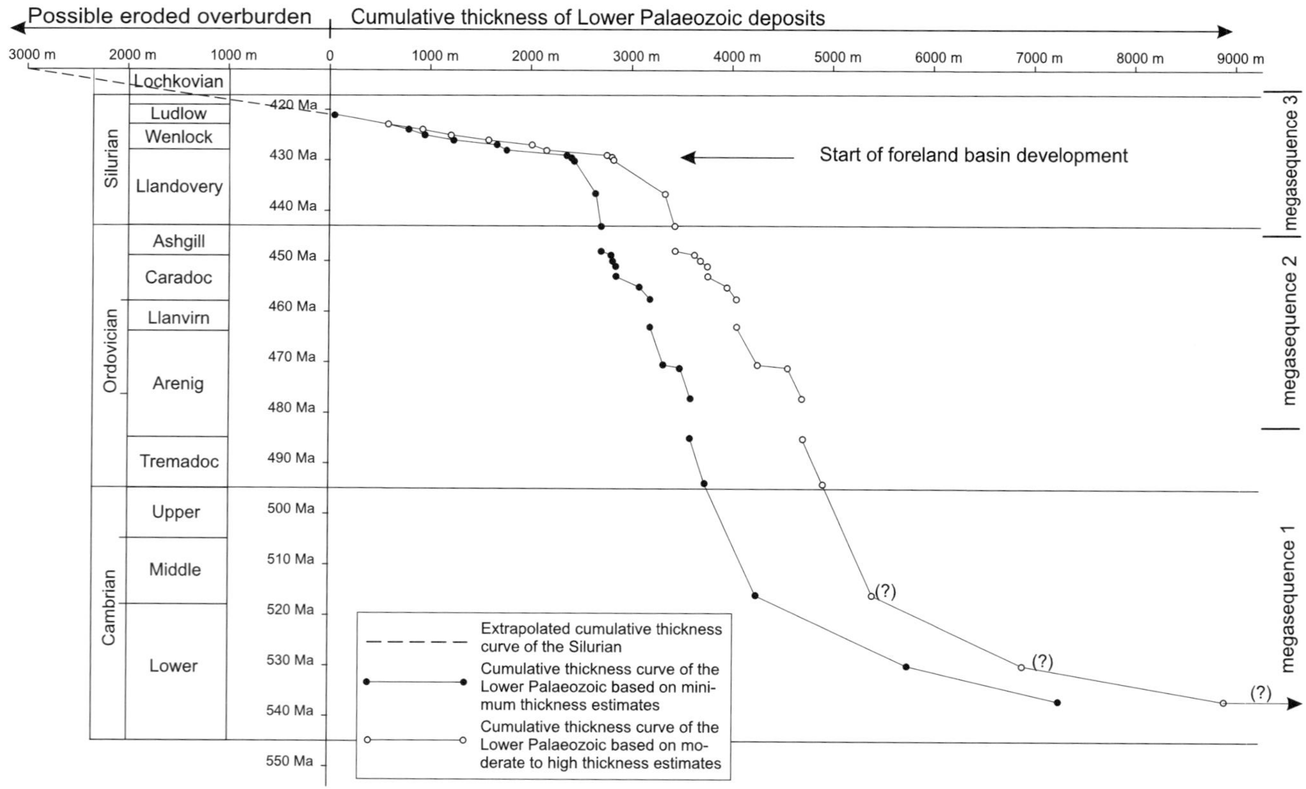

Fig. 3. Cumulative thickness curve of the Lower Palaeozoic sediments of the Brabant Massif, central and north Brabant Basin, taken from Verniers *et al.* (2001), plotted against the absolute time-scale of Gradstein & Ogg (1996) (Debacker 2001*a*). The high rates of sediment thickness against time resulting from Silurian foreland basin development (Van Grootel *et al.* 1997) are extrapolated to higher stratigraphic levels in order to get a minimum value for the age of compressive deformation along the southern, Silurian rim of the Brabant Massif, taking into account the overburden necessary for cleavage development and metamorphism at moderate geothermal gradients.

Oxhe Inlier, deep shelf deposits occur with benthic macrofauna (trilobites and ostracodes) and occasional storm beds (Oxhe Formation; Dean 1991; Schallreuter *et al.* 2000).

After a probable hiatus from the mid Caradoc to the mid Ashgill (about 9 Ma) megasequence 3 starts with a (calcareous) mudstone sedimentation of the Fosses Formation (Upper Ashgill). At the top of this formation two thin (<2 m sandstone or conglomerate strata contain lower to middle Ashgill limestone boulders indicative of a nearby carbonate platform (Tourneur *et al.* 1993; Herbosch *et al.* 2001). These two coarse sand or conglomerate levels are tentatively dated as Hirnantian using chitinozoans, and interpreted as resulting from the eustatic sea-level drop caused by Hirnantian glaciation (Herbosch *et al.* 2001). In the Silurian, the rather thick and continuous mostly argillaceous sedimentation continues until the late Ludlow. No turbidites have been observed. Some levels contain calcareous nodules: they occur in the upper Wenlock part of the Jonquoi Formation in the same stratigraphical level as the carbonate-rich levels in the Welsh Borderland (Wenlock Limestone) (Maes *et al.* 1979). The numerous levels with graptolites and occasional benthic fauna such as brachiopods or trilobites (Michot 1954) indicate a deeper shelf environment of deposition. Due to the poor state of the outcrops and the shortage of detailed stratigraphical studies neither subsidence nor provenance studies can be made, but it seems likely that the subsidence rate did not accelerate or slow down. However, the presence of an often graptolitic mudstone facies lacking turbidites, and several long-lived hiatuses or disconformities, shows clearly that most of the Condroz Inlier was situated on a shelf.

Ardennes inliers and Bolland borehole (Figs 1 & 2, column 10)

There are four Ardennes inliers. In the two larger inliers, Stavelot-Venn and Rocroi, only megasequences 1 and 2 are present. The two smaller ones, Serpont and Givonne, contain only parts of megasequence 1 and will not be discussed further. Megasequence 3 (late Ordovician and Silurian) is absent due to the deformation of the inliers during the Ardennian Phase (Michot 1980). After erosion of the deformation belt, the Rocroi, Serpont and Givonne inliers were covered by Přídolí or lower Lochkovian strata, and the Stavelot-Venn Inlier, situated further to the north, was covered by upper Lochkovian strata (Borremans & Bultynck 1986; Steemans 1989; Godefroid *et al.* 1994; Godefroid 1995; Godefroid & Cravatte 1999).

The Stavelot-Venn Inlier contains at the base, above an inferred Neoproterozoic basement (Von Hoegen *et al.* 1990), pale coloured quartz-arenite and mudstone deposited on a relatively near-shore to intertidal shallow shelf, and later on a more open and deep shelf (Deville Group; Von Hoegen *et al.* 1985, 1990; Geukens, 1999). Higher up, a coarse turbiditic facies alternates with mudstone-siltstone dominated siliciclastics, probably deposited as turbidites on the lower slope/upper fan region of a submarine fan-valley system (Revin Group; Von Hoegen *et al.* 1985; Geukens, 1999; Verniers *et al.* 2001). No significant gaps are observed, although the transition from the Deville to the Revin Groups was never observed without faulted contacts. The top of the Cambrian terminates with black mudstone (La Gleize Formation). The Tremadocian (Jalhay Formation) starts with Bouma-type turbidites (Solwaster Member), is followed by low-density turbidites (Spa Member), and ends with mudstone and sandstone deposited on a platform (Lierneux Member) (Lamens 1985; Lamens & Geukens 1985). Hence the Tremadocian is thought to represent the filling of an epicontinental basin by a northward prograding clastic wedge (Lamens 1985, 1986). The fine-grained turbidites of the Spa Member are identical in facies and age to the Chevlipont Formation of the Condroz Inlier and the Brabant Massif (Herbosch, unpublished data). The late Tremadocian upper member (Lierneux) (Catot 1992; Vanguestaine 1989, 1992) or its time-equivalents are not found elsewhere in Belgium. All units described above are mostly deposited in a deep environment and are grouped here in megasequence 1. Its thickness, difficult to evaluate, is estimated to exceed 2400 m, with only 300–400 m for the Lower Cambrian shelf deposits (Verniers *et al.* 2001).

According to Vanguestaine (1992) a gap of uncertain duration exists between megasequences 1 and 2. A drastic change in depositional environment is observed, from a shallow platform at the top of megasequence 1 (Lierneux Member of Jalhay Formation) to a probable deep water depositional environment of the mudstone-siltstone dominated siliciclastics with turbiditic levels of the Ottré Formation (?Arenig) at the base of megasequence 2 (Lamens *et al.* 1986). These pass into fine siliciclastic pelagic to turbiditic sediments of late Arenig or Llanvirn age (Bihain Formation) deposited in a deep-sea environment (Lamens 1988). The thickness of megasequence 2 is estimated at more than 330 m.

In the Rocroi Inlier a similar megasequence 1 is present but apparently less thick than in the Stavelot-Venn Inlier (1135–2020 m). Megasequence 2 is only partly represented (Vieux-Moulins de Thilay Formation, late Arenig to Llanvirn) (Verniers *et al.* 2001).

Thus, although in the Ardennes inliers megasequence 1 is similar to that of the Brabant Massif, the Lower Cambrian part in the Ardennes was deposited on a shelf, and in deeper water in the Brabant Massif. The Middle and Upper Cambrian and Tremadocian strata are about the same thickness in the Ardennes inliers and the Brabant Massif and are similar in depositional environment. However, sedimentation continued in the Stavelot-Venn Inlier until the late Tremadocian, later than in the other Belgian areas. Megasequence 2, well represented in the Stavelot-Venn Inlier, has a different composition in the Brabant Massif, where deposition was in a different basin.

The 3001 m deep Bolland borehole (Figs 1 & 2, column 9), in its lowest 126 m, contains folded mudstone of Arenig–Llanvirn age (Michot 1979; Vanguestaine *et al.* unpublished *in* Verniers & Van Grootel 1991), whose affinity to the Brabant Massif or Stavelot-Venn Inlier has not been established. The borehole is situated within the northernmost part of the Variscan deformation front (Hance *et al.* 1999). According to Michot (1979) the borehole may prove the presence of two successive deformation phases: (1) Lochkovian–Pragian unconformably covers cleaved Ordovician mudstone with a Llanvirn to early Lochkovian hiatus in between (corresponding to the Ardennian Phase); (2) gently dipping upper Frasnian covers folded Lochkovian–Pragian, indicating deformation during the Pragian–late Givetian, which he named the Bollandian Phase. This is now questioned by Vanbrabant (2001) in the Prayon outcrop, near Liège, where a similar discontinuity is visible, but which can be explained by disharmonic folding, excluding a tectonic unconformity.

From the Belgian data it becomes apparent that megasequence 1 is developed similarly in the three main areas, the Brabant Massif, the Condroz Inlier and the Ardennes inliers, and that megasequence 2, although similar in the Brabant Massif and the Condroz Inlier, differs substantially in the Ardennes inliers. The separation into three different areas is even more evident during megasequence 3. Shelf deposits occur in the Brabant Massif, in the southwestern Brabant Shelf, which is similar to the Condroz Shelf (Condroz Inlier). There is a turbiditic basin in the north and central Brabant Basin, and in the Ardennes inliers uplift and deformation occurs instead of sedimentation.

Krefeld High (Figs 1 & 2, column 11)

A 100 by 25 km large, NNW–SSE oriented subcrop high has been postulated under the city of Krefeld and between the cities of Düsseldorf and Nijmegen. It is supposed to contain Lower Palaeozoic and Proterozoic rocks. Thick Givetian conglomerates near Düsseldorf testify to a nearby sediment source. Boreholes proved Eifelian shallow-marine strata below these conglomerates, indicating a Givetian uplifted area, the Krefeld High, to the west, at present situated below the Lower Rhine Embayment (Neumann-Mahlkau 1982; Neumann-Mahlkau & Ribbert 1998). The high was uplifted from the Givetian until the end of the Famennian and was the source area of the Upper Famennian feldspar-rich littoral sandstone of the Dinant Synclinorium (Thorez 1969 unpublished, in Paproth *et al.* 1986; Paproth *et al.* 1986; Thorez *et al.* 1988 and Thorez pers. comm.). Their geochemical analysis indicates the (Middle Netherlands-) Krefeld High as the nearby gneissic source, which by unroofing became increasingly feldspathic and micaceous during the late Famennian. A Scandinavian source is excluded because of the distance and the tropical weathering in that area, which does not fit the presence of abundant and fresh feldspar in the Famennian in Belgium. This is the only indirect indication of a nearby Proterozoic basement, close to the surface, north of the Brabant Massif.

Ebbe and Remscheid inliers of the Rhenish Massif (Figs 1 & 2, columns 12 & 13)

Ordovician and Silurian rocks, structurally surrounded by Lower Devonian (Lochkovian) rocks, occur in the Ebbe and the Solingen-Remscheid-Altena anticlines in Sauerland, east of Cologne, Germany. These sediments are usually portrayed as belonging to Avalonia, together with most of the parautochthonous parts of the Rheno-Hercynian Belt (e.g. Pharaoh 1999). The strongly tectonized Herscheider Schichten are a monotonous 800 m thick clastic succession of dark blue, grey and black mudstone and siltstone without any carbonate content. The lithological monotony of the succession makes it difficult to distinguish between the four different lithological units. The up to 65 m thick Plettenberger Bänderschiefer is a dark compact mudstone with abundant thin pyritous silty layers

(Eiserhardt *et al.* 1981). The overlying 150–200 m thick Unterer (Kiesberter) Tonschiefer is a dark mudstone with infrequent sandy layers. The (Rahlenberger) Grauwackenschiefer is about 300 m thick, and contains true thin-bedded greywacke, and siliceous concretions in the upper part. The youngest unit is the Oberer (Solinger) Tonschiefer which reaches some 200 m thickness in the Ebbe Inlier. The Herscheider Schichten have been dated using chitinozoans (Samuelsson *et al.* 2002*b*), graptolites and acritarchs (Maletz and Servais 1993), but trilobites, phyllocarids, foraminifers, ostracods and trace fossils have also been recorded (Koch 1999). Their inferred age-ranges are as follows (Samuelsson *et al.* 2002): Plettenberger Bänderschiefer (early Abereiddian, earliest Llanvirn); Unterer (Kiesberter) Tonschiefer (early to mid Abereiddian, early Llanvirn); (Rahlenberger) Grauwackenschiefer (Aurelucian, earliest Caradoc); and Oberer (Solinger) Tonschiefer (late Caradoc). The presence or a hiatus of the late Arenig could not be established. The highest units were only poorly dated until the study of Samuelsson *et al.* (2002*b*).

Several lines of evidence led Samuelsson *et al.* (2002*b*) to support the suggestion of Jaeger (1967) that the Herscheider Schichten were deposited in the same basin as the Rügen rocks. The Nd-isotope studies show a remarkable similarity between the Ebbe Inlier, the Stavelot-Venn Inlier, the Brabant Massif, the Lake District and the Welsh Basin (Gerdes *et al.* 2000, 2001*a*, *b*). This conclusion supported the idea of a continuation of Eastern Avalonia east of the Brabant Massif into northern Germany (e.g. Maletz & Servais 1993; Pharaoh 1999). However, although both areas have a similar deeper shelf depositional environment as shown by lithology, sedimentary structures and trace fossils, they could have been deposited on two different but neighbouring microplates of Avalonia which may perhaps be better termed the Avalonian Terrane Assembly.

The Herscheider Schichten are separated by a sedimentation gap from the overlying uppermost Silurian rocks of the Köbbinghäuser Schichten (Timm *et al.* 1981). This 100 m thick unit consists mainly of dark shales with carbonate bands (Timm 1981) dated by macrofossils as latest Silurian (post-Ludlow, i.e. Přídolí). The overlying *c.* 200 m thick Ockrige Kalke has a similar lithology to the Köbbinghäuser Schichten, but with more frequent carbonaceous bands. It belongs mainly in the early Devonian (early Lochkovian) which implies a further hiatus between the Ockrige Kalke and the overlying Hüinghäuser Schichten (Timm *et al.* 1981).

Southern North Sea–Lüneburg Terrane (Far Eastern Avalonia) (Fig. 1)

The crust below the *Anglo-Dutch Basin* (see Pharaoh 1999) is generally considered as part of Eastern Avalonia or an Avalonia-related microplate, but for 250 km to the NE of the Dowsing-South Hewett Fault Zone and north of the Rhenish Massif, pre-Devonian basement is not encountered by deep boreholes (Pharaoh 1999). Strong geophysical contrasts in the crust either side of this lineament led Pharaoh *et al.* (1995) to infer the presence of an Ordovician terrane boundary here, with the Southern North Sea Terrane (D. Franke 1995*a*, *b*; British Geological Survey 1996) extending as far NE as the Elbe Lineament (EL, Fig. 1). Boreholes only reach the basement at the Mid-North Sea/Ringkøbing-Fyn High and penetrated schists which underwent prograde greenschist-amphibolite metamorphism at 450–425 Ma, and retrogression at 415–400 Ma (Frost *et al.* 1981). The former is considered to mark docking of Avalonia to Baltica in late Ordovician time, contemporary with the Shelveian Phase in onshore UK (Toghill 1992; Pharaoh *et al.* 1995) and the Ardennian Phase in Belgium. Further evidence for the timing of docking comes from the G14 borehole near Rügen Island, located on the Baltica side of the Thor Suture, which began to receive reworked Gondwanan fossils of Llanvirn age and Neoproterozoic micas in early–mid Ashgill time (Samuelsson *et al.* 2001; Vecoli & Samuelsson 2001*a*,*b*). The detritus was fed across the suture zone from the Heligoland–Pomerania Deformation Belt in northern Germany and Poland (Fig. 1), which formed at the leading edge of Far Eastern Avalonia during the collision. Further deformation occurred during development of a foredeep on the East European Craton margin throughout the Silurian, culminating in the Scandian Phase. Gneisses outcropping in the Rheno-Hercynian Zone of Germany have yielded U–Pb ages in the range 575–560 Ma (Baumann *et al.* 1991; Molzahn *et al.* 1998) (see above) which suggest that the protolith of the crystalline basement of the Southern North Sea Terrane is of late Neoproterozoic age. Deep seismic reflection data indicate that this crust has a rather flat Moho, almost uniformally at 30 km depth, and a brightly reflective lower crust, comparable to that of Variscan Europe (Meissner & Bortfeld

1990). This apparently rules out the involvement of much older, thicker crust e.g. sourced from the East European Craton.

Danish and northern Germany subsurface, Rügen and Pomerania, NW Poland (Baltica and Heligoland–Pomerania Deformation Belt) (Figs 1 & 2, numbers and columns 15, 16, 17 & 18)

In the Danish subsurface, the Lower Palaeozoic consists of barely deformed, predominantly pelitic rocks (Sorgenfrei & Buch 1964; Poulsen 1969, 1974; Frost *et al.* 1981; Michelsen & Nielsen 1991; Vejbaek *et al.* 1994). The stratigraphy of the area can be exemplified by the relatively continuous, 400 m thick Cambrian to Lower Silurian subsurface section cut by the Slagelse-1 borehole of western Sjaelland. A comprehensive description of the lithology and biostratigraphy of this section can be found in Sorgenfrei & Buch (1964) and Poulsen (1969, 1974). The Lower Cambrian consists of basal quartzitic sandstone (2972–2966 m), followed by a dark-grey shaly succession between 2966 and 2944 m, rich in trilobites and graptolites (Poulsen 1969). An essentially black shale succession ('Alum Shales') between 2944 and 2917 m has been dated as Middle–Upper Cambrian. According to Poulsen (1974), the Lower Silurian (Llandovery) is developed between 2637 and 2811 m and consists of finely laminated black to grey silts and shales. The interval between 2811 and 2917 m, of similar lithology, is devoid of fossils and may be attributable to undifferentiated Ordovician – lowermost Silurian. Similar, but generally incomplete Lower Palaeozoic successions are encountered in several other wells drilled in the Norwegian-Danish Basin (e.g. Nøvling-1, Terne-1, and Pernille-1) and in borehole G-14, located in the southernmost Baltic Sea, offshore of Rügen Island (Franke *et al.* 1994). These successions are closely correlatable, both lithologically and biostratigraphically, with the classical platform, undeformed Lower Palaeozoic successions of Scania (Sweden) and Bornholm (Poulsen 1969, 1974; Michelsen & Nielsen 1991; Vejbaek *et al.* 1994; Samuelsson *et al.* 2001). The Lower Palaeozoic sediments of the Norwegian–Danish Basin as well as those of the Bornholm area indicate a long period of tectonic stability, with absence of any significant synsedimentary tectonic activity between the early Cambrian and the latest Ordovician/earliest Silurian. This was followed by a period of foreland basin-type rapid sedimentation starting in the Llandovery and continuing up to Wenlock times (Vejbaek *et al.* 1994).

New biostratigraphic (palynological) and geochronological ($^{40}Ar/^{39}Ar$) datings have been obtained from the lower Silurian sediments of boreholes Slagelse-1 and G-14 (e.g. Vecoli & Samuelsson 2001*a*). These new studies, together with other recent investigations of the Cambrian successions (e.g. Giese & Köppen 2001) confirmed that the Lower Palaeozoic sedimentary sequences of the Danish subsurface were deposited on the southwestern passive margin of Baltica, as the main detrital input was derived from the stable cratonic regions of the Baltic Shield. However, $^{40}Ar/^{39}Ar$ dating of numerous white mica grains from the Llandovery of the Slagelse-1 borehole, has yielded a mixed age spectrum with two distinct age populations clustered around the values 500–600 Ma, and 900–1000 Ma. The isotopic age values of 500–600 Ma were interpreted as indicating detrital provenance from a Cadomian-age domain, in addition to a clastic input from a Sveconorwegian consolidated basement (Baltic Shield). A similar mixed age spectrum was obtained by the isotopic dating of single white mica grains separated from Llandovery sediments of the G-14 borehole (Vecoli *et al.* 2000). From the isotopic dating of the white micas, it was inferred that detrital grains with a Cadomian age were transported from a Peri-Gondwana-related microplate onto the southwestern margin of Baltica. More precise provenance information can be derived from the interpretation of reworked microfossil associations of mixed age (late Cambrian to Llanvirn) in the uppermost Ordovician to Lower Silurian sediments of the Danish–North German Basin. The study of microfloral and microfaunal affinity of these reworked microfossils (acritarch and chitinozoans) permitted identification of a Peri-Gondwanan microplate as the source area with its related detritus encroaching on the southwestern border of Baltica since late Ordovician times. These findings support a late Ordovician docking/collision of 'Far Eastern Avalonia', with Baltica resulting in the Southern North Sea–Lüneburg Terrane.

In the Rügen area, Caledonian deformed units of early Palaeozoic age, thrust onto the Baltic Shield, were encountered and extensively cored by several boreholes both onshore and in adjacent offshore areas (e.g. boreholes Rügen-5, Rügen-3, Arkona-101, Lohme-2, Binz-1, H2, K5, L2–1). In this region, the stratigraphic cover of the borehole section is much more complete than in southern Jylland. Numerous studies, including provenance, structural and isotopic analysis, confirmed that the Lower Palaeozoic (mainly Ordovician) of Rügen and adjacent areas can be considered as nappe piles of Peri-Gondwana origin (e.g. Giese *et al.* 1994, 1997*a*, *b*; Katzung *et al.* 1993; Dallmeyer *et al.* 1999; Mc Cann 1998). Microfloral and microfaunal analyses permit assignment of the Rügen succession to a Peri-Gondwana derived microplate, probably Avalonia or Far Eastern Avalonia (Servais & Katzung 1993; Cocks & Verniers 2000; Samuelsson *et al.* 2000; Vecoli & Samuelsson 2000*b*; Cocks 2002).

The Ordovician sediments of West Pomerania, NW Poland, are closely comparable with the coeval strata present in the Rügen subsurface, and show clear similarities in their stratigraphical, and petrographical composition (e.g. Franke 1994). These rocks are well-known from several boreholes in the area (Bednarczyk 1974; Samuelsson *et al.* 2002*a*) and show evidence of folding and faulting. However, most of the sedimentary sequence is virtually unmetamorphosed and uncleaved (Franke 1994). The entire Ordovician succession, consisting of siltstone with sparse carbonaceous intercalations and concretions, is subdivided into five lithological units (Bednarczyk 1974; Samuelsson *et al.* 2002*a*) and is overlain by thick Devonian, Carboniferous and Permian successions.

The pioneering biostratigraphic work on the deformed Ordovician units by Bednarczyk (1974) has recently been refined. The recorded ages range from late Darriwillian (Llandeilian) to late Caradoc (Samuelsson *et al.* 2002*a*). Correlation with the succession underlying Rügen is straightforward, and quantitative assessment of microfossil assemblages of Pomerania and adjacent areas interpreted as belonging to Avalonia demonstrates the temporal and spatial association of Pomerania (and the Heligoland–Pomerania Deformation Belt) with Avalonia or Far Eastern Avalonia. A number of studies have demonstrated that onset of foreland basin sedimentation began in mid Caradoc to late Ashgill times in the Rügen area (e.g. Vecoli & Samuelsson 2001*b*). In Pomerania, clastic sedimentation was active at least to the mid Caradoc, further constraining the lower time limit for the start of the deformation of the Heligoland–Pomerania Deformation Belt (Samuelsson *et al.* 2002*a*).

Rheno-Hercynian zones of Avalonia in Cornwall (Cornubia) and central Germany (Giessen nappes) (Fig. 1, numbers 20 & 23)

The Rheno-Hercynian nappes of Cornubia, and central Germany as far south as the Northern Phyllite Belt, represent a part of Avalonia displaced by thrusting during the Variscan Orogeny, as they contain Avalonian faunas and therefore lie to north of the Rheic Suture (Cocks *et al.* 1997). Deep seismic profiles reveal that the internal structure of the Variscan Orogen changes in style from thick-skinned in the west (Cheadle *et al.* 1987) to thin-skinned, with ramp-flat geometry, in northern France, Belgium and Germany (Cazes *et al.* 1986; Bois *et al.* 1990; Meissner & Bortfeld 1990; Meissner *et al.* 1994). The Rheic Suture (Cocks & Fortey 1982; Ziegler 1990) lies between the Avalonian margin of Laurussia and the Armorican part of the Saxothuringian Zone, and other possible components of the Armorican Terrane Assembly (ATA) (Tait *et al.* 1997). During the opening phase of the Rheic Ocean, Avalonia rifted away from Gondwana, during the Arenig. By late Silurian time the ocean was closing rapidly, with a magmatic arc generated by southward subduction in the Mid-German Crystalline High (Dallmeyer *et al.* 1995). The northern margin of the ocean briefly experienced back-arc extension with initiation of the Rheno-Hercynian Basin, and formation of the Lizard-Giessen ophiolites (W. Franke 1995) in a marginal basin at about 397 Ma (Clark *et al.* 1998). These basins were inverted/obducted during overthrusting by the Normannian Terrane approximately 20 Ma later (Sandeman *et al.* 1997, 2000), during the Ligerian Phase of Ziegler (1990). Subsequently the Rheno-Hercynian Basin was transformed to a foreland basin and received flysch from the eroded, uplifted Saxothuringian terranes (Franke 1998; Franke *et al.* 1999), deformation continuing until late Carboniferous time (Oncken *et al.* 1995, 1999, 2000).

Magmatism

England and Wales

In eastern England a compositionally diverse assemblage of calc-alkaline magmatic rocks is present. The exposed South Leicestershire Diorite Suite is interpreted as a differentiated plutonic suite (Le Bas 1972), which seismic evidence suggests was emplaced as a tabular intrusive body (Allsop 1987). This body, and the related Mountsorrel Granodiorite, were emplaced at about 450 Ma (Pidgeon & Aftalion 1978; Noble *et al.* 1993), contemporaneous with arc-related granitic plutonism in the Lake District (Hughes *et al.* 1996; Evans *et al.* 1994). Further to the NE, numerous boreholes prove volcanic rocks in the pre-Carboniferous basement (Pharaoh *et al.* 1991), e.g. Cox's Walk Borehole, which proved 234 m of andesite and dacite lavas (Pharaoh *et al.* 1991). Unfortunately few of the boreholes penetrated as deeply, and the volcanic rocks are rarely associated with sedimentary strata, so that none is well constrained by biostratigraphic data. Virtually all occurrences have suffered hydrothermal alteration and metamorphism, thus whole-rock Rb–Sr isochron ages cannot be interpreted reliably in terms of age of igneous crystallization (Pharaoh *et al.* 1993*a*). U–Pb zircon ages are less severely affected by subsequent alteration and are considered to represent more reliable ages for crystallization of these suites. Welded felsic ash-flow tuffs from three of the provings (Glinton, Orton, Oxendon Hall boreholes) yielded precise concordant ages of about 615 Ma (Noble *et al.* 1993). These are comparable in age to the lithologically identical Padarn Tuff in North Wales (Reedman *et al.* 1984), but geochemically distinct from the Charnian Supergroup, with which they are tectonically juxtaposed (Pharaoh & Carney 2000). They are believed to represent the Neoproterozoic crustal substrate to this part of Avalonia. εNd values = –5 (at 615 Ma) support the presence of older crust, perhaps as old as 1400 Ma (Tucker & Pharaoh 1991; Noble *et al.* 1993). Other borehole provings, e.g. of welded ash-flow tuff at North Creake, Norfolk, and subvolcanic microgranite at Moorby, Lincolnshire (Pharaoh *et al.* 1997), have yielded imprecise but reliable concordant ages of about 450 Ma (Caradoc) (Noble *et al.* 1993). These rocks represent the latest stage of subduction magmatism associated with closure of the Iapetus Ocean and Tornquist Sea (Pharaoh *et al.* 1993*a*).

The products of the earliest, Tremadocian, phase of calc-alkaline magmatism are preserved at Rhobell Fawr, North Wales (Kokelaar 1986). Avalonia may still have been attached to Gondwana at this time. Thick piles of volcanic rocks preserved in North Wales and the Lake District record subduction underflow to the south beneath Avalonia in Llanvirn to Caradoc time (Cooper *et al.* 1993), although a considerable part of these rocks represent crustal melts.

Deep seismic profiling in the southern North Sea (Blundell *et al.* 1991) has enabled the mapping of a prominent SW-dipping zone of mantle reflectivity below the Dowsing-South Hewett Fault Zone. This reflectivity has been inferred to result from subduction underflow beneath Avalonia (Lee *et al.* 1993) and may well be that associated with the preserved record of Ordovician volcanic rocks (Pharaoh *et al.* 1993*a*, 1995). If true, it means that a small marginal oceanic basin may have lain outboard of Avalonia proper, separating it from the Southern North Sea–Lüneburg Terrane (Pharaoh *et al.* 1995; D. Franke 1995*a*); Winchester *et al.* in press), which lay peripheral to the main Thor-Tornquist Suture. Lamprophyric sills and dykes were emplaced within the back-arc region of the Midlands Microcraton until latest Ordovician time (Thorpe *et al.* 1993; Noble *et al.* 1993).

Volcanic rocks of Llandovery age occur in southern Britain remote from the Ordovician calc-alkaline suites, and are inferred to be associated with opening of the Rheic Ocean to south of Avalonia (Pharaoh *et al.* 1991).

Brabant Massif

In the Brabant Massif there were three periods of volcanism between the Cambrian and the mid Devonian: two minor ones intercalated in the Lower Cambrian and in the lower Tremadocian of megasequence 1 and one major period at the top of megasequence 2 and lingering into the base of megasequence 3, from mid Caradoc to late Llandovery in the Brabant Massif and continuing into the early Wenlock in the Condroz Inlier (Neuville-sous-Huy; Michot 1934; Maes *et al.* 1979).

Interstratified magmatic rocks in the Cambrian Tubize Formation have been briefly mentioned in the archives of the Geological survey south of Tubize and by Van Grootel *et al.* (1997), while Corin (1965) mentioned some feldspar-rich layers that can be interpreted as metatuffite in the Brussels area. These rocks have so far not been studied in detail, especially as their interstratified or possible intrusive position and the epizonal metamorphic grade makes interpretation rather difficult.

In recent boreholes in the Marcq valley, 25 km WSW of Brussels, porphyritic volcanic rocks were found intercalated between Tremadocian turbiditic metasediments of the Chevlipont Formation. Their apparent thickness amounts to 12 m in one horizon and 8 m in another, separated by an interval of about 20 m (De Vos, unpublished report Geol. Surv. Belg. 1997, 1998, 2000; Debacker 1998 unpublished report to ANRE; Piessens 1998 unpublished report to ANRE; Purvis, unpublished report to ANRE 1999; Vanguestaine & Verniers, pers. comm.). As all the rocks are altered, metamorphosed and deformed, their precise initial position is difficult to determine. They are described as acid volcanic rocks of ignimbrite origin or as a submarine mass-flow equivalent, with lapilli, tuff or ignimbrite in a chlorite–quartz–sulphide (pyrite) matrix and sericite-rich fiamme, occasionally also with carbonates. They are totally altered, with no fresh primary mineralogy apart from quartz phenocrysts (Purvis unpublished report to ANRE 1999). Although the occurrence of these volcanic rocks might be important, because they could indicate magmatic activity before or during the rifting episode that caused Avalonia to drift away from Gondwana in the early Ordovician, their origin is still uncertain. However, this magmatism is contemporaneous with that in the Tremadocian at Rhobell Fawr in North Wales (see below).

Late Ordovician to early Silurian magmatism is well documented in the Anglo-Brabant Deformation Belt. It occurs along the southern part of the main axis of the Brabant Massif. The magmatism is clearly calc-alkaline, and probably caused by subduction of oceanic lithosphere beneath the Avalonia microcontinent (André *et al.* 1986*b*; Pharaoh *et al.* 1993*a*; Van Grootel *et al.* 1997). It is mostly dacitic in composition with some rhyolites and rare andesitic and basaltic rocks, mostly observed as inclusions within the more felsic rocks.

In the Belgian part the volcanic rocks are mostly ash-flow tuffs and volcanic breccias, with some sporadic lava flows (e.g. at Fauquez and Voroux-Goreux). Nevertheless, several associated subvolcanic sill or neck-like bodies have been recognized. The best-exposed are the Quenast neck and the Lessines and Bierghes sills, respectively dated at 433 ± 10 Ma and 414 ± 16 Ma (André & Deutsch 1984). Most original magmatic textures have been severely altered by low-temperature secondary parageneses during several successive events: (1) in some cases spilitization immediately after the submarine extrusion (Hertogen & Verhaeren unpublished report to the ANRE 1999); (2) the hydrothermal activity linked to subvolcanic injection of dacitic magmas within water-saturated Ordovician sediments (André & Deutsch 1986); (3) the metamorphism associated with the Lower Devonian to Eifelian deformation phase (André *et al.* 1981); (4) a late Givetian to late Devonian (375 Ma) Sr isotopic resetting (André & Deutsch 1985); (5) finally subaerial weathering following uplift and erosion (Van Grootel *et al.* 1997). Biostratigraphic dating of volcano-sedimentary sequences in the Brabant Massif (outcrops and boreholes) indicate an early Ashgill age for the peak of extrusive magmatism. Some extrusive volcanism continued into the Llandovery in the eastern part of the Brabant Massif, and into the lower Wenlock in the Condroz Inlier. Hence the period of extrusion is concentrated in the lower Ashgill (Corin 1965; Verniers & Van Grootel 1991; Van Grootel *et al.* 1997). The decreasing age of magmatism towards the east of the belt was commented on by Pharaoh *et al.* (1995) and related to the oblique convergence of Avalonia with Baltica, involving significant anticlockwise rotation of the Avalonia microplate.

Three main steps have been recognized in the crustal magmatic differentiation of the late Ordovician magmatism: stage 1: a high-pressure magmatic crystal fractionation balanced by some granulite-facies rock contamination that gave rise to the basaltic andesites; stage 2: a low-pressure magmatic differentiation of these basaltic andesites that produced a series of dacitic residual melts; stage 3: a Plinian-like mechanically driven, crystal versus glass eruptive segregation that generated the rhyolitic tuffs. The characteristic features of the stage 1 were detailed in André (1991) and will not be discussed further here. Only the last two stages are reviewed briefly below.

Stage 2 is very well expressed throughout a series of cognate andesitic inclusions found within the Quenast dacitic neck (André 1983). This series of inclusions helps to define a regional liquid line of descent, which runs from the parental basaltic andesites to the residual dacites. Figures 4 and 5 illustrate some geochemical features of this liquid line (detailed analytical data are listed in a file that can be consulted from the British Library Document Supply Centre and the Society Library). Most andesites and dacites from the Brabant Massif (except those from the far eastern part of the massif, e.g. Voroux-Goreux) fit in this liquid line that turns out to be a characteristic feature of the magmatic differentiation in the area. The spectacular decrease in the TiO_2 content at the transition between basaltic andesites and andesites (Fig. 4B) clearly demonstrates that Ti-oxide played a leading role in the differentiation of those rocks. This feature, joined to very pronounced Ta, Cr, Ni depletions of the basaltic andesites, and their moderate light rare earth elements (REE) La–Sm enrichment, coupled to a limited heavy REE (Tb–Lu) fractionation, pinpoint the calc-alkaline character of this liquid line of descent (André *et al.* 1986*a*). The leading role played by plagioclase settlings during the crystal fractionation process is proved by the gradual deepening of the negative Eu anomaly along the liquid line of descent. This has been confirmed by the discovery of several plagioclase cumulate inclusions (André 1983) that display a pronounced positive Eu anomaly (Fig. 5A). The most likely fractionating mineral assemblage has been estimated (André 1983) as plagioclase (75 wt %), clinopyroxene (10 wt %), amphibole (10 wt %) and Ti-oxide (5 wt %).

Stage 3 has been recognized within a series of pyroclastic tuff sequences discovered in the boreholes of Deerlijk and Harelbeke. At both places, the thick pyroclastic layers begin with

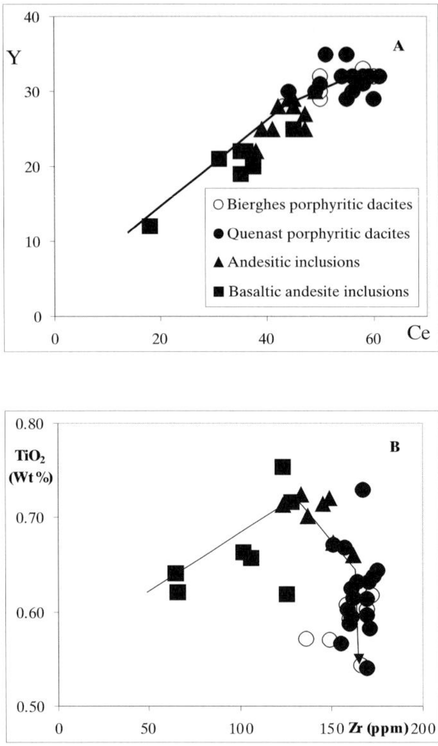

Fig. 4. A. Linear Y–Ce variation in the Quenast basaltic andesites, andesites and dacites that illustrates the gradual increase of the hygromagmatophile elements during the differentiation of the Quenast liquid line of descent. B. TiO_2–Zr contents in the Quenast basaltic andesites, andesites and dacites that demonstrates the change of behaviour of TiO_2 from incompatible to compatible behaviour in the course of the differentiation of the Quenast liquid line of descent.

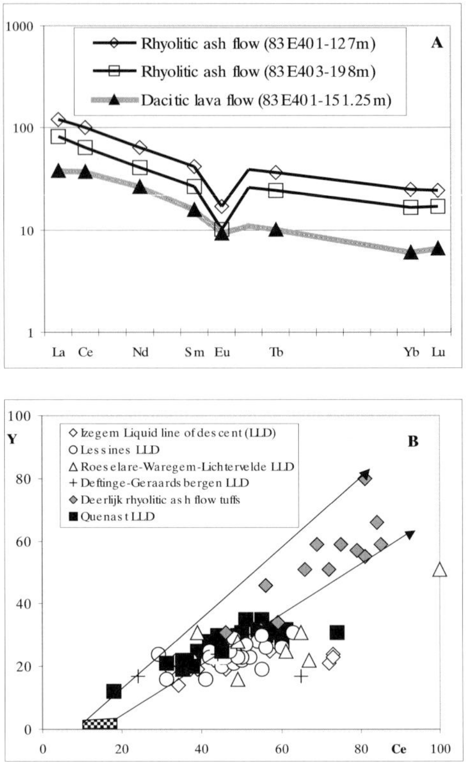

Fig. 5. A. REE patterns from the Deerlijk rhyolitic ash flow tuffs compared to the local andesitic–dacitic lava flow. B. Y–Ce variations in the Deerlijk rhyolitic ash flow tuffs compared to those of the Quenast, Izegem, Lessines, Roeselare and Deftinge liquid lines of descent. From the strong geochemical contrast, the mechanism of differentiation of those ash flow tuffs is concluded to be different from the crystal fractionation processes that controlled the differentiation of the various liquid lines of descent. The strong Y enrichment of the Deerlijk tuffs pinpoint the role of Plinian gravitational segregation (symbolized by arrow lines) of plagioclase with low Y contents (black and white square pattern box) in their genesis.

coarse-grained feldspar-rich levels of Plinian-like fall deposits at the base that grade upward to glassy tuffs showing a normal grading of lithic fragments and an inverse grading of pumice fragments. The upper part of the ash flow is made up of a pyroclastic air-fall deposit of fine-grained ash depleted in lithic and crystal fragments. Compared to the local dacitic lava flows, the fine ash exhibits a deep negative Eu anomaly that corroborates the strong role of the gravity driven plagioclase separation during the eruptive activity (Fig. 5A). The parental magma of these tuffs has been defined comparing their geochemistry to the regional liquid lines of descent involving plagioclase as the main crystal separated from the magma (Fig. 5B). This points to an andesitic composition for the most likely parental magma, supported by the similarity of initial Nd isotopic composition between the ash flow tuffs ($\varepsilon^{420} = -2.0$, André 1983) and the andesites and dacites ($-0.8 < \varepsilon^{420} < -2.9$, André 1991).

Tectonometamorphic evolution

For the name of deformation phase present in each area we refer to Figure 2.

Welsh Basin

The earliest recorded tectonic deformation is dextral strike-slip along the Welsh Borderland

Fault System e.g. Pontesford–Linley Fault (Lynas 1988), and the Tywi Anticline, reflecting Ashgill age, Shelveian deformation (Toghill 1992), prior to development of a regional unconformity beneath megasequence 3. All three megasequences were deformed by a penetrative cleavage development during the later (early Devonian) Acadian Phase, most spectacularly developed in the Cambrian slate belts of North Wales. Cleavage transection is mainly clockwise (Soper *et al.* 1987). A reactivation of all the more ancient crustal lineaments occurred, with sinistral strike-slip, e.g. on the Menai Straits, Bala and Welsh Borderland Fault Systems. Major southward-directed thrusting occurred in Anglesey (Carmel Head Thrust) and elsewhere. Metamorphism is up to epizone (greenschist facies), in contrast to the Midlands Microcraton, typified by the diagenetic zone (Merriman *et al.* 1993).

Northern England

Evidence of Shelveian deformation here is equivocal, unless the pre-megasequence 3 unconformity is a consequence of this deformation. However, Branney & Soper (1988) have argued for a volcanotectonic origin for structures in the Borrowdale volcanic sequence (megasequence 2). The strongly penetrative deformation is Acadian in age, as in North Wales. A slaty cleavage is ubiquitously developed in pelitic rocks, axial planar to upright tectonic folds, whose development is inferred to have occurred in a southward-migrating foreland thrust belt (Hughes *et al.* 1993) following closure of the Iapetus Ocean, in late Silurian time. Metamorphic grade reaches epizonal values (Fortey *et al.* 1993). Extensive post-folding southward-directed thrusting affects the northern edge of the Lake District (Hughes *et al.* 1993; Fortey *et al.* 1993). High heat production granites such as the Shap intrusion were emplaced towards the end of cleavage development, at about 395 Ma (Wadge *et al.* 1978; Soper & Kneller 1990; Merriman *et al.* 1995). These relationships suggest that crustal thickening associated with deformation was an essential prerequisite to granite generation in early Devonian time.

Eastern England

Not much is known on this mostly subcrop area except that rocks as young as Silurian have a penetrative cleavage and metamorphism up to epizone (greenschist facies), which is therefore likely to be of Acadian age (Pharaoh *et al.* 1987; Merriman *et al.* 1993). The Precambrian to Cambrian age rocks of the Charnian Supergroup, which crop out just beyond the NE limit of the Midlands Microcraton, contain greenschist facies schistosity with a probable Acadian age (Pringle *et al.* in prep.; Pharaoh unpublished data). It is inferred, but not proven, that most lineaments in the subcrop region suffered dextral reactivation, associated with anticlockwise cleavage transection (Pharaoh *et al.* 1987). Further north, the dominant NW–SE structural grain in the Carboniferous basins of the East Midlands region is inferred to have been inherited from the early Palaeozoic basement by the reactivation of Acadian thrusts and strike-slip faults. The region lies closer to the Thor-Tornquist Suture than the rest of onshore Britain, and might therefore be expected to have suffered deformation during the earlier, late Ordovician, Shelveian Phase. However, there is at present no clear evidence for this.

Brabant Massif

The Brabant Massif forms the southeastern part of the Anglo-Brabant Deformation Belt, situated just north of the Variscan deformation front (cf. Van Grootel *et al.* 1997; Oncken *et al.* 2000). Several outcrop areas along the southern rim of the massif allow a study of its structural architecture, with four structural areas (Fig. 6).

The Brabant Massif is characterized by an overall NW–SE structural grain (Fourmarier 1921; De Vos *et al.* 1993; Van Grootel *et al.* 1997; Sintubin 1997*a*,*b*, 1999; Debacker 2001*a*), curving into a more ENE–WNW direction (Fourmarier 1921; Sintubin 1997*a*,*b*, 1999; Debacker 2001*a*) in the east. The massif shows an apparently symmetrical disposition with a Cambrian core flanked on both sides by Ordovician–Silurian strata (Fig. 6). Weakly metamorphosed (Geerkens & Laduron *Etude du métamorphisme du Massif du Brabant.* Unpublished BNRE-report 1996; Van Grootel *et al.* 1997) and deformed Lower Palaeozoic metasediments are unconformably overlain by undeformed, diagenetic Givetian deposits (Legrand 1967; Verniers & Van Grootel 1991; Louwye *et al.* 1992; De Vos *et al.* 1993; Van Grootel *et al.* 1997; Debacker *et al.* 1999). The deformation of the Lower Palaeozoic strata is mainly reflected by the cogenetic development of folds and cleavage (Fourmarier 1921; Debacker 1997; Sintubin 1997*a*, 1999; Sintubin *et al.* 1998; Debacker *et al.* 1997; Debacker 1999, 2001*a*). There is evidence for only one main progressive deformation event (Debacker 1997, cf. Sintubin 1997*a*, 1999; Debacker 2001*a*).

The Lower to Middle Cambrian core of the

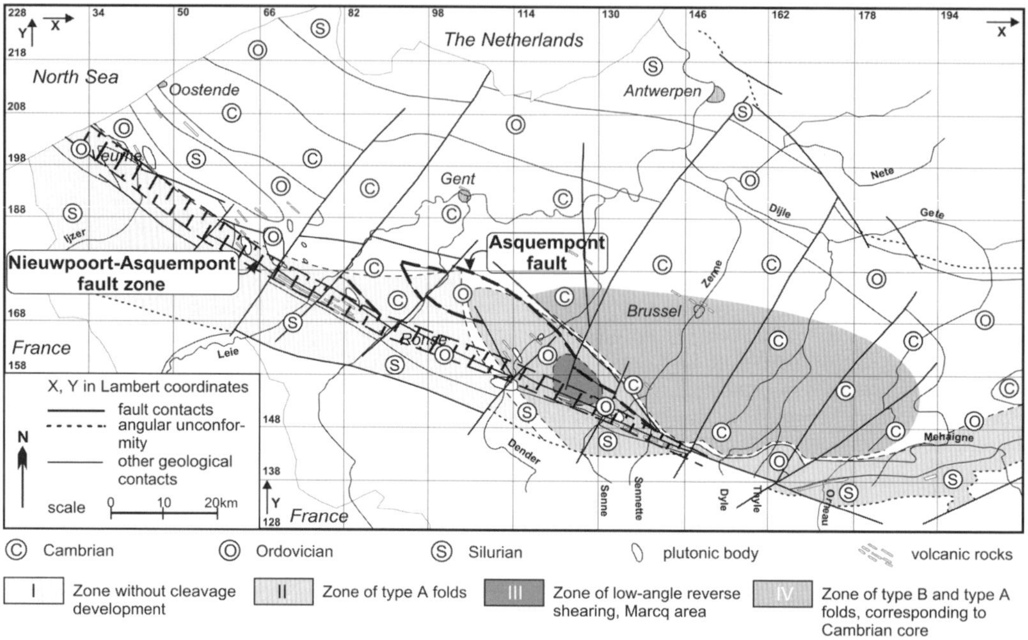

Fig. 6. Geological subcrop map of the Brabant Massif after De Vos *et al.* (1993) and Van Grootel *et al.* (1997), showing the four structural areas distinguished on the basis of deformation style and deformation intensity (Debacker 2001*a*). Structural area II roughly corresponds to the southern subdomain of Sintubin (1997*a*), structural area IV approximately coincides with the northern subdomain of Sintubin (1997*a*) and structural area III falls within the intermediate domain of Sintubin (1997*a*). The limits between the different structural areas are of a gradual nature. Two relatively important fault structures, redefined by Debacker (2000, 2001*a*) are added. The Asquempont fault is a pre-cleavage and pre-folding low-angle extensional décollement or detachment, forming the limit between the Lower to lower Middle Cambrian and the Ordovician. The Nieuwpoort-Asquempont fault zone is redefined as a WNW–ESE-trending post-cleavage fault zone, essentially consisting of north- and south-dipping normal faults, locally deforming the Lower Palaeozoic basement into a horst-and-graben geometry. The Ronse–Veurne line on the map coincides approximately with the southern limit of the Nieuwpoort-Asquempont fault zone.

massif is characterized by an overall steepness of bedding and cleavage (cf. Legrand 1968) and the presence of open, upright to steeply inclined, steeply plunging to reclined, sub-angular folds (cf. Sintubin 1997*a*, 1999; Sintubin *et al.* 1998; Debacker 2001*a*). These folds have NW–SE- to NNE–SSW-trending long limbs and NE–SW- to WNW–ESE-trending short limbs, apparently reflecting a Z-fold geometry (cf. Sintubin *et al.* 1998). The NW–SE- to NNE–SSW-trending fold limbs are subparallel to the pronounced NW-SE-trending aeromagnetic lineaments in the Cambrian core of the massif (Sintubin *et al.* 1998; Debacker 2001*a*). Open to close, moderately inclined to upright, sub-horizontal to gently plunging fold trains are present between these aeromagnetic lineaments and a gradual transition exists between the steeply and gently plunging fold types (Debacker 2001*a*, *b*). The domainal occurrence of both fold types in the Cambrian core, the particular presence of fold trains with a Z-fold geometry and the apparent dextral displacement along the aeromagnetic lineaments are all interpreted as the expression of a dextral transpressional shear (cf. Sintubin *et al.* 1998; Sintubin 1999; Debacker 2001*a*). To the SW, the aeromagnetic high of the Cambrian core of the massif is delimited by a steep NE–SW-trending aeromagnetic gradient (cf. Sintubin 1999; Sintubin & Everaerts 2002). This gradient coincides with the northeastern limit of an important NE-SW-trending gravity low (Fig. 7 and figs 7 & 8 in Sintubin & Everaerts 2002) (cf. De Meyer 1983; Everaerts *et al.* 1996), indicative of a low-density body at depth (see below). Above its roof a low-angle reverse shear zone has been identified, of which the development and propagation occurred cogenetically with folding and cleavage (Debacker 1999). Synkinematic metamorphic fluid flow along this shear zone caused mineralization and an intense alteration (Piessens *et al.* 2000, 2002; Piessens 2001). The Ordovician–Silurian strata, south of the Cambrian core, show open to tight, moderately inclined to upright, sub-horizontal to gently plunging, weakly south-verging fold trains. A small axial cleavage transection (*sensu* Johnson 1991) often occurs, the sense and amount of which is to a large extent related to the periclinal nature of the folds (Debacker *et al.* 1997, 1999; Debacker 2001*a*; cf. Treagus & Treagus 1981). Finally, to the SW of the low-density body, uncleaved subhorizontal to gently dipping low-anchizonal to diagenetic Silurian strata occur (Verniers & Van Grootel 1991; Van Grootel *et al.* 1997).

Several authors have advocated the importance of thrust faults in the Brabant Massif (e.g. Mortelmans 1955). However, recent studies (Giese *et al.* 1997*b*; Debacker 2000, 2001*a*) have shown no evidence for large-scale thrusting. The Asquempont fault (Fig. 6), juxtaposing the Lower Ordovician and the Lower to lower Middle Cambrian and previously considered to be an important reverse fault (e.g. Legrand 1967), is currently interpreted as a low-angle extensional detachment or décollement, formed prior to folding and cleavage development (Debacker 2000, 2001*a*). The Nieuwpoort-Asquempont fault zone, running along the southern limit of the low-density body at depth, is not an important strike-slip fault zone (cf. André & Deutsch 1985) but essentially consists of steep north- and south-dipping normal faults, formed after cleavage development (Debacker 2000, 2001*a*).

As becomes apparent from the particular distribution of the deformation, the low-density body at depth had a significant influence on the structural architecture of the Brabant Massif (De Vos 1997; Sintubin 1999; Sintubin & Everaerts 2002; Debacker 2001*a*). During deformation this low-density body behaved as a rigid object. The pervasive deformation north of this low-density body resulted in a steep belt, predominantly composed of Cambrian metasediments. Along the northeastern side of the low-density body a dextral transpressional shear occurred, resulting in a lateral, eastward escape of core material. Locally, some material was thrust over the roof of the low-density body along low-angle reverse shear zones. As deformation proceeded, also the Ordovician and Silurian strata in the southern part of the massif became deformed. However, to the SW of the low-density body, the Silurian strata were shielded from deformation. Possibly this low-density body also controlled the position of the Nieuwpoort-Asquempont fault zone.

The Brabant Massif is characterized by low-grade to very low-grade metamorphism, ranging from the lower anchizone to the epizone (Geerkens & Laduron unpublished BNRE-report 1996; Van Grootel *et al.* 1997). Maximum temperature in the core of the massif is estimated at 350 °C in the lowermost Cambrian and 250 °C in the Lower Cambrian Tubize Formation (André & Deutsch 1985; Giese *et al.* 1997*b*). Microscopic observations indicate that metamorphism occurred prior to and during cleavage development and occasionally outlasted the main compression (Giese *et al.* 1997*b*; Debacker & Sintubin unpublished data). Metamorphism may be interpreted as essentially resulting from burial (cf. Giese *et al.* 1997*b*). If so, a linear relationship should exist between grade of metamorphism and sediment age, as favoured by Giese *et al.* (1997*b*). As in Wales or in the Lake District contrasting burial depths can be explained by differential thicknesses of upper Silurian and Lower Devonian, since removed by erosion. One would expect that in the SE prolongation of the large Silurian Anglia basin, thick Silurian deposits are present all over the Brabant Massif, except for its SW shelf area. A re-interpretation by Debacker (2001*a*) of the results of a large-scale illite crystallinity study by Geerkens & Laduron (unpublished report to BNRE, 1996) and Van Grootel *et al.* (1997) shows that the situation is slightly more complex. With the exception of the thin Silurian shallow deposits along the southwestern rim, illite crystallinity values in the thick Silurian turbidite deposits along the southern, southeastern and northern rim are commonly similar to or higher than those obtained in the Cambrian and Ordovician in the more central parts of the massif. The most feasible explanation for this apparent anomaly is to invoke an extra load on top of the rims of the massif, which was much thinner or absent above the Cambrian and Ordovician of the more central parts. The inferred extra load along the rims with respect to the core can be explained by a relative uplift of the latter with respect to the former during the late Ordovician or early Silurian and/or by different thicknesses of missing Devonian.

A comparison of stratigraphic thickness and degree of metamorphism also points in this direction. In order to explain the low anchizonal metamorphism of the Gorstian deposits below the angular unconformity along the southern rim of the massif, one has to invoke the presence of at least 5 km of eroded overburden at average geothermal gradients (approximately 36 °C km^{-1}, cf. Anglia Basin, Merriman *et al.* 1993; cf. Van Grootel *et al.* 1997; Debacker 2001*a*). Expressed in stratigraphic thickness, the Lower Cambrian Tubize Formation, having experienced temperatures up to 250 °C, is situated at least 6 km below the Gorstian deposits (Fig. 3; Verniers *et al.* 2001). As such, in the assumption of burial metamorphism, one would necessitate a very low geothermal gradient (approximately 20 °C km^{-1}) in order to explain the low temperatures in the Lower Cambrian. However, this low geothermal gradient does not seem compatible with the anchizonal metamorphism of the Gorstian deposits, since it would necessitate an overburden along the southern rim of at least 8 km, which is not likely (cf. Van Grootel *et al.* 1997). Alternatively, a higher geothermal gradient (> 36 °C km^{-1}), more compatible with the temperatures measured in the Silurian along the rims, would result in a temperature in the Lower Cambrian well above the measured 250 °C (Debacker 2001*a*). The hypothesis that the uplift of the central core would be responsible for the Silurian and supposed Lower Devonian foreland basins at the northern and southern sides of that core, as suggested by Debacker (2001*a*) and Sintubin & Everaerts (2002), needs still to be tested by modelling. It also does not offer the mechanism to explain the foreland basin development of the Silurian Anglia basin, situated in the NW prolongation of the Brabant Massif.

Ardennes inliers

The Ardennes inliers with Cambrian to Middle Ordovician siliciclastic rocks, are situated in the High Ardennes Slate Belt (Sintubin *et al.* 2000), forming a part of the Variscan fold and thrust belt, the Ardennes Allochthon (Meilliez & Mansy 1990). The overall architecture of the inliers is very similar, with a predominantly east–west-trending structural grain and strongly north-verging fold trains. The dominant, pervasive cleavage is south-dipping. The overall structure of the inliers is that of a north-verging overturned antiform, shown also by the marked difference between the southern and northern unconformity. The southern unconformity is always weakly south-dipping. The dominant cleavage, both above and below the unconformity gradually becomes subparallel to the unconformity. The northern unconformity is subvertical to slightly overturned. The dominant, weakly south-dipping to subhorizontal, cleavage crosscuts the unconformity. The Lower Palaeozoic inliers are segmented by Variscan. Because these thrusts are commonly related to major facies transitions within the Devonian sedimentary sequence, they are interpreted as being inverted normal faults determining the Variscan basin development (Meilliez & Mansy 1990; Meilliez *et al.* 1991; Oncken *et al.* 2000). Along the southern unconformity, Variscan epizonal and rare mesozonal metamorphism occur. These metamorphic zones are clearly visible as highs on the aeromagnetic anomaly map (Sintubin & Everaerts unpublished data).

Most of the research has been based on the two largest inliers (Rocroi and Stavelot-Venn). The small Serpont Inlier can be considered as the eastern continuation of the Rocroi Inlier, forming the backbone of the culmination zone of the Ardennes Anticlinorium. The structural relationship of the Givonne Inlier (Fig. 1) with the Rocroi Inlier is unknown.

The Rocroi Inlier has an overall east–west-trending structural grain. Its internal structural architecture consists of two narrow antiformal areas, composed of Lower Cambrian units (Deville Group), separated by a very broad synformal area, composed of Middle to Upper Cambrian units (Revin Group) and some Middle Ordovician (Salm Group).

The Stavelot-Venn Inlier is situated close to the Variscan thrust front. The main thrusts crosscutting the inlier actually form part of the Variscan front thrust complex (Xhoris-Monschau, Eupen, Venn and Theux thrusts) (Hance *et al.* 1999). The inlier itself consists of two distinct lithostructural domains, separated by the Xhoris-Monschau thrust (Fielitz 1992; Sintubin 1994). The southern domain is dominated by two dome-like structures, composed of Lower Cambrian, predominantly quartzitic strata (Deville Group). The overall internal structural grain is east–west-trending, comparable to the other three inliers. The northern domain consists of a number of Variscan thrust sheets (Hance *et al.* 1999; Sintubin & Matthijs 1998) in the footwall of the Xhoris-Monschau thrust. The overall structural grain has a NE–SW trend (N60 °E), which coincides with the structural grain outside the inlier. Superposed on the border zone between both domains, a post-Variscan graben structure is found. The infill is considered Permian in age (Renier 1902; Smolderen 1987). Finally, significant uplift occurred since the Chattian in the northern part of the Stavelot-Venn Inlier as part of the rift shoulder of the Roer Valley Graben System (Gullentops 1997).

The exact determination of the Caledonian tectonometamorphic event in the Ardennes inliers (Ardennian phase) is obscured by the subsequent Variscan tectonometamorphic event during the Carboniferous.

Strata on both sides of the southern unconformities suffered Variscan epizonal high temperature–low pressure metamorphism. This burial to early synkinematic metamorphism reached 400–450 °C and 100–300 MPa in the Stavelot-Venn Inlier and up to 500 °C and 200–300 MPa in the Rocroi-Serpont Inlier (see Fielitz & Mansy 1999). This metamorphic event probably occurred between 349 and 308 Ma (see Fielitz & Mansy 1999). The metamorphism at the northern unconformity reached the anchizone (up to 230 °C) (see Fielitz & Mansy 1999).

The issue of the presence of Ardennian metamorphism (traditionally called 'Caledonian' in the Belgian literature) has been resolved by an extensive fluid inclusion and mineralogical study of the quartz veins in and around the Stavelot-Venn Inlier (Ferket *et al.* 1998; Schroyen 2000). Variscan veins in the southern domain give pressure–temperature conditions similar to the host-rocks (385–435 °C/175–275 MPa). In the northern domain Variscan fluids have much higher pressure and temperature conditions (350–400 °C / 200–250 MPa) than those for the host rocks (about 240 °C). This anomaly within the northern domain of the Stavelot-Venn Inlier, but also north of the Stavelot-Venn Inlier, is explained by the expulsion of hot metamorphic fluids to the foreland during Variscan thrust stacking.

Also for the Ardennian metamorphism a northwards (to the foreland?) gradient could be deduced. In the southern domain pressure–temperature conditions (200–300 °C/205–295 MPa at 30 °C/km) are lower than the Variscan metamorphic conditions whereas in the northern domain the conditions (180–270 °C/80–130 MPa at 50 °C/km) are higher than the Variscan metamorphic conditions. Also striking is the large difference in geothermal gradient between both domains.

The very existence of an Ardennian deformation event has long been disputed. The discussion focused on both Rocroi and Stavelot-Venn inliers, because of their large extent and high degree of exposure. The main argument against the existence of a significant Ardennian deformation event is the apparent presence of only one pervasive cleavage throughout the inliers and the apparent continuity of cleavage trajectory across both southern and northern unconformities. This led a number of authors (e.g. Hugon 1983; Hugon & Le Corre 1979; Le Gall 1992) to postulate that the Lower Palaeozoic inliers only suffered a Variscan tectonometamorphic event. Le Gall (1992) even considered a decoupling level at the unconformity. Based on the overwhelming evidence of the presence of an angular unconformity this model can definitely not be retained. Other authors (e.g. Meilliez 1989; Meilliez *et al.* 1991; Meilliez & Mansy 1990; Lacquement 2001) suggested that an Ardennian deformation occurred, but without any cleavage development. Major P_1 fold structures were interpreted as synsedimentary slump folds. Still other authors (e.g. Klein 1977, 1980; Delvaux de Fenffe & Laduron 1984, 1991; Belanger *et al.* 1997; Belanger 1998; Piessens & Sintubin 1997) considered the internal structural architecture of the inliers to be the reflection of an important tectonometamorphic event, called the Ardennian phase by Michot (1980). The subsequent Variscan deformation is very heterogeneous and only expressed locally in the previously deformed slaty basement inliers. The current authors favour the latter model. To date a detailed study of the deformation history has only been performed in the southern part of the Rocroi Inlier (Belanger 1998). An Ardennian, polyphase deformation history resulted in a complexly structured slaty basement. NW–SE trending, SE-plunging, NE-verging kilometre-scale recumbent P_1 folds are associated with the development of a pervasive, south-dipping, axial-planar, slaty cleavage. Superposed on these P_1 folds, north–south to NW–SE trending, upright P_2 folds are responsible for a type 3 fold interference pattern. No cleavage development is related to the P_2 folding. Only towards the west, where P_2 folding gradually intensifies, is the primary cleavage intensively crenulated, resulting in a pervasive linear fabric. Although not yet demonstrated, a similar, complex, polyphase deformation history can be assumed in the other basement inliers.

In the framework of the Variscan tectonics the Lower Palaeozoic inliers are bounded by structures which can be considered as footwall shortcuts of a crustal ramp (see Oncken *et al.* 2000). The development of the Ardennes Allochthon led to fault-assisted antiformal stacking (Le Gall 1992), generating the current basement inliers within the High Ardennes Slate Belt. Variscan deformation is strongly partitioned within the previously deformed slaty basement inliers. Primarily the pre-existing, Ardennian, slaty cleavage has been reactivated (see Belanger 1998; Klein 1980) because the Variscan and Ardennian kinematic framework showed a similar orientation (especially in the Rocroi and Serpont inliers). The result of this reactivation is that on a regional scale the inliers are apparently affected by only one pervasive cleavage and that the unconformities are apparently cross-cut by a single pervasive cleavage. At a smaller scale different features (e.g. asymmetric boudinage, small-scale folds, multiple crenulation cleavage) indicate a heterogeneous and strongly localized Variscan deformation (e.g. Delvaux de Fenffe & Laduron 1984, 1991; Fielitz 1992; Piessens & Sintubin 1997; Belanger 1998). Due to the previously deformed nature of the slaty basement inliers, leading to a very limited and localized internal Variscan deformation within the inliers, the Devonian cover sequence, suffering significant Variscan shortening due to cleavage development, is detached from its basement. It is therefore fair to assume that shear strain is localized in a zone along the unconformity (cf. Le Gall 1992), as suggested by the sigmoidal cleavage trajectory at the southern unconformities. However, no shear deformation occurred on the unconformity itself, where Le Gall (1992) placed it, but in a zone on both sides of the unconformity. With respect to the Variscan deformation the basement inliers acted as low-strain zones bordered by high-strain zones in which shear deformation was strongly localized (cf. Burg 1999).

The tectonometamorphic event in the Ardennes inliers was called the Ardennian Phase by Michot (1980) and is constrained in time by the hiatus at the unconformity, which ranges from the Llanvirn (Verniers *et al.* 2001) to the Přídolí (Godefroid & Cravatte 1999) and spans about 45 Ma (see Fig. 2, column 10). Previously, this event was considered to occur during the mid Caradoc (Martin *et al.* 1970; Michot 1980). The presence of a thin conglomerate (Cocriamont member) in the Condroz Inlier at the base of the Ashgill Fosses Formation above a hiatus of part of the Upper Ordovician was considered as evidence for the Caradoc age of the deformation. Recent fieldwork places the conglomeratic level high in the Fosses Formation or even above it and links it to the late Ashgill Hirnantian glaciation sea level fall (Herbosch *et al.* 2001, see above). Reworked Ordovician acritarchs in the Silurian deposits of the Condroz Inlier were interpreted by Martin (1969) as indicative of the unroofing of the 'Ardennes Deformation Belt' from the earliest Silurian onwards.

Krefeld High and Ebbe and Remscheid inliers

In these three areas structural studies are lacking. A recent study on epizonal phyllitic clasts in the Givetian conglomerate from the Viersen 1001 borehole in the Krefeld High, dated them using acritarchs as upper Tremadocian (equivalent to the Lierneux Member, Stavelot-Venn Inlier). The pebbles are thought to have been derived from a nearby cliff of the Krefeld High. K–Ar dating on the phyllitic clasts indicate an age of 394 ± 8 Ma, reflecting a tectonic imprint, coeval with the Brabantian (= Acadian) Phase, on the Tremadocian of the Krefeld High and an age of 315 ± 7 Ma in the matrix of the conglomerate, a Variscan imprint (Ahrendt *et al.* 2001*a*, *b*). No observations are available yet to indicate the time range of the sedimentation gap caused by the deformation.

In the Ebbe and Remscheid inliers the tectonics are apparently very complex, and this, together with the limited access to measurable outcrops, has led to only limited data being published. The Ebbe Inlier and covering Devonian strata are part of a NE–SW trending structure consisting of two anticlines and a syncline. Generally, the anticline limbs plunge gently towards both NE and SW (Timm *et al.* 1981). The tectonics of the northern flank of the Ebbe Inlier were briefly discussed by Eiserhardt *et al.* (1981) who suggested that the anticline might have formed tectonically in three phases, a so-called 'Caledonian', a Variscan and a post-Variscan. The time range of the 'Caledonian' phase here, is in between the late Caradoc and the Ludlow (Fig. 2).

Southern North Sea–Lüneburg Terrane

Only along the northeastern rim have boreholes penetrated this terrane and information is scanty. Boreholes on the Mid-North Sea High and Ringkøbing-Fyn High penetrated schists which underwent prograde greenschist–amphibolite metamorphism at 450–425 Ma, and retrogression at 415–400 Ma (Frost *et al.* 1981). The former is considered to mark the docking of Avalonia with Baltica in the late Ordovician, contemporaneous with the Shelveian Phase in the onshore UK (Toghill 1992; Pharaoh *et al.* 1995) and with the Ardennian phase in the Ardennes inliers.

Recent palynological studies and geochronological ($^{40}Ar/^{39}Ar$) datings of detrital micas of the Lower Silurian sediments of boreholes Slagelse-1 and Pernille-1 (e.g. Vecoli & Samuelsson 2001*a*) corroborate previous models (e.g. Frost *et al.* 1981) which included most of the Danish subsurface within the southwestern passive margin of Baltica. However, the low-grade metamorphic, pelitic to psammitic sequences of unknown or uncertain stratigraphic position occurring in southernmost Jylland and in the adjacent areas of northern Germany (e.g. boreholes Løgumkloster-1 and Flensburg-Z1) were considered parts of a separate terrane (the so-called 'South Jylland Terrane') by Franke (1994). Petrographic examination of pelitic and arenitic samples from the borehole Løgumkloster-1 showed a very low metamorphic grade for these rocks, and has revealed an intense recrystallization of the rock-matrix, possibly due to hydrothermal alteration. New single-grain $^{40}Ar/^{39}Ar$ datings of detrital micas from the same borehole (between 2721 and 2724 m) show main age patterns between 500 and 450 Ma and between 350 and 390 Ma. This complex age pattern clearly reflects a main 'Caledonian' imprint, together with a younger event at around 360 Ma which is more difficult to interpret but could represent a further thermal event, maybe also responsible for the strong recrystallization of the rock matrix: e.g. a hydrothermal event. The $^{40}Ar/^{39}Ar$ age spectrum of single grains of detrital micas from the nearby Brøns-1a borehole (sample taken at *c.* 2530 m) is completely different, showing a late Neoproterozoic signature (ages in the range 517–578 Ma with a main concentration at around 570 Ma). The absence of 'Caledonian' ages may signify that in this particular area the intensity of deformation was not high enough to cause a rejuvenation of the primary isotopic information.

New single grain $^{40}Ar/^{39}Ar$ analyses have also been performed on detrital white micas separated from arenitic strata in the Flensburg-Z1 borehole, northern Germany, yielding a main age cluster around 870 Ma, with some scattered younger (*c.* 640 Ma) and older (*c.* 1000 Ma) ages. Comparable Sveconorwegian–Grenvillian ages are known from the southwestern Baltic Shield (e.g. SW Sweden: Johansson *et al.* 1991; northern Denmark: Larsen 1971), supporting a Baltica-related detrital source area for the sediments of the Flensburg-Z1 borehole. On the basis of the above data, it is not possible to support the assignment of the sequences cut by the Løgumkloster-1, Brøns-1A, and Flensburg-Z1 boreholes to a single homogeneous tectonostratigraphic terrane, as suggested by previous authors (e.g. the Southern Jylland Terrane of Franke 1994). The present heterogeneous isotopic data are more in agreement with the hypothesis of the EUGENO-S Working Group (1988) which assigned the 'Caledonian' deformed pre-Devonian of South Jylland to a nappe pile of a marginal thrust belt at the northern front of the Danish–North German–Polish 'Caledonian' mountain chain. The variable intensity of deformation and of tectonothermal rejuvenation of the isotopic ages in the Løgumkloster-1 and Brøns-1A boreholes is consistent with the latter hypothesis. Unfortunately, however, there are no precise fossil fauna/floral data to support a Peri-Gondwana origin of these allochtonous thrust-sheet units of southern Denmark.

The presence of apparently Baltica-related sediments in the Flensburg-Z1 borehole, south of the 'Caledonian' deformed nappe piles of South Jylland could result from an uplifted block of the Baltic Shield basement together with its metasedimentary cover. A similar situation was envisaged for the area of the mid Rügen depression by Dallmeyer *et al.* (1999), where a sedimentary succession (Loissin-1 borehole) of Baltica provenance is present south of the Avalonia-derived Rügen allochtonous units (Giese *et al.* 1994, 1997*a*). In agreement with the hypothesis of Dallmeyer *et al.* (1999), and with various lines of geophysical evidence (e.g. Thybo *et al.* 1990; Aichroth *et al.* 1992; Rabbel *et al.* 1995) the present data seem to support a more southerly extension of the Baltica basement beneath northern Germany than previously thought. However, a serious problem for this interpretation arises from the unknown age of the metasedimentary sequences cored by the Flensburg-Z1 and Loissin-1 boreholes.

The crustal structure beneath the Polish 'Caledonides' is still controversial because of the relative scarcity of geophysical and borehole data (Schlüter *et al.* 1997; Dadlez 2000). The Lower Palaeozoic Pomeranian 'Caledonides' clearly resemble the Lower Palaeozoic of Rügen, both from a stratigraphical, structural, and palaeontological point of view (Bednarczyk 1974; Dadlez 2000; Samuelsson *et al.* 2002*a*). The presence of a Tornquist Sea suture near the 'Caledonian' Front in Pomerania is commonly accepted in prevailing opinion (e.g. Tanner & Meissner 1996). However, Dadlez (2000) claimed that seismic data suggest the presence of 'several crustal blocks' in the basement of the German–Polish lowlands, and that if it is accepted that the crustal structure of the Rügen area is typical for East Avalonia, then it must follow that 'Avalonia does not extend to the east beyond the Odra river' (Dadlez 2000, p. 233).

Potential field data of Belgium and surrounding area

The potential field maps of Belgium and surrounding area (cf. Chacksfield et al. 1993; Mansy *et al.* 1999; Everaerts 2000) and calculated derivatives (see Everaerts 2000; Sintubin & Everaerts 2002) have helped to clarify the structural architecture of the predominantly concealed Anglo-Brabant Deformation Belt

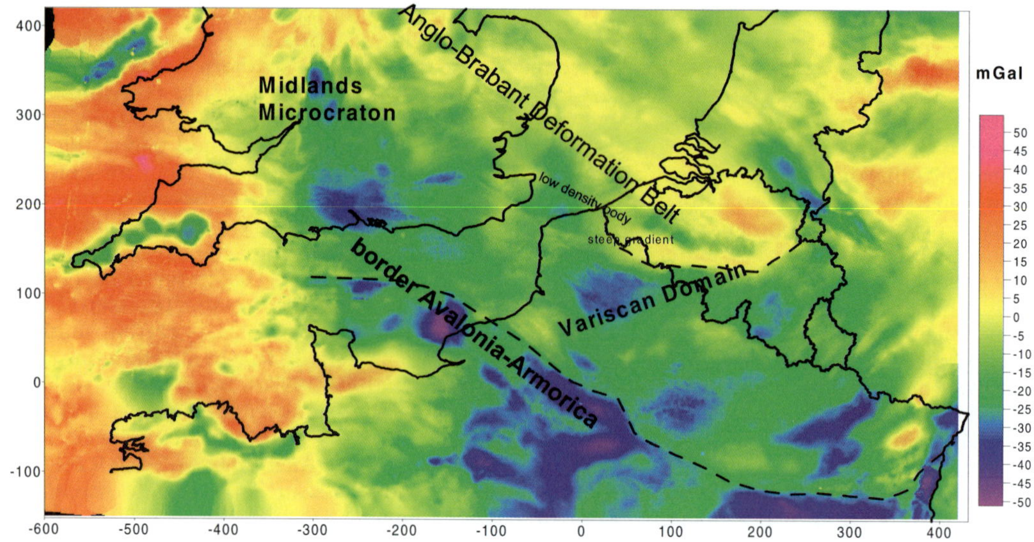

Fig. 7. Gravity field images, indicating the cratonic area(s) in Belgium, northern France, Midlands (UK), the position of the low density body interpreted as a granitic batholith, and the relation to the Midlands Microcraton (modified after Everaerts 2000).

(Lee *et al.* 1993), particularly in its southeastern extremity (Brabant Massif) where it could be correlated with field data (Sintubin 1997*b*, 1999; Debacker 2001*a*; Sintubin & Everaerts 2002). A discussion of these detailed analyses can be found in the cited works. In this paper we will only discuss the gravity anomaly maps on a larger scale to provide a more regional picture of the subsurface east of the Midlands Microcraton. The Bouguer gravity map (Fig. 7) has recently been prepared using the gravity data base of the Royal Observatory of Belgium and contains more than 250 000 points of measurement. All gravity data are referenced to the gravity datum of Uccle (1976) (IGSN71–0.048 mGal). The density reduction used for the calculation of the Bouguer anomaly on land is 2.67 g/cm^3. On sea the free air anomaly was used. The theoretical gravity was computed using the 1980 geodetic reference system formula (International Association of Geodesy (1980). The geodetic co-ordinates (φ, λ) were converted into the planar Belgian Lambert projection. After combination of the offshore and onshore data, a gridding procedure was performed using the kriging method with Surfer software using a cell size of 1 km, a search radius of 30 km, and a linear weighting coefficient to avoid aliasing in poorly covered areas. The data outside Belgium were provided for France by the BRGM, for Great Britain by the BGS, for The Netherlands from the Meetkundige Dienst in Delft and for Germany from the European Geotraverse (Blundell *et al.* 1992).

In Belgium the main feature is the density contrast between the northern part and the southern part (De Vos *et al.* 1992; Chacksfield *et al.* 1993; Mansy *et al.* 1999). The positive anomaly (in yellow and red on Fig. 7) in the northern part (north of the Condroz Inlier) continues below the North Sea and East Anglia (cf. Lee *et al.* 1993). It coincides largely with the Anglo-Brabant Deformation Belt (Lee *et al.* 1993), but also includes the southwestern Brabant Shelf. This positive anomaly and the steep gradient observed in the south of the Brabant Massif, and the mass excesses it represents, have been explained by the pelitic composition of the supposed Cambrian to Silurian sediments (De Vos *et al.* 1992). However, it can be explained better by a combination of two phenomena south of the break: (1) the 'Brabant Massif', or its prolongation, is situated at progressively deeper levels towards the south, and (2) a coal basin is present. A gravity model testing the impact of both the phenomena to explain the anomaly confirms this possibility (Everaerts 2000). SW of the positive anomaly, an elongate, NW–SE trending negative anomaly (in green on Fig. 7) is interpreted as a low-density body at depth (De Meyer 1983, 1984; Chacksfield *et al.* 1993; Everaerts *et al.* 1996). The latter authors modelled this anomaly with very steeply dipping limits as a steep-sided low-density body (2.63 g cm^{-3}), possibly a granite. Because of its spatial concordance with the Upper Ordovician and Llandovery magmatic province (see above), this granitic body has been interpreted as a batholith (see also Mansy *et al.* 1999). Sintubin (1997*b*, 1999) and Debacker (1999, 2001*a*) demonstrated that this body, oriented according to the structural grain of the Brabant Massif, had an important control on the Brabantian deformation (see above), and on the sedimentation pattern from the Middle Devonian to Lower Carboniferous (Hennebert 1993). Because it has not yet been reached by boreholes, its composition, formation and age can only be speculated upon. Hypotheses include: (1) a sedimentary basin (De Meyer 1983, 1984); (2) an intrusion genetically related to the Upper Ordovician rhyolitic activity in outcrop and subcrop (De Vos 1997); (3) a granitoid Proterozoic basement block (Sintubin 1999; Sintubin & Everaerts **201**, 00–01); or (4) a Variscan batholith (Rabae & Kearey 1997). The latter hypothesis is contradicted by observations indicating its inferred presence before the Brabantian deformation event (see above).

South of the Brabant Massif, in the Ardennes and below the Paris Basin lies a large gravity low with some more pronounced anomalies. The light anomaly below the Ardennes can be explained by the presence of Devonian–Carboniferous rocks. Gravity modelling supports this hypothesis (Everaerts 2000). The Mesozoic rocks overlying the Variscan belt below the Paris Basin are also very light and so reinforce the anomaly in this region. The southern border of the Rheno-Hercynian Zone and of some smaller subterranes in Germany attached to Avalonia–Baltica during the Devonian (see overview in Franke 2000) is marked as the Rheic suture on Figure 1. This suture is also represented in France by the Bray–Vitel fault zone and can be seen on the gravity anomaly map (Fig. 7). It shows that the latter fault zone in the Channel area trends east–west. South of this fault zone the anomaly belongs to the Armorican Terrane Assembly or the Moldanubian zone. Some negative anomalies in this area have been interpreted as granitic bodies by the ECORS team, who modelled a different crustal density between both sides of the fault (Galdeano & Guillon 1988). The southern limit of Avalonia is hence placed at the Bray Fault. Furthermore the Cornubian Variscan granites can also easily be seen on the map by their corresponding negative anomalies in green.

On this map the Midlands Microcraton is characterized by a gravity low. North of the Bray fault in the Paris Basin and in the Ardennes some more pronounced lows are also present. There is a large gravity low with similar orientation on the Bouguer anomaly map from the Midlands Microcraton under the Straits of Dover below northern France, north of the Bray–Vital fault, below the Ardennes, as far as the area below the Rhenish Massif. The northern limit in Belgium can be traced south of the elongated positive anomaly of central Belgium or north of the elongated negative anomaly. Proterozoic basement can be postulated below this large area (Sintubin & Everaerts 2002). In the Ardennes, gravimetric and magnetic modelling by Chacksfield *et al.* (1993) also suggested the presence of higher magnetic and lower density rock in the lower crust, unknown at the surface.

Discussion

With the observations and data described above, we will discuss successively: (1) the basin evolution of Eastern Avalonia around and east of the Midlands Microcraton and (2) the evolution of the Avalonia-related microplate, Far Eastern Avalonia; (3) to which of the two microplates the Krefeld High and the Ebbe and Remscheid inliers belong; (4) the direct and indirect evidence for Proterozoic basements in both terranes; (5) the possible models which could explain the observed deformation in the Brabant Massif (and the Ardennes inliers); (6) the constraints controlling a larger model for all observed phenomena in Eastern Avalonia east of the Midlands Microcraton; and (7) an outline of a possible new model.

Basin evolution of Eastern Avalonia around and east of the Midlands Microcraton

From the palaeontological and palaeomagnetic evidence shown above, it is clear that the Cambrian to mid Devonian history of Eastern Avalonia has to be divided into three or four periods, each with a very different geodynamic context: (1) an early Cambrian to Tremadocian period, when Avalonia was still attached to Gondwana at a high latitude, close to the northern part of South America or West Africa (McKerrow *et al.* 1992; Cocks & Torsvik 2002); (2) an Arenig to mid Ashgill period when Avalonia was isolated as a separate microcontinent, drifting away from Gondwana towards Baltica and (3) a late Ashgill to early mid Devonian period when Avalonia docked with Baltica and together both collided with Laurentia by the Wenlock; (4) early to mid

Devonian extensional rifting, forming the Rheno-Hercynian Basin at about the same time that compression occurs in the Anglo-Brabant Deformation Belt. The exact relationship needs to be analysed more and will not be discussed further. For comparison with the surrounding palaeocontinents the ranges of sedimentation, hiatus or deformation phases of Saxothuringia, Tepla-Barrandian and Central Armorican Massif are also indicated on Figure 2. They show very different histories of basin evolution in relation to Eastern Avalonia.

The three megasequences defined in the Welsh Basin (Woodcock 1990*b*) have important tectonic controls, while shorter cyclicities may have a eustatic control. The Lower Cambrian to Upper Tremadocian megasequence 1 can be equated with the period when Avalonia was attached to Gondwana; the Lower Arenig to middle Ashgill megasequence 2, with the period that Avalonia detached from Gondwana, creating (part of) the Rheic Ocean in its wake and the closure of the Iapetus Ocean and the Tornquist Sea in front of it; the upper Ashgill to Lower Devonian megasequence 3 with the period of collision with Baltica and Laurentia, when subduction magmatism of the Iapetus Ocean beneath Avalonia terminated and collision with Laurentia initiated. During megasequence 1 the stratigraphy and sedimentology indicates different environments (shelf, slope or basin), which changed though time. During the early Cambrian a shelf is present on the Midlands Platform and in the Stavelot-Venn and Rocroi inliers, on an unknown and possibly later detached basement, while a slope is present in the central and northern Brabant Basin. No Cambrian sediments are observed in the southwestern part of the Brabant Massif at the future site of the southwestern Brabant Shelf with deposition of megasequence 3. During the mid and late Cambrian, the slope environment remained in Brabant, but Stavelot-Venn and Rocroi experienced a deepening with turbiditic basin deposits, reverting to a slope environment in the Tremadocian. Similar environmental changes, associated with the development of localized rifts within the Midlands Microcraton, have been described by Smith & Rushton (1993). During the deposition of megasequence 2 a shelf environment persists in all areas (Brabant, Condroz, Stavelot-Venn) with a rather thin sequence reflecting the reduced source area of the Avalonian microcontinent, surrounded by oceans. In the Midlands Microcraton, uplift in the back-arc region means that megasequence 2 strata are almost completely absent. During the deposition of megasequence 3 a shelf environment is present in the microcraton, southwestern Brabant Shelf and the Condroz Shelf, while in the central and north Brabant Basin a deep shelf changes in the late Llandovery to a slope or basin from the latest Llandovery to the end of the Silurian.

In terms of subsidence, megasequence 1 in the Lower Cambrian in Wales and the Brabant Massif records a rifting event, possibly indicating a first (failed?) separation from Gondwana *contra* the model of Murphy *et al.* (1999). Megasequence 2 records an uninformative subsidence curve and megasequence 3 shows shelf sedimentation on the Midlands Platform and on the southwestern Brabant Shelf and Condroz Shelf; further away from the shelves, turbiditic basins develop in Wales, the Lake District and the central and north Brabant Basin. This is explained by the shape of the subsidence curves as respectively a pull-apart basin in Wales and foreland basin in the latter two areas.

Provenance studies using Nd geochemistry show that the overall pattern of the Nd isotope evolution curve is similar in the Stavelot-Venn Inlier, the Brabant Massif, the Lake District and Wales (Thorogood 1990; André 1991; Gerdes *et al.* 2000, 2001*a*, *b*). It suggests that throughout the three periods all four basins have a similar basin evolution history and that they belong to one and the same microcontinent or part of a larger palaeocontinent.

If a pre-Variscan restoration is made of the Condroz Inlier, allowing for Variscan transport of 20–30 km (Adams & Vandenberghe 1999), this latter area would be situated SW of the prolongation of the slope or basin of the central and northern Brabant Basin. From the new sedimentological observations presented above, we can deduce the presence of a shelf during megasequences 2 and 3 in the southwestern Brabant Shelf and on the Condroz Shelf, a basin in the central and northern Brabant Basin, and (during megasequence 2 only) a shelf in the Ardennes inliers.

Pharaoh *et al.* (1995), Van Grootel *et al.* (1997), Pharaoh (1999) and Winchester *et al.* (in press) present different lines of evidence to separate the Eastern Avalonia microcontinent from an Avalonia-related microplate now represented in the Southern North Sea–Lüneburg Terrane. The stratigraphic evidence presented here is another line of evidence. The presence of the three megasequences on the northeastern side of the Midlands Microcraton, in the Brabant Massif, Condroz Inlier and of two of the three megasequences in the Ardennes inliers indicates an attachment of these areas to the Midlands Microcraton.

Basin evolution of the Avalonia-related microplate (now present in the Southern North Sea–Lüneburg Terrane).

From present knowledge it is still uncertain if the Rügen and Pomerania subcrops also belong to the basins marginal to the Midlands Microcraton. The top of megasequence 1 is present in Rügen. The Middle Ordovician part of megasequence 2 contains turbidites in its upper part, earlier than in Belgium. Hence it is not certain whether the threefold division of megasequences can be extended to Rügen and Pomerania. The Middle Ordovician high-latitude planktonic fauna and flora of the latter areas have well been established by Servais *et al.* (2001), Samuelsson *et al.* (2001) and Vecoli & Samuelsson (2001*b*) for Rügen, and Samuelsson *et al.* (2002) for Pomerania, and are clearly different from the low-latitude planktonic fossils in nearby boreholes (e.g. borehole G14) on Baltica (*contra* Cocks 2000). The absence, however, of endemic benthic fauna that could determine the palaeocontinental affinity of Rügen and Pomerania, and hence of the northern part of Far Eastern Avalonia, means that a clear link with microcratonic Avalonia cannot be established. A similar facies and age is the only poor evidence. Planktonic microfossils, however, indicate high-latitude affinity, supposedly Peri-Gondwanan and clearly different from the present-day juxtaposed Baltica. Neither palaeomagnetic studies, nor complete stratigraphic logs are available from the cores to determine the palaeolatitude or a sequential stratigraphy (presence or absence of the three megasequences). Hence the attribution of the Rügen and Pomerania subcrop to an Avalonia-related microplate (Pharaoh 1999) is justified. New evidence from Nd isotope studies indicate a different signal for the Middle Ordovician of Rügen and Pomerania than from the mutually similar signals obtained in Wales, the Lake District, the Brabant Massif and the Ardennes inliers (Gerdes *et al.* 2000, 2001*a*, *b*). This would point to the presence of a different microplate adjacent to Rügen-Pomerania at the time of sedimentation.

However, from various benthic fossil groups (see above) it is established that by late Caradoc and certainly Ashgill, contacts between Avalonia east and west of the Midlands Microcraton, were established with Baltica, allowing shallow marine fauna to cross the Thor-Tornquist Suture. This indicates that Eastern Avalonia and Far Eastern Avalonia were not very distant from each other at that time.

To which terrane do the Krefeld High and Ebbe and Remscheid inliers belong?

The boundary between the Southern North Sea–Lüneburg Terrane (SNSLT, Fig. 1) and Eastern Avalonia around the Midlands Microcraton was placed in previous Palaeozoic terrane boundary maps of central and NW Europe along the Dowsing-South Hewett Fault Zone (DSHFZ, Fig. 1) and Lower Rhine Lineament (LRL, Fig. 1) (Pharaoh 1999; Pharaoh *et al.* 1995; Winchester *et al.* in press). The Krefeld High and the Ebbe and Remscheid inliers on these maps are situated north of the LRL and hence in the SNSLT. Balanced cross-sections through the eastern Rhenish Massif (see Oncken *et al.* 2000) indicate that only a small amount of northwestward Variscan thrusting occurred there; less than a few kilometres northwards for the Ebbe and Remscheid inliers and no displacement for the Krefeld High, which is situated at the Variscan deformation front. Hence their present position is close to their pre-Variscan position.

In the Ebbe Inlier, the top of megasequence 1 and much of megasequence 2 occur in a shelf facies, very similar to the Condroz Shelf, and have a depositional hiatus due to deformation, with the same age as the Ardennes inliers (Ardennian Phase?). A structural study is needed to investigate the impact of this deformation. The benthic fauna in megasequence 2 of the Ebbe Inlier indicates a Peri-Gondwanan affinity; however a clear distinction between Avalonia or an Avalonia-related microplate (SNSLT), based on fauna, is not yet possible.

In the Krefeld High, below the Mesozoic / Cenozoic Roer Valley Graben, the sediments exhibit a facies and age most closely comparable to those of the Stavelot-Venn Inlier, while an isotopic age determination of 394 Ma for the cleavage formation, indicates a Brabantian Phase or related (Acadian) orogenic imprint. In the poorly constrained northern part of the Krefeld High, it is inferred that Proterozoic gneisses served as the source for Upper Famennian (late Devonian) feldspar-rich sediments in Belgium. The boundary between Eastern Avalonia and Far Eastern Avalonia can hence be postulated to fall within the Krefeld High, north of the Krefeld borehole of Ahrendt *et al.* (2001*a*, *b*) and south of the supposed Proterozoic subcrop (north limit ABDB, Fig. 1).

The ε^{Nd} studies in the Ebbe Inlier indicate an evolutionary curve through the Ordovician very similar to that of the Stavelot-Venn Inlier, the Brabant Massif, the Lake District and Wales. It indicates similarity of provenances during deposition of megasequence 2 for all areas mentioned above, but is insufficient to prove affinity of the Ebbe Inlier to Eastern Avalonia.

The similarity in facies and benthic fauna alone is also insufficient proof for the affinity of the Krefeld High and the Ebbe and Remscheid inliers to Eastern Avalonia, but, combined with the presence of similar tectonic history to the Brabant Massif and Ardennes inliers, it strongly indicates their affinity to the Brabant, Condroz and Ardennes area, and hence Eastern Avalonia. The margin of Eastern Avalonia should therefore be located north of the two inliers. In this hypothesis the Anglo-Brabant Deformation Belt has its continuation into the Krefeld High.

The Anglo-Brabant Deformation Belt (ABDB) may continue east of the Rhine. The gravity image (Fig. 7) indicates that the signature of the ABDB may extend east of the Roer Valley graben, in the northern part of the eastern Rhenish Massif. The ABDB may therefore extend into the Remscheid and Ebbe inliers. Three deep seismic lines (MOBIL 6 and 7, Belcorp) through the Midlands Microcraton and the ABDB, image a rather transparent, poorly reflective crust in Avalonia, and as this character is also observed under the eastern Rhenish Massif (Bouckaert *et al.* 1988; DEKORP Research Group 1991) it is possible that the Avalonian crust extends under most of the Rhenish Massif.

Direct and indirect evidence for Proterozoic basement under both terranes

For Eastern Avalonia there is direct evidence for Neoproterozoic basement under the Midlands Microcraton in outcrops and boreholes (see above). Indirect sedimentological evidence comes from the presence of a shelf in the Silurian in the SE Brabant Massif with a turbiditic basin north of it. If the Silurian in the central part of the Condroz Inlier is restored for Variscan shortening it is also situated in the prolongation of the shelf in the SW Brabant Massif and to the south of the turbiditic basin. The presence of a persistent shelf from the Ordovician to the Silurian might indicate the presence of a stable block below it. From the field evidence in the Midlands Microcraton mentioned above, it seems that the whole region is underlain by Neoproterozoic crust. This Neoproterozoic basement possibly extends beneath the southwestern part of the Brabant Massif (cf. Sintubin & Everaerts 2002).

For the Far Eastern Avalonia microplate there is indirect sedimentological evidence of a Proterozoic basement by the postulated presence of a source area, situated in the northern area of the Krefeld High, for the Upper Famennian sandstones in the Dinant Synclinorium (see above).

Possible models for the observed deformation in the Brabant Massif and the Ardennes inliers

The sudden appearance of turbidites during mid Caradoc time (Verniers *et al.* 2001) and the inferred large-scale slumping (Debacker *et al.* 2001) record tectonic instability of the Brabant Massif at the top of megasequence 2. Extrusive calc-alkaline magmatism at the top of megasequence 2 and the base of megasequence 3 records subduction in the vicinity of the Brabant Massif. This relatively short period of magmatism was followed by foreland basin development during the main period of megasequence 3 deposition, starting in the early Wenlock (Van Grootel *et al.* 1997) and maybe in the latest Llandovery when the first thick turbidites were deposited (Verniers *et al.* 2001), implying an onset of tectonic loading within or near the Brabant Massif (Debacker 2001*a*). The higher metamorphic grades in the Silurian rim are probably directly related to this increased subsidence. The inferred absence of overburden above the core of the massif, combined with the anomalous metamorphic grade in the northern and southern Silurian rims, indicates that separate Silurian foreland basins developed on both sides of the core. One explanation could be that the core itself was partly responsible for the Silurian foreland basin development. As such, it appears that, during the latest Llandovery, the Cambrian and Ordovician strata in the core of the massif were already experiencing shortening, simultaneous with deposition in foreland basins north and south of the core. The progressive uplift of the core may have been associated with the development of pre-cleavage low-angle extensional detachments or décollements such as the Asquempont Fault (Debacker 2001*a*). This is compatible with the progressive nature of the deformation. The symmetrical disposition of the Silurian foreland basins with respect to the core of the massif may reflect a symmetrical deformation style, postulated as north-vergent on the north-side and, as observed, south-vergent on the south-side (Sintubin & Everaerts 2002). This matches the steep belt-nature of the Cambrian core of the massif, flanked to the south, and possibly also to the north, by Ordovician–Silurian flat belts (Sintubin & Everaerts 2002). Considering the constant northward-directed Silurian turbidite influx along the southern rim

(Verniers & Van Grootel 1991), the gradually rising core was still submerged during the Gorstian. Well-preserved, non-metamorphic Silurian and early Lochkovian reworked acritarchs in upper Lower Lochkovian deposits along the northern rim of the Dinant synclinorium are considered to be derived from the Brabant Massif (Steemans 1989) and if so, it indicates that during the late early Lochkovian the core of the massif had emerged. However, deposition continued along the southern (and northern?) rim. Taking into account the anchizonal degree of metamorphism, and the presence of quartz veins and a well-developed cleavage in the Gorstian deposits below the angular unconformity, an eroded overburden of at least 5 km is inferred at average geothermal gradients (~36 °C km^{-1}). Cumulative thickness curves of the Ordovician–Silurian, based on Verniers *et al.* (2001) and Van Grootel *et al.* (1997) indicate that this thickness of overburden could not have been reached before the late Pragian, and, considering a likely decrease in sedimentation rates as the foreland basin was progressively consumed by deformation, this overburden was probably only attained during the Emsian. The extensive late Emsian to earliest Eifelian conglomerates along the north-rim of the Dinant Synclinorium are considered to mark the end of compression (cf. Michot 1980; Verniers & Van Grootel 1991; Van Grootel *et al.* 1997). The Givetian conglomerates above the angular unconformity along the southern rim of the Brabant Massif probably represent the last erosion products. In short, the deformation history of the Brabant Massif can be considered to be the progressive inversion of a Cambrian rift basin during a period of at least 35 Ma (Debacker 2001*a*).

Much less is known about the pre-Variscan deformation history of the Ardennes inliers. Considering the stratigraphic gap between the Middle Ordovician and the Přídolí (about 40 Ma), the Ardennian Phase of deformation in the Ardennes inliers definitely occurred earlier than the Brabantian Phase in the Brabant Massif. Possibly the Ardennes inliers experienced a similar long-lived deformation event, consisting of the progressive inversion of Cambrian basins. In this respect, although having a different timing, the Brabantian Phase and the Ardennian Phase might even have a similar cause (Debacker 2001*a*; Sintubin & Everaerts 2002).

Constraints for a model

Only in the Anglo-Brabant Deformation Belt are there sufficient data, reviewed in the light of published data (see above) for a model to be proposed that is consistent with the above observations and the following constraints.

(1) Palaeomagnetic inclination results from Wales indicate a shift from 60°south in the Tremadocian to 13°south in the Wenlock.
(2) The model has to fit earlier models of the Avalonia–Baltica collision closing the Tornquist Sea, and of the termination of Iapetus subduction magmatism by the mid Caradoc, with sediments crossing the Iapetus Suture from mid/late Wenlock onwards (King 1994).
(3) It has to comply with a proposed anticlockwise rotation, which could explain the diachronous onset of turbiditic sedimentation and deformation from southern Ireland to the Lake District (Soper & Woodcock 1990; Soper *et al.* 1992) and the decrease in age of peak magmatic activity from west to east along the belt (Pharaoh *et al.* 1995).
(4) It has also to integrate a supposed anticlockwise rotation, deduced from palaeomagnetism, of 55° between the Lake District and North Wales from the Caradoc to the mid Devonian (Piper 1997). The east of the microcraton and hence the area of the Anglo-Brabant Deformation Belt would also be subjected to this rotation.
(5) It has to explain the calc-alkaline subduction-related magmatism in the Anglo-Brabant Deformation Belt in a more or less 500 km long belt from the Lake District to Quenast, from Caradoc to Ashgill (about 10 Ma), at a different timing from the one in Wales and Lake District. Subduction started before the mid Caradoc, but extrusive magmatism was only produced in the mid Caradoc, and definitely in the early Ashgill.
(6) It has to explain the high-temperature and low-pressure deformation of the four Ardennes inliers and the Remscheid and Ebbe inliers (Ardennian Phase).
(7) Foreland basin development in the Brabant Massif occurred from the late Llandovery to early Devonian (cf. Lee *et al.* 1993; Van Grootel *et al.* 1997), and the Anglian Basin is also assumed to have thick Silurian strata (Woodcock & Pharaoh 1993).
(8) Finally it should explain the high-temperature and low-pressure progressive deformation of the Brabant Massif, north of the Brabantian deformation front (Sintubin 1999), which was accompanied by substantial shortening (Debacker 2001*a*) in the late Silurian to Eifelian, during the Brabantian Phase (Michot 1980; Van Grootel *et al.*

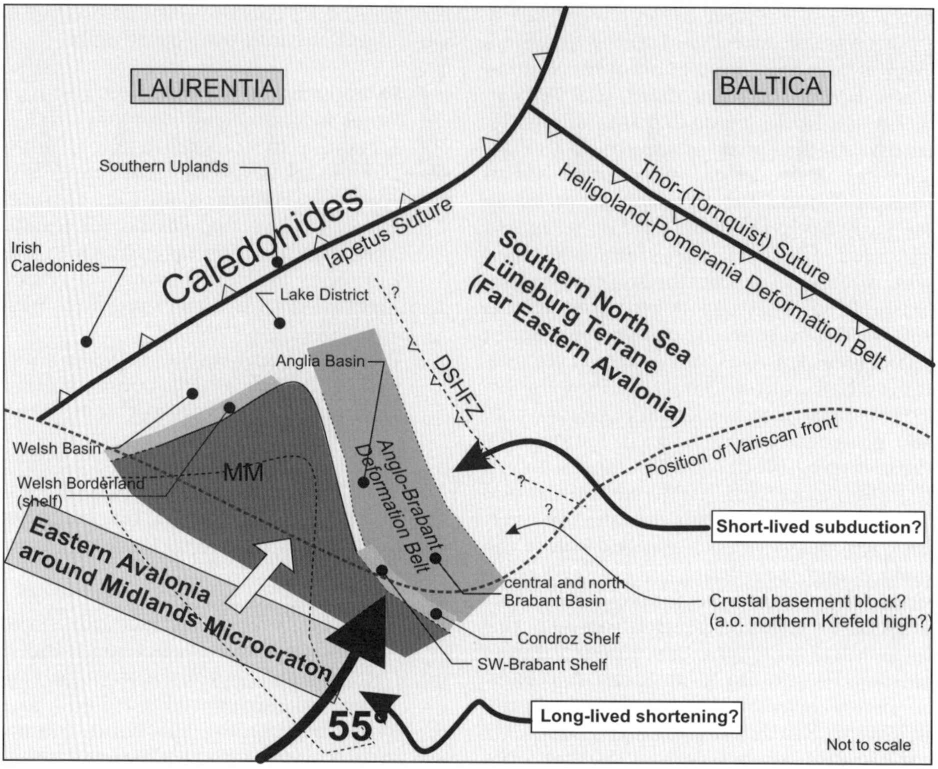

Fig. 8. Schematic map (after Debacker 2001*a*) showing the 55° anticlockwise rotation of the Lake District with respect to Wales and Midlands Microcraton and adjacent areas of Eastern Avalonia, from Caradoc to Emsian (after Piper 1997). Shown is also the implied movement for the area east of the Midlands Microcraton and the Anglo-Brabant Deformation Belt (see text for explanation).

1997). This deformation belt extends northeastwards of the Stavelot-Venn Inlier as far as the Krefeld High (Ahrendt *et al.* 2001*a*, *b*) and possibly farther east; it is present in East Anglia, in the Brabant Massif (except for its undeformed southwestern part) and possibly in a small part of the Condroz Inlier. The Ebbe and Remscheid inliers are not affected because sedimentation of the Rheno-Hercynian Basin is already occurring there. Contemporaneous with the long-lived and slow deformation in the ABDB, extension occurred in the Rheno-Hercynian Zone, the large and wide area south of the ABDB, where a rift basin was active from uppermost Silurian (Upper Ludlow) to the Emsian (Lower Devonian). By the end of the Emsian continental rifting ceased with the formation of the Rheno-Hercynian Ocean farther south (see synthesis in Oncken *et al.* 2000).

A tentative model

The new data presented here allow the following model for the basin development, subduction magmatism and deformation history of Eastern Avalonia east of the Midlands Microcraton to be proposed (Fig. 8).

An extensional basin or series of basins developed during early Cambrian time, forming a rift along the Anglo-Brabant–Ardennes area (cf. Crowley *et al.* 2000; Winchester *et al.* in press), soon after the Neoproterozoic Panafrican Orogeny or even during it (Murphy *et al.* 1999). Hence a zone of crustal weakness was established which formed a template controlling structural evolution from the Cambrian to the Middle Devonian.

The Far Eastern Avalonia and the Eastern Avalonia microcontinents are both considered to be Gondwana-derived, but in view of the arguments above, the former may have left Gondwana earlier than the latter. A narrow ocean is inferred to have separated them during their northward transit. The Eastern Avalonia microcontinent should end north of the present day location of the Anglo-Brabant Deformation Belt and of the Rhenish Massif.

When Avalonia detached from Gondwana during the late Tremadocian to early Arenig (Prigmore *et al.* 1997), subduction underflow occurred beneath the Irish, Welsh and Lake District segments of Eastern Avalonia. The Midlands Microcraton and the Anglo-Brabant area were rotated approximately 55° anticlockwise in relation to the Lake District (Piper 1997). The Anglo-Brabant area was in this scenario oriented parallel to transport and remained a passive margin, with on its eastern side, an oceanic basin separating it from Far Eastern Avalonia.

When the Iapetus Ocean subduction stopped during Caradoc–Ashgill time, tectonic instability and turbidites were generated simultaneously in the Brabant Massif, and subduction magmatism was initiated in the same area. When Avalonia docked with Baltica in the late Caradoc or early Ashgill, an anticlockwise rotation started, provoking subduction of oceanic crust in the basin lying between Eastern Avalonia and the Far Eastern Avalonia microplate, giving rise to the calc-alkaline subduction magmatism from The Wash to Quenast.

The anticlockwise rotation of Wales and the Midlands Microcraton, relative to the Lake District in Caradoc to mid Devonian time, is invoked as the driving mechanism for the further evolution of the Anglo-Brabant Deformation Belt. The rotation implies for the ABDB a scissors-like closure leading to a shortening of about 500 km in its southern part (Brabant Massif) (Fig. 8). This movement can explain the short-lived late Ordovician–earliest Silurian subduction of oceanic crust revealed by subduction-related magmatism. Volcanism ended in the Brabant Massif in the early Llandovery, possibly because the oceanic crust had subducted completely and the continental crust of the SNSLT had approached the SW limit of Baltica. Continuing rotation created the Silurian foreland basin development of the central and northern Brabant Basin (Van Grootel *et al.* 1997) and, supposedly in the Anglian Basin (Woodcock & Pharaoh 1993). It may also have created the conditions for the long-lived Wenlock to early Eifelian Brabantian shortening and inversion of the central steep belt of the Brabant Massif. Some areas were shielded from deformation, such as the southwestern Brabant Shelf and the Condroz Shelf, possibly because they were situated on top of a Neoproterozoic basement, or a more rigid, low density body at depth. The end of rotation by Eifelian time terminated the Brabantian deformation.

How the Ardennes, and the Ebbe and Remscheid inliers were deformed in the Ardennian Phase between the Caradoc and the Silurian, is still unclear. One possibility would be that this deformation was caused by the first collision between this part of Avalonia and Baltica; in this case, the Ardennes, Ebbe and Remscheid inliers would have been in a more easterly position during the early period of the anticlockwise rotation, reaching Baltica first. A similar scenario may be assumed calling upon rotations of cratonic cores within the Avalonian Terrane Assemblage.

Conclusion

The three megasequences from Cambrian to mid Devonian described in Wales and the Welsh Borderland by Woodcock (1990*b*) are also present, or parts of them, on Avalonia east of the Midlands Microcraton (Brabant Massif, Condroz, Ardennes, Remscheid and Ebbe inliers, Krefeld high). The three megasequences are caused by a tectonic driving mechanism and can be explained as the result of three different geodynamic contexts: an early phase with extensional basins or rifting and rather thick sequences, a second phase with a shelf basin with moderately thin sequences and a later phase with a shelf or foreland basin development and thick sequences.

Deposition of megasequence 1 started shortly after the Cadomian orogeny in the earliest Cambrian as rift basins opened along the position of the Anglo-Brabant Deformation Belt and along other directions in the Ardennes and Remscheid inliers and the Krefeld High. The extensional basin development is indicated by the concave up subsidence curve in the Brabant Massif. Megasequence 1 was deposited in a palaeogeographical situation when Avalonia as a separate microcontinent did not yet exist and where the study area was still attached to Gondwana, situated off the north coast of South America or West Africa (according to two different types of reconstruction). The megasequence 1 ends at the widespread hiatus in the study area from the late Tremadocian to the latest mid Arenig.

Megasequence 2 was deposited on a deeper shelf with rather thin sequences present in most of the areas. The upper part of the megasequence 2 is disturbed by a short-lived deeper basin interval with turbidites and locally large-scale slides in the late Caradoc. No clear increasing or decreasing subsidence can be deduced from subsidence curves. The microcontinent Avalonia, and with it the study area, had detached from Gondwana as a separate microcontinent during deposition of megasequence 2. The limited source area of the microcontinent explains the thin sequences.

Megasequence 3 from late Caradoc till latest Silurian contains shallow shelf deposits which continue on some of the areas in thick shelf deposits and on other areas in a thick foreland basin development, clearly indicated by the convex up subsidence curves. Subduction magmatism occurs from The Wash and East Anglia as far as the Brabant Massif. Well dated as lower Ashgill in the Brabant Massif it is short-lived and hence indicative of a short period of subduction. The subduction magmatism implies a subduction of oceanic crust somewhere parallel to the magmatic subcrop area. This implies an oceanic crust which would be situated within what is assumed to be the Avalonia microcontinent (see below).

Deformation of the megasequences 1 and 2 or 1 to 3 differs from area to area. North and NW of the Midlands microcraton the late Ordovician Shelveian unconformity (see Fig. 2) is interpreted by some as a deformation phase. In Wales, away from the microcraton, and in the Lake District the Acadian phase is long-lived and active from early to mid Devonian. In the Ardennes inliers a deformation is active between the late Ordovician and the Silurian (Ardennian Phase), with a similar intensity as the core of the Brabant Massif, when present erosion levels are compared. The Brabant Massif is partly deformed by the long-lived Brabantian Phase from late Silurian till early mid Devonian. Both the Ardennes inliers and the Brabant Massif are not classic orogenic belts, only slate belts where no more than the epizone is reached, at present erosion levels. Areas supposedly close to the microcraton or basement are nearly undeformed (SW Brabant Massif and central Condroz).

A model of anticlockwise rotation of Avalonia of about 55° from Caradoc to Emsian (Piper 1997) is proposed to explain the setting of deposition of megasequence 3 and the subsequent Acadian and Brabantian deformation. As soon as Avalonia reached Baltica in Caradoc time it created a short-lived subduction magmatism from the Wash to the Brabant Massif and, soon after the magmatism ended, a foreland basin developed. Possibly during and after that development a long-lived and slow shortening and compression occurred leading to the deformation of the Anglo-Brabant Deformation Belt. Locally the deformation was shielded by the presence of basement structures. Note that the contemporaneity in the early Devonian between the shortening of the Anglo-Brabant Deformation Belt and the extension in the Rheno-Hercynian Zone, situated adjacent to the south. Possibly the extensional Rheno-Hercynian Basin was caused by the same slow rotation of Eastern Avalonia.

From the review of recently published and new data presented and discussed above, it appears that the study area, Avalonia east of the Midlands Microcraton, probably consists of two terranes. The first is the remnant of the palaeocontinent Avalonia, in which the history of its eastern part has been described above. The second terrane can be called the palaeocontinent Far Eastern Avalonia. Its relics are only rarely observed in the few deep boreholes into the Heligoland–Pomerania Deformation Belt, in southern Denmark, NE Germany and NE Poland. In between the two areas, the latter and the Anglo-Brabant Deformation Belt only few data are available. Absolute age determinations on inherited zircons in younger magmatic rocks indicate only Proterozoic basement (the Ecke gneiss, the huge xenolith in the Variscan Bröcken granite, Harz mountains (Berthelsen 1992); Flechtingen and Penkun boreholes (Breitkreuz & Kennedy 1999); a nearby northern provenance area of feldspars in the Upper Famennian sandstone of the Dinant Synclinorium (Thorez 1969 cited in Paproth *et al.* 1986; see above). For Far Eastern Avalonia a palaeogeographical history is postulated similar to Avalonia, with rifting from Gondwana in Arenig or earlier times and collision with Baltica before mid Ashgill (Samuelsson *et al.* 2001) and deformation between the late Ordovician and latest Silurian. The Avalonia concept might need to be expanded to an 'Avalonian Terrane Assemblage' with cratonic cores and small short-lived oceans as in the Armorican Terrane Assemblage.

The numerous comments of the editor J. Winchester and the referees N. Woodcock and P. Lane are much appreciated. T.C. Pharaoh publishes with the approval of the Executive Director, British Geological Survey. This study contains the results of the 'Palaeozoic Amalgamation of Central Europe' (PACE) TMR network project (ERBFMRXCT97 0136) and the EUROPROBE TESZ project. Part of the work was also sponsored by the FWO research project 'Tectonics of the early Palaeozoic basin development in NW Europe: basin analysis and magnetic fabric analysis in the Belgian Caledonides' (N° G.0094.01) to M. Sintubin & J. Verniers and by FNRS grant FRFC N°2.4506.97 to A. Herbosch.

Prof. J. Thorez (Université de Liège) is acknowledged for the detailed information on the Upper Famennian of Belgium and Dr Y. Vanbrabant (Université de Liège) for sharing his new data on the

Prayon outcrop. Mrs N. Reynaert is thanked for typing parts of several versions. Joakim Samuelsson acknowledges a postdoctoral scholarship from the Swedish Foundation for International Cooperation in Research and Higher Education (STINT). Manuel Sintubin and Tim Debacker were, at the time of this study, respectively postdoctoral fellow and research assistant of the Fund for Scientific Research, Flanders (Belgium). This is a publication in the EUROPROBE series.

References

ADAMS, R. & VANDENBERGHE, N. 1999. The Meuse valley section across the Condroz-Ardennes (Belgium) based on a predeformational sediment wedge. *In:* SINTUBIN, M., VANDYCKE, S. & CAMELBEECK, T. (eds) Palaeozoic to Recent tectonics in the NW European Variscan Front Zone. *Tectonophysics,* **309**, 179–195.

AHRENDT, H., RIBBERT, K.-H., VANGUESTAINE, M. & WEMMER, K. 2001*a*. K-Ar and acritarch dating of phyllite clasts from a resedimented Middle Devonian conglomerate in the northwestern part of the Rhenish Slate Mountains. *Abstracts of the Meuse-Rhine Euregio Geologists Meeting in the Hautes-Fagnes/Hohes Venn (Belgium, May, 4–5, 2001)*, p. 3.

AHRENDT, H., RIBBERT, K.-H., VANGUESTAINE, M. & WEMMER, K. 2001*b*. K-Ar and acritarch dating of phyllite clasts from a resedimented Middle Devonian conglomerate in the northwestern part of the Rhenish Slate Mountains. *Zeitschrift der deutschen geologischen Gesellschaft*, **152**, 365–377.

AICHROTH, B., PRODEHL, C. & THYBO, H. 1992. Crustal structure along the Central segment of the EGT from seismic refraction studies. *Tectonophysics,* **207**, 43–64.

ALLSOP, J. M. 1987. Patterns of late Caledonian intrusive activity in eastern and northern England from geophysics, radiometric dating and basement geology. *Proceedings of the Yorkshire Geological Society*, **46**, 335–353.

ANDRÉ, L. 1983. *Origine et évolution des roches éruptives du Massif du Brabant (Belgique). Implications au niveau de l'étude des roches magmatiques transformées et de géologie calédonienne du nord-ouest de l'Europe.* Unpublished PhD thesis, Université Libre de Bruxelles.

ANDRÉ, L. 1991. The concealed crystalline basement in Belgium and the 'Brabantia' microplate concept: constraints from the Caledonian magmatic and sedimentary rocks. *In:* ANDRÉ, L., HERBOSCH, A., VANGUESTAINE, M. & VERNIERS, J. (eds) Proceedings of the international meeting on the Caledonides of the Midlands and the Brabant Massif, 1989. *Annales de la Société Géologique de Belgique*, **114**, 117–139.

ANDRÉ, L. & DEUTSCH, S. 1984. Les porphyres de Quenast et de Lessines: géochronologie, géochimie isotopique et contribution au problème de l'âge du socle précambrien du Massif du Brabant (Belgique). *Bulletin de la Société belge de Géologie,* **93**, 375–384.

ANDRÉ, L. & DEUTSCH, S. 1985. Very low-grade metamorphic Sr isotopic resettings of magmatic rocks and minerals: Evidence for a late Givetian strike-slip division of the Brabant Massif (Belgium). *Journal of the Geological Society, London,* **142**, 911–923.

ANDRÉ, L. & DEUTSCH, S. 1986. Magmatic $^{87}Sr/^{86}Sr$ relicts in hydrothermally altered quartz diorites (Brabant Massif, Belgium) and the role of epidote as a Sr filter. *Contributions to Mineralogy and Petrology,* **92**, 104–112.

ANDRÉ, L., DEUTSCH, S. & MICHOT, J. 1981. Données géochronologiques concernant le développement tectono-métamorphique du segment Calédonien Brabançon. *Annales de la Société Géologique de Belgique,* **104**, 241–253.

ANDRÉ, L., DEUTSCH, S. & HERTOGEN, J. 1986*a*. Trace element and Nd isotopes in shales as indexes of provenance and crustal growth: the early Paleozoic from the Brabant Massif (Belgium). *Chemical Geology,* **57**, 101–115.

ANDRÉ, L., HERTOGEN, J. & DEUTSCH, S. 1986*b*. Ordovician-Silurian magmatic provinces in Belgium and the Caledonian orogeny in middle Europe. *Geology,* **14**, 879–882.

BAUMANN, A., GRAUERT, B., MECKLENBERG, S. & VINX, R. 1991. Isotopic age determinations of crystalline rocks of the Upper Harz Mountains, Germany. *Geologische Rundschau*, **80**, 669–690.

BEDNARCZYK, W. 1974. The Ordovician in the Koszalin-Chojnice region (Western Pomerania). *Acta Geologica Polonica*, **24,** 581–600.

BELANGER, I. 1998. *Effets d'une déformation hétérogène sur un ensemble mécaniquement anisotrope: le cas du Massif de Rocroi (Ardenne franco-belge).* Unpublished PhD thesis, Université Catholique de Louvain.

BELANGER, I., SINTUBIN, M. & LADURON, D. 1997. Strain analysis along the southern border of the Rocroi Massif (Belgian-French Ardennes). *Aardkundige Mededelingen,* **8**, 11–14.

BELGIAN GEOLOGICAL SURVEY 1994. Aeromagnetic map of the Brabant Massif: residual total field reduced to the pole. Scale 1/100000.

BERTHELSEN, A. 1992. From Precambrian to Variscan Europe. *In:* BLUNDELL, D., FREEMAN, R. & MUELLER, S. (eds) *A continent revealed – the European Geotraverse.* Cambridge University Press, 153–164.

BLUNDELL, D., FREEMAN, R. & MUELLER, S. (eds) 1992. *A continent revealed – the European Geotraverse.* Cambridge University Press.

BLUNDELL, D. J., HOBBS, R. W., KLEMPERER, S. L., SCOTT-ROBINSON, R., LONG, R. E., WEST, T. E. & DUIN, E. 1991. Crustal structure of the central and southern North Sea from BIRPS deep seismic reflection profiling. *Journal of the Geological Society, London,* **148**, 445–458.

BOIS, C. & ECORS SCIENTIFIC PARTY 1990. Major geodynamic processes studied from the ECORS deep seismic profiles in France and adjacent areas. *Tectonophysics*, **173**, 397–410.

BOUCKAERT, J., FOCK, W.& VANDENBERGHE, N. 1988. First results of the Belgian Geotraverse 1986 (Belcorp). *Annales de la Société Géologique de Belgique,* **111**, 279–290.

BORREMANS, G. & BULTYNCK, P. 1986. Conodontes du calcaire de Naux-Gedinnien Inférieur au sud immédiat du Massif de Rocroi (Ardenne française). *Aardkundige Mededelingen*, **3**, 45–58.

BRANNEY, M. J. & SOPER, N. J. 1988. Ordovician volcano-tectonics in the English Lake District. *Journal of the Geological Society, London,* **145**, 367–376

BRASIER, M. D., INGHAM, J. K. & RUSHTON, A. W. A. 1992. Cambrian. *In:* COPE, J. C. W., INGHAM, J. K. & RAWSON, P. F. (eds) *Atlas of Palaeogeography and Lithofacies*. Geological Society, London, Memoir **13**, 13–18.

BREITKREUZ, C. & KENNEDY, A. 1999. Magmatic flare-up at the Carboniferous/Permian boundary in the NE German basin revealed by SHRIMP zircon ages. *Tectonophysics*, **302**, 307–326.

BRIDGE, D. MCC., CARNEY, J. N., LAWLEY, R. S. & RUSHTON, A. W. A. 1998. Geology of the country around Coventry and Nuneaton. Memoir for 1: 50 000 Geological Sheet 169 (England and Wales). *British Geological Survey*, 1–185.

BRITISH GEOLOGICAL SURVEY 1996. *Tectonic map of Britain, Ireland and adjacent areas.* PHARAOH, T. C., MORRIS, J. H., LONG, C. B. & RYAN, P. D. (compilers) 1: 1 500 000. Keyworth, Nottingham, British Geological Survey.

BURG, J.-P. 1999. Ductile structures and instabilities: their implication for Variscan tectonics in the Ardennes. *In:* SINTUBIN, M., VANDYCKE, S. & CAMELBEECK, T. (eds) Palaeozoic to Recent tectonics in the NW European Variscan Front Zone. *Tectonophysics*, **309**, 1–25.

CATOT, E. 1992 Contribution à l'étude des acritarches du Salmien Inferieur du massif de Stavelot. *Annales de la Société Géologique de Belgique*, **115**, 371.

CAZES, M., MASCLE, A., TORREILLES, X., BOIS, C., DAMOTTE, X., MATTE, P., RAOULT, X., PHAM, V. N., HIRN, A. & GALDEANO, X. 1986. Large Variscan overthrusts beneath the Paris Basin. *Nature,* **323**, 144–147.

CHACKSFIELD, B., DE VOS, W., D'HOOGE, L., DUSAR, M., LEE, M., POITEVIN, C., ROYLES, C. & VERNIERS, J. 1993. A new look at Belgian aeromagnetic and gravity data through image-based display and integrated modelling techniques. *Geological Magazine,* **130**, 583–591.

CHEADLE, M. J., MCGEARY, S., WARNER, M. R. & MATHEWS, D. H. 1987. Extensional structures on the western UK continental shelf: a review of evidence from deep seismic profiling. *In*: COWARD, M. P., DEWEY, J. K. & HANCOCK, P. I. (eds) *Continental extensional tectonics*. Geological Society, London, Special Publications, **28**, 445–465.

CLARK, A. H., SCOTT, D. J., SANDEMAN, H. A., BROWLEY, A. V. & FARRAR, E. 1998. Siegenian generation of the Lizard ophiolite: U-Pb zircon age data for plagiogranite, Porthkerris, Cornwall. *Journal of the Geological Society, London*, **155**, 595–598.

CHLUPAC, I., HAVLICEK, V., KRIZ, J., KUKAL, Z. & STORCH, P. 1998. Palaeozoic of the Barrandian (Cambrian to Devonian). Czech Geological Survey, Prague, 183 p.

COCKS, L. R. M. 2000. The Early Palaeozoic geography of Europe. *Journal of the Geological Society, London,* **157**, 1–10.

COCKS, L. R. M. 2002. Key Lower Palaeozoic faunas from near the Trans-European Suture Zone. *In:* WINCHESTER, J. A., PHARAOH, T. C. & VERNIERS, J. (eds) *Palaeozoic Amalgamation of Central Europe.* Geological Society, London, Special Publications, **201**, 37–46.

COCKS, L. R. M. & FORTEY, R. A. 1982. Faunal evidence for ocean separations in the Palaeozoic of Britain. *Journal of the Geological Society, London,* **139**, 465–478.

COCKS, L. R. M. & FORTEY, R. A. 1990. Biogeography of Ordovician and Silurian faunas. *In:* MCKERROW, W. S. & SCOTESE, C. R. (eds) *Palaeozoic palaeogeography and biogeography*. Geological Society, London, Memoirs, **12**, 97–104.

COCKS, L. R. M., HOLLAND, C. H. & RICKARDS, R. B. 1992. A revised correlation of Silurian rocks in the British Isles. *Geological Society, London, Special Report,* N° **21**, 1–32.

COCKS, L. R. M., MCKERROW, W. S. & VAN STAAL, C. R. 1997. The margins of Avalonia. *Geological Magazine,* **134**, 627–636.

COCKS, L. R. M. & TORSVIK, T. H. 2002. Earth geography from 500 to 400 million years ago: a faunal and palaeomagnetic review. *Journal of the Geological Society, London*, **159**, 631–644.

COCKS, L. R. M. & VERNIERS, J. 2000. Applicability of planktonic and nektic fossils to palaeogeographical reconstructions. *Acta Universitatis Carolinae – Geologica*, **42**(1998), 399–400.

COOPER, A. H., MILLWARD, D., JOHNSON, E. W. & SOPER, N. J. 1993. The early Palaeozoic evolution of northwest England. *Geological Magazine,* **130**, 711–724.

CORIN, F. 1965. Atlas des roches éruptives de Belgique. *Mémoires pour servir à l'Explication des Cartes géologiques et Minières de la Belgique,* **4**, 1–190.

CROWLEY, Q. G., FLOYD, P. A., WINCHESTER, J. A., FRANKE, W. & HOLLAND, J. G. 2000. Early Palaeozoic rift-related magmatism in Variscan Europe: fragmentation of the Armorican Terrane Assemblage. *Terra Nova,* **12**, 171–180.

DADLEZ, R. 2000. Pomeranian Caledonides (NW Poland), fifty years of controversies: a review and a new concept. *Geological Quarterly*, **44**, 221–236.

DALLMEYER, R. D., FRANKE, W. & WEBER, K. (eds) 1995. *The Pre-Permian geology of Central and Eastern Europe*. Springer, Berlin, 1–604.

DALLMEYER, R. D., GIESE, U., GLASMACHER, U. & PICKEL, W. 1999. First $^{40}Ar/^{39}Ar$ age constraints for the Caledonian evolution of the Trans-European Suture Zone in NE Germany. *Journal of the Geological Society, London*, **156**, 279–290.

DEAN, T. W. 1991. Ordovician trilobites from the inlier at Le Petit Fond d'Oxhe, Belgium. *Bulletin de l'Institut Royal des Sciences Naturelles de Belgique, Sciences de la Terre*, **61**, 135–155.

DEBACKER, T. N. 1997. A structural study of the

Silurian turbidite deposits in the Ronquières area, Brabant Massif. *Annales de la Société Géologique de Belgique,* **120**, 104–105.

DEBACKER, T. N. 1999. Folds trending at various angles to the transport direction in the Marcq area, Brabant Massif, Belgium. *Geologica Belgica,* **2**, 159–172.

DEBACKER, T. N. 2000. The Asquempont fault, southeastern Anglo-Brabant fold belt, Belgium. *Abstracts Volume Joint meeting of EUROPROBE (TESZ) and PACE* Projects (September 16–23, 2000, Zakopane, Poland), 18–19.

DEBACKER, T. N. 2001*a*. *Palaeozoic deformation of the Brabant Massif within eastern Avalonia: how, when and why?* Unpublished Ph. D. thesis, Ghent University.

DEBACKER, T. N. 2001*b*. Gently plunging steeply to moderately inclined folds versus steeply plunging upright to steeply inclined folds in the southeastern Anglo-Brabant fold belt, Belgium. *Abstracts Volume Tectonic Studies Group annual meeting (January 3–6, 2001, Leeds, UK).*

DEBACKER, T. N., SINTUBIN, M. & VERNIERS, J. 1997. The Ronquières Section revisited (Brabant Massif, Belgium). *Aardkundige mededelingen,* **8**, 53–56.

DEBACKER, T. N., SINTUBIN, M. & VERNIERS, J. 1999. Cleavage/fold relationships in the Silurian metapelites, southeastern Anglo-Brabant fold belt (Ronquières, Belgium). *Geologie & Mijnbouw,* **78**, 47–56.

DEBACKER, T. N., SINTUBIN, M. & VERNIERS, J. 2001. Large-scale slumping deduced from structural relations and sedimentary features: a study in the Lower Palaeozoic Anglo-Brabant fold belt, Belgium. *Journal of the Geological Society, London,* **158**, 341–352.

DEKORP RESEARCH GROUP 1991. Results of the DEKORP 1 (BELCORP-DEKORP) deep seismic reflection studies in the western part of the Rhenish Massif. *Geophysical Journal International,* **106**, 203–227.

DELVAUX de FENFFE, D. & LADURON, D. 1984. Analyse structurale au bord sud du Massif de Rocroi (Ardennes françaises). *Bulletin de la Société belge de Géologie,* **93**, 11–26.

DELVAUX de FENFFE, D. & LADURON, D. 1991. Caledonian and Variscan structures in the Rocroi-Ardenne Lower Paleozoic basement (Belgium and adjacent countries). *Annales de la Société Géologique de Belgique,* **114**, 141–162.

DE MEYER, F. 1983. Gravity interpretation of the western flank of the Brabant Massif. *Koninklijk Meteorologisch Instituut van België, Publicaties,* Serie A/*Institut Royal Météorologique de Belgique, Publications*, Série A, **111**, 1–50.

DE MEYER, F. 1984. Two structural models for the western flank of the Brabant Massif. *Geophysical Prospecting,* **32**, 37–50.

DE VOS, W. 1997. Influence of the granitic batholith of Flanders on Acadian and later deformation (Brabant Massif, Belgium). *Aardkundige Mededelingen,* **8**, 49–52.

DE VOS, W., POOT, B., HUS, J. & EL KHAYATI, M. 1992. Geophysical characterization of lithologies from the Brabant Massif as a contribution to gravimetric and magnetic modelling. *Bulletin de la Société belge de Géologie,* **101**, 173–180.

DE VOS, W., VERNIERS, J., HERBOSCH, A. & VANGUESTAINE, M. 1993. A new geological map of the Brabant Massif, Belgium. *Geological Magazine,* **130**, 605–611.

D'OMALIUS D'HALLOY, J. J. 1842. *Coup d'oeil sur la géologie de la Belgique*. Hayez, Bruxelles,1–132.

DUMONT, A. 1847. Mémoire sur les terrains ardennais et rhénan de l'Ardenne, du Rhin, du Brabant et du Condroz. I. Terrain ardennais. *Mémoires de l'Académie Royale de Belgique, Classe des Sciences,* **20**, 1–163.

EISERHARDT, K.-H., HEYCKENDORF, K. & THOMBANSEN, E. 1981. Zur Stratigraphie und Tektonik des nördlichen Ebbe-Teilsattels (Sauerland, Rheinisches Schiefergebirge). *Mitteilungen aus dem Geologisch-Paläontologischen Institut der Universität Hamburg,* **50**, 199–238.

EUGENO-S WORKING GROUP 1988. Crustal structure and tectonic evolution of the transition between the Baltic Shield and the North German Caledonides (the EUGENO-S Project). *Tectonophysics*, **150**, 253–348.

EVANS, D., ROWLEY, W., CHADWICK, R., KIMBELL, G. & MILLWARD, D. 1994. Seismic reflection data and the internal structure of the Lake District batholith, Cumbria, northern England. *Proceedings of the Yorkshire Geological Society,* **50**, 11–24.

EVERAERTS, M. 2000. *L'interpretation structurale de la Manche au Rhin: apport du filtrage des champs de potentiel.* Unpublished PhD thesis, Université Catholique de Louvain.

EVERAERTS, M., POITEVIN, C., DE VOS, W. & STERPIN, M. 1996. Integrated geophysical/geological modelling of the western Brabant Massif and structural implications. *Bulletin de la Société belge de Géologie*, **105**, 41–59.

FERKET, H., MUCHEZ, P., SCHROYEN, K. & SINTUBIN, M. 1998. Metamorphism in the Stavelot Massif: a study of quartz veins in the basal Devonian conglomerates (Lochkovian). *Aardkundige Mededelingen*, **9**, 7–16.

FIELITZ, W. 1992. Variscan transpressive inversion in the northwestern central Rhenohercynian belt of western Germany. *Journal of Structural Geology,* **14**, 547–564.

FIELITZ, W. & MANSY, J-L. 1999. Pre- and synorogenic burial metamorphism in the Ardenne and neighbouring areas (Rhenohercynian zone, central European Variscides). *In:* SINTUBIN, M., VANDYCKE, S. & CAMELBEECK, T. (eds) Palaeozoic to Recent tectonics in the NW European Variscan Front Zone. *Tectonophysics*, **309**, 227–256.

FORTEY, N. J., ROBERTS, B. & HIRONS, S. R. 1993. Relationship between metamorphism and structure in the Skiddaw Group, English Lake District. *Geological Magazine*, **130**, 631–638.

FORTEY, R. A., HARPER, D. A. T., INGHAM, J. K., OWEN, A. W. & RUSHTON, A. W. A. 1995. A revision of Ordovician series and stages from the historical type area. *Geological Magazine*, **132**, 15–30.

FORTEY, R. A., HARPER, D. A. T., INGHAM, J. K., OWEN, A. W., PARKES, M. A., RUSHTON, A. W. A. & WOODCOCK, N. H. 2000. A revised correlation of Ordovician rocks in the British Isles. *Geological Society, London, Special Report,* **24**, 1–83.

FOURMARIER, P. 1921. La tectonique du Brabant et des régions voisines. *Mémoires de l'Académie Royale de Belgique, Classe des Sciences,* 2ème série, **4**, 1–95.

FRANKE, D. 1994. The deformational history of the Caledonian terranes at Baltica's southwestern margin. *Zeitschrift für Geologische Wissenschaften*, **22**, 67–80.

FRANKE, D. 1995a. The Caledonian terranes along the southwestern border of the East European platform – evidence, speculation and open questions. *In:* GEE, D. G. & BECKHOLMEN, M. (eds) The Trans-European Suture Zone: EUROPROBE in Liblice 1993. *Studia Geophysica et Geodaetica*, **39**, 241–256.

FRANKE, D. 1995b. The North Variscan Foreland. *In:* DALLMEYER, D., FRANKE, W. & WEBER, K. (eds) *The Pre-Permian geology of Central and Eastern Europe.* Springer, Berlin, 554–566.

FRANKE, D., GRÜNDEL, J., LINDERT, W., MEISNER, B., SCHULZ, E. ZAGORA, I. & ZAGORA, K. 1994. Die ostseebohrung G14 – eine Profilübersicht. *Zeitschrift Geologische Wissenschaften*, **22**, 235–240.

FRANKE, W. 1995. Rhenohercynian foldbelt: autochton and nonmetamorphic nappe units- stratigraphy *In:* DALLMEYER, D., FRANKE, W. & WEBER, K. (eds) *The Pre-Permian geology of Central and Eastern Europe*. Springer, Berlin, 33–49.

FRANKE, W. 1998. Tectonic and plate tectonic units at the North Gondwana margin: evidence from the Central European Variscides. *Schriften des Staatlichen Museums für Mineralogie und Geologie zu Dresden,* **9**, 132–134.

FRANKE, W. 2000. The mid-European segment of the Variscides: tectonostratigraphic units, terrane boundaries and plate tectonic evolution. *In:* FRANKE, W., HAAK, V., ONCKEN, O. & TANNER, D. (eds) *Orogenic processes: Quantification and modelling in the Variscan Belt.* Geological Society, London, Special Publications, **179**, 35–61.

FRANKE, W., ALTHERR, R., HAAK, V. & ONCKEN, O. 1999. Orogenic processes – quantification and modelling in the Variscides: summary of summaries. Geologische Vereinigung 89 th Annual Meeting, Freiberg, Saxony. *Terra Nostra*, **99**, 7–10.

FROST, R. T. C., FITCH, F. J. & MILLER, J. A. 1981. The age and nature of the crystalline basement of the North Sea Basin. *In*: ILLING, L. V. & HOBSON, G. D. (eds) *Petroleum geology of the continental shelf of north-west Europe.* Institute of Petroleum, London, 43–57.

GALDEANO, A. & GUILLON, J. C. 1988. Interprétation de données magnétiques et gravimétriques dans l'étude de la croûte terrestre par sismique profonde (profil Nord de la France). *Editions Technip*, 1–260.

GERDES, A., PHARAOH, T. C., DEBACKER, T. & VERNIERS, J. 2000. Nd-isotope systematics of Lower Palaeozoic sediments from the Brabant Massif. *Abstracts volume of the 'Joint Meeting of EUROPROBE TESZ and PACE projects' in Zakopane and Holy Cross Mountains, Poland, 16 to 23 September 2000,* pp. 23–24.

GERDES, A., PHARAOH, T. C., DEBACKER, T. & VERNIERS, J. 2001*a*. Nd-isotope systematics of Lower Palaeozoic sediments from Eastern Avalonia: precise monitoring of magmatic and orogenic activity. *Abstracts volume of the 'European Union of Geosciences' in Strasbourg, France, 8–12 April 2001*, p. 535.

GERDES, A., PHARAOH, T. C., TIMMERMANN, H., VERNIERS, J., VECOLI, M., NOBLE, S. & PARRISH, R. 2001*b*. Nd-isotope record of clastic sediments in response to orogenic processes: from Gondwana break-up to Palaeozoic amalgamation of Central Europe. *Abstract for the Symposium 'GV & DGG, 2001 Margins Meeting' October 2–6, 2001, Kiel, Germany,* 1 p.

GEUKENS, F. 1999. Notes accompagnant une révision de la carte structurale du Massif de Stavelot. *Aardkundige Mededelingen*, **9**, 183–190.

GIBBONS, W. 1990. Transcurrent ductile shear zones and the dispersal of the Avalon Superterrane. *In:* D'LEMOS, R. S., STRACHAN, R. A. & TOPLEY, C. G. (eds) *The Cadomian Orogeny.* Geological Society, London, Special Publications, **51**, 407–423.

GIESE, U., & KÖPPEN, S. 2001. Detrital record of Early Palaeozoic clastic sediments at the southern border of the Baltic Shield – Provenance signals for a Caledonian geodynamic evolution. *Neues Jahrbuch für Geologie und Paläontologie, Abhandlungen*, **222**, 215–251.

GIESE, U., KATZUNG, G. & WALTER, R. 1994. Detrital composition of Ordovician sandstones from the Rügen boreholes: implications for the evolution of the Tornquist Ocean. *Geologische Rundschau*, **93**, 293–308.

GIESE, U., DALLMEYER, R. D., KRAMM, U. & TSCHERNOSTER, R. 1997*a*. Provenance analysis across a Caledonian suture, NE Germany. *Terra Nova*, **9**, 595.

GIESE, U., DALLMEYER, R. D., KRAMM, U. & WEBER, J. 1997*b*. The Caledonian deformation of the Brabant Massif and the Early Palaeozoic in north-east Germany: compared. *Geological Magazine*, **134**, 637–652.

GODEFROID, J. 1995. *Dayia shirleyi* ALVAREZ & RACHEBOEUF, 1986. Un brachiopode silurien dans les 'Schistes de Mondrepuis' à Muno (sud de la Belgique). *Bulletin de l'Institut Royal des Sciences Naturelles de Belgique, Sciences de la Terre,* **65**, 269–272.

GODEFROID, J. & CRAVATTE, T. 1999. Les brachiopodes et la limite Silurien/Dévonien à Muno (sud de la Belgique). *Bulletin de l'Institut Royal des Sciences Naturelles de Belgique, Sciences de la Terre,* **69**, 5–26.

GODEFROID, J., BLIECK, A., BULTYNCK, P., DEJONGHE, L., GERRIENNE, P., HANCE, L., MEILLIEZ, F., STAINIER, P. & STEEMANS, P. 1994. Les formations du Dévonien Inférieur du Massif de la Vesdre, de la fenêtre de Theux et du Synclinorium de Dinant (Belgique, France). *Mémoires pour servir à l'Explication des Cartes Géologiques et Minières de la Belgique*, **38**, 1–144.

GRADSTEIN, F. & OGG, J. 1996. A Phanerozoic time scale. *Episodes*, **19**, annexe.

GRAULICH, J. M. 1961. Le sondage de Wépion. *Mémoire Explicatif des Cartes Géologiques et Minières de la Belgique*, **2**,1–102.

GULLENTOPS, F. 1997. Uplift of the Hautes Fagnes. *Aardkundige Mededelingen*, **8**, 95–98.

HANCE, L., GHYSEL, P., LALOUX, M., DEJONGHE, L. & MANSY, J-L. 1999. Influence of a heterogeneous lithostructural layering on orogenic deformation of the Variscan Front Zone (eastern Belgium). *In:* SINTUBIN, M., VANDYCKE, S. & CAMELBEECK, T. (eds) Palaeozoic to Recent tectonics in the NW European Variscan Front Zone. *Tectonophysics,* **309**, 161–177.

HENNEBERT, M. 1993. Rôle possible des structures profondes du Massif Cambro-Silurien du Brabant dans l'évolution des bassins sédimentaires post-Calédoniens (Belgique et Nord de la France). *Annales de la Société Géologique de Belgique,* **116**, 147–162.

HERBOSCH, A., VANGUESTAINE, M., DEGARDIN, J. M., DEJONGHE, L., FAGEL, N. & SERVAIS, T. 1991. Etude lithostratigraphique, biostratigraphique et sédimentologique du sondage de Lessines (bord méridional du Massif du Brabant, Belqique). *Annales de la Société Géologique de Belgique*, **114**, 195–212.

HERBOSCH, A., VERNIERS, J., DEBACKER, T., BILLIAERT, B., DE SCHEPPER, S. & BELMANS, M. 2001. The Lower Palaeozoic stratigraphy and sedimentology of the Brabant Massif in the Dyle and Orneau valleys and of the Condroz Inlier at Fosses: an excursion guidebook. *In: Pre-symposium excursion of the International Symposium on 'Early Palaeozoic Palaeogeographies and Biogeographies of Western Europe and North Africa'.* Université des Sciences et Techniques de Lille (September 22–26, 2001), 1–77.

HOWELLS, M. F., REEDMAN, A. J. & CAMPBELL, S. D. G. 1990. *Ordovician Marginal Basin volcanism in Snowdonia (N.W. Wales).* Book Series, British Geological Survey.

HUGHES, R. A., COOPER, A. H. & STONE, P. 1993. Structural evolution of the Skiddaw Group (English Lake District) on the northern margin of eastern Avalonia. *Geological Magazine*, **130**, 621–629.

HUGHES, R. A., EVANS, J. A., NOBLE, S. R. & RUNDLE, C. C. 1996. U-Pb chronology of the Ennerdale and Eskdale intrusions supports sub-volcanic relationships with the Borrowdale Volcanic Group (Ordovician, English Lake District). *Journal of the Geological Society, London*, **153**, 33–38.

HUGON, H. 1983. Structures et déformation du Massif de Rocroi (Ardennes). *Bulletin de la Société Géologique et Minéralogique de Bretagne*, **15**, 109–143.

HUGON, H. & LE CORRE, C. 1979. Mise en évidence d'une déformation hercynienne en régime cisaillant progressif dans le Massif de Rocroi (Ardenne). *Comptes Rendus de l'Académie des Sciences, Paris*, **289**, 615–618.

JAEGER, H. 1967. Ordoviz auf Rügen. Datierung und Vergleich mit anderen Gebieten. *Berichte der deutschen Geschellschaft geologischer Wissenschaften A,* **12,** 165–176.

JOHANSSON, A., LINDH, A. & MÖLLER, C. 1991. Late Sveconorwegian (Grenvillian) high pressure granulite facies metamorphism in southwest Sweden. *Journal of Metamorphic Geology*, **9**, 283–292.

JOHNSON, T. E. 1991. Nomenclature and geometric classification of cleavage-transected folds. *Journal of Structural Geology*, **13**, 261–274.

KATZUNG, G. 2001. The Caledonides at the southern margin of the East European Craton. . *Neues Jahrbuch für Geologie und Paläontologie, Abhandlungen*, **222**, 3–53.

KATZUNG, G., GIESE, U., WALTER, R. & VON WINTERFELD, C. 1993. The Rügen Caledonides, northeast Germany. *Geological Magazine*, **130**, 725–730.

KING, L. M. 1994. Subsidence analysis of Eastern Avalonian sequences: implications for Iapetus closure. *Journal of the Geological Society, London*, **151**, 647–657.

KIRBY, G. A., BAILY, H., CHADWICK, R. A., EVANS, D. J., HOLLIDAY, D. W., HOLLOWAY, S., HULBERT, A. G., PHARAOH, T. C., SMITH, N. J., AITKENHEAD, N. & BIRCH, B. 2000. The structure and evolution of the Craven Basin and adjacent areas. Keyworth, Nottingham, British Geological Survey, 1–114.

KLEIN, C. 1977. L'ardennotype. *Comptes Rendus de l'Académie des Sciences,* Série D, **284**,1263–1266.

KLEIN, C. 1980. L'intérêt tectonogénétique de la discordance post-calédonienne en Ardenne. Les notions d'hérédité mécanique et d'induction tectonique. *Bulletin de la Société belge de Géologie,* **89**, 1–54.

KOCH, L. 1999. Die Familie Cyclopygidae (Trilobita) im Ordovizium des Ebbe- und Remscheider Sattels (Rheinisches Schiefergebirge, Deutschland). *Neues Jahrbuch für Geologie und Paläontologie, Abhandlungen*, **213**, 375–431.

KOKELAAR, B. P. 1986. Petrology and geochemistry of the Rhobell Volcanic Complex: amphibole dominated fractionation at an early Ordovician arc volcano in North Wales. *Journal of Petrology*, **27**, 887–914.

KOKELAAR, B. P., HOWELLS, M. F., BEVINS, R. E., ROACH, R. A. & DUNKEY, P. N. 1984. The Ordovician marginal basin of Wales. *In:* KOKELAAR, B. P. & HOWELLS, M. F. (eds) *Marginal basin geology*. Geological Society, London, Special Publications, **16**, 245–269.

LACQUEMENT, F. 2001. *L'Ardenne Varisque. Déformation progressive d'un prisme sédimentaire préstructuré, de l'affleurement au modèle de chaîne.* Société Géologique du Nord Publications, **29**.

LAMENS, J. 1985. Transition from turbidite to shallow-water sedimentation in the lower Salmian (Tremadocian-lower Ordovician) of the Stavelot Massif, Belgium. *Sedimentary Geology*, **44**, 121–142.

LAMENS, J. 1986. Depositional history of Salmian (Lower Ordovician sediments in Belgium). *Aardkundige Mededelingen*, **3**,125–138.

LAMENS, J. 1988. Sedimentology of the Cambro-Ordovician of the Caledonian Stavelot Massif. Excursion A-5. *In:* HERBOSCH, A. (ed.) *IAS, 9th European Regional Meeting, Excursion Guidebook Leuven-Belgium*. Belgian Geological Survey, 67–76.

LAMENS, J. & GEUKENS, F. 1985. Sedimentary structures in the lower Salmian of the Stavelot Massif (Belgium) as indicators of turbidite sedimentation. *Bulletin de la Société belge de Géologie*, **941**, 29–40.

LAMENS, J., GEUKENS, F. & VIAENE, W. 1986. Geological setting and genesis of coticules (spessartine metapelites) in the Lower Ordovician of the Stavelot Massif (Belgium). *Journal of the Geological Society, London*, **143**, 253–258.

LARSEN, O. 1971. K/Ar determinations from the Precambrian of Denmark. *Danmarks Geologiske Undersøgelse,* Series II, **97**, 1–37.

LE BAS, M. J. 1972. Caledonian igneous rocks beneath central and eastern England. *Proceedings of the Yorkshire Geological Society*, **39**, 71–86.

LEE, M. K., PHARAOH, T. C., WILLIAMSON, J. P., GREEN, C. A. & DE VOS, W. 1993. Evidence on the deep structure of the Anglo-Brabant Massif from gravity and magnetic data. *Geological Magazine,* **130**, 575–582.

LE GALL, B. 1992. The deep structure of the Ardennes Variscan thrust belt from structural and ECORS seismic data. *Journal of Structural Geology,* **14**, 531–546.

LEGRAND, R. 1967. Ronquières. Documents géologiques. *Mémoires pour servir à l'Explication des Cartes Géologiques et Minières de la Belgique,* Mémoire, **6**, 1–60.

LEGRAND, R. 1968. Le Massif du Brabant. *Mémoires pour servir à l'Explication des Cartes Géologiques et Minières de la Belgique,* Mémoire, **9**, 1–148.

LESPÉRANCE, P. J. & SHEEHAN, P. M. 1988. Trilobites et Brachiopodes ashgilliens (Ordovicien supérieur) de l''Assise' de Fosse, Bande de Sambre-Meuse (Belgique). *Bulletin de l'Institut Royal des Sciences Naturelles de Belgique, Sciences de la Terre,* **57**(1987), 91–123.

LINNEMANN, U., GEHMLICH, M., TICHOMIROWA, M., BUSCHMANN, B. & BOMBACH, K. 1998. Tectonostratigraphic Events of the Peri-Gondwanan Basement of the Saxo-Thuringian Composite Terrane (Central European Variscides). *Schriften des Staatlichen Museums für Mineralogie und Geologie zu Dresden,* **9**, 159–161.

LOUWYE, S., VAN GROOTEL, G. & VERNIERS, J. 1992. The stratigraphy of the type locality of the ?late Wenlock/ early Ludlow Mont Godart Formation and the early Ludlow Ronquières Formation, Brabant Massif, Belgium. *Annales de la Société Géologique de Belgique*, **115**, 307–331.

LYNAS, B. D. T. 1988. Evidence for dextral oblique-slip faulting in the Shelve Ordovician inlier, Welsh Borderland: implications for the south British Caledonides. *Geological Journal*, **23**, 39–57.

MAES, G., RICKARDS, R. B., ROMBOUTS, L. & VANDEVELDE, N. 1979. Silurian formations between Neuville-sous-Huy and Ombret: their correlation, age and structure. *Annales de la Société Géologique de Belgique*, **101**(1978), 31–36.

MALETZ, J. & SERVAIS, T. 1993. Acritarchs and graptolites from the Early Llanvirn of the Herscheider Schichten (Rheinisches Schiefergebirge, Germany). *Neues Jahrbuch für Paläontologie und Geologie, Abhandlungen*, **190**, 125–142.

MALETZ, J. & SERVAIS, T. 1998. Upper Ordovician graptolites from the Brabant Massif, Belgium. *Geobios*, **31**, 21–37.

MANSY, J-L., EVERAERTS, M. & DE VOS, W. 1999. Structural analysis of the adjacent Acadian and Variscan fold belts in Belgium and northern France from geophysical and geological evidence. *Tectonophysics,* **309**, 99–116.

MARTIN, F. 1969. Ordovicien et Silurien belges; données nouvelles apportées par l'étude des Acritarches. *Bulletin de la Société belge de Géologie*, **77**(1968), 175–181.

MARTIN, F., MICHOT, P. & VANGUESTAINE, M. 1970. Le Flysch Caradocien d'Ombret. *Annales de la Société Géologique de Belgique*, **93**, 337–362.

MCCANN, T. 1998. Lower Palaeozoic evolution of the northeast German Basin/Baltica borderland. *Geological Magazine*, **135**, 129–142.

MCKERROW, W. S., SCOTESE, C. R. & BRASIER, M. D. 1992. Early Cambrian reconstructions. *Journal of the Geological Society, London,* **149**, 599–606.

MCKERROW, W. S., MACNIOCAILL, C. & DEWEY, J. F. 2000. The Caledonian Orogeny redefined. *Journal of the Geological Society, London,* **157**, 1149–1154.

MEILLIEZ, F. 1989. Tectonique distensive et sédimentation à la base du Devonien, en bordure NE du Massif de Rocroi (Ardenne). *Annales de la Société Géologique du Nord*, **107**, 281–295.

MEILLIEZ, F. & MANSY, J-L. 1990. Déformation pelliculaire différenciée dans une série lithologique hétérogène: le Dévono-Carbonifère de l'Ardenne. *Bulletin de la Société géologique de France,* **8**, 177–188.

MEILLIEZ, F., ANDRÉ, L., BLIECK, A., FIELITZ, W., GOFFETTE, O., HANCE, L., KHATIR, A., MANSY, J-L., OVERLAU, P. & VERNIERS, J. 1991. Ardenne – Brabant. *Sciences Géologiques, Bulletin*, **441**, 3–29.

MEISSNER, R., & BORTFELD, R. K. (eds) 1990. *DEKORP-Atlas. Results of Deutsches Kontinentales Reflexionsseismisches Programm*. Springer, Berlin, 1–19.

MEISSNER, R., SADOWIAK, P. & THOMAS, S. A. 1994. East Avalonia, the third partner in the Caledonian collisions: evidence from deep seismic reflection data. *Geologische Rundschau*, **83**, 186–196.

MERRIMAN, R. J., PHARAOH, T. C., WOODCOCK, N. H. & DALY, P. 1993. The metamorphic history of the concealed Caledonides of eastern England and their foreland. *In:* PHARAOH, T. C., MOLYNEUX, S. G., MERRIMAN, R. J., LEE, M. K. & VERNIERS, J. (eds) *Proceedings of the International Conference on the Caledonides of the Anglo-Brabant Massif (Nottingham, September 1992)*. Special Issue on the Caledonides of the Anglo-Brabant Massif. Geological Magazine, **130**, 613–620.

MERRIMAN, R. J., REX, D. C., SOPER, N. J. & PEACOR, D. R. 1995. The age of the Acadian cleavage in northern England, UK: K-Ar and TEM analysis of a Silurian metabentonite. *Proceedings of the Yorkshire Geological Society*, **50**, 255–265.

MICHELSEN, O. & NIELSEN, L. H. 1991. Well records on the Phanerozoic stratigraphy in the Fennoscandian Border Zone, Denmark. *Danmarks Geologiske Undersøgelse,* Series A, **29**, 1–37.

MICHOT, P. 1934. La stratigraphie du Silurien de la bande Sambre-et-Meuse. *Mémoires de l'Académie Royale de Belgique, Classe des Sciences,* 2ème série, **13**, 1–108.

MICHOT, P. 1954. Le Silurien. *In:* FOURMARIER, P. (ed.) *In: Prodrôme d'une description géologique de la Belgique.* Imprimerie Vaillant Carmanne, 39–82.

MICHOT, P. 1979. La faille mosane et la phase hyporogénique bollandienne, d'âge emsien dans le rameau calédonien condruso-brabançon. *Annales de la Société Géologique de Belgique*, **101**, 321–335.

MICHOT, P. 1980. Le segment tectogénique calédonien belge. *Mémoires de l'Académie Royale de Belgique, Classe des Sciences,* Collection in –8°, 2ème série, **43**, 1–61.

MOLYNEUX, S. G. 1991. The contribution of palaeontological data to an understanding of the Early Palaeozoic framework of eastern England. *Annales de la Société Géologique de Belgique*, **114,** 93–105.

MOLZAHN, M., ANTHES, G. & REISCHMANN, T. 1998. Single zircon Pb/Pb age geochronology and isotope systematics of the Rhenohercynian basement. *Terra Nostra* **98/1**, 67–68.

MORTELMANS, G. 1952. Observations nouvelles sur les 'porphyroïdes' caradociens de la gare d'Hennuyères. *Bulletin de la Société belge de Géologie, de Paléontologie et d'Hydrologie,* **61**, 176–197.

MORTELMANS, G. 1955. Considérations sur la structure tectonique et la stratigraphie du Massif du Brabant. *Bulletin de la Société belge de Géologie, de Paléontologie et d'Hydrologie*, **64**, 179–218.

MURPHY, J. B., KEPPIE, J. D., DOSTAL, J. & NANCE, R. D. 1999. Neoproterozoic-early Paleozoic evolution of Avalonia. *In:* RAMOS, V. A. & KEPPIE, J. D. (eds) *Laurentia-Gondwana Connections before Pangea.* Geological Society of America Special Papers, **336**, 253–266.

MURPHY, J. B., STRACHAN, R. A., NANCE, R. D. PARKER, K. D., FOWLER, M. B. 2000. Prota-Avalonia: A 1.2–1.0 Ga tectonothermal event and constraints for the evolution of Rodinia. *Geology,* **28**, 1071–1074.

NANCE, R. D. & MURPHY, J. B. 1994. Contrasting basement isotopic signatures and the palinspastic restoration of peripheral orogens: Example from the Neoproterozoic Avalonian-Cadomian belt. *Geology*, **22**, 617–620.

NEUMANN-MAHLKAU, P. N. 1982. Die Gerölle in der Schwarzbachtall-Konglomeraten bei Ratingen und ihre paläogeographische Aussage. *Senckenbergiana lethaea*, **63**, 79–95.

NEUMANN-MAHLKAU, P. N. & RIBBERT, K.-H. 1998. Die Konglomerate der Givet-Stufe östlich des Brabanter Massivs. *Fortschritte in der Geologie von Rheinland und Westfalen*, **37**, 393–421.

NOBLE, S. R., TUCKER, R. D. & PHARAOH, T. C. 1993. Lower Palaeozoic and Precambrian igneous rocks from eastern England, and their bearing on late Ordovician closure of the Tornquist Sea: constraints from U-Pb and Nd isotopes. *Geological Magazine,* **130**, 738–747.

ONCKEN, O., FRANZKE, H. J., DITMAR, U. & KLÜGEL, T. 1995. Rhenohercynian foldbelt: metamorphic units-structure. *In:* DALLMEYER, D., FRANKE, W. & WEBER, K. (eds) *The Pre-Permian geology of Central and Eastern Europe.* Springer, Berlin, 108–117.

ONCKEN, O., VON WINTERFELD, C. & DITTMAR, U. 1999. Accretion of a rifted passive margin: The Late Paleozoic Rhenohercynian fold and thrust belt (Middle European Variscides). *Tectonics,* **18**, 75–91.

ONCKEN, O., PLESCH, A., WEBER, J., RICKEN, W. & SCHRADER, S. 2000. Passive margin detachment during arc-continent collision (Central European Variscides). *In:* FRANKE, W., HAAK, V., ONCKEN, O. & TANNER, D. (eds) *Orogenic processes: Quantification and modelling in the Variscan Belt.* Geological Society, London, Special Publications, **179**, 199–216.

PAPROTH, E., DREESEN, R. & THOREZ, J. 1986. Famennian paleogeography and event stratigraphy of northwestern Europe. *In:* BLESS, M. J. M. & STREEL, M. (eds) *Late Devonian events around the Old Red Continent.* Annales de la Société géologique de Belgique, **109**, 175–186.

PARIS, F. & LE HÉRISSÉ, A. 1992. Palaeozoic in Western Brittany (Outline of the Armorican geological history and geological itinerary in the Crozon Peninsula). *Cahiers de Micropaléontologie*, **7**, 5–28.

PHARAOH, T. C. 1999. Palaeozoic terranes and their lithospheric boundaries within the Trans-European Suture Zone (TESZ): a review. *Tectonophysics,* **314**, 17–41.

PHARAOH, T. C. & CARNEY, J. 2000. *Introduction* Chapter 1. *In:* CARNEY, J., HORAK, J. M., PHARAOH, T. C., GIBBONS, W., WILSON, D., BARCLAY, W. J., BEVINS, R. E., COPE, J. C. W. & FORD, T. D. *Precambrian rocks of England and Wales*, p. 1–18. Geological Conservation Review, Joint Nature Conservation Committee – Chapman & Hall, 1–239.

PHARAOH, T. C. & GIBBONS, W. 1994. Precambrian rocks in England and Wales south of the Menai Strait Fault System. *In:* GIBBONS, W. & HARRIS, A. L. (eds) *A revised correlation of Precambrian rocks in the Britisch Isles.* Geological Society, London, 85–97.

PHARAOH, T. C., GIBBONS, W., WEBB, P. C. & BECKINSALE, R. D. 1987. The concealed Caledonides of eastern England: preliminary results of a multidisciplinary study. *Proceedings of the Yorkshire Geological Society,* **46**, 355–364.

PHARAOH, T. C., MERRIMAN, R. J., EVANS, J. A., BREWER, T. S., WEBB, P. C. & SMITH, N. J. P. 1991. Early Palaeozoic arc-related volcanism in the concealed Caledonides of southern Britain. *Annales de la Société Géologique de Belgique,* **114**, 63–92.

PHARAOH, T. C., BREWER, T. S. & WEBB, P. C. 1993*a*. Subduction-related magmatism of late Ordovician age in eastern England. *Geological Magazine,* **130**, 647–656.

PHARAOH, T. C., MOLYNEUX, S. G., MERRIMAN, R. J., LEE, M. K. & VERNIERS, J. 1993*b*. The Caledonides of the Anglo-Brabant Massif reviewed. *In:* PHARAOH, T. C., MOLYNEUX, S. G., MERRIMAN, R. J., LEE, M. K. & VERNIERS, J. (eds) *Special Issue on the Caledonides of the Anglo-Brabant Massif.* Geological Magazine, **130**, 561–562.

PHARAOH, T. C., ENGLAND, R. & LEE, M. K. 1995. The concealed Caledonide basement of eastern England and the southern North Sea – a review. *In:* GEE, D. G. & BECKHOLMEN, M. (eds) The Trans-European Suture Zone. Europrobe in Liblice 1993. *Studia Geophysica et Geodaetica*, **39**, 330–346.

PHARAOH, T. C., ALLSOP, J. M., HOLLIDAY, D. W., MERRIMAN, R. J., KIMBELL, G. S., RUNDLE, C. C., BREWER, T. S., NOBLE, S. R. & EVANS, C. J. 1997. The Moorby Microgranite: a deformed high level intrusion of Ordovician age in the concealed Caledonian basement of Lincolnshire. *Proceedings of the Yorkshire Geological Society,* **51**, 329–342.

PIDGEON, R. T. & AFTALION, M. 1978. Cogenetic and inherited zircon U-Pb systems in Palaeozoic granites from Scotland and England. *In:* BOWES, D. R. & LEAKE, B. E. (eds) *Crustal evolution in northwest Britain and adjacent regions.* Geological Journal Special Issue, **10**, 183–220.

PIESSENS, K. 2001. *Metallogenese van de gemineraliseerde schuifzone te Sint-Pieters-Kapelle (Brabantmassief, België).* Unpublished PhD thesis, Katholieke Universiteit Leuven.

PIESSENS, K. & SINTUBIN, M. 1997. Partitioning of Variscan strain in the southern part of the Caledonian Stavelot-Venn Inlier in the Ardenne Allochthon (Belgium). *Aardkundige Mededelingen*, **8**, 135–138.

PIESSENS, K., MUCHEZ, P., VIAENE, W., BOYCE, A., DE VOS, W., SINTUBIN, M. & DEBACKER, T. 2000. Alteration and fluid characteristics of a mineralised shear zone in the Lower Palaeozoic of the Anglo-Brabant belt, Belgium. *Journal of Geochemical Exploration*, **69–70**, 317–321.

PIESSENS, K., MUCHEZ, P., DEWAELE, S., BOYCE, A., DE VOS, W., SINTUBIN, M., DEBACKER, T., BURKE, E. A. J. & VIAENE, W. 2002. Fluid flow, alteration and polysuphide mineralisation associated with a low-angle reverse shear zone in the Lower Palaeozoic of the Anglo-Brabant fold belt, Belgium. *Tectonophysics*, **348**, 73–92.

PIPER, J. D. A. 1997. Tectonic rotation within the paratectonic British Caledonides and Early Palaeozoic location of the orogen. *Journal of the Geological Society, London*, **154**, 9–14.

POULSEN, C. 1969. The Lower Cambrian from Slagelse no. 1, Western Sealand. *Danmarks Geologiske Undersøgelse, Series II*, **93**.

POULSEN, C. 1974. Further contributions to the knowledge of the Palaeozoic of Slagelse no. 1, Western Sealand. *Danmarks Geologiske Undersøgelse, Series II,* **101.**

PRIGMORE, J. K., BUTLER, A. J. & WOODCOCK, N. H. 1997. Rifting during separation of Eastern Avalonia from Gondwana: Evidence from subsidence analysis. *Geology*, **25**, 203–207.

RABAE, A. M. & KEAREY, P. 1997. Granites of the western Anglo-Brabant Massif. *Annales de la Société Géologique de Belgique,* **120**, 217–226.

RABBEL, W., FÖRSTE, K., SCHULZE, A., BITTNER, R., RÖHL, J. & REICHERT, J. C. 1995. A high-velocity layer in the lower crust of the North German Basin. *Terra Nova*, **7**, 327–337.

REEDMAN, A. J., LEVERIDGE, B. E. & EVANS, R. B. 1984. The Arfon Group ('Arvonian') of North Wales. *Proceedings of the Geologists' Association*, **95**, 313–321.

REGNELL, G. 1951. Revision of the Caradocian-Ashgillian cystoid fauna of Belgium with notes on isolated pelmatozoan stem fragments. *Mémoires de l'Institut Royal des Sciences Naturelles de Belgique*, **120**, 1–47.

RENIER, A. 1902. Le Poudingue de Malmédy. Essai géologique. *Annales de la Société Géologique de Belgique,* **29**, M145–223.

SAMUELSSON, J. & VERNIERS, J. 2000. Ordovician Chitinozoa Biozonation of the Brabant Massif, Belgium. *Review of Palaeobotany and Palynology,* **113**, 105–129.

SAMUELSSON, J., VERNIERS, J. & VECOLI, M. 2000. Chitinozoan faunas from the Rügen Ordovician (Rügen 5/66 and Binz 1/73 wells), NE Germany. *Review of Palaeobotany and Palynology,* **113,** 131–143.

SAMUELSSON, J., VECOLI, M. & BEIER, H. 2001. Ordovician – Silurian palynostratigraphy (Chitinozoa and acritarchs) of the G 14–1/86 borehole, southern Baltic Sea. *Neues Jahrbuch für Geologie und Paläontologie, Abhandlungen*, **222**, 259–290.

SAMUELSSON, J., VECOLI, M., BEDNARCZYK, W. S. & VERNIERS, J. 2002*a*. Timing of the Avalonia-Baltica plate convergence as inferred from palaeogeographic and stratigraphic data of chitinozoan assemblages in West Pomerania, northern Poland. *In:* WINCHESTER, J. A., PHARAOH, T. C. & VERNIERS, J. (eds) *Palaeozoic Amalgamation of Central Europe.* Geological Society, London, Special Publications, **201**, 95–114.

SAMUELSSON, J., GERDES, A., KOCH, L., SERVAIS, T. & VERNIERS, J. 2002*b*. Chitinozoa and Nd isotope stratigraphy of the Ordovician rocks in the Ebbe Anticline, NW Germany. *In:* WINCHESTER, J. A., PHARAOH, T. C. & VERNIERS, J. (eds) *Palaeozoic Amalgamation of Central Europe.* Geological Society, London, Special Publications, **201**, 115–134.

SANDEMAN, H. A. I., CLARK, A. H., STYLES, M. T., SCOTT, D. J., MALPAS, J. G. & FARRAR, E. 1997. Geochemistry and U-Pb and ^{40}Ar-^{39}Ar geochronology of the Man of War Gneiss, Lizard Complex, SW England: pre-Hercynian arc-type crust with a Sudeten-Iberian connection. *Journal of the Geological Society, London*, **154**, 403–447.

SANDEMAN, H. A. I., CLARK, A. H., SCOTT, D. J. & MALPAS, J. G. 2000. The Kennack Gneiss of the Lizard Peninsula, Cornwall, SW England:

commingling and mixing of mafic and felsic magmas accompanying Givetian continental incorporation of the Lizard ophiolite. *Journal of the Geological Society, London*, **157**, 1127–1242.

SCHALLREUTER, R., VERNIERS, J. & DE GEEST, P. 2000. An Ordovician ostracode from Belgium. *Neues Jahrbuch für Geologie und Paläontologie, Monatshefte*, **2000**, 570–576.

SCHLÜTER, H. U., BEST, G., JÜRGENS, U. & BINOT, F. 1997. Interpretation reflexionsseismischer Profile zwischen baltischer Kontinentalplatte und kaledonischem Becken in der südlichen Ostsee-erste Ergebnisse. *Zeitschrift der deutschen geologischen Gesellschaft*, **148**, 1–32.

SCHROYEN, K. 2000. *Caledonische en Varistische Fluïda en de metamorfe evolutie van het Stavelot-Venn-massief.* Unpublished PhD thesis, Katholieke Universiteit Leuven.

SERVAIS, T. 1991. Discovery of turbiditical levels in the late Ordovician of the Sennette valley (Brabant Massif, Belgium). *Annales de la Société Géologique de la Belgique*, **114**, 247–251.

SERVAIS, T. 1994. The Ordovician acritarchs from Rügen, (NE-Germany) palaeobiogeographical evidence for the attribution to Eastern Avalonia. *Neues Jahrbuch für Geologie und Paläontologie, Abhandlungen*, **9**, 566–580.

SERVAIS, T. & FATKA, O. 1997. The recognition of the Trans-European-Suture Zone by the palaeogeographic distribution pattern of Early to Middle Ordovician acritarchs. *Geological Magazine*, **134**, 617–625.

SERVAIS, T. & KATZUNG, G. 1993. Acritarch dating of Ordovician sediments of the island of Rügen (NE Germany. *Neues Jahrbuch für Geologie und Paläontologie, Monatshefte*, **1993**, 713–723.

SERVAIS, T. & MALETZ, J. 1992. Lower Llanvirn (Ordovician) graptolites and acritarchs from the 'Assise de Huy', Bande de Sambre-et-Meuse, Belgium. *Annales de la Société Géologique de Belgique*, **115**, 265–285.

SERVAIS, T., HERBOSCH, A. & VANGUESTAINE, M. 1993. Review of the stratigraphy of the Ordovician in the Brabant Massif, Belgium. *Geological Magazine*, **130**, 699–710.

SERVAIS, T., SAMUELSSON, J., SEHNERT, M., VECOLI, M. & GIESE, U. 2001. Ordovician palynomorphs from the subsurface of Rügen (NE-Germany): review and perspectives. *Neues Jahrbuch für Geologie und Paläontologie, Abhandlungen*, **222**, 291–307.

SHEEHAN, P. M. 1987. Late Ordovician (Ashgillian) Brachiopods from the region of the Sambre and Meuse rivers, Belgium. *Bulletin de l'Institut Royal des Sciences Naturelles de Belgique, Sciences de la Terre*, **57**, 5–81.

SINTUBIN, M. 1994. Phyllosilicate preferred orientation in relation to strain path determination in the Lower Paleozoic Stavelot-Venn Massif (Ardennes, Belgium). *Tectonophysics*, **237**, 215–231.

SINTUBIN, M. 1997*a*. Cleavage-fold relationships in the Lower Paleozoic Brabant Massif (Belgium). *Aardkundige Mededelingen*, **8**, 161–164.

SINTUBIN, M. 1997*b*. Structural implications of the aeromagnetic lineament geometry in the Lower Paleozoic Brabant Massif (Belgium). *Aardkundige Mededelingen*, **8**, 165–168.

SINTUBIN, M. 1999. Arcuate fold and cleavage patterns in the southeastern part of the Anglo-Brabant Fold Belt (Belgium): tectonic implications. *In:* SINTUBIN, M., VANDYCKE, S. & CAMELBEECK, T. (eds) Palaeozoic to Recent tectonics in the NW European Variscan Front Zone. *Tectonophysics*, **309**, 81–97.

SINTUBIN, M. & EVERAERTS, M. A. 2002. A compressional wedge model for the Lower Palaeozoic Anglo-Brabant Belt (Belgium), based on the potential field data. *In:* WINCHESTER, J. A., PHARAOH, T. C. & VERNIERS, J. (eds) *Palaeozoic Amalgamation of Central Europe.* Geological Society, London, Special Publications, **201**, 327–344.

SINTUBIN, M. & MATTHIJS, J. 1998. Structural implications of the geometry of the western margin of the Lower Paleozoic Stavelot Massif in the Ninglinspo area (Nonceveux, Belgium). *Aardkundige Mededelingen*, **9**, 97–110.

SINTUBIN, M., BRODKOM, F. & LADURON, D. 1998. Cleavage-fold relationships in the Lower Cambrian Tubize Group, southeast Anglo-Brabant Fold Belt (Lembeek, Belgium). *Geological Magazine*, **135**, 217–226.

SINTUBIN, M., KENIS, I., SCHROYEN, K., MUCHEZ, PH. & BURKE, E.A.J. 2000. 'Boudinage' in the High-Ardenne Slate Belt (Belgium), reconsidered from the perspective of the 'interboudin' veins. *Journal of Geochemical Exploration*, **69–70**, 511–516.

SMITH, N. J. P. & RUSHTON, A. W. A. 1993. Cambrian and Ordovician stratigraphy related to structure and seismic profiles in the western part of the English Midlands. *Geological Magazine*, **130**, 665–671.

SMOLDEREN, A. 1987. Age range and origin of the limestone clasts of the Malmédy conglomerate (Permian, Ardennes, Belgium). *Bulletin de la Société belge de Géologie*, **961**, 55–80.

SOPER, N. J. 1986. The Newer Granite problem: a geotectonic view. *Geological Magazine*, **123**, 227–236.

SOPER, N. J. & HUTTON, D. H. W. 1984. Late Caledonian sinistral displacements in Britain: Implications for a three-plate collision model. *Tectonics*, **3**, 781–794.

SOPER, N. J. & KNELLER, B. C. 1990. Cleaved microgranite dykes of the Shap swarm in the Silurian of NW England. *Geological Journal*, **5**, 161–170.

SOPER, N. J. & WOODCOCK, N. H. 1990. Silurian collision and sediment dispersal patterns in southern Britain. *Geological Magazine*, **127**, 527–542.

SOPER, N. J., WEBB, B. C. & WOODCOCK, N. H. 1987. Late Caledonian (Acadian) transpression in north-west England: timing, geometry and geotectonic significance. *Proceedings of the Yorkshire Geological Society*, **46**, 175–192.

SOPER, N. J., STRACHAN, R. A., HOLDSWORTH, R. E., GAYER, R. A. & GREILING, R. O. 1992. Sinistral transpression and the Silurian closure of Iapetus. *Journal of the Geological Society, London*, **149**, 871–880.

SORGENFREI, T. & BUCH, A. 1964. Deep tests in

Denmark 1935–1959. *Danmarks Geologiske Undersøgelse*, Series III, **36**.

STEEMANS, PH. 1989. Paléogéographie de l'Eodévonien Ardennais et des régions limitrophes. *Annales de la Société Géologique de Belgique*, **112**, 103–119.

TAIT, J. A., BACHTADSE, V., FRANKE, W. & SOFFELL, H. C. 1997. Geodynamic evolution of the European Variscan Foldbelt: Palaeomagnetic and geological constraints. *Geologische Rundschau*, **86**, 585–598.

TAIT, J. A., SCHÄTZ, M., BACHTADSE, V. & SOFFEL, H. 2000. Palaeomagnetism and Palaeozoic palaeogeography of Gondwana and European terranes. *In:* FRANKE, W., HAAK, V., ONCKEN, O. & TANNER, D. (eds) *Orogenic processes: Quantification and modelling in the Variscan Belt.* Geological Society, London, Special Publications, **179**, 21–34.

TANNER, B. & MEISSNER, R. 1996. Caledonian deformation upon southwest Baltica and its tectonic implications: alternatives and consequences. *Tectonics*, **15**, 803–812.

THOREZ, J., GOEMAERE, E. & DREESEN, R. 1988. Tide-and wave-influenced depositional environments in the Psammites du Condroz (Upper Famennian) in Belgium. *In:* DE BOER, P. L. *et al.* (eds) *Tide-influenced sedimentary environments and facies*. Reidel Publ. Co., 389–415.

THOROGOOD, E. J. 1990. Provenance of the Pre-Devonian sediments of England and Wales. *Journal of the Geological Society, London*, **147**, 591–94.

THORPE, R. S., GASKARTH, J. W. & HENNEY, P. J. 1993. Composite Ordovician lamprophyre (spessartite) intrusions around the Midlands Microcraton in central Britain. *Geological Magazine*, **130**, 657–663.

THYBO, H., KIORBOE, L. L., MOELLER, C., SCHONHARTING, G. & BERTHELSEN, A. 1990. Geophysical and tectonic modelling of EUGENO-S profiles. *In:* FREEMAN, R. & MUELLER, S. (eds) *Sixth EGT workshop: Data compilations and synoptic interpretation*. European Science Foundation, Strasbourg, 93–104.

TIMM, J. 1981. Die Faziesentwicklung der ältesten Schichten des Ebbe-Antiklinoriums. *Mitteilungen aus dem Geologisch-Paläontologischen Institut der Universität Hamburg*, **50**, 147–173.

TIMM, J., DEGENS, E. T. & WIESNER, M. G. 1981. Erläuterungen zur Geologischen Karte des zentralen Ebbe-Antiklinoriums 1: 25 000. *Mitteilungen aus dem Geologisch-Paläontologischen Institut der Universität Hamburg*, **50**, 59–75.

TOGHILL, P. 1992. The Shelveian event, a Late Ordovician tectonic episode in southern Britain (Eastern Avalonia). *Proceedings of the Geologists' Association, London*, **103**, 31–35.

TORSVIK, T. H. 1998. Palaeozoic palaeogeography: A North Atlantic viewpoint. *Geologiska Föreningens i Stockholm Förhandlingar*, **120**, 109–118.

TORSVIK, T. H., SMETHURST, M. A., VAN DER VOO, R., TRENCH, A., ABRAHAMSEN, N. & HALVORSEN, E. 1992. Baltica. A synopsis of Vendian-Permian paleomagnetic data and their palaeotectonic implications. *Earth Science Reviews*, **33**, 133–152.

TORSVIK, T. H., TRENCH, A., SVENSSON, I. & WALDERHAUG, H. 1993. Mid Silurian palaeomagnetic results from the East Mendips Inlier, Southern Britain: Palaeogeographic significance and major revision of the apparent polar wander path for Eastern Avalonia. *Geophysical Journal International*, **113**, 651–668.

TORSVIK, T. H., SMETHURST, M. A., MEERT J. G., VAN DER VOO, R., MCKERROW W. S., BRASIER, M. D., STURT, B. A. & WALDERHAUG, H. J. 1996. Continental break-up and collision in the Neoproterozoic and Palaeozoic – a tale of Baltica and Laurentia. *Earth Science Reviews*, **40**, 229–258.

TOURNEUR, F., VANGUESTAINE, M., BUTTLER, C., MAMET, B., MOURAVIEFF, N., POTY, E. & PRÉAT, A. 1993. A preliminary study of Ashgill carbonate beds from the lower part of the Fosses Formation (Condroz, Belgium). *Geological Magazine*, **130**, 673–679.

TREAGUS, J. E. & TREAGUS, S. H. 1981. Folds and the strain ellipsoid: a general model. *Journal of Structural Geology*, **3**, 1–17.

TRENCH, A. & TORSVIK, T. H. 1991. A revised Paleozoic apparent polar wander path for south Britain (eastern Avalonia). *Geophysical Journal International*, **104**, 227–233.

TRENCH, A. & TORSVIK, T. H. 1992. The closure of the Iapetus Ocean and Tornquist Sea: new palaeomagnetic constraints. *Journal of the Geological Society London*, **149**, 867–870.

TRENCH, A., TORSVIK, T. H. & MCKERROW, W. S. 1992. The palaeogeographic evolution of Southern Britain during early Palaeozoic times: a reconciliation of palaeomagnetic and biogeographic evidence. *In:* PERROUD, H. & BYLUND, G. (eds) Palaeomagnetism of old European rocks. *Tectonophysics,* **201**, 75–82.

TUCKER, R. D. & MCKERROW, W. S. 1995. Early Palaeozoic chronology: a review in light of new U-Pb zircon ages from Newfoundland and Britain. *Canadian Journal of Earth Sciences*, **32**, 368–379.

TUCKER, R. D. & PHARAOH, T. C. 1991. U-Pb zircon ages for Late Precambrian igneous rocks in southern Britain. *Journal of the Geological Society, London*, **148**, 435–443.

TURNER, J. S. 1949. The deeper structure of central and northern England. *Proceedings of the Yorkshire Geological Society*, **27**, 280–297.

VANBRABANT, Y. 2001. *Evolution géodynamique de la partie orientale de l'Allochtone de l'Ardennais: Observations Structurales et Modélisations Numériques*. Unpublished PhD thesis, Université de Liège.

VANDER AUWERA, J. & ANDRÉ, L. 1985. Sur le milieu de dépôt, l'origine des matériaux et le faciès métamorphique de l'Assise de Tubize (Massif du Brabant, Belgique). *Bulletin de la Société belge de Géologie,* **94**, 171–184.

VANGUESTAINE, M. 1989. Late Tremadocian acritarchs in the Chevron boreholes (Stavelot Massif, Belgium). *Abstract International Meeting on the Caledonides of the Midlands and the Brabant Massif*, **46**.

VANGUESTAINE, M. 1992. Biostratigraphie par

acritarches du Cambro-Ordovicien de Belgique et des regions limitrophes: synthèse et perspectives d'avenir. *Annales de la Société Géologique de Belgique*, **115**, 1–18.

VAN GROOTEL, G., VERNIERS, J., GEERKENS, B., LADURON, D., VERHAEREN, M., HERTOGEN, J. & DE VOS, W. 1997. Timing of magmatism, foreland basin development, metamorphism and inversion in the Anglo-Brabant fold belt. *Geological Magazine*, **134**, 607–616.

VECOLI, M. & SAMUELSSON, J. 2001*a*. Reworked acritarchs as provenance indicators in the Lower Palaeozoic of Denmark. *Comptes Rendus de l'Académie des Sciences*, Série IIa, **332**, 465–471.

VECOLI, M. & SAMUELSSON, J. 2001*b*. Quantitative evaluation of microplankton palaeobiogeography in the Ordovician-Early Silurian of the northern TESZ (Trans-European Suture Zone): implications for the timing of the Avalonia-Baltica collision. *Review of Palaeobotany and Palynology*, **115**, 43–68.

VECOLI, M., GIESE, U. & SAMUELSSON, J. 2000. Closure of the Tornquist Ocean and Avalonia-Baltica collision: chronostratigraphic constraints from provenance analysis and micropalaeontology. *In*: Variscan-Appalachian dynamics: the building of the Upper Paleozoic basement. *Basement Tectonic program and abstracts*, **15**, A Coruña, Spain, 173–175.

VEJBAEK, O. V., STOUGE, S. & POULSEN, D. K. 1994. Palaeozoic tectonic and sedimentary evolution and hydrocarbon prospectivity in the Bornholm area. *Danmarks Geologiske Undersøgelse,* Series A, **34**.

VERNIERS, J. 1983. The Silurian of the Mehaigne area (Brabant Massif, Belgium), lithostratigraphy and features of the sedimentary basin. *Professional Paper of the Geological Survey of Belgium*, **203**, 1–117.

VERNIERS, J. & VAN GROOTEL, G. 1991. Review of the Silurian in the Brabant Massif, Belgium. *In*: ANDRÉ, L., HERBOSCH, A., VANGUESTAINE, M. & VERNIERS, J. (eds) *Proceedings of the International Meeting on the Caledonides of the Midlands and the Brabant Massif, Brussels, 20–23 September 1989*. Annales de la Société géologique de Belgique, **114**, 163–193.

VERNIERS, J. & DE VOS, W. 1995. Recent research on the Brabant Massif. *Studia Geophysica et Geodaetica,* **39**, 347–353.

VERNIERS, J., ANDRÉ, L., HERBOSCH, A. & VANGUESTAINE, M. 1991. Introduction to the International Meeting on the Caledonides of the Midlands and the Brabant Massif. *In*: ANDRÉ, L., HERBOSCH, A., VANGUESTAINE, M. & VERNIERS, J. (eds) *Proceedings of the International Meeting on the Caledonides of the Midlands and the Brabant Massif, Brussels, 20–23 september 1989*. Annales de la Société géologique de Belgique, **114**, 1–3.

VERNIERS, J., HERBOSCH, A., VAN GUESTAINE, M., GEUKENS, F., DELCAMBRE, B., PINGOT, J. L., BELANGER, I., HENNEBERT, M., DEBACKER, T., SINTUBIN, M. & DE VOS, W. 2001. Cambrian-Ordovician-Silurian lithostratigraphical units (Belgium). *In:* BULTYNCK, P. & DEJONGHE, L. (eds) *Guide to a revised lithostratigraphic scale of Belgium. Geological Belgica*, **4**, 5–38.

VON HOEGEN, J., LEMME, B., ZIELINSKI, J. & WALTER, R. 1985. Cambrian and Lower Ordovician in the Stavelot-Venn Massif. A model for depositional history. *Neues Jahrbuch für Geologie und Paläontologie, Abhandlungen*, **171**, 217–235.

VON HOEGEN, J., KRAMM, U. & WALTER, R. 1990. The Brabant Massif as part of Armorica/Gondwana: U-Pb isotopic evidence from detrital zircons. *Tectonophysics*, **185**, 37–50.

WADGE, A. J., GALE, N. H., BECKINSALE, R. D. & RUNDLE, C. C. 1978. A Rb-Sr isochron age for the Shap granite. *Proceedings of the Yorkshire Geological Soc*iety, **42**, 297–305.

WINCHESTER, J. A. & THE PACE TMR NETWORK TEAM, including BELKA, Z., PATOCKA, F. & KACHLIK, V. in press. Palaeozoic Amalgamation of Central Europe: new results from recent geological and geophysical investigations. *Tectonophysics*.

WOODCOCK, N. H. 1990*a*. Transpressive Acadian deformation across the Central Wales Lineament. *Journal of Structural Geology*, **12**, 329–337.

WOODCOCK, N. H. 1990*b*. Sequence stratigraphy of the Palaeozoic Welsh Basin. *Journal of the Geological Society, London,* **147**, 537–547.

WOODCOCK, N. H. 1991. The Welsh, Anglian and Belgian Caledonides compared. *Annales de la Société Géologique de Belgique*, **114**, 5–17

WOODCOCK, N. H. & GIBBONS, W. 1988. Is the Welsh Borderland Fault System a terrane boundary? *Journal of the Geological Society, London*, **145**, 915–933.

WOODCOCK, N. H. & PHARAOH, T. C. 1993. Silurian facies beneath East Anglia. *In:* PHARAOH, T. C., MOLYNEUX, S. G., MERRIMAN, R. J., LEE, M. K. & VERNIERS, J. (eds) *Proceedings of the International Conference on the Caledonides of the Anglo-Brabant Massif (Nottingham, September 1992)* Special Issue on the Caledonides of the Anglo-Brabant Massif. *Geological Magazine*, **130**, 681–690.

WOODCOCK, N. H. & STRACHAN, R. 2000. *Geological History of Britain and Ireland.* Blackwell Science Ltd, Oxford.

ZIEGLER, P. A. 1982. *Geological Atlas of Western and Central Europe*. Shell International Petroleum Maatschappij BV, The Hague.

ZIEGLER, P. A. 1990. *Geological Atlas of Western and Central Europe.* Shell International Petroleum Maatschappij BV, The Hague/ Geological Society Publishing House, Bath/ Elsevier, Amsterdam.

Timing of the Avalonia–Baltica plate convergence as inferred from palaeogeographic and stratigraphic data of chitinozoan assemblages in west Pomerania, northern Poland

JOAKIM SAMUELSSON[1,2,3], MARCO VECOLI[4], WIESŁAW S. BEDNARCZYK[5] & JACQUES VERNIERS[1]

[1]*Research Unit Palaeontology, Department of Geology and Pedology, Ghent University, Krijgslaan 281 / S 8, B-9000 Ghent, Belgium*

[2]*Géosciences-Rennes, UMR 6118 du CNRS, Université de Rennes I, F-350 42 Rennes-cedex, France*

[3]*Present address: Institute of Earth Sciences, Historical Geology and Palaeontology, Norbyvägen 22, S-752 36 Uppsala, Sweden (e-mail: joakim.samuelsson@geo.uu.se)*

[4]*Institut für Geologische Wissenschaften, Martin-Luther Universität Halle/Wittenberg, Domstrasse 5, D-06108 Halle (Saale), Germany*

[5]*Polish Academy of Sciences, Institute of Geological Sciences, ul. Twarda 51/55, Pl-00–818 Warszaw, Poland*

Abstract: Tectonically disturbed Ordovician rocks penetrated by deep drillholes in Pomerania, NW Poland (Koszalin–Chojnice Zone) belong to the Heligoland–Pomerania Deformation Belt. Earlier data demonstrate that the Avalonia–Baltica collision occurred in Late Ordovician times, but in Pomerania, the timing of convergence has not been ascertained, and it is uncertain if the rocks underneath the Koszalin–Chojnice Zone belong to Avalonia or Baltica. Data from chitinozoans, organic-walled Palaeozoic microfossils with applications in biostratigraphy and palaeobiogeography, were assessed from ten boreholes (Brda 2; Brda 3; Chojnice 5; Karsina 1; Kościernica 1; Nowa Wieś 1; Okunino 1; Sarbinowo 1; Skibno 1; Wyszebórz 1) to address these problems. The results improve the biostratigraphy of the cores and demonstrate that the youngest Ordovician rocks are of a Burrellian (early mid Caradoc) to Cheneyan (late mid Caradoc) age. Because these rocks are interpreted as forming part of the deformation belt, the obtained ages put a lower age limit on the initiation of foreland basin sedimentation on the foreland of the orogeny, i.e. the Baltic platform. Quantitative comparison of chitinozoan assemblages demonstrates a high level of similarity between Pomerania and Avalonia. Together, Pomerania and Avalonia show greater similarity to Baltoscandia than to North Gondwana, supporting the idea that the Tornquist Ocean had narrowed significantly in early Caradoc times.

Deformed rocks of Early Palaeozoic age are known from drillings in southern Denmark (Jylland), northern Germany (Schleswig and Rügen) and northern Poland (west Pomerania). These tectonically disturbed units are part of the so-called 'Danish–North German–Polish Caledonides', or the 'Heligoland–Pomerania Deformation Belt' (HPDB), an ancient fold belt interpreted as being created during the collision and subsequent amalgamation of the palaeocontinents Avalonia and Baltica during Late Ordovician to early Silurian times (Brochwicz-Lewiński *et al.* 1981; Ziegler 1988; Katzung *et al.* 1993; Samuelsson *et al.* 2001; Vecoli & Samuelsson 2001*a*; Winchester *et al.* 2002).

The Trans-European Suture Zone (TESZ; Pharaoh *et al.* 1997) is an important lithospheric boundary separating the Archaean and Proterozoic crust of the East European Craton (EEC) from the younger crustal domains of western Europe (Gee & Zeyen 1996; Pharaoh *et al.* 1997; Pharaoh 1999). The TESZ consists of the boundaries of a number of smaller terranes, including the Teisseyre–Tornquist Zone (TTZ) in western Poland, and is concealed over most of its length by deep sedimentary basins of Permian to Cenozoic age (Pharaoh 1999). As a consequence, the exact geometry of the HPDB and Baltica contact in NW Poland (west Pomerania) is unclear (e.g. Franke 1995).

In the vicinity of the German island of Rügen (southern Baltic Sea), deep boreholes encounter

From: WINCHESTER, J. A., PHARAOH, T. C. & VERNIERS, J. 2002. *Palaeozoic Amalgamation of Central Europe.* Geological Society, London, Special Publications, **201**, 95–113. 0305-8719/02/$15.00

thrust and deformed anchizonal greywackes of Ordovician age in the hanging wall of the Thor Suture, formerly known as the Caledonian Deformation Front (Berthelsen 1998). Contrary to the opinion of Cocks *et al.* (1997), sediment provenance studies and quantitative palynological analyses of the Rügen Ordovician successions support the inclusion of these rocks within the eastern extension of Avalonia (Giese *et al.* 1994; Dallmeyer *et al.* 1999; Vecoli & Samuelsson 2001*a*).

There are convincing arguments that the buried Lower Palaeozoic rocks of west Pomerania were also part of Avalonia (e.g. Franke 1967, 1978, 1990; Teller 1974; Dadlez 1978; Pożaryski & Kotański 1978; Pożaryski 1987; Franke & Znosko 1988; Berthelsen 1992). Supporting data come partly from the similarity in overall geological development between Pomerania and those rocks underlying Rügen, and partly because directly to the NW along the Thor Suture in the southern Baltic Sea, seismic data indicate the existence of a highly deformed Caledonian accretionary wedge, detached on Cambrian and Ordovician black shales, and thrust northeastwards onto the foreland of the orogeny (Schlüter *et al.* 1997). However, the exact location of the northeastern border of the microcontinent of Avalonia, which thus probably forms part of present-day Poland south of the Thor Suture, is still debatable (see e.g. Pharaoh 1999 and his fig. 1; Winchester *et al.* 2001), especially whether the Lower Palaeozoic rocks underlying the TTZ itself should also be regarded as Avalonian.

The timing of the collision between Avalonia and Baltica is now well established as restricted to the Late Ordovician, especially in SW Baltoscandia (e.g. Frost *et al.* 1981; McKerrow & Cocks 1986; Pharaoh *et al.* 1995; Torsvik 1998; Samuelsson *et al.* 2001; Vecoli & Samuelsson 2001*a*), but the constraints on the timing of the deformation events in rocks underlying Pomerania have not yet been evaluated in detail. Chitinozoans can provide palaeolatitudinal and biostratigraphical data, and here we discuss the chitinozoan faunas in 49 samples from Ordovician levels in the boreholes Brda 2; Brda 3; Chojnice 5; Karsina 1; Kościernica 1; Nowa Wieś 1; Okunino 1; Sarbinowo 1; Skibno 1; and Wyszebórz 1. These were drilled in the narrow NW–SE zone in NW Poland usually referred to as the Koszalin–Chojnice Structural Zone. The aim of this paper is to provide a more precise biostratigraphy for the drilled rocks, shed more light on the position of the Avalonia–Baltica suture in this region, and assess the timing of deformation of the HPDB.

Results from earlier stratigraphic and quantitative analyses of acritarch and chitinozoan assemblages on both sides of the TESZ have demonstrated that foreland basin sedimentation commenced in mid Caradoc to late Ashgill times in the Rügen area (Samuelsson *et al.* 2001; Vecoli & Samuelsson 2001*a*), and not later than Llandovery times in southern Denmark (Vecoli & Samuelsson 2001*b*). In Pomerania, sedimentation of the fold belt itself was active at least up to the Burrellian (middle Caradoc), constraining the lower time limit for the formation of the belt. In the Llandeilian (uppermost Llanvirn/uppermost Darriwilian) to Aurelucian (lowermost Caradoc) stratigraphic interval, the chitinozoan fauna from Pomerania is most readily compared to those from other areas located on east Avalonia.

Previous Palaeozoic stratigraphic research in northern Poland

In the 1960s, Ordovician rocks were drilled in west Pomerania by oil and gas companies, and by the Polish Geological Institute. Early papers dealing with the stratigraphy of these cores include e.g. those by Dadlez (1967) and Modliński (1968).

Bednarczyk (1974) described the drill cores investigated in this paper, analysed their graptolite faunas, and provided the first biostratigraphic dating. However, only recently has the full potential of these drillings for understanding the geological evolution of the area been explored. Bednarczyk *et al.* (1999) presented preliminary results from integrated graptolite, acritarch and chitinozoan biostratigraphy of the borehole Skibno 1, and Wrona *et al.* (2001) give the full account of these data. Based on acritarch biostratigraphy, Szczepanik (2000) correlated subsurface Ordovician successions from the Koszalin–Chojnice Structural Zone (i.e. Chojnice 5; Jamno IG 1; Karsina 1; Brda 3; and Nowa Wieś 1 boreholes) with the better studied Ordovician successions of the East European Platform, and suggested some refinements to the stratigraphy originally proposed by Bednarczyk (1974).

Chitinozoans were previously investigated from cores drilled in the Łeba Region, NE of the Koszalin–Chojnice Zone, on the other side of the TTZ. Eisenack (1972) described Silurian chitinozoans and acritarchs from the Łeba borehole drilled in 1936/38. Podhalańska (1979) reported the occurrence of chitinozoans, including *Laufeldochitina stentor* (Eisenack) of early Caradoc age, in five boreholes (Białogóra 1 and 2; Piaśnica 2; Dębki 3; and Mieroszyno 8). Wrona (1980) briefly mentioned chitinozoans

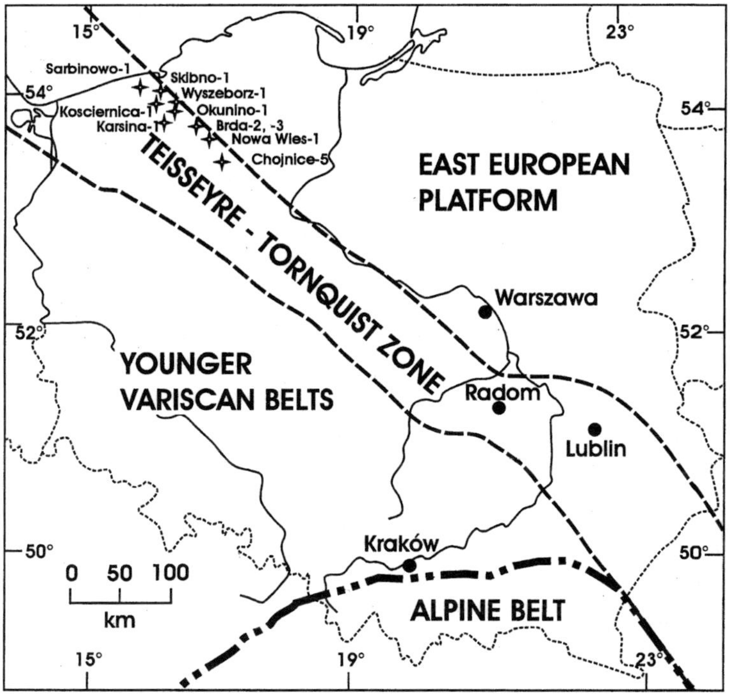

Fig. 1. Simplified tectonic map of Poland showing the geographic positions of the boreholes investigated in this paper.

from northeastern Poland, and Bednarczyk (1999) discussed Arenig to lower Ashgill chitinozoan assemblages in the Sokolica 1 and Klewno 1 boreholes of the Kętrzyn region, north of the TTZ.

Geology and tectonics

General overview

Folded Ordovician and Silurian (up to Pridoli) strata appear along the SW margin of the EEC (Fig. 1; Teller 1974; Dadlez 1978) and are different in their overall geological development from rocks north of the Thor Suture (see e.g. review by Pharaoh 1999). As in the Rügen area (NE Germany), the Lower Palaeozoic deformed rocks in the Koszalin–Chojnice and Toruń–Radom tectonic zones were thrust over the crystalline basement of the EEC. Accordingly, the Pomeranian deformed rocks are included in the fold belt usually called the 'Danish–North German–Polish Caledonides', and collectively interpreted as belonging to the Pomerania (or Pomorze) terrane together with the Lower Palaeozoic rocks of the subsurface of Rügen. The term 'Caledonian' should properly be restricted to the areas affected by the development and closure of the Iapetus Ocean more to the west (McKerrow *et al.* 2001). Thus, acknowledging that this fold belt is palaeogeographically separated from the almost contemporaneous Anglo-Brabant Deformation Belt to the SW we refer to it as the HPDB (Winchester *et al.* 2002). The Pomerania terrane was lodged against a restraining bend of the TTZ in sinistral kinematics of the closure of the Tornquist Ocean (Unrug *et al.* 1999). SE of Pomerania Palaeozoic rocks are as yet unknown.

The investigated drill cores

Ordovician rocks in Pomerania (NW Poland), deformed during pre-Devonian times, have been known from drillcores for more than 30 years. The Ordovician successions were penetrated by 14 boreholes situated in an area extending from Chojnice in the SE to Koszalin in the NE (Bednarczyk 1974). The area, known in Polish geological literature as the Koszalin–Chojnice Structural Zone, is a fault-bounded narrow band situated between the SW margin of the EEC to the NE, and the younger Variscan mobile belts to the SW (Fig. 1). The

Ordovician deposits are buried beneath an overburden of Late Permian, Carboniferous, and Devonian age, 1727 m thick in the Skibno 1 drillcore in the NW part of the region, and up to 4890 m in the Chojnice 5 drillcore in the SE. Strong folding and tectonic deformation are the main features which distinguish the Ordovician sediments from their contemporaneous counterparts in the EEC. All Ordovician sediments show steep dips, ranging from 10° to 90°, numerous dislocations, and slickensides. Nevertheless, most of the successions are unmetamorphosed and uncleaved (Franke 1994). Graptolites and brachiopods are present, albeit with a limited number of taxa, except for the graptolite fauna from Chojnice 5 which yielded a moderately varied assemblage (Bednarczyk 1974). Lithologically, the Ordovician succession in the boreholes consists of strata that have been subdivided into five different units.

(1) Dark grey siltstone with abundant fine muscovite flakes and pyrite. Pyrite is dispersed or occurs in concentrations (Skibno 1: 1882.6–2807.0 m; Karsina 1: 3142.5–3203.0 m). In the central part of the region, thin anhydrite veins occur (Okunino 1: 1873.2–2009.5 m).
(2) Dark grey siltstone with dolomitic mudstone (Skibno 1: 1802.3–1882.6 m; Sarbinowo 1: 2802.5–3000.0 m) or with greyish-beige dolomite laminae (Wyszebórz 1: 2753.2–3021.4 m) or dolomitic concretions (Kościernica 1: 2818.4–2853.0 m) and with dark brown limestone intercalations (Brda 2: 2602.0–3000.0 m).
(3) Dark grey greenish-tinted dolomitic siltstone intercalated with red dolomitic siltstone (Skibno 1: 1727.0–1802.3 m).
(4) Dark grey siltstone infrequently bluish-tinted with light-grey bluish-tinted tuffite intercalations (Brda 3: 2153.0–2901.5 m; Nowa Wieś 1: 2417.5–2900.0 m).
(5) Dark grey nearly black graptolite bearing siltstone, with numerous pyrite concretions in the SE part of the Koszalin–Chojnice structural zone (Chojnice 5: 4890.0–5055.5 m).

The coprolite *Tomaculum problematicum* Groom, was recovered from units (1), (2) and (3) in the form of unbranched, faecal-like burrows. Previously, this trace fossil was regarded as typical of Perigondwana, but recent interpretation and revision (Orr 1996) indicate that it might be more widely distributed, both spatially and temporally.

Bednarczyk (1974) subdivided the Pomerania rocks into two local graptolite assemblage biozones, which can be correlated with the *Hustedograptus teretiusculus* to lower *Diplograptus multidens*, and the upper *D. multidens* to *Dicranograptus clingani* biozones respectively, of the standard British graptolite biozonation (Fortey *et al.* 1995). Where available, new fossil data have confirmed these ages (Bednarczyk *et al.* 1999; Szczepanik 2000; Wrona *et al.* 2002).

Material and methods

The studied material consists of 49 core samples from the Brda 2; Brda 3; Chojnice 5; Karsina 1; Kościernica 1; Nowa Wieś 1; Okunino 1; Sarbinowo 1; Skibno 1; and Wyszebórz 1 boreholes in west Pomerania south of the TTZ (Fig. 1). Samples which are referred to in the text are indicated with the borehole name followed by the sampling depth in metres. Full taxonomic treatment of the chitinozoan faunas is published elsewhere (Samuelsson 2002). Laboratory extraction and concentration of palynomorphs were carried out at the Research Unit Palaeontology, Department of Geology and Pedology, Ghent University, Belgium by using the standard palynological method described by Paris (1981).

Quantitative assessment of the chitinozoan assemblages

Palynomorphs, e.g. acritarchs and chitinozoans, can provide palaeolatitudinal as well as high resolution biostratigraphic data when carefully analysed (e.g. Paris 1981, 1993; Achab 1988, 1991; Playford *et al.* 1995; Tongiorgi & Di Milia 1999; Vecoli 1999; Vecoli & Samuelsson 2001*a*). In order to quantify the observed taxonomic similarities and differences between the present chitinozoan assemblages with those of contemporaneous settings in the palaeocontinents of Baltica, North Gondwana and Avalonia, we used multivariate statistical cluster analysis, and calculated a coefficient of similarity (Clark & Harteberg 1983) by using our own and previously published data.

In short, cluster analysis evaluates the overall similarity between objects mathematically, in this case the chitinozoan taxa, which are then displayed in the form of a dendrogram. Comparison between objects is based on an agglomerative clustering algorithm which allows combining the most similar objects in clusters of successively higher ranks until all objects are arranged in a single, hierarchical group (Hazel 1970). Our approach to this method is discussed in more detail in Vecoli & Samuelsson (2001*a*).

A number of previous studies have successfully employed the coefficient of similarity (CS) *sensu* Clark & Harteberg (1983) for bioprovincialism evaluation. This simple and straightforward approach is expressed by the formula CS = $2v/(a+b)$ where v is the number of species in common between two compared assemblages, and a and b are the total number of species in each assemblage, respectively.

Some problems exist with these two approaches: in a few reported sections, although the chitinozoan assemblages are excellently illustrated and described, the authors did not include the complete assemblages in their account of the chitinozoan content of the investigated section, but only the more representative and morphologically more interesting taxa. In order to overcome this obvious drawback, we have included as many sections as possible in the analyses. Another problem is the possible effect of oceanic palaeocurrents which do not necessarily correlate to geographic distance. A third problem is the specific determination of the chitinozoans: some species, such as *Rhabdochitina magna* Eisenack, *Euconochitina conulus* (Eisenack) or *Euconochitina* cf. *communis* (Taugourdeau) possess few distinctive features. It is therefore possible that these taxa are of a 'waste basket' character, and that a number of biological species are included in such taxa. Furthermore, for cluster and similarity index analysis to be reliable, it is important that the data matrices employed are of approximately the same size. Thus, the obtained results should be regarded as rough approximates only.

In the investigated area, three index fossils of the widely used Baltoscandian chitinozoan biozonation (Nõlvak & Grahn 1993; Nõlvak 1999*a*) were identified, namely *Laufeldochitina stentor* (Eisenack), *Lagenochitina dalbyensis* (Laufeld) and *Spinachitina cervicornis* (Eisenack), corresponding to the late Llandeilian (late Darriwilian) to Aurelucian (*stentor* Biozone), early Harnagian (*dalbyensis* Biozone), and late Harnagian to Longvillian (*cervicornis* Biozone) times. However, we made cluster and CS analyses for the *stentor* biozone only, because impoverished assemblages were recovered from the other two biozones. In addition, corresponding strata in other areas identified as belonging to east Avalonia have not yet been analysed for their chitinozoan faunas, or as is the case for published data from the Ebbe Anticline (Samuelsson *et al.* 2002), or the Brabant Massif (Samuelsson & Verniers 2000), are not suitable for analysis because of the poor preservation and low diversity of the faunas.

Results

Palynological results

Forty-three out of 49 samples (88%), from the ten investigated drillcores yielded moderately well-preserved chitinozoans. In total, more than 13 200 chitinozoan specimens were observed. The microfossils are generally dark-brown in colour, although a few thin-walled forms appear yellowish-brown in transmitted light. All observed chitinozoans are flattened. They occur in low to high numbers, i.e. from less than one to 232 chitinozoans per gram of rock. From the borehole Okunino 1, only a single indeterminable chitinozoan was recovered. Summaries of the total taxonomic composition of the observed faunas and quantitative abundance values are given in Figures 2 to 10. For detailed taxonomy and illustrations of the chitinozoans, see Samuelsson (2002).

The taxonomic diversity is quite high, with an average of six species determined per fossiliferous sample. Thus it is comparable to Middle and Upper Ordovician chitinozoan biodiversity reported from elsewhere (e.g. Paris 1981, 1996; Nõlvak & Grahn 1993; Miller 1996; Paris & Nõlvak 1999).

Brda 2 For the two uppermost samples the quality of chitinozoan preservation is good, with for example spines at the base of *Spinachitina bulmani* (Jansonius) in fair condition. The lowermost sample (Brda 2/2930.6–2930.7) yielded chitinozoans that are very poorly preserved, in most cases with the entire vesicle penetrated by holes. Their deep dark-brown colour is in contrast to that of the other samples from this drilling. Taxonomically, the chitinozoan assemblage from all three investigated samples is dominated by *S. bulmani*. In sample Brda 2/2603–2606, *Desmochitina cocca* (Eisenack) is also abundant. In this sample, a few specimens kept under open nomenclature as *Cyathochitina* sp. A and *Hercochitina* sp. B appear. Among some of the specimens described as *Desmochitina minor* Eisenack in Brda 2/2603–2606, there is a tendency towards a shape similar to that of *Desmochitina amphorea* Eisenack.

Brda 3 One sample of the three investigated was barren. For the fossiliferous samples, preservation is good, with long fragile specimens, such as *Rhabdochitina magna* Eisenack and *Conochitina minnesotensis* (Stauffer) recovered nearly intact. Delicate details such as spines and chamber ornamentation are clearly visible. *S. bulmani* dominates in the two fossiliferous

Brda 2	*Cyathochitina* sp. A	*Hercochitina* sp. B	*Belonechitina capitata*	*Conochitina minnesotensis?*	*Desmochitina minor*	*Desmochitina cocca*	*Conochitina homoclaviformis?*	*Spinachitina bulmani*	*Belonechitina micracantha*	*Cyathochitina campanulaeformis*	*Conochitina chydaea*	*Cyathochitina* spp.	*Desmochitina* spp.	*Ancyrochitina?* sp.	**undetermined**	**total number of Chitinozoa**	**sample weight (g)**	**Chitinozoa / g sample**
2603 - 2606 (1)	○	◎	◎	◆	□	□	◎	●				◎	◎		○	221	11.0	20
2732 - 2733 (3)						◆		■	◎	◎	◎	◆			○	2408	10.4	232
2930.6 - 2930.7 (4)								■			◆			◆	○	496	10.0	50

Fig. 2. Chitinozoan taxa recovered from the Brda 2 borehole. Numbers within parenthesis after sample depth are working sample numbers. Legend: ◆ = <1%, ◎ = 1–5%, ○ = 6–10%, □ = 11–25%, ● = 26–50%, ■ = 51–100%.

Brda 3	*Belonechitina robusta*	*Conochitina chydaea*	*Conochitina dolosa*	*Desmochitina cocca*	*Desmochitina erinacea*	*Desmochitina nodosa*	*Spinachitina bulmani*	*Spinachitina cervicornis?*	*Fungochitina actonica?*	*Rhabdochitina magna*	*Conochitina minnesotensis*	*Conochitina homoclaviformis*	*Pistillachitina elegans?*	*Euconochitina primitiva*	*Rhabdochitina gracilis*	*Rhabdochitina* spp.	*Cyathochitina?* spp.	*Conochitina* spp.	**undetermined**	**total number of Chitinozoa**	**sample weight (g)**	**Chitinozoa / g sample**
2355 - 2361 (5)	◎	◎	○	○	◆	◎	■	◎								◎			□	342	11.0	31
2646 - 2647 (6)																				0	10.5	0
2733 - 2738 (7)	◎					◎	●		◎	◎	□	◎	◆	◎	○		◎	◎	□	186	11.5	16

Fig. 3. Chitinozoan taxa recovered from the Brda 3 borehole. Numbers within parenthesis after sample depth are working sample numbers. Grey field indicates barren sample. Legend: ◆ = <1%, ◎ = 1–5%, ○ = 6–10%, □ = 11–25%, ● = 26–50%, ■ = 51–100%.

samples, but specimens identified as *Belonechitina robusta* (Eisenack), and possibly *Spinachitina cervicornis* (Eisenack) occur in sample Brda 3/2355–2361 (Fig. 3). *B. robusta* was also identified in Brda 3/2733–2738.

Chojnice 5 Six out of seven investigated samples yielded chitinozoans in a sound state of preservation, for example with delicate features well preserved (Fig. 4). Throughout the sampled interval, specimens described as *Euconochitina* cf. *communis* (Taugourdeau) are a characteristic component. In the samples from the deeper parts of the borehole, *Spinachitina bulmani* dominates (Fig. 4). Specimens belonging to the genus *Desmochitina* are also common, and a few *Cyathochitina calix* (Eisenack) occur in samples Chojnice 5/5053–5054 and 5052–5053.

Karsina 1 Three samples from this borehole yielded a reasonably diverse and well preserved assemblage (Fig. 5). Specimens attributed to *Belonechitina* cf. *micracantha* (Eisenack) were recovered from all three samples. A number of specimens kept under open nomenclature such as *Cyathochitina* sp. A, *Armoricochitina* sp. A, *Hercochitina* sp. A, and *Hercochitina* sp. B, were also retrieved.

Kościernica 1 Two samples were investigated (Fig. 6), and the quality of preservation of chitinozoans was good. In sample Kościernica

Chojnice 5	*Rhabdochitina gracilis*	*Cyathochitina campanulaeformis*	*Desmochitina nodosa*	*Desmochitina erinacea?*	*Euconochitina* cf. *communis*	*Rhabdochitina gallica*	*Conochitina* cf. *homoclaviformis*	*Armoricochitina* sp. A	*Desmochitina cocca*	*Belonechitina micracantha*	*Spinachitina bulmani*	*Desmochitina typica*	*Cyathochitina calix*	*Cyathochitina* sp. A	*Conochitina chydaea*	*Rhabdochitina* spp.	*Cyathochitina* spp.	*Belonechitina* spp.	*Conochitina* spp.	*Desmochitina* spp.	*Pistillachitina* spp.	*Spinachitina* spp.	CONOCHITININAE indet.	*Spinachitina* spp.	*Lagenochitina* sp.	**undetermined**	**total number of Chitinozoa**	**sample weight (g)**	**Chitinozoa / g sample**
4895.1 - 4896.1 (8)																											0	10.8	0
4956.5 - 4957.5 (9)	○	□	●	◎	◎											◎	◎	○	◎							○	76	10.8	7
5000 - 5001 (10)			■		□	◆	◎	◎									◆		◆	◎	◆	◎				○	587	11.9	49
5005 - 5006 (14)		◎			●				●									◎					◎			□	99	12.0	8
5006.75 - 5006.8 (17)			○		□	◎		◎	□	◎	□	○							◎	◎		◆				□	904	11.0	82
5052 - 5053 (19)					□				□	○	□	○	□	◎			◎		◎	○				◎		□	97	10.5	9
5053 - 5054 (21)					□				○		●		◎		○		◎				◆				◆	□	488	11.8	41

Fig. 4. Chitinozoan taxa recovered from the Chojnice 5 borehole. Numbers within parenthesis after sample depth are working sample numbers. Grey field indicates barren sample. Legend: ◆ = <1%, ◎ = 1–5%, ○ = 6–10%, □ = 11–25%, ● = 26–50%, ■ = 51–100%.

Karsina 1	*Cyathochitina* sp. A	*Conochitina chydaea*	*Cyathochitina* cf. *kuckersiana*	*Desmochitina cocca*	*Belonechitina* cf. *micracantha*	*Belonechitina micracantha*	*Desmochitina minor*	*Armoricochitina* sp. A	*Cyathochitina campanulaeformis*	*Conochitina* cf. *dolosa*	*Euconochitina* cf. *communis*	*Conochitina homoclaviformis?*	*Rhabdochitina gracilis*	*Hercochitina* sp. A	*Rhabdochitina gallica*	*Hercochitina* sp. B	*Rhabdochitina magna*	*Conochitina* spp.	*Euconochitina* spp.	*Cyathochitina* spp.	*Belonechitina* spp.	*Lagenochitina* sp.	*Desmochitina* spp.	**undetermined**	**total number of Chitinozoa**	**sample weight (g)**	**Chitinozoa / g sample**
3141 - 3143 (22)	◎	◎	◎	◎	□													◎	◎	◎	●		◎	□	206	12.5	16
3166.4 - 3167.4 (23)			□		●			◆	◎	◎	◎	◆	◆						◎	○		◎	◎	●	406	10.4	39
3167.4 - 3168.4 (24)				□	□		◎		○		□		◆	◆	◎	◆	◎				●		○	○	116	11.2	10

Fig. 5. Chitinozoan taxa recovered from the Karsina 1 borehole. Numbers within parenthesis after sample depth are working sample numbers. Legend: ◆ = <1%, ◎ = 1–5%, ○ = 6–10%, □ = 11–25%, ● = 26–50%, ■ = 51–100%.

1/2850.0–2852.0, *Euconochitina* cf. *communis* (Taugourdeau) dominates the assemblage. In Kościernica 1/2820.0–2823.15 a possible *Lagenochitina dalbyensis* (Laufeld) occurs, as do a number of smooth, cono- to claviform-shaped specimens here identified as *Conochitina brevis* Benoît & Taugourdeau.

Nowa Wieś 1 Sample Nowa Wieś 1/2710–2711 revealed a well-preserved assemblage, the other two samples yielded poor assemblages, both in terms of preservation and in taxonomic abundance. *Spinachitina cervicornis* Eisenack is an important constituent of the assemblage from the well-preserved sample, accompanied by specimens belonging to the genus *Rhabdochitina* (Fig. 7).

Okunino 1 A sample from the interval 1877.6 to 1878.5 m proved to be barren (10.5 g of rock was dissolved). Another sample from the interval 1878.5 to 1878.8 m yielded a single poorly preserved chitinozoan, indeterminable even to the generic level (10.3 g of rock was dissolved). No further attempts at finding chitinozoans in this borehole were made, and we treated the two investigated samples as barren.

Sarbinowo 1 The six investigated samples proved to be fossiliferous (Fig. 8) and yielded well-preserved chitinozoans e.g. *Belonechitina robusta* (Eisenack), *Euconochitina* cf. *communis* (Taugourdeau), and *Belonechitina* cf. *micracantha* (Eisenack) in the three uppermost samples. Also noteworthy are specimens attributed to *Hercochitina* sp. A, *Lagenochitina* cf. *deunffi* (Paris), and *Eisenackitina*? sp. The latter shows a vague similarity to *Eisenackitina rhenana* (Eisenack), but preservation of these few specimens is too poor to allow definite attribution to that taxon.

Skibno 1 Two out of 17 investigated samples were barren, but the remaining 15 samples yielded a highly diverse chitinozoan assemblage with up to 72 chitinozoans per gram of sample

Kościernica 1	*Conochitina brevis*	*Lagenochitina dalbyensis?*	*Belonechitina* cf. *micracantha*	*Cyathochitina campanulaeformis*	*Desmochitina minor*	*Conochitina minnesotensis*	*Euconochitina conulus*	*Belonechitina robusta*	*Euconochitina* cf. *communis*	*Pistillachitina?* sp.	*Lagenochitina* sp.	*Spinachitina?* spp.	*Conochitina* spp.	*Belonechitina* spp.	*Cyathochitina* spp.	*Desmochitina* spp.	undetermined	total number of Chitinozoa	sample weight (g)	Chitinozoa / g sample
2820.0 - 2823.15 (25)	□	◎	◎	□	◆					◎	◎	◆	◎	○	□		●	180	11.6	15
2850.0 - 2852.0 (26)			□	◎		○	◎	◆	●				◎	◎	◎	◎	□	680	11.4	60

Fig. 6. Chitinozoan taxa recovered from the Kościernica 1 borehole. Numbers within parenthesis after sample depth are working sample numbers. Legend: ◆ = <1%, ◎ = 1–5%, ○ = 6–10%, □ = 11–25%, ● = 26–50%, ■ = 51–100%.

Nowa Wieś 1	*Cyathochitina* sp. A	*Desmochitina minor*	*Spinachitina bulmani*	*Spinachitina cervicornis*	*Rhabdochitina gracilis?*	*Rhabdochitina gallica*	*Conochitina chydaea*	*Cyathochitina* spp.	*Rhabdochitina* sp.	*Euconochitina?* sp.	*Spinachitina?* spp.	*Cyathochitina?* sp.	*Fungochitina* spp.	*Lagenochitina* sp.	*Desmochitina* spp.	*Belonechitina* spp.	CONOCHITININAE indet.	undetermined	total number of Chitinozoa	sample weight (g)	Chitinozoa / g sample
2488.2 - 2490.2 (27)	◎	◎						○	◎	◎	□							■	25	10.8	2
2492.3 - 2496.3 (28)											□	□						■	7	10.6	0.7
2710 - 2711 (31)			□	○	□	□	◎	◆					○	◆	◎	◆	◎	●	206	11.0	19

Fig. 7. Chitinozoan taxa recovered from the Nowa Wieś 1 borehole. Numbers within parenthesis after sample depth are working sample numbers. Legend: ◆ = <1%, ◎ = 1–5%, ○ = 6–10%, □ = 11–25%, ● = 26–50%, ■ = 51–100%.

Sarbinowo 1	*Hercochitina* sp. A	*Conochitina chydaea*	*Belonechitina robusta*	*Conochitina brevis*	*Desmochitina nodosa*	*Rhabdochitina gracilis*	*Lagenochitina* cf. *deunffi*	*Rhabdochitina gallica*	*Desmochitina minor*	*Euconochitina* cf. *communis*	*Belonechitina* cf. *micracantha*	*Desmochitina cocca*	*Conochitina dolosa?*	*Desmochitina* spp.	*Belonechitina* spp.	*Cyathochitina* spp.	*Conochitina* spp.	*Euconochitina?* spp.	*Eisenackitina?* sp.	undetermined	total number of Chitinozoa	sample weight (g)	Chitinozoa / g sample
2845 - 2846 (34)	●	○	○	◎	○									◎	□					□	139	10.1	14
2846 - 2847 (35)	◆	○	□	◎		○	◆	◎	◆	◆					□	□	◆	◆		□	141	11.4	12
2888 - 2889 (37)										□	□	●	◎	□	◆		◎			□	256	10.3	25
2912.1 - 2912.2 (39)	□	○	◎			◎			○			○		○	◎			◎		●	76	10.9	7
2951 - 2952 (41)		□									□			□			□			●	8	12.1	0.7
2952 - 2953 (42)	○	◎		◎		◎				◎					◎	□			□	●	202	11.3	18

Fig. 8. Chitinozoan taxa recovered from the Sarbinowo 1 borehole. Numbers within parenthesis after sample depth are working sample numbers. Legend: ◆ = <1%, ◎ = 1–5%, ○ = 6–10%, □ = 11–25%, ● = 26–50%, ■ = 51–100%.

Skibno 1	1722.3 - 1725.18 (43)	1730.92 - 1731.84 (46)	1745 - 1746 (48)	1795 - 1796 (50)	1838.4 - 1839.4 (51)	1880.1 - 1881.1 (52)	1913 - 1914 (53)	1950.22 - 1951.14 (54)	2037 - 2038 (55)	2110.5 - 2110.7 (57)	2330.7 - 2331.7 (59)	2435.2 - 2435.3 (60)	2528 - 2528.1 (61)	2592.36 - 2592.96 (64)	2653.2 - 2653.3 (66)	2716.87 - 2716.97 (69)	2802.7 - 2802.8 (70)
Lagenochitina dalbyensis			○														
Conochitina dolosa			○														
Cyathochitina campanulaeformis			●					◎	○	◎							
Desmochitina cocca			◎	□	○	◎	●		□	◎	◎		●	◎			
Desmochitina nodosa?			◎														
Euconochitina conulus			◎				◎							○			
Armoricochitina sp. A				◎													
Conochitina minnesotensis				○		○		◎	□	◎	●			◎	◆		
Euconochitina cf. *communis*				◎	◎	●	◎	◎		◎	◎			○			●
Conochitina cf. *homoclaviformis*				□	○	◎											
Desmochitina minor				◎	○		◎		□		◎		○				
Cyathochitina sp. A				□	□		◎										
Spinachitina bulmani					○								□	□	■	■	
Belonechitina robusta					◆	◎											
Rhabdochitina magna							◎										
Cyathochitina cf. *calix*							◎		◎								
Belonechitina micracantha							□		◎	●	○		◎		◆	◎	○
Laufeldochitina stentor								◆		◎							
Eisenackitina rhenana								■									
Lagenochitina sp. A								◎		○				◎			
Rhabdochitina gracilis								◎									
Lagenochitina deunffi								◆									
Belonechitina cf. *micracantha*								◆	◎								
Belonechitina capitata									◎								
Rhabdochitina gallica										○				□			
Calpichitina lata?														◎			
Conochitina brevis														□			
Rhabdochitina spp.			◎	◎								●	◎		◎		
Cyathochitina spp.			◎														
Desmochitina spp.			◎	◎	●	○	○	◆	◎	◎	◎		◎	◎			
CONOCHITININAE indet.			◆					◎	◎							◆	
Conochitina spp.				◎		○	○				◎						◎
Pistillachitina? spp.				◎				◆									
Cyathochitina? spp.					◎						○		◎				
Belonechitina spp.						◎				□	◎		◎	◆		○	□
Euconochitina spp.												■					◎
undetermined			●	□	□	●	□	○	●	●	□	◆	□	●	○	□	□
total number of Chitinozoa	0	0	720	266	226	67	264	704	96	106	238	3	146	222	254	328	24
sample weight (g)	10.3	10.8	10.0	11.9	10.3	11.4	13.8	11.0	10.3	10.7	10.6	11.7	11.0	10.6	11.1	10.7	11.5
Chitinozoa / g sample	0	0	72	22	22	6	19	64	9	10	22	0.3	13	21	23	31	2

Fig. 9. Chitinozoan taxa recovered from the Skibno 1 borehole. Numbers within parenthesis after sample depth are working sample numbers. Grey fields indicate barren samples. Legend: ◆ = <1%, ◎ = 1–5%, ○ = 6–10%, □ = 11–25%, ● = 26–50%, ■ = 51–100%.

(Fig. 9) in a good state of preservation. Almost all species recovered from the other Pomeranian drillholes investigated by us are also present in Skibno 1. Three index species of the Baltoscandian chitinozoan biozonation (Nõlvak & Grahn 1993) were observed: Skibno 1/1745–1746 yielded abundant well-preserved *Lagenochitina dalbyensis* (Laufeld); Skibno 1/2110.5–2110.7 and Skibno 1/1950.22–1951.14 yielded *Laufeldochitina stentor* (Eisenack); the latter sample also yielded *Eisenackitina rhenana* (Eisenack) which is the index fossil of a subzone in the *stentor* biozone. The index fossils occur in the same stratigraphic order as in the biozonation. In addition, three important species, namely *Spinachitina bulmani* (Jansonius), *Belonechitina*

Wyszebórz 1	*Pistillachitina elegans*	*Desmochitina cocca*	*Lagenochitina dalbyensis?*	*Euconochitina cf. communis*	*Spinachitina bulmani*	*Cyathochitina campanulaeformis*	*Belonechitina robusta?*	*Conochitina minnesotensis*	*Belonechitina capitata*	*Cyathochitina sp. A*	*Euconochitina conulus*	*Conochitina spp.*	*Belonechitina spp.*	*Cyathochitina spp.*	*Desmochitina spp.*	CONOCHITININAE indet.	**undetermined**	**total number of Chitinozoa**	**sample weight (g)**	**Chitinozoa / g sample**
2863.6 - 2864.5 (71)	□	□	○	○								○	●				□	**33**	**10.0**	**3**
2866.33 - 2867.26 (73)			□	□								○	□				●	**14**	**12.9**	**1**
2884.4 - 2885.4 (76)			◎		◎	◎	○	◎	○	◎	◎	◎	□	○	◎	◎	●	**1024**	**13.1**	**78**

Fig. 10. Chitinozoan taxa recovered from the Wyszebórz 1 borehole. Numbers within parenthesis after sample depth are working sample numbers. Legend: ◆ = <1%, ◎ = 1–5%, ○ = 6–10%, □ = 11–25%, ● = 26–50%, ■ = 51–100%.

micracantha (Eisenack) and *Lagenochitina deunffi* Paris were observed, and some less conspicuous forms such as *Euconochitina conulus* (Eisenack), *Conochitina minnesotenis* (Stauffer), *Rhabdochitina gracilis* Eisenack, as well as specimens kept under open nomenclature, e.g. *Armoricochitina* sp. A, *Cyathochitina* sp. A, and *Lagenochitina* sp. A.

Wyszebórz 1 All three investigated samples contain chitinozoans, the most fossiliferous being Wyszebórz 1/2884.4–2885.4 which yielded 78 chitinozoans per gram of rock (Fig. 10). Specimens here referred to as *Lagenochitina dalbyensis*? (Laufeld) occur in all three samples. They are kept in open nomenclature because their distorted and wrinkled chamber makes it impossible to place them confidently as *L. dalbyensis*.

Cluster and similarity coefficient analyses results

The dendrogram produced by cluster analysis (Fig. 11) demonstrates the high level of similarity between the Pomeranian chitinozoans of the *stentor* biozone (borehole Skibno 1) and contemporary Avalonian assemblages (Shropshire, Ebbe Anticline and Rügen 5/66 drillcore). In turn, the chitinozoans of Pomerania and Avalonia show a greater similarity to time-equivalent assemblages from Baltica than those from North Gondwana.

The coefficients of similarity are very similar for all four chitinozoan assemblages (Fig. 12). Coefficients of similarity calculated for Pomerania, Avalonia, Baltoscandia and North Gondwana fall within a relatively narrow range from 0.38 to 0.54. The highest similarity (CS = 0.54) is between Pomerania and Avalonia, and the weakest link is between Pomerania and North Gondwana (CS = 0.38).

Discussion

Relative ages of the sediments in the boreholes

In the following evaluation of the relative ages of the investigated borehole intervals, stratigraphic data has been obtained from the present as well as earlier published work. Figure 13 gives a summary outline of the correlations discussed below. References given to the lower, middle and upper (or early, mid and late) Caradoc respectively, refer to the Velfreyan, Costonian and Harnagian substages for the lower (and early) Caradoc, the Soudleyan, Longvillian and Woolstonian substages for the middle (or mid) Caradoc, and to the Marshbrookian, Actonian and Onnian substages for the upper (or late) Caradoc (Fortey *et al.* 1995).

Brda 2 The chitinozoans from the investigated interval belong to the same biostratigraphic assemblage. In itself, this assemblage gives a broad stratigraphic range, with *Belonechitina capitata* (Eisenack), ranging from the late Abereiddian to the late Caradoc (Paris 1981; Grahn 1982), and *Spinachitina bulmani* (Jansonius), ranging from the early Caradoc to the late Ashgill (Jansonius 1964; Elaouad-Debbaj 1984), being the most biostratigraphically significant species. Thus, a Caradoc age can be assigned to the investigated interval (2603–2930.7 m) on

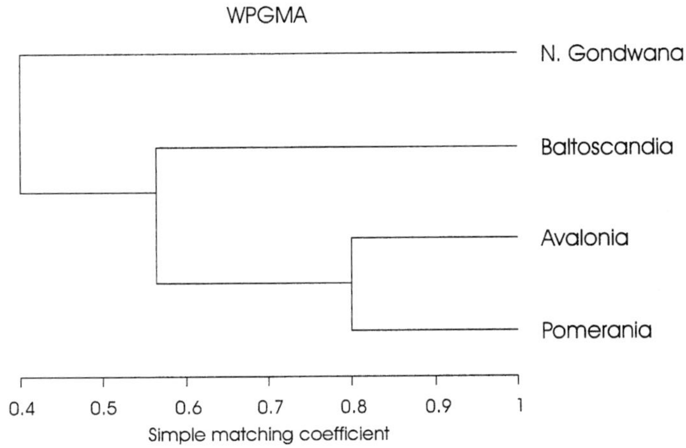

Fig. 11. Resulting dendrogram from the cluster analysis of the chitinozoan assemblages of the *stentor* chitinozoan biozone (latest Llanvirn to early Caradoc age; Nõlvak & Grahn 1993) from Pomerania, east Avalonia, Baltoscandia, and North Gondwana using the average linking technique and the Weighted Pair Group Median Average algorithm (Vecoli & Samuelsson 2001*a*). Data obtained herein was used for Pomerania; Eisenack (1939), Jenkins (1967), and Samuelsson *et al.* (2000) for Avalonia; Laufeld (1967), Grahn (1981*a*, *b*, 1982), Podhalańska (1979), Bauert & Bauert (1998), Nõlvak (1999*b*), and Nõlvak *et al.* (1999), for Baltoscandia; Paris (1981, 1987, 1990), Soufiane & Achab (1993), Al-Hajri (1995), Oulebsir & Paris (1995), Paris *et al.* (2000) for North Gondwana.

grounds of chitinozoans alone. Based on the presence of recovered graptolites, among them *Glyptograptus* cf. *teretiusculus*, Bednarczyk (1974) assigned the interval 2602 to 3000 m to his local *teretiusculus–acutus* graptolite biozone with a latest Llanvirn (Llandeilian)/latest Darriwilian to early Caradoc age. We suggest an early Caradoc age based on chitinozoan and graptolite data.

Brda 3 The two fossiliferous samples yielded four significant chronostratigraphic marker species: *Belonechitina robusta* (mid to late Caradoc in Baltoscandia, Nõlvak & Grahn 1993), *S. bulmani* (early Caradoc to late Ashgill, Jansonius 1964; Elaouad-Debbaj 1984), *Fungochitina actonica*? (Burrellian to late Ashgill, Elaouad-Debbaj 1984) and possibly also *Spinachitina cervicornis* (mid Caradoc, Nõlvak & Grahn 1993), recovered in sample Brda 3/2355–2361 and referred to as *S. cervicornis*? A mid Caradoc age, based on the recovered chitinozoan assemblage, is suggested for the interval 2355–2738 m. Bednarczyk (1974) assigned the interval 2153–2901.5 m to his local *bicornis–truncatus* graptolite assemblage biozone, with a mid to late Caradoc age. Szczepanik (2000) ascribed the interval 2153.4–2900.5 m to his local acritarch biozone B, with a latest Llanvirn to mid Caradoc age, although he also stated that evaluation of the stratigraphic succession is difficult here, probably because of overturning. In conclusion, we suggest a mid Caradoc age for the interval investigated for its chitinozoans, with the possibility that deeper parts of the core may be of latest Llanvirn age.

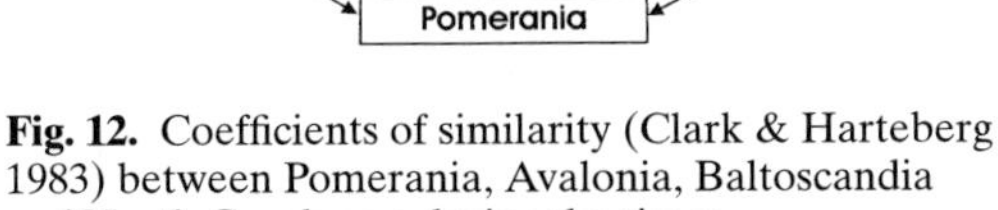

Fig. 12. Coefficients of similarity (Clark & Harteberg 1983) between Pomerania, Avalonia, Baltoscandia and North Gondwana during the times corresponding to *stentor* chitinozoan biozone (latest Llanvirn to early Caradoc age; Nõlvak & Grahn 1993). Data is the same as for the cluster analysis (see legend to Fig. 11 for references).

Chojnice 5 The presence of *Cyathochitina calix* (Eisenack) in the lowest two samples distinguishes that interval, 5052–5054 m, from the overlying interval, 4956.5–5006.75 m. *Cyathochitina calix* is known from upper Arenig to lower Caradoc strata in Baltoscandia (Grahn

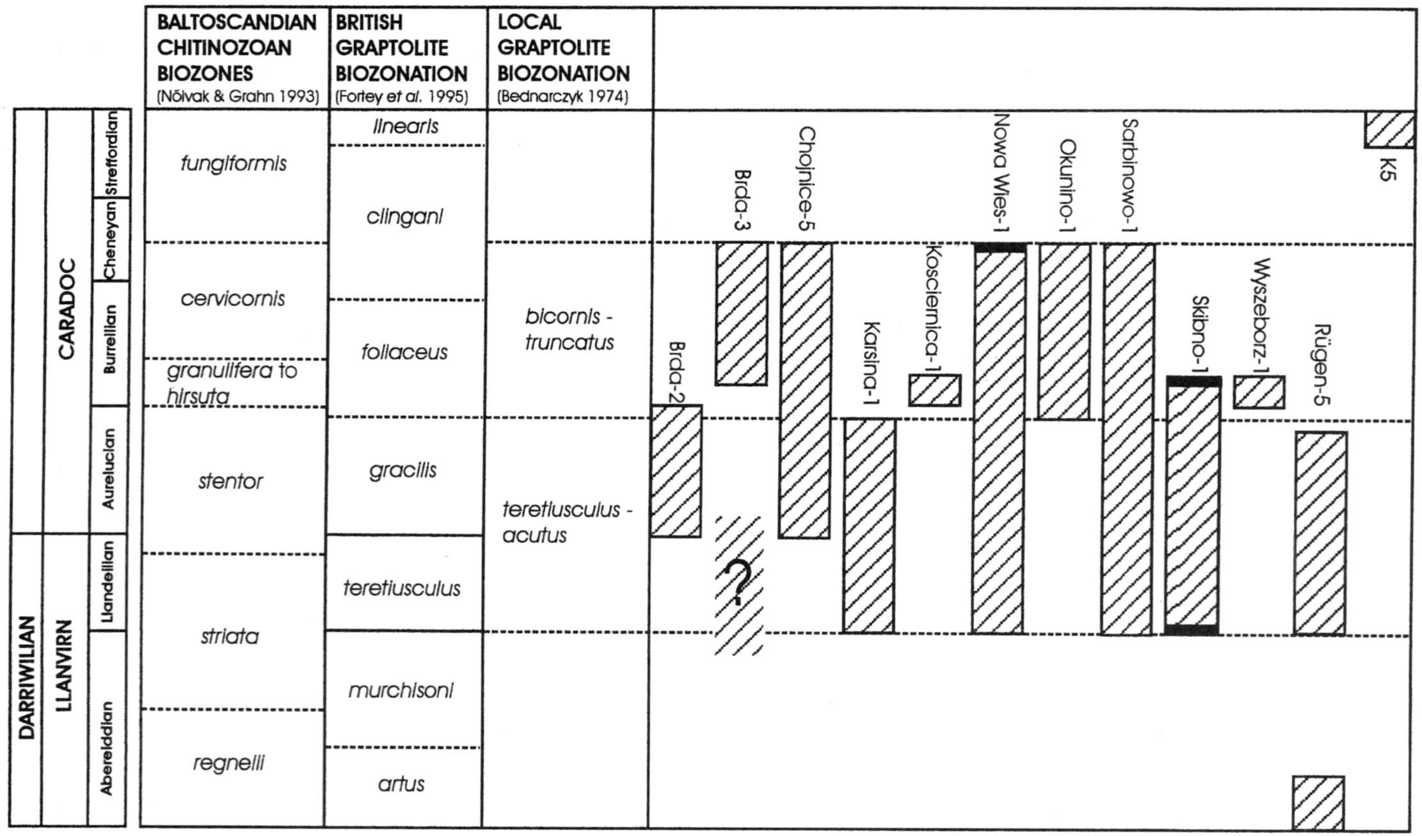

Fig. 13. Simplified stratigraphic outline and the ages as suggested in this paper for the ten west Pomeranian boreholes, and the Rügen 5/66 and K5 boreholes (Maletz 1998; Samuelsson *et al.* 2000; Vecoli & Samuelsson 2001*a*) in relation to the British Series (and the single defined global Stage), British Stages, Baltoscandian chitinozoan biozonation (Nõlvak & Grahn 1993), British graptolite biozonation (Fortey *et al.* 1995), and the local graptolite biostratigraphy (Bednarczyk 1974), respectively. Vertical thicknesses not drawn to scale. The possible age-ranges are drawn; so far more delimited evidence is only available for the boreholes Nowa Wieś 1, and Skibno 1, denoted by thick lines in this diagram.

1982; Nõlvak & Grahn 1993). Other biostratigraphically significant chitinozoans from the interval above, 4956.5–5006.8 m, e.g. *Desmochitina nodosa* Eisenack, point to a mid Caradoc age (Grahn *et al.* 1994). Bednarczyk (1974) assigned the interval 4890.0–5055.5 m to his *bicornis–truncatus* graptolite biozone (mid to late Caradoc age), although no truly biostratigraphically significant graptolite specimens were recovered from the interval 5048.0–5055.5 m.

Based on the abundance of acritarchs belonging to the genus *Veryhachium*, Szczepanik (2000) suggested that the youngest deposits investigated by him in the Pomeranian fold belt occur in the Chojnice 5 borehole, and he proposed a Caradoc to Ashgill age for the interval 4713–5043 m. However, the *Veryhachium* plexus, comprising either quadrangular and triangular forms, is very difficult to treat taxonomically even when well-preserved specimens are available, and it has an uncertain stratigraphic value (e.g. Fensome *et al.* 1990; Sarjeant & Stancliffe 1994). Triangular forms of *Veryhachium* that can be attributed to *V. trispinosum* were present since early Arenig times in South China (Yin Leiming 1995) and the Baltic area (C. Ribecai, pers. comm. 2001). *Veryhachium* is abundantly present with triangular and quadrangular forms in the lower Llanvirn of Öland, Sweden (Ribecai & Tongiorgi 1995). The abundance of *Veryhachium* in low-diversity acritarch assemblages in post-early Arenig times has probably a palaeoecological rather than a chronostratigraphic significance. For these reasons, we consider that the Caradoc–Ashgill chronostratigraphic attribution of the borehole interval 4713–5043 m by Szczepanik (2000) is highly unreliable. Consequently, we suggest that the interval 5052–5054 m is of early Caradoc age, and the overlying interval, 4713–5048 m is of mid Caradoc age.

Karsina 1 The three investigated samples yielded only chitinozoans with broad stratigraphic ranges, indicating an early Llanvirn to late Caradoc age. Graptolites recovered from the interval 3142–3203 m belong to the local *teretiusculus–acutus* biozone of latest Llanvirn (Llandeilian) to early Caradoc age (Bednarczyk 1974). Szczepanik (2000) recovered acritarchs from the interval 3140–3201.5 m which he interpreted as indicative of Llandeilian to early Caradoc ages. However, none of the 14 acritarch species listed by Szczepanik (2000; table 4, p. 282) are characteristic of the Llandeilian–Caradoc interval. The tentative dating by Szczepanik (2000) is apparently based on some general feature of the impoverished acritarch assemblage (e.g. prevalence of large acanthomorphic acritarchs, presence of species of *Multiplicisphaeridium* and *Ordovicidium*) and on resemblance to previously described acritarch assemblages dated as Llandeilian on purely palynological grounds (Górka 1980). The present chitinozoan data are consistent with the late Llanvirn (Llandeilian)–Caradoc age suggested by Bednarczyk (1974) for the studied part of the Karsina 1 borehole.

Kościernica 1 The presence of *Belonechitina robusta* and the possible *Lagenochitina dalbyensis* (referred to as *Lagenochitina dalbyensis*?) are indicative of an early mid Caradoc age (Nõlvak & Grahn 1993) for the interval 2820–2852 m. Bednarczyk (1974) assigned the interval 2818–2853 m to the *bicornis–truncatus* biozone (middle to upper Caradoc). Considering these data, we suggest a mid Caradoc, and probably an early mid Caradoc (Harnagian) age.

Nowa Wieś 1 Only the sample from the deepest part of the drillcore (Nowa Wieś 1/2710–2711) yielded a biostratigraphic marker fossil: *Spinachitina cervicornis* which is known from late Idavere to early Oandu times in Baltoscandia (Nõlvak & Grahn 1993), corresponding to mid Caradoc (Burrellian) time. A sample higher in the borehole (Nowa Wieś 1/2488.2–2490.2) yielded *Desmochitina minor* and *Cyathochitina* sp. A, both with long biostratigraphic ranges, and a few other specimens determined to the generic level. Bednarczyk (1974) assigned the interval 2832–2900 m to the local *teretiusculus–acutus* graptolite biozone (Llandeilian to early Caradoc age), and the overlying 2488.3–2832 m to the *bicornis–truncatus* biozone (mid to late Caradoc age). Szczepanik (2000) recovered acritarchs at 2893 m which he interpreted as typical of Middle Ordovician (upper Arenig to Llanvirn) strata, although this interpretation is highly debatable. At the other two levels investigated by the latter author (i.e. at 2404 m and 2757.5 m) acritarch assemblages are dominated by the genus *Veryhachium*, which, contrary to Szczepanik (2000), are not indicative of any particular age (Fensome *et al.* 1990; Sarjeant & Stancliffe 1994). In conclusion, we propose a Llandeilian age for the interval 2832–2900 m based mainly on graptolite data, and a mid Caradoc age for the interval 2404–2832 m, on the basis of the chitinozoan and graptolite data.

Okunino 1 No chitinozoan data were obtained. The assignment by Bednarczyk (1974) of the investigated 1873.2–1975.9 m to the *bicornis–truncatus* graptolite Biozone (mid to late Caradoc age) is here accepted.

Sarbinowo 1 The available chitinozoan data suggest that the investigated part of the core might belong to two different biostratigraphic intervals. The lower interval (2951–2953 m), characterized by the absence of *Belonechitina robusta*, and the presence of *Eisenackitina*? sp. (in sample Sarbinowo 1/2952–2953 only), yielded chitinozoans that in the absence of any good biostratigraphic marker point to an open early Llanvirn to late Caradoc age (from the presence of *Conochitina chydaea* Jenkins; Paris 1981). The presence of *Belonechitina robusta* in the upper interval, 2845–2912.2 m, indicates a mid to late Caradoc age (Nõlvak & Grahn 1993). Bednarczyk (1974) also recognized two different intervals in his evaluation of the graptolite faunas. He ascribed the interval 2950–3000 m to the local *teretiusculus–acutus* graptolite biozone (Llandeilian to early Caradoc age), and the overlying 2802–2950 m to the *bicornis–truncatus* biozone (mid to late Caradoc age). The biostratigraphic data from the graptolites are thus corroborated by the chitinozoans.

Skibno 1 This borehole yielded the taxonomically most diverse assemblage. Two recovered chitinozoan species are index species in the Baltoscandian chitinozoan biozonation: *Laufeldochitina stentor* and *Lagenochitina dalbyensis* (Nõlvak & Grahn 1993). The lowest sample (Skibno 1/2802.7–2802.8) yielded *Belonechitina micracantha*, with a stratigraphic range spanning the middle Arenig to the Ashgill series (Paris 1981; Elaouad-Debbaj 1984). In samples from the interval above, 2528–2716.97 m, *Spinachitina bulmani* was observed, with a stratigraphic range from the lower Caradoc to the upper Ashgill (Jansonius 1964; Elaouad-Debbaj 1984). Two samples (Skibno 1/2435.2–2435.3, and Skibno 1/2330.7–2331.7) did not yield any biostratigraphically significant taxa. In the interval 1950.22 to 2110.7 m *Laufeldochitina stentor* was observed, indicating a latest Llanvirn to early Caradoc age (Nõlvak & Grahn 1993). In sample Skibno 1/1950.22–1951.14, *L. stentor* was accompanied by abundant and well-preserved *Eisenackitina rhenana*, the index species of a Baltoscandian subzone of Aurelucian age (Nõlvak & Grahn 1993). Specimens referred to *Cyathochitina* cf. *calix* in sample Skibno 1/1913–1914 are indicative of an early Caradoc age, and we suggest that this sample belongs to the *stentor* biozone, and possibly the *rhenana* subzone. The presence of *Belonechitina robusta* in two samples from the interval 1839.4–1880.1 m are indicative of a mid Caradoc age, and we place these two samples as well as the one immediately above (Skibno 1/1795–1796, which only yielded e.g. *Conochitina* cf. *homoclaviformis*), in the same stratigraphic unit as the sample Skibno 1/1745–1746 which yielded abundant *L. dalbyensis*. The latter species has a known biostratigraphic range limited to Harnagian, i.e. lower Burrellian, lower middle Caradoc strata in Baltoscandia (Nõlvak & Grahn 1993). Bednarczyk *et al.* (1999) in their preliminary investigation of the Skibno 1 chitinozoans and acritarchs, recovered *L. stentor* from the interval 2038.6 to 2112.5 m, later confirmed by Wrona *et al.* (2001). Bednarczyk (1974) subdivided Skibno 1 into two stratigraphic intervals, 1802.3–2807 m which he assigned to the local *teretiusculus–acutus* graptolite biozone (Llandeilian to early Caradoc age), and the overlying 1727–1802.3 m to the *bicornis–truncatus* biozone (mid to late Caradoc age), respectively. Acritarchs from the drillcore were interpreted as supporting the more detailed graptolite and chitinozoan biostratigraphy, and so were a few isolated findings of brachiopods and conodonts (Bednarczyk *et al.* 1999; Wrona *et al.* 2001). Based on the new detailed chitinozoan data and supported by earlier, especially graptolite work (Bednarczyk 1974), a broad late Llanvirn (Llandeilian) to early Caradoc age is suggested for the 2330.7–2807 m interval. A latest Llanvirn to early Caradoc age for the 1913–2112.5 m interval, with at least firm evidence for an Aurelucian age at the 1950.22–1951.14 m interval. An early Harnagian age is suggested for the interval 1745 to 1881.1 m.

Wyszebórz 1 Specimens referred to *Lagenochitina dalbyensis*? because of their poor preservation are indicative of an early mid Caradoc age for the entire investigated interval (2863.6–2885.4 m). This age determination is corroborated by the presence of other poorly preserved specimens, attributed to *Belonechitina robusta*? in sample Wyszebórz 1/2884.4–2885.4. Representatives of *Spinachitina bulmani* and *Belonechitina capitata* in the same sample suggest a Caradoc age. Bednarczyk (1974) assigned the entire interval 2753.2–3021.4 m to his *bicornis–truncatus* graptolite biozone (mid to late Caradoc age) and the chitinozoans thus delimit this age determination.

From the available evidence, the youngest Ordovician rocks in the investigated boreholes are of a mid Caradoc age. Firm evidence for this age was recovered in a sample at a depth of 2710–2711 m in the Nowa Wieś 1 borehole, where the Baltoscandian index species *Spinachitina cervicornis* (Eisenack) was observed, having a late Idavere to early Oandu chronostratigraphic range in Baltoscandia (Nõlvak & Grahn 1993), corresponding to the British mid Caradoc (Burrellian) time.

Correlation with other parts of the Heligoland–Pomerania Deformation Belt

It is possible to correlate the rocks penetrated by the Pomerania drillings with other rock units belonging to the HPDB. The most complete Lower Palaeozoic succession in the Rügen area is recovered from the borehole Rügen 5/66, drilled onshore in the northern part of the island (Servais & Molyneux 1997; Maletz 1998; Samuelsson *et al.* 2000). The deepest part could be assigned to the latest Tremadoc (Servais & Molyneux 1997; Samuelsson *et al.* 2000) and no equivalent rocks have been recorded in Pomerania. Higher in the Rügen 5/66 borehole, sediments of an early Abereiddian age are present (Maletz 1998; Samuelsson *et al.* 2000), also without any counterparts in the Pomeranian rocks (Fig. 13). At about 2700–3100 m depth, sediments belonging to the *teretiusculus* graptolite biozone (Llandeilian) occur (Maletz 1998), and these as well as rocks placed within the *gracilis* graptolite biozone, 1558.8–2610.6 m, seem to have direct counterparts in the Pomerania drillings (Brda 2; ?Brda 3; Karsina 1; Nowa Wieś 1; Sarbinowo 1; and Skibno 1, Fig. 13). Other drillings in the Rügen area, both inshore and offshore (Binz 1/73; Lohme 2/70; H2), yielded Ordovician rocks attributable to the *Siphonochitina formosa* chitinozoan biozone of Llanvirn/Darriwilian age (Samuelsson *et al.* 2000; Samuelsson & Servais 2001; Vecoli & Samuelsson 2001*a*), which do not seem to have any documented counterparts in Pomerania. Another borehole, K5, drilled offshore of Rügen to the NW of the Pomerania drillings investigated in this paper, yielded chitinozoans of a mid to late Caradoc age (Vecoli & Samuelsson 2001*a*). The K5 succession thus possibly represents the youngest rocks within the HPDB (Fig. 13).

In South Jylland (Denmark) and Schleswig-Holstein (northwestern Germany), deep boreholes have come upon folded rocks that, although different from those of the Rügen and Pomerania areas and the southern North Sea, have been incorporated in the Danish–North German–Polish Caledonides. Unfortunately, no biostratigraphic data have been obtained (Schleswig-Holstein) or made available (South Jylland). Correlations with the Pomeranian rocks are therefore impossible at present, as radiometric ages only provide an indication of the times of metamorphism and deformation (Franke 1995; Pharaoh 1999).

Presence of potentially coeval strata in Pomerania and the German part of the HPDB is certainly not evidence for any generic relationship between the two areas. However, it demonstrates that stratigraphic correlation between these two areas is possible, and indirectly substantiates the earlier assumption that they belong to the same orogeny.

Palaeobiogeographic evaluation of the Pomerania chitinozoan assemblages

Palaeomagnetic data shows that Avalonia drifted away from North Gondwana towards Baltica during the course of the Ordovician. At the same time Baltica was rotating slowly (Torsvik 1998) while drifting towards the palaeoequator at a more reduced velocity than that of Avalonia. Palaeogeographic reconstructions of southern latitudes show Avalonia at a palaeolatitude of about 60°south during the Arenig and Llanvirn times (e.g. Tait *et al.* 1997) and at about 30°south at the Late Ordovician (Tait *et al.* 1997), with a widening Rheic Ocean separating Avalonia from North Gondwana to the south. There is good evidence for close proximity of Avalonia to Baltica by mid Caradoc times, when faunal and microplankton interchange between the two areas had become significant (Cocks & Fortey 1982, 1990; Samuelsson & Verniers 2000; Vecoli & Samuelsson 2001*a*). Thus, palaeogeographic reconstructions for the Caradoc usually show the easternmost parts of Avalonia close to the southwestern parts of Baltica (e.g. Torsvik 1998).

With the possible exception of those taxa treated under open nomenclature, the Pomerania assemblages yielded chitinozoans well-known from especially Baltoscandia, North Gondwana, and other parts of Avalonia such as Shropshire (UK). The cluster and CS analyses quantify this observation, and show that during *stentor* times (uppermost Llandeilian and Aurelucian stages, i.e. upper Llanvirn/upper Darriwilian to lower Caradoc), the west Pomeranian successions can be identified as Avalonian (Figs 11 and 12). The chitinozoan faunas preserved in Pomerania and the rest of Avalonia are more similar to those of Baltoscandia than to those of North Gondwana at these times (Fig. 11). This supports the palaeogeographical scenario outlined above, and the position of the Avalonia–Baltica suture thus lies to the north of the Koszalin–Chojnice structural zone and the TTZ (Winchester *et al.* 2001).

Timing of the Avalonia–Baltica collision

In the Rügen area to the NW of Pomerania and south of the Thor Suture, the youngest preserved sediments in the deformed Lower Palaeozoic successions are of a mid Caradoc and possibly early Ashgill age (Vecoli & Samuelsson 2001*a*). The presence of reworked acritarchs of Llanvirn age and Avalonian affinity in the Ashgill successions of the southwestern border of Baltica (G-14 borehole) proves that erosion of the HPDB had commenced before these times (Samuelsson *et al.* 2001). Thus, the timing of the closure of the Tornquist Ocean can with a high degree of confidence be assigned to an interval from mid Caradoc to late Ashgill times (Vecoli & Samuelsson 2001*a*). The presence of reworked Middle Ordovician acritarchs of clear Perigondwanan palaeobiogeographical affinity in Lower Silurian strata north of the Thor Suture (the boreholes Slagelse 1 and Pernille 1) in southern Denmark suggests that foreland basin sedimentation was active since at least early Silurian times in this area (Vecoli & Samuelsson 2001*b*). Unfortunately, lack of data from stratigraphically significant intervals makes any evaluation of the timing and large-scale geometry of the Avalonia/Baltica collision impossible.

The demonstrated close proximity of Pomerania, representing the easternmost part of Avalonia, to Baltica in late Llanvirn/late Darriwilian to early Caradoc times corroborates the earlier conclusions that closure of the Tornquist Ocean took place in the Late Ordovician.

Conclusions

A refined biostratigraphy of the investigated boreholes situated in the Koszalin–Chojnice structural Zone, south of the Thor Suture has been possible, in most cases narrowing down the stratigraphic results from the pioneering work of Bednarczyk (1974) on the same boreholes. Straightforward correlation with coeval strata in especially the Rügen area is possible, which is in line with the other evidence favouring the deposition of the Pomerania strata in the same fold belt as the Rügen rocks. Cluster and coefficient of similarity analyses of chitinozoans belonging to the Baltoscandian *stentor* chitinozoan biozone from the Skibno 1 borehole, shows the close similarity of these with contemporary assemblages from other parts of Avalonia (i.e. Shropshire, the Ebbe Anticline and the Brabant Massif). If this is interpreted in terms of palaeolatitudinal distance, the observed pattern supports the Avalonian affinity of the Pomeranian host sediments, and the Avalonia/Baltica suture consequently lies to the north of the Koszalin–Chojnice structural zone. The closer similarity of Avalonian (including Pomeranian) chitinozoan assemblages to those of Baltoscandia than North Gondwana support earlier observations that the Tornquist Ocean separating Avalonia and Baltica had started to narrow drastically in early Caradoc times. The youngest tectonically disturbed sediments encountered in Pomerania (HPDB) are of a Burrellian to Cheneyan (mid Caradoc) age, which puts a lower age limit on the initiation of foreland basin sedimentation on the Baltic platform to the north of present-day Pomerania.

This project was initiated during the postdoctoral studies of J. Samuelsson and M. Vecoli at the universities of Ghent (Belgium), and Halle (Germany), respectively, while recipients of research grants from the EU funded TMR program PACE (Palaeozoic Amalgamation of Central Europe). The project was finished by J.S. at the University of Rennes I (France) with fiscal support of a postdoctoral fellowship from the Swedish government through STINT. F. Paris (Rennes) kindly hosted J.S. during his stay, and read an earlier version of the manuscript. S. Van Cauwenberghe (Ghent) processed the samples for microfossils. J. Verniers was research director of the Fund for Scientific Research (Flanders), Belgium until 2000. S. Molyneux and R. Wrona carefully reviewed this paper.

References

ACHAB, A. 1988. Mise en évidence d'un provincialisme chez les Chitinozoaires ordoviciens. *Canadian Journal of Earth Sciences,* **25**, 635–638.

ACHAB, A. 1991. Biogeography of Ordovician Chitinozoa. *In*: BARNES, C. R. & WILLIAMS, S. H. (eds) *Advances in Ordovician Geology*. Geological Survey of Canada Professional Papers, **90**, 135–142.

AL-HAJRI, S. 1995. Biostratigraphy of the Ordovician Chitinozoa of northwestern Saudi Arabia. *Review of Palaeobotany and Palynology,* **89**, 27–48.

BAUERT, G. & BAUERT, H. 1998. Distribution of chitinozoans in the Tartu (453) core. *In*: PÕLDVERE, A., KLEESMENT, A., PAALITS, I., MEIDLA, T., BAUERT, G., BAUERT, H., STOUGE, S., VALIUKEVICIUS, J. & SAADRE, T. (authors) *Tartu (453) drill core.* Geological Journal of the Geological Survey of Estonia **1**, Appendix 14.

BEDNARCZYK, W. 1974. The Ordovician in the Koszalin-Chojnice region (Western Pomerania). *Acta Geologica Polonica,* **24**, 581–600.

BEDNARCZYK, W. 1999. Chitinozoa-bearing Horizons in the Ordovician of Northern Poland in the Light of Conodont Stratigraphy. *Bulletin of the Polish Academy of Sciences Earth Sciences,* **47**, 1–13.

BEDNARCZYK, W., STEMPIEŃ-SAŁEK, M. & WRONA, R. 1999. Integrated biostratigraphy (graptolite, acritarch and chitinozoan) of the subsurface Caradocian in Pomerania, NW Poland. *Acta Universitatis Carolinae Geologica,* **43**, 53–54.

BERTHELSEN, A. 1992. Mobile Europe. *In*: Blundell, D., Freeman, R., Mueller, S (eds) *A continent revealed: the European geotraverse*. Cambridge University Press, Cambridge, 11–32.

BERTHELSEN, A. 1998. The Tornquist Zone northwest of the Carpathians: an intraplate pseudosuture. *Geologiska Föreningens i Stockholm Förhandlingar,* **120**, 223–230.

BROCHWICZ-LEWIŃSKI, W., POŻARYSKI, W. & TOMCZYK, H. 1981. Mouvements coulisants de grande ampleur au Paleozoique inférieur le long de la marge sud ouest de la plate-forme est-Europeenne. *Comptes Rendus de l'académie des Sciences, A. Sciences de la Terre et des Planètes,* **293**, 855–858.

CLARK, D. L. & HARTEBERG, E. W. 1983. Palaeoenvironmental factors and the distribution of conodonts in Lower Triassic of Svalbard and Nepal. *Fossils and Strata,* **15**, 171–175.

COCKS, L. R. M. & FORTEY, R. A. 1982. Faunal evidence for oceanic separations in the Palaeozoic of Britain. *Journal of the Geological Society, London,* **139**, 465–478.

COCKS, L. R. M. & FORTEY, R. A. 1990. Biogeography of Ordovician and Silurian Faunas. *Geological Society of London Memoir,* **12,** 97–104.

COCKS, L. R. M., MCKERROW, W. S. & VAN STAAL, C. R. 1997. The Margins of Avalonia. *Geological Magazine,* **134**, 627–636.

DADLEZ, R. 1967. New sections of the Zechstein substratum in North-West Poland. *Kwartalnik Geologiczny,* **11** (in Polish with English summary).

DADLEZ, R. 1978. Sub-Permian rock complexes in the Koszalin-Chojnice Zone. *Kwartalnik Geologiczny,* **22**, 269–299 (in Polish with English summary).

DALLMEYER, R. D., GIESE, U., GLASMACHER, U. & PICKEL, W. 1999. First $^{40}Ar/^{39}Ar$ age constraints for the Caledonian evolution of the Trans-European Suture Zone in NE Germany. *Journal of the Geological Society, London,* **156**, 279–290.

EISENACK, A. 1939. Chitinozoen und Hystrichosphaeriden im Ordovicium des Rheinischen Schiefergebirges. *Senckenbergiana,* **21**, 135–152.

EISENACK, A. 1972. Chitinozoen und andere Mikrofossilien aus der Bohrung Łeba, Pommern. *Palaeontographica A,* **139**, 64–87.

ELAOUAD-DEBBAJ, Z. 1984. Chitinozoaires Ashgilliens de l'Anti-Atlas (Maroc). *Geobios,* **17**, 45–68.

FENSOME, R. A., WILLIAMS, G. L., BARSS, M. S., FREEMAN, J. M. & HILL, J. M. 1990. Acritarchs and fossil prasinophytes: an index to genera, species and intraspecific taxa. *Contribution Series of the American Association of Stratigraphic Palynologists Foundation,* **25**, 1–771.

FORTEY, R. A., HARPER, D. A. T., INGHAM, J. K, OWEN, A. W. & RUSHTON, A. W. A. 1995. A revision of Ordovician series and stages from the historical type area. *Geological Magazine,* **132**, 15–30.

FRANKE, D. 1967. Zu den Varisziden und zum Problem der Kaledoniden im nördlichen Mitteleuropa. *Berichte der Deutschen Gesellschaft für Geologische Wissenschaften A,* **12**, 83–140.

FRANKE, D. 1978. Entwicklung und Bau der Paläozoiden im nördlichen Mitteleuropa. Teil 1. *Zeitschrift für Geologische Wissenschaften,* **6**, 5–32.

FRANKE, D. 1990. Der präpermische Untergrund der Mitteleuropäischen Senke. Fakten und Hypothesen. *Niedersächsische Akademie der Wissenschaften, Veröffentlichungen,* **4**, 19–75.

FRANKE, D. 1994. The deformational history of the Caledonian terranes at Baltica's southwest margin. *Zeitschrift für Geologische Wissenschaften,* **22**, 67–80.

FRANKE, D. 1995. The Caledonian terranes along the south-western border of the East European Platform – evidence, speculation and open questions. *In*: GEE, D.G. & BECKHOLMEN, M. (eds) The Trans-European Suture Zone: EUROPROBE in Liblice 1993. *Studii Geophysica et Geodetica,* **39**, 241–256.

FRANKE, D. & ZNOSKO, J. 1988. Einige Fragen der baikalisch-kaledonischen Entwicklung im Gebiet südwestlich der Tornquist-Teisseyre Zone. *Zeitschrift für angewandte Geologie,* **34**, 33–36.

FROST, R. T. C., FITCH, F. J. & MILLER, J. A. 1981. The age and nature of the crystalline basement of the North Sea Basin. *In*: ILLING, L. V. & HOBSON, G. D. (eds) *Petroleum Geology of the Continental Shelf of North-West Europe.* Institute of Petroleum, London, 43–57.

GEE, D. G. & ZEYEN, H. 1996. GEOPROBE 1996 – Lithosphere dynamics – origin and evolution of continents. EUROPROBE Secretariat, Uppsala.

GIESE, U., KATZUNG, G. & WALTER, R. 1994. Detrital composition of Ordovician sandstones from the Rügen boreholes: implications for the evolution of the Tornquist Ocean. *Geologische Rundschau,* **83**, 293–308.

GRAHN, Y. 1981*a*. Ordovician Chitinozoa from the Stora Åsbotorp boring in Västergötland south-central Sweden. *Sveriges Geologiska Undersökning Serie C,* **787**, 1–40.

GRAHN, Y. 1981*b*. Middle Ordovician Chitinozoa from Öland. *Sveriges Geologiska Undersökning Serie C,* **784**, 1–51.

GRAHN, Y. 1982. Caradocian and Ashgillian Chitinozoa from the subsurface of Gotland. *Sveriges Geologiska Undersökning Serie C,* **788**, 1–66.

GRAHN, Y., IDIL, S. & ØSTVEDT, A. M. 1994. Caradocian and Ashgillian chitinozoan biostratigraphy of the Oslo-Asker and Ringerike districts, Oslo Region, Norway. *Geologiska Föreningens i Stockholm Förhandlingar,* **116**, 147–160.

GÓRKA, H. 1980. Le microplancton de l'Ordovicien moyen de Strabla (Pologne). *Acta Paleontologica Polonica,* **25**, 261–277.

HAZEL, J. E. 1970. Binary Coefficient and clustering in biostratigraphy. *Geological Society of America, Bulletin,* **81**, 3237–3252.

JANSONIUS, J. 1964. Morphology and classification of some Chitinozoa. *Bulletin of Canadian Petroleum Geology,* **12**, 901–918.

JENKINS, W. A. M. 1967. Ordovician Chitinozoa from Shropshire. *Palaeontology,* **10**, 436–488.

KATZUNG, G., GIESE, U., WALTER, R. & VON WINTERFELD, C. 1993. The Rügen Caledonides, northeast Germany. *Geological Magazine,* **130**, 725–730.

LAUFELD, S. 1967. Caradocian Chitinozoa from Dalarna, Sweden. *Geologiska Föreningens i Stockholm Förhandlingar,* **89**, 275–349.

MALETZ, J. 1998. Die Graptolithen des Ordoviziums von Rügen (Norddeutschland, Vorpommern). *Paläontologisches Zeitschrift,* **72,** 351–372.

MCKERROW, W. S. & COCKS, L. R. M. 1986. Oceans, island arcs, and olistostromes: the use of fossils in distinguishing sutures, terranes and environments around the Iapetus Ocean. *Journal of the Geological Society, London,* **143**, 185–191.

MCKERROW, W. S. & MAC NIOCAILL, C. & DEWEY, J. F. 2001. The Caledonian Orogeny redefined. *Journal of the Geological Society, London,* **157**, 1149–1154.

MILLER, M. A. 1996. Chitinozoa. *In*: JANSONIUS J. & MCGREGOR D. C. (eds) Palynology: principles and applications. *American Association of Stratigraphic Palynologists Foundation,* **1**, 307–336.

MODLIŃSKI, Z. 1968. Ordovician in West Pomerania. *Kwartalnik Geologiczny,* **12**, 488–492 (In Polish).

NÕLVAK, J. 1999*a*. Ordovician chitinozoan biozonation of Baltoscandia. *Acta Universitatis Carolinae Geologica,* **43**, 287–290.

NÕLVAK, J. 1999*b*. Distribution of chitinozoans in the Taga-Roostoja (25A) core. *In*: PÕLDVERE, A., MENS, K., KOPPELMAA, H., VIIRA, V., MÄNNIK, P., NÕLVAK, J, & KIVISILLA, J. (authors) *Taga-Roostoja (25A) drill core.* Geological Journal of the Geological Survey of Estonia **2**, Appendix 6.

NÕLVAK, J. & GRAHN, Y. 1993. Ordovician Chitinozoa zones from Baltoscandia. *Review of Palaeobotany and Palynology,* **79**, 245–269.

NÕLVAK, J., GRAHN, Y. & STURKELL, E. F. F. 1999. Chitinozoan biostratigraphy of the Middle Ordovician Dalby Limestone in the Fjäcka section, Siljan District, Sweden. *Proceedings of the Estonian Academy of Sciences,* **48**, 75–85.

ORR, P. J. 1996. The ichnofauna of the Skiddaw Group (early Ordovician) of the Lake District, England. *Geological Magazine,* **133**, 193–216.

OULEBSIR, L. & PARIS, F. 1995. Chitinozoaires ordoviciens du Sahara algérien: biostratigraphie et affinités paléogéographiques. *Review of Palaeobotany and Palynology,* **86**, 49–68.

PARIS, F. 1981. Les Chitinozoaires dans le Paleozoïque du sud-ouest de l'Europe. *Mémoires de la Société géologique et minéralogique de Bretagne,* **26**, 1–412.

PARIS, F. 1987. Succession lithostratigraphique du Paléozoïque de l'Aquitaine nord. *Géologie profonde de la France. Thème 7, Bassins Paléozoïques cachés sous l'Aquitaine,* **144**, 3–21.

PARIS, F. 1990. The Ordovician Chitinozoa biozones of the Northern Gondwana Domain. *Review of Palaeobotany and Palynology,* **66**, 181–209.

PARIS, F. 1993. Evolution paléogéographique de l'Europe au Paléozoïque inférieur: le test des chitinozoaires. *Comptes Rendus de l'académie des Sciences, A. Sciences de la Terre et des Planètes,* **316**, 273–280.

PARIS, F. 1996. Chitinozoan biostratigraphy and palaeoecology. *In*: JANSONIUS, J. & MCGREGOR, D. C. (eds) *Palynology: principles and applications.* American Association of Stratigraphic Palynologists Foundation, **2**, 531–552.

PARIS, F. & NÕLVAK, J. 1999. Biological interpretation and paleobiodiversity of a cryptic fossil group: the 'chitinozoan animal'. *Geobios,* **32**, 315–324.

PARIS, F., VERNIERS, J. & AL-HAJRI, S. 2000. Ordovician Chitinozoans from Central Saudi Arabia. *In*: AL-HAJRI, S. & OWENS, B. (eds) *Stratigraphic Palynology of the Palaeozoic of Saudi Arabia.* GeoArabia Special Publication, **1**, 42–56.

PHARAOH, T. C. 1999. Palaeozoic terranes and their lithospheric boundaries within the Trans-European Suture Zone (TESZ): a review. *Tectonophysics,* **314**, 17–41.

PHARAOH, T. C., ENGLAND, R. W. & LEE, M. K. 1995. The concealed Caledonide basement of Eastern England and the southern North Sea – a review. *In*: GEE, D. G. & BECKHOLMEN, M. (eds) The Trans-European Suture Zone: EUROPROBE in Liblice 1993. *Studii Geophysica et Geodetica,* **39**, 85–97.

PHARAOH, T. C., ENGLAND, R. W., VERNIERS, J. & ŻELAŹNIEWICZ, A. 1997. Introduction: Geological and geophysical studies in the Trans-European Suture Zone. *Geological Magazine,* **134**, 585–590.

PLAYFORD, G., RIBECAI, C. & TONGIORGI, M. 1995. Ordovician genera *Peteinosphaeridium, Liliosphaeridium,* and *Cycloposphaeridium*: morphology, taxonomy, biostratigraphy, and palaeogeographic significance. *Bollettino della Società Paleontologica Italiana,* **34**, 3–54.

PODHALAŃSKA, T. 1979. The correlative significance of the Chitinozoa-bearing horizon in the Caradocian profiles of the Łeba region (NW Poland). *Acta Geologica Polonica,* **29**, 441–448.

POŻARYSKI, W. 1987. Geology of the sub-Permian trough and its basement. *In*: RACZYNSKA, A. & POŻARYSKI, W. (eds) *Budowa Geologiczna Polski, Tektonika, Warszawa.* Prace Instytutu Geologicznego, **119**, 174–186. (In Polish).

POŻARYSKI, W. & KOTAŃSKI, Z. 1978. Baikalian, Caledonian and Variscan events in the forefield of the East-European Platform. *Zeitschrift der Deutschen Geologischen Gesellschaft,* **129**, 391–402.

RIBECAI, C. & TONGIORGI, M. 1995. Arenigian acritarchs from Horns Udde (Öland, Sweden): a preliminary report. *Review of Palaeobotany and Palynology,* **86**, 1–11.

SAMUELSSON, J. 2002. Chitinozoans from deep drillholes in west Pomerania, Poland. *Review of Palaeobotany and Palynology,* in press.

SAMUELSSON, J. & SERVAIS, T. 2001. Chitinozoa biostratigraphy of subsurface Ordovician sediments from the Lohme 2/70 well, Island of Rügen (NE-Germany). *Neues Jahrbuch für Geologie und Paläontologie, Abhandlungen,* **222**, 73–90.

SAMUELSSON, J. & VERNIERS, J. 2000. Ordovician chitinozoan biozonation of the Brabant Massif, Belgium. *Review of Palaeobotany and Palynology,* **113**, 131–143.

SAMUELSSON, J., VERNIERS, J. & VECOLI, M. 2000. Chitinozoan faunas from the Rügen Ordovician (Rügen 5/66 and Binz 1/73 wells), NE Germany. *Review of Palaeobotany and Palynology,* **133**, 105–129.

SAMUELSSON, J., VECOLI, M. & BEIER, H. 2001. Ordovician–Silurian palynostratigraphy (Chitinozoa and acritarchs) of the G 14–1/86 borehole, southern Baltic Sea. *Neues Jahrbuch für Geologie und Paläontologie, Abhandlungen,* in press.

SAMUELSSON, J., GERDES, A., KOCH, L., SERVAIS, T. & VERNIERS, J. 2002. Chitinozoa and Nd isotope stratigraphy of the Ordovician rocks in the Ebbe Anticline, NW Germany. *In*: WINCHESTER, J. A., PHARAOH, T. C. & VERNIERS, J. (eds) *Palaeozoic Amalgamation of Central Europe*. Geological Society, London, Special Publications, **201**, 115–131.

SARJEANT, W. A. S. & STANCLIFFE, R. P. 1994. The *Micrhystridium* and *Veryhachium* complexes (Acritarcha: Acanthomorphitae and Polygonomorphitae): a taxonomic reconsideration. *Micropaleontology,* **40**, 1–77.

SCHLÜTER, H. U., BEST, G., JÜRGENS, U. & BINOT, F. 1997. Interpretation reflexionsseismischer Profile zwischen baltischer Kontinentalplatte und kaledonischem Becken in der südlichen Ostsee – erste Ergebisse. *Zeitschrift der deutschen geologischen Gesellschaft,* **148**, 1–32.

SERVAIS, T. & MOLYNEUX, S. G. 1997. The *messaoudensis-trifidum* acritarch assemblage (Ordovician: late Tremadoc – early Arenig) from the subsurface of Rügen (Baltic Sea, NE Germany). *Palaeontographia Italica,* **84**, 113–161.

SOUFIANE, A. & ACHAB, A. 1993. Quelques assemblages de chitinozoaires de l'Ordovicien du Maroc, bassin de Tadla. *Geobios,* **26**, 535–553.

SZCZEPANIK, Z. 2000. The Ordovician acritarchs of the Pomeranian Caledonides and their foreland – similarities and differences. *Geological Quarterly,* **44**, 275–295.

TAIT, J. A., BACHTADSE, V., FRANKE, W., SOFFEL, H. C. 1997. Geodynamic evolution of the European Variscan Foldbelt: Palaeomagnetic and geological constraints. *Geologische Rundschau,* **86**, 585–598.

TELLER, L. 1974. The Silurian of the margin of the East European platform in the region of Miastko-Chojnice (NW Poland). *Acta Geologica Polonica,* **24**, 563–579.

TONGIORGI, M. & DI MILIA, A. 1999. Differentiation and spread of the Baltic Acritarch Province (Arenig-Llanvirn). *In*: TONGIORGI, M. & PLAYFORD, G. (eds) Studies in Palaeozoic palynology. Selected papers from the CIMP Symposium at Pisa, 1998. *Bollettino della Società Paleontologica Italiana,* **38,** 297–312.

TORSVIK, T. H. 1998. Palaeozoic palaeobiogeography: a North Atlantic viewpoint. *Geologiska Föreningens i Stockholm Förhandlingar,* **120**, 109–118.

UNRUG, R., HARAŃCZYK, C. & CHOCYK-JAMIŃSKA, M. 1999. Easternmost Avalonian and Armorican-Cadomian terranes of central Europe and Caledonian-Variscan evolution of the polydeformed Kraków mobile belt: geological constraints. *Tectonophysics,* **302**, 133–157.

VECOLI, M. 1999. Cambro-Ordovician palynostratigraphy (acritarchs and prasinophytes) and palaeogeography of the Hassi-R'Mel area and northern Rhadames Basin, North Africa. *Palaeontographia Italica,* **86**, 1–112.

VECOLI, M. & SAMUELSSON, J. 2001*a*. Quantitative evaluation of microplankton palaeobiogeography in the Ordovician-Early Silurian of the northern TESZ (Trans European Suture Zone): implications for the timing of the Avalonia-Baltica collision. *Review of Palaeobotany and Palynology,* **115**, 43–68.

VECOLI, M. & SAMUELSSON, J. 2001*b*. Reworked acritarchs as provenance indicators in the Lower Palaeozoic of Denmark. *Comptes Rendus de l'académie des Sciences, A. Sciences de la Terre et des Planètes,* **332**, 465–471.

WINCHESTER, J. A. & THE PACE TMR NETWORK TEAM. 2002. Palaeozoic amalgamation of central Europe: New results from recent geological and geophysical investigations. *Tectonophysics,* in press.

WRONA, R. 1980. Microarchitecture of the chitinozoan vesicles and its paleobiological significance. *Acta Palaeontologica Polonica,* **25**, 123–163.

WRONA, R., BEDNARCZYK, W. S. & STEMPIEŃ-SAŁEK, M. 2001. Chitinozoans and acritarchs from the Ordovician of Skibno 1 borehole, Pomerania, Poland: implications for stratigraphy and palaeogeography. *Acta Geologica Polonica,* **51**, 317–331.

YIN LEIMING. 1995. Early Ordovician acritarchs from the Hunjiang, Jilin and Yichang regions, Hubei, China. *Palaeontologica Sinica A,* **185**, 107–170.

ZIEGLER, P. A. 1988. Geodynamic model for the Palaeozoic crustal consolidation of western and central Europe. *Tectonophysics,* **126**, 303–328.

Chitinozoa and Nd isotope stratigraphy of the Ordovician rocks in the Ebbe Anticline, NW Germany

J. SAMUELSSON[1,2,3], A. GERDES[4], L. KOCH[5], T. SERVAIS[6] & J. VERNIERS[1]

[1]*Research Unit Palaeontology, Department of Geology and Pedology, Ghent University, Krijgslaan 281 / S 8, B-9000 Ghent, Belgium*

[2]*Géosciences-Rennes, UMR 6118 du CNRS, Université de Rennes I, F-350 42 Rennes-cedex, France*

[3]*Present address: Institute of Earth Sciences, Historical Geology and Palaeontology, Norbyvägen 22, S-754 36 Uppsala, Sweden (e-mail: joakim.samuelsson@geo.uu.se)*

[4]*NERC Isotope Geoscience Laboratory, Kingsley Dunham Centre, Keyworth, Notts NG 12 5GG, UK*

[5]*Heinrich-Heine-Straße 5, D-58256 Ennepetal, Germany*

[6]*USTL, UPRESA 8014 du CNRS, Bât. SN5, F-59655 Villeneuve d'Ascq Cedex, France*

Abstract: Strongly tectonized Ordovician rocks appear in the Ebbe Anticline (Rheinisches Schiefergebirge), West Germany. These fine-grained detrital rocks of the Herscheider Schichten are divided into the Plettenberger Bänderschiefer, Unterer (Kiesberter) Tonschiefer, (Rahlenberger) Grauwackenschiefer, and the Oberer (Solinger) Tonschiefer. The scope of this investigation was to improve the dating of the entire Ordovician succession, but especially the Oberer (Solinger) Tonschiefer. We used chitinozoans, Palaeozoic microfossils of high biostratigraphic value, and Nd isotopes, which previously have been used for correlation and terrane affinity analysis. Chitinozoan preservation is poor, but some taxa could be confidentially identified to the species level. The εNd(t) values obtained from the Ordovician succession range from –8.0 to –9.2. Joint evaluation of chitinozoan and Nd isotope data together with previously known age-ranges suggest the following ages for the Herscheider Schichten: Plettenberger Bänderschiefer (early Abereiddian, earliest Llanvirn), Unterer (Kiesberter) Tonschiefer (early to mid Abereiddian, early Llanvirn), (Rahlenberger) Grauwackenschiefer (Aurelucian, earliest Caradoc), and Oberer (Solinger) Tonschiefer (late Caradoc). The Ebbe εNd(t) values are most readily compared with εNd(t) values from Avalonia, and we therefore support the inclusion of the Ordovician rocks of the Ebbe Anticline in that palaeocontinent.

The Rheno-Hercynian belt to the south and east of the Ardennes has few outcrops in its geographic extension from the Rheinisches Schiefergebirge through the Harz Mountains in central Europe. Ordovician rocks appear in small isolated outcrops located in the northeastern part of the Rheinisches Schiefergebirge, West Germany in the Ebbe and Solingen–Remscheid–Altena anticlines in the Sauerland region, east of Cologne (Fig. 1). Near the towns of Plettenberg and Herscheid, about 25 km SE of Hagen, in the Ebbe Anticline a strongly tectonized succession of monotonous black and grey mudstones and siltstones occur. These rocks were referred to the Herscheider Schichten by Fuchs (1912, 1935). Similar rocks crop out 3 km SW of Solingen (Solingen–Remscheid–Altena Anticline). In the Ordovician successions that constitute the Ebbe Anticline e.g. trilobite faunas comparable to the Mediterranean faunal province occur (Siegfried 1969; Koch 1999*a*). The Ebbe Anticline sediments are usually interpreted as belonging to the palaeocontinent Avalonia, together with most of the autochthonous parts of the Rheno-Hercynian Belt (e.g. Pharaoh 1999).

The Herscheider Schichten of the Ebbe Anticline are subdivided in four lithological units, from bottom to top: the Plettenberger Bänderschiefer, the Unterer (Kiesberter) Tonschiefer, the (Rahlenberger) Grauwackenschiefer and the Oberer (Solinger) Tonschiefer (Beyer 1941*a*,*b*,*c*). The ages of the two lower units, the Plettenberger Bänderschiefer and the Unterer (Kiesberter) Tonschiefer, are fairly well-constrained by graptolites, and the age of the (Rahlenberger) Grauwackenschiefer is constrained by previously known chitinozoan data,

From: WINCHESTER, J. A., PHARAOH, T. C. & VERNIERS, J. 2002. *Palaeozoic Amalgamation of Central Europe*. Geological Society, London, Special Publications, **201**, 115–131. 0305-8719/02/$15.00

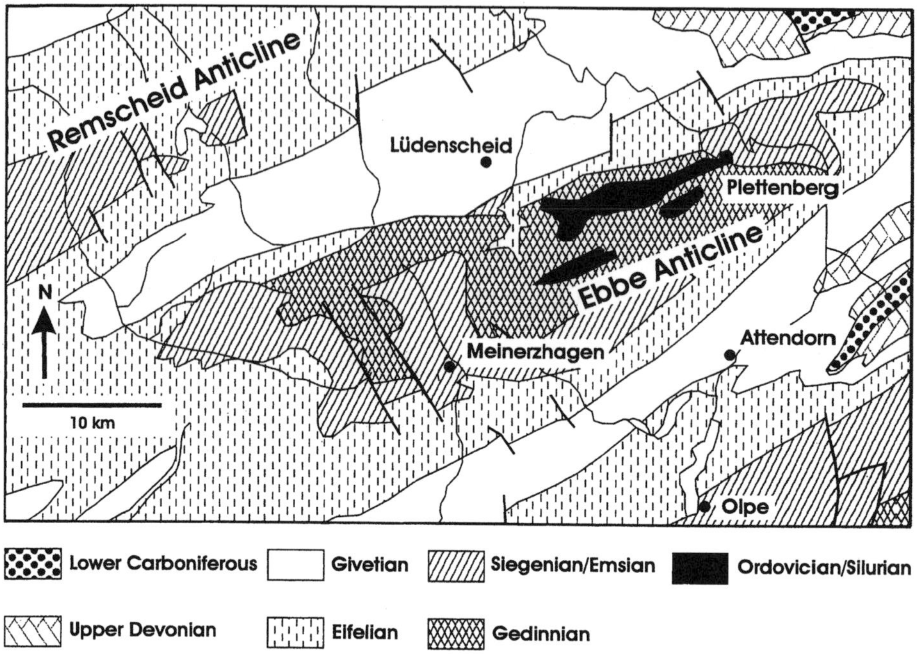

Fig. 1. Geological overview map of the Ebbe Anticline and its surroundings in the Rheinisches Schiefergebirge. Modified from Koch *et al.* (1990).

but the age of the Oberer (Solinger) Tonschiefer is only broadly assumed to be Late Ordovician.

Chitinozoans are organic-walled Palaeozoic microfossils of unknown biological affinity that are extremely useful for biostratigraphy, but also palaeobiogeography (e.g. Miller 1996; Paris 1996). A number of studies carried out in recent years has demonstrated the applicability of this fossil group for detailed stratigraphy and palaeobiogeography also for successions deposited on Eastern Avalonia (Samuelsson & Verniers 2000; Samuelsson *et al.* 2000, 2002; Samuelsson & Servais 2001; Vecoli & Samuelsson 2001).

Results from Nd-isotope studies can, under certain circumstances, act as a powerful tool for correlation and dating of clastic sedimentary rocks. Furthermore, Nd-isotope ratios of fine-grained clastic sedimentary rocks reflect their average detritus composition and can help to identify potential source regions from which they were originally derived (e.g. Thorogood 1990; Evans 1992). Changes in the supply of detritus due to orogenic and magmatic activity will cause variation of the sedimentary Nd-isotope composition through time. These changes, if significant enough, can be traced over wide areas, e.g. a micro-continent (Gerdes *et al.* 2001) or even a major continent (Patchett *et al.* 1999). Avalonia and some Variscan crustal blocks, where the latter are sometimes grouped together as the Armorican Terrane Assemblage (ATA; Tait *et al.* 1997), are a result of the Early Palaeozoic break-up of northern Gondwana. The geological record indicates differences in their rift, drift and subsequent amalgamation history and their crustal provenance (e.g. Nance & Murphy 1996; Tait *et al.* 1997; Winchester *et al.* 2002). Distinct supply of juvenile detritus to the marginal basins and distinct basement lithologies yield characteristic Nd-isotope trends of these terranes. Eastern Avalonia probably has the best studied and dated sediment successions and the most complete Nd-isotope data set. New and published analyses of about 200 Cambrian to Silurian sediments from the Brabant Massif, the Ardennes, the Welsh Basin, the Welsh Borderland and the English Lake District show that Nd isotopes can be used as a powerful tool to correlate different well-dated sedimentary successions of a specific palaeogeographic unit (Gerdes *et al.* 2001). Local variation can be seen usually only in the amount of juvenile, volcanogenic detritus supplied during the Ordovician drift stage. Thus, when a specific isotope trend from a terrane is known, Nd isotopes may even give some constraints for the deposition time of poorly dated terrigenous sediments.

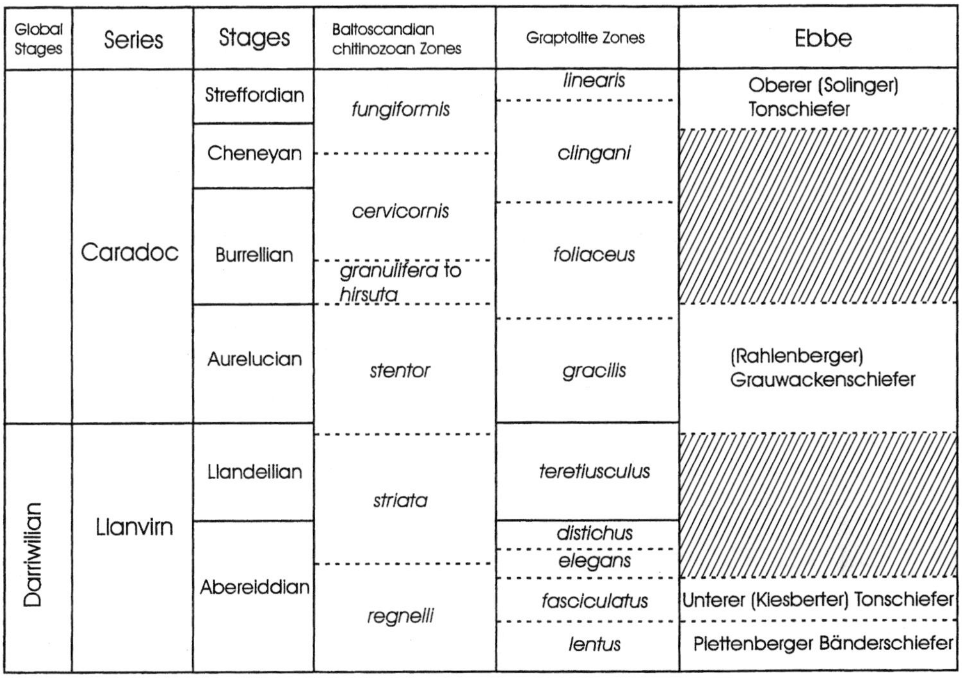

Fig. 2. Simplified stratigraphic outline and the ages as suggested in this paper of the four units of the Herscheider Schichten in relation to British Series (and the single defined global Stage), British Stages, Baltoscandian chitinozoan biozonation (Nõlvak & Grahn 1993), and graptolite biozonation (Fortey *et al.* 1995; Maletz 1995), respectively. Stratigraphical terminology follows Eiserhardt *et al.* (2001). Vertical thickness not drawn to scale. Oblique lines indicate lack of evidence for deposition.

The scope of the present study is to place age constraints on the Oberer (Solinger) Tonschiefer, and also to provide further stratigraphic constraints on the previously suggested ages of the other Ordovician units of the Ebbe Anticline. By doing this, we will obtain further indirect evidence for the deposition of these sediments on the microcontinent Avalonia. Based on chitinozoan and Nd-isotope data, we suggest a late Caradoc age for the Oberer (Solinger) Tonschiefer, and confirm earlier suggested ages for the other units.

Previous palaeontological studies

Whereas a number of papers have reported on fossils from the Ebbe Anticline, e.g. trilobites (e.g. Richter & Richter 1937, 1954; Siegfried 1969; Koch & Lemke 1994, 1995*a*,*b*, 1997, 1998*a*,*b*, 2000; Koch 1999*a*,*b*), graptolites (e.g. Jentsch & Stein 1961; Maletz & Servais 1993), phyllocarids (Beyer 1941*a*,*b*,*c*; Koch & Brauckmann 1998), foraminiferans (Riegraf & Niemeyer 1996), ostracodes (Schallreuter 1996; Schallreuter & Koch 1999), and trace fossils (Richter & Richter 1939*a*,*b*, 1941; Beyer 1941*a*; Eiserhardt *et al.* 2001) very few papers deal with the organic-walled microfossil assemblages. Eisenack (1939) demonstrated the presence of acritarchs and chitinozoans from siliceous concretions ('Kieselknollen', or 'Kieselgallen') in an unknown level in the (Rahlenberger) Grauwackenschiefer, that has now been built over, in the town of Rahlenberg (Herscheid). His paper is the only previous publication dealing in detail with chitinozoans from the Ebbe Anticline and he reported and originally described a number of chitinozoan species in that study. Acritarchs from the Plettenberger Bänderschiefer and the Unterer (Kiesberter) Tonschiefer were dealt with in Maletz & Servais (1993).

Geological overview and previously obtained ages

A graphic summary of the stratigraphy of the Herscheider Schichten can be found in Figure 2. After the original studies by Beyer in the early

Ebbe	unit	? *Belonechitina* ? *capitata*	*Belonechitina capitata*	*Cyathochitina* sp. A	*Belonechitina* ? *micracantha*	*Desmochitina cocca*	*Conochitina* ? *minnesotensis*	*Desmochitina minor*	*Pistillachitina* ? *pistillifrons*	*Rhabdochitina* ? *magna*	*Euconochitina* ? *conulus*	*Desmochitina* cf. *piriformis*	*Conochitina* sp. A	*Conochitina* ?sp. A	*Cyathochitina calix*	*Rhabdochitina* ? *gracilis*	*Laufeldochitina* ? *clavata*	*Cyathochitina* ? *campanulaeformis*	*Cyathochitina campanulaeformis*	*Lagenochitina* ? *ponceti*	CONOCHITININAE indet.	? *Lagenochitina* spp.	*Conochitina* spp.	*Desmochitina* spp.	? *Rhabdochitina* spp.	*Belonechitina* spp.	*Cyathochitina* spp.	? *Conochitina* spp.	? *Cyathochitina* spp.	*Euconochitina* spp.	*Lagenochitina* spp.	? *Laufeldochitina* spp.	*Rhabdochitina* spp.	? *Eisenackitina* spp.	? *Belonechitina* spp.	? *Pistillachitina* sp.	? *Euconochitina* sp.	undetermined	total number of Chitinozoa	sample weight (g)	Chitinozoa / g sample
EA99-1	Oberer (Sol.)	4	1	3	1	2	2														2	2	2	2														15	36	144	0.2
EA99-2	Tonschiefer		1				1	4													3		2		1	1	1											9	23	101	0.2
EA99-5																																							0	12	0
EA99-6	(Rahlenberger)							2	3														1					1										1	8	58	0.1
EA99-15	Grauwacken				1																							1	1									3	6	54	0.1
EA99-16	Schiefer																																						0	12	0
EA99-17						1				3	1										4		3				2											10	24	54	0.4
EA99-3A	U. Tonsch./																										2											4	6	54	0.1
EA99-4	Grauwackens.	1										2													1			6										2	12	52	0.2
EA99-9			23			4			3				5	74	8	16					21		100			1				10	1							27	293	13	22
EA99-10	Unterer																												2		1	3						5	11	13	0.8
EA99-12	(Kiesberter)																					2	24					1				1						10	38	12	3.3
EA99-13	Tonschiefer					4											5				12		3							7	3							15	49	12	4.0
EA99-14						3											13				27		10				1			2	12		1	3				48	120	14	8.5
EA99-18																						2	4												1			2	9	13	0.7
EA99-7									12						1			1				1	6		1					1								7	30	12	2.5
EA99-8	Plettenberger																		4			1	1		2											1	1	1	11	12	0.9
EA99-19	Bändersch.	1			1	3					2						1				11		14				1		2		2			2				16	56	12	4.5
EA99-20			1								2									5	10		11	4						7	3			10				13	66	13	5.1

Fig. 3. Chitinozoan taxa recovered from the Herscheider Schichten, Ebbe Anticline, in absolute numbers.

1940s, a multidisciplinary group of geologists from the University of Hamburg investigated the Ebbe Anticline in the late 1970s (Degens *et al.* 1981). The history of Ebbe Anticline research and the geology of the Ebbe Anticline Ordovician successions have recently been reviewed by Maletz (2000) and Eiserhardt *et al.* (2001). In the latter paper, the terminology is reviewed according to the modern usage of stratigraphical nomenclature. This new stratigraphical terminology is also adopted here (see Fig. 2).

The Ebbe Anticline rocks, referred to as the Herscheider Schichten, consist of a monotonous thick clastic succession of shales and siltstones without any carbonate content in total some 800 m thick (Timm 1981). The lithological monotony of the succession makes it difficult to distinguish between the four different lithological units that Beyer (1941*a,b,c*) established (Degens *et al.* 1981). The base of the Ordovician rocks is not known, and the contact with the overlying late Silurian and Devonian rocks is obscure (Degens *et al.* 1981).

The Plettenberger Bänderschiefer is up to 65 m thick and is a monotonous dark grey to bluish compact shale with abundant thin silty layers where pyrite is common (Eiserhardt *et al.* 1981). The overlying 150–200 m thick Unterer (Kiesberter) Tonschiefer incorporate dark grey to bluish and black shale with infrequent sandy layers. The (Rahlenberger) Grauwackenschiefer is about 300 m thick, and consists of massive black to grey and bluish-grey silty shale with sandy layers and true thin-bedded greywacke. In the upper part, siliceous concretions are common in places. Presumably the youngest of the Herscheider Schichten units is the Oberer (Solinger) Tonschiefer, a dark grey to bluish and black shale with coarser silty or sandy bands, which reaches some 200 m thickness in the Ebbe Anticline.

Graptolites from the Plettenberger Bänderschiefer belong to the *Holmograptus lentus* Zone and indicate an early Abereiddian (earliest Llanvirn/mid Darriwilian) age. This age assignment is corroborated by the presence of early Llanvirn acritarchs (Maletz & Servais 1993). The overlying Unterer (Kiesberter) Tonschiefer yielded graptolites typical of the *Nicholsonograptus fasciculatus* Zone, which is stratigraphically positioned immediately above the *lentus* Zone, also indicative of an Abereiddian age, although somewhat younger. In the Unterer (Kiesberter) Tonschiefer, early Llanvirn/mid Darriwilian acritarchs occur as well (Maletz & Servais 1993). Hitherto, the youngest confidentially dated strata in the Herscheider Schichten are the (Rahlenberger) Grauwackenschiefer, which yielded the biostratigraphically significant chitinozoan species *Laufeldochitina stentor* (Eisenack) and *Eisenackitina rhenana* (Eisenack) from siliceous concretions within the Grauwackenschiefer (Eisenack 1939). In Baltoscandia, *E. rhenana* is the index species of a subzone within the *L. stentor* biozone which has an age limited to the Kukruse Stage (Nõlvak & Grahn 1993), corresponding to the Aurelucian Stage (lowermost Caradoc) in Britain. Ages obtained from poorly preserved graptolites in the (Rahlenberger) Grauwackenschiefer are less precise, and indicate a Llandeilian (latest Llanvirn/latest Darriwilian) or younger age (Maletz 2000). For the Oberer (Solinger) Tonschiefer an undiagnosed post-Llanvirn Ordovician age was suggested, based on the presence of a few trilobites (Koch 1999*a*; Maletz 2000).

Material and methods

The studied material consists of 19 outcrop samples collected by the authors in November 1999 from the four lithological units that together constitute the Herscheider Schichten in the Ebbe Anticline (Fig. 1). The samples were not taken in strict stratigraphic order, but can be described as spot samples. Samples are referred in the text with the field sample numbers (Fig. 3). For the exact location of the samples, see the Appendix. For the Nd-isotope analyses, two samples from the Oberer (Solinger) Tonschiefer were analysed and one sample each from the other three units (Table 1).

Palynological extraction

Laboratory extraction and concentration of palynomorphs were carried out using the standard palynological method described by Paris (1981). Slides for optical microscopy were prepared using Canada Balsam as the embedding medium. Scanning electron microscope (SEM) study and photography of chitinozoans was carried out at the Department of Geology and Pedology, Ghent University, Belgium, with a JEOL scanning microscope 6400 at 10 kV. All illustrated specimens are deposited in the collection of the Research Unit Palaeontology, Department of Geology and Pedology, Ghent University, Belgium.

Nd-isotopes

Approximately 200 mg of powdered sample were combined with a ^{149}Sm–^{150}Nd mixed spike, digested at *c.* 120 °C in a HF-HNO_3 mixture over 4 days, evaporated to dryness, attacked with

Table 1. *Nd-isotope values obtained from the Herscheider Schichten, Ebbe Anticline, Germany*

Sample	Formation	Age* (Ma)	Sm (ppm)	Nd (ppm)	^{147}Sm/^{144}Nd	^{143}Nd/^{144}Nd ($\pm 2\sigma_m$)	^{143}Nd/^{144}Nd Initial	εNd Initial
EA99-1	Oberer (Solinger) Tonschiefer	452	6.752	40.55	0.1007	0.511911 ± 0.000003	0.511613	–8.7
EA99-2	Oberer (Solinger) Tonschiefer	452	6.768	40.67	0.1006	0.511893 ± 0.000003	0.511595	–9.0
EA99-5	(Rahlenberger) Grauwackenschiefer	457	8.244	46.75	0.1066	0.511956 ± 0.000003	0.511637	–8.0
EA99-9	Unterer (Kiesberter) Tonschiefer	465	9.787	50.65	0.1168	0.511927 ± 0.000003	0.511571	–9.1
EA99-19	Plettenberger Bänderschiefer	465	9.307	49.19	0.1144	0.511917 ± 0.000003	0.511569	–9.2

*Estimated ages (see text) using time scale proposed by McKerrow & van Staal (2000).

HNO_3 and after evaporation re-dissolved overnight in 6 M HCl, once again evaporated to dryness and finally dissolved in 2.5 M HCl. Sm and Nd concentrations and Nd isotope ratios were analysed in static mode on a Finnegan MAT 262 multicollector mass spectrometer at the NERC Isotope Geosciences Laboratory (NIGL) after standard chromatographic separation. ^{143}Nd/^{144}Nd was normalized to a ^{146}Nd/^{144}Nd ratio of 0.7219. Nd blanks were less than 0.5 ng and remained negligible.

The NIGL in-house J&M standard gave a ^{143}Nd/^{144}Nd of 0.511185 ± 12 (2σ, n = 15) during the course of this study. Sample data are reported relative to a J&M value of 0.511126, which corresponds to the accepted ^{143}Nd/^{144}Nd of 0.511860 for La Jolla. Nine analyses of the rock standard BCR-1 at NIGL over a period of 8 months yielded ^{147}Sm/^{144}Nd of 0.13867 ± 0.2 (2σ) and ^{143}Nd/^{144}Nd of 0.512638 ± 15 (2σ). Results are reported as εNd-values, the derivations in parts per 10^4 from chondritic Nd at time of sedimentation.

Results

Palynological results

Almost all recovered specimens are broken and in a poor state of preservation, but some are nevertheless preserved in three dimensions. Thus, the abundance of chitinozoans in each processed sample as indicated by a number of specimens per gram of rock must be regarded only as approximate. These numbers vary between 0.1 and 22 (EA99–9) chitinozoans per gram of rock (Fig. 3). Two samples from the (Rahlenberger) Grauwackenschiefer proved to be barren. Because of the poor state of preservation, a large number of specimens remain undetermined or are determined to the generic level only; sometimes the classification even at this level proved difficult. The chitinozoans which are reasonably well preserved are all attributed to well-known taxa, hence, no formal taxonomic descriptions have been made, and no synonymy lists are provided. In the following section, abbreviations for the biometric measurements are as follows: L = total length; Lp = chamber length; Dp = chamber diameter, Dc = diameter of oral tube, and n = number of specimens included in the calculation. The terminology and the generic assignments of the observed chitinozoans correspond to the new generic classification proposed by Paris *et al.* (1999). Because of the poor preservation of the chitinozoans, open nomenclature has been used for numerous taxa.

Plettenberger Bänderschiefer Four investigated samples from around the village of Herscheid and the small town of Plettenberg yielded few chitinozoans in a poor state of preservation that nevertheless allowed recognition of some chitinozoan species, i.e. *Belonechitina capitata* (Eisenack), *Cyathochitina calix* (Eisenack), *Cyathochitina campanulaeformis* (Eisenack) and *Desmochitina cocca* Eisenack. A number of other chitinozoans are uncertainly determined to the specific level, i.e. *Euconochitina ?conulus*, *Pistillachitina ?pistillifrons*, *Laufeldochitina ?clavata*, *Belonechitina ?micracantha* and *Lagenochitina ?ponceti* (Fig. 3). The specimen attributed to *C. calix* (in EA99–7) is reasonably well-preserved (L = 185 μm; Dp = 65 μm; Dc = 40 μm; Lp = 135 μm) and a weakly developed carina is clearly visible (Fig. 4c). Twelve incomplete chitinozoans (average: L > 135 μm; Dp = 63.3 μm, n = 12; Dc = 41.2 μm, n = 9) are similar to *P. pistillifrons* (Eisenack), with long weakly conical necks, and pronounced aboral swellings, here attributed to *P. ?pistillifrons* (Fig. 4c). Five specimens in EA99–20 are attributed to *L. ?ponceti* (Rauscher). The general morphology and the biometry (average: Dp = 90 μm, n = 5; Dc = 66.5 μm, n = 4; Lp = 114.5 μm, n = 4) fits well with that of those illustrated by Paris (1981) from the Domfront Synclinal, Orne, France, but the specific assignment is far from certain as these specimens are flattened, and their necks are missing (Fig. 4i).

Unterer (Kiesberter) Tonschiefer Six samples yielded a few complete, three-dimensionally preserved specimens. The most fossiliferous sample that we encountered from the entire Ebbe Anticline comes from the Unterer (Kiesberter) Tonschiefer (EA99–9). This particular sample produced a great number of specimens which belong to the subfamily Conochitininae. However most of these specimens were very difficult to determine even to the generic level, as only very general characters can be observed. Some fossils were attributed to *Rhabdochitina ?gracilis*, and five other are treated as *Conochitina* sp. A. The latter are characterized by a short vesicle, a flat base with a sharp margin and a short mucron-like structure (average: L = 154 μm, n = 3; Dp = 53 μm, n = 5; Dc = 33.5 μm, n = 2). One hundred specimens in the same sample (EA99–9) are only determined as *Conochitina* spp. These specimens can be compared to poorly diagnostic species such as *Conochitina claviformis* Eisenack and *Conochitina incerta* Eisenack. Several specimens of *Belonechitina capitata* (Eisenack) were also observed in this sample. Two other samples (EA99-13 and -14) yielded some specimens similar to *Laufeldochitina clavata* (Jenkins). Because all of them are incomplete (Fig. 4h), we formally refer to them as *L. ?clavata* (average in EA99–14: Dp = 95.5 μm, n = 4; Dcarina > 80 μm).

(Rahlenberger) Grauwackenschiefer In the three fossiliferous samples from Rahlenberg, the Hangweg of Kiesbert-Höh and the village of Frehlinghausen (see Appendix) a few fragmentary chitinozoans in a very poor state of preservation were observed. These were attributed to *Belonechitina ?micracantha*, *Desmochitina minor* Eisenack, *Euconochitina ?conulus*, *Pistillachitina ?pistillifrons* (Fig. 4b) and *Rhabdochitina ?magna*, but most were determined to the generic level only, or could not be attributed even to that level (Fig. 3).

Two samples from a temporary excavation at Herscheid (see Appendix), EA99-3A and -4, come from an outcrop mapped as Unterer (Kiesberter) Tonschiefer (Degens *et al.* 1981), but because of the difficulty in separating the Unterer (Kiesberter) Tonschiefer from the overlying (Rahlenberger) Grauwackenschiefer, the possibility exists that they belong to the latter unit. From these samples, only three specimens could be tentatively determined to two species: *?Belonechitina ?capitata* and *Desmochitina* cf. *piriformis* (Fig. 4a), both in EA99–4 (Fig. 3).

Oberer (Solinger) Tonschiefer The two samples that were obtained close to the village of Hohl (see Appendix) yielded inadequately preserved and broken chitinozoans only. In addition, the small number of reasonably preserved chitinozoans possess few discriminate characters that permit a definite attribution to the specific level of these fossils. The following taxa were determined to the specific level: *Belonechitina capitata* (Eisenack), *Desmochitina cocca* Eisenack and *Desmochitina minor* Eisenack. Attribution of specimens to the following taxa were less certain: *Cyathochitina* sp. A, *Belonechitina ?micracantha* (Figs 4l, 4m), and *Conochitina ?minnesotensis*. The two specimens of *B. capitata* (average: L > 117 μm; Dp = 65 μm, n = 2; Dc = 52 μm, n = 2) show clear although worn aboral spines (Figs 4j, 4k), and the shapes of the incomplete vesicles are identical to the holotype (Eisenack 1962). The specimens called *Cyathochitina* sp. A herein (Fig. 4n) have a conspicuous flexure and a short neck (average: L = 127.5 μm, n = 2; Dp = 101.7 μm, n = 3; Dc = 46.5 μm, n = 2; Lp = 95 μm, n = 2) and are very similar to those also referred to *Cyathochitina* sp. A in deep drillholes in Pomerania (Samuelsson *et al.* 2002). Two specimens are designated to *?Lagenochitina* spp., and one of them, although even the generic attribution is uncertain due to the poor preservation, is similar to *Lagenochitina dalbyensis* (Laufeld), both in terms of shape, e.g. the flat base typical for *L. dalbyensis* which is present, and size (L > 130 μm; Dp = 65 μm). Unfortunately, no further *L. dalbyensis*-like specimens were observed despite several renewed macerations and investigation of additional sample material.

Nd-isotope results

Nd-isotope data are listed in Table 1 and plotted in comparison to Lower Palaeozoic sediments from Avalonia and the ATA in Figure 5. The obtained εNd(t) values from the Ebbe Anticline range from –8.0 to –9.2 and thus show less variation than e.g. the Ordovician Brabant Massif sediments. However, samples from the Plettenberger Bänderschiefer and the Unterer (Kiesberter) Tonschiefer have εNd(t) values of –9.1 to –9.2, which is comparable to values from the

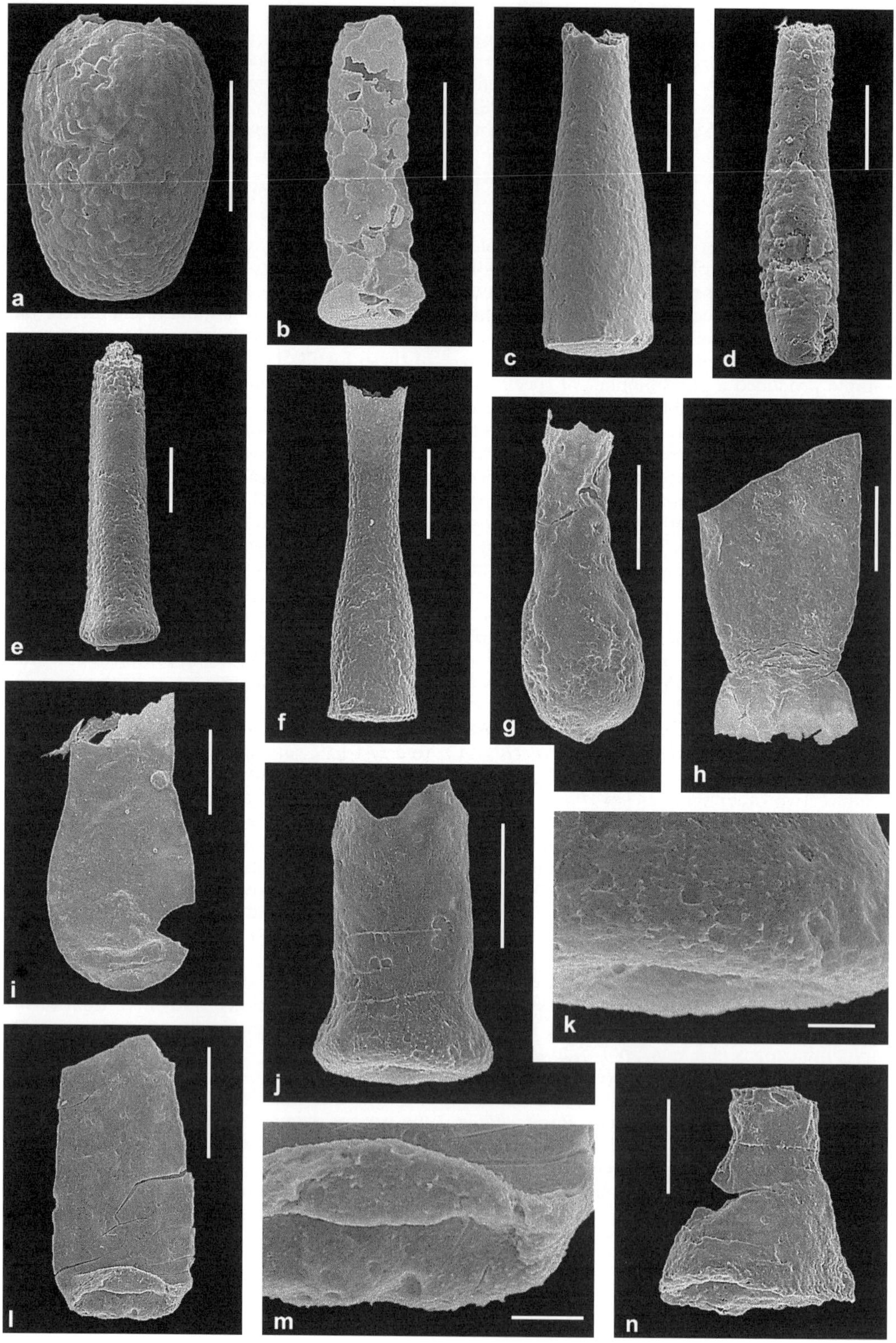
a
b
c
d
e
f
g
h
i
j
k
l
m
n

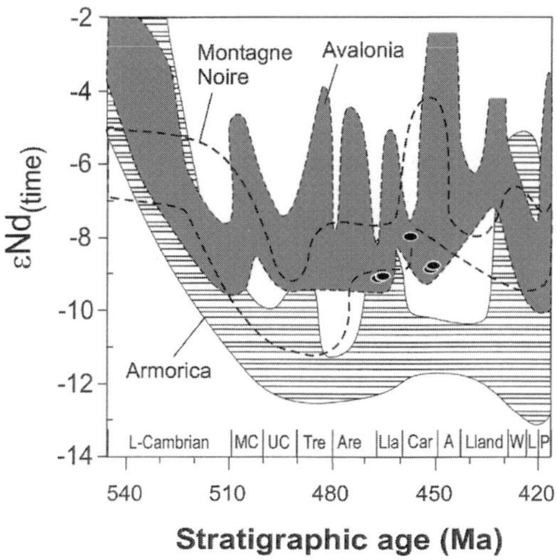

Fig 5. εNd(t) v. stratigraphic ages for sediments from the Ebbe Anticline (black circles) in comparison to well dated sediment successions from Perigondwanan terranes. Avalonia, sediments from Welsh basin, Welsh borderland, English Lake District, Stavelot Massif and Brabant Massif (*c.* 200 analyses); Armorica, sediments from Brittany, Cantabria and Central Iberia (*c.* 70 analyses); Montagne Noire, sediments from the Montagne Noire and the Pyrenees (*c.* 35 analyses). Nd isotopic evolution curves constructed from data in Michard *et al.* (1985); Thorogood (1990); Evans (1992) and references therein; McCaffrey (1994); Leng & Evans (1994); Stone & Evans (1997); Nagler *et al.* (1995); Beetsma (1995); Simien *et al.* (1999); André *et al.* (1986); André (1991); Gerdes *et al.* (unpublished data). Abbreviations: MC = Middle Cambrian, UC = Upper Cambrian, Tre = Tremadoc, Are = Arenig, Lla = Llanvirn, Car = Caradoc, A = Ashgill, Lland = Llandovery, W = Wenlock, L = Ludlow, P = Pridoli.

Brabant Massif, Wales and Lake District shales from upper Arenig to mid Llanvirn/late Darriwilian, which range from –8.7 to –9.4 (Thorogood 1990; Stone & Evans 1997; Gerdes *et al.* unpublished data). Comparable values are also the –8.0 εNd(t) of the (Rahlenberger) Grauwackenschiefer and those of lower Caradocian sediments from the Brabant Massif, which range from –8.0 to –8.2 (Gerdes *et al.* unpublished data). For the Oberer (Solingen) Tonschiefer no exact depositional age is previously known, however, an εNd(t) as low as –9.0 can be found in the Brabant Massif and the Welsh sediments above the lower Llanvirn Stage (middle Darriwilian) only in middle to upper Caradoc (Burrellian to Streffordian) and Upper Silurian strata.

Discussion

Stratigraphy

Apart from *Cyathochitina* sp. A and *Conochitina* sp. A, the chitinozoans recovered from the Ebbe Anticline are well-known from other localities, and their stratigraphic ranges have been established with high confidence, especially in North Gondwana (e.g. Paris 1981, 1990, 1996) and Baltoscandia (e.g. Nõlvak & Grahn 1993). With few exceptions, the observed chitinozoan taxa have broad stratigraphic ranges. Most of them, however, are confined to the Middle and Upper Ordovician.

Plettenberger Bänderschiefer The presence of *Belonechitina capitata* suggests a late Abereiddian (early Llanvirn/mid Darriwilian) to late Caradoc age (Paris 1981; Grahn 1982). The presence of *Cyathochitina calix* support that wide age assignment, as this taxon is known from upper Arenig to lower Caradoc strata elsewhere (Grahn 1982; Nõlvak & Grahn 1993). Some of the taxa kept in open nomenclature, i.e. especially *Euconochitina ?conulus*, *Pistillachitina ?pistillifrons* and *Lagenochitina ?ponceti* also support this time-range, provided they are conspecific with the original species.

Fig. 4. Selected chitinozoans from the Ordovician units of the Herscheider Schichten, Ebbe Anticline, W Germany. Number in parenthesis after field sample name refers to specimen on SEM preparate. Scale bar 50 μm in all illustrations except in (f), (k) and (m). (**a**) *Desmochitina* cf. *piriformis* (Laufeld 1967). EA99-4 (3). Unterer (Kiesberter) Tonschiefer/(Rahlenberger) Grauwackenschiefer. (**b**) *Pistillachitina ?pistillifrons* (Eisenack 1939). EA99-6 (7). (Rahlenberger) Grauwackenschiefer. (**c**) *Cyathochitina calix* (Eisenack 1931). EA99-7 (13). Plettenberger Bänderschiefer. (**d**) *?Rhabdochitina* sp. EA99-7 (28). Plettenberger Bänderschiefer. (**e**) *Pistillachitina ?pistillifrons* (Eisenack 1939). EA99-7 (30). Plettenberger Bänderschiefer. (**f**) *Cyathochitina calix* (Eisenack 1931). EA99-9 (26). Unterer (Kiesberter) Tonschiefer. Scale bar 100 μm. (**g**) *Lagenochitina* sp. EA99-14 (4). Unterer (Kiesberter) Tonschiefer. (**h**) *Laufeldochitina ?clavata* (Jenkins 1967). EA99-14 (13). Unterer (Kiesberter) Tonschiefer. (**i**) *Lagenochitina ?ponceti* (Rauscher 1973). EA99-20 (19). Plettenberger Bänderschiefer. (**j**) *Belonechitina capitata* (Eisenack 1962). EA99-1 (00-699: 2). Oberer (Solinger) Tonschiefer. (**k**) Detail of chamber surface at base of specimen in (**j**) showing small spines. Scale bar 10 μm. (**l**) *Belonechitina ?micracantha* (Eisenack 1931). EA99-1 (01-761: 8). Oberer (Solinger) Tonschiefer. (**m**) Detail of chamber surface at base of specimen as in (l), showing very small and probably worn spines. Scale bar 10 μm. (**n**) *Cyathochitina* sp. A. EA99-1 (01-761: 11). Oberer (Solinger) Tonschiefer.

The supply of volcanogenic detritus was probably the reason for a general increase of the εNd(t) values (–8 to –5) in middle Llanvirn/middle Darriwilian to lower Caradoc sediments from Brabant and Wales (Thorogood 1990; Evans 1992; Gerdes *et al.* unpublished data). Episodic magmatic activity in Ordovician time seems to have delivered juvenile detritus contemporaneously over a more than 700 km wide area, causing a specific Nd-isotope evolution trend for Avalonian sedimentation (Thorogood 1990; Gerdes *et al.* 2001; Fig. 5). However, Middle Cambrian to Upper Ordovician (Caradoc) sedimentation is usually dominated by detritus derived from the exposed Gondwanan/Avalonian crust. Its average composition is defined by the relatively uniform εNd(t) value (–8.9 ± 0.4; n = 41) of Avalonian sediments lacking juvenile detritus. Correlating the Plettenberger *Bänderschiefer* εNd(t) value with those obtained from other Avalonian areas suggests deposition not after mid Llanvirn/late Darriwilian or at late Caradoc times for this unit. Thus, the Abereiddian (early Llanvirn/mid Darriwilian) age given by graptolites is supported both by the recovered chitinozoans and the Nd-isotope data.

Unterer (Kiesberter) Tonschiefer As in the Plettenberger Bänderschiefer, the biostratigraphically most important species is *Belonechitina capitata* which indicates a late Abereiddian (early Llanvirn/mid Darriwilian) to late Caradoc age (Paris 1981; Grahn 1982). Specimens referred to *Laufeldochitina ?clavata* hint at a middle to late Llanvirn age for the Unterer (Kiesberter) Tonschiefer, as true *L. clavata* are known from upper Abereiddian to Llandeilian (upper Darriwilian) strata elsewhere (Paris 1990). The recovered graptolites are indicative of an Abereiddian age (Maletz & Servais 1993), and that assignment is thus confirmed by the sparse chitinozoan data.

Independently, the Nd isotopes also support this age attribution, as sediments with an εNd(t) value of about –9 are unknown from the ATA or from Avalonia between late Llanvirn/late Darriwilian to mid Caradoc times.

(Rahlenberger) Grauwackenschiefer No biostratigraphically unequivocal taxa were recovered, but the presence of *Pistillachitina ?pistillifrons* in EA99-6 gives a hint about the age of the (Rahlenberger) Grauwackenschiefer as true *P. pistillifrons* (Eisenack) are known from lower Caradoc to lower Ashgill strata (Elaouad-Debbaj 1986). The chitinozoans reported by Eisenack (1939) from siliceous concretions within the (Rahlenberger) Grauwackenschiefer include (taxonomy as currently understood) *Cyathochitina campanulaeformis* (Eisenack), possible *Cyathochitina calix* (Eisenack), *Laufeldochitina stentor* (Eisenack), *Euconochitina primitiva* (Eisenack), *Belonechitina micracantha* (Eisenack), *Desmochitina minor* Eisenack, *Eisenackitina rhenana* (Eisenack), *Rhabdochitina magna* Eisenack, *Pistillachitina pistillifrons* (Eisenack), *Conochitina minnesotensis* (Stauffer) and *Conochitina* aff. *claviformis* (Eisenack). Eisenack (1939) did not give precise information from where his siliceous concretion sample of the Grauwackenschiefer came from. However, Richter & Richter gave the material to him (Eisenack 1939), and these two authors mention the locality, 'Fundpunkt 04', in a later publication describing the trilobite '*Cyclopyge*' *illaenoides* (Richter & Richter 1954, p. 54; see also Koch 1999*a*, p. 385, 'Fundpunkt 8'). This locality has now been built over, and our locality close to that of Eisenack (1939), behind the Herscheid school (Turnhalle) was excavated much later. Unfortunately, our siliceous concretion (EA99–6) yielded only a few poor chitinozoans. Therefore, the chitinozoan assemblage reported by Eisenack (1939) still gives the most precise biostratigraphic age for the (Rahlenberger) Grauwackenschiefer, i.e. Aurelucian (earliest Caradoc).

The εNd(t) values in the Brabant Massif sediments seem to decrease continuously from about –8.0 to –8.8 in lower to upper Caradoc strata (Gerdes *et al.* 2001). Correlation of the (Rahlenberger) Grauwackenschiefer with the Brabant Massif sediments shows that the Nd-isotope data support the more precise biostratigraphic age.

The two samples belonging either to the Unterer (Kiesberter) Tonschiefer or the (Rahlenberger) Grauwackenschiefer might be of an early Caradoc age, as *Desmochitina* cf. *piriformis*, provided it is a true *D. piriformis,* indicates this age (Paris 1981). The samples therefore possibly derive from the (Rahlenberger) Grauwackenschiefer.

Oberer (Solinger) Tonschiefer The species *Belonechitina capitata* is known from strata of late Abereiddian (early Llanvirn/mid Darriwilian) to late Caradoc ages (Paris 1981; Grahn 1982). Because the Oberer (Solinger) Tonschiefer is observed to be overlying the (Rahlenberger) Grauwackenschiefer (Timm *et al.* 1981) and the latter unit has been assigned an earliest Caradoc age, the Oberer (Solinger) Tonschiefer must be of the same age, or younger, than the

(Rahlenberger) Grauwackenschiefer. We therefore suggest an early to late Caradoc age for the Oberer (Solinger) Tonschiefer based on chitinozoans only. If we also correlate the Nd-isotope data of the Oberer (Solinger) Tonschiefer with that of sediments from the Brabant Massif, the obtained εNd(t) values of –8.7 to –9 indicate a late Caradoc or late Silurian age (Fig. 5). Accordingly, based on the combined chitinozoan and Nd-isotope study, a late Caradoc age is suggested for the Oberer (Solinger) Tonschiefer.

Attribution of the Ebbe Anticline (Herscheider Schichten) to Avalonia

The Ebbe Anticline belongs to the Rheinisches Schiefergebirge which in turn forms part of the Rheno-Hercynian Zone of central Europe north of the so-called Rheic Suture and the Saxo-Thuringian Zone (e.g. Erdtmann 1991; Cocks *et al.* 1997; Pharaoh 1999). The Rheno-Hercynian nappes of Cornubia (SW England) and central Europe are interpreted as representing parts of Avalonia separated from the parent continent during the Variscan Orogeny (Dallmeyer *et al.* 1995). In addition, most of the autochthonous parts of the Rheinisches Schiefergebirge were also part of Avalonia, because fossil faunas in these areas are most closely comparable to other Avalonian faunas (e.g. Siegfried 1969; Anderle 1987; Koch 1999*a*).

Individual chitinozoan taxa typical for a higher latitude area (here identified as Avalonia or Perigondwana) were recovered neither by the present investigation nor by Eisenack (1939). A possible exception would be those specimens herein attributed to *Lagenochitina ?ponceti* from the Plettenberger Bänderschiefer, as this taxon was previously never described from low-latitude Baltoscandian sediments. However, the poor level of preservation and the uncertain taxonomic attribution of these specimens do not provide evidence for the inclusion of Ebbe Anticline in Avalonia. Instead, the total taxonomic composition of the chitinozoan assemblage itself appears to be of some palaeobiogeographic significance. Cluster analysis and coefficient of similarity calculations on the Eisenack (1939) chitinozoan assemblage from the chitinozoan *stentor* biozone (i.e. the (Rahlenberger) Grauwackenschiefer) together with other contemporary Avalonian assemblages from the Rügen 5/66 borehole (North Germany) and the Brabant Massif, Belgium, show that the thus postulated Avalonian fauna is different from contemporary faunas recovered from both North Gondwana and Baltoscandia (Samuelsson *et al.* 2002). This indirectly supports the inclusion of the Ebbe Anticline succession in Avalonia. In addition, acritarchs in the Ebbe Anticline are typical cold-water, high-latitude species, and thus provide additional evidence for that attribution (Maletz & Servais 1993).

Lower Palaeozoic sediments from the ATA show different εNd(t) trends compared to eastern Avalonian sediments (Fig. 5). The ATA is characterized by at least two distinct regions in terms of their sedimentary Nd-isotope composition, each with similar εNd(t) trends over a wide depositional area (Gerdes *et al.* 2001). Sediments from the Montagne Noire and the Pyrenees show less negative εNd(t) values in comparison to the Ebbe Anticline sediments, and sediments from Brittany, Cantabria and Central Iberia usually more negative εNd(t) values during Llanvirn and Caradoc times (Michard *et al.* 1985; Nagler *et al.* 1995; Beetsma 1995; Simien *et al.* 1999). Thus the Nd-isotope signatures support the idea that the Ordovician succession of the Ebbe Anticline belongs to the same palaeogeographic unit as the Lower Palaeozoic of the Brabant Massif and the Welsh Basin, i.e. Avalonia.

Stratigraphic correlations

A much clearer picture of the stratigraphic relationship between the rock successions interpreted as forming part of Eastern Avalonia has emerged over the last years (e.g. Verniers *et al.* 2002; Fig. 6).

To the west of the Ebbe Anticline, Ordovician rocks are known from Belgium, where Lower Palaeozoic sediments occur in three distinct tectonostratigraphical units, from north to south the Brabant Massif (which is underlying most of Belgium), the Condroz Inlier (also called 'Bande de Sambre-et-Meuse'), and the Ardenne inliers (Stavelot, Rocroi, Serpont, and Givonne inliers) in southern Belgium. The Ebbe succession is most similar to the succession of dark shale and sandstone of the Condroz Inlier. The Huy Formation of an early Llanvirn age in the Condroz Inlier has a lithology and a fossil content (acritarchs, graptolites, trilobites, and ichnofossils) similar to those of the Plettenberger Bänderschiefer and the Unterer (Kiesberter) Tonschiefer. The Huy Formation is overlain by the Sart-Bernard, Vitrival-Bruyère and Oxhe formations, of which only the latter is accurately dated by trilobites (Dean 1991) as Longvillian (latest Burrellian, mid Caradoc) in age. The two other formations, which only occur locally, are probably of an Abereiddian to Aurelucian age (Servais & Maletz 1992) and

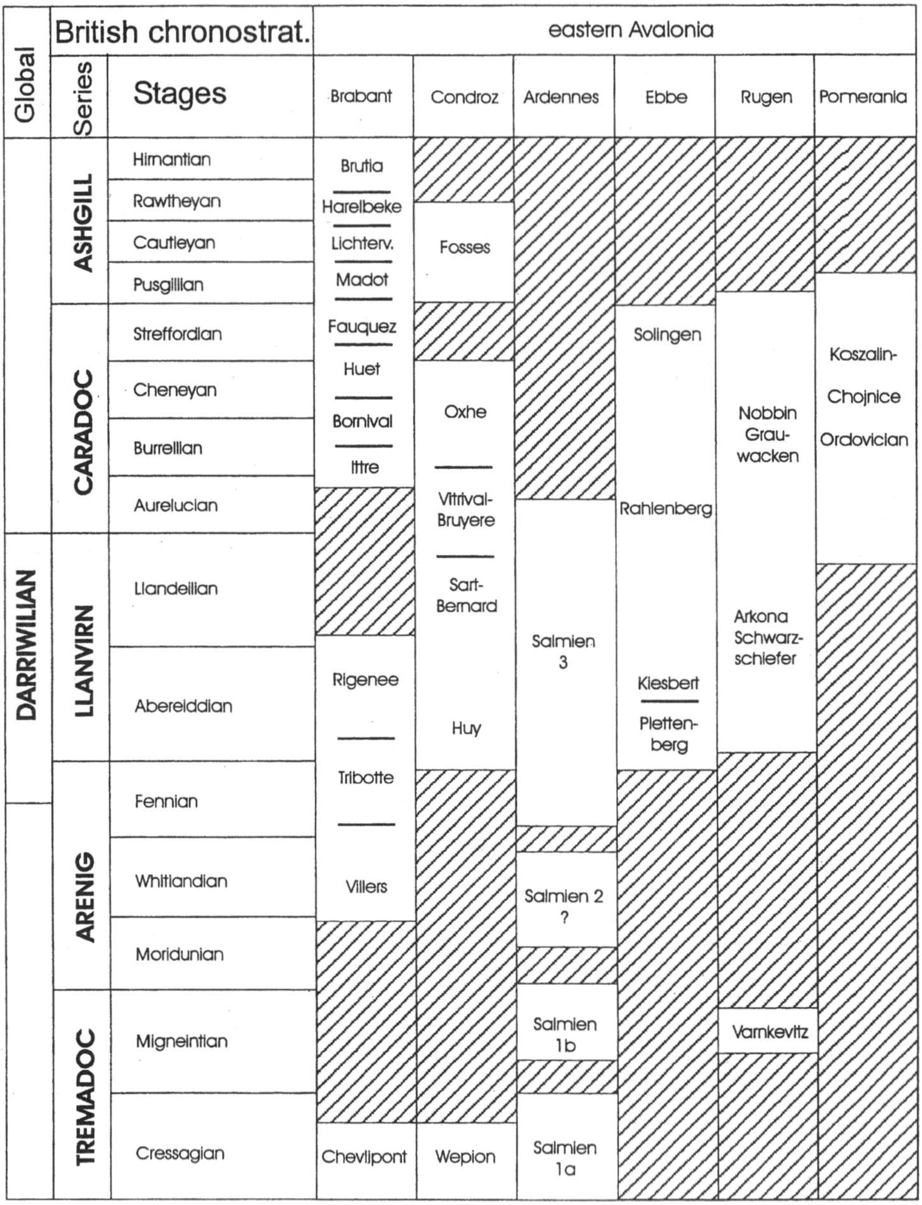

Fig. 6. Correlation between Ordovician units of the Brabant Massif (Samuelsson & Verniers 2000), the Condroz Inlier (Servais & Maletz 1992), the Stavelot Inlier (Verniers *et al.* in press.), the Ebbe Anticline, the Rügen successions (Servais *et al.* 2001), and Pomerania (Samuelsson *et al.* 2002). Chronostratigraphy after Fortey *et al.* (1995). Oblique lines indicate lack of evidence for deposition.

thus represent possible equivalents of the (Rahlenberger) Grauwackenschiefer. The other formations in the Condroz Inlier, the Wépion Formation of a Tremadoc age and the Fosses Formation of an Ashgill (pre-Hirnantian) age, have no equivalents in the Ebbe area.

Although the Plettenberger Bänderschiefer and the Unterer (Kiesberter) Tonschiefer may be correlated with the Rigenée Formation of the Brabant Massif, dated as Abereiddian (early Llanvirn/mid Darriwilian) in age (Maletz & Servais 1998; Samuelsson & Verniers 2000), the

(Rahlenberg) Grauwackenschiefer does not appear to have any time-equivalent counterparts in the Brabant Massif. However, the suggested late Caradoc age of the Oberer (Solinger) Tonschiefer may have its parallel in the late Caradoc or earliest Ashgill Huet and Fauquez formations of the Brabant Massif (Samuelsson & Verniers 2000). Unfortunately, no hitherto observed chitinozoan taxa are common between Oberer (Solinger) Tonschiefer and the Madot Formation.

In the Stavelot Inlier, a succession of shale and siltstone forming the so-called 'Salm 3' unit of the Salm Group were attributed to the Arenig–Llandeilo series by Vanguestaine (1986), i.e. Arenig to upper Llanvirn/upper Darriwilian strata (Verniers *et al.* 2001), and can thus be compared with the two lowermost Ebbe Anticline units. In the Rocroi Inlier, corresponding strata are attributed to the 'Revin 5' unit of the Revin Group (Roche *et al.* 1986). Correlations between the Ardenne inliers and the Ebbe Anticline therefore remain problematical.

Jaeger (1967), who first described the lithology and the palaeontology (graptolites and phyllocarids) of the Ordovician rocks underlying the island of Rügen (NE Germany), pointed out the close relationship of Rügen with the Ordovician of the Rheinisches Schiefergebirge. He suggested that both areas are related palaeogeographically, possibly belonging to the same sedimentation area, with Rügen being located at the margin and the Rheinisches Schiefergebirge in the central part of the Rheno-Hercynian Belt. This view, which is similar to the modern interpretation of the Rügen Ordovician rocks being part of Eastern Avalonia, is sustained by recent studies which indicate that the Ordovician succession of Rügen is of a Llanvirn to Caradoc (and possibly early Ashgill) age (Servais *et al.* 2001). Furthermore, both lithology, macrofossils (e.g. graptolites and phyllocarids), and microfossils (acritarchs and chitinozoans) of Rügen and the Ebbe anticline are very similar.

The Ordovician succession of the Koszalin–Chojnice area in western Pomerania (northwestern Poland) can be considered as an eastward extension of Eastern Avalonia, i.e. the Rügen succession. Samuelsson *et al.* (2002) evaluate the stratigraphy of this Polish succession, which in part can be correlated with the uppermost unit of the Rügen Ordovician. Therefore, it corresponds to the Upper Ordovician units of the Ebbe Anticline (Fig. 6).

This project was started during the postdoctoral studies of J. Samuelsson and A. Gerdes while recipients of research grants from the EU funded TMR program PACE (Palaeozoic Amalgamation of Central Europe) at the University of Ghent (Belgium) and the NERC Isotope Geoscience Laboratory (UK), respectively. The project was finished by J.S. at the University of Rennes I (France) with a postdoctoral fellowship from the Swedish government through STINT. F. Paris (Rennes) kindly hosted J.S. during his latter stay. S. Van Cauwenberghe (Ghent) processed the samples for microfossils. O. Paris (Rennes) drew the chitinozoan figure. K.-H. Eiserhardt (Hamburg) discussed Ebbe Anticline stratigraphy. J. Verniers was research director of the Fund for Scientific Research (Flanders), Belgium until 2000. Constructive reviews were made by S. Noble and J. Maletz. This paper is a contribution to IGCP 410 'The Great Ordovician Biodiversification Event'.

Appendix

Sample	Locality	Formation	Notes
EA99-1	Hohl	Oberer (Solinger) Tonschiefer	Road to Hohl, *c.* 75 m from road cross
EA99-2	Hohl	Oberer (Solinger) Tonschiefer	2.3 m S of EA99–1
EA99-3A	Herscheid, Neubaugebiet	Unterer (Kiesberter) Tonschiefer or (Rahlenberger) Grauwackenschiefer	1–1.5 m from NNW corner of Weissdornweg 13
EA99-4	Herscheid, Neubaugebiet	Unterer (Kiesberter) Tonschiefer or (Rahlenberger) Grauwackenschiefer	'Kieselgallen', Schlehenweg
EA99-5	Herscheid, Rahlenberg Schule	(Rahlenberger) Grauwackenschiefer	Behind sports hall (Turnhalle), 4 m from left corner of house, 3.6 m from wall, *c.* 1.1 m height

Appendix – *continued*

Sample	Locality	Formation	Notes
EA99-6	Herscheid, Rahlenberg Schule	(Rahlenberger) Grauwackenschiefer	'Kieselgallen' to the left of EA99-5, *c.* 1.5 m above base of sports hall
EA99-7	Herscheid Umgehungstrasse	Plettenberger Bänderschiefer	1.40 m above road, 4 m to the right of road sign
EA99-8	Herscheid Umgehungstrasse	Plettenberger Bänderschiefer	1.90 m above road, 19 m to the left of road sign
EA99-9	Brenscheider Fahrweg	Unterer (Kiesberter) Tonschiefer	200 m east of Waldminer Kreuz, in a 'sunken' dirt road, 0.70 m above road
EA99-10	Brenscheider Fahrweg	Unterer (Kiesberter) Tonschiefer	0.90 m above road, 5 m above EA99-9
EA99-12	Hangweg NE Kiesbert	Unterer (Kiesberter) Tonschiefer	Maletz & Servais (1993) loc. A, 2 m above road
EA99-13	Hangweg NE Kiesbert	Unterer (Kiesberter) Tonschiefer	Maletz & Servais (1993) loc. A
EA99-14	Hangweg NE Kiesbert	Unterer (Kiesberter) Tonschiefer	Maletz & Servais (1993) loc. C, 2.5 m above road
EA99-15	Hangweg NE Kiesbert	(Rahlenberger) Grauwackenschiefer	37 m west of oak, 1.7 m above road, base of outcrop
EA99-16	Frehlinghausen	(Rahlenberger) Grauwackenschiefer	'Kieselgallen', base of outcrop, not exactly positioned
EA99-17	Frehlinghausen	(Rahlenberger) Grauwackenschiefer	'Kieselgallen', 30 cm above ground
EA99-18	Plettenberg – Hechmecker Weg	Unterer (Kiesberter) Tonschiefer	*c.* 1 m above ground
EA99-19	Ziegelei Loos (Plettenberg)	Plettenberger Bänderschiefer	Behind supermarket, *c.* 1 m above ground
EA99-20	Ziegelei Loos (Plettenberg)	Plettenberger Bänderschiefer	1 m from previous sample, 1 m above ground

References

ANDERLE, H.-J. 1987. Entwicklung und Stand der Unterdevon-Stratigraphie im südlichen Taunus. *Geologisches Jahrbuch Hessen,* **115**, 81–98.

ANDRÉ, L. 1991. The concealed crystalline basement in Belgium and the 'Brabantia' microplate concept: constraints from the Caledonian magmatic and sedimentary rocks. *In*: ANDRÉ, L., HERBOSCH, A., VANGUESTAINE, M. & VERNIERS, J. (eds) *Proceedings of the international meeting on the Caledonides of the Midlands and the Brabant Massif,* 117–139.

ANDRÉ, L., DEUTSCH, S. & HERTOGEN, J. 1986. Trace-element and Nd isotopes in shales as indexes of provenance and crustal growth – the Early Paleozoic from the Brabant Massif (Belgium). *Chemical Geology,* **57**, 101–115.

BEETSMA, J. J. 1995. The late Proterozoic / Palaeozoic and Hercynian crustal evolution of the Iberian Massif, N Portugal, as traced by geochemistry and Sr-Nd-Pb isotope systematics of pre-Hercynian terrigenous sediments and Hercynian granitoids, Ph. D. thesis, *University of Amsterdam.*

BEYER, K. 1941*a*. Zur Kenntnis des Silurs im Rheinischen Schiefergebirge. 1. Das Auftreten von *Tomaculum problematicum* Groom im Ebbe-Sattel und die Bedeutung der Kotpillen-Schnur für die Gliederung des Sauerländischen Ordoviziums. *Jahrbuch der Reichsstelle für Bodenforschung,* **61**, 198–221.

BEYER, K. 1941*b*. Zur Kenntnis des Silurs im Rheinischen Schiefergebirge. 2. Die Plettenberger Bänderschiefer, das älteste Ordovizium im Rechtsrheinischen Schiefergebirge. *Jahrbuch der Reichsstelle für Bodenforschung,* **61**, 222–253.

BEYER, K. 1941*c*. Zur Kenntnis des Silurs im Rheinischen Schiefergebirge. 3. Die Gliederung des Ordoviziums im Kern des Remscheider Sattels. *Jahrbuch der Reichsstelle für Bodenforschung,* **61**, 254–266.

COCKS, L.R.M., MCKERROW, W.S. & VAN STAAL, C.R. 1997. The Margins of Avalonia. *Geological Magazine,* **134**, 627–636.

DALLMEYER, R.D., FRANKE, W. & WEBER, K. (eds) 1995. *The Pre-Permian geology of Central and Eastern Europe*. Springer, Berlin.

DEAN, W. T. 1991. Ordovician trilobites from the inlier at Le Petit Fond d'Oxhe, Belgium. *Bulletin de l'Institut Royal des Sciences naturelles de Belgique, Sciences de la Terre,* **61**, 135–155.

DEGENS, E. T., TIMM, J. & WONG, H. K. (eds) 1981. Rheinisches Schiefergebirge: Ebbe-Anticlinorium. Fazies, Stratigraphie, Tektonik. *Mitteilungen aus dem Geologisch-Paläontologischen Institut der Universität Hamburg,* **50**, 1–282.

EISENACK, A. 1931. Neue Mikrofossilien des baltischen Silurs. I. *Paläontologische Zeitschrift,* **13**, 74–118.

EISENACK, A. 1939. Chitinozoen und Hystrichosphaeriden im Ordovicium des Rheinischen Schiefergebirges. *Senckenbergiana,* **21**, 135–152.

EISENACK, A. 1962. Neotypen baltischer Silur-Chitinozoen und neue Arten. *Neues Jahrbuch für Geologie und Paläontologie, Abhandlungen,* **114**, 291–316.

EISERHARDT, K.-H., HEYCKENDORF, K. & THOMBANSEN, E. 1981. Zur Stratigraphie und Tektonik des nördlichen Ebbe-Teilsattels (Sauerland, Rheinisches Schiefergebirge). *Mitteilungen aus dem Geologisch-Paläontologischen Institut der Universität Hamburg,* **50**, 199–238.

EISERHARDT, K.-H., KOCH, L. & EISERHARDT, W. L. 2001. Revision des Ichnotaxon *Tomaculum* Groom, 1902. *Neues Jahrbuch für Geologie und Paläontologie, Abhandlungen,* **221**, 325–358.

EISERHARDT, K.-H., KOCH, L. & MALETZ, J. 2001. Nördliches Rheinisches Schiefergebirge. *In*: STRATIGRAPHISCHE KOMMISSION DEUTSCHLANDS (Hrsg.) *Stratigraphie von Deutschland II: Ordovizium, Kambrium; Vendium, Riphäikum, Teil II.* Courier Forschungs-Institut Senckenberg, **235**, 90–108.

ELAOUAD-DEBBAJ, Z. 1986. Chitinozoaires de la Formation inférieure du Ktaoua, Ordovicien supérieur de l'Anti-Atlas (Maroc). *Hercynica,* **2**, 35–55.

EVANS, J. A. 1992. Geochemical and isotope composition of pebbles from the Caban Conglomerate Formation and their bearing on the source of Welsh Palaeozoic sedimentary rocks. *Geological Magazine,* **129**, 581–587.

ERDTMANN, B.-D. 1991. The post-Cadomian early Palaeozoic tectonostratigraphy of Germany. *Annales de la Société Géologique de Belgique,* **114**, 19–43.

FORTEY, R. A., HARPER, D. A. T., INGHAM, J. K., OWEN, A. W. & RUSHTON, A. W. A. 1995. A revision of Ordovician series and stages from the historical type area. *Geological Magazine,* **132**, 15–30.

FUCHS, A. 1912. Zur Stratigraphie der Lenneschiefer und des jüngeren Unterdevons im östlichen Taunus. *Jahrbuch Preußische Geologische Landesanstalt,* **33**, 474–477.

FUCHS, A. 1935. Geologische Karte von Preußen und benachbarten deutschen Ländern. Erläuterungen zu Blatt Solingen. *Preußische Geologische Landesanstalt LFG,* **316**, 1–65.

GERDES, A., PHARAOH, T. C., TIMMERMANN, H., VERNIERS, J., VECOLI, M., NOBLE, S. R. & PARRISH, R. R. 2001. Nd-isotope record of clastic sediments in response to orogenic processes: from Gondwana break-up to Palaeozoic amalgamation of Central Europe. *In*: ROTH, S. & RUGGEBERG, A. *2001 Margins Meeting,* Schriftenreihe der Deutschen Geologischen Gesellschaft, **14**, 67–69.

GRAHN, Y. 1982. Caradocian and Ashgillian Chitinozoa from the subsurface of Gotland. *Sveriges Geologiska Undersökning Serie C,* **788**, 1–66.

JAEGER, H. 1967. Ordoviz auf Rügen. Datierung und Vergleich mit anderen Gebieten. *Berichte der deutschen Geschellschaft für geologische Wissenschaften A,* **12**, 165–176.

JENKINS, W. A. M. 1967. Ordovician Chitinozoa from Shropshire. *Palaeontology,* **10**, 436–488.

JENTSCH, S. & STEIN, V. 1961. Neue Fossilfunde im Ordovizium des Ebbe-Sattels. *Paläontologische Zeitschrift,* **35**, 200–208.

KOCH, L. 1999*a*. Die Familie Cyclopygidae (Trilobita) im Ordovizium des Ebbe-Sattels und Remscheider Sattels (Rheinisches Schiefergebirge, Deutschland). *Neues Jahrbuch für Geologie und Paläontologie, Abhandlungen,* **213**, 375–431.

KOCH, L. 1999*b*. *Corrugatagnostus* (Metagnostidae, Trilobita) aus dem Ordovizium des Ebbe-Sattels (Rheinisches Schiefergebirge, Deutschland). *Geologica et Palaeontologica,* **33**, 9–19.

KOCH, L. & BRAUCKMANN, C. 1998. Phyllocariden (Crustacea) aus dem Unteren Llanvirn (Ordovizium) des Ebbe-Sattels (Rheinisches Schiefergebirge). *Neues Jahrbuch für Geologie und Paläontologie, Monatshefte,* **1998**, 55–64.

KOCH, L. & LEMKE, U. 1994. *Waldminia spinigera* n.g., n.sp. ein neuer Trilobit aus dem Ordovizium des Ebbe-Sattels. *Dortmunder Beiträge zur Landeskunde, naturwissenschaftliche Mitteilungen,* **28**, 67–74.

KOCH, L. & LEMKE, U. 1995*a*. Trilobiten aus dem Unteren Tonschiefer (Unteres Llanvirn, Ordovizium) von Kiesbert (Ebbe-Sattel, Rheinisches Schiefergebirge). Teil 1. *Geologie und Paläontologie in Westfalen,* **39**, 15–55.

KOCH, L. & LEMKE, U. 1995*b*. Neue Trilobiten und Graptholiten-Funde aus dem Unteren Tonschiefer (Unteres Llanvirn, Ordovizium) von Kiesbert (Ebbe-Sattel). *Dortmunder Beiträge zur Landeskunde, naturwissenschaftliche Mitteilungen,* **29**, 7–19.

KOCH, L. & LEMKE, U. 1997. *Corrugatagnostus magnodosus* n. sp. ein neuer Trilobit aus dem Unteren Llanvirn (Ordovizium) von Kiesbert (Ebbe-Sattel, Rheinisches Schiefergebirge). *Neues Jahrbuch für Geologie und Paläontologie, Monatshefte,* **1997**, 297–307.

KOCH, L. & LEMKE, U. 1998*a*. Die Gattungen *Girvanopyge* Kobayashi 1960 und *Waldminia* Koch & Lemke 1994 (Remopleurididae, Trilobita) im Unteren Llanvirn (Ordovizium) des Ebbe-Sattels und des Remscheider Sattels (Rheinisches Schiefergebirge, Deutschland). *Neues Jahrbuch für Geologie und Paläontologie, Monatshefte,* **1998**, 494–512.

KOCH, L. & LEMKE, U. 1998*b*. *Dionide* Barrande 1847 und *Dionidella* Prantl & Pribyl 1949 (Dionididae, Trilobita) aus dem Ordovizium des Ebbe-Sattels (Rheinisches Schiefergebirge, Deutschland). *Neues Jahrbuch für Geologie und Paläontologie, Monatshefte,* **1998**, 613–625.

KOCH, L. & LEMKE, U. 2000. Seltene Trilobiten-Arten aus dem Ordovizium des Ebbe-Sattels (Rheinisches Schiefergebirge, Deutschland). *Neues Jahrbuch für Geologie und Paläontologie, Monatshefte,* **2000**, 513–544.

KOCH, L., LEMKE, U. & BRAUCKMANN, C. 1990. *Vom Ordovizium bis zum Devon: Die fossile Welt des Ebbe-Gebirges*, **198 pp**. Hagen (v.d. Linnepe).

LAUFELD, S. 1967. Caradocian Chitinozoa from Dalarna, Sweden. *Geologiska Föreningens i Stockholm Förhandlingar,* **89**, 275–349.

LENG, M. J. & EVANS, J. A. 1994. Provenance of Late Ashgill (Hirnantian) Fine-Grained Sediments and Pebbles in the Welsh Basin – a Nd and Sr Isotope Study. *Geological Journal*, **29**, 1–9.

MALETZ, J. 1995. The Middle Ordovician (Llanvirn) graptolite succession of the Albjära core (Scania, Sweden) and its implication for a revised biozonation. *Zeitschrift für geologische Wissenschaften,* **23**, 249–259.

MALETZ, J. 2000. Review of the Ordovician biostratigraphy of the Herscheid Schichten (Rheinisches Schiefergebirge, Germany). *Neues Jahrbuch für Geologie und Paläontologie, Abhandlungen,* **218**, 45–60.

MALETZ, J. & SERVAIS, T. 1993. Acritarchs and graptolites from the Early Llanvirn of the Herscheider Schichten (Rheinisches Schiefergebirge, Germany). *Neues Jahrbuch für Geologie und Paläontologie, Abhandlungen,* **190**, 125–142.

MALETZ, J. & SERVAIS, T. 1998. Upper Ordovician graptolites from the Brabant Massif, Belgium. *Geobios,* **31**, 21–37.

MCCAFFREY, W. D. 1994. Sm-Nd isotopic characteristics of sedimentary provenance – the Windermere Supergroup of NW England. *Journal of the Geological Society*, **151**, 1017–1021.

MCKERROW, W. S. & VAN STAAL, C. R. 2000. The Palaeozoic time scale reviewed. *In*: FRANKE, W., HAAK, V., ONCKEN, O. & TANNER, D. (eds) *Orogenic processes: Quantifications and modelling in the Variscan belt*, Geological Society, London, Special Publications, **179**, 5–8.

MICHARD, A., GURRIET, P., SOUDANT, M. & ALBARÉDE, F. 1985. Nd isotopes in French Phanerozoic shales: External vs. internal aspects of crustal evolution. *Geochimica et Cosmochimica Acta*, **49**, 601–610.

MILLER, M. A. 1996. Chitinozoa. *In*: JANSONIUS J. & MCGREGOR D. C. (eds) Palynology: principles and applications. *American Association of Stratigraphic Palynologists Foundation,* **1**, 307–336.

NAGLER, T. F., SCHAFER, H. J. & GEBAUER, D. 1995. Evolution of the Western-European Continental-Crust – Implications from Nd and Pb Isotopes in Iberian Sediments. *Chemical Geology*, **121**, 345–357.

NANCE, R. D. & MURPHY, J. B. 1996. Basement isotopic signatures and Neoproterozoic palaeogeography of Avalonian-Cadomian and related terranes in the circum-North Atlantic. *In*: NANCE, R. D. & THOMPSON, M. D. (eds) *Avalonian and related peri-Gondwanan terranes of the circum-North Atlantic*. Geological Society of America Special Paper, **304**, 333–346.

NÕLVAK, J. & GRAHN, Y. 1993. Ordovician Chitinozoa zones from Baltoscandia. *Review of Palaeobotany and Palynology,* **79**, 245–269.

PARIS, F. 1981. Les Chitinozoaires dans le Paleozoïque du sud-ouest de l'Europe. *Mémoires de la Société géologique et minéralogique de Bretagne,* **26**, 1–412.

PARIS, F. 1990. The Ordovician Chitinozoa biozones of the Northern Gondwana Domain. *Review of Palaeobotany and Palynology,* **66**, 181–209.

PARIS, F. 1996. Chitinozoan biostratigraphy and palaeoecology. *In*: JANSONIUS, J. & MCGREGOR, D. C. (eds) Palynology: principles and applications. *American Association of Stratigraphic Palynologists Foundation*, **2**, 531–552.

PARIS, F., GRAHN, Y., NESTOR, V. AND LAKOVA, I. 1999. A revised chitinozoan classification. *Journal of Paleontology,* **73**, 549–570.

PATCHETT, P. J., ROSS, G. M. & GLEASON, J. D. 1999. Continental drainage in North America during the Phanerozoic from Nd isotopes. *Science*, **283**, 671–673.

PHARAOH, T. C. 1999. Palaeozoic terranes and their lithospheric boundaries within the Trans-European Suture Zone (TESZ): a review. *Tectonophysics,* **314**, 17–41.

RAUSCHER, R. 1973. Recherches micropaléontologiques et stratigraphiques dans l'Ordovicien et le Silurien en France. Etude des Acritarches, des Chitinozoaires et des spores. *Sciences géologiques, Université Louis Pasteur de Strasbourg, Institut de Géologie,* **38**, 1–224.

RICHTER, R. & RICHTER, E. 1937. Die Herscheider Schiefer, ein zweites Vorkommen von Ordovicium im Rheinischen Schiefergebirge, und ihre Beziehung zu den wiedergefundenen *Dayia*-Schichten. *Senckenbergiana,* **19**, 289–313.

RICHTER, R. & RICHTER, E. 1939*a*. Eine Lebens-Spur (*Syncoprulus pharmaceus*), gemeinsam dem rheinischen und böhmischen Ordovicium. *Senckenbergiana,* **21**, 152–167.

RICHTER, R. & RICHTER, E. 1939*b*. Die Kot-Schnur *Tomaculum* Groom (= *Syncoprulus* Rud. & E. Richter), ähnliche Scheitel-Platten und beider stratigraphische Bedeutung. *Senckenbergiana,* **21**, 278–291.

RICHTER, R. & RICHTER, E. 1941. Das stratigraphische Verhalten von *Tomaculum* als Beispiel für die Bedeutung von Lebensspuren. *Senckenbergiana,* **23**, 133–135.

RICHTER, R. & RICHTER, E. 1954. Die Trilobiten des Ebbe-Sattels und zu vergleichende Arten (Ordovizium, Gotlandium/Devon). *Abhandlungen der senckenbergischen naturforschenden Gesellschaft,* **488**, 1–76.

RIEGRAF, W. & NIEMEYER, J. 1996. Agglutinierte Foraminiferen aus Graptolithen-Schwarzschiefern des Llanvirniums (Ordovizium) von Plettenberg im Sauerland (Nordrhein-Westfalen, NW-Deutschland). *Paläontologische Zeitschrift,* **70**, 19–36.

ROCHE, M., SABIR, M., STEEMANS, P. & VANGUESTAINE, M. 1986. Palynologie du sondage et de la région de Willerzie. *Aardkundige Mededelingen,* **3**, 149–190.

SAMUELSSON, J. & SERVAIS, T. 2001. Chitinozoa biostratigraphy of subsurface Ordovician sediments from the Lohme 2/70 well, Island of Rügen (NE-Germany). *Neues Jahrbuch für Geologie und Paläontologie, Abhandlungen*, **222**, 241–258.

SAMUELSSON, J. & VERNIERS, J. 2000. Ordovician chitinozoan biozonation of the Brabant Massif, Belgium. *Review of Palaeobotany and Palynology,* **113**, 131–143.

SAMUELSSON, J., VERNIERS, J. & VECOLI, M. 2000. Chitinozoan faunas from the Rügen Ordovician (Rügen 5/66 and Binz 1/73 wells), NE Germany. *Review of Palaeobotany and Palynology,* **133**, 105–129.

SAMUELSSON, J., VECOLI, M., BEDNARCZYK, W. S. & VERNIERS, J. 2002. Timing of the Avalonia-Baltica plate convergence as inferred from palaeogeographic and stratigraphic data of chitinozoan assemblages in West Pomerania, northern Poland. *In*: WINCHESTER, J. A., PHARAOH, T. C. & VERNIERS, J. (eds) *Palaeozoic Amalgamation of Central Europe*, Geological Society, London, Special Publications, **201**, 95–113.

SCHALLREUTER, R. 1996. Die ersten ordovizischen Ostrakoden aus Westfalen. *Geologie und Paläontologie in Westfalen,* **42**, 61–71.

SCHALLREUTER, R. & KOCH, L. 1999. Ostrakoden aus dem Unteren Llanvirn (Ordoviz) von Kiesbert (Ebbe-Sattel, Rheinisches Schiefergebirge). *Neues Jahrbuch für Geologie und Paläontologie, Monatshefte,* **1999**, 477–489.

SERVAIS, T. & MALETZ, J. 1992. Lower Llanvirn (Ordovician) graptolites and acritarchs from the 'assise de Huy', bande de Sambre-et-Meuse, Belgium. *Annales de la Société Géologique de Belgique,* **115**, 265–285.

SERVAIS, T., SAMUELSSON, J., SEHNERT, M., VECOLI, M., GIESE, U. & VERNIERS, J. 2001. Ordovician palynomorphs from the subsurface of Rügen (NE-Germany): review and perspectives. *Neues Jahrbuch für Geologie und Paläontologie, Abhandlungen,* **222**, 291–307.

SIEGFRIED, P. 1969. Trilobiten aus dem Ordovizium des Ebbe-Sattels im Rheinischen Schiefergebirge. *Paläontologische Zeitschrift,* **43**, 148–168.

SIMIEN, F., MATTAUER, M. & ALLEGRE, C. J. 1999. Nd isotopes in the stratigraphic record of the Montagne Noire (French Massif Central): No significant Paleozoic juvenile inputs, and pre-Hercynian paleogeography. *Journal of Geology,* **107**, 87–97.

STONE, P. & EVANS, J. A. 1997. A comparison of the Skiddaw and Manx groups (English Lake District and Isle of Man) using neodymium isotopes. *Proceedings of the Yorkshire Geological Society*, **51**, 343–347.

THOROGOOD, E. J. 1990. Provenance of the Pre-Devonian Sediments of England and Wales – Sm-Nd Isotopic Evidence. *Journal of the Geological Society*, **147**, 591–594.

TAIT, J. A., BACHTADSE, V., FRANKE, W. & SOFFEL, H. C. 1997. Geodynamic evolution of the European Variscan fold belt: palaeomagnetic and geological constraints. *Geologische Rundschau*, **86**, 585–598.

TIMM, J. 1981. Die Faziesentwicklung der ältesten Schichten des Ebbe-Antiklinoriums. *Mitteilungen aus dem Geologisch-Paläontologischen Institut der Universität Hamburg,* **50**, 147–173.

TIMM, J., DEGENS, E. T. & WIESNER, G. M. 1981. Erläuterungen zur Geologischen Karte des zentralen Ebbe-Antiklinoriums 1: 25 000. *Mitteilungen aus dem Geologisch-Paläontologischen Institut der Universität Hamburg,* **50**, 59–75.

VANGUESTAINE, M. 1986. Progrès recents de la stratigraphie par acritarchs du Cambro-Ordovicien d'Ardenne, d'islande, d'Angleterre, du Pays de Galles et de Terre Neuve orientale. *Annales de la Société Géologique du Nord,* **105**, 65–76.

VECOLI, M. & SAMUELSSON, J. 2001. Quantitative evaluation of microplankton palaeobiogeography in the Ordovician-Early Silurian of the northern TESZ (Trans European Suture Zone): implications for the timing of the Avalonia-Baltica collision. *Review of Palaeobotany and Palynology,* **115**, 43–69.

VERNIERS, J., HERBOSCH, A., VANGUESTAINE, M., GEUKENS, F., DELCAMBRE, B., PINGOT, J. L., BELLANGER, I., HENNEBERT, M., DEBACKER, T., SINTUBIN, M. & DE VOS, W. 2001. The lower Palaeozoic formations in Belgium. *Geologica Belgica* **4**, 5–38.

VERNIERS, J., PHARAOH, T. C., ANDRÉ, L., DEBACKER, T., DE VOS, W., EVERAERTS, M., HERBOSCH, A., SAMUELSSON, J., SINTUBIN, M. & VECOLI, M. 2002. Lower Palaeozoic Basin Development and collision history of eastern Avalonia. *In*: WINCHESTER, J. A., PHARAOH, T. C. & VERNIERS, J. (eds) *Palaeozoic Amalgamation of Central Europe*, Geological Society, London, Special Publications, **201**, 47–94.

WINCHESTER, J. A. & THE PACE TMR NETWORK TEAM. 2002. Palaeozoic amalgamation of central Europe: New results from recent geological and geophysical investigations. *Tectonophysics*, in press.

The $^{40}Ar/^{39}Ar$ ages from the West Sudetes (NE Bohemian Massif): constraints on the Variscan polyphase tectonothermal development

D. MARHEINE[1], V. KACHLÍK[2], H. MALUSKI[1], F. PATOČKA[3] & A. ŻELAŹNIEWICZ[4]

[1]*ISTEEM-CNRS, Laboratoire de Géochronologie, CC 58, Place Eugène Bataillon, Université Montpellier 2, 34095 Montpellier cedex 5, France (e-mail: maluski@dstu.univ-montp2.fr)*

[2]*Department of Geology & Palaeontology, Charles University, Albertov 6, 128 43 Prague 2, Czech Republic*

[3]*Institute of Geology, Academy of Sciences of the Czech Republic, Rozvojová 135, 165 00 Prague 6, Czech Republic*

[4]*Institute of Geology, Polish Academy of Sciences, Podwale 75, 50449 Wroclaw, Poland*

Abstract: The West Sudetes (NE margin of the Bohemian Massif) consist of a complex mosaic of several tectonometamorphic units juxtaposed during the Variscan orogeny. The polyphase Variscan tectonothermal development of the West Sudetes was determined by $^{40}Ar/^{39}Ar$ ages of single grains and mineral concentrates. Late Famennian (359 Ma) mica ages from the high-grade Góry Sowie Block suggest continuous uplift after a Late Devonian high temperature–low pressure (HT–LP) event contemporaneous with the end of subduction-related high pressure–low temperature (HP–LT) metamorphism in the East Krkonoše Complex.

Mid-Late Devonian high pressure events in the Krkonoše-Jizera Terrane and Orlica-Śnieżnik Dome are followed by coeval high temperature events between 345 and 335 Ma (Viséan). The latter are interpreted as consequence of uplift, and decompression during overthrusting of both complexes on their forelands. Subsequent small- to large-scale shear movements dated at around 325–320 Ma (early Namurian) affected the Orlica-Śnieżnik Dome, Krkonoše-Jizera Terrane, including the Intra-Sudetic Fault, and also the eastern Lusatian Granitoid Complex. They were accompanied by contemporaneous emplacement of the Krkonoše-Jizera pluton. The upper limit of the tectonometamorphic and magmatic activity is dated at 314–312 Ma (Namurian/Westphalian boundary). The final juxtaposition of the diversified tectonometamorphic units, which constitute the West Sudetes, took place in early Namurian times.

The Bohemian Massif occupies a key position as the largest exposed part of the Variscan orogen in Central Europe (e.g. Matte *et al.* 1990). Recently it has been presented as a complex mosaic of terranes, with each one showing independent protolith and tectonometamorphic development. The terrane amalgamation of Central Europe was a result of the Variscan multiple interactions during the Variscan orogeny between the Gondwana-derived Armorican Terrane Assemblage with Baltica and East Avalonia which were already attached to extraneous (also peri-Gondwanan?) fragments accreted in pre-Variscan cycles. The subsequent late Variscan large-scale thrust and horizontal shear movements created the dominant architecture of the Bohemian Massif. Numerous attempts have been made to identify individual terranes in the Bohemian Massif, define them regionally and describe their evolution (e.g. Franke 1989; Matte *et al.* 1990; Oczlon 1992; Cymerman *et al.* 1997; Tait *et al.* 1997; Pharaoh 1999).

The West Sudetes are the easternmost part of the Saxothuringian Zone of the European Variscan orogen (e.g. Franke *et al.* 1993; Narębski 1994; Franke & Żelaźniewicz 2000), and form the northern and northeastern margins of the Bohemian Massif (Fig. 1). There they have a unique position facing the NW–SE oriented Trans-European Suture Zone (TESZ) separating Palaeozoic Europe from the Precambrian East European Craton (e.g. Pharaoh 1999). The succession of Palaeozoic tectonothermal events is recorded in the Cambrian to Upper Carboniferous meta-igneous and meta-sedimentary rocks of the West Sudetes.

This paper presents new information on the West Sudetic metamorphic and igneous rocks as

From: WINCHESTER, J. A., PHARAOH, T. C. & VERNIERS, J. 2002. *Palaeozoic Amalgamation of Central Europe.* Geological Society, London, Special Publications, **201**, 133–155. 0305-8719/02/$15.00

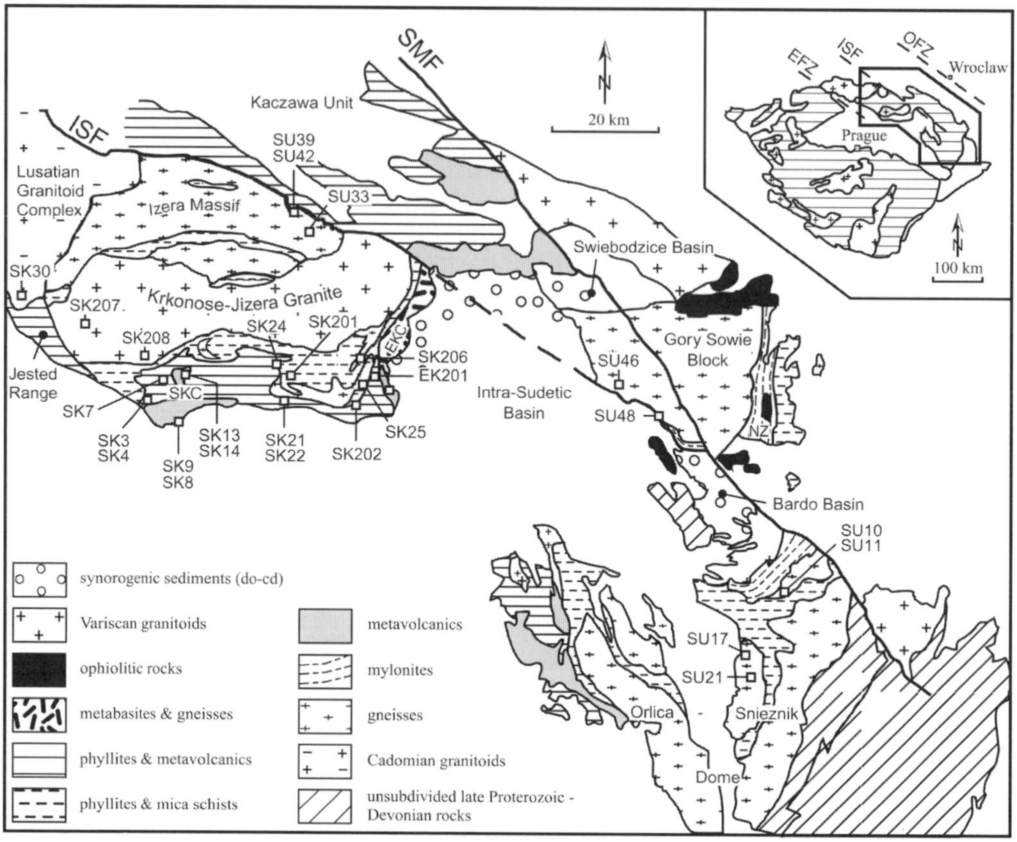

Fig. 1. Simplified geological map of the West Sudetes (modified after Aleksandrowski *et al.* 1997) with sample localities. EKC, East Krkonoše Complex; ISF, Intra-Sudetic-Fault; NZ, Niemcza Shear Zone; SKC, South Krkonoše Complex; SMF, Sudetic Marginal Fault. Black squares, sample localities. Inset: schematic map of the Bohemian Massif. EFZ, Elbe Fault Zone; OFZ, Odra Fault Zone.

well as shear zones and mylonites developed within them. It is based on the interpretation of dating using the $^{40}Ar/^{39}Ar$ step-wise heating technique. The resulting ages were evaluated in order to determine the sequence of prominent events, i.e. timing of terrane collisions, amalgamations and strike-slip movements in the West Sudetes and their significance for the Palaeozoic evolution of the Central European part of the Variscan orogen.

Geological setting

The West Sudetes are composed of Upper Proterozoic to Lower Carboniferous low- to medium-grade metamorphosed sedimentary and volcanosedimentary sequences, with local tectonic insertions of high pressure rocks, that were intruded by latest Proterozoic as well as early and late Palaeozoic granitoids (Svoboda & Chaloupský 1966; Teisseyre 1973; Żelaźniewicz 1997). The West Sudetes are a collage of the differentiated lithotectonic units which are interpreted as terranes, defined according to autonomous stratigraphic, igneous and tectonometamorphic records (Narębski 1994; Cymerman *et al.* 1997; Franke & Żelaźniewicz 2000). The assembly of the West Sudetic terrane mosaic is interpreted as a result of (early?) Variscan (Maluski & Patočka 1997) collision of Gondwana-derived terranes with Baltica (± East Avalonia) (e.g. Franke 2000) and late Variscan large-scale shear movements along prominent strike-slip faults parallel to the TESZ (e.g. Aleksandrowski *et al.* 1997).

The Krkonoše-Jizera Terrane

An important role in the terrane evolution of the West Sudetes is attributed to the

Krkonoše-Jizera Terrane (KJT, after Narębski 1994). Three lithotectonic units are currently distinguished in the Krkonoše-Jizera Terrane sequence and are described in structurally upwards succession (e.g. Kachlík & Patočka 1998).

(1) The basal autochthonous unit includes the Cadomian Lusatian granitoids, which intrude the late Proterozoic flysch sequence (Chaloupský *et al.* 1989). The unit is exposed along the NW margin of the Ještěd Range (the westernmost part of the KJT at the boundary with the Lusatian Granitoid Complex, Fig. 1) as a foreland of overlying lithotectonic units. It experienced greenschist facies metamorphism of Cadomian age (between *c.* 560 and 545 Ma) and has a non-penetrative Variscan overprint (e.g. Kröner *et al.* 1994*a*).

(2) The overlying parautochthonous to allochthonous unit contains a very low-grade metamorphosed early to late Palaeozoic volcanosedimentary sequence with features typical of the Saxothuringian Zone (e.g. Chlupáč 1993). This unit, which experienced only late Variscan lower greenschist facies metamorphism forms several imbricated slices in the central and possibly also the eastern parts of the Ještěd Range (Kachlík & Patočka 2001).

(3) The uppermost allochthonous composite unit comprises a large antiform of the Izera and Krkonoše (Kowary) gneisses, which include metamorphosed Cambrian to Ordovician granitoids (e.g. Borkowska *et al.* 1980; Kryza & Pin 1997; Białek 1998). The core is intruded by the late Variscan Krkonoše-Jizera granite pluton. Its southern and eastern rims consist of the Lower Palaeozoic metamorphosed volcanosedimentary sequences of the South and East Krkonoše Complexes (e.g. Fajst *et al.* 1998; Kachlík & Patočka 1998; Patočka *et al.* 2000; Dostal *et al.* 2001). The East Krkonoše Complex (EKC), exposed on the east of the KJT, was defined by Berg (1912), Oberc (1960) and Teisseyre (1968). It comprises the Rýchory Mountains (Czech Republic), the Lasocki Range and Rudawy Janowickie Mountains (Poland) (Patočka & Smulikowski 1998). The South Krkonoše Complex (SKC) is situated on the SW margin of the KJT (Fajst *et al.* 1998). The complexes show considerable diversity both in metamorphic grade and protolith composition, and are mostly tectonically bounded. They underwent early Variscan blueschist facies metamorphism, followed by a widespread greenschist facies overprint (Patočka *et al.* 1996) which is related to the Early Carboniferous tectonic uplift of the previously subducted crustal slices.

Interpretations of the tectonometamorphic development of the KJT vary widely, but encompass the effects of the Cadomian, Caledonian (?) and Variscan orogenies (see review by Chlupáč 1993). Lithostratigraphic studies on the South Krkonoše Complex (Chaloupský 1963, 1966; Kachlík 1997; Chlupáč 1997, 1998), and the ^{40}Ar/^{39}Ar dating from the East Krkonoše Complex (Maluski & Patočka 1997) show that the KJT structure is a result of Variscan tectonometamorphic processes. The KJT volcanosedimentary successions were deposited during a protracted period of intracontinental rifting of the Cadomian basement, and (as suggested by some of the EKC metabasites) the formation of an oceanic basin of limited extent (Kryza *et al.* 1995; Winchester *et al.* 1995; Patočka *et al.* 2000; Dostal *et al.* 2000, 2001).

Collision of peri-Gondwanan microplates with Baltica (± East Avalonia) in Middle to Late Devonian closed the above-mentioned basin, and produced progressive stacking of the basin fill. The early Variscan subduction-related blueschist facies metamorphism of estimated peak conditions T = 400–450 °C and P = 10–12 kbar (Patočka *et al.* 1996) affected the rocks of the subducted plate. Its ending is dated at 365–360 Ma (Maluski & Patočka 1997). The subsequent greenschist facies retrogression (345–340 Ma; Maluski & Patočka 1997) was followed by the late tectonic Krkonoše–Jizera pluton intrusions (Pin *et al.* 1987; Mierzejewski *et al.* 1994) and major late Variscan shearing and thrusting which produced the NW–SE directed linear fabric of the KJT.

Góry Sowie Block

The Góry Sowie Block (GSB) is a tectonostratigraphic unit composed of the gneiss–migmatite Góry Sowie Complex (GSC) with minor occurrences of felsic granulite and basite–ultrabasite rocks (Żelaźniewicz 1990, 1995). The triangular block is bounded by steep fault zones with records of polyphase ductile and then brittle deformation. Magnetic and gravimetric data suggest that its eastern part is underlain by the Sudetic ophiolite (Fig. 1) dated by the U–Pb zircon method at *c.* 420–400 Ma (Oliver *et al.* 1993; Żelaźniewicz *et al.* 1998). During the Late Devonian the GSB together with the ophiolite were rapidly exhumed and both delivered clasts to adjacent sedimentary Late Devonian–Early Carboniferous Bardo and Świebodzice basins (e.g. Żelaźniewicz 1997; Hladil *et al.* 1999; Kryza *et al.* 1999).

Based on relationships between deformational structures and successive stages of migmatite formation, five phases (deformation phases D_1–D_5) of evolution of the gneiss–migmatite portion of the GSC have been discerned, with peak metamorphic conditions during migmatization at 3–6 kbar and 700–730 °C (Żelaźniewicz 1990, 1995). Several isotopic dates on migmatites formed in different phases show that the peak conditions were attained at Mid–Late Devonian boundary (van Breemen *et al.* 1988; Bröcker *et al.* 1998; Timmermann *et al.* 2000). Age constraints on post-peak metamorphism are documented by Rb–Sr ages between 375 ± 4 and 360 ± 7 Ma (van Breemen *et al.* 1988; Bröcker *et al.* 1998) for D_2–D_5 migmatite phases. They suggest a rapid uplift, which was contemporaneous with blueschist-facies metamorphism in Palaeozoic rocks of the East Krkonoše Complex (Maluski & Patočka 1997).

Minor occurrences of felsic granulites and mantle-derived garnet peridotite, yielding Early Devonian isotopic ages (O'Brien *et al.* 1997; Brueckner *et al.* 1996) were tectonically inserted into migmatites (Żelaźniewicz 1995) and were retrogressively metamorphosed contemporaneously with the progressive migmatization under upper amphibolite facies conditions which obliterated the boundary shear zones.

Orlica-Śnieżnik Dome

The Orlica-Śnieżnik Dome (OSD) is the southeasternmost tectonostratigraphic unit of the West Sudetes (Fig. 1) and the Saxothuringian Zone, and it has a complicated fault zone contact with the Moravo-Silesian zone further east. The OSD generally consists of a mainly orthogneissic core, locally embracing (ultra) high pressure eclogites and granulites, enveloped by a variegated middle–lower amphibolite facies series of mica schists and paragneisses with marbles and amphibolites, which is surrounded in turn by greenschist-facies metapelites and metabasites. Metabasites are geochemically linked with an intraplate rift setting (Floyd *et al.* 1996). Palaeontological data, although controversial, for metasediments (Gunia 1996 and references therein) point to the late Proterozoic–Early Cambrian age. Pb/Pb dating of zircon from a leptynite yielded an age of *c.* 520 Ma (Kröner *et al.* 1997).

The orthogneiss core of the OSD consists of (1) porphyritic coarse- to medium-grained metagranites (Śnieżnik augen gneisses) of calc-alkaline affinity and (2) laminated, variably grained, two-mica migmatitic alkaline gneisses (Gierałtów gneisses), with lensoid bodies of augen gneisses, amphibolites, eclogites and felsic granulites (Borkowska *et al.* 1990). A structural inventory of the migmatitic gneisses differs profoundly from that of the augen metagranites, whereas both the gneiss variants are zonally sheared and deformed to mylonites and ultramylonites.

Eclogite bodies have amphibolitized margins and occur in narrow discontinuous belts (Dumicz 1993). Their protoliths range from MORB, through calc-alkaline to bimodal volcanic rocks (Bakun-Czubarow 1998), which excludes single source of the original mafic rocks and obscures their relationships to the gneissic hosts. Most of the eclogites show signs of UHP metamorphism under conditions of 660–800 °C and above 27 kbar, followed by marked decompression under amphibolite facies conditions (Bakun-Czubarow 1992, 1998; Bröcker & Klemd 1996), with isothermal decompression in the pressure range 9–5 kbar at ≤ 600 °C (Kozłowski & Bakun-Czubarow 1997), or 11–4 kbar at 650–600 °C (Bröcker & Klemd 1996). Strong metamorphic contrasts between neighbouring eclogites, granulites and some gneisses and amphibolites on one hand and other gneisses and amphibolites, mica schists and phyllites on the other hand, point to large tectonic displacements between and within these rock units.

Review of geochronology

Lusatian Granitoid Complex (LGC)

Constraints on the sedimentation age of the Lusatian greywackes, representing the oldest rocks of the LGC, are given by magmatic zircons of synsedimentary pyroclastic intercalations dated at 562 ± 4 Ma using the Pb/Pb evaporation method (Gehmlich *et al.* 1997). These late Proterozoic greywackes were later intruded by the voluminous Lusatian pluton granitoids.

The time of zircon crystallization in the western Lusatian granodiorites is given by Pb/Pb zircon ages between 550 and 535 Ma (mean age 542 Ma) for the biotite-granodiorite near Kindisch reported by Tichomirova *et al.* (1997) and for the muscovite-bearing biotite-granodiorite (near Kubschütz) with 542 ± 9 Ma (Kröner *et al.* 1994*a*).

Further east, on the northwestern flank of the Krkonoše-Jizera region, the Lusatian granodiorites, referred to as the Zawidów granodiorite and the Leśna gneisses (foliated variant of granodiorite) which also yielded Early Cambrian intrusion ages (U–Pb zircon lower intercept) of 540 +6/−7 Ma and 540 +19/−21 Ma, respectively (Korytowski *et al.* 1993). Different older emplacement Pb/Pb single zircon evaporation ages for the various granitoids of the LGC between 587 and 560 Ma were reported by Kröner *et al.* (1994*a*) which coincides with

K-feldspar Pb/Pb model ages between 589 and 563 Ma obtained by Bielicki *et al.* (1989).

The so-called Rumburk granite of East Lusatia is considered as a member of the most differentiated and youngest granitoid generation which intruded the LGC. For the undeformed Rumburk granite a late Middle Cambrian Rb–Sr whole rock age of 501 ± 32 Ma was reported by Borkowska *et al.* (1980). More recently, it yielded latest Cambrian to Early Ordovician Pb/Pb zircon emplacement ages between 494 ± 12 Ma and 480 ± 12 Ma (Hammer *et al.* 1997). An older emplacement age of 571 ± 16 Ma is given by Kröner *et al.* (1994*a*).

Small bodies of hornblende monzogranite (Wiesa) and amphibole-bearing granodiorite (Kleinschweidnitz) were dated by the Pb/Pb single zircon evaporation method at 304 ± 14 Ma and 312 ± 10 Ma (Late Carboniferous), respectively (Kröner *et al.* 1994*a*; Hammer *et al.* 1997).

A compilation of published and unpublished Rb–Sr and K–Ar age data (up to 1992) of the pre-Variscan Lusatian granitoids is given by Kröner *et al.* (1994*a*).

Krkonoše-Jizera Terrane (KJT)

From the KJT (situated to the east of the LGC), the oldest available ages were provided by the Izera Gneisses derived from the Cambrian Izera granites. In the western part of the KJT, the gneissic samples from Frýdlant yielded Pb/Pb single zircons ages of 515 ± 8 Ma and 504 ± 10 Ma (Kröner *et al.* 1994*b*). Likewise in the eastern part of the KJT, the weakly foliated granitic samples from Perła Zachodu yielded the U–Pb zircon lower intercept ages of 515 +5/–7 Ma (Korytowski *et al.* 1993) and 514 +5/–6 Ma (Philippe *et al.* 1995). A further U–Pb lower intercept age of 493 ± 2 Ma for a Rumburk-type metagranite was regarded as the minimum emplacement age due to moderate U-content in the investigated zircons (Oliver *et al.* 1993).

The Rb–Sr whole rock ages of weakly deformed Izera Gneiss (462 ± 15 Ma) and a leucogneiss variety (473 ± 16 Ma) dated by Borkowska *et al.* (1980) were possibly affected by (partial) homogenization during the Variscan metamorphic overprint. The Late Carboniferous Rb–Sr ages of 320–310 Ma (using one muscovite and two biotite grains) were interpreted as the products of isotopic resetting due to the Krkonoše-Jizera pluton intrusion (see below) (Borkowska *et al.* 1980).

The Kowary gneiss (Poland), petrographically equivalent of the Izera Gneiss from the northern part of the KJT yielded compatible U–Pb zircon lower intercept ages between 492 and 481 Ma (Oliver *et al.* 1993). Published Pb/Pb single zircon ages on the Krkonoše gneiss (on the Czech territory corresponding to the Kowary gneiss) date its magmatic origin between 509 and 490 Ma (Kröner *et al.* 1994*b*, 1997).

East and South Krkonoše Complexes (EKC; SKC) In the EKC the geochronological data were reported from the Rýchory Mountains (Czech Republic), and their northern continuation, the Rudawy Janowickie Mountains (Poland). The porphyroids from the Rýchory Mountains provided the Rb–Sr whole rock age of 495 ± 9 Ma and in combination with associated greenschists, an age of 501 ± 8 Ma. Both of the ages were interpreted to date the Cambro-Ordovician rift-related magmatism (Bendl & Patočka 1995). Recent U–Pb data on zircons from the Rýchory Mountains mafic blueschists yielded a protolith age of 485 ± 4 Ma (Timmermann *et al.* 1999). In the Rudawy Janowickie Mountains, Oliver *et al.* (1993) obtained by U–Pb method (on zircons) an age of 505 ± 5 Ma from a felsic volcanic rock boudin as well as an age of 494 ± 2 Ma from a rock described as 'hornblende gabbro', that in fact corresponds to a Paczyn gneiss variety (Patočka & Smulikowski 1998).

The Bítouchov metagranite from the South Krkonoše Complex (SKC) was dated at *c.* 540 Ma (U–Pb zircon; Dörr unpublished). This metagranite yields a homogeneous age pattern recording neither inherited components nor younger zircons (Dörr pers. comm.), and thus may indicate the presence of a Cadomian basement.

The early Variscan high pressure–low temperature metamorphism of the EKC in the eastern Rýchory Mountains was dated by the single grain Ar/Ar technique on phengites in the mafic blueschists yielding Late Devonian plateau ages of 364 ± 2 Ma and 359 ± 2 Ma, respectively and were interpreted as the end of subduction-related blueschist metamorphism. Other phengites were resetted by the subsequent retrogressive greenschist facies overprint and provided plateau ages between 345 and 340 Ma (Maluski & Patočka 1997).

Krkonoše-Jizera pluton The late-tectonic Krkonoše-Jizera pluton intruded central sectors of the KJT. It consists essentially of several different granite types representing at least two distinct (major) magmatic events: The medium grained aphyric Tanvald two-mica granite, cropping out only on the SW and west margin of the Krkonoše-Jizera pluton, is assumed to be the oldest type (Klomínský 1969). The medium to coarse grained porphyritic granite (the 'Liberec-type' granite), and a younger fine to medium grained facies, the 'Krkonoše ridge-type granite' (e.g. Mierzejewski *et al.* 1994) constitute the main part of the pluton. The emplacement of the porphyritic Liberec-type granite was dated by Rb–Sr whole rock isochrons at around 330–325 Ma (Pin *et al.* 1987; Duthou *et al.* 1991) whereas the finer grained Krkonoše ridge-type granite yielded Rb–Sr whole rock isochron ages about 310 ± 5 Ma (Mierzejewski *et al.* 1994). However, the coarse grained porphyritic monzogranite near Liberec provided a Pb/Pb zircon evaporation age of 304 ± 14 Ma (Kröner *et al.* 1994*a*).

Góry Sowie Block (GSB)

Pb/Pb single zircon ages between 487 ± 2 and 482 ± 2 Ma of gneisses in the GSB have been interpreted as crystallization ages of Ordovician granitoids (Kröner & Hegner 1998). The layered migmatitic gneiss from Zagórze and the diatexitic migmatite from Potoczek yielded the U–Pb monazite ages of 381 ± 2 Ma (lower intercept; van Breemen *et al.* 1988) and of 383–379 Ma (nearly concordant; Bröcker *et al.* 1998), respectively. They are identical with concordant xenotime ages of 384–380 Ma obtained for the augen gneiss from Sokolec (Bröcker *et al.* 1998) and correspond well with recent U–Pb monazite and xenotime ages of 378 ± 2 Ma and 383–370 Ma, for the anatectic granite and the pegmatite, respectively (Timmermann *et al.* 2000). These highly consistent results of rocks deformed during the D_2 to D_5 stages contrast with the Pb/Pb single zircon evaporation data published by Kröner & Hegner (1998), which range from 473 to 440 Ma, and the U–Pb zircon lower intercept age of 461 +50/–2 Ma of the diatexite (Oliver *et al.* 1993). After studying the internal morphology of zircons from these rocks the apparent Ordovician ages were reinterpreted as inherited or mixed ages dominated by an older core component (Timmermann *et al.* 2000). This interpretation is supported by several well-documented Rb–Sr thin-slab and mineral isochron ages of the Góry Sowie migmatites varying from 375 ± 4 Ma to 362 ± 8 Ma (Bröcker *et al.* 1998) and 372 ± 7 Ma to 360 ± 7 Ma (van Breemen *et al.* 1988). These results together point to Late Devonian high temperature metamorphism (cf. Tucker *et al.* 1998).

Somewhat older Early Devonian U–Pb and Pb/Pb metamorphic zircon ages of 401 ± 10 Ma and 402 ± 1 Ma, respectively were reported for felsic high pressure granulites (O'Brien *et al.* 1997). For a mantle-derived garnet–peridotite associated with the granulites, Brueckner *et al.* (1996) obtained an identical age of 402 ± 3 Ma using the Sm–Nd method (Cpx–Opx–Grt–whole rock isochron). These ages were interpreted as mineral growth ages during high pressure-granulite facies pressure and temperature conditions.

A single post-tectonic granitoid body in the eastern part of the GSB (Piława), dated at 334 ± 2 Ma (Pb/Pb zircon; Kröner & Hegner 1998), is identical to an adjacent syn-tectonic granodiorite of the Niemcza Zone to the east (Fig. 1), which yielded a Pb/Pb single zircon age of 334 ± 2 Ma (Kröner & Hegner 1998) and a U–Pb lower intercept zircon age of 338 +2/–3 Ma (Oliver *et al.* 1993) from the same quarry at Koźmice.

Late Variscan activity in the GSB is documented by the Ar/Ar total fusion ages of two to three grains of muscovite and biotite (Oliver & Kelley 1993) which show a range of ages of mylonitized gneisses between 337 ± 13 Ma to 319 ± 17 Ma. These are thought to provide a very rough time estimate for movements on mylonite zones bordering the GSB.

Orlica-Śnieżnik Dome (OSD)

The protolith ages of orthogneisses in the OSD have been determined on many zircons by the U–Pb and Pb/Pb methods. Cambrian Pb/Pb single zircon ages of *c.* 520 Ma were obtained for felsic metavolcanic rocks in the Stronie formation and for the Gierałtów and Śnieżnik orthogneisses (Kröner *et al.* 1997), which intruded the Stronie Formation. However, Pb/Pb zircon ages of 507 ± 10 Ma, 503 ± 4 Ma and 499 ± 15 Ma were also reported as apparent emplacement ages for their granitic protolith (Kröner *et al.* 1994*b*). The differences are possibly caused by an influence of inherited components, as revealed by U–Pb dating of abraded zircons from the Śnieżnik gneiss dated between 540 and 500 Ma (Borkowska & Dörr 1998). These results were confirmed by SHRIMP U–Pb and Pb/Pb analyses on zircons yielding ages of *c.* 500 Ma reflecting the age of magmatic crystallization of the protolith and 540–530 Ma in a few inherited zircon cores for both Śnieżnik and Gierałtów gneisses (Turniak *et al.* 2000). U–Pb zircon lower intercept ages between 504 ± 3 Ma and 488 +4/–7 Ma of mylonitized Śnieżnik gneiss support a Cambro-Ordovician emplacement of the gneiss protolith (Oliver *et al.* 1993; Kröner *et al.* 1994*b*). The Rb–Sr whole rock analyses of the Gierałtów gneiss yielded ages of 487 ± 11 Ma (van Breemen *et al.* 1982) as well as 465 ± 35 Ma and 464 ± 18 Ma (Borkowska *et al.* 1990), both interpreted as protolith emplacement ages. Some data record a Devonian metamorphic event in the OSD; for example the Rb–Sr whole rock age of 395 ± 35 Ma on the Śnieżnik augengneiss (Borkowska *et al.* 1990), the Rb–Sr thin slab whole rock isochron age of 396 ± 17 Ma and the U–Pb zircon lower intercept age of 372 ± 7 Ma both on Gierałtów gneisses associated with eclogites, and nearly concordant U–Pb zircon ages between 369 and 360 Ma on an omphacite granulite assumed as the date of high pressure–high temperature metamorphism (Bröcker *et al.* 1997).

Carboniferous ages of metamorphism were documented in all the above-mentioned lithologies or varieties respectively. Eclogite lenses yielded Sm–Nd garnet–whole rock–(clinopyroxene) isochron ages, grouped between 352 and 326 Ma (Brueckner *et al.* 1991; Bröcker *et al.* 1997) and the U–Pb zircon lower intercept age of 337 ± 3 Ma was considered by Bröcker *et al.* (1997) to date the late stage of high pressure-metamorphism. However, the SHRIMP analyses on rims of the Gierałtów gneiss zircons yielded concordant age of 342 ± 6 Ma (Turniak *et al.* 2000) which is considered to record the high temperature–low pressure metamorphism peak. The Rb–Sr biotite–muscovite–whole rock (Śnieżnik gneiss) and phengite–whole rock (eclogite; Gierałtów gneiss) isochron ages of 335 ± 5 Ma (Borkowska *et al.* 1990) and 333–329 Ma (Bröcker *et al.* 1997), respectively, fit into this range.

The late Variscan history of the OSD was dated by the ^{40}Ar/^{39}Ar method. The ^{40}Ar/^{39}Ar plateau ages of 329 ± 2 Ma (muscovite) and 328 ± 2 Ma (biotite) were reported for the migmatitic Gierałtów gneiss; a muscovite age of 328 ± 2 Ma for the mylonite Śnieżnik augengneiss and a hornblende age of 327 ± 2 Ma for the Lewin Kłodzki amphibolites have also been recorded (Steltenpohl *et al.* 1993). An identical biotite plateau age of 328 ± 3 Ma for a migmatitic gneiss in the Śnieżnik unit is given by Maluski *et al.* (1995). All these late Viséan Ar/Ar ages were interpreted as uplift-related cooling ages.

The youngest age in the OSD (313 ± 3 Ma) was obtained by the ^{40}Ar/^{39}Ar method on undeformed primary muscovite. It reflects a cooling under static conditions after a late increase in temperature (Maluski *et al.* 1995).

Analytical techniques

^{40}Ar/^{39}Ar measurements

The mineral concentrates were prepared by crushing, sieving, Frantz magnetic separation and handpicking. Finally the minerals were cleaned using ultrasonic treatment successively in alcohol, acetone and distilled water.

According to the grain sizes, either mineral concentrates (≤ 250 μm) or single grains (> 250 μm) were analysed. In the case of the mineral concentrates, *c.* 100 mg of material encapsulated in evacuated quartz vials were irradiated for 58 hours in the Osiris reactor in Saclay (France). The single grain samples, wrapped in aluminium-foil packets were irradiated in the McMaster reactor in Ontario (Canada) for nearly 70 hours. The monitors were the MMhb-1 hornblende standard (Alexander *et al.* 1978) with a recommended age of 520.4 ± 1.7 Ma (Samson & Alexander 1987) and an internal hornblende standard with an intralaboratory age of 344.5 ± 3 Ma.

The ^{40}Ar/^{39}Ar analyses were completed at the ^{40}Ar/^{39}Ar Laboratoire de Géochronologie of the Université Montpellier II. The mineral concentrates were incrementally heated at first with a resistance furnace followed by a Mo crucible coupled with a high-frequency inductor for the high-temperature steps. The isotopic measurements were carried out on a highly modified THN 205E noble-gas mass spectrometer. Details of the analytical procedure are given in Maluski *et al.* (1993) and Monié *et al.* (1994).

The single grain analyses were performed using a LEXEL 3500 continuous 6W argon-ion laser for step-wise heating and a MAP 215-50 noble gas mass spectrometer equipped with a Nier source and a Johnston MM1 electron multiplier for the mass analyses. A detailed description is given by Monie *et al.* (1994, 1997).

The measured isotope ratios were corrected for total system blanks, atmospheric contamination, effects of mass discrimination, irradiation induced mass interference due to Ca and Cl and radioactive decay of ^{37}Ar and ^{39}Ar isotopes. The age calculation is based on the constants recommended by the 'IUGS subcommission on geochronology' quoted in Steiger & Jäger (1977) and cited by McDougall & Harrison (1999). The reported 1σ-errors for plateau and total ages include the uncertainties of the monitors and their ^{40}Ar/^{39}Ar ratios. The detailed results can be obtained from the Society Library or the British Library Document Supply Centre, Boston Spa, Wetherby, West Yorkshire LS23 7BQ, UK as Supplementary Publication No. SUP 18179 (14 pages).

Electron microprobe analyses

Polished thin-sections was used for electron microprobe studies. Analyses of mineral chemistry were prepared on the Cameca SX100 electron microprobe at the University of Montpellier II. The operating conditions were a 20 kV acceleration voltage, a 10 nA beam current and a counting time of 30 s per element (peak and background).

Results

$^{40}Ar/^{39}Ar$ dating

Twenty-eight samples of crystalline rocks from the West Sudetes were dated by the $^{40}Ar/^{39}Ar$ incremental step-wise heating technique (Figs 2, 3, 4). Depending of the mineral grain sizes, either mineral concentrates (conventional technique) or single grains (laser technique) were analysed.

From the East Lusatia (west of Bílý Kostel) a cataclastic granodiorite (the Zawidów-type) was analysed (sample SK30). It contains sericite flakes (pseudomorphs after muscovites) and relicts of chloritized biotite. A small (~ 300 μm) sericite–muscovite single grain (SK30 #1) displayed an irregular complex age spectrum whereas the second analysis of two sericite-muscovite grains heated together (SK30 #2) showed an age spectrum with low-temperature steps beginning at 152 ± 23 Ma and increasing regularly up to a plateau of 323 ± 6 Ma (87.3% of ^{39}Ar released).

From the Krkonoše-Jizera Terrane (KJT) 21 samples were analysed, including rocks of the Izera Gneiss, Intra-Sudetic Fault Zone, Krkonoše-Jizera pluton, South and East Krkonoše Complexes.

Two samples of biotite and muscovite rich Izera gneiss near Siedlecin (Perła Zachodu) were investigated: an undeformed metagranite (sample SU35) and a sheared gneiss (SU33). Single muscovite of the former sample displayed a completely disturbed age spectrum lacking any clear features. Muscovite from the sheared gneiss (SU33) revealed a disturbed spectrum as well, but, beside the scattering low-temperature steps and one exceptional step at intermediate temperatures of 366 Ma (representing the interval of 39 to 55% of the released ^{39}Ar) an age of *c.* 335–328 Ma is quite evident. Two distinct 'plateau ages' were calculated, excluding the step of 366 Ma, which yield ages of 335 ± 3 Ma and 328 ± 3 Ma corresponding to 27.5% and 44.9% of the released ^{39}Ar, respectively. A coexisting sheared biotite yielded an overall plateau age of 294 ± 3 Ma, but with two recognizable plateaus for the low- and high-temperature steps, which are 289 ± 3 Ma (54.8% released ^{39}Ar) and 301 ± 4 Ma (42.8% released ^{39}Ar), respectively.

Two mylonites (samples SU39 and SU42) from the Intra-Sudetic Fault (Pilchowice dam) belonging to the KJT were analysed. A single grain of muscovite from the sample SU39 yielded the plateau age of 324 ± 3 Ma (93% released ^{39}Ar) followed by two younger steps and a final high-temperature step of 326 ± 3 Ma. The muscovite in sample SU42 displayed an age spectrum beginning with seven scattered low-temperature steps (9.4% of the released ^{39}Ar) followed by an age plateau at 333 ± 3 Ma, corresponding to 68.6% of the ^{39}Ar release and ending with two high-temperature steps with an integrated age of 339 ± 3 Ma.

From the Krkonoše-Jizera pluton two granite varieties, the Liberec-type granite and Tanvald-type granite were dated. The former is a porphyritic biotite–monzogranite (sample SK207) which provided a single biotite for analysis. The obtained age spectrum displayed increasing low-temperature steps (representing 20% of the released ^{39}Ar) from 50 Ma up to 316 Ma. The following age plateau comprising 16 heating steps (70% of ^{39}Ar released) yielded age of 320 ± 2 Ma. Two last high-temperature steps reveal a slight decrease in ages to 315 Ma and 314 Ma. Single muscovite of the Tanvald-type two-mica granite (sample SK208) records a well-defined plateau age at 312 ± 2 Ma (94.8% of the released ^{39}Ar).

The following data were obtained on the set from the South Krkonoše Complex (SKC). From the phyllonitized metagranite (sample SK3) near Železný Brod eight very small grains (≤ 200 μm) of muscovite were analysed together displaying regular increment of ages until the total fusion age of 350 ± 4 Ma, representing 62.5% of released ^{39}Ar. Zoned phlogopite from an altered olivine–pyroxene minette (SK4) cross-cutting the SKC metasediments after young apparent ages in the first four low-temperatures steps revealed ages between 321 and 327 Ma followed by an age plateau at 314 ± 6 Ma corresponding to 71.5% released ^{39}Ar.

From a metagabbro (SK7) sample from Loužnice an amphibole (crossite overgrown by aggregates of fibrous pale actinolitic hornblende and actinolite) revealed a well-defined age plateau at 321 ± 6 Ma after the first scattered low-temperature steps corresponding to 10.9% of the released ^{39}Ar.

Muscovite from a mylonitized Bitouchov metagranite (SK8) showed a plateau age of 352 ± 6 Ma (93.4% released ^{39}Ar) whereas the muscovite concentrate of a nearby greenschist sample (SK9) from the Jizera River Valley yielded a concordant plateau age of 344 ± 3 Ma (94.7% released ^{39}Ar).

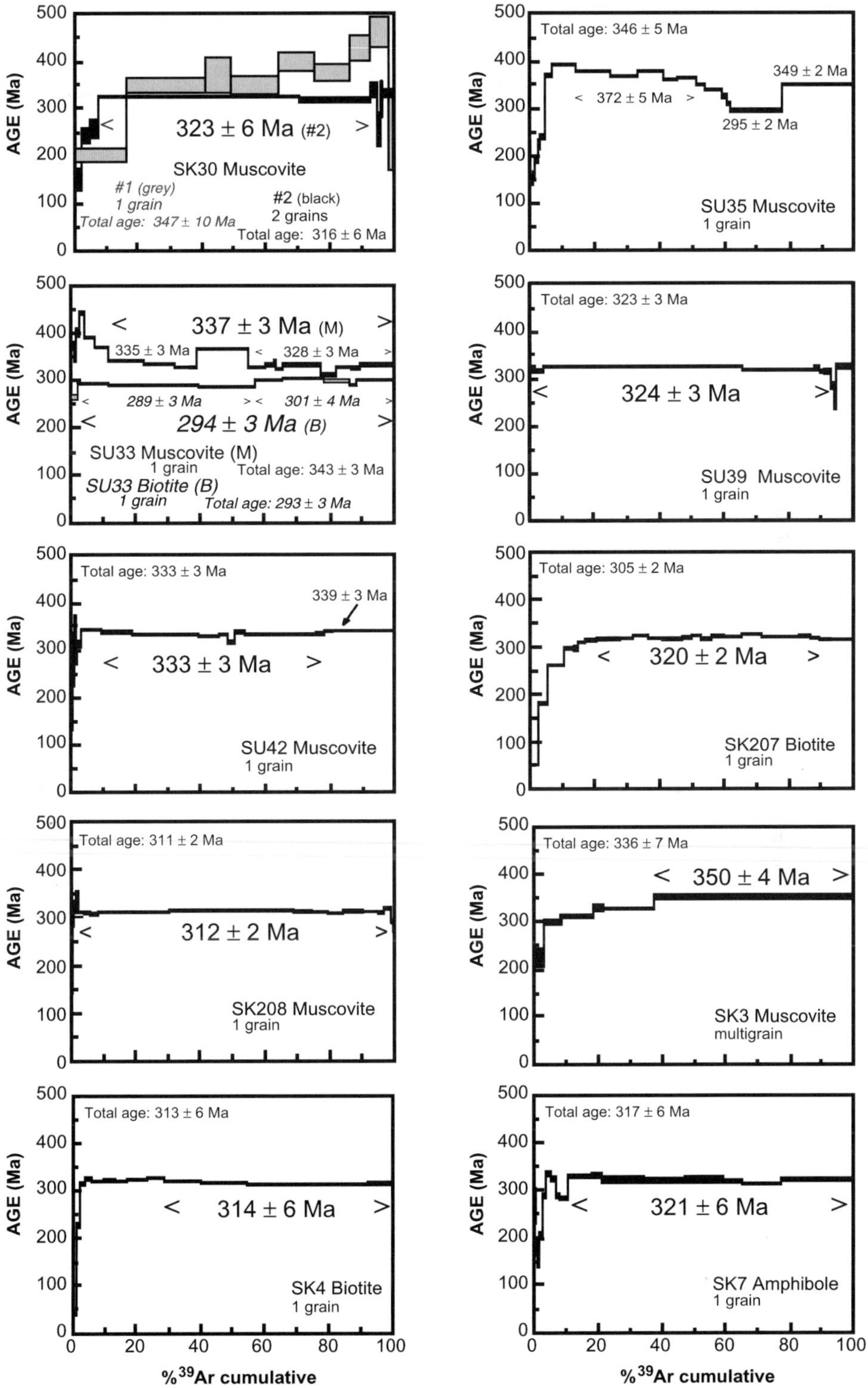

Fig. 2. ^{40}Ar/^{39}Ar age spectra from the Lusatian Granitoid Complex and the Krkonoše-Jizera Terrane.

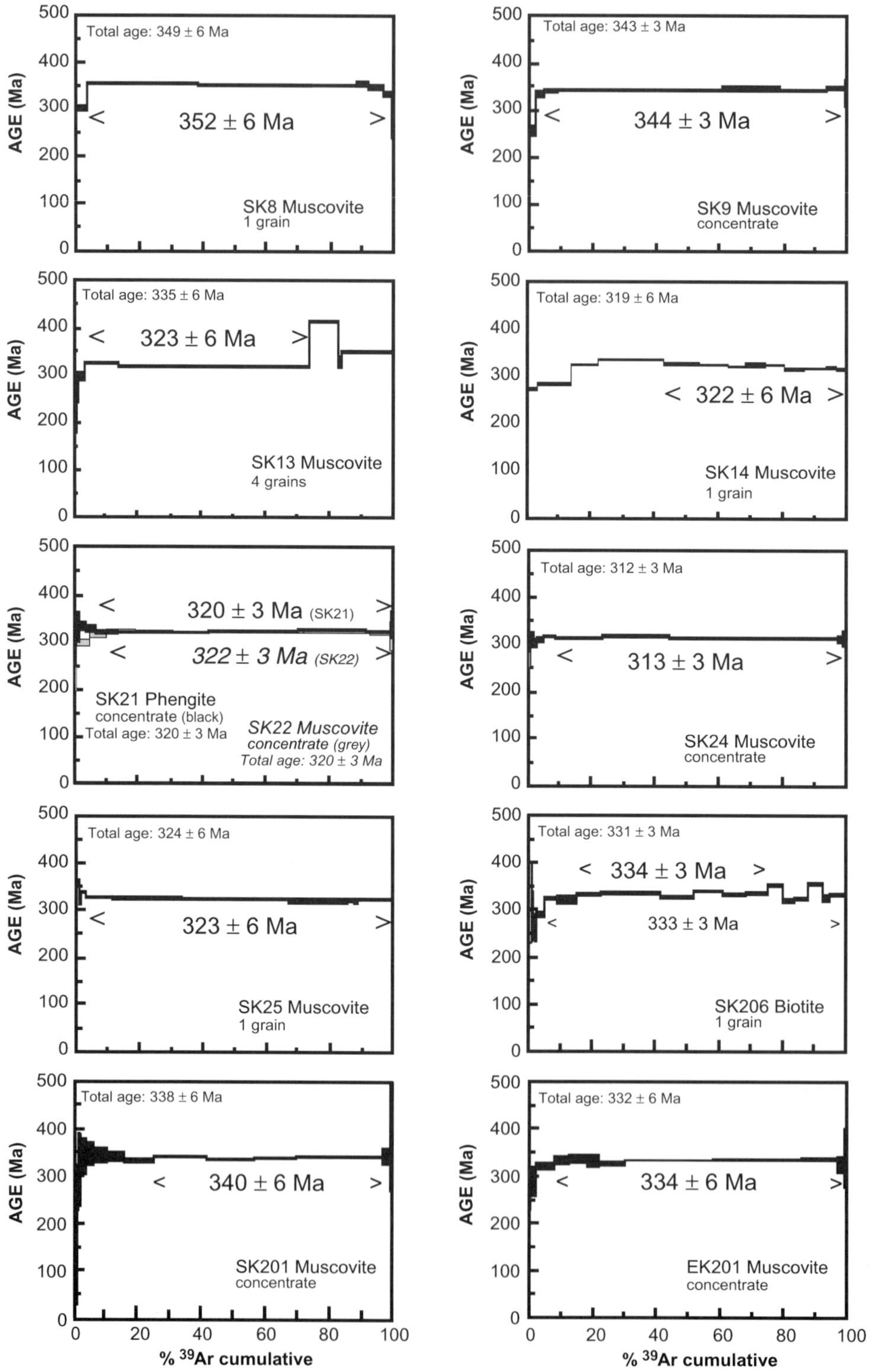

Fig. 3. $^{40}Ar/^{39}Ar$ age spectra from the Krkonoše-Jizera Terrane.

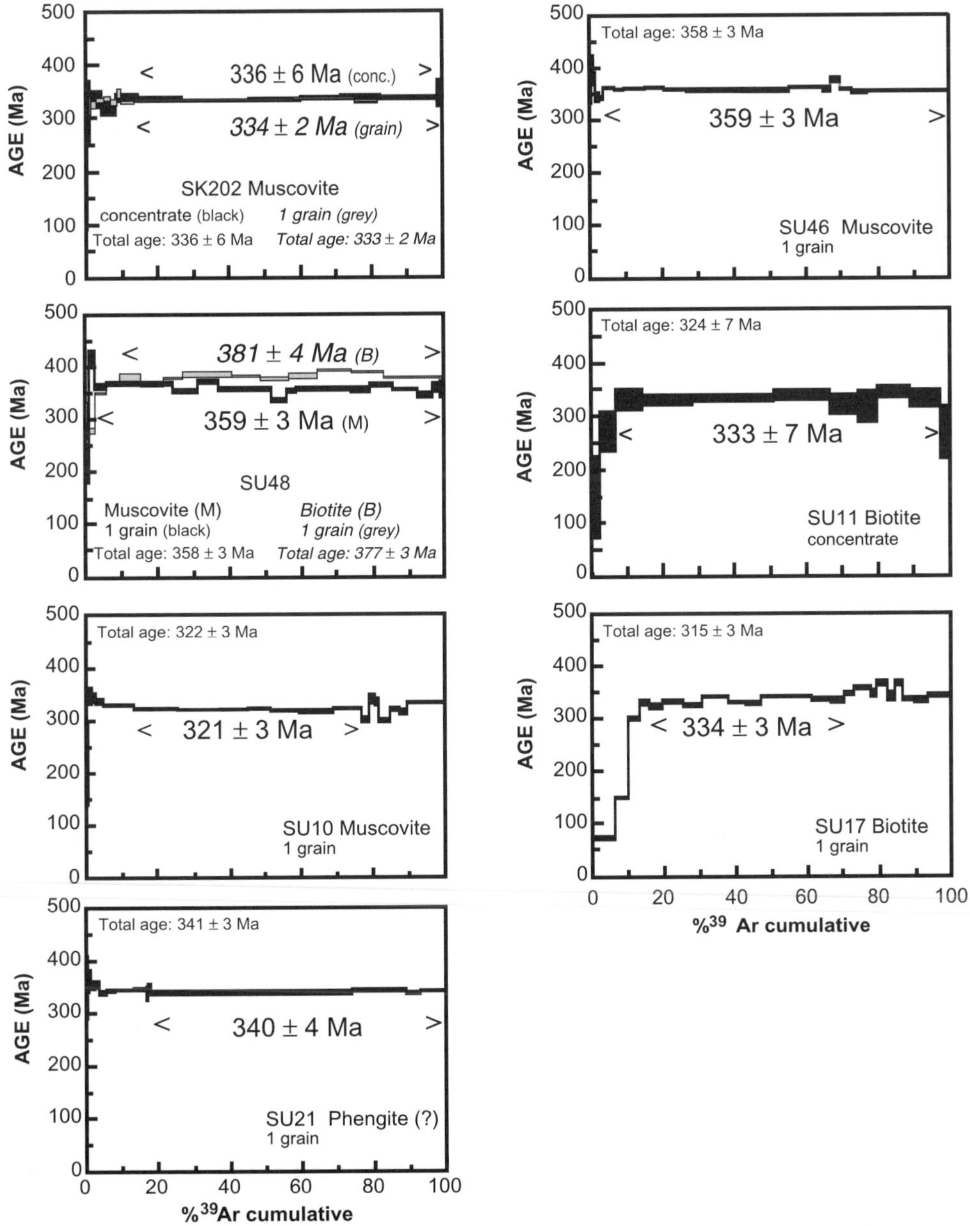

Fig. 4. ^{40}Ar/^{39}Ar age spectra from the Krkonoše-Jizera Terrane, Góry Sowie Block and the Orlica-Śnieżnik Dome.

From a banded metatuffite greenschist sampled near Návarov Castle (SK13) four small muscovite grains (160–250 μm) were dated together, and after the first low-temperature steps, displayed an age plateau at 323 ± 6 Ma, representing 70% of released ^{39}Ar. This is followed by an exceptionally high value of 418 Ma after which the ages decreased abruptly and ended in a total fusion age of 352 ± 1 Ma.

From the western part of the SKC (close to the sample site SK13) a large sericite flake provided by a Ordovician–Silurian sericite–quartz metarhyolite tuff (SK14) from the Kamenice River was analysed. The age spectrum showed increasing low- to intermediate-temperature steps up to 335 Ma continuing in the age plateau of 322 ± 6 Ma equal to 57.1% of the released ^{39}Ar. Mineral concentrates of phengite from the sheared blueschist and phengitic muscovite in sheared phyllite (SK21 and SK22, respectively) near Vrchlabí yielded well-defined plateau ages of 320 ± 3 Ma (92.4% ^{39}Ar release) and 322 ± 3 Ma (89% released ^{39}Ar), respectively.

From the sheared Krkonoše gneiss sampled at the Elbe Dam, the muscovite concentrate (SK24) was dated giving a plateau age of 313 ± 3 Ma, corresponding to 92.5% released ^{39}Ar. A single grain of muscovite from the sheared Krkonoše gneiss (SK25) sampled in the Úpa Valley near the Leszczyniec shear zone provided a plateau age of 323 ± 6 Ma calculated for 96.6% of the ^{39}Ar release. A single biotite separated from massive coarse-grained porphyritic metagranite (SK206) from the northeastern part of the Krkonoše gneiss body displayed considerable age discordance between the low- and high-temperature steps. The intermediate temperature gas fractions give an integrated plateau age of 334 ± 3 Ma representing 60.1% of the released ^{39}Ar.

From deformed Ordovician–Silurian conglomeratic quartzite (SK201) in the vicinity of the Krkonoše orthogneiss in the Malé Labe Valley, a muscovite concentrate was separated and analysed. The obtained age spectrum reveals a distinct age plateau at 340 ± 6 Ma calculated for 72.1% ^{39}Ar released.

Muscovite concentrate was analysed from the strongly sheared and mylonitized porphyroid (sample EK201) located in the Leszczyniec shear zone in the East Krkonoše Complex (EKC). The concentrate records the well-defined plateau age of 334 ± 6 Ma, corresponding to 89.2% of the ^{39}Ar release. Another Ordovician–Silurian quartzite (SK202) near Janské Lázně was dated on single muscovite grain (~1mm) and muscovite concentrate, respectively. The latter yielded an age spectrum with irregular low-temperature steps, terminated in a flat age plateau at 336 ± 6 Ma (83.9% of the released ^{39}Ar). The single grain records the age plateau at 334 ± 2 Ma, corresponding to 90.1% of the released ^{39}Ar.

From the SW, part of the Góry Sowie Block (GSB) muscovite from the D_3 anatectic granite (sample SU46, south of Walim) with sillimanite lineation (Żelaźniewicz 1990) yielded the plateau age of 359 ± 3 Ma representing 96.3% of released ^{39}Ar. The muscovite of sheared gneiss (SU48) from a sinistral fault zone of the SW margin of the GSB near Przygórze revealed similar spectrum with the identical plateau age of 359 ± 3 Ma (97.6% of ^{39}Ar release). However, the coexisting biotite displays the age spectrum with irregularly increasing low-temperature steps up to apparent ages slightly oscillating around 380 Ma. The integration of 90.5% released ^{39}Ar give the 'plateau age' of 381 ± 4 Ma.

From the eastern part of the Orlica-Śnieżnik Dome (OSD) four samples were dated: the migmatitic Gierałtów gneiss (SU11), mylonitic Gierałtów gneiss (SU10), mylonitic Śnieżnik gneiss (SU17) and orthogneiss (SU21) associated with eclogites from the transition between Gierałtów and Śnieżnik gneisses.

Biotite concentrate of migmatitic Gierałtów gneiss (SU11) NE of Radochów displays an age spectrum with considerable errors, due to very small amounts of Ar released during the analysis. The integrated plateau age of 333 ± 7 Ma was calculated for 90.5% of the released ^{39}Ar. The apparent younger ages in the initial low-temperature and final high-temperature steps matched higher Ca/K ratios, which suggest intra-sample inhomogeneity. The muscovite of the mylonitic Gierałtów gneiss from a sharply limited sinistral shear zone (SU10) in the migmatitic gneisses (SU11) displays an age plateau at 321 ± 3 Ma, representing 63.7% of the released ^{39}Ar, followed by scattered high-temperature steps finishing at 333 ± 1 Ma.

A large biotite flake (500 μm) of the mylonitized Śnieżnik gneiss at Idzikow (SU17) records an age spectrum with increasing low-temperature steps from 73 to 329 Ma, succeeded by the age plateau at 334 ± 3 Ma, corresponding to 54.6% released ^{39}Ar. The following high-temperature steps scatter between ages of 366 and 333 Ma terminating with a total fusion age of 343 ± 2 Ma. A single white mica (phengite) of the high pressure-orthogneiss at the SE end of Międzygórze (SU21) displays the age plateau of 340 ± 4 Ma, corresponding to 82.2% released ^{39}Ar.

Additional electron microprobe investigations

The blueschist white mica samples occurring on foliation planes or as inclusions in carbonates (from the vicinity of Vrchlabí, SK21) were analysed by electron microprobe (Table 1). The results are plotted in the Si v. Al diagram (see Fig. 5). The foliation plane white micas are phengites with high Si contents of 3.37 to 3.49 per formula unit (f.u.). In comparison, the white mica inclusions in carbonate yielded somewhat lower Si (f.u.) contents of 3.31 to 3.40; although, they are also regarded as phengites. Two white

Table 1. *Microprobe analyses of white micas from sample SK21*

Analysis No.	2	4	5	6	7	8	9	10	11	12	13	14	15	16	18	19	20	21	22	23
Position*	Fol.	Fol.	Fol.	Fol.	Fol.	Fol.	Fol.	Fol.	Fol.	Fol.	Fol.	Fol.	Incl.	Incl.	Incl.	Incl.	Incl.	Incl.	Incl.	Incl.
SiO_2	50.94	50.17	50.99	49.89	49.79	49.61	49.58	49.88	50.30	49.75	51.28	50.81	47.72	49.38	49.33	47.37	49.76	49.61	50.90	49.94
TiO_2	0.18	0.17	0.16	0.20	0.18	0.21	0.18	0.20	0.18	0.18	0.15	0.18	0.09	0.17	0.23	0.05	0.16	0.11	0.13	0.24
Al_2O_3	26.62	27.01	26.74	27.88	27.28	27.09	27.83	28.03	26.46	27.44	24.97	25.80	36.76	28.24	30.21	37.60	27.81	27.93	27.87	28.26
Cr_2O_3	0.02	0.01	0.01	0.03	0.07	0.02	0.00	0.08	0.02	0.04	0.04	0.00	0.04	0.01	0.12	0.04	0.15	0.01	0.06	0.14
FeO^T	2.83	2.74	2.81	3.20	2.81	3.62	2.73	2.92	2.80	2.92	3.05	3.10	1.18	2.80	2.01	0.60	2.76	2.81	2.60	2.73
MnO	0.06	0.03	0.00	0.01	0.01	0.02	0.02	0.01	0.00	0.00	0.00	0.00	0.00	0.00	0.00	0.00	0.02	0.00	0.01	0.00
MgO	3.22	3.13	3.18	2.72	2.92	3.11	2.79	2.76	3.08	2.84	3.69	3.43	0.89	2.76	2.26	0.33	2.65	2.86	3.09	2.52
CaO	0.02	0.00	0.00	0.00	0.03	0.03	0.01	0.00	0.03	0.02	0.00	0.03	0.13	0.05	0.08	0.12	0.04	0.01	0.01	0.02
Na_2O	0.26	0.28	0.21	0.42	0.34	0.39	0.54	0.47	0.28	0.49	0.17	0.16	5.23	0.42	0.69	6.11	0.46	0.50	0.43	0.50
K_2O	9.41	9.66	9.74	9.28	9.02	9.81	9.11	9.00	8.95	8.80	10.16	9.20	2.85	8.93	8.99	1.44	9.13	9.05	9.22	9.04
H_2O	4.43	4.41	4.44	4.42	4.38	4.41	4.40	4.42	4.37	4.39	4.40	4.39	4.63	4.40	4.47	4.66	4.40	4.40	4.48	4.43
Sum	97.99	97.61	98.28	98.07	96.82	98.31	97.19	97.78	96.46	96.88	97.92	97.10	99.51	97.16	98.38	99.32	97.34	97.30	98.80	97.83
Atomic proportions on the basis of 11 oxygen atoms																				
Si	3.443	3.411	3.441	3.377	3.402	3.373	3.378	3.377	3.445	3.396	3.489	3.466	3.089	3.362	3.307	3.046	3.385	3.375	3.405	3.377
Ti	0.009	0.009	0.008	0.010	0.009	0.011	0.009	0.010	0.009	0.009	0.008	0.009	0.004	0.008	0.012	0.002	0.008	0.006	0.006	0.012
Al	2.121	2.165	2.127	2.225	2.197	2.171	2.235	2.237	2.136	2.208	2.002	2.075	2.805	2.266	2.387	2.925	2.230	2.240	2.197	2.252
Cr	0.001	0.001	0.001	0.002	0.004	0.001	0.000	0.004	0.001	0.002	0.002	0.000	0.002	0.001	0.006	0.002	0.008	0.001	0.003	0.008
Fe	0.160	0.156	0.158	0.181	0.160	0.206	0.156	0.165	0.160	0.167	0.173	0.177	0.064	0.160	0.113	0.032	0.157	0.160	0.146	0.154
Mn	0.003	0.002	0.000	0.001	0.001	0.001	0.001	0.001	0.000	0.000	0.000	0.000	0.000	0.000	0.000	0.000	0.001	0.000	0.001	0.000
Mg	0.325	0.318	0.320	0.275	0.298	0.315	0.283	0.278	0.314	0.289	0.374	0.348	0.086	0.280	0.226	0.032	0.269	0.290	0.308	0.254
Ca	0.001	0.000	0.000	0.000	0.002	0.002	0.001	0.000	0.002	0.001	0.000	0.002	0.009	0.003	0.005	0.008	0.003	0.001	0.001	0.001
Na	0.034	0.037	0.027	0.055	0.045	0.051	0.071	0.062	0.037	0.065	0.022	0.022	0.657	0.056	0.090	0.762	0.060	0.067	0.055	0.065
K	0.812	0.838	0.838	0.802	0.786	0.851	0.792	0.778	0.782	0.766	0.882	0.800	0.235	0.775	0.769	0.118	0.792	0.786	0.787	0.780
OH	2.000	2.000	2.000	2.000	2.000	2.000	2.000	2.000	2.000	2.000	2.000	2.000	2.000	2.000	2.000	2.000	2.000	2.000	2.000	2.000
Sum	8.909	8.935	8.920	8.928	8.904	8.982	8.926	8.912	8.887	8.905	8.953	8.899	8.950	8.912	8.915	8.928	8.914	8.925	8.910	8.904

* Fol. - on foliation plane; Incl. - as inclusion

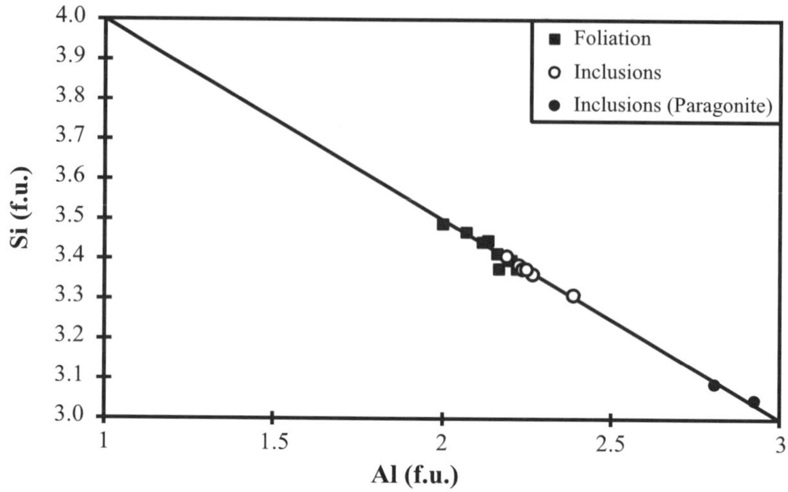

Fig. 5. Si/Al (f.u.) diagram for white micas of the blueschist sample SK21 showing the microprobe results of micas occurring in the foliation plane or as inclusions in carbonate, respectively.

mica inclusions show paragonitic composition that, as regards metabasites, indicate an earlier blueschist facies metamorphic episode (Guidotti 1984). Considering a temperature range of 300–500 °C for the blueschist facies in the South Krkonoše Complex (Patočka *et al.* 1996; Kryza 1998) the pressure-conditions are estimated at more than 10 kbar.

Discussion and interpretation of the $^{40}Ar/^{39}Ar$ ages

The mineral plateau ages are interpreted to date last cooling through appropriate temperatures necessary for intracrystalline retention of argon. Although these so-called closure temperatures are variable and hard to define (e.g. Villa 1997; McDougall & Harrison 1999), 'nominal' values of closure temperatures are used in most geochronological studies. The applied temperatures are typically *c.* 500 °C for hornblende (Harrison 1981), 350 °C for muscovite (e.g. Purdy & Jäger 1976) and 300 °C for biotite (e.g. Hodges 1991).

Lusatian Granitoid Complex (LGC)

The Early Cambrian cataclastic Zawidów-type granodiorite (540 Ma, U–Pb zircon; Korytowski *et al.* 1993) from the southeastern part of the LGC reveals highly disturbed $^{40}Ar/^{39}Ar$ age spectra on recrystallized muscovites (SK30#1) due to partial resetting of their K–Ar systems during sericitization related to the later cataclasis. However, the completely reset sericite flakes (SK30#2) date cataclastic deformation at 323 ± 6 Ma (Viséan/Namurian).

Krkonoše-Jizera Terrane (KJT)

The results from the KJT distinguish several prominent age intervals: 352–350 Ma, 344–340 Ma, 336–333 Ma, 324–320 Ma and 314–312 Ma (Fig. 6). The following discussion is subdivided into these age groups.

352–350 Ma and 344–340 Ma The oldest $^{40}Ar/^{39}Ar$ ages were obtained from strongly mylonitized metagranites of the South Krkonoše Complex (SKC). In the analysed metagranite samples (SK3 and SK8) the box shaped quartz ribbons and plastic recrystallization of quartz indicate elevated temperatures in the range of 400–450 °C during deformation corresponding to higher greenschist facies metamorphic conditions. The samples provided ages of about 352–350 Ma, interpreted as cooling after mylonitization during tectonic exhumation and thrusting after the high pressure event. While SK8 showed a concordant age plateau at 352 ± 6 Ma, SK3 yields younger apparent ages in low- to intermediate-temperature steps due to partial diffusion after the tectonothermal event at 350 ± 4 Ma. This partial diffusion probably took place during uplift-related greenschist facies metamorphism which ceased at *c.* 340 Ma as indicated by $^{40}Ar/^{39}Ar$ ages of 344 ± 3 Ma provided by the metavolcanic greenschist (SK9) of the SKC and 340 ± 6 Ma measured on the ductile deformed conglomeratic quartzite (SK201) overlying the Krkonoše gneiss. These ages are

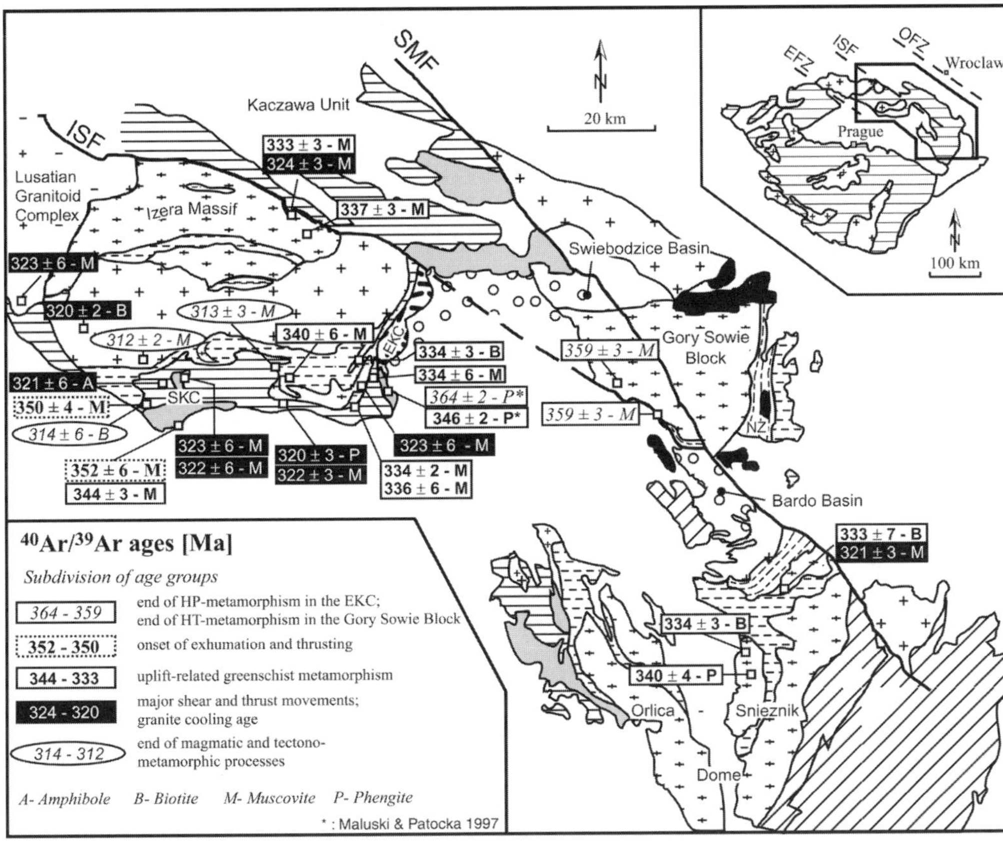

Fig. 6. ^{40}Ar/^{39}Ar plateau ages in the West Sudetes. For the legend and explanations see Figure 1.

consistent with the ^{40}Ar/^{39}Ar age span of 345–340 Ma from the EKC (Rýchory Mountains), which is interpreted as the age of the greenschist facies retrogression after the high pressure–low temperature event (Maluski & Patočka 1997).

336–333 Ma Ages between 336 and 334 Ma were obtained from the slightly deformed Krkonoše gneiss sample (SK206) with only poorly developed foliation as well as from rocks of the East Krkonoše Complexes involving the strongly sheared quartzite sample (SK202) with a prominent NW–SE oriented lineation dipping moderately to the SE, and the mylonitized porphyroid (EK201) of the Leszczyniec shear zone. The similar ages of biotite and muscovite indicate relatively rapid cooling through their appropriate closure temperatures. As both nearly undeformed and mylonitized samples yielded the same ages, some local shearing and deformation under greenschist facies conditions may be assumed as results of the same event. These ages are interpreted to date nappe thrusting in the South and East Krkonoše Complexes.

In the eastern part of the KJT, the Izera gneiss (sheared variety) provided the mica ages reflecting the Viséan tectonothermal overprint. However, the seriously disturbed age spectrum of the sheared gneiss sample (SU33), suggests that an excess of Ar may have been incorporated. Provided that the calculated ages of 335 ± 3 Ma and 328 ± 3 Ma represent the timing of a distinct tectonothermal event, it has to be decided whether they correspond to either post-deformational cooling throughout widespread greenschist facies metamorphism or subsequent local shearing in the Izera gneiss body. The interpretation of this age has to remain not unequivocal because both microfabric studies and the data on the coexisting sheared biotites failed to give any further information. The latter revealed a disturbed age spectrum indicating partial loss of Ar. The apparent ages of 301 ± 3 Ma and 289 ± 3 Ma are therefore regarded as minimum ages which probably post-date the latest granite intrusion of the Krkonoše-Jizera pluton (310 ± 5 Ma: Mierzejewski *et al.* 1994; 304 ± 14 Ma: Kröner *et al.* 1994*a*; 312 ± 2 Ma: this work).

In the northern part of the KJT two mylonites from the Intra-Sudetic Fault (ISF) (Pilchowice dam) provided muscovites yielding distinct plateau ages of 333 ± 3 Ma (SU42) and 324 ± 3 Ma (SU39). Taking into account more pronounced shear strain features of the sample SU39 in a comparison with SU42, the age difference may indicate that during the shearing SU39 experienced also higher temperatures sufficient for opening/resetting of the muscovite K–Ar system at 324 ± 3 Ma.

However, the above described ages may reflect several different stages of (re)activation of the ISF (e.g. Aleksandrowski *et al.* 1997). The latter interpretation is supported by the apparent age given by the last two high-temperature steps in the age spectrum of sample SU42 (339 ± 3 Ma) that is comparable to the age of the greenschist overprint in the EKC and SKC.

According to the Ar/Ar data from the SKC, EKC, Krkonoše and Izera gneisses, ubiquitous early Viséan (344–333 Ma) greenschist facies metamorphism associated with local shearing and thrusting dominated in the KJT during exhumation processes following the high pressure–low temperature event.

324–320 Ma Most of the ages are grouped between 324 Ma and 320 Ma. The example is the previously mentioned mylonite age from the internal part of the ISF at the eastern border of the Izera gneisses of 324 ± 3 Ma. In the SKC the sheared metatuffite greenschist (SK13) and weakly deformed metagabbro (SK7) reveal plateau ages of 323 ± 6 Ma and 321 ± 6 Ma, respectively. The metatuffite underwent a greenschist facies overprint with subsequent strong shearing. The plateau age of 323 ± 6 Ma dates the time of shearing whereas the last high-temperature-steps of the Ar-release spectrum reveal an age of 352 ± 1 Ma which may represent a record of the onset of exhumation and thrusting according to the ages from the metagranites of the SKC. The metagabbro contains indicators (Na-amphiboles) of a former high pressure event which was overprinted under greenschist facies conditions. By comparison with the age of the deformed metatuffite the plateau age of the slightly deformed metagabbro (SK7) is 320 ± 6 Ma and must date either shearing or some late deformation.

In the connecting belt between the South and East Krkonoše Complexes the samples of glaucophane blueschist (SK21) and encompassing graphite–sericite phyllite (SK22) were taken. Both samples retrogresed under greenschist facies conditions, and experienced later shearing, too. The blueschist and phyllite white micas are phengites and phengitic muscovites, respectively (see above). The ages of 320 ± 3 Ma and 322 ± 3 Ma provided by the micas are interpreted as records of late shearing. From the SKC also the sericite–quartz metarhyolite tuff sample (SK14) revealed an identical age of 322 ± 6 Ma. The strongly sheared metagranite sample (SK25) from the southeastern part of the Cambro-Ordovician Krkonoše gneiss (509–490 Ma; Kröner *et al.* 1994*b*, 1997) adjacent to the Leszczyniec shear zone yielded a well-defined age plateau of 323 ± 6 Ma, which is evidence of complete resetting during the shear and fault movements.

An identical $^{40}Ar/^{39}Ar$ age (320 ± 2 Ma) was obtained on the sample SK207 (biotite) from the Krkonoše-Jizera pluton (Liberec-type granite). The biotite age, representing the closure of the K–Ar system at a temperature of *c.* 300 ± 50 °C (e.g. Hodges 1991), constrains an emplacement and cooling of the granite throughout the late Variscan shearing mentioned above and corresponds to results on the magnetic fabric of the Krkonoše-Jizera pluton (Diot *et al.* 1995) as well as to the description of WNW–ESE trending feldspar lineation by Cloos (1925). These structural features indicate that both the granite and in its metamorphic envelope shared deformations. The biotite cooling age fits with the lack of contact metamorphism along the ISF (e.g. Mierzejewski & Oberc-Dziedzic 1990). The contact metamorphism with conditions of temperature above 600 °C and pressure below 2 kbar (Aleksandrowski *et al.* 1997) was older than the biotite cooling at *c.* 320 Ma whereas the semi-brittle and brittle left-lateral displacements along the ISF were the youngest process.

According to the above results, small- to large-scale shear and thrust movements took place in the KJT between 324–320 Ma close to the Viséan/Namurian boundary.

314–312 Ma The youngest ages were obtained on the Tanvald-type granite (Krkonoše-Jizera pluton), the minette dyke cross-cutting the SKC metasediments, and sheared metagranite from the northwestern part of the Krkonoše gneiss nearby the Krkonoše-Jizera pluton. The ideal age plateau of 312 ± 2 Ma of a large muscovite (>2mm) of the Tanvald-type granite (SK208) from the SW border of the Krkonoše-Jizera pluton dates the cooling after a later magmatic pulse than the Liberec-type granite intrusion.

This age matches with the Rb–Sr whole rock ages of 310 ± 5 Ma obtained for the fine-grained variety (ridge-type granite) of the Krkonoše-Jizera pluton (Mierzejewski *et al.* 1994). The muscovite of the Tanvald-type granite, which is considered to be the oldest granite generation of the Krkonoše-Jizera pluton (e.g. Klomínský 1969), may be reset as the result of the late magmatic pulse dated at 312 ± 2 Ma. The late or final magmatic pulse generated the dykes crosscutting the SKC metasediments (Pták 1962). The postmetamorphic and postdeformational dykes are represented by the minette (SK4: phlogopite) dated at 314 ± 6 Ma. They correspond to the final stage of Variscan magmatic and tectonometamorphic processes.

The sample SK24 presents a complex history. Despite its belonging to sheared Krkonoše gneiss, the age of its muscovite is much younger (313 ± 3 Ma) than the defined age of shearing in the KJT (324–320 Ma). Its proximity to the youngest fine grained granite facies may explain a complete resetting of the small muscovites (160–250 μm) which occurred within temperature conditions high enough for such a process, but not sufficient to create contact metamorphic mineral growth.

Góry Sowie Block (GSB)

The syntectonic granite mobilisate (SU46) yielded a muscovite age of 359 ± 3 Ma. This cooling age supports the conclusion of a late Devonian high temperature–medium pressure metamorphism occurring in the GSB, as proposed by Bröcker *et al.* (1998) and Timmermann *et al.* (2000) who found the ages of 372 ± 3 Ma (Rb–Sr isochron) and 378 ± 2 Ma (U–Pb monazite), respectively. Our result, linked to these ages contradicts the interpretation of Kröner & Hegner (1998) based on a Pb/Pb zircon evaporation age of 473 ± 2 Ma.

The sheared gneiss (SU48) from an oblique sinistral shear zone on the SW margin of the GSB near Przygórze yielded the identical muscovite age of 359 ± 3 Ma identical to that of the syntectonic granite mobilisate. The shearing, which overprinted the earlier amphibolite facies dextral strike-slip event, took place under low-temperature greenschist conditions (biotite was substituted by chlorite and quartz behaved plastically) which were insufficient to reset the K–Ar system of the muscovite. Therefore this age is also interpreted as the date of an uplift related cooling postdating the peak metamorphism. The coexisting biotite yielded the 'older plateau age' of 381 ± 4 Ma. The calculation of the corresponding inverse isotope correlation age yielded 372 ± 4 Ma with an inverse ordinate intercept (^{40}Ar/^{36}Ar ratio) of 564 ± 224. This could suggest, although the isotope correlation is not well defined (MSWD = 8.8), an intracrystalline contamination with excess argon components. An alternative explanation of the apparent plateau age (which is even older than the U–Pb monazite ages) would be the K-loss during chloritization of the biotite leading to erroneous higher 'age' values.

Orlica-Śnieżnik Dome (OSD)

The ^{40}Ar/^{39}Ar age results from the Śnieżnik Massif, in the eastern part of the OSD, range between 340 and 321 Ma.

The early Viséan white mica (probably phengite) age (340 ± 4 Ma) obtained on a high pressure orthogneiss (SU21) (associated with eclogites) corresponds with a SHRIMP result on zircon rims of a migmatitic Gierałtów gneiss (342 ± 6 Ma) which is interpreted to be close to the high temperature–low pressure peak metamorphism (Turniak *et al.* 2000). These coincidental zircon and white mica ages imply fast cooling, and represent the time of uplift and decompression after older high pressure and high temperature events.

Biotites separated from the samples of the mylonitic Śnieżnik and migmatitic Gierałtów gneisses (SU17 and SU11) yielded ages of 334 ± 3 Ma and 333 ± 7 Ma, respectively. Both ages represent the continuation of uplift-related cooling since biotites are less Ar-retentive (i.e. lower closing temperatures) than the white micas dated at 340 Ma (SU21). Therefore the minimum age of the Śnieżnik gneiss mylonitization is 334 ± 3 Ma. The equivalent age shown by the migmatitic Gierałtów gneiss (SU11) confirms this mid-Viséan uplift-related cooling, and is consistent with previously published data from the OSD (see Review of geochronology).

The sample of mylonitic Gierałtów gneiss (SU10) from a sinistral shear zone (lineation 240/15) exposed about ten metres from the sample SU11 yielded a muscovite age of 321 ± 3 Ma. Regarding the older biotite age of the mylonite country rock, this age is considered as the time of movements in this discrete shear zone. Consequently it is evidence for the very limited extent of the early Namurian shearing.

Several samples show low-temperature extraction ages (samples: SK4, SK7, SK30, SK207, SU11, SU17, SU35) suggesting a post-Palaeozoic overprint in the northeastern Bohemian Massif. Comparable ^{40}Ar/^{39}Ar ages from this region have been interpreted as thermal pulse related to the Alpine–Carpathian orogeny (Maluski *et al.* 1995).

Conclusions: Variscan tectonothermal development in the West Sudetes

The West Sudetes are a complex mosaic of peri-Gondwanan crustal fragments or terranes that are considered to be members of the recently termed Armorican Terrane Assemblage (Tait *et al.* 1997). Its components were welded together by closing intervening narrow basins (seaways) and oceans due to series of mutual collisions as well as collisions with Baltica (± East Avalonia) culminating in the Variscan orogeny.

The closure of the intervening seaways by oceanic lithosphere subduction was associated with high pressure metamorphism. The oldest high pressure–high temperature event (15–20 kbar, 900–1000 °C) in the West Sudetes was dated at 402 Ma (Early Devonian) on felsic granulites and garnet peridotites of the Góry Sowie Block (Brueckner *et al.* 1996; O'Brien *et al.* 1997). Subsequently, an amphibolite-facies (i.e. high temperature–medium pressure) metamorphism associated with widespread migmatization at temperatures of about 700–730 °C and pressures of around 4–5 kbar affected the granulites and peridotites as well as the surrounding migmatite gneisses (Żelaźniewicz 1995). The peak of the later metamorphic event is dated around 380 Ma (van Breemen *et al.* 1988; Bröcker *et al.* 1998; Timmermann *et al.* 2000), i.e. to the earliest Late Devonian (Frasnian) (cf. Tucker *et al.* 1998). Rb–Sr ages between 374 and 360 Ma (van Breemen *et al.* 1988; Bröcker *et al.* 1998) and $^{40}Ar/^{39}Ar$ data of 359 Ma (SU46, SU48) indicate a post-peak metamorphic cooling down to temperatures of around 350 °C (nominal muscovite blocking temperature for Ar, e.g. Purdy & Jäger 1976), and evidence of a rapid Famennian uplift of the GSB after the preceding early Frasnian amphibolite facies event. This is the final stage of metamorphism in the GSB. Following the Late Devonian the GSB became the source area of conglomerate sequences in the adjacent sedimentary basins (Żelaźniewicz 1997; Kryza *et al.* 1999).

The Famennian uplift of the GSB was contemporaneous with the subduction-related blueschist facies metamorphism in the EKC terminating at about 365–360 Ma (Maluski & Patočka 1997). Eclogite facies metamorphism in the OSD where U–Pb zircon ages in the span between 372 and 360 Ma may define the time of high pressure–high temperature metamorphic episode (Bröcker *et al.* 1997).

The pressure and temperature conditions in the high pressure units of the EKC (Rýchory Mountains) and OSD (Międzygórze and Złote units) are quite different. In the Rýchory Mountains the blueschist facies metamorphism was estimated at T = 300–500 °C and P = 7–10 kbar (Patočka *et al.* 1996; Kryza 1998) whereas the eclogites in the OSD underwent UHP metamorphism in a continent–continent collisional regime at around 660–800 °C and above 27 kbar (Bakun-Czubarow 1989, 1992; Bröcker & Klemd 1996). The ultra high pressure event in the OSD was followed by isothermal decompression and cooling from eclogite to amphibolite facies conditions of T = 600–650 °C and P = 4–11 kbar (Bröcker & Klemd 1996; Kozlowski & Bakun-Czubarow 1997).

The U–Pb and Sm–Nd data that cluster around 340 Ma (early Viséan) are considered to be the final stage of high pressure and temperature metamorphism (Bröcker *et al.* 1997) or as close to the peak of a distinct high temperature–low pressure overprint (Turniak *et al.* 2000), respectively. In a combination with the identical $^{40}Ar/^{39}Ar$ result of 340 Ma on the sample SU21 these ages are interpreted to be a consequence of rapid uplift and decompression during thrusting in east to NE direction (e.g. Turniak *et al.* 2000). This deformation continued through middle Viséan times under greenschist facies conditions as documented by cooling ages of 334–333 Ma on the samples SU11, SU17 and the earlier published Rb–Sr and Ar/Ar data grouped between 335 and 328 Ma (Borkowska *et al.* 1990; Steltenpohl *et al.* 1993; Maluski *et al.* 1995; Bröcker *et al.* 1997).

In the KJT the blueschist facies metamorphism of the latest Devonian age was rapidly followed by tectonic exhumation and thrusting of once deeply buried crustal slices at around 350 Ma (early to middle Tournaisian) producing the uplift-related greenschist facies metamorphism dated at 345–340 Ma (Maluski & Patočka 1997; Marheine *et al.* 1999). Thrusting and deformation under greenschist facies metamorphism produced by propagation of the Variscan orogenic front generally from east to west direction in the KJT (e.g. Kachlík & Patočka 1998) was dated between *c.* 345 and 335 Ma. The propagation of the orogenic wedge is shown by the decrease of metamorphism from a garnet zone on the east to a chlorite zone on the NW and by diminishing ages of flysch sedimentation onsets (Kachlík & Patočka 1998, 2001). The younger ages (of *c.* 335 Ma) occurring in the eastern part of the KJT may be also interpreted as out-of-sequence-stacking at the back of a thrust belt (e.g. Plesch & Oncken 1999), where still deeply buried and therefore hot material was uplifted and thrust over an already cooled nappe-pile. During this stage the dextral strike-slip ISF zone was activated. In a more general interpretation the KJT, comprising the Izera gneiss, Krkonoše gneiss, EKC and SKC underwent a pervasive early to middle Viséan greenschist metamorphism with localized shearing and thrusting between 345 and 335 Ma.

The comparison of $^{40}Ar/^{39}Ar$ ages from the KJT and OSD reveal coeval uplift and decompression in both complexes. The KJT is interpreted as a pile of parautochthonous to allochthonous slices thrust to the NW on the Saxothuringian foreland in Early Carboniferous times, while the OSD is stacked in an east to NE direction on the Moravo-Silesian nappe-pile (e.g. Schulmann & Gayer 2000; Turniak *et al.* 2000). The overthrusting of both complexes on their forelands is related to microcontinent collisions in the West Sudetes (e.g. Maluski & Patočka 1997; Cymerman *et al.* 1997; Kachlík & Patočka 2001).

Subsequent early Namurian (325–320 Ma) small- to large-scale shear movements including thrusting, strike-slip and normal faulting, affected the OSD and KJT as well as the ISF zone and the eastern part of the LGC. These major late Variscan processes modified the dominant composed NW–SE directed linear fabric of the KJT, reactivated the ISF, and generated the contemporaneous emplacement of the late-tectonic Krkonoše-Jizera pluton (biotite $^{40}Ar/^{39}Ar$ cooling age: 320 ± 2 Ma).

The latest magmatic and tectonometamorphic processes in the KJT are dated at 314–312 Ma (Namurian/Westphalian boundary). This limit is set both by the intrusion of the latest Krkonoše-Jizera pluton related dyke (represented by the post-tectonic minette) and the age of the contact aureole of the pluton (within the adjacent contact metamorphosed gneiss). An identical $^{40}Ar/^{39}Ar$ age obtained on post-deformational muscovites from the OSD rocks reflects cooling under static conditions (Maluski *et al.* 1995).

In conclusion, the prominent early Namurian (325–320 Ma) localized shear and thrust movements accompanied (and followed) by NW–SE oriented extension (e.g. Mazur & Kryza 1996) resulted in juxtaposition of the distinct KJT tectonometamorphic units. The corresponding ages of the shear movements in the OSD give evidence for coeval tectonic processes over large parts of the West Sudetes. Thus the juxtaposition of the diversified West Sudetic terranes took place during early Namurian times and the termination of tectonometamorphic processes is constrained at the Namurian/Westphalian boundary.

The distribution of $^{40}Ar/^{39}Ar$ ages in the West Sudetes reflects the complexity of the Variscan polyphase deformation and metamorphism ranging from very low-grade to eclogite facies. The Famennian subduction-related eclogite/blueschist metamorphism may be interpreted as recording closure of narrow seaway(s) between fragments of the Armorican Terrane Assemblage (ATA). That would have required an amalgamation of members of the ATA prior to their final collision with East Avalonia and Baltica, respectively, which commenced in the Tournaisian (e.g. Stoppel & Zscheked 1971; Marheine 1997; Franke 2000). The interpretation is supported by the presence of widespread Famennian subduction-related high pressure–low temperature metamorphism along the Armorican Terranes from Malpica-Tuy in the northwestern Iberian Massif (e.g. Santos Zalduegui *et al.* 1995; Rodriguez Aller *et al.* 1997) through the Ile de Groix and Champtoceaux in the Armorican Massif (e.g. Ballèvre *et al.* 1999, 2000; Bosse *et al.* 2000, 2001) to the East Krkonoše Complex in the Bohemian Massif. All these parts of the ATA subsequently underwent broadly synchronous deformation and thrusting during Viséan exhumation processes provoked by joint accretion and the docking with East Avalonia and Baltica.

Support and funding of this research by the PACE TMR network (contract number ERBFMRX-CT97-0136) from the European Union is gratefully acknowledged. Many thanks for reviews, constructive comments and suggestions to F. Neubauer and W. Dörr.

References

Aleksandrowski, P., Kryza, R., Mazur, S. & Żaba, J. 1997. Kinematic data on major Variscan strike-slip faults and shear zones in the Polish Sudetes, northeast Bohemian Massif. *Geological Magazine*, **134**, 727–739.

Alexander, E. C. Jr., Mickelson, G. M. & Lanphere, M. A. 1978. MMhb-1: A new $^{40}Ar/^{39}Ar$ dating standard. *In:* Zartman, R. E. (eds) *Short Paper, 4th International Conference on Geochronology, Cosmochronology and Isotope Geology. U.S. Geological Survey Open File Report,* ***78–701***, 6–8.

Bakun-Czubarow, N. 1989. On the Sudetic eclogites, SW Poland. *Terra Cognita (Third Int. Eclogite Conf., Würzburg, Germany)* **1**, 2.

Bakun-Czubarow, N. 1992. Quartz pseudomorphs after coesite and quartz exsolutions in the eclogitic clinopyroxenes of the Złote Mountains in the Sudetes (SW Poland). *Archiwum Mineralogiczne*, **48**, 3–25.

Bakun-Czubarow, N. 1998. Ilmenite-bearing eclogites of the West Sudetes – their geochemistry and mineral chemistry. *Archiwum Mineralogiczne*, **51**, 29–110.

Ballèvre, M., Bosse, V. & Capdevila, R. 1999. Large-scale correlations of the Variscan metamorphic terranes across the Ibero-Armorican

Arc (Western Europe). *Journal of Conference Abstracts*, **4**, 88.

BALLÈVRE, M., BOSSE, V., CAPDEVILA, R., DE JONG, K., FÉRAUD, G., GUERROT, C., PEUCAT, J. J. & RUFFET, G. 2000. Structure and thermochronology of the Champtoceaux Complex (France): a key for the Ibero-Armorican correlations. *In: Variscan-Appalachian dynamics: the building of the Upper Paleozoic basement*. Basement Tectonics 15, A Coruña, Spain, Program and Abstracts, 52–55.

BENDL, J. & PATOČKA, F. 1995. The ^{87}Rb-^{86}Sr isotope geochemistry of the metamorphosed bimodal volcanic association of the Rýchory Mts crystalline complex, West Sudetes, Bohemian Massif. *Geologia Sudetica*, **29**, 3–18.

BERG, G. 1912. Die Kristallinen Schiefer des Östlichen Riesengebirges. *Abhandlungen der Preussischen Landesanstanstalt*, **68**, 1–184.

BIAEK, D. 1998. Aspects of geochemistry of the Zawidów Granodiorite and Izera Granite-arc to rift transition? *Geolines*, **6**, 11.

BIELICKI, K. H., HAASE, G., EIDAM, J., KEMNITZ, H., SCHUST, F., GERSTENBERGER, H., HILLER, H. & HABEDANK, M. 1989. Pb-Pb and Rb-Sr dating of granitoids from the Lusatian Block. *ZfI-Mitteilungen, 5th Working Meeting Isotopes in Nature Proceedings II, Leipzig*, 23–45.

BORKOWSKA, M. & DÖRR, W. 1998. Some remarks on the age and mineral chemistry of orthogneisses from the Lądek-Śnieżnik metamorphic massif – Sudetes, Poland. *Terra Nostra*, **98/2**, 27–30.

BORKOWSKA, M., HAMEURT, J. & VIDAL, P. 1980. Origin and age of Izera gneisses and Rumburk granites in the West Sudetes. *Acta Geologica Polonica*, **30**, 121–145.

BORKOWSKA, M., CHOUKROUNE, P., HAMEURT, J. & MARTINEAU, F. 1990. A geochemical investigation of the age, significance, and structural evolution of the Caledonian-Variscan granite-gneisses of the Śnieżnik metamorphic area, central Sudetes, Poland. *Geologia Sudetica*, **25**, 1–27.

BOSSE, V., BALLÈVRE, M., FÉRAUD, G. & PEUCAT, J. J. 2000. Petrological and gochronological constraints in the Ile de Groix blueschists (Armorican Massif, France). *In: Variscan-Appalachian dynamics: the building of the Upper Paleozoic basement*. Basement Tectonics 15, A Coruña, Spain, Program and Abstracts, 63–66.

BOSSE, V., FÉRAUD, G., RUFFET, G., BALLÈVRE, M., PEUCAT, J. J. & DE JONG, K. 2001. Evidence for a late Devonian continental subduction in the Variscan belt: geochronology of eclogite-facies rocks from the Champtoceaux Complex (Armorican Massif, France). *Geological Journal*, **35**, 297–325.

BRÖCKER, M. & KLEMD, R. 1996. Ultrahigh-Pressure metamorphism in the Śnieżnik Mountains (Sudetes, Poland): P-T constraints and geological implications. *Journal of Geology*, **104**, 417–433.

BRÖCKER, M., COSCA, M. & KLEMD, R. 1997. Geochronologie von Eklogiten und assoziierten Nebengesteinen des Orlica-Śnieżnik Kristallins (Sudeten, Poland): Ergebnisse von U-Pb, Sm-Nd, Rb-Sr und Ar-Ar Untersuchungen. *Terra Nostra*, **97/5**, 29–30.

BRÖCKER, M., ŻELAŹNIEWICZ, A. & ENDERS, M. 1998. Rb-Sr and U-Pb geochronology of migmatic gneisses from the Góry Sowie (West Sudetes, Poland): the importance of Mid-Late Devonian metamorphism. *Journal of the Geological Society, London*, **155**, 1025–1036.

BRUECKNER, H. K., MEDARIS, L. G. & BAKUN-CZUBAROW, N. 1991. Nd and Sr age and isotope patterns from Variscan eclogites of the eastern Bohemian Massif. *Neues Jahrbuch für Mineralogie, Abhandlungen*, **163**, 169–196.

BRUECKNER, H. K., BLUSZTAJN, J. & BAKUN-CZUBAROW, N. 1996. Trace element and Sm-Nd 'age' zoning in garnets from peridotites of the Caledonian and Variscan Mountains and tectonic implications. *Journal of Metamorphic Geology*, **14**, 61–73.

CHALOUPSKÝ, J. 1963. Conglomerates in the Krkonoše Crystalline Complex (in Czech, English abstract). *Sborník Ústředního ústavu geologického*, **28**, 143–190.

CHALOUPSKÝ, J. 1966. Caledonian and Variscan orogenies in the Ještěd Crystalline Complex (in Czech, English abstract). *Sborník geologických věd, ř. Geologie*, **10**, 7–37.

CHALOUPSKÝ, J., ČERVENKA, J., JETEL, J., KRÁLÍK, F., LÍBALOVÁ, J., PÍCHOVÁ, E., POKORNÝ, J., POŠMOURNÝ, K., SEKYRA, J., SHRBENÝ, O., ŠALANSKÝ, J., ŠRÁMEK, J. & VÁCL, J. 1989. *Geology of the Krkonoše and Jizerské hory Mts.* Geological Survey Czechoslovakia, Prague.

CHLUPÁČ, I. 1993. Stratigraphic evaluation of some metamorphic units in the north part of the Bohemian Massif. *Neues Jahrbuch für Geologie und Paläontologie, Abhandlungen*, **188**, 363–388.

CHLUPÁČ, I. 1997. Palaeozoic ichnofossils in phyllites near Železný Brod, northern Bohemia. *Journal of the Czech Geological Society*, **42**, 75–94.

CHLUPÁČ, I. 1998. New paleontological finds in the western part of the Krkonoše-Jizerské hory Mts. metamorphics and their impact on stratigraphical and tectonical concepts. *Geolines*, **6**, 29–30.

CLOOS, H. 1925. *Einführung in die tektonische Behandlung magmatischer Erscheinungen (Granittektonik). I Teil. Das Riesengebirge in Schlesien, Bau, Bildung und Oberflächengestaltung*. Berlin.

CYMERMAN, Z., PIASECKI, M. A. J. & SESTON, R. 1997. Terranes and terrane boundaries in the Sudetes, northeast Bohemian Massif. *Geological Magazine*, **134**, 717–725.

DIOT, H., MAZUR, S. & PIN, C. 1995. Karkonosze batholith (NE Bohemian Massif): The evidence for pluton emplacement during transtensional-extensional collapse. *Journal of the Czech Geological Society*, **40**, 62.

DOSTAL, J., PATOČKA, F. & PIN, C. 2000. Early Palaeozoic intracontinental rifting and early seafloor spreading in the Central West Sudetes (Bohemian Massif): Geochemical and Nd-Sr isotopic study on metavolcanic rocks of the East Krkonoše Complex. *Geolines*, **10**, 19–20.

DOSTAL, J., PATOČKA, F. & PIN, C. 2001. Middle/Late Cambrian intracontinental rifting in the central West Sudetes, NE Bohemian Massif (Czech

Republic): geochemistry and petrogenesis of the bimodal metavolcanic rocks. *Geological Journal*, **36**, 1–17.

DUTHOU, J. L., COUTURIE, J. P., MIERZEJEWSKI, M. P. & PIN, C. 1991. Rb/Sr age of the Karkonosze granite on the base of the whole rock method. *Przegląd geologiczny*, **2**, 75–79.

DUMICZ, M. 1993. The history of eclogites in the geological evolution of the Śnieżnik crystalline complex based on mesostructural analysis. *Geologica Sudetica*, **27**, 21–64.

FAJST, M., KACHLÍK, V. & PATOČKA, F. 1998. Geochemistry and petrology of the Early Palaeozoic Železný Brod Volcanic Complex (W Sudetes, Bohemian Massif): geodynamic interpretations. *Geolines*, **6**, 14–15.

FLOYD, P. A., WINCHESTER, J. A., CIESIELCZUK, J., LEWANDOWSKA, A., SZCZEPAŃSKI, J. & TURNIAK, K. 1996. Geochemistry of early Palaeozoic amphibolites from the Orlica-Śnieżnik dome, Bohemian massif: petrogenesis and palaeotectonic aspects. *Geologische Rundschau*, **85**, 225–238.

FRANKE, W. 1989. Variscan plate tectonics in Central Europe – current ideas and open questions. *Tectonophysics*, **169**, 221–228.

FRANKE, W. 2000. The mid-European segment of the Variscides: tectono-stratigraphic units, terrane boundaries and plate tectonic evolution. *In:* FRANKE, W., HAAK, V., ONCKEN, O. & TANNER, D. (eds) *Orogenic processes: Quantification and modelling in the Variscan Belt.* Geological Society, London, Special Publications, **179**, 35–62.

FRANKE, W. & ŻELAŹNIEWICZ, A. 2000. The eastern termination of the Variscides terrane correlation and kinematic evolution. *In:* FRANKE, W., HAAK, V., ONCKEN, O. & TANNER, D. (eds) *Orogenic processes: Quantification and modelling in the Variscan Belt.* Geological Society, London, Special Publications, **179**, 63–86.

FRANKE, W., ŻELAŹNIEWICZ, A., A. POREBSKI, S. J. & WAJSPRYCH, B. 1993. Saxothuringian Zone in Germany and Poland: differences and common features. *Geologische Rundschau*, **82**, 583–599.

GEHMLICH, M., LINNEMANN, U., TICHOMIROVA, M., LÜTZNER, H. & BOMBACH, K. 1997. Die Bestimmung des Sedimentationsalters cadomischer Krustenfragmente im Saxothuringikum durch die Einzelzirkon-Evaporationsmethode. *Terra Nostra*, **97/5**, 46–49.

GUIDOTTI, C. V. 1984. Micas in metamorphic rocks. *In:* BAILEY, S. W. (eds) Micas. *Reviews in Mineralogy*, **13**, 357–456.

GUNIA, T. 1996. Problem wieku marmurów okolicy Stronia Śląskiego na podstawie ikroskamieniałości (Sudety) *Acta Universitatis Wratislaviensis, Prace Geologiczno-Mineralogiczne*, **62**, 5–48.

HAMMER, J., BRÖCKER, M. & KRAUSS, M. 1997. Alter und geologische Signifikanz von Deformationszonen im östlichen Teil des Lausitzer Granitoidkomplexes. *Terra Nostra*, **97/5**, 62–63.

HARRISON, T. M. 1981. Diffusion of ^{40}Ar in hornblende. *Contributions to Mineralogy and Petrology*, **78**, 324–331.

HLADIL, J., MELICHAR, R., OTAVA, J., GALLE, A., KRS, M., MAN, O., PRUNER, P., ČEJCHAN, P. & OREL, P. 1999. The Devonian in the easternmost Variscides, Moravia: a holistic analysis directed towards comprehension of the original context. *Abhandlungen der Geologischen Bundesanstalt*, **54**, 27–47.

HODGES, K. V. 1991. Pressure–temperature–time paths. *Annual Review of Earth Planetary Sciences*, **19**, 207–236.

KACHLÍK, V. 1997. Litostratigrafie a stavba železnobrodského krystalinika: výsledek variských tektodeformačních procesů. *Zprávy o geologických výzkumech v roce*, **1996**, 30–31.

KACHLÍK, V. & PATOČKA, F. 1998. Lithostratigraphy and tectonomagmatic evolution of the Železný Brod Crystalline Unit: Some constraints for the palaeotectonic development of the W Sudetes (NE Bohemian Massif). *Geolines*, **6**, 34–35.

KACHLÍK, V. & PATOČKA, F. 2001. Late Devonian to Early Carboniferous bimodal volcanic rocks of the Ještěd Range Unit: constraints on the development of the Variscan orogenic wedge. *Geolines*, **13**, 74–75.

KLOMÍNSKÝ, J. 1969. Krkonošsko-Jizerský granitoidní masív. *Sborník geologických věd, ř. Geologie*, **15**, 1–134.

KORYTOWSKI, A., DÖRR, W. & ŻELAŹNIEWICZ, A. 1993. U-Pb dating of (meta)granitoids in the NW Sudetes (Poland) and their bearing on tectonostratigraphic correlation. *Terra Nova*, **5**, 331–332.

KOZŁOWSKI, A. & BAKUN-CZUBAROW, N. 1997. Fluidy w minerałach asocjacji eklogitowych z rejonu Śnieżnika. *Polskie Towarzystwo Mineralogiczne – Prace Specjalne*, **9**, 114–115.

KRÖNER, A. & HEGNER, E. 1998. Geochemistry, single zircon ages and Sm-Nd systematics of granitoid rocks from the Góry Sowie (Owl Mts), Polish West Sudetes: evidence for early Palaeozoic arc-related plutonism. *Journal of the Geological Society, London*, **155**, 711–724.

KRÖNER, A., HEGNER, E., HAMMER, J., HAASE, G., BIELICKI, K.-H., KRAUSS, M. & EIDAM, J. 1994*a*. Geochronology and Nd-Sr systematics of Lusatian granitoids: significance for the evolution of the Variscan orogen in east-central Europe. *Geologische Rundschau*, **83**, 357–376.

KRÖNER, A., JAECKEL, P. & OPLETAL, M. 1994*b*. Pb-Pb and U-Pb zircon ages for orthogneisses from eastern Bohemia: further evidence for a major Cambro-Ordovician magmatic event. *Journal of the Czech Geological Society*, **39**, 61.

KRÖNER, A., JAECKEL, P. & JAECKEL, P. 1997. Cambrian to Ordovician granitoid orthogneisses in the Polish and Czech West Sudetes Mts. and their geodynamic significance. *Terra Nostra*, **97/11**, 67–68.

KRYZA, R. 1998. High pressure-low temperature metamorphism in the West Sudetes: Tectonic implications. *Geolines*, **6**, 38–39.

KRYZA, R. & PIN, C. 1997. Cambrian/Ordovician magmatism in the Polish Sudetes: no evidence for subduction-related setting. *Terra Nova*, **7**, 144.

KRYZA, R., MAZUR, S. & PIN, C. 1995. The Lesczyniec meta-igneous complex in the eastern part of the

Karkonosze-Izera Block, Western Sudetes: trace element and Nd isotope study. *Neues Jahrbuch für Mineralogie, Abhandlungen*, **170**, 59–74.

KRYZA, R., MAZUR, S. & ALEKSANDROWSKI, P. 1999. Pre-Late Devonian unconformity in the Kłodzko area excavated: a record of Eo-Variscan metamorphism and exhumation in the Sudetes. *Geologica Sudetica*, **32**, 127–137.

MALUSKI, H. & PATOČKA, F. 1997. Geochemistry and $^{40}Ar/^{39}Ar$ geochronology of the mafic metavolcanics from the Rýchory Mts. complex (west Sudetes, Bohemian Massif): palaeotectonic significance. *Geological Magazine*, **134**, 703–716.

MALUSKI, H., RAJLICH, P. & MATTE, P. 1993. ^{40}Ar-^{39}Ar dating of the Inner Carpathians Variscan basement and Alpine mylonitic overprinting. *Tectonophysics*, **223**, 313–337.

MALUSKI, H., RAJLICH, P. & SOUCEK, J. 1995. Prevariscan, Variscan and Early Alpine thermo-tectonic history of the north-eastern Bohemian Massif: An $^{40}Ar/^{39}Ar$ study. *Geologische Rundschau*, **84**, 345–358.

MARHEINE, D. 1997. Zeitmarken im variszischen Kollisionsbereich des Rhenoherzynikums-Saxothuringikums zwischen Harz und Sächsischem Granulitmassiv – Ergebnisse von K/Ar-Altersbestimmungen. *Göttinger Arbeiten zur Geologie und Paläontologie*, **75**, 1–97.

MARHEINE, D., KACKLÍK, V., PATOČKA, F. & MALUSKI, H. 1999. The Variscan polyphase tectonothermal development in the South Krkonoše Complex (W-Sudetes, Czech Republic). *Journal of Conference abstracts*, **4**, 95.

MATTE, P., MALUSKI, H., RAJLICH, P. & FRANKE, W. 1990. Terrane boundaries in the Bohemian Massif: Result of large-scale Variscan shearing. *Tectonophysics*, **177**, 151–170.

MAZUR, S. & KRYZA, R. 1996. Superimposed compressional and extensional tectonics in the Karkonosze-Izera Block, NE Bohemian Massif. *In:* ONCKEN, O., LANSSEN, C. (eds) *Basement tectonics 11*, Kluwer, Dordrecht, 51–66.

MCDOUGALL, I. & HARRISON, T. M. 1999. *Geochronology and thermochronology by the $^{40}Ar/^{39}Ar$ method*. Second edition. Oxford Monographs on Geology and Geophysics, **9**, Oxford.

MIERZEJEWSKI, M. P. & OBERC-DZIEDZIC, T. 1990. The Izera-Karkonosze Block and its tectonic development (Sudetes, Poland). *Neues Jahrbuch für Geologie und Paläontologie, Abhandlungen*, **179**, 197–222.

MIERZEJEWSKI, M. P., PIN, C., DUTHOU, J. L. & COUTURIE, J. P. 1994. Sr-Nd isotopic study of the Karkonosze granite (Western Sudetes). *In:* KRYZA, R. (ed.) *Igneous activity and metamorphic evolution of the Sudetes area. Abstracts*, Wroclaw University, 82.

MONIÉ, P., SOLIVA, J., BRUNEL, M. & MALUSKI, H. 1994. Les cisaillements mylonitiques du granite de Millas (Pyrénées, France). Âge Crétacé $^{40}Ar/^{39}Ar$ et interprétation tectonique. *Bulletin de la Société Géologique de France*, **165**, 559–571.

MONIÉ, P., CABY, R. & ARTHAUD, M.-H. 1997. The Neoproterozoic Brasiliano orogeny in Northeast Brazil: $^{40}Ar/^{39}Ar$ and petrostructural data from Ceara. *Precambrian Research*, **81**, 241–264.

NARĘBSKI, W. 1994. Lower to Upper Paleozoic tectonomagmatic evolution of NE part of the Bohemian Massif. *Zentralblatt für Geologie und Paläontologie*, **9/10**, 961–972.

OBERC, J. 1960. Tectonics of the eastern Karkonozsy and their relation to the structure of Sudetes (in Polish, English abstract). *Acta Geologica Polonica*, **10**, 1–48.

O'BRIEN, P. J., KRÖNER, A., JAECKEL, P., HEGNER, E., ŻELAŹNIEWICZ, A. & KRYZA, R. 1997. Petrological and isotopic studies on Paleozoic high pressure granulites, Góry Sowie Mts, Polish Sudetes. *Journal of Petrology*, **38**, 433–456.

OCZLON, M. S. 1992. Gondwana and Laurussia before and during the Variscan orogeny in Europe and related areas. *Heidelberger Geowissenschaflichet Abhandlungen*, **53**, 1–56.

OLIVER, G. J. H. & KELLEY, S. 1993. ^{40}Ar-^{39}Ar fusion ages from the Polish Sudetes: Variscan tectonothermal reworking of Caledonian protoliths. *Neues Jahrbuch für Geologie und Paläontologie, Monatshefte*, **1993**, 321–344.

OLIVER, G. J. H., CORFU, F. & KROGH, T. E. 1993. U-Pb ages from SW Poland: evidence for a Caledonian suture zone between Baltica and Gondwana. *Journal of the Geological Society, London*, **150**, 355–369.

PATOČKA, F. & SMULIKOWSKI, W. 1998. The geochemical correlation of the East Krkonoše Complex metabasites (West Sudetes, Czech Republic and Poland) – relation to the Cambrian-Ordovician northern Gondwana breakup: A review. *Acta Universitatis Carolinae, Geologica*, **42**, 485–488.

PATOČKA, F., PIVEC, E. & OLIVERIOVA, D. 1996. Mineralogy and petrology of mafic blueschists from the Rýchory Mts. crystalline complex (Western Sudetes, Bohemian Massif). *Neues Jahrbuch für Mineralogie, Abhandlungen*, **170**, 313–330.

PATOČKA, F., FAJST, M. & KACHLÍK, V. 2000. Mafic-felsic to mafic-ultramafic Early Palaeozoic magmatism of the West Sudetes (NE Bohemian Massif): the South Krkonoše Complex. *Zeitschrift für geologische Wissenschaften*, **28**, 177–210.

PHARAOH, T. C. 1999. Palaeozoic terranes and their lithospheric boundaries within the Trans-European Suture Zone (TESZ): a review. *Tectonophysics*, **314**, 17–41.

PHILIPPE, S., HAACK, U., ŻELAŹNIEWICZ, A., DÖRR, W. & FRANKE, W. 1995. Preliminary geochemical and geochronological results on shear zones in the Izera-Karkonosze Block (Sudetes, Poland). *Terra Nostra*, **95/8**, 122.

PIN, C., MIERZEJEWSKI, M. P. & DUTHOU, J. L. 1987. Rb/Sr isochron age of the Karkonosze granite from quarry Szklarska Poręba Huta and determination of its $^{87}Sr/^{86}Sr$ initial ratio. *Przegląd geologiczny*, **35**, 512–517.

PLESCH, A. & ONCKEN, O. 1999. Orogenic wedge growth during oblique convergence – constraints on the mechanics from its kinematic record (Rhenohercynian thrust belt, Central Germany). *Tectonophysics*, **309**, 117–139.

PTÁK, J. 1962. Lamprophyre dykes of the Jizera granite massif and their relation to the granite tectonics (in Czech). *Zprávy geologického výzkumu v roce*, **1962**, 71–75.

PURDY, J. W. & JÄGER, E. 1976. K/Ar ages on rock-forming minerals from the Central Alps. *Memorie degli Instituti di Geologia e Mineralogia dell'università di Padova*, **30**, 1–31.

RODRIGUEZ ALLER, J., COSCA, M. & GIL IBARGUCHI, J. I. 1997. ^{40}Ar/^{39}Ar dating of Hercynian metamorphism in the HP Malpica Tuy allochthon and LP parautochthon, Iberian Massif, NW Spain. *Terra Nova*, **9**, 497.

SAMSON, S. D. & ALEXANDER, E. C. JR. 1987. Calibration of the interlaboratory ^{40}Ar/^{39}Ar dating standard, MMhb-1. *Chemical Geology*, **66**, 27–34.

SANTOS ZALDUEGUI, J. F., SCHÄRER, U. & GIL IBARGUCHI, J. I. 1995. Isotope constraints on the age and origin of magmatism and metamorphism in the Malpica-Tuy allochthon, Galicia, NW Spain. *Chemical Geology*, **121**, 91–103.

SCHULMANN, K. & GAYER, R. 2000. A model for a continental accretionary wedge developed by oblique collision: the NE Bohemian Massif. *Journal of the Geological Society, London*, **157**, 401–416.

STEIGER, R. H. & JÄGER, E. 1977. Subcommission on geochronology: Convention on the use of decay constants in geo- and cosmochronology. *Earth Planet Science Letters*, **36**, 359–362.

STELTENPOHL, M. J., CYMERMAN, Z., KROGH, E. & KUNK, M. J. 1993. Exhumation of eclogitized continental basement during Variscan lithospheric delamination and gravitational collapse, Sudety Mountains, Poland. *Geology*, **21**, 1111–1114.

STOPPEL, D. & ZSCHEKED, J. G. 1971. Zur Biostratigraphie und Fazies des höheren Mitteldevons und Oberdevons im Westharz mit Hilfe der Conodonten- und Ostracodenchonologie. *Beiheft zum Geologischen Jahrbuch*, **108**, 1–84.

SVOBODA, J. & CHALOUPSKÝ, J. 1966. The West Sudeten Crystalline. *In:* SVOBODA, J., DVOŘÁK, J., HAVLENA, V., HAVLÍČEK, V., HORNÝ, R., CHLUPÁČ, I., KLEIN, V., KOPECKÝ, L., MALECHA, A., MALKOVSKÝ, M., SOUKUP, J., TÁSLER, R., VÁCL, J. & ŽEBERA, K. (eds) *Regional geology of Czechoslovakia I.* Geological Survey, Prague, 215–278.

TAIT, J. A., BACHTADSE, V., FRANKE, W. & SOFFEL, H. C. 1997. Geodynamic evolution of the European Variscan fold belt: Palaeomagnetic and geological constraints. *Geologische Rundschau*, **86**, 585–598.

TEISSEYRE, J. H. 1968. On the Old-Palaeozoic initial volcanism in the West Sudetes. *Acta Geologica Polonica*, **18**, 239–256.

TEISSEYRE, J. H.. 1973. Metamorphic complex of Rudawy Janowickie and Lasocki Grzbiet ridge. *Geologia Sudetica*, **8**, 7–129.

TICHOMIROVA, M., LINNEMANN, U. & GEHMLICH, M. 1997. Zircon ages as magmatic time marks – Comparison of the crustal evolution in different units of the Saxothuringian zone. *Terra Nostra*, **97/11**, 137–141.

TIMMERMANN, H., PARRISH, R. R., NOBLE, S. R., KRYZA, R. & PATOČKA, F. 1999. Single cycle Variscan orogeny inferred from New U-(Th-)Pb data from the Sudetes Mountains in Poland and the Czech Republic. *In:* 4th PACE TMR meeting, Geological Institute, Copenhagen, World Wide Web Address: http://www.pangaea.esci.keele.ac.uk/pace/.

TIMMERMANN, H., PARRISH, R. R., NOBLE, S. R. & KRYZA, R. 2000. New U-Pb monazite and zircon data from the Sudetes Mountains in SW Poland: evidence for a single-cycle Variscan orogeny. *Journal of the Geological Society, London*, **157**, 265–268.

TUCKER, R. D., BRADLEY, D. C., VER STRAETEN, C. A., HARRIS, A. G., EBERT, J. R. & MCCUTCHEON, S. R. 1998. New U-Pb zircon ages and the duration and division of Devonian time. *Earth and Planetary Science Letters*, **158**, 175–186.

TURNIAK, K., MAZUR, S. & WYSOCZAŃSKI, R. 2000. SHRIMP zircon geochronology and geochemistry of the Orlica-Śnieżnik gneisses (Variscan belt of Central Europe) and their tectonic implications. *Geodinamica Acta*, **13**, 293–312.

VAN BREEMEN, O., AFTALION, M., BOWES, D. R., DUDEK, A., MIŠAŘ, Z., POVONDRA, P. & VRÁNA, S. 1982. Geochronological studies of the Bohemian Massif, Czechoslovakia and their significance in the evolution of Central Europe. *Transactions of the Royal Society of Edinburgh, Earth Science*, **73**, 89–108.

VAN BREEMEN, O., BOWES, D. R., AFTALION, M. & ŻELAŹNIEWICZ, A. 1988. Devonian tectonothermal activity in the Góry Sowie gneissic block, Sudetes, southwestern Poland: evidence from Rb-Sr and U-Pb isotopic studies. *Annales Societatis Geologorum Poloniae*, **58**, 3–19.

VILLA, I. M. 1997. Isotopic closure. *Terra Nostra*, **10**, 42–47.

WINCHESTER, J. A., FLOYD, P. A., CHOCYK, M., HORBOWY, K. & KOZDROJ, W. 1995. Geochemistry and tectonic environment of Ordovician meta-igneous rocks in the Rudawy Janowickie Complex, SW Poland. *Journal of the Geological Society, London*, **152**, 105–115.

ŻELAŹNIEWICZ, A. 1990. Deformation and metamorphism in the Góry Sowie gneiss complex, Sudetes, SW Poland. *Neues Jahrbuch für Geologie und Paläontologie, Abhandlungen*, **179**, 129–157.

ŻELAŹNIEWICZ, A. 1995. Część przedsudecka bloku sowiogórskiego. *Przewodnik LXVI Zjazdu PTG, Wroclaw*, 85–109.

ŻELAŹNIEWICZ, A. 1997. The Sudetes as a Palaeozoic orogen in Central Europe. *Geological Magazine*, **134**, 691–702.

ŻELAŹNIEWICZ, A., DÖRR, W. & DUBIŃSKA, E. 1998. Lower Devonian oceanic crust from U-Pb zircon evidence and Eo-Variscan event in the Sudetes. *Terra Nostra*, **98/2**, 174–176.

Palaeozoic terrane amalgamation in Central Europe: a REE and Sm–Nd isotope study of the pre-Variscan basement, NE Bohemian Massif

QUENTIN G. CROWLEY[1,2], HILKE TIMMERMANN[2,4], STEPHEN R. NOBLE[2] & J. GRENVILLE HOLLAND[3]

[1]*School of Earth Sciences and Geography, Keele University, Staffordshire ST5 5BG, UK*

[2]*NERC Isotope Geosciences Laboratory, British Geological Survey, Keyworth, Nottingham NG12 5GG, UK (e-mail: qcrowley@bgs.ac.uk)*

[3]*Department of Geological Sciences, University of Durham, Durham DH1 3LE, UK*

[4]*Present address: Institute für Geowissenschaften und Lithosphärenforschung, JLU, Giessen, D-35390, Germany*

Abstract: The West Sudetes, NE Bohemian Massif, comprises several suspect terranes accreted to the margins of Laurussia during Variscan orogenesis. Whole rock REE and Sm–Nd isotope data for seven separate provinces (Izera, Kaczawa, Rudawy Janowickie and Kłodzko complexes; Fore-Sudetic and Góry Sowie Blocks; Ślęża Ophiolite) suggest involvement of a variety of crustal and mantle sources. Felsic metasedimentary rocks ($\varepsilon Nd(t)$ = –8.3 to –5.0) have two stage T_{DM} ages of 1.9 to 1.5 Ga, whereas acidic metavolcanic rocks and granite gneisses ($\varepsilon Nd(t)$ = –5.4 to +0.8) have two stage T_{DM} ages of 1.5 to 1.0 Ga. A range of sources is implicated: predominantly Archaean and Palaeoproterozoic sources for the metasedimentary rocks, and Archaean, Palaeoproterozoic and Neoproterozoic to early Palaeozoic sources for the meta-igneous felsic lithologies. LREE depleted tholeiitic metabasites ($(Ce/Yb)_N$ = 0.8 to 3.4) generally have $\varepsilon Nd(t)$ = +4.0 to +9.1, indicating derivation from depleted mantle asthenosphere. LREE enriched meta-alkali basalts ($(Ce/Yb)_N$ = 4.6 to 10.1) with $\varepsilon Nd(t)$ between +3.1 and +7.0 implicate utilization of enriched mantle asthenosphere.

Analogous lithologies from elsewhere in the Sudetes, North Bohemian Massif and the Armorican Terrane Assemblage have similar REE abundances, εNd values and T_{DM} ages. Complexes previously considered to have had disparate Neoproterozoic to early Palaeozoic histories may be integrated into a unifying geodynamic model of derivation from the North Gondwanan (North African) margin during a widespread episode of continental margin break-up.

The terrane status, tectonostratigraphic affinity and provenance of several crustal blocks in the West Sudetes (NE Bohemian Massif) of Poland and the Czech Republic is widely disputed. Eastward correlation of the Central European Variscides beyond the Elbe Zone into the NE Czech Republic and SW Poland is problematic due to major strike-slip offset, more intense deformation and a thick Cenozoic metasedimentary cover to the east. Correlation of tectonic units in Central Europe is ultimately dependent on an understanding of large-scale crustal features and delineation of distinct tectonostratigraphic units in the Variscides.

Within the Sudetes a number of crustal blocks have previously been assigned terrane status (e.g. Matte *et al.* 1990; Oliver *et al.* 1993; Cymerman *et al.* 1997). Furthermore, significant crustal lineaments have been classified previously as terrane boundaries (Don 1990; Cymerman & Piasecki 1994). This paper presents new whole rock REE and Sm–Nd isotope data for the West Sudetes and evaluates the significance of the terrane concept to individual crustal blocks from this sector of the NE Bohemian Massif. Previous studies of this kind have tended to focus on a single province (Furnes *et al.* 1994; Bendl & Patočka 1995; Kryza *et al.* 1995; O'Brien *et al.* 1997; Kröner & Hegner 1998; Dostal *et al.* 2000, 2001). In contrast, we compare analogous lithologies from separate fault-bounded blocks within the West Sudetes, as well as analogous lithologies from elsewhere in the Bohemian Massif (Erzgebirge orthogneisses, Kröner *et al.* 1995; Schmadicke *et al.* 1995; Tichomirowa *et al.* 2001; Fichtelgebirge orthogneisses and metavolcanic rocks, Siebel *et al.* 1997; Lusatian granitoids, NE Bohemian Massif granitoids, Kröner *et al.* 1994; KTB metabasites, Teichmann & Basu 1996; Harms *et al.* 1997; von Quadt 1997) and

From: WINCHESTER, J. A., PHARAOH, T. C. & VERNIERS, J. 2002. *Palaeozoic Amalgamation of Central Europe.* Geological Society, London, Special Publications, **201**, 157–176. 0305-8719/02/$15.00

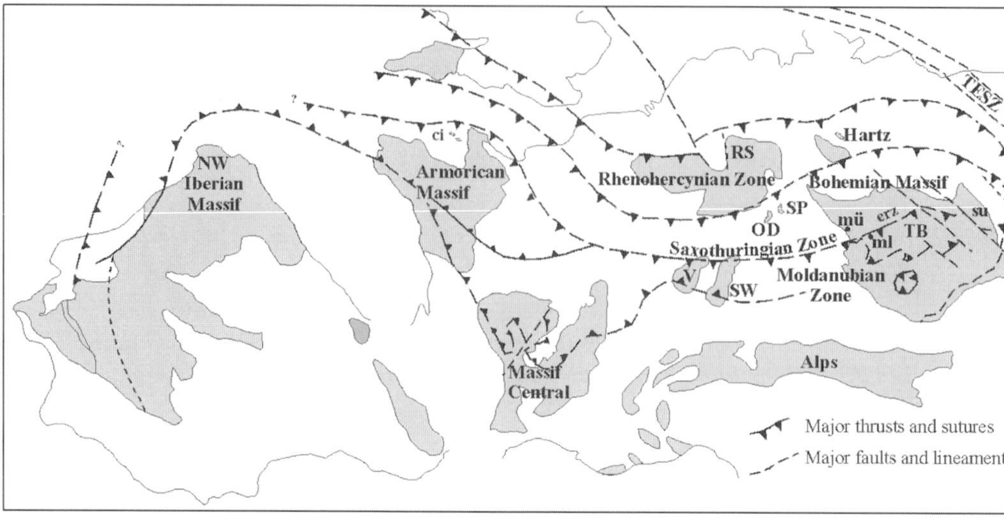

Fig. 1. Sketch map of the Variscan Belt in Europe (pre-Permian configuration). Adapted from Franke 1989; Winchester *et al.* 2002. TESZ, Trans-European Suture Zone; su, Sudetes; mü, Münchberg; erz, Erzgebirge; ml, Mariánské-Lázně; RS, Rheinische Schiefer; SP, Spessart; OD, Odenwald; SW, Scharzwald, V, Vosges; ci, Channel Islands.

also elsewhere in Europe (Massif Central, Armorican Massif, NW Iberian Massif, Pin 1981; Bernard-Griffiths *et al.* 1985; Pin & Carme 1987; Liew & Hofmann 1988; Guerrot *et al.* 1989; Peucat *et al.* 1990; Dallmeyer *et al.* 1991; D'Lemos & Brown 1993; Pin & Marini 1993; Nagler *et al.* 1995; Santos Zalduegi *et al.* 1995, 1996; Dabard *et al.* 1996; Samson & D'Lemos 1998; Fernández-Suárez *et al.* 1999, 2000; Simien *et al.* 1999). REE and Sm–Nd isotope chemistry is used to elucidate further the crustal history (T_{DM} – depleted mantle model ages) and magmatic evolution of crustally derived or dominated lithologies as well as to constrain the sources, mantle evolution and chemistry of mantle-derived magmatic rocks.

An understanding of the tectonic evolution of the West Sudetes provides a basis for a correlative geodynamic model over a wide area of the European Variscides. Whereas a northern North African (West African Craton) origin is widely favoured for the various fault-bounded crustal blocks of the West Sudetes (e.g. McKerrow *et al.* 1992; Cocks 2000; Crowley *et al.* 2000*b*; Franke 2000), inherited components with peri-Amazonian affinities have also been reported from this and other areas of the Bohemian Massif (Friedl *et al.* 2000; Hegner & Kröner 2000). In terms of Neoproterozoic continental reconstructions (e.g. Nance & Murphy 1996) both these interpretations indicate a Gondwanan provenance, however they disagree on whether a North African or South American source was used for this part of the North Bohemian Massif. The new geochemical and isotopic evidence presented here not only serves to aid in a re-evaluation of the affinity of several fault-bounded blocks exposed in the West Sudetes, it also further constrains their tectonic setting, provenance and Neoproterozoic to early Palaeozoic geodynamic evolution.

Geological background

Variscan Europe comprises a number of separate crustal blocks and tectono-stratigraphic units with differing affinities (Kossmat 1927; Matte 1986; Franke 1989, 2000). In Central Europe these blocks include the Armorican Terrane Assemblage (ATA), the high-grade Moldanubian Zone, the Mid-German Crystalline High, and to the north, the Rheno-Hercynian Zone and southern margins of eastern Avalonia (Fig. 1). The geodynamic and tectonothermal history of each of these areas is variably constrained. The ATA is the collective term for a number of distinct micro-plates (Saxothuringia, Teplá-Barrandia, Moldanubia, Armorica, Iberia) that lie south of the southeastern active margin of the Rheno-Hercynian Zone (Mid-German Crystalline High) and form the inner part of the orogen. The root of the Moldanubian Zone is of uncertain origin, whereas relicts of the Cadomian Orogen are preserved in the Teplá-Barrandian Unit and areas of the Saxothuringian basement. These constituent units of the Central European Variscides form a belt of mobile Phanerozoic terranes that amalgamated with Baltica along the Trans-European Suture Zone during the Palaeozoic (e.g. Ziegler 1990; Dallmeyer *et al.* 1995; Pharaoh 1999). Accretion of Avalonia to Baltica and Laurentia was initiated in the early Ordovician. It was accomplished by closure of the Tornquist Sea in the Ashgill and of the Iapetus Ocean by mid to late Silurian times (Cocks & Fortey 1982; Harper *et al.* 1996; Cocks *et al.* 1997; Van Staal *et al.* 1998). Members of the ATA experienced a separate subduction–collision event prior to their late Devonian to early Carboniferous amalgamation along the southern margins of the newly formed Laurussia (Franke 2000; Marheine *et al.* 2000; Marheine *et al.* 2002).

The West Sudetes are amongst the northernmost exposures of the Bohemian Massif; farther to the east lies the East European Craton that developed as part of Baltica. This area of the Bohemian Massif therefore represents a NE termination of exposed terranes affected by Variscan orogenesis and widely considered to have a predominant peri-Gondwanan affinity (Cocks 2000; Crowley *et al.* 2000*a*,*b*; Floyd *et al.* 2000; Franke 2000). Geochronological studies on the timing of tectono-thermal episodes within the Sudetes have highlighted the importance of Cadomian (*c.* 900–500 Ma), pre-Variscan (*c.* 500–400 Ma) and Variscan (*c.* 383–340 Ma) phases of metamorphism or magmatism (Oliver *et al.* 1993; O'Brien *et al.* 1997; Hegner & Kröner 2000; Marheine *et al.* 2000; Timmermann *et al.* 2000). In the West Sudetes in general, the prolonged Palaeozoic magmatic and tectonothermal history that culminated in Variscan Orogenesis may be interpreted in terms of a complex tectonic evolution from continental margin, continental rift, oceanic rift, destructive plate margin and arc–continent to continent–continent collisional tectonic settings. Partial preservation of earlier tectonic cycles and crustal accretion episodes as recorded by inherited or detrital phases in some lithologies attest to the involvement of several generations of relict components from previous orogenies.

Seven areas within the West Sudetes were targeted for sampling (complexes of Izera, Kaczawa, Rudawy-Janowickie, Fore-Sudetic Block, Ślęża Ophiolite, Góry-Sowie Block and the Kłodzko metamorphic complex). Thirty-three samples were selected for Sm–Nd isotope and REE analysis (Fig. 2). As these were chosen from a larger major and trace element geochemical database comprising more than 600 separate whole rock analyses, they are considered representative of the granitic, felsic schistose and metabasic lithologies from each of the complexes.

The Izera Complex consists of variably deformed, greenschist to lower-amphibolite facies, granitic gneisses, meta-rhyolites, schistose felsic metasedimentary rocks (mica schists) and minor metabasite intrusives. Several geochemically similar textural varieties of Izera gneiss exist (Klimas-August 1989; Borkowska *et al.* 1980; Seston *et al.* 2000). U–Pb and $^{207}Pb/^{206}Pb$ zircon ages of *c.* 500 Ma have been interpreted as emplacement ages (Oliver *et al.* 1993; Kröner *et al.* 1994), but inherited 546–916, 1105 and 1176 Ma $^{207}Pb/^{206}Pb$ zircon ages have also been documented (Hegner & Kröner 2000). The post-collisional Karkonosze granite that divides the complex into northern and southern regions intruded these lithologies at 328±12 Ma (Pin *et al.* 1988*a*). Structural evidence indicates that Izera lithologies were probably subjected to both Cadomian and Variscan deformation events (Chaloupski 1988; Seston *et al.* 2000), whereas $^{40}Ar/^{39}Ar$ ages of *c.* 335 Ma place an upper age limit on Variscan greenschist metamorphism in the Izera Complex (Marheine *et al.* 2000). The Izera schists are considered to represent crustally derived rocks into which protoliths of the Izera Gneiss were emplaced, with the metabasite bodies being intruded at some later 'pre-Variscan' stage. Izera quartz–mica schists occur in three main areas; the northern, central and southern schist belts. Three mica schist samples were taken from the northern schist belt (Fig. 2). Metabasites from the northern Izera province are dominantly alkaline in character, but a minor group of metatholeiites also occurs (Seston *et al.* 2000). Two Izera metabasites were also sampled to the north of the Central Schist Belt.

Relict blueschist assemblages have been described from mafic lithologies belonging to the Rudawy-Janowickie Complex, SE of the Izera Complex (Kryza *et al.* 1995; Patočka *et al.* 1996; Smulikowski 1995). A U–Pb zircon age of 485±4 Ma from mafic blueschists (Timmermann *et al.* 1999) represents the protolith age, whereas *c.* 360 Ma and 340 Ma $^{40}Ar/^{39}Ar$ ages have been obtained for the blueschist and greenschist metamorphic events respectively (Maluski & Patočka 1997). Geochemical and Sm–Nd isotope data for rocks of the Leszczyniec region, including the Rudawy-Janowickie Complex (Kryza *et al.* 1995) indicate that the metabasites represent N-MORBs extracted from strongly depleted mantle sources (εNd_{500} +7.0 to +8.0). Also present in the region are two separate types of metarhyolite; those formed by fractionation of mafic magmas (εNd_{500} +6.2 to +6.8) and a

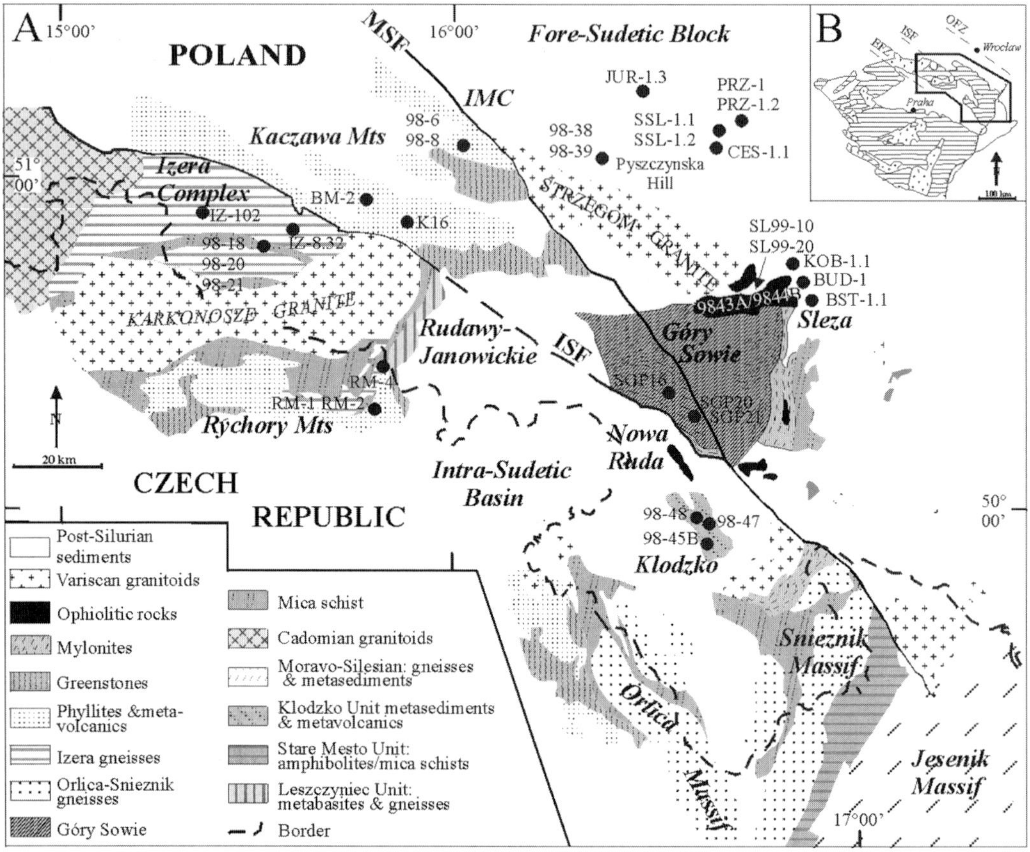

Fig. 2. (A) Simplified geological map of the NE Bohemian Massif (adapted from Aleksandrowski *et al.* 1997) illustrating sample localities of this study. (B) Schematic map of the Bohemian Massif. EFZ, Elbe Fault Zone; ISF, Intra-Sudetic Fault; OFZ, Odra Fault Zone; MSF, Marginal Sudetic Fault; IMC, Imbramowice Metamorphic Complex.

second minor group formed by extreme crustal contamination of mafic magmas (εNd_{500} +2.8 to +1.6). Both mafic and felsic protoliths were considered by Kryza *et al.* (1995) to have been intruded in an extensional setting. Dostal *et al.* (2000, 2001) described similar Cambro-Ordovician bimodal rift-related magmatism from the same region where greenschist, blueschist and amphibolite facies metabasites were found to be characterized by εNd(t) +3.1 to +6.6, +6.0 to +7.7 and +7.0 to +8.0 respectively. Progressive partial melting of upwelling mantle asthenosphere was attributed as a plausible process for the genesis of these metabasites. Highly fractionated metarhyolites gave εNd(t) of –6.0 to –4.0, implying protoliths contain a significant continental crust component. Three samples were taken from this region: two mafic blueschists from Trutnov, and an orthogneiss from Temny Dul.

The Kaczawa Complex is situated NE of Izera between the Intra-Sudetic and Marginal Sudetic faults (Fig. 2). Jerzmański (1965) and Baranowski *et al.* (1990) recognized seven separate tectonic units within the Kaczawa Complex. More recently, Seston *et al.* (2000) have used lithological, structural and geochemical criteria to define two separate thrust sheet units (Świerzawa and Dobromierz) bounded by ductile shear zones. Furnes *et al.* (1989) produced a 511 ± 39 Ma U–Pb zircon age from a metatrachyte, whereas constraints from fossil evidence indicate the presence of mid-Ordovician conodont bearing flysch (Rzeszówek Unit, Świerzawa thrust sheet, Baranowski & Urbanek 1972), Silurian graptolite shales (Chełmiec Unit, Świerzawa thrust sheet, Jerzmański 1965) and Devonian conodont and radiolarian bearing shales and metacherts (Świerzawa thrust sheet, Urbanek 1978). The

Nd isotope characteristics of meta-tholeiitic, transitional and meta-alkali basalts from Kaczawa reported by Furnes *et al.* (1994) were εNd(t) +8.7 to +3.2, 0.0 to +7.2 and +2.7 to +3.7 respectively. These data were taken as evidence for distinctly different mantle sources being used; an N-MORB depleted source, a less depleted mantle source and an enriched asthenosphere plume-like component. Rhyodacite lavas and acidic volcaniclastics with εNd(t) –3.6 to –4.8 were considered to have formed by partial melting of continental crust. Five samples were taken from the Kaczawa Complex for this study; three felsic schistose metasediments from the Dobromierz thrust sheet, a metatholeiitic basalt from the Dobromierz thrust sheet and an alkali metabasalt from the Świerzawa thrust sheet.

The Fore-Sudetic Block is situated east of the Marginal Sudetic Fault and forms the northernmost limit of the Bohemian Massif (Fig. 2). However, it is largely buried under Tertiary cover sequences; outcrops of basement are scarce and it is primarily known from borehole samples. The Fore-Sudetic block consists of variably deformed and predominantly weakly metamorphosed (upper greenschist–lower amphibolite facies) Cambro-Ordovician to early Carboniferous lithologies and is intruded by the Carboniferous post-orogenic Strzegom Granite. Three distinct structural units separated by mylonite zones were recognized by Seston *et al.* (2000) and these are characterized by the following lithological associations: (1) the Wądroże granitoid gneiss, (2) phyllonitic schists and scarce meta-alkali basalts and (3) metatholeiites. Crowley *et al.* (2000*c*) examined the geochemistry of metasedimentary, metabasic and granitic basement lithologies from thirteen separate borehole sites within the Fore-Sudetic Block. In addition to the structural units of Seston *et al.* (2000), the geochemical data indicated that both metabasic and granitic samples from the Borek Strzeliński borehole have affinities with the Ślęża ophiolite. The geological boundaries of this ophiolite body are thus extended, placing it at a structurally higher level in the tectonically imbricated nappe stack of the Fore-Sudetic Block. Eleven samples, nine of which come from seven separate borehole sites (Borek Strzeliński, granitoids; Budziszow, metabasite, metasediment; Cesarzowice, metasediments; Jurcz, metabasites; Kobierzyce, metasediment; Przedmoście, granitoids; Środa-Śląska, metasediments) are included in this study. A further two metabasite samples, from the Imbramowice metamorphic complex, were taken from outcrop at Pyszczyńska Hill. These samples cover a large area of the Fore Sudetic Block between the Marginal Sudetic Fault and the Middle Odra Fault zone further to the east.

A south to north traverse of the Ślęża ophiolite reveals a pseudostratigraphy of deformed serpentinized peridotites, layered gabbros, cumulates, isotropic gabbros, sheeted dykes, fine-grained metabasites and deformed pillow lavas (Narębski & Majerowicz 1985; Floyd *et al.* 2000, 2002). These lithologies have been hornfelsed by intrusion of the post-collisional Strzegom granite. Proximal to the Ślęża ophiolite is the Nowa Ruda ophiolite; both were the focus of a Sm–Nd, Rb–Sr isotope and trace element study by Pin *et al.* (1988*b*). They were dated as 353 ± 16 Ma (whole rock Sm–Nd) and 351 ± 21 Ma (plagioclase–whole rock Sm–Nd) respectively. Żelaźniewicz *et al.* (1998) however, obtained a *c.* 440–400 Ma U–Pb zircon age for Ślęża, suggesting that the younger Sm–Nd ages reflect resetting during a later event. The ophiolite metabasites display isotopic characteristics attributable to their derivation from a depleted mantle source (Ślęża, εNd(t) +7.9 to +8.7; Nova Ruda, εNd(t) +7.6 to +9.1, Pin *et al.* 1988*b*) consistent with their development at normal ocean ridges. Four samples were taken from the Ślęża ophiolite; three metabasalts from Mt Ślęża and one from Kunow, which Narębski & Majerowicz (1985) considered to represent a sheeted dyke complex associated with the ophiolite.

The Góry Sowie block occupies the highest structural position in the Sudetes. It is cross-cut by the Marginal Sudetic Fault, so that its NE part is situated in the Fore-Sudetic block and its SW part in the Sudetes Mountains (Fig. 2). The Góry Sowie block is dominantly composed of amphibolite facies paragneisses containing partly retrogressed high-pressure eclogite to granulite lenses, migmatized pelitic and psammitic metasediments and variably deformed granitoids. These lithologies have been intruded by small metabasite bodies. Rare serpentinites have also been documented from the west of the block (Hentschel 1943; Cymerman 1987; Winchester *et al.* 1998). Winchester *et al.* (1998) attributed intrusion of the metabasites to a widespread rift-related magmatic event prevalent in the Sudetes and the North Bohemian Massif in early Palaeozoic times. Although the Góry Sowie block is considered anomalous in the Sudetes in terms of its relict eclogite to granulite metamorphic assemblages and rare Archaean detrital zircon components (Oliver *et al.* 1993; Kröner & Hegner 1998; Hegner & Kröner 2000), Timmermann *et al.* (2000) argued for a single Variscan orogenic cycle. In a petrological and isotopic study of the high-pressure felsic granulites,

O'Brien *et al.* (1997) reported εNd(t) of –11 to –4 and Nd model ages of 1.4 to 1.8 Ga, interpreted to represent use of highly heterogeneous sources incorporating both Archaean and Proterozoic crustal material. Three granitoids were sampled from the west and SW Góry Sowie block (Fig. 2).

The Kłodzko metamorphic complex is situated to the south of the Góry Sowie block and west of the Intra-Sudetic Fault (Fig. 2). Narębski *et al.* (1988) recognized three separate tectonic units: (1) a lower sedimentary complex containing flysch capped by limestones with an Upper Silurian (Ludlovian) coral fauna; (2) a middle complex dominantly composed of phyllites, metabasites and meta-rhyolites; and (3) an upper magmatic complex consisting predominantly of metabasites and meta-rhyolites. Metamorphic grade varies from mid- to upper-greenschist facies in the north and central parts of the complex to lower amphibolite facies in the south. Narębski *et al.* (1988) considered the products of bimodal magmatism to represent fragmented remnants of the circum-Góry Sowie ophiolites. Three samples were taken from the Kłodzko metamorphic complex, a metabasite from Ksiazek quarry, a metagabbro from Gorzuchow and a metabasite from Święcko, all within the northern upper magmatic complex.

Analytical methods

Whole rock samples were trimmed to discard any weathered material, jaw crushed and ground to a fine powder in an agate tema-mill. All samples were analysed for major and trace elements by XRF spectrometry (ARL 8420) at Keele University, England. Details of analytical techniques can be found in Floyd and Castillo (1992) and Winchester *et al.* (1992). These analyses were used to select representative samples for the subsequent REE and isotope investigations. Cs, Hf, Sc, Ta, U and REE determinations were obtained by ICP-MS at the University of Durham, England. Details of analytical techniques, accuracy and precision can be found in Turner *et al.* (1999).

Samples analysed for their Nd isotope compositions and Sm and Nd concentrations (by isotope dilution) were spiked with a mixed $^{150}Nd/^{149}Sm$ isotope tracer solution and digested in PFA Teflon vessels using HF-HNO_3 at 120 °C for a minimum of sixty-three hours. The samples were then evaporated to dryness and the resultant residue dissolved in distilled HNO_3 and again evaporated to dryness. Residues were re-dissolved in 6 M HCl for 12 hours at 120 °C to ensure complete conversion of the samples to chloride form. The REE were first separated by eluting distilled 2.5 M and 6.2 M HCl through AG50-X8 cation exchange resin in quartz columns. Purification of Sm and Nd was accomplished using quartz columns containing Bio-beads coated with HDEHP-di- (2-ethylexyl) orthophosphoric acid (Richard *et al.* 1976), eluting 0.3 M and 0.6 M HCl to recover Nd and Sm fractions respectively. Sm and Nd concentration and Nd isotope data were obtained on the NERC Isotope Geoscience Laboratory's (NIGL) MAT262 mass spectrometer employing static multi-collection. During the course of analysis, the La Jolla international standard yielded $^{143}Nd/^{144}Nd = 0.511893 \pm 12$ (1σ, n = 13). Repeated analysis of a NIGL house standard prepared from Johnson and Matthey Nd_2O_3 gave $^{143}Nd/^{144}Nd = 0.511163 \pm 11$ (1σ, n = 23). Total analytical blanks for Nd were between 85 and 105 pg for three separate runs. Corrections for mass fractionation were made relative to $^{146}Nd/^{144}Nd = 0.7219$. The ^{147}Sm decay constant used here is 6.54×10^{-12} a^{-1} (Lugmair & Marti 1978). Calculations of εNd(t), T_{CHUR} and T_{DM}; $(^{143}Nd/^{144}Nd)_{CHUR} = 0.512638$, $(^{143}Nd/^{144}Nd)_{DM} = 0.5131$, $(^{147}Sm/^{144}Nd)_{CHUR} = 0.1967$ and $(^{147}Sm/^{144}Nd)_{DM} = 0.225$ (McCulloch *et al.* 1983), were calculated following DePaolo (1981) whereas the two-stage T_{DM} was calculated following the model of Liew and Hofmann (1988).

Results

Rare earth element geochemistry

REE analyses for representative samples in this study are presented in Table 1. Felsic metasediments from the Izera Complex, Kaczawa Complex and the Fore-Sudetic Block display remarkably similar chondrite normalized REE profiles (Fig. 3a), with LREE enrichment ($(Ce/Yb)_N$ = 4.3 to 8.9) and moderate negative Eu anomalies (Eu/Eu* = 0.4 to 0.7). This is consistent with their derivation from a felsic igneous source. Granitoids from the Fore-Sudetic Block and the Góry Sowie Block are also characterized by LREE enrichment ($(Ce/Yb)_N$ = 8.6 to 10.1) and similar negative Eu anomalies ((Eu/Eu* = 0.5 to 0.7) (Fig. 3b). A meta-rhyolite from the Rudawy Janowickie Complex, however, (sample RM-4) with much lower ΣREE (Fig. 3b), only slight LREE enrichment ($(Ce/Yb)_N$ = 2.0) and a significantly negative Eu anomaly (Eu/Eu* = 0.2). This is consistent with extreme fractional crystallization of plagioclase feldspar in a REE depleted magma. In this respect, it is chemically distinct from the other felsic magmatic rocks sampled in this study. The range of REE from

Table 1. *REE data for selected samples from the West Sudetes, NE Bohemian Massif*

Sample	Description	Locality	La	Ce	Pr	Nd	Sm	Eu	Gd	Tb	Dy	Ho	Er	Tm	Yb	Lu	Eu/Eu*
9818	Acidic schist	Izera Complex	39.00	75.11	9.32	43.18	6.19	1.22	4.97	0.74	4.04	0.78	2.02	0.35	2.18	0.36	0.67
98-20	Acidic schist	Izera Complex	21.37	43.64	5.49	20.31	3.70	0.78	2.86	0.48	3.02	0.72	2.28	0.44	2.69	0.42	0.73
98-21	Acidic schist	Izera Complex	20.68	45.58	5.97	22.34	4.34	0.85	3.82	0.72	4.67	1.00	2.69	0.46	0.73	0.44	0.64
IZ-8.32	Alkali metabasite	Izera Complex	11.94	25.61	3.54	15.46	3.68	1.22	4.16	0.64	3.57	0.69	1.68	0.26	1.45	0.22	0.95
IZ-102	Tholeiitic metabasite	Izera Complex	2.45	7.12	1.29	7.18	2.45	0.93	3.76	0.67	4.21	0.94	2.59	0.44	2.62	0.44	0.94
K16	Tholeiitic metabasite	Kaczawa Complex	8.55	21.47	3.27	15.38	3.94	1.43	4.50	0.70	4.00	0.79	2.02	0.32	1.82	0.28	1.04
BM-2	Alkali metabasite	Kaczawa Complex	35.81	76.96	10.41	42.51	8.18	2.36	7.25	1.00	5.20	0.95	2.28	0.35	1.97	0.29	0.94
98-6	Acidic schist	Kaczawa Complex	37.62	75.18	9.50	36.27	6.90	1.45	5.97	0.87	4.77	0.96	2.68	0.45	2.80	0.46	0.69
98-8	Acidic schist	Kaczawa Complex	33.16	67.69	8.24	31.43	5.84	1.22	4.98	0.70	3.88	0.79	2.18	0.38	2.37	0.39	0.69
RM-1	Tholeiitic metabasite	Rýchory Mts	36.90	73.89	9.64	39.35	8.23	2.49	8.68	1.37	8.03	1.65	4.46	0.76	4.47	0.72	0.90
RM-2	Tholeiitic metabasite	Rýchory Mts	30.59	77.55	11.64	53.24	13.12	3.73	14.81	2.38	13.77	2.79	7.09	1.07	5.94	0.76	0.82
RM-4	Orthogneiss	Rýchory Mts	6.28	19.76	1.89	6.62	2.03	0.12	2.33	0.56	3.57	0.81	2.30	0.42	2.56	0.36	0.17
BST-1.1	Plagio-granite	Fore-Sudetic Block	80.19	139.61	15.16	46.87	5.87	1.22	2.68	0.30	1.30	0.21	0.49	0.07	0.43	0.07	0.94
PRZ-1.2	Orthogneiss	Fore-Sudetic Block	36.74	72.91	9.27	36.26	7.31	1.10	6.40	0.89	4.71	0.90	2.31	0.37	2.19	0.35	0.49
CES-1.1	Acidic schist	Fore-Sudetic Block	31.92	60.96	7.45	27.46	4.55	0.93	3.87	0.57	3.10	0.61	1.63	0.28	1.77	0.29	0.68
JUR-1.3	Alkali metabasite	Fore-Sudetic Block	20.23	46.55	6.00	24.94	5.10	1.69	4.93	0.75	4.29	0.84	2.26	0.35	2.11	0.33	1.03
PRZ-1	Tholeiitic metabasite	Fore-Sudetic Block	0.17	0.55	0.12	0.82	0.42	0.26	0.69	0.14	0.93	0.20	0.58	0.10	0.58	0.09	1.47
BUD 1-1	Low-Ti metabasite	Fore-Sudetic Block	0.33	0.68	0.09	0.39	0.12	0.12	0.14	0.03	0.16	0.04	0.10	0.02	0.12	0.02	2.98
KOB-1.1	Acidic schist	Fore-Sudetic Block	28.61	58.92	7.57	28.70	6.26	0.85	6.16	0.99	5.85	1.22	3.44	0.59	3.54	0.56	0.42
SSL-1.1	Low-Ti metabasite	Fore-Sudetic Block	2.69	6.90	1.18	6.46	2.17	0.95	3.00	0.55	3.64	0.80	2.32	0.38	2.29	0.36	1.14
SSL-1.2	Tholeiitic metabasite	Fore-Sudetic Block	3.77	10.01	1.80	9.49	3.08	0.93	4.33	0.75	4.80	1.08	3.13	0.54	3.36	0.57	0.78
98-38	Tholeiitic metabasite	Fore-Sudetic Block	3.10	9.46	1.76	9.60	3.21	1.42	4.65	0.80	4.88	1.05	2.83	0.46	2.63	0.41	1.12
98-39	Tholeiitic metabasite	Fore-Sudetic Block	4.69	11.99	2.00	10.24	3.14	1.21	4.17	0.72	4.38	0.93	2.48	0.40	2.28	0.36	1.02
98-43A	Tholeiitic metabasite	Ślęża Ophiolite	4.39	13.89	2.60	14.09	4.61	1.40	6.71	1.17	7.15	1.54	4.17	0.70	4.02	0.62	0.77
98-44B	Tholeiitic metabasite	Ślęża Ophiolite	9.67	29.35	5.37	28.55	9.08	2.92	12.60	2.19	13.34	2.82	7.50	1.22	7.01	1.09	0.83
98-45B	Tholeiitic metabasite	Kłodzko	6.50	16.63	2.80	14.76	4.90	1.78	7.47	1.34	8.56	1.87	5.20	0.88	5.15	0.82	0.90
98-47B	Tholeiitic metabasite	Kłodzko	4.71	11.78	1.96	10.15	3.11	1.18	4.26	0.74	4.49	0.96	2.56	0.42	2.41	0.38	0.99
98-48	Tholeiitic metabasite	Kłodzko	4.42	13.11	2.48	13.65	3.85	1.33	4.30	0.64	3.61	0.74	1.93	0.30	1.74	0.27	1.00
SGP-16	Granitoid	Sowie Góry Block	36.24	74.41	9.32	34.24	6.25	1.24	5.16	0.78	0.84	0.84	2.32	0.41	2.58	0.43	0.67
SGP-20	Granitoid	Sowie Góry Block	25.88	57.14	7.49	29.82	6.38	1.12	5.62	0.81	3.98	0.69	1.66	0.26	1.46	0.23	0.57

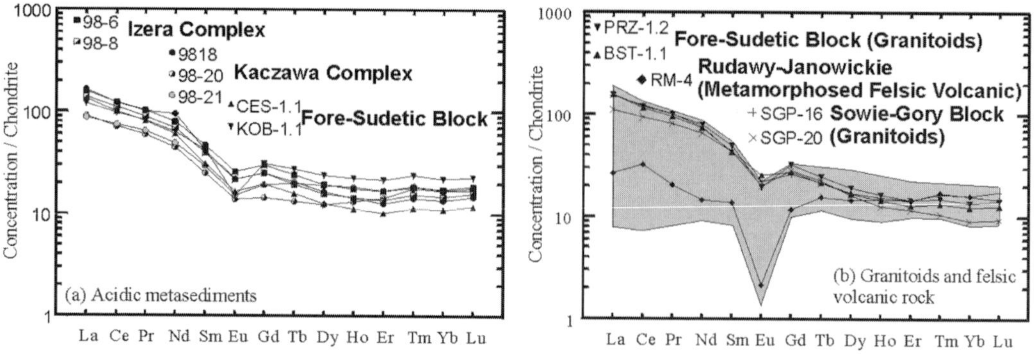

Fig. 3. Chondrite normalized REE diagrams. (a) felsic metasediments, (b) granitoids and metamorphosed felsic volcanic rock, shaded area indicates range of previously published REE values for analogous lithologies in the West Sudetes (Furnes *et al.* 1994; Kryza *et al.* 1995; Dostal *et al.* 2001).

analogous felsic lithologies in the Sudetes is also indicated in Fig 2b; REE abundances reported here fall within this expected range.

Metabasites are divided according to their trace element chemistry into low-Ti tholeiitic, tholeiitic to transitional and alkalic types. A ratio of V/Ti = 20 discriminates between low Ti tholeiitic and tholeiitic varieties whereas a ratio of Nb/Y = 0.67 discriminates between the tholeiitic and alkalic fields. Alkali metabasites from the complexes of Rudawy Janowickie, Kaczawa, Izera and the Fore-Sudetic Block typically exhibit $(Ce/Yb)_N$ = 4.3 to 10.1, generally with higher ΣREE than other lithologies. Comparison with typical REE abundances of meta-alkali basalts from elsewhere in the Sudetes indicates that these samples fall within the expected range of values. Only HREE abundances in a sample from the Rudawy Janowickie Complex (RM-1) deviate slightly from this range (Fig. 4a). Tholeiitic metabasites from the complexes of Rudawy Janowickie, Kaczawa, Kłodzko and the Fore-Sudetic Block display relatively flat REE profiles, with $(Ce/Yb)_N$ generally within the range of 0.8 to 2.0. Sample RM-2 from the Rudawy Janowickie Complex is an enriched metatholeiite and has higher ΣREE than more depleted metatholeiite varieties (Fig. 4b). Metatholeiites from the Ślęża Ophiolite, however, are distinct in their REE abundances; these tend to display slightly negative Eu anomalies (Eu/Eu* = 0.8) and HREE enrichment ($(Ce/Yb)_N$ = 0.9 to 1.1), quite distinct from other tholeiitic metabasites in the Sudetes (Fig. 4c). Two low-Ti metabasites from the Fore-Sudetic Block show extreme depletion in REE. Whereas sample PRZ-1 falls within the expected range, BUD-1 however has low, near chondrite REE values and a positive Eu anomaly (Eu/Eu* = 3.0) indicating a cumulate origin (Fig. 4d).

Sm–Nd Isotope geochemistry

All Sm–Nd isotope data and model ages are presented in Table 2. Two stage T_{DM} ages (Liew & Hofmann 1988) were calculated for continental crust derived or dominated lithologies. Felsic metasedimentary rocks and continental granitoids have εNd(t) ranging from –8.3 to –5.0, –5.4 to –2.2 and two stage T_{DM} ages of 1.3 to 1.8 Ga and 1.0 to 1.5 Ga respectively. A granite sample (BST-1.1) from the Borek Strzeliński borehole (Fore-Sudetic Block) that is closely associated with mafic and ultramafic lithologies yielded εNd(t) –0.3 and a T_{DM} age of 1.4 Ga. A Góry-Sowie granitoid (SGP-20) displays an εNd(t) value of +0.8 and a T_{DM} age of 1.0 Ga. The meta-rhyolite sample (RM-4) is characterized by an εNd(t) value of –5.4 and a two-stage T_{DM} age of 1.3 Ga (using a typical crustal $^{147}Sm/^{144}Nd$ ratio of 0.12). The low-Ti metatholeiites, metatholeiites and meta-alkali basalts display εNd(t) between +5.4 and +6.9, +6.1 and +8.9 and +3.1 and +7.0 respectively. Metatholeiites from the Ślęża ophiolite are characterized by a narrow range of εNd(t) values (+8.2 to +9.1). Ranges of εNd(t) values with respect to Sm–Nd ratios for all the samples lithologies in this study are illustrated in Figure 5.

Comparison within the North Bohemian Massif

Polish and Czech Sudetes

Previous Sm–Nd isotope studies of lithologies from individual fault bounded crustal blocks in the West Sudetes (see Figs 5 and 6a for comparison with new data) generally agree that the early Palaeozoic bimodal magmatism was rift-related (e.g. Furnes *et al.* 1994; Dostal *et* al. 2001). Mafic

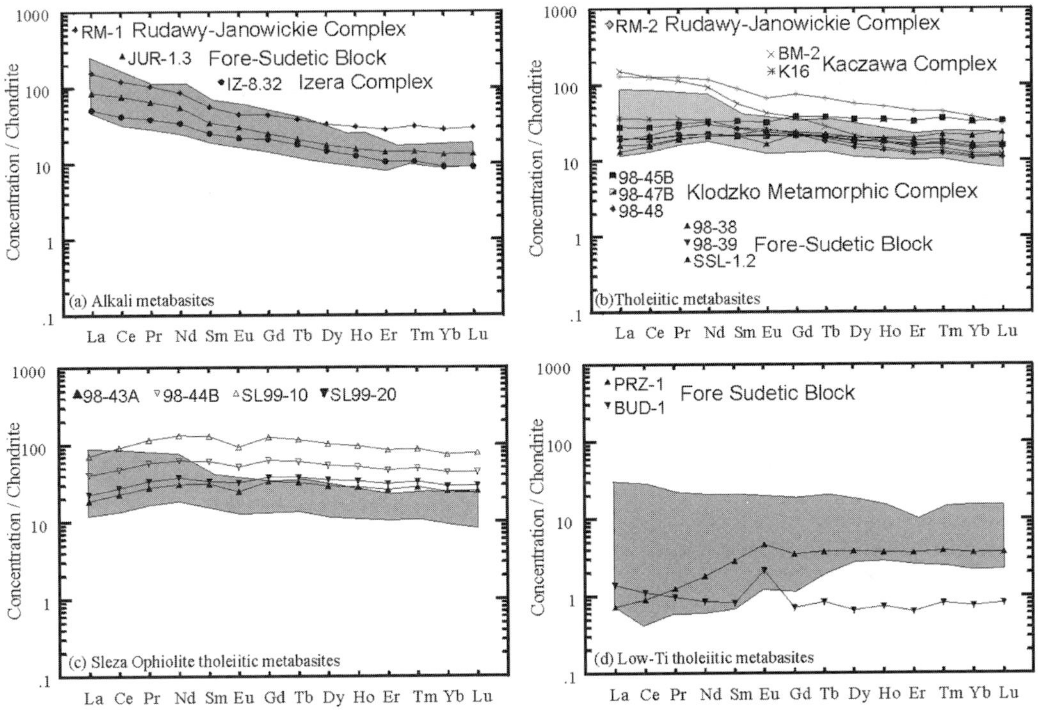

Fig. 4. Chondrite normalized (normalizing values from McDonough & Sun 1995) REE diagrams of metabasites from the study area. (a) alkali metabasites; (b) tholeiitic metabasites; (c) low-Ti tholeiitic metabasites; (d) tholeiitic metabasites from the Ślęża Ophiolite, shaded areas indicate range of previously published REE values for analogous lithologies in the West Sudetes (Pin *et al.* 1988*b*; Furnes *et al.* 1994; Kryza *et al.* 1995; Seston *et al.* 2000; Dostal *et al.* 2001).

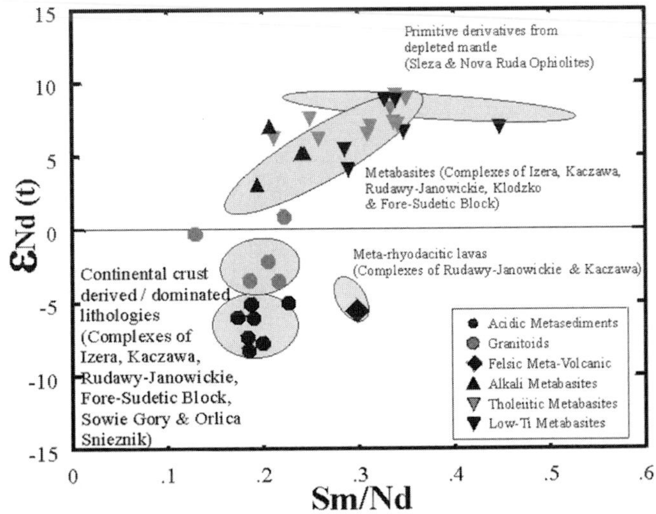

Fig. 5. Comparison of εNd(t) versus Sm/Nd for the data presented in this study (data points illustrated) with previously published Nd-isotope data from the North Bohemian Massif (Liew & Hofmann 1988; Pin *et* al. 1988*b*; Furnes *et al.* 1994; Bendl & Patočka 1995; Kröner *et al.* 1994, 1995; Beard *et* al. 1995; Kryza *et al.* 1995; O'Brien *et al.* 1997; Schmadicke *et al.* 1995; Teichmann & Basu 1996; Harms *et al.* 1997; von Quadt 1997; Siebel *et al.* 1997; Kröner & Hegner 1998; Dostal *et al.* 2000; 2001; Hegner & Kröner 2000; Tichomirowa *et al.* 2001). Range of previously published data represented by shaded fields.

Table 2. *Sm–Nd isotope data for samples from the West Sudetes, NE Bohemian Massif*

Sample	Description	Locality	Assumed age (Ma)	Sm (ppm)	Nd (ppm)	$^{147}Sm/^{144}Nd$	$^{143}Nd/^{144}Nd$*		$^{143}Nd/^{144}Nd$†	εNd(0)	εNd(t)	T_{DM} (Ga) (2-stage)§
98-18	Acidic schist	Izera Complex	500	6.24	33.81	0.1115	0.511981	±6	0.511616	−12.8	−7.4	1.80
98-20	Acidic schist	Izera Complex	500	3.62	19.54	0.1121	0.511937	±8	0.511570	−13.7	−8.3	1.86
98-21	Acidic schist	Izera Complex	500	4.50	22.49	0.1210	0.511991	±6	0.511594	−12.6	−7.8	1.78
IZ-8.32	Alkali metabasite	Izera Complex	500	3.73	15.48	0.1458	0.512738	±6	0.512260	1.9	5.2	
IZ-102	Tholeiitic metabasite	Izera Complex	500	2.43	6.94	0.2117	0.513143	±8	0.512449	9.8	8.9	
K16	Tholeiitic metabasite	Kaczawa Complex	500	3.55	13.70	0.1566	0.512821	±6	0.512308	3.6	6.1	
BM-2	Alkali metabasite	Kaczawa Complex	500	8.45	43.49	0.1175	0.512536	±6	0.512151	−2.0	3.1	
98-6	Acidic schist	Kaczawa Complex	500	5.91	31.44	0.1137	0.512106	±6	0.511734	−10.4	−5.1	1.61
98-8	Acidic schist	Kaczawa Complex	500	5.91	30.99	0.1152	0.512061	±6	0.511683	−11.3	−6.1	1.67
RM-1	Tholeiitic metabasite	Rýchory Mts	500	8.54	40.09	0.1289	0.512750	±6	0.512328	2.2	6.5	
RM-2	Tholeiitic metabasite	Rýchory Mts	500	12.16	48.57	0.1513	0.512888	±6	0.512392	4.9	7.8	
RM-4	Orthogneiss	Rýchory Mts	500	2.12	7.09	0.1803	0.512309	±8	0.511718	−6.4	−5.4	1.30
BST-1.1	Plagio-granite	Fore-Sudetic Block	500	5.44	42.03	0.0782	0.512234	±6	0.511978	−7.9	−0.3	1.41
PRZ-1.2	Orthogneiss	Fore-Sudetic Block	500	7.27	35.26	0.1246	0.512290	±6	0.511882	−6.8	−2.2	1.32
CES-1.1	Acidic schist	Fore-Sudetic Block	500	4.26	24.46	0.1052	0.512030	±6	0.511686	−11.9	−6.0	1.72
JUR-1.3	Alkali metabasite	Fore-Sudetic Block	500	4.90	23.63	0.1254	0.512764	±6	0.512353	2.5	7.0	
PRZ-1	Tholeiitic metabasite	Fore-Sudetic Block	500	0.43	0.97	0.2707	0.513235	±6	0.512348	11.6	6.9	
BUD-1	Low-Ti metabasite	Fore-Sudetic Block	500	0.11	0.38	0.1730	0.512839	±10	0.512272	3.9	5.4	
KOB-1.1	Acidic schist	Fore-Sudetic Block	500	6.44	28.39	0.1372	0.512186	±10	0.511736	−8.8	−5.0	1.48
SSL-1.1	Low-Ti metabasite	Fore-Sudetic Block	500	2.14	6.16	0.2103	0.513023	±10	0.512334	7.5	6.6	
SSL-1.2	Tholeiitic metabasite	Fore-Sudetic Block	500	3.20	9.49	0.2041	0.513036	±10	0.512367	7.8	7.3	
98-38	Tholeiitic metabasite	Fore-Sudetic Block	500	3.16	9.33	0.2051	0.513133	±8	0.512461	9.7	9.1	
98-39	Tholeiitic metabasite	Fore-Sudetic Block	500	2.98	9.61	0.1872	0.512952	±8	0.512339	6.1	6.7	
98-43A	Tholeiitic metabasite	Ślęża Ophiolite	420	4.70	13.89	0.2044	0.513113	±6	0.512550	9.3	8.9	
98-44B	Tholeiitic metabasite	Ślęża Ophiolite	420	8.15	24.38	0.2021	0.513074	±6	0.512518	8.5	8.2	
SL99-10	Tholeiitic metabasite	Ślęża Ophiolite	420	18.60	56.71	0.1983	0.513095	±8	0.512549	8.9	8.8	
SL99-20	Tholeiitic metabasite	Ślęża Ophiolite	420	5.59	16.46	0.2054	0.513113	±6	0.512548	9.3	8.8	
98-45B	Tholeiitic metabasite	Kłodzko	500	5.01	14.67	0.2064	0.513036	±6	0.512360	7.8	7.1	
98-47B	Tholeiitic metabasite	Kłodzko	500	3.11	9.91	0.1894	0.512973	±8	0.512352	6.5	7.0	
98-48	Tholeiitic metabasite	Kłodzko	500	3.95	13.63	0.1753	0.512775	±8	0.512201	2.7	4.0	
SGP16	Granitoid	Sowie Góry Block	500	4.93	26.45	0.1128	0.512185	±6	0.511815	−8.8	−3.5	1.48
SGP20	Granitoid	Sowie Góry Block	500	6.53	29.30	0.1346	0.512478	±6	0.512037	−3.1	0.8	1.04
SGP21	Granitoid	Sowie Góry Block	500	7.48	34.47	0.1312	0.512241	±8	0.511811	−7.7	−3.6	1.40

* measured ratio, corrected for spike. Errors are 2σ, refer to least significant digit(s) and are ± two standard errors of the mean
† initial ratio
§ Two-stage depleted mantle age following the model of Liew and Hofmann (1988)

magmatism involved three main sources: (1) a depleted mantle source with low time-integrated Nd–Sm ratios (MORB-like) which experienced small amounts of crustal contamination; (2) a less depleted mantle source; and (3) an enriched mantle asthenosphere source (upwelling mantle plume) which interacted with the depleted mantle source. Mixing of the enriched and depleted mantle source end-members may account for the transitional tholeiitic series. Although the Sm–Nd isotope evidence is also compatible with derivation of the alkali metabasites by crustal contamination of partial melts of a depleted mantle source, this is not substantiated by their trace element chemistry. Trace element modelling (e.g. Zr v. Th/Ta, Ti/Y v. Zr/Y, Ta/Yb v. Th/Yb, Zr/Nb v. La/Sm) indicates that the alkali metabasites were produced by partial melting of an enriched mantle source that experienced little or no crustal contamination; moreover they are similar in composition to both modern day and ancient plume-related basalts (Floyd *et al.* 2000). Felsic volcanic rocks contain a significant crustal component, of which two separate varieties have been described: (1) those with negative εNd(t) which formed by partial melting of continental crust; and (2) those displaying slightly positive εNd(t) values and significant negative Eu anomalies which formed by crustal contamination of extremely fractionated mafic magmas. The MORB-like series displays the greatest variation in εNd(t) (+4.2 to +8.7), whereas the alkali metabasites are characterized by a more tightly constrained range of εNd(t) values (+1.8 to 3.7).

Metabasites from the Ślęża Ophiolite analysed in this study display a narrow range of positive εNd(t) values (+8.2 to +8.9), indicating that they were derived from a strongly depleted source. No alkali metabasalts have been identified in the Ślęża Ophiolite. The data presented here for the Ślęża Ophiolite metabasites agree with the samples analysed by Pin *et al.* (1998*b*). Moreover samples from Pyszczyńska Hill (98–38, 98–39), previously considered to belong to the Imbramowice Metamorphic Complex (Majerowicz & Pin 1994), have affinities with the Ślęża metabasites (εNd(t) = +9.1 and +6.7). This suggests a greater extension of the Ślęża ophiolite outcrop limit within the Fore-Sudetic Block.

The low initial εNd values of felsic metasediments analysed in this study are comparable to those analysed by Hegner and Kröner (2000) from an Orlica-Snieznik Dome metagreywacke (εNd = –8.6, average crustal residence age 1.9 Ga), which is consistent with derivation from either a predominantly Mesoproterozoic provenance, or a mixture of Archaean to Palaeoproterozoic and Neoproterozoic source material. The εNd values and T_{DM} ages exhibited by late Neoproterozoic to early Palaeozoic granitoids in the Sudetes (Kröner *et al.* 1994; O'Brien *et al.* 1997; Kröner & Hegner 1998; Hegner & Kröner 2000) suggest crustal sources with similar compositional ranges to those of the felsic metasediments, but with either predominantly Mesoproterozoic source material, or Archaean to Palaeoproterozoic and Neoproterozoic to early Palaeozoic source components. The granitoids analysed here yield comparable results to other granitoids in the Sudetes, again indicating a relatively homogenous crustal reservoir. Two granitoids (BST-1.1 and SGP-20) are characterized by εNd(t) values (–0.3 and +0.8 respectively) indicating that they contain a significant proportion of mantle-derived material. Although not common in the Sudetes, this is not unexpected in an extensional regime where considerable volumes of mafic magmas have been emplaced or have resided for extended periods in an attenuated continental crust setting. Sample RM-4, a meta-rhyolite from the Rýchory Mts, is characterized by εNd = –5.4 and a T_{DM} age of 1.3 Ga. Similar ranges of values have been reported for analogous lithologies from the Kaczawa Complex (Furnes *et al.* 1994).

Germany and NW Czech Republic

Correlation between tectonic klippen occurring in parts of Saxothuringian Germany, in the NW Czech Republic (Mariánské-Lázně Complex) and the Polish and Czech Sudetes indicates several similarities in the Neoproterozoic to mid-Palaeozoic geotectonic settings between these areas (Franke *et al.* 1993; Patočka *et al.* 1993; Crowley *et al.* 1999, 2000*a*,*b*; Seston *et al.* 2000). The Münchberg klippe, Zone of Erbendorf-Vohenstrauss, and the Mariánské-Lázně Complex all feature a phase of early Palaeozoic mafic dominated magmatism (see discussion in Crowley *et al.* 2002). εNd values for mafic magmatism from these areas range from +10.2 (Mariánské-Lázně Complex, Beard *et al.* 1995) to –5.0 (KTB drill hole in the Zone of Erbendorf-Vohenstrauss; Harms *et al.* 1997), but generally fall within the range of +8.6 to +4.0 (see Fig. 6b for comparison with new data). Three main sources for this mafic magmatism are suggested from Sm–Nd isotope systematics: (1) MORB-like asthenospheric mantle; (2) enriched 'within-plate' plume asthenospheric mantle; and (3) pelagic and upper continental crustal contaminants (Teichmann & Basu 1996; Harms *et al.* 1997; von Quadt 1997).

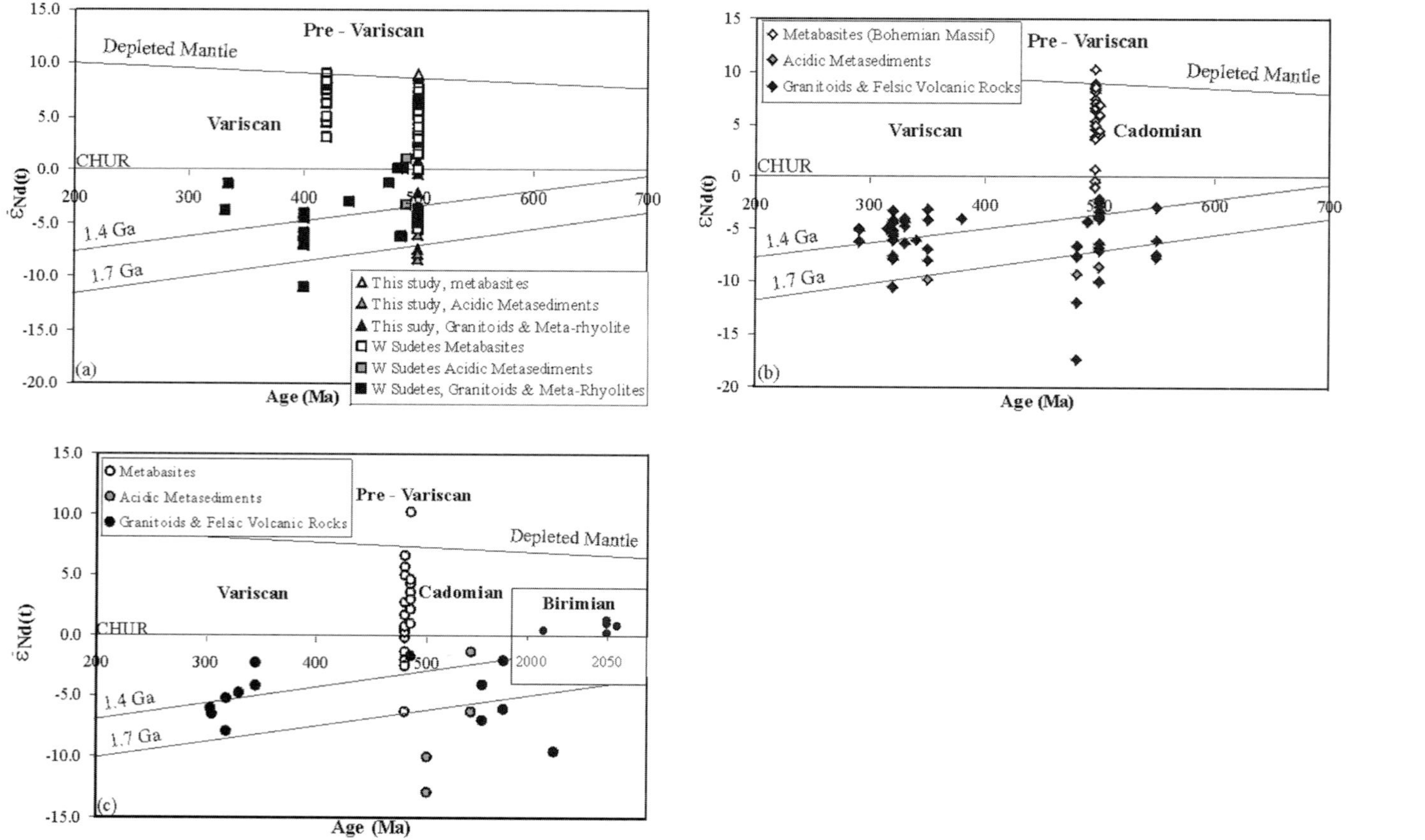

Fig. 6. εNd(t) versus geological age for (a) the data presented in this study and elsewhere in the West Sudetes (Pin *et* al. 1988*b*; Furnes *et al.* 1994; Bendl & Patočka 1995; Kryza *et al.* 1995; O'Brien *et al.* 1997; Kröner & Hegner 1998; Dostal *et al.* 2000, 2001; Hegner & Kröner 2000); (b) Bohemian Massif, Odenwald, Spessart, Schwarzwald and Vosges (Liew & Hofmann 1988; Liew *et al.* 1989; Kröner *et al.* 1994, 1995; Beard *et al.* 1995; Schmadicke *et al.* 1995; Teichmann & Basu 1996; Harms *et al.* 1997; von Quadt 1997; Siebel *et al.* 1997; Tichomirowa *et al.* 2001); (c) Massif Central, Armorican Massif (including Channel Islands) and NW Iberian Massif (Pin 1981; Bernard-Griffiths *et al.* 1985; Pin & Carme 1987; Liew & Hofmann 1988; Peucat *et al.* 1990; Dallmeyer *et al.* 1991; D'Lemos & Brown 1993; Pin & Marini 1993; Nagler *et al.* 1995; Santos Zalduegi *et al.* 1995, 1996; Dabard *et al.* 1996; Samson & D'Lemos 1998; Fernández-Suárez *et al.* 1999, 2000; Simien *et al.* 1999). Reference T_{DM} age evolution lines for 1.4 and 1.7 Ga were calculated assuming a typical crustal $^{147}Sm/^{144}Nd$ ratio = 0.12 (after Liew & Hofmann 1988).

As in the West Sudetes, early Palaeozoic magmatism represented by the para-autochthonous Saxothuringian Fichtelgebirge orthogneisses and metarhyolites is considered to have developed in an extensional geotectonic environment (Siebel *et al.* 1997). εNd(t) values of –2.9 to –6.4 and T_{DM} ages of 1.5 to 1.7 Ga indicate that this acid magmatism formed by partial melting of a mixture of continental crustal material from either predominantly Mesoproterozoic or Archaean to Palaeoproterozoic and Neoproterozoic to early Palaeozoic sources.

High pressure and ultra-high pressure lithologies that crop out in Germany include the Saxonian Granulites and Erzgebirge. The origin of these granulite protoliths is disputed; it is uncertain whether they have affinities with the Mid-German Crystalline High, or with Saxothuringian lithologies (Molzahn *et al.* 1998; Franke & Stein 2000). U–Pb SHRIMP dating of zircons from high-grade felsic gneisses of the eastern Erzgebirge has constrained an emplacement age of *c.* 540 Ma for an orthogneiss and revealed a dominant *c.* 575 Ma crustal source for a paragneiss (Tichomirowa *et al.* 2001). Other gneisses from the Erzgebirge area have yielded $^{207}Pb/^{206}Pb$ zircon protolith ages of *c.* 495 to 480 Ma (Kröner *et al.* 1995; Kröner & Willner 1995). Furthermore, U–Pb SHRIMP dating of inherited or xenocrystic zircons from these felsic gneisses indicates the presence of Neoproterozoic (*c.* 600 to 700 Ma), Palaeoproterozoic (*c.* 2100 to 2200 Ma) and Archaean (*c.* 2700 to 2800 Ma) populations (Tichomirowa *et al.* 2001). Other U–Pb xenocrystic zircon ages from the Erzgebirge gneisses are *c.* 2464, 1910 and 850 Ma (Kröner *et al.* 1995). These age ranges are consistent with utilization of pre-existing Gondwanan (North African) continental crust. In terms of the Sm–Nd isotope systematics associated high pressure mafic eclogites from the Erzgebirge region are characterized by εNd(t) from +4.4 to +6.9 suggesting that the protoliths were derived from a depleted mantle source (Schmadicke *et al.* 1995).

Comparison with pre-Variscan basement in Europe

Inliers of pre-Variscan basement of the ATA that have previously been correlated with either the Saxothuringian or Moldanubian zones occur in the Armorican Massif (northern France and the Channel Islands), and Massif Central (central and southeastern France). The relationship between εNd(t) and geological age of representative areas of these other areas of pre-Variscan basement in Europe is presented in Figure 6c. Liew and Hofmann (1988) report T_{DM} ages of up to 3.0 Ga for a hornblende gneiss from northern Austria (Moldanubian Zone), although T_{DM} ages of 1.4 to 1.7 Ga are considered to be more typical of lithologies from the Polish, Czech, German, Austrian and French (Massif Central) sectors of the Saxothuringian and Moldanubian zones (including the Odenwald, Spessart, Schwarzwald and Vosges as well as parts of the southern, central and eastern Bohemian Massif). These T_{DM} ages were attributed to mixing between Palaeoproterozoic (*c.* 2.0 Ga) crustal components and early Palaeozoic juvenile additions. Furthermore, it was proposed by Liew and Hofmann (1988) that the relatively large Palaeozoic contributions did not significantly affect these T_{DM} ages that are considered to reflect 'regional basement values'.

The antiquity of areas of western European pre-Variscan basement determined directly by U–Pb dating of zircons and also suggested by whole rock T_{DM} ages indicates the presence of 2.0 to 2.7 Ga sources (Samson & D'Lemos 1998; Guerrot *et al.* 1989; Dallmeyer *et al.* 1991). These studies have identified Archaean to Palaeoproterozoic components in granulites from the Bay of Biscay (1.9 Ga and 2.7 Ga) and Palaeoproterozoic gneisses from the Channel Islands (e.g. Icart gneiss: 2061 ± 2 Ma). In the NW Iberian Massif, *c.* 1812 and 2763 Ma inherited ages (U–Pb, zircon) have been detected in early Ordovician eclogitic and granulitic felsic gneisses (Peucat *et al.* 1990; Santos Zalduegui *et al.* 1995). Moreover, early Palaeozoic metasediments from the same area have yielded detrital U–Pb zircon ages up to 2000 Ma (Fernández-Suárez *et al.* 1999, 2000) and whole rock T_{DM} ages of 1.7 to 2.1 Ga (Peucat *et al.* 1990; Nagler *et al.* 1995), also implying the input of Palaeoproterozoic to Archaean source material in this region. Fernández-Suárez *et al.* (2000) detected a Mesoproterozoic detrital and xenocrystic zircon component in the Neoproterozoic to early Palaeozoic sedimentary and volcanosedimentary rocks of the NW Iberian Massif. Due to an apparent lack of Mesoproterozoic ages in the West African Craton, these ages have been taken as evidence for a peri-Amazonian location, (proximal to West Avalonia) for the original sedimentary basin.

Early Palaeozoic metabasites (*c.* 480 to 500 Ma) from the Massif Central are characterized by εNd(t) values with two distinct ranges, consistent with the use of both depleted and enriched mantle asthenosphere sources, with the latter experiencing a limited amount of crustal contamination (Pin & Carme 1987; Pin &

Marini 1993). Subsequent synkinematic Variscan granitic intrusions from the southern Massif Central contain inherited xenocrystic zircon components dated at *c.* 1800 Ma (Pin 1981), indicating that a Palaeoproterozoic source contributed to Variscan crustal formation in this area.

Discussion and conclusions

The new geochemical and Nd isotopic data presented here indicate that, regardless of metamorphic grade and geographic locality within the West Sudetes, each of the separate rock types defines distinct ranges in εNd(t). Felsic metasediments display the oldest T_{DM} ages, whereas the meta-rhyolite and crustally dominated granitoids are characterized by a relatively narrow range of T_{DM} ages (Table 2).

In the West Sudetes, normalized REE patterns of felsic metasediments are consistent with their derivation from an acid igneous source. Nd isotopes suggest that there was a lack of newly differentiated continental crust in the provenance. Granitoids typically display LREE enriched chondrite normalized patterns with moderately negative Eu anomalies signifying fractional crystallization (plagioclase) of crustally-derived or dominated magmas. Nd systematics indicate that most granitoids were formed by partial melting of either a predominantly Mesoproterozoic aged source, or a mixture of Archaean to Palaeoproterozoic and Neoproterozoic sources with the addition of early Palaeozoic material. Two granitoids, however, (BST-1.1 and SGP-20), display slightly negative to slightly positive εNd(t) values, and hence slightly younger T_{DM} ages due to the incorporation of a significantly larger proportion of juvenile material. The metarhyolite sample RM-4 is unusual in that it displays only slight LREE enrichment and a pronounced negative Eu anomaly.

The metabasites may be divided into three separate categories according to their REE chemistry and Sm–Nd isotope ratios. These divisions support the metabasite categories as defined by selected HFSE ratios (e.g. Nb/Y and Ti/V). This geochemical evidence indicates that alkali metabasites were derived from an enriched mantle source (plume), whereas the metatholeiites originated from a depleted mantle asthenosphere (N-MORB to P-MORB), and the low-Ti metatholeiites from the most depleted mantle source. Although trace element modelling indicates that some metabasites were subjected to a limited degree of sediment contamination (Kryza *et al.* 1995; Winchester *et al.* 1998; Crowley *et al.* 2000*b*; Floyd *et al.* 2000), Nd isotopes illustrate that if this has occurred, it must have been minor, at least in the samples examined in this study. Ślęża ophiolite metabasites that have flat chondrite-normalized REE profiles with MREE and HREE enrichment relative to other Sudeten metatholeiites. Nd isotopes suggest the Ślęża metatholeiites used a relatively homogeneous depleted mantle asthenosphere source. Taking a late Silurian to early Devonian (*c.* 420–400 Ma) age (Żelaźniewicz *et al.* 1998) for the Ślęża ophiolite, as opposed to the early Palaeozoic ages for most of the basic magmatism now preserved in the West Sudetes, the difference in nature of the sources could be taken as evidence for migration of the ATA and associated developing seaways away from a source of enriched upwelling mantle asthenosphere.

It had been proposed that pre-Permian rocks of the Sudetes might be divided into five or seven distinct terranes (Cymerman & Piasecki 1994; Cymerman *et al.* 1997; Cymerman 1999). This terrane concept was based mainly on the presence of zones of ductile shearing and in some instances on differences in metamorphic grades between juxtaposed blocks. The new geochemical and isotope data presented here does not substantiate this theory. Furthermore, it not only discounts the importance of the Intra-Sudetic and Marginal Sudetic faults as terrane boundaries, but also questions the validity of the Elbe Fault as such a dividing structural boundary. Rather than acting as primary crustal dislocations between disparate blocks, it is proposed that these lineaments formed during the final stages of Variscan collisional orogenesis. However, it is also likely that the position of main crustal lineaments, such as the Elbe Fault, were strongly influenced by pre-existing deep basement structures (Winchester *et al.* 2002).

Buried basement of the Fore-Sudetic block had been tentatively correlated with the SE active margin of the Rheno-Hercynian Zone (Grocholski 1986; Bankwitz *et al.* 1990), parts of the Saxothuringian Zone (Franke & Żelaźniewicz 2000), or as representing part of a separate terrane (Cymerman *et al.* 1997). The evidence presented here indicates that the Fore-Sudetic Block located between the Marginal Sudetic Fault and the Middle Odra Fault Zone is geochemically very similar to the Saxothuringian Zone to the west. Indeed all samples examined in this study indicate that this part of the NE Bohemian Massif has affinities with the Saxothuringian Zone. Franke and Żelaźniewicz (2000) considered that the portion of the Fore-Sudetic Block situated further to the east between the Middle Odra Fault Zone and the Odra Fault Zone represents a continuation of the Mid-German Crystalline High, thus delineating an oroclinal bend at the eastern termination of the Variscides.

Comparison of Nd data over a wide area of the European Variscides reveals that the West Sudetes conform with much of central and western Europe. Only three of the continental crust dominated samples (three Izera schist samples; 98–18, 98–20, 98–21) fall outside the T_{DM} age range of 1.4 to 1.7 Ga defined by Liew and Hofmann (1988). These have T_{DM} ages *c.* 1.8 Ga, which is identical to that of the Mömbris schist from Spessart that forms part of the Saxothuringian Zone (Liew & Hofmann 1988). This is also the age of inherited zircons in Variscan granitoids from the Massif Central (Pin 1981) and it represents the age of a magmatic event within the proto-ATA. The only known regions of basement in the European Variscides where T_{DM} ages outside the above range occur are (1) the North Armorica–NW Iberian Massif and its possible extension into the Bay of Biscay (T_{DM} ages of up to 2.8 Ga), which is similar to ages from the West African Craton (Nance & Murphy 1996), and (2) the Moravo-Silesian Unit, where SHRIMP Mesoproterozoic and Palaeoproterozoic inherited zircon ages of 1.2, 1.5 and 1.65 to 1.8 Ga suggest that it originated as a peri-Gondwanan terrane proximal to the Amazonian cratonic province (Friedl *et al.* 2000).

The origin of pre-Variscan basement in Europe remains controversial. Although a North African Gondwanan realm is widely accepted as the provenance for the ATA, other theories have been proposed. Hegner and Kröner (2000) favour a peri-Amazonian origin for the NE Bohemian Massif (West Sudetes and Erzgebirge) and Silesian Domain based on T_{DM} ages of 1.4 to 1.7 Ga and 1.1 to 1.3 Ga from Neoproterozoic to early Palaeozoic granitoid gneisses in the respective areas. These data, together with Mesoproterozoic detrital and xenocrystic $^{207}Pb/^{206}Pb$ zircon ages, were taken by Hegner and Kröner (2000) as evidence for a crustal component apparently not evident in North Africa. Mesoproterozoic T_{DM} ages however do not necessarily signify a crustal component of this age and only provide a lower age limit of the older components involved (Arndt & Goldstein 1987; DePaolo *et al.* 1991). Such T_{DM} ages may also result from a heterogeneous crustal source comprised of Archaean to Palaeoproterozoic and juvenile source material. The Mesoproterozoic $^{207}Pb/^{206}Pb$ zircon ages of Hegner and Kröner (2000) were determined using the Kober (1987) technique, which is unable to reveal either lead loss or discordance. These ages could therefore represent mixed ages between Archaean to Palaeoproterozoic cores and Neoproterozoic overgrowths, with additional early Palaeozoic zircon growth occurring in the Cambro-Ordovician granitoids. Significantly, Neoproterozoic sediments now occurring as schists (Izera schists) display Palaeoproterozoic T_{DM} ages, whereas early Palaeozoic continental crust dominated granitoids and felsic volcanic rocks from the same area display Mesoproterozoic T_{DM} ages. This difference can be accounted for by the Neoproterozoic sediments using a predominantly Archaean to Palaeoproterozoic provenance whereas the Cambro-Ordovician felsic magmatic rocks could contain inherited Archaean, Palaeoproterozoic and Neoproterozoic components with an addition of early Palaeozoic material. Further evidence against the involvement of a Mesoproterozoic component has come from a recent U–Pb SHRIMP study of zircons from early Palaeozoic (*c.* 540 Ma) felsic gneisses of the Erzgebirge (Tichomirowa *et al.* 2001). This unequivocally confirms the presence of inherited Neoproterozoic (*c.* 600 to 700 Ma), Palaeoproterozoic (*c.* 2.1 to 2.2 Ga) and Archaean (*c.* 2.7 to 2.8 Ga) zircon components. Detailed U–Pb zircon dating is needed to investigate this scenario for the extension of the Saxothuringian Zone in the West Sudetes.

The new data presented here are compatible with other geological evidence in supporting a North African, Gondwanan origin for the West Sudetes crustal blocks in this study. This new evidence further sustains a viable geodynamic model that is applicable across a wide area of the European Variscides. Crustal melting of the pre-existing Cadomian crust due to a phase of plume incubation resulted in genesis of granitoids with abundant inherited zircon phases. Widespread tensional forces along the continental margins together with the presence of an upwelling mantle plume initiated development of a magmatically active rift system. Formation of seaway spreading centres ensued where both enriched (plume) and depleted (MORB-like) mantle sources were utilised. Progressive development of this tensional regime assisted initial dispersion of the ATA. The late Devonian to early Carboniferous incorporation of the Armorican Terrane Assemblage along the southern margins of Laurussia therefore resulted in a re-assembly of related crustal blocks predominantly derived from a North African Gondwanan source.

The authors wish to acknowledge EU funding of the Palaeozoic Amalgamation of Central Europe (PACE) research network (CORDIS TMR contract number: ERBFMRXCT970136). Thanks to J.A. Winchester, P.A. Floyd, W. Franke, Z. Cymerman, R. Kryza, S. Mazur, P. Aleksandrowski, F. Patočka and V. Kachlík for an introduction to the geology of the West Sudetes and to the PGI for providing access to core material from the Fore-Sudetic block. H. Timmermann thanks A. Gerdes for helpful discussion of the isotope data. R.A. Strachan and J. Menuge provided thorough reviews of the manuscript.

References

ARNDT, N. & GOLDSTEIN, S. L. 1987. Use and abuse of crust-formation ages. *Geology*, **15**, 893–895.

ALEKSANDROWSKI, P., KRYZA, R., MAZUR, S. & ŻABA, J. 1997. Kinematic data on major Variscan strike-slip faults and shear zones in the Polish Sudetes, northeast Bohemian Massif. *Geological Magazine*, **134**, 727–737.

BANKWITZ, Z., BANKWITZ, E. & WASTERNACK J. 1990. The Mid-German Crystalline Zone and its supposed continuation into SW-Poland. *Terra Nostra*, **99**, 59–60.

BARANOWSKI, Z. & URBANEK, Z. 1972. Ordovician conodonts from the epimetamorphic complex from Rzeszowek in the Kaczawa Mountains (Western Sudetes). *Bulletin Academy Polska Science Ser de la Terre* **20**, 211–216.

BARANOWSKI, Z., HAYDUKIEWICZ, A., KRYZA, R., LORENC, S., MUSZYŃSKI, A., SOLECKI, A. & URBANEK, Z. 1990. Outline of the geology of the Góry Kaczawskie (Sudetes, Poland). *Neues Jahrbuch für Geologie und Paläontologie, Abhandlungen,* **179**, 223–257.

BEARD, B. L., MEDARIS, L. G., JOHNSON, C. M., JELÍNEK, E., TONIKA, J. & RICIPUTI, L. R. 1995. Geochronology and geochemistry of eclogites from the Mariánské Lázně Complex, Czech Republic: implications for Variscan orogenesis. *Geologische Rundschau*, **84**, 552–567.

BENDL, J. & PATOČKA F. 1995. The Rb–Sr isotope geochemistry of the metamorphosed bimodal volcanic association of the Rýchory Mts Crystalline Complex, West Sudetes, Bohemian Massif. *Geologia Sudetica*, **29**, 3–18.

BERNARD-GRIFFITHS, J., PEUCAT, J. J., SHEPPARD, S. & VIDAL, P. 1985. Petrogenesis of Hercynian leucogranites from the southern Armorican Massif: contribution of REE and isotopic (Sr, Nd, Pb and O) geochemical data to the study of source rock characteristics and ages. *Earth and Planetary Science Letters*, **74**, 235–250.

BORKOWSKA, M, HAMEURT, J. & VIDAL P. 1980. Origin and age of the Izera gneisses and Rumburk granites in the Western Sudetes. *Acta Geologica Polonica*, **30**, 121–149.

CHALOUPSKI, J. 1988. Lugicum: Precambrian and Lower Palaeozoic in the Krkonoše-Jizerskie Hory Metamorphic Complex. *In*: ZOUBEK, Q. (ed.) *Precambrian in younger fold belts: European, Variscides, The Carpathians and Balkans.* Wiley and Sons, Chichester, 153–160.

COCKS, L. R. M. 2000. The Early Palaeozoic geography of Europe. *Journal of the Geological Society, London*, **157**, 1–10.

COCKS, L. R. M. & FORTEY, R. A. 1982. Faunal evidence for oceanic separations in the Palaeozoic of Britain. *Journal of the Geological Society, London*, **139**, 465–478.

COCKS, L. R. M., MCKERROW, W. S. & VAN STAAL, C. R. 1997. The margins of Avalonia. *Geological Magazine*, **134**, 627–636.

CROWLEY, Q. G., SESTON, R. & WINCHESTER, J. A. 1999. Palaeozoic volcanism and nappe structure in the Bohemian Massif: Münchberg and the Polish Sudetes compared. *In*: *Old crust – new problems: geodynamics and utilization 89th annual meeting of the Geologische Vereinigung e. V. Terra Nostra*, **1**, 71–72.

CROWLEY, Q. G., FLOYD, P. A. & WINCHESTER, J. A. 2000*a*. Early Palaeozoic magmatism in the European Variscides: evidence for plume related continental break-up. *Geological Society, London, GEOSCIENCE 2000 abstracts*, 46.

CROWLEY, Q. G., FLOYD, P. A., WINCHESTER, J. A., FRANKE, W. & HOLLAND, J. G. 2000*b*. Early Palaeozoic rift-related magmatism in Variscan Europe: fragmentation of the Armorican Terrane Assemblage. *Terra Nova*, **12**, 171–180.

CROWLEY, Q. G., KOZDROJ, W., FLOYD, P. A., WINCHESTER, J. A., CWOJDZINSKI, S. & HOLLAND, J. G. 2000*c*. Tectono-Stratigraphic Affinities of the Buried Fore-Sudetic Block: an Insight from Geochemical Studies. *Joint meeting of EUROPROBE (TESZ) and PACE projects, Zakopane, Holy Cross Mountains, Poland (abstract supplement)*, 15–16.

CROWLEY, Q. G., KOZDROJ, W., ŠTĚDRÁ, V., WINCHESTER, J. A., KACHLIK, V. & HOLLAND, J. G. 2002. The Mariánské-Lázně Complex, NW Bohemian Massif: development and destruction of an early Palaeozoic seaway. *In*: WINCHESTER, J. A., PHARAOH, T. C. & VERNIERS, J. (eds) *Palaeozoic Amalgamation of Central Europe*. Geological Society, London, Special Publications, **201**, 177–195.

CYMERMAN, Z. 1987. Connection of the Slęża Ophiolite with the Variscan structure of the Sowie Góry metamorphic rocks. *Przeglad Geologiczny*, **6**, 304–312.

CYMERMAN, Z. 1999. Terranes and terrane boundaries in the Sudetes. *Geolines*, **8**, 15–16.

CYMERMAN, Z. & PIASECKI, M. A. J. 1994. The terrane concept in the Sudetes, Bohemian Massif. *Geological Quarterly*, **38**, 191–210.

CYMERMAN, Z., PIASECKI, M. A. J. & SESTON, R. 1997. Terranes and terrane boundaries in the Sudetes, northeast Bohemian Massif. *Geological Magazine*, **134**, 717–725.

DABARD, M. P., LOI, A. & PEUCAT, J. J. 1996. Zircon typology combined with Sm–Nd whole rock isotope analysis to study Brioverian sediments from the Armorican Massif. *Sedimentary Geology*, **101**, 243–260.

DALLMEYER, R. D., D'LEMOS, R. S., STRACHAN, R. A. & MUELLER, P. A. 1991. Tectonothermal chronology of early Cadomian arc development in Guernsey and Sark, Channel Islands. *Journal of the Geological Society*, **148**, 691–702.

DALLMEYER, R. D., FRANKE, W. & WEBER, K. (eds) 1995. The Variscan belt of Central and Eastern Europe. Springer, Berlin, Heidelberg, NewYork.

DEPAOLO, D. J. 1981. Trace element and isotopic effects of combined wall rock assimilation and fractional crystallisation. *Earth and Planetary Letters*, **53**, 198–202.

DEPAOLO, D. J., LINN, A. M. & SCHUBERT, G. 1991. The continental crust distribution: methods of determining mantle separation ages from Sm–Nd isotopic data and application to the southwestern United States. *Journal of Geophysical Research*, **96**, 2071–2088.

D'LEMOS, R. S. & BROWN, M. 1993. Sm–Nd isotope characteristics of late Cadomian granite magmatism in northern France and the Channel Islands. *Geological Magazine*, **130**, 797–804.

DON, J. 1990. The differences in Palaeozoic facies-structural evolution of the West Sudetes. *Neues Jahrbuch für Geologie und Paläontologie Abhandlungen*, **179**, 307–328.

DOSTAL, J., PATOČKA, F. & PIN, C. 2000. Early Palaeozoic intracontinental rifting and early sea floor spreading in the central west Sudetes (Bohemian Massif): geochemical and Nd-Sr isotopic study on metavolcanic rocks of the east Krkonoše Complex. *Geolines*, **10**, 19–20.

DOSTAL, J., PATOČKA, F. & PIN, C. 2001. Middle/Late Cambrian intracontinental rifting in the central West Sudetes, NE Bohemian Massif (Czech Republic): geochemistry and petrogenesis of the bimodal metavolanic rocks. *Geological Journal*, **36**, 1–17.

FERNÁNDEZ-SUÁREZ, J., GUTIÉRREZ-ALONSO, G. G, JENNER, G. A. & TUBRETT, M. N. 1999. Crustal sources in Lower Palaeozoic rocks from NW Iberia: insights from laser ablation U–Pb ages of detrital zircons. *Journal of the Geological Society*, **156**, 1065–1068.

FERNÁNDEZ-SUÁREZ, J., GUTIÉRREZ-ALONSO, G. G, JENNER, G. A. & TUBRETT, M. N. 2000. New ideas on the Proterozoic-Early Palaeozoic evolution of NW Iberia: insights from U–Pb detrital zircon ages. *Precambrian Research*, **102**, 185–206.

FLOYD, P. A. & CASTILLO, P. R. 1992. Geochemistry and petrogenesis of Jurassic ocean crust basalts, ODP led 129, Site 801. *In*: LARSON, R., LAUNCELOT, Y. *et al.* (eds) *Proceedings of ODP, scientific results*, **46**, 137–150.

FLOYD, P. A., WINCHESTER, J. A., SESTON, R., KRYZA, R. & CROWLEY, Q. G. 2000. Review of geochemical variation in Lower Palaeozoic metabasites from the NE Bohemain Massif: intracratonic rifting and plume-ridge interaction. *In*: FRANKE, W., ALTHERR, R., HAAK, V. & ONCKEN, O. (eds) *Orogenic processes: Quantification and modelling in the Variscan Belt of Central Europe*, Geological Society, London, Special Publications, **179**, 155–174.

FLOYD, P. A., CROWLEY, Q. G., WINCHESTER, J. A., KOZDROJ, W. & CYMERMAN, Z. 2002. Pillow lavas in the Ślęża Ophiolite and their geochemical relationship to other Sudetic mafic components. *In*: WINCHESTER, J. A., PHARAOH, T. C. & VERNIERS, J. (eds) *Palaeozoic Amalgamation of Central Europe*. Geological Society, London, Special Publications, **201**, 197–215.

FRANKE, W. 1989. Tectonostratigraphic units in the Variscan belt of Europe. *Geological Society of America*, **230**, 67.

FRANKE, W. 2000. The mid-European segment of the Variscides: tectono-stratigraphic units, terrane boundaries and plate tectonic evolution. *In*: FRANKE, W., ALTHERR, R., HAAK, V. & ONCKEN, O. (eds) *Orogenic processes: Quantification and modelling in the Variscan Belt of Central Europe*, Geological Society, London, Special Publications, **179**, 35–61.

FRANKE, W. & STEIN, E. 2000. Exhumation of high-grade rocks in the Saxothuringian Belt: geological constraints and geodynamic concepts. *In*: FRANKE, W., ALTHERR, R., HAAK, V. & ONCKEN, O. (eds) *Orogenic processes: Quantification and modelling in the Variscan Belt of Central Europe*, Geological Society, London, Special Publications, **179**, 337–354.

FRANKE, W. & ŻELAŹNIEWICZ, A. 2000. The eastern termination of the Variscides: terrane correlation and kinematic evolution. *In*: FRANKE, W., ALTHERR, R., HAAK, V. & ONCKEN, O. (eds) *Orogenic processes: Quantification and modelling in the Variscan Belt of Central Europe*, Geological Society, London, Special Publications, **179**, 63–86.

FRANKE, W., ŻELAŹNIEWICZ, A., PORĘBSKI, S. J. & WAJSPYRCH, B. 1993. Saxothuringian zone in Germany and Poland: differences and common features. *Geologische Rundschau*, **83**, 583–599.

FRIEDL, G., FINGER, F., MCNAUGHTON, N. J. & FLETCHER, I. R. 2000. Deducing the ancestry of terranes: SHRIMP evidence for South America-derived Gondwana fragments in Central Europe. *Geology*, **28**, 1035–1038.

FURNES, H., KRYZA, R. & MUSZYNSKI, A. 1989. Geology and geochemistry of early Palaeozoic volcanics of the Sierzawa Unit, Kaczawa Mts, W. Sudetes, Poland. *Neues Jahrbuch für Geologie und Paläontologie, Monatshefte*, **3**, 136–154.

FURNES, H., KRYZA, R., MUSZYNSKI, A., PIN, C. & GARMANN, L. B. 1994. Geochemical evidence of progressive, rift-related early Palaeozoic volcanism in the western Sudetes. *Journal of the Geological Society, London*, **151**, 91–109.

GROCHOLSKI, A. 1986. Proterozoic and Palaeozoic of southwestern Poland in a light of new data. *Biuletyn Instytutu Geologicznego*, **335**, 7–29.

GUERROT, C., PEUCAT, J. J, CAPDEVILA, R. & DOSSO, L. 1989. Archaean protoliths within Early Proterozoic granulitic crust of the west European Hercynian belt: possible relics of the west African Craton. *Geology*, **17**, 241–244.

HARMS, U., CAMERON, K.L., SIMON, K. & BRATZ, H. 1997. Geochemistry and petrogenesis of metabasites from the KTB ultradeep borehole, Germany. *Geologische Rundschau*, **86**, 155–166.

HARPER, D. A. T., MACNIOCAILL, C. & WILLIAMS, S. H. 1996. The palaeogeography of early Ordovician Iapetus terranes: an integration of faunal and palaeomagnetic data. *Palaeogeography, Palaeoclimatology, Palaeoecology*, **121**, 297–312.

HEGNER, E., & KRÖNER, A. 2000. Review of Nd data and xenocrystic and detrital ages from the pre-Variscan basement in the Eastern Bohemian Massif: speculations on palinspastic reconstructions. *In*: FRANKE, W., ALTHERR, R., HAAK, V. & ONCKEN, O. (eds) *Orogenic processes: Quantification and modelling in the Variscan Belt of Central Europe*, Geological Society, London, Special Publications, **179**, 113–129.

HENTSCHEL, H. 1943. Die kalksilikatischen Bestandmassen in den Gneisen des Eulen-Gebirges (Schlesien). *Mineralogische Petrolosche Mitteilungen*, **55**, 1–121.

JERZMAŃSKI, J. 1965. Geology of north-eastern part of the Kaczawa Mts and their eastern extent. *Biuletyn Instytut Geologniczny,* **185**, 109–193. In Polish with English summary.

KLIMAS-AUGUST, K. 1989. Genesis of gneisses and granites from the eastern part of the Izera metamorphic complex in the light of study of zircons from selected geological profiles. *Geologia Sudetica*, **24**, 1–71 [in Polish with English summary].

KOBER, B. 1987. Single-zircon evaporation combined with Pb^+ emitter bedding for $^{207}Pb/^{206}Pb$-age investigation using thermal ion mass spectrometry and implications to zirconology. *Contributions to Mineralogy and Petrology*, **96**, 63–71.

KOSSMAT, F. 1927. Gliederung des varistischen Gebirgsbaues. *Abhandlungen des Sächsischen Geologischen Landesamtes*, **1**, 1–39.

KRÖNER, A. & HEGNER, E. 1998. Geochemistry, single grain zircon ages and Sm–Nd systematics for granitoid rocks from the Góry Sowie (Owl Mts), Polish west Sudetes: evidence for early Palaeozoic arc-related plutonism. *Journal of the Geological Society, London*, **155**, 711–724.

KRÖNER, A. & WILLNER, A. P. 1995. Magmatische und metamorphe Zirkonalter für Quarz-Feldspat-Gesteine der Gneis-Eklogit-Einheit des Erzgebirges. *Terra Nostra*, **95**, 112.

KRÖNER, A., WILLNER, A. P., HAMMER, J., HASSE, G., BIELICKI, K. H., KRAUSS, M. & EIDAM, J. 1994. Geochronology and Nd-Sr systematics of Lusatian granitoids – significance for the evolution of the Variscan orogen in East-Central-Europe. *Geologische Rundschau*, **82**, 357–376.

KRÖNER, A., WILLNER, A. P., HEGNER, E., FRISCHBUTTER, A., HOFMANN, J. & BERGNER, R. 1995. Latest Precambrian (Cadomian) zircon ages, Nd isotopic systematics and P-T evolution of granitoid orthogneisses of the Erzgebirge, Saxony and Czech-Republic. *Geologische Rundschau*, **84**, 437–456.

KRYZA, R., MAZUR, S. & PIN, C. 1995. The Leszczyniec meta-igneous complex in the eastern part of the Karkonosze-Izera block, Western Sudetes: trace elements and Nd isotope study. *Neues Jahrbuch Mineral Abhandlungen*, **170**, 59–74.

LIEW, T. C. & HOFMANN, A. W. 1988. Precambrian crustal components, plutonic associations, plate environment of the Hercynian Fold Belt of Central Europe: indications from a Nd and Sr isotopic study. *Contributions to Mineralogy and Petrology*, **98**, 129–138.

LIEW, T. C., FINGER, F. & HOCK, V. 1989. The Moldanubian granitoid plutons of Austria: chemical and isotopic studies bearing on their environmental setting. *Chemical Geology*, **76**, 41–55.

LUGMAIR, G. W. & MARTI, K. 1978. Lunar initial $^{143}Nd/^{144}Nd$: differential evolution of the lunar crust and mantle. *Earth and Planetary Science Letters*, **39**, 349–357.

MAJEROWICZ, A. & PIN, C. 1994. Mafic and ultramafic rocks of the Imbramowice area, north of the Ślęża ophiolite: geochemical and Sm–Nd isotope characteristics. *In*: KRYZA, R. (ed.): *Magmatisme et evolution metamorphique des Sudetes*. Wyd University Wrocław, 71–75.

MALUSKI, H. & PATOČKA, F. 1997. Geochemistry and ^{40}Ar-^{39}Ar geochronology of the mafic metavolcanic rocks from the Rýchory Mountains complex (west Sudetes, Bohemian Massif); palaeotectonic significance. *Geological Magazine*, **134**, 703–716.

MARHEINE, D., KACHLÍK, V., PATOČKA, F., MALUSKI, H. & ŻELAŹNIEWICZ, A. 2000. Variscan polyphase tectonothermal record in the west Sudetes (Bohemian Massif) – deduced from Ar-Ar ages. *In*: *Variscan-Appalachian dynamics: the building of the Upper Palaeozoic basement*. Basement Tectonics 15, A Coruna, Spain, Program and Abstracts, 254–257.

MARHEINE, D., KACHLÍK, V., MALUSKI, H., PATOČKA, F. & ŻELAŹNIEWICZ, A. 2002. The $^{40}Ar/^{39}Ar$ ages from the West Sudetes (NE Bohemian Massif): constraints on the Variscan polyphase tectonothermal development. *In*: WINCHESTER, J. A., PHARAOH, T. C. & VERNIERS, J. (eds) *Palaeozoic Amalgamation of Central Europe*. Geological Society, London, Special Publications, **201**, 133–155.

MATTE, P. 1986. Tectonic and plate tectonic model for the Variscan belt of Europe. *Tectonophysics*, **126**, 329–374.

MATTE, P., MALUSKI, H., RAJLICH, P. & FRANKE, W. 1990. Terrane boundaries in the Bohemian Massif: Results of large-scale Variscan shearing. *Tectonophysics*, **177**, 151–170.

MCCULLOCH, M. T., JACQUES, A. L., NELSON, D. R., & LEWIS, J. D. 1983. Nd and Sr isotopes in kimberlites and lamproites from Western Australia: enriched mantle origin. *Nature*, **302**, 400–403.

MCKERROW, W. S., SCOTESE, C. R. & BRASIER, M. D. 1992. Early Cambrian continental reconstructions. *Journal of the Geological Society, London*, **149**, 599–606.

MOLZAHN, M., ONCKEN, O. & REISCHMANN, T. 1998. Isotopengeochemische Charakterisierung von Krustenblocken innerhalb der Varisziden Mitteleuropas. *Terra Nostra*, **98**, 108–109.

NAGLER, T. F., SCHAFER, H. J. & GEBAUER, D. 1995. Evolution of the Western European continental crust – implications from Nd and Pb isotopes in Iberian sediments. *Chemical Geology,* **121**, 345–357.

NANCE, R. D. & MURPHY, J. B. 1996. Basement isotopic signatures and Neoproterozoic palaeogeography of Avalonian-Cadomian and related terranes in the circum-North Atlantic. *In*: NANCE, R. D. & THOMPSON, M. D. (eds) *Avalonian and related peri-Gondwanan terranes of the circum-North Atlantic*. Geological Society of America Special Paper **304**, 333–346.

NARĘBSKI, W. & MAJEROWICZ, A. 1985. Ophiolites of the surrounding Sowie Góry Mts. Block and lower Palaeozoic initialites of the Polish Sudetes. *In:* DOBRETSOV, N. L. & ZONENSHEIN, L. (eds) *Riphean-Lower Paleozoic Ophiolites of Northern Eurasia*. Nauka Novosibirsk, 86–105.

NARĘBSKI, W., WOJCIECHOWSKA, I. & DOSTAL, J. 1988. Initial rifting volcanics in the Kłodzko Metamorphic Complex (Polish Middle Sudetes), evidenced by chemical data. *Bulletin of the Polish Academy of Sciences, Earth Sciences*, **36**, 261–269.

O'BRIEN P. J., KRÖNER, A., JAECKEL, P., HEGNER, E., ŻELAŹNIEWICZ, A. & KRYZA, R. 1997. Petrological and isotopic studies on Palaeozoic high-pressure granulites, Góry Sowie Mts, Polish Sudetes. *Journal of Petrology*, **38**, 433–456.

OLIVER, G. J. H., CORFU, F. & KROGH, T. E. 1993. U–Pb ages from SW Poland: evidence for a Caledonian suture zone between Baltica and Gondwana. *Journal of the Geological Society, London*, **150**, 366–369.

PATOČKA, F., VLAŠMINSKÝ, P. & BLECHOVÁ, K. 1993. Geochemistry of Early Paleozoic volcanics of the Barrandian Basin (Bohemian Massif, Czech Republic): implications for paleotectonic reconstructions, *Jahrbuch für Geologie und Palaontologie Abhandlungen*, **136**, 873–896.

PATOČKA, F., PIVEC, E. & OLIVERIOVÁ, D. 1996. Mineralogy and petrology of mafic blueschists from the Rýchory Mts. Crystalline complex (West Sudetes, Bohemian Massif). *Neues Jahrbuch für Mineralogy, Abhandlungen*, **170**, 313–330.

PEUCAT, J. J., BERNARD-GRIFFITHS, J., GIL IBARGUCHI, J. I., DALLMEYER, R. D., MENOT, R. P., CORNICHET, J. & IGLESIAS PONCE DE LEON, M. 1990. Geochemical and geochronological cross section of the deep Variscan crust: The Cabo Ortegal high-pressure nappe (north western Spain). *Tectonophysics*, **177**, 263–292.

PHARAOH, T. C. 1999. Palaeozoic terranes and their lithospheric boundaries within the Trans-European Suture Zone (TESZ): a review. *Tectonophysics*, **314**, 17–41.

PIN, C. 1981. Old inherited zircons in two synkinematic Variscan granitoids: the 'granite du Pinet' and the 'orthogneiss de Marvejois' (southern French Massif Central). *Neues Jahrbuch für Mineralogie, Abhandlungen*, **142**, 27–48.

PIN, C. & CARME, F. 1987. A Sm–Nd isotopic study of 500 Ma old oceanic crust in the Variscan belt of western Europe: the Chamrousse ophiolite complex, Western Alps (France), *Contributions to Mineralogy and Petrology*, **96**, 406–413.

PIN, C. & MARINI, F. 1993. Early Ordovician continental break-up in Variscan Europe: Nd-Sr isotope and trace element evidence from bimodal igneous associations of Southern Massif Central, France. *Lithos*, **29**, 177–196.

PIN, C., MIERZEJEWSKI, M., DUTHOU, J. L. & COUTURIE, J. P. 1988*a*. Etude isotopique Rb–Sr du granite de Karkonosze. *In*: LORENC, S. & MAJEROWICZ, A. (eds) *Petrologie et Geologie du Socle Varisque de Sudetes Polonaise*. Resultats de la cooperation entre les Universités de Wrocław et Clermont-Ferrand. Université Clermont-Ferrand, 8–44.

PIN, C., MIERZEJEWSKI, M. & WOJCIECHOWSKA, I. 1988*b*. Upper Paleozoic oceanic crust in the Polish Sudetes: Nd-Sr isotope and trace element evidence. *Lithos*, **21**, 195–209.

RICHARD, P., SHIMIZU, N. & ALLÈGRE, C. J. 1976. $^{143}Nd/^{144}Nd$, a natural tracer: an application to oceanic basalts. *Earth and Planetary Letters*, **31**, 269–278.

SAMSON, S. D. & D'LEMOS, R. S. 1998. U–Pb geochronology and Sm–Nd isotopic composition of Proterozoic gneisses, Channel Islands, UK. *Journal of the Geological Society, London*, **155**, 609–618.

SANTOS ZALDUEGUI, J. F., SCHÄRER, U. & GIL IBARGUCHI, J. I. 1995. Isotope constraints on the age and origin of magmatism and metamorphism in the Malpica-Tui allochthon, Galicia, NW Spain. *Chemical Geology*, **121**, 91–103.

SANTOS ZALDUEGUI, J. F., SCHÄRER, U., GIL IBARGUCHI, J. I. & GIRARDEAU, J. 1996. Origin and evolution of the Palaeozoic Cabo Ortegal ultramafic-mafic complex (NW Spain): U–Pb, Rb–Sr and Pb–Pb isotope data. *Chemical Geology*, **129**, 281–304.

SCHMADICKE, E., MEZGER, K., COSCZ, M. A. & OKRUSCH, M. 1995. Variscan Sm–Nd and Ar–Ar ages of eclogite-facies rocks from the Erzgebirge, Bohemian Massif. *Journal of Metamorphic Geology*, **13**, 537–552.

SESTON, R., WINCHESTER, J. A., PIASECKI, M. A., CROWLEY, Q. G. & FLOYD, P. A. 2000. A structural model for the western-central Sudetes: a deformed stack of Variscan thrust sheets. *Journal of the Geological Society, London*, **157**, 1155–1167.

SIEBEL, W., RASCHKA, H. IRBER, W., KREUZER, H., LENZ, K. L., HOHNDORF, A. & WENT, I. 1997. Early Palaeozoic acid magmatism in the Saxothuringian belt: new insights from a geochemical and isotopic study of orthogneisses and metavolcanic rocks from the Fichtelgebirge, SE Germany. *Journal of Petrology*, **38**, 203–230.

SIMIEN, F., MATTAUER, M. & ALLEGRE, C. J. 1999. Nd isotopes in the stratigraphic record of the Montagne Noire (French Massif Central): no significant Palaeozoic juvenile inputs, and pre-Hercynian paleogeography. *Journal of Geology*, **107**, 87–97.

SMULIKOWSKI, W. 1995. Evidence of glaucophane-schist facies metamorphism in the East Karkonosze Complex, West Sudetes, Poland. *Geologische Rundschau*, **84**, 720–737.

TEICHMANN F. & BASU, A. R. 1996. Nd-Sr isotopic and trace element study of rocks and fluids from the continental deep drilling Project (KTB), Germany. *Geologische Rundschau*, **85**, 162–171.

TICHOMIROWA, M., BERGER, H. J., KOCH, E. A., BELYATSKI, B. V., GOTZE, J., KEMPE, U., NASDALA, L. & SCHALTEGGER, U. 2001. Zircon ages of high-grade gneisses in the Eastern Erzgebirge (Central European Variscides) – constraints on origin of the rocks and Precambrian to Ordovician magmatic events in the Variscan foldbelt. *Lithos*, **56**, 303–332.

TIMMERMANN, H., PARRISH, R. R., NOBLE, S. R., KRYZA, R. & PATOČKA, F. 1999. Single cycle Variscan Orogeny inferred from new U-(Th)-Pb data from the Sudetes mountains in Poland and the Czech Republic. Abstracts from the 4th PACE TMR meeting, Copenhagen, Denmark, October 9th–10th 1999.

TIMMERMANN, H., PARRISH, R. R., NOBLE, S. R., KRYZA, R. & PATOČKA, F. 2000. New U–Pb monazite and zircon data from the Sudetes Mountains in southwest Poland: evidence for a single cycle Variscan orogeny. *Journal of the Geological Society, London*, **152**, 265–268.

TURNER, S. P., PLATT, J. P., GEORGE, R. M. M., KELLEY, S. P., PEARSON, D. G. & NOWELL, G. M. 1999. Magmatism associated with orogenic collapse of the Betic-Alboran domain, SE Spain. *Journal of Petrology*, **40**, 1011–1036.

URBANEK, Z. 1978. The significance of Devonian conodont faunas for the stratigraphy of epimetamorphic rocks of north-eastern part of Góry Kaczawskie. Geologia Sudetica, 13, 7–26.

VAN STAAL, C. R., DEWEY, J. F., MACNIOCAILL, C. & MCKERROW, W. S. 1998. The Cambrian-Silurian tectonic evolution of the northern Appalachians and British Caledonides: history of a complex, west and southwest Pacific-type segment of Iapetus. *In*: BLUNDELL, D. J. & SCOTT, A. C. (eds) *Lyell: the past is the key to the present*. Geological Society, London, Special Publication, **143**, 199–242.

VON QUADT, A. 1997. U–Pb zircon and Sr-Nd-Pb whole rock investigations from the continental deep drilling (KTB). *Geologische Rundschau*, **86**, 258–271.

WINCHESTER, J. A., VAN STAAL, C. R. & LANGTON, J. P. 1992. The Ordovician volcanics of the northern Miramichi Highlands, New Brunswick. *Canadian Journal of Earth Sciences*, **29**, 1430–1447.

WINCHESTER, J. A., FLOYD, P. A., AWDANKIEWICZ, M. PIASECKI, M. A. J. AWDANKIEWICZ, H. GUNIA, P. & GLIWICZ, T. 1998. Geochemistry and tectonic significance of metabasite suites in the Góry Sowie Block, SW Poland. *Journal of the Geological Society, London*, **155**, 155–164.

WINCHESTER, J. A. & PACE TMR NETWORK. 2002. Palaeozoic amalgamation of Central Europe: new results from recent geological and geophysical investigations. *Tectonophysics*, in press.

ŻELAŹNIEWICZ, A., DÖRR, W. & DUBIŃSKA, E. 1998. Lower Devonian oceanic crust from U–Pb zircon evidence and Eo-Variscan events in the Sudetes. *Terra Nostra*, **98**, 174–176.

ZIEGLER, P. A. 1990. *Geological atlas of western and central Europe*. 2nd edition, Shell Internationale Petroleum Maatschappij B.V./Geological Society of London/Elsevier, Amsterdam.

The Mariánské-Lázně Complex, NW Bohemian Massif: development and destruction of an early Palaeozoic seaway

QUENTIN G. CROWLEY.[1,5], PETER A. FLOYD[1], VERONIKA ŠTĚDRÁ[2], JOHN A. WINCHESTER[1], VACLAV KACHLÍK[3], J. GRENVILLE HOLLAND[4]

[1]*School of Earth Sciences and Geography, Keele University, Staffordshire ST5 5BG, UK*

[2]*Czech Geological Survey, Klarov 3/131, 118 21 Prague, Czech Republic*

[3]*Natural Science Faculty, Section of Geology, Charles University, Albertov 6, Prague, Czech Republic*

[4]*Department of Geological Sciences, University of Durham, Durham DH1 3LE, UK*

[5]*Present address: NERC Isotope Geosciences Laboratory, British Geological Survey, Kingsley Dunham Centre, Keyworth NG12 5GG, UK (e-mail: qcrowley@bgs.ac.uk)*

Abstract: The Mariánské-Lázně Complex is a Cambro-Ordovician terrane of oceanic affinity tectonically emplaced between the Saxothuringian Zone and Teplá-Barrandian Unit, NW Czech Republic. It forms a SE-dipping allochthonous body that comprises the largest contiguous exposure of metamorphosed basic and ultrabasic lithologies in the Bohemian Massif. Petrographic evidence indicates that a significant proportion of protoliths underwent eclogite facies metamorphism (570 to 720 °C, 1.44 to 2.10 GPa), followed by an increase in temperature (up to around 800 °C) and a subsequent widespread retrograde amphibolite facies event (550 to 680 °C, 0.75 to 1.20 GPa). New major and trace element geochemical analyses of metamorphosed basic and ultrabasic lithologies indicate that they exhibit geochemical characteristics attributable to a sea floor origin. The metabasites were generated at a spreading centre that interacted with deep-seated upwelling mantle asthenosphere. Separate, independently fractionating basic melt batches existed: these were derived from depleted and enriched asthenosphere and depleted sub-continental lithosphere sources. Geochemical correlation of the Mariánské-Lázně Complex with other early Palaeozoic metabasic provinces facilitates comparison of metabasic lithologies occurring in tectonically dislocated nappe pile thrust sheets, and allows delineation of important suture zones in the European Variscides.

The Variscides of Central Europe are a complex tectonic mosaic composed of several terranes and associated tectono-stratigraphic units of varied affinities (Kossmat 1927; Matte 1986; Franke 1989, 2000). It includes the southern margin of eastern Avalonia, the Rheno-Hercynian Zone and to the SE the Mid-German Crystalline High, the Armorican Terrane Assemblage (ATA), which incorporates fragmented remnants of the Cadomian orogen partly preserved in the Saxothuringian Zone and Teplá-Barrandian Unit, and the high-grade Moldanubian Zone of uncertain origin. These components of the Central European Variscides amalgamated with Baltica along the Trans-European Suture Zone (e.g. Ziegler 1990; Dallmeyer *et al.* 1995). Accretion of Avalonia was a diachronous event; it collided first with Baltica in the Ashgill and with Laurentia in the Llandovery or Wenlock (Cocks *et al.* 1997; Cocks 2000). Tectonic units now occurring south of the Rheno-Hercynian Zone experienced a separate subduction–collision related event prior to their late Devonian to early Carboniferous amalgamation along the southern margins of the newly-formed 'Old-Red Continent', Laurussia (e.g. Pharaoh 1999; Franke 2000; Marheine *et al.* 2002).

Ocean sutures mark the location of notable terrane boundaries; the Rheic suture separates the Rheno-Hercynian Zone from the Saxothuringian Zone and associated Mid-German Crystalline High to the south (Cocks & Fortey 1982; Franke & Engel 1986). Closure of early Palaeozoic seaways that separated members of the ATA have resulted in a bilateral symmetry of suture zones on the northern and southern flanks of the Moldanubian Zone and associated Teplá-Barrandian Unit (Matte 1986; Franke 1989, 1995; Matte *et al.* 1990). With the exception of the later Rheno-Hercynian Basin, which acted as a successor basin opening in response to subduction and closure of the Rheic Ocean, all major Central European depocentres

From: WINCHESTER, J. A., PHARAOH, T. C. & VERNIERS, J. 2002. *Palaeozoic Amalgamation of Central Europe.* Geological Society, London, Special Publications, **201**, 177–195. 0305-8719/02/$15.00

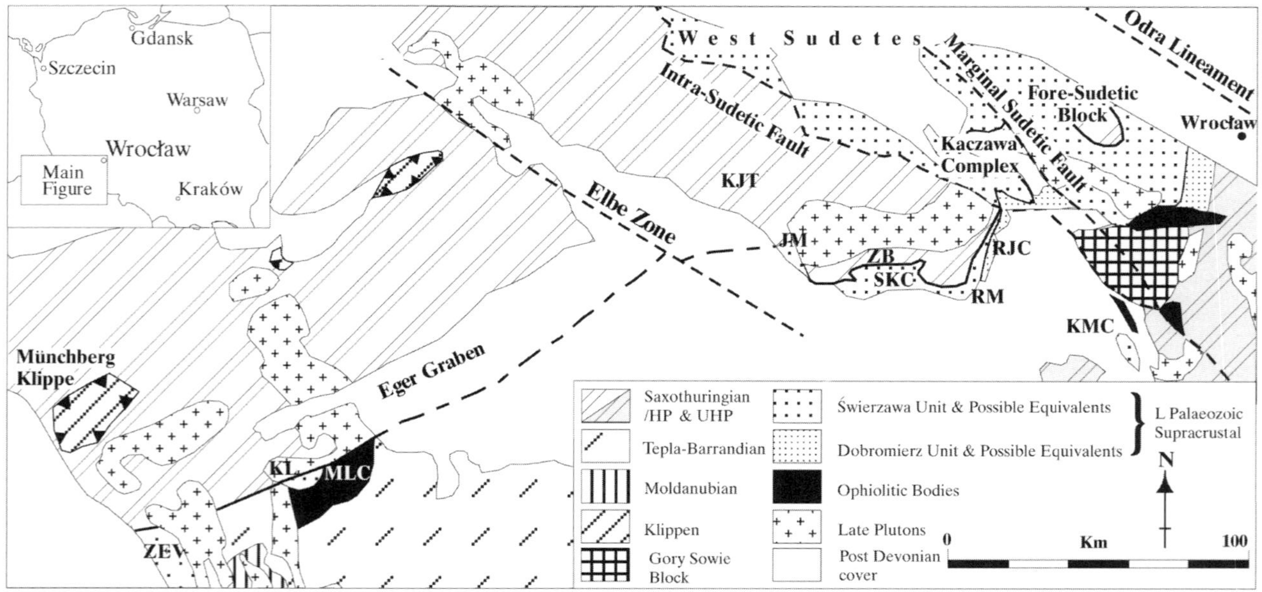

Fig. 1. Simplified geological map of the northern Bohemian Massif (adapted from Seston *et al.* 2000). KMC, Kłodsko Metamorphic Complex; RJC. Rudawy Janowickie Complex; RM, Rýchory Mountains; SKC, South Karkonše Complex; ZB, Železný Brod; JM, Jested Mountains; KJT, Karkonše-Izera Terrane; MLC, Mariánské-Lázně Complex; KL, Kladská Unit; ZEV, Zone of Erbendorf Vohenstrauss.

which eventually became involved in the Variscan Orogeny initially developed in late Cambrian to early Ordovician times (Franke 2000; Tait *et al.* 2000). This was a period of substantial crustal attenuation and rifting, often accompanied by the development of rift-related magmatism, and in some instances, generation of oceanic crust (Furnes *et al.* 1994; Briand *et al.* 1995; Winchester *et al.* 1995, 1998; Floyd *et al.* 1996, 2000; Patočka *et al.* 1997; Kachlík & Patočka 1998). Correlation of early Palaeozoic bimodal magmatism across Europe attests to the large-scale fragmentation of Gondwanan-derived terrane assemblages (Pin 1990; Pin & Marini 1993; Crowley *et al.* 2000*a*,*b*).

Large metabasic provinces and true ophiolites are particularly important in elucidating the complex early Palaeozoic tectonic evolution of the European Variscides. The detailed study presented here focuses on the petrogenetic and metamorphic development of one such province; the Mariánské-Lázně Complex (MLC), NW Czech Republic. This study aims to ascertain the palaeotectonic setting of magmatism, constrain its metamorphic history, and establish a geodynamic framework for comparison with other early Palaeozoic metabasic provinces within the European Variscides.

Geological setting

The total area of the MLC is approximately 225 km^2, although gravity and magnetic surveys, together with the occurrence of eclogite xenoliths in the Doupov Cenozoic volcanic complex suggest that it continues some 100 km to the ENE (Jelinek *et al.* 1997; Štědrá 1997). It is situated on a tectonic boundary that separates the Saxothuringian Zone to the NW and Teplá-Barrandian Unit to the SE (Fig. 1). Typically, the Saxothuringian parautochthon is described as consisting of Cadomian basement intruded and overlain by Cambro-Ordovician rift-related bimodal magmatism, overlain by Ordovician to late Devonian pelagic sediments and late Devonian to early Carboniferous flysch sediments (Falk *et al.* 1995; Franke 1989, 2000; Linnemann *et al.* 2000). The Saxothuringian allochthon, as typified by the Münchberg tectonic klippe, features a series of thrust sheets in which the metamorphic grade and sequence of protolith and metamorphic ages are broadly inverted. At its base is Carboniferous flysch while MORB-type eclogites with *c.* 500 Ma protolith ages (Gebauer & Grüenfelder 1979; Söllner *et al.* 1981) occur at the top of the klippe. Franke (1989) suggested that the Münchberg allochthonous sequence may represent an accretionary wedge complex containing fragments of the Teplá-Barrandian and metamorphosed tectonic slices of an early Palaeozoic basin. Thermobarometry calculations, $^{147}Sm/^{144}Nd$ and $^{40}Ar/^{39}Ar$ dating of eclogite metamorphism indicate that subduction of the Saxothuringian Cambro-Ordovician volcanosedimentary sequence was initiated in early Devonian times (*c.* 400 Ma) and proceeded to depths of more than 70km (Franz *et al.* 1986; Kreuzer *et al.* 1989; Stosch & Lugmair 1990; Okrusch *et al.* 1991; Beard *et al.* 1995). $^{40}Ar/^{39}Ar$ dating of mafic blueschists in the Czech Sudetes constrains a late Devonian age (*c.* 360 Ma) for cessation of subduction induced metamorphism (Maluski & Patočka 1997).

The Kladská Unit over which the MLC is thrust has been correlated with Ordovician to Silurian Saxothuringian lithologies. According to Kachlík (1997*a*) it may represent a local expression of the parautochthon. It comprises epidote–amphibolite facies metasediments and metabasites occurring below a thrust contact to the NW of the MLC, and as isolated rafts intruded by the late Variscan post-orogenic Karlovy Vary and Bor granite plutons.

The Teplá-Barrandian Unit (Perunica microcontinent of Havlíček *et al.* 1994) features a well preserved section of the Cadomian Orogen in Central Europe and is considered to have behaved as a separate microplate within the ATA in early Palaeozoic times (Franke 1989; Matte *et al.* 1990). It was subsequently thrust to the NW over the MLC. The Teplá-Barrandian Unit contains Proterozoic metasedimentary and metavolcanic lithologies at its base, overlain by Cambrian molasse derived from the Cadomian Orogen and late Cambrian volcanics indicative of an intraplate tectonic setting (Kukal 1971; Patočka *et al.* 1993; Chaloupský *et al.* 1995). An Ordovician to Devonian volcanosedimentary sequence disconformably overlies the older cover of the Teplá-Barrandian basement and is considered to have developed in an extensional passive margin setting (Chlupáč *et al.* 1992; Waldhausrová 1997). Latest Cambrian to early Ordovician granitoids of the Teplá-Barrandian Unit are also envisaged to have been generated and emplaced in a tensional tectonic regime (Zulauf *et al.* 1997). Metamorphic grade in much of the Teplá-Barrandian Unit is low (anchimetamorphic to lower greenschist facies), but increases to upper amphibolite facies (kyanite–sillimanite) along much of its western margins and in the NW adjacent to the MLC (Cháb *et al.* 1997). Zulauf *et al.* (1997) considered this Barrovian isograd geometry to have resulted from two separate Cadomian deformation events that

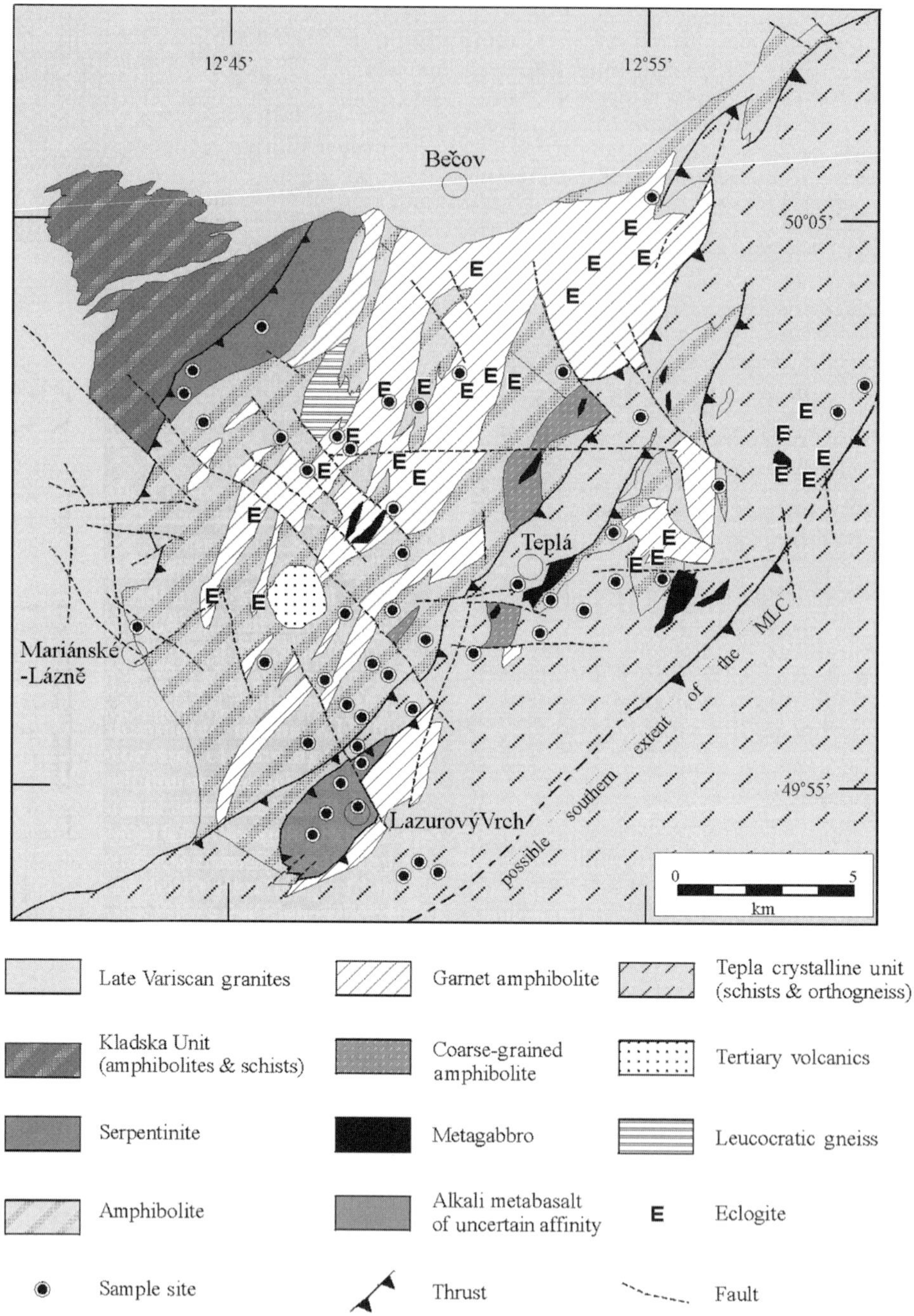

Fig. 2. Geological map of the Mariánské-Lázně Complex and surrounding area (adapted from Jelínek *et al.* 1997). Note proposed southern extent of the MLC.

occurred prior to emplacement of the post-Cadomian, pre-Variscan granitoids. Kachlík (1997*b*) however, states that growth of kyanite in the aureole of the Lestkov pluton postdates intrusion of Cambrian granitoids and hence is Variscan rather than Cadomian-related. Emplacement of end-Variscan granitoids, such as the Kladruby pluton, Sedmihoří stock and

Štěnovice stock, at *c.* 340 Ma (Cháb 1997) may have resulted from late Devonian subduction under the SE margins of the microplate. 'Variscan' collision in the Teplá-Barrandian Unit is recorded by the deformed and weakly metamorphosed upper mid-Devonian (Givetian) sediments and has been confirmed by $^{40}Ar/^{39}Ar$ mineral plateau ages of *c.* 376 to 362 Ma (Dallmeyer & Urban 1998).

Mariánské-Lázně Complex

A NE–SW trending body of serpentinites approximately 1.5 km thick with minor occurrences of tremolite–actinolite schists forms the base of the structural succession (Fig. 2). It overthrusts low-grade metasediments and transitional to alkali metabasites of the Kladská Unit to the NW, on which isolated tectonic klippen of serpentinite also occur. The main body of the MLC consists of metagabbros, garnet-bearing and garnet-free amphibolites, mafic and intermediate eclogites, rare mafic and intermediate granulites and isolated bodies of orthogneiss. Mega-boudins of eclogite surrounded by amphibolite, and occasionally shear zones containing leucocratic partial melts, are also recognized at outcrop scale.

A metagabbro body in the south of the MLC yielded a late Cambrian (496 ± 1 Ma) U–Pb zircon protolith age (Bowes & Aftalion 1991), whereas eclogite metamorphism has been constrained, using Sm–Nd geochronology of garnet–omphacite pairs (Beard *et al.* 1995), as Givetian to late Frasnian–early Famennian in age (377 ± 7 and 367 ± 4 Ma).

Petrography and geobarometry

Serpentinization of the ultramafic lithologies preferentially affected the outer margins of the body and was accompanied by dynamic recrystallization to form actinolite and tremolite schists. Aggregates of metamorphic dark green spinel surrounded by prismatic and acicular orthoamphiboles, tremolite, and Mg-chlorite in rocks forming discontinuous boudinaged bodies in host serpentinites may indicate that the ultrabasic sole also underwent high-grade metamorphism.

Petrologically, the eclogitic rocks may be divided into three main groups: garnet–omphacite eclogites (garnet, omphacite, plagioclase, kyanite ± quartz ± hornblende), low-alkali eclogites (clinopyroxene, garnet, quartz, cummingtonite, hornblende) and garnet pyroxenite eclogites (garnet, pyroxene, globular symplectite of hornblende and plagioclase). Common eclogite accessory minerals include rutile, ilmenite, monazite, and apatite. Rare mafic granulites are characterized by the presence of Na-clinopyroxene, with an assemblage of garnet, rutile, kyanite, sapphirine, spinel, quartz, anorthite, actinolite and coarse-grained plagioclase aggregates.

In general, the amphibole-rich garnetiferous rocks present in the MLC represent retrogressed high pressure assemblages (hornblende, plagioclase, garnet, clinopyroxene, quartz, epidote, rutile, titanite). Two groups of amphibolites however, do not appear to show any obvious relict eclogite assemblages; these are the Lazurový Vrch amphibolites and a group of metagabbros with primary pyroxenes and relict magmatic textures; both are found along the southern margins of the MLC.

Anatectic rocks that crystallized from hydrous melts occur in a wide range of high temperature facies from upper amphibolite and high pressure eclogite to granulite lithologies. It is likely that felsic melts were generated in more than one stage; individual generations differ in content of mafic minerals, grade of hydration of mafic minerals, and in deformational regime. Injections of migmatitic leucocratic assemblages (quartz, plagioclase–quartz–garnet, plagioclase-hornblende±garnet) are common in the northern part of the MLC.

Results obtained from thermobarometric studies of garnet amphibolite assemblages (Stedra 1996) were calculated according to the calibration by Blundy and Holland (1994), Graham and Powell (1984) and Kohn and Spear (1990). Additional constraints were obtained by jadeite content in clinopyroxenes, and by comparison with experimentally determined stability fields of sapphirine, spinel and corundum with reaction curves in P–T space (Spear 1995). Geothermobarometry calculations indicate that conditions of eclogite metamorphism were in the range of 570 to 720 °C and 1.44 to 2.10 GPa whereas rare granulite assemblages were formed in conditions between 680 and 800 °C and 0.9 to 1.4 GPa. Temperature estimates calculated for a group of garnetiferous amphibolites are relatively high, at the granulite–amphibolite facies transition (630–680 °C). Pressure estimates for the garnet amphibolites are between 0.75 and 1.2 GPa. Late mineral assemblages record the decrease in pressure and temperature through the field of stability of lower amphibolite and greenschist facies assemblages. The intense synmetamorphic shear deformation substantially enhanced these later changes.

Table 1. *Major, trace and REE geochemical analyses of representative lithologies from the MLC*

	Low-Ti tholeiites							Metatholeiites												Alkali metabasites					
	LREE depleted																								
MLC99-	21	32	51	23	25	39	67	14	26	34	41	47	61	69	70	75	85	95	111	92	93	94	MLQ-1	MLQ-2	MLQ-3
Major elements (wt.%)																									
SiO2	43.89	49.44	55.02	45.60	58.62	49.32	48.05	53.81	52.09	52.22	45.88	63.79	48.48	48.55	49.80	50.00	49.02	52.39	47.86	46.06	47.42	44.85	46.69	46.18	44.06
TiO2	1.18	0.56	0.27	0.55	0.61	1.04	0.81	1.17	1.83	1.83	3.11	0.72	2.93	1.52	1.90	1.97	2.09	2.23	1.60	2.26	1.90	3.20	2.95	2.90	2.77
Al2O3	16.81	15.64	13.58	18.44	16.34	14.63	16.46	18.31	14.49	14.44	15.33	14.90	19.98	17.26	14.15	14.48	18.81	17.25	14.41	14.92	13.53	16.99	16.46	15.49	16.03
Fe2O3	17.24	16.27	11.78	11.27	9.27	10.29	9.35	6.20	13.59	11.51	15.01	8.19	11.59	11.64	14.03	11.30	10.32	10.58	13.42	13.07	13.57	13.16	10.93	11.84	10.64
MnO	0.21	0.29	0.22	0.20	0.18	0.18	0.14	0.09	0.21	0.19	0.23	0.18	0.19	0.22	0.22	0.19	0.15	0.18	0.18	0.21	0.17	0.10	0.20	0.16	0.18
MgO	7.66	6.57	7.05	7.80	3.99	9.66	7.70	6.15	6.18	6.77	5.96	2.07	4.42	8.31	6.91	7.88	5.24	5.26	7.76	4.44	3.99	2.30	4.35	5.20	5.26
CaO	10.98	9.36	9.36	13.68	7.26	11.99	13.48	9.43	8.42	9.41	9.40	5.82	7.79	9.55	9.69	10.15	8.22	6.15	9.79	12.77	13.72	9.23	12.17	10.60	14.05
Na2O	1.05	1.98	2.62	0.96	3.02	2.21	2.30	4.02	2.46	2.88	3.09	3.53	3.73	2.87	2.55	2.53	4.43	3.84	2.95	1.75	2.86	1.43	3.17	3.85	2.62
K2O	0.47	0.12	0.01	0.86	0.71	0.11	0.09	0.27	0.31	0.53	0.86	0.36	0.10	0.45	0.03	0.70	0.78	1.18	0.52	1.23	0.95	6.03	1.27	0.82	0.63
P2O5	0.03	0.06	0.01	0.03	0.14	0.07	0.07	0.18	0.21	0.16	0.48	0.29	0.68	0.20	0.12	0.17	0.64	0.29	0.12	1.20	1.38	0.66	0.88	0.82	1.16
LOI	0.97	0.13	0.14	1.23	0.76	0.47	1.70	1.08	0.47	0.31	0.57	0.38	0.18	0.37	0.31	1.05	0.83	1.11	0.73	1.37	1.02	2.04	1.34	0.71	1.79
Total	100.49	100.42	100.06	100.62	100.90	99.97	100.15	100.71	100.26	100.25	99.92	100.23	100.07	100.94	99.71	100.42	100.53	100.46	99.34	99.28	100.51	99.99	100.41	98.57	99.19
Trace elements (ppm)																									
Ba	89	91	30	254	269	46	31	93	78	146	333	374	76	169	1304	157	280	501	117	266	181	878	304	126	255
Cl	4	0	0	0	0	0	1	10	0	0	4	5	9	33	0	17	63	20	1	42	81	0	16	0	0
Cr	117	143	330	241	45	277	309	99	225	205	77	33	48	123	185	226	96	166	186	41	39	157	105	130	119
Cs	2.16	0.24	0.08	4.38	2.46	0.18	0.33	2.59	0.41	0.76	1.40	0.20	0.40	0.89	0.08	0.91	1.02	2.00	0.74	0.60	1.23	1.00	0.00	0.00	0.00
Cu	210	53	34	193	65	84	77	206	83	41	72	37	30	38	66	47	39	37	47	40	40	47	27	29	35
Ga	12	12	13	16	15	13	14	12	18	19	20	16	18	18	19	17	22	20	16	29	25	23	23	22	19
Hf	0.36	0.41	0.26	0.67	0.51	1.46	1.15	3.46	2.82	2.77	1.82	1.28	0.59	2.82	3.03	3.41	6.92	1.21	2.39	10.21	9.26	7.21	7.82	6.74	6.77
Nb	12	10	1	6	3	6	2	3	10	9	2	2	3	7	3	8	8	9	0	55	47	54	50	47	45
Ni	61	24	59	53	10	120	88	70	49	32	15	8	6	67	41	51	25	26	57	7	9	117	87	86	100
Pb	6	0	6	4	8	5	5	0	6	3	9	4	5	8	5	5	11	10	5	11	18	10	4	4	3
Rb	13	5	5	24	14	5	5	6	8	15	26	8	5	14	4	16	20	40	15	20	22	74	24	16	1
S	484	9	0	57	102	257	89	113	25	122	469	3	12	355	213	32	32	33	50	40	31	4	0	5	0
Sc	40	61	43	43	31	34	37	41	40	42	38	23	12	27	47	45	22	35	50	17	15	21	0	0	0
Sr	96	54	30	416	233	106	110	416	110	118	272	403	249	271	115	209	457	261	138	324	237	250	715	706	465
Ta	0	0	0	0	0	0	0	0	0	1	1	0	1	1	1	1	1	1	0	4	3	3	4	3	3
Th	0.16	0.02	0.05	0.31	0.12	0.05	0.04	2.47	0.34	2.12	0.27	0.25	0.11	1.19	0.17	1.28	0.58	0.84	0.26	5.36	4.64	3.87	5.02	4.46	4.18
U	0.13	0.01	0.02	0.22	0.03	0.02	0.01	0.73	0.18	0.77	0.13	0.12	0.09	0.43	0.10	0.47	0.46	0.39	0.11	1.47	1.23	0.93	1.33	1.12	1.10
V	536	338	268	342	243	233	195	323	349	293	459	102	160	189	413	327	182	268	300	62	55	192	203	206	212
Y	10	13	8	9	21	25	22	23	50	39	32	20	20	26	47	38	38	42	34	71	63	35	37	32	36
Zn	109	115	120	84	106	77	73	34	112	91	99	87	75	85	108	87	131	110	103	174	190	123	107	109	102
Zr	14	16	10	26	20	57	45	135	110	108	71	50	23	110	118	133	270	47	93	398	361	281	305	263	264
La	0.29	0.34	0.38	1.95	2.71	1.34	1.15	11.50	5.37	10.33	9.98	6.83	5.97	9.55	4.55	7.53	29.00	13.80	4.05	54.20	43.30	35.00	44.59	40.75	40.34
Ce	0.66	0.44	0.78	3.16	6.51	4.55	3.95	21.10	10.71	24.19	25.58	16.75	14.85	22.35	9.64	19.10	68.54	33.50	11.40	116.00	93.00	74.00	91.27	81.27	82.60
Pr	0.09	0.12	0.14	0.57	1.06	0.95	0.79	2.70	2.20	3.72	4.22	2.63	2.50	3.32	2.17	3.15	10.38	5.40	2.04	16.40	12.90	10.00	11.36	10.03	10.47
Nd	0.44	0.57	0.69	2.98	5.64	5.61	4.74	11.40	11.08	17.29	21.07	13.17	13.15	15.31	11.33	15.53	45.10	24.90	10.90	71.10	56.60	42.30	47.21	41.66	43.63
Sm	0.16	0.19	0.30	1.00	1.70	2.15	1.85	2.86	3.67	4.88	5.14	3.66	3.58	4.02	3.78	4.64	9.34	7.33	3.63	16.15	13.10	9.15	9.47	8.45	9.08
Eu	0.10	0.10	0.13	0.47	0.66	0.80	0.72	0.89	1.20	1.48	1.62	1.31	2.15	1.37	1.39	1.44	2.49	1.30	1.19	4.20	3.79	2.96	2.96	2.62	2.84
Gd	0.36	0.44	0.79	1.30	2.22	3.23	2.74	3.66	6.10	5.94	5.53	4.11	4.34	4.62	5.55	5.54	8.56	8.42	4.87	16.00	13.25	8.60	9.01	7.88	8.73
Tb	0.11	0.14	0.22	0.21	0.42	0.60	0.52	0.62	1.20	1.05	0.86	0.65	0.64	0.77	1.12	0.99	1.26	1.40	0.89	2.42	2.05	1.24	1.30	1.14	1.28
Dy	1.00	1.42	1.56	1.27	2.94	3.73	3.37	3.83	7.70	6.54	5.05	3.88	3.37	4.59	7.46	6.05	6.69	7.85	5.60	13.11	11.25	6.52	6.95	6.07	6.86
Ho	0.23	0.40	0.33	0.27	0.69	0.79	0.75	0.82	1.66	1.36	1.03	0.82	0.75	0.95	1.63	1.25	1.28	1.54	1.20	2.50	2.17	1.20	1.27	1.12	1.28
Er	0.62	1.26	0.86	0.71	1.95	2.19	2.11	2.18	4.59	3.69	2.57	2.27	2.11	2.54	4.45	3.35	3.18	3.82	3.24	6.23	5.48	2.89	3.34	2.94	3.33
Tm	0.10	0.23	0.14	0.12	0.35	0.36	0.36	0.35	0.77	0.62	0.39	0.38	0.36	0.42	0.74	0.55	0.51	0.60	0.54	0.98	0.86	0.44	0.50	0.44	0.50
Yb	0.54	1.42	0.86	0.68	2.04	2.13	2.09	2.03	4.54	3.60	2.22	2.33	2.09	2.45	4.43	3.15	2.83	3.32	3.10	5.48	4.85	2.41	2.90	2.49	2.81
Lu	0.09	0.24	0.15	0.11	0.35	0.35	0.34	0.32	0.75	0.59	0.34	0.39	0.20	0.40	0.73	0.50	0.45	0.52	0.50	0.85	0.75	0.37	0.43	0.37	0.42

Geochemistry of meta-ultrabasic and metabasic lithologies

In this study 149 samples were analysed (see Table 1 for representative analyses); these were collected from all main outcrops to provide good sampling cover and ensure that the lithologies are representative of the area. The data set mainly comprises MLC metabasic lithologies, but also includes a number of samples of metabasites from the Teplá-Barrandian Unit in order to compare chemical variations and also to attempt to constrain the poorly exposed intervening tectonic boundaries. All samples were analysed for major and trace elements by XRF spectrometry (ARL 8420) at Keele University, England, whereas Cs, Hf, Sc, Ta, U and rare earth element (REE) determinations were accomplished using ICP-MS at the University of Durham, England. Details of analytical techniques, accuracy and precision can be found in Floyd and Castillo (1992) and Turner *et al.* (1999).

Metamorphism and element mobility

As discussed above, many MLC lithologies have been subjected to polyphase metamorphism attaining, in some instances, eclogite and granulite facies. All lithologies have therefore experienced some degree of element mobility, particularly involving the large ion lithophile (LIL) elements. In order to overcome the inherent problems of element mobility, the relatively immobile REE and high field strength (HFS) elements have been used in the geochemical characterization of metabasic protolith lithologies (e.g. Pearce & Cann 1973; Smith & Smith 1976; Floyd & Winchester 1978). Changes in fluid composition at high metamorphic grades are known to affect some of the relatively immobile elements (Janardhan *et al.* 1982). The MLC metabasite groups have been geochemically defined independently of metamorphic grade, indicating that there is no correlation between chemistry and metamorphic grade and thereby suggesting that no significant REE or HFS element mobility has occurred. Even small amounts of partial melting can deplete LIL elements significantly, and to a certain degree REE and HFS element abundances (e.g. James *et al.* 1987). To avoid the effects of element mobility due to partial melting, metabasites were not sampled from localities where partial melting was known to have occurred. Screening for cumulates was accomplished using the criteria defined by Pearce (1996); samples with compositions considered to deviate significantly from a true melt composition were not used to determine petrogenetic features or mantle source characteristics in this study.

Chemical variations and structural position

Serpentinized peridotites at the base of the complex, display FeO_{tot}/MgO ratios and other geochemical features similar to lithologies considered to have been derived from oceanic mantle (Jelinek & Stedra 1997). Most of the MLC ultrabasic lithologies are typical of suboceanic upper mantle, although high Ni concentrations (>2000ppm) indicate a cumulate origin for some.

Geochemical variation of metabasic lithologies within the MLC permits classification of three broad compositional groups (Fig. 3): low-Ti tholeiitic metabasites, tholeiitic metabasites and alkali metabasites. Nb–Y greater than 0.65 distinguishes alkali from tholeiitic metabasites (Floyd & Winchester 1978), whereas Ti–V greater than 20 distinguishes high-Ti from low-Ti tholeiitic metabasites (Cox 1988; Floyd *et al.* 2000). MORB-normalized multi-element diagrams emphasize the differences between these three chemical groups (Fig. 4). Relative to MORB chemistry, the low-Ti metabasites are characterized by depletion of the HFS elements, the metatholeiites display a relatively flat HFS element distribution, whereas the meta-alkali basalts display the steepest enrichment patterns. Certain features are common to both the tholeiitic and low-Ti tholeiitic metabasites, these include positive K and negative Nb anomalies and slight LREE enrichment relative to adjacent HFS elements. Chondrite normalized REE patterns again confirm this three-fold division of the MLC metabasites (Fig. 5). The low-Ti tholeiites display lower ΣREE than other groups. Two separate subgroups of low-Ti metatholeiites are recognized; those with strongly LREE depleted patterns with average $(Ce–Yb)_N = 0.21$ and those displaying relatively flat REE profiles with average $(Ce–Yb)_N = 0.75$. The former subgroup is previously unrecognized in early Palaeozoic metabasites from the North Bohemian Massif. The metatholeiites also display relatively flat to mildly enriched REE profiles (average $(Ce–Yb)_N = 2.23$), but are more enriched than the low-Ti group and with the exception of one sample (ML99–85) display a continuous REE enrichment. The meta-alkali basalts display the greatest LREE enrichment average ($Ce–Yb_N = 6.02$).

Enrichment in the LIL elements (e.g. Rb, K) may be a consequence of selective enrichment due to element mobility during amphibolization,

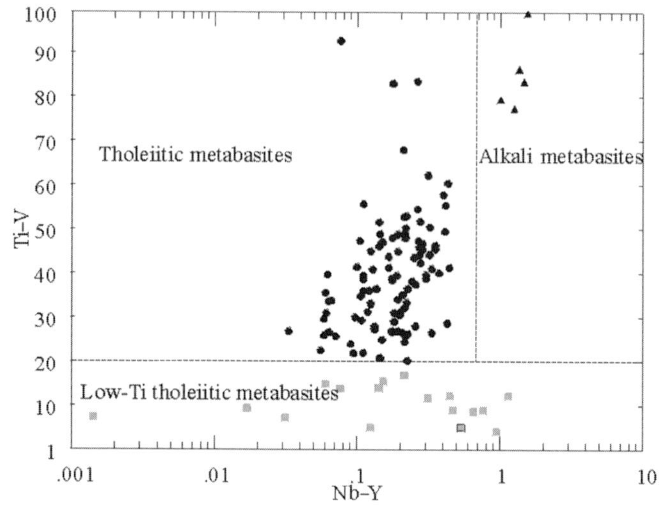

Fig. 3. Nb–Y v. Ti–V plot of Mariánské-Lázně Complex metabasites. Delineation of compositional fields from Floyd and Winchester (1978) and Floyd *et al.* (2000).

whereas the anomalies in HFS elements signify a particular petrogenetic process. For example, high Th and low Ti–Y or Ta–Yb ratios can be caused by crustal or pelagic contamination of a mantle source, this may be due to relative late-stage assimilation of sedimentary material, or it may be an inherent geochemical characteristic of the source area. The REE depletion and enrichment patterns exhibited by the three metabasite groups are difficult to reconcile fully with any one singular magmatic process. The possible causes of these geochemical variations are now evaluated by means of further geochemical investigations.

The greatest geochemical diversity occurs in the NW of the complex, to the SE of the serpentinites; this zone features variably metamorphosed (amphibolite to eclogite facies) low-Ti tholeiitic and tholeiitic metabasites. There is no direct correlation between textural features, metamorphic grade and chemical composition. The observed chemical variations and structural position of this part of the complex suggests that it may originally have comprised feeder dykes intruded into cumulates, relatively primitive gabbros and intermediate lithologies. Similar structures and sequences of large-scale igneous layering are well documented in undeformed ophiolites (e.g. Troodos, Gass 1980; Oman, Hopson *et al.* 1981). Meta-alkali basalts are restricted to Lazurový Vrch in the southern part of the complex (Fig. 2); these are the only occurrence of such composition within the MLC.

Magmatic processes and sources

Fractional crystallization trends can account for Zr variations (Figs 6–8), with the most evolved rocks of the meta-alkali basalt suite (mugearite) being extremely Zr-enriched (>300 ppm Zr). Although some of this chemical variation may be attributed to fractional crystallization, both within each of the separate chemical groups and in the metabasites as a whole, it cannot explain significant deviation from a given Nb–Y or Zr–Y ratio. A combination of open system fractionation and contamination of heterogeneous sources may account for some of these deviations (e.g. Rehkamper & Hofmann 1997), and must be considered as contributing factors in the context of magmatism generated at a rifted continental margin setting. Certain geochemical signatures can be attributed to contamination of mafic magmas; some geochemical anomalies (e.g. low Ti–Y or Ta–Yb ratios) displayed by the low-Ti and a proportion of the tholeiitic metabasites are considered indicative of a certain degree of contamination in these magmas. To quantify the degree of this possible crustal contamination, geochemical modelling of assimilation and fractional crystallization (AFC) processes was undertaken (DePaolo 1981). Pelagic sediment (post-Archean sediment composite (PASC); Taylor & McLennan 1985) and upper continental crust (UCC; Gromet *et al.* 1984; Taylor & McLennan 1985) were taken as possible contaminants. Closed system fractional crystallization cannot account for the observed

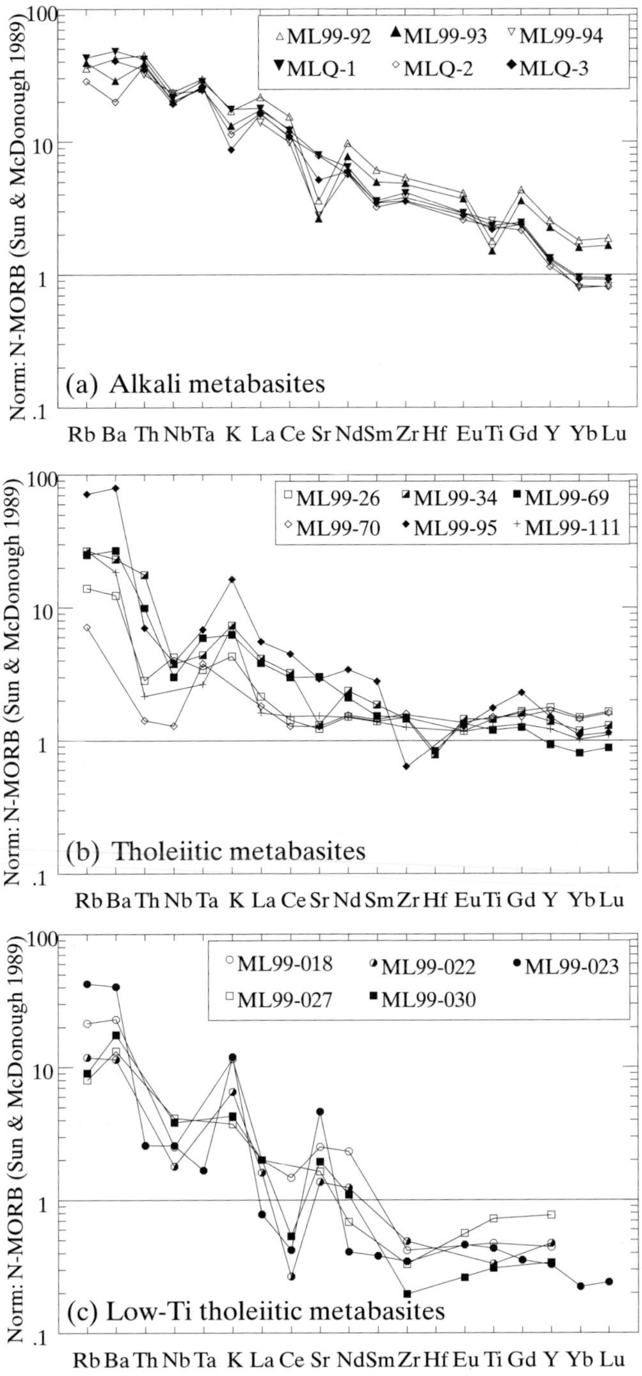

Fig. 4. MORB normalized multi-element diagrams of Mariánské-Lázně Complex metabasites. (a) alkali metabasites, (b) metatholeiites, (c) low-Ti metatholeiites. Normalizing values from Sun & McDonough (1989).

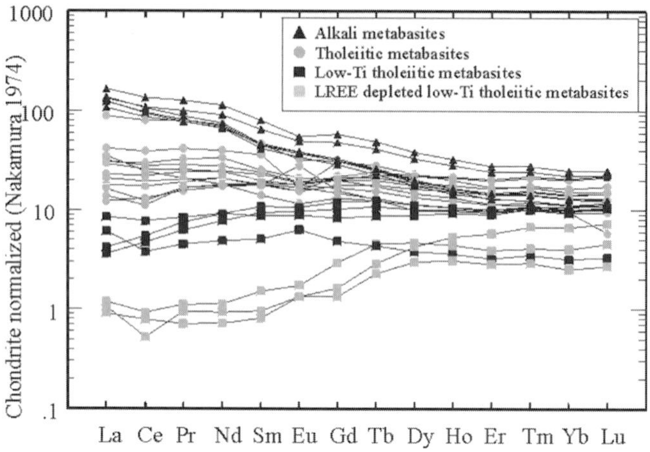

Fig. 5. Chondrite normalized REE plot (normalizing values from Nakamura 1974). Note that two separate categories of low-Ti tholeiites are distinguishable.

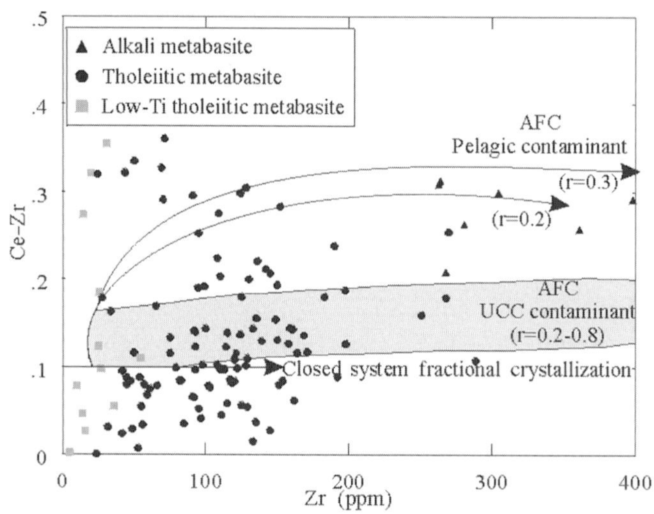

Fig. 6. Geochemical modelling of assimilation and fractional crystallization (AFC) trends based on the model of DePaolo (1981). Curves are constructed for a fractionating basaltic magma with an initial composition of 20 ppm Zr and 2 ppm Ce. Bulk K_D of Zr and Ce for $Ol_{10}Cpx_{40}Pl_{50}$ fractionation are 0.0652 and 0.0929 respectively. Values for upper continental crust (UCC) and pelagic contaminants are taken from Gromet *et al.* (1984), McLennan (1989) and Taylor & McLennan (1981).

variation in Ce–Zr ratios, AFC with UCC as a contaminant also fails to explain the observed metabasite compositional diversity (Fig. 6). The lack of distinct curvilinear trends in this diagram can be accounted for by a variety of parent magma compositions (i.e. related to source heterogeneity). In general AFC with a pelagic contaminant produces a trend close to that of many of the MLC metabasites; however unrealistically high proportions of pelagic contaminant (20 to 30%) had to be used in an attempt to account for the observed geochemical variations. Similar evidence has suggested that AFC processes with a pelagic contaminant cannot solely account for the compositional variation of other early Palaeozoic metabasic provinces in the northern Bohemian Massif (e.g. Floyd *et al.* 1996, 2000).

Pearce and Norry (1979) attributed Zr–Y variations of basaltic lithologies, from a variety of tectonic settings, to source-related processes

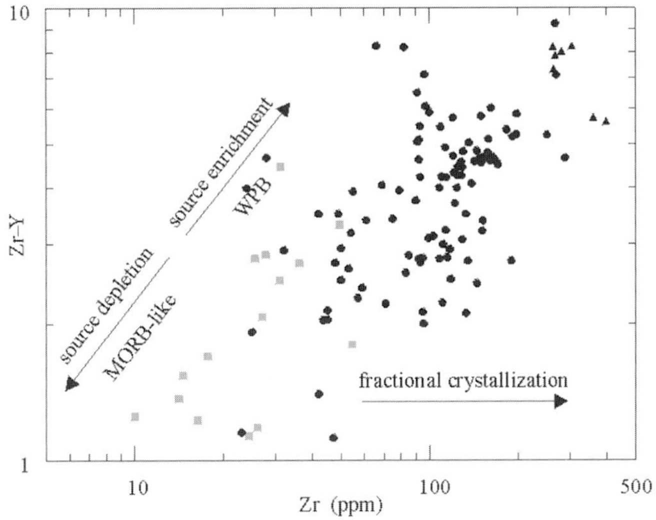

Fig. 7. Zr v. Zr–Y plot (adapted from Pearce and Norry 1979) of Mariánské-Lázně Complex metabasites indicating geochemical trends resulting from source heterogeneities and fractional crystallization processes.

and characteristics. Fig. 7 illustrates the compositional trends of Pearce and Norry (1979) in relation to the MLC metabasites; although fractional crystallization accounts for the Zr variations at a given Zr–Y ratio, changes in this ratio are essentially governed by source composition. In Figure 7 the low-Ti and 'depleted' metatholeiites correspond to a MORB-like depleted source. Floyd *et al.* (2000) characterized similar compositional fields of metabasites from elsewhere in the North Bohemian Massif as having been sourced in depleted lithospheric subcontinental and depleted asthenospheric mantle respectively. Most MLC metatholeiites and all MLC alkali metabasalts plot in the WPB-enriched field of Figure 7, thus indicating that they used an enriched asthenosphere (plume) source component. A number of possible petrogenetic settings are considered in order to elucidate the nature of the source further. The influence of depleted (MORB-like) and enriched (OIB or plume) sources together with PASC contaminants on the geochemical evolution of the MLC metabasites is shown in Figure 8. Most data points are dispersed around the average MORB source composition (Sun & McDonough 1989) and trend towards both the average OIB-plume composition (Sun & McDonough 1989) and PASC (Taylor & McLennan 1985). The trends of separate, variably melted, source compositions are represented by two magmatic arrays: spinel–lherzolite and garnet–lherzolite (Pearce 1996). With the exception of the low-Ti and 'depleted' metatholeiites, the MLC lithologies tend to plot along the compositional trend of the variably melted spinel–lherzolite array, but also trend towards the PASC. The alkali metabasites were generated by low degrees of partial melting of an enriched mantle source (such as an upwellling mantle plume). The composition of most metatholeiites, however, may be explained in terms of different degrees of partial melting of depleted (sublithospheric mantle and MORB-like) and variably enriched (plume) mantle sources, or mixtures of the end members. A limited amount of pelagic contaminant was also involved in the generation of these basaltic magmas. Fractional crystallization vectors are also illustrated and indicate how the resultant melt changes in composition (in relation to Nb–Y and Zr–Ti) according to the fractionating mineral assemblage. Again, this confirms that fractional crystallization can account for some of the compositional variations, although it is evident that separate, independently fractionating, partial melt batches existed and that variably enriched and depleted sources were used.

Comparison with adjacent tectonic units

Both the Kladská Unit and the Teplá-Barrandian Unit contain metabasic lithologies. The main part of the Kladská Unit, to the NW, is isolated from the MLC by a zone of ductile thrusting, whereas Jelínek *et al.* (1997) proposed that the Teplá-Barrandian Unit is tectonically interleaved with the MLC along its SE margins.

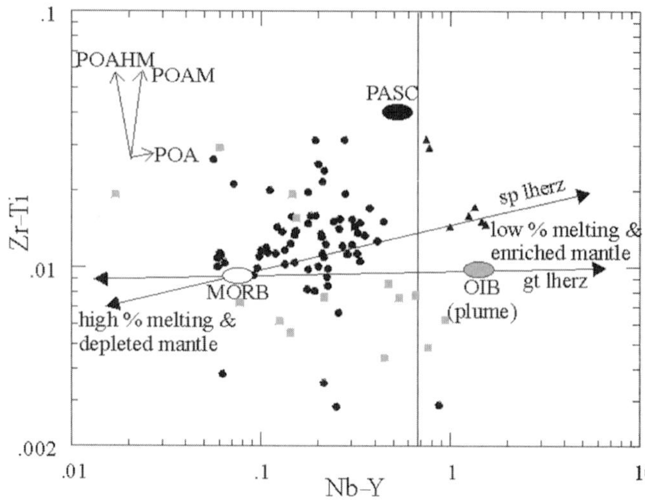

Fig. 8. Nb–Y v. Zr–Ti plot (adapted from Pearce 1996) of Mariánské-Lázně Complex metabasites. Approximate fractional vectors for plagioclase (P), olivine (O), augite (A), hornblende (H) and magnetite (M) are indicated. Compositions of N-MORB, OIB and PASC are also included.

Within the MLC itself, it is likely that thrust slices with contrasting P–T–t paths have been juxtaposed and that a certain degree of structural repetition has taken place, as deduced from differing peak eclogite and amphibolite metamorphic assemblages (Jelínek 1997; Štědrá 1997). In the North Bohemian Massif, metabasite chemistry has previously been successfully used as criteria for delineating thrust slice boundaries in areas that have experienced a complex deformational history, or indeed in areas of poor exposure (e.g. Winchester *et al.* 1995; Kachlík 1997*b*; Seston *et al.* 2000). On this basis, the boundaries between the MLC and adjacent units can be geochemically evaluated.

Metabasites of the Kladská Unit are distinct in that they experienced only low-grade (epidote–amphibolite facies) metamorphism. Both meta-alkali and metatholeiitic basalts are present; Kachlík (1997*b*) reports that the Kladská Unit metabasites have lower MgO and higher $FeO_{(t)}$ than their MLC counterparts. Another distinct difference is that the Kladská suite contains a higher proportion of meta-alkali basalts, moreover these meta-alkali basalts are chemically less evolved than those occurring at Lazurový Vrch, near the southern margins of the MLC (Fig. 9). The affinity of the Lazurový Vrch meta-alkali basalts is problematic, as they differ in chemistry and metamorphic grade from most of the MLC metabasites to the north, are chemically distinct and spatially separate from the Kladská Unit and are of contrasting metamorphic grade to the adjacent Teplá-Barrandian Unit lithologies. Kachlík (1997*b*) considered these metabasites to represent part of the Kladská Unit, but given their chemical differences and structurally higher position, to the south of the MLC, it is proposed that these lithologies define a tectonically separate thrust sheet.

A number of metabasites previously assigned to the Teplá-Barrandian Unit were analysed (Table 1). Although the Teplá-Barrandian Unit is considered to represent a distinct microplate and is unique in the context of the Bohemian Massif, the composition of the sampled metabasites is remarkably similar to those of the adjacent MLC. Geochemically, metabasites from this unit (Cháb *et al.* 1997) are indistinguishable in their chemistry from metatholeiitic MLC lithologies. Moreover, geophysical surveys in this area (Pokorný 1993; Šrámek 1994) have detected pronounced magnetic and gravity anomalies parallel, but south of the boundary between the MLC and Teplá-Barrandian Unit as it appears in Figure 2. It is likely that Variscan thrusting produced structural repetition and tectonic interleaving of the MLC and Teplá-Barrandian Unit. Important lineaments in this region evident from geophysical and field investigations include: (1) a high-strain zone at the SE margin of the MLC where metagreywackes of TBU affinity and the Tepla orthogneiss crop out; (2) a geophysical lineament that corresponds to the southeasternmost exposures of garnet–amphibolites and eclogites; and (3) further to the south, a separate thrust juxtaposes biotite and kyanite

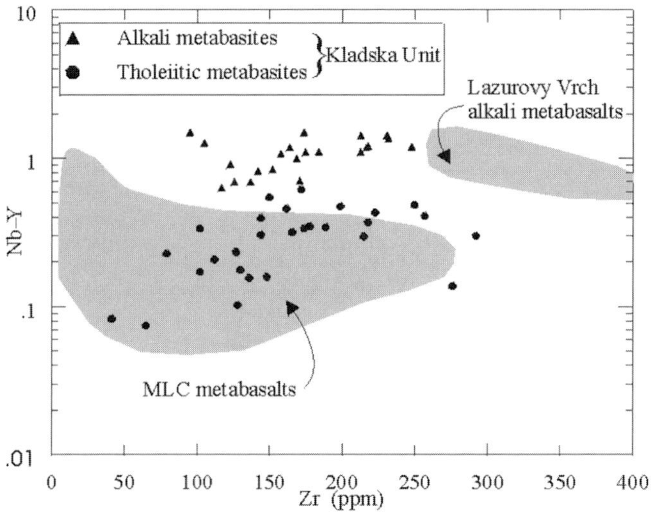

Fig. 9. Comparative plot (Zr v. Nb–Y) plot of Mariánské-Lázně Complex (shaded area) and Kladská Unit metabasites (unpublished data). Note the Zr enrichment unique to the Lazurový Vrch alkali metabasites.

metamorphic zones at the southern margin of the Lestkov pluton.

Correlation within the Central European Variscides

Early Palaeozoic rift-related magmatism may be correlated across much of the North Bohemian Massif, from the West Sudetes (occurring in, or bounding, the complexes of Izera, Kaczawa, Rudawy Janowickie, Rýchory Mountains, Železný Brod, Góry-Sowie, Klodzko, Orlica-Śnežnik) in the east, through to the Mariánské-Lázně Complex, Kladská Unit, Zone of Erbendorf-Vohenstrauß (ZEV) and Münchberg Massif. (Borkowska *et al.* 1980; von Quadt 1990; Bosbach *et al.* 1991; Furnes *et al.* 1994; Bendl & Patočka 1995; Franke *et al.* 1995; Winchester *et al.* 1995; Floyd *et al.* 1996, 2000; Jelínek & Štědrá 1997; Kachlík 1997*b*; Patočka & Smulikowski 1997; Patočka & Kachlík 1998; Crowley *et al.* 2000*a*,*b*). This correlation is based on magmatic ages, geochemical similarities of metabasic lithologies and also their relative position in the structurally imbricated Variscan nappe pile sequence. Figure 10 outlines similarities of protolith ages, metamorphic grade, and imbrication sequence of the main tectonic slices across different sectors of the North Bohemian Massif. Although the thrust sheet distribution and geometry is a consequence of the final stages of Variscan collisional orogenesis, the overall similarity of these features attests to a common early Palaeozoic magmatic and tectonic evolution over a wide area of the North Bohemian Massif. Differences in metamorphic grade may be explained by differing depths of burial and rates of exhumation. In general, as exemplified by Münchberg, metamorphic grade increases from anchimetamorphic at the base of the nappe pile sequence, to eclogite facies at the top. This is considered an effect of sequential back-stripping and exhumation of deeper lithospheric levels as collisional orogenesis proceeded. An exception to this correlation between metamorphic grade and position in the nappe pile sequence occurs in the West Sudetes of Poland where the low grade Devonian Ślęża Ophiolite resides at a high structural level, immediately beneath the high grade Góry-Sowie block. Development of Devonian arc/back arc volcanism in the Jeseníky Mts, NE Czech Republic, (Patočka 1987; Patočka & Valenta 1996) is considered to have developed as a consequence of closure of the intervening seaway between the re-amalgamated ATA and the southern margins of the 'Old Red Continent' Laurussia. It is envisaged that the Ślęża Ophiolite was obducted in response to such a tectonic setting (see Floyd *et al.* 2002).

Previous correlation between the MLC and Münchberg–ZEV have been made on the basis of lithological associations, metamorphic grade and to a certain extent, metabasite chemistry (Franke *et al.* 1995; Štědrá & Cháb 1997; Crowley *et al.* 1999). Protolith ages of 490–495 Ma (von Quadt 1990) for metagabbros from the KTB drill site in the northern part of the ZEV

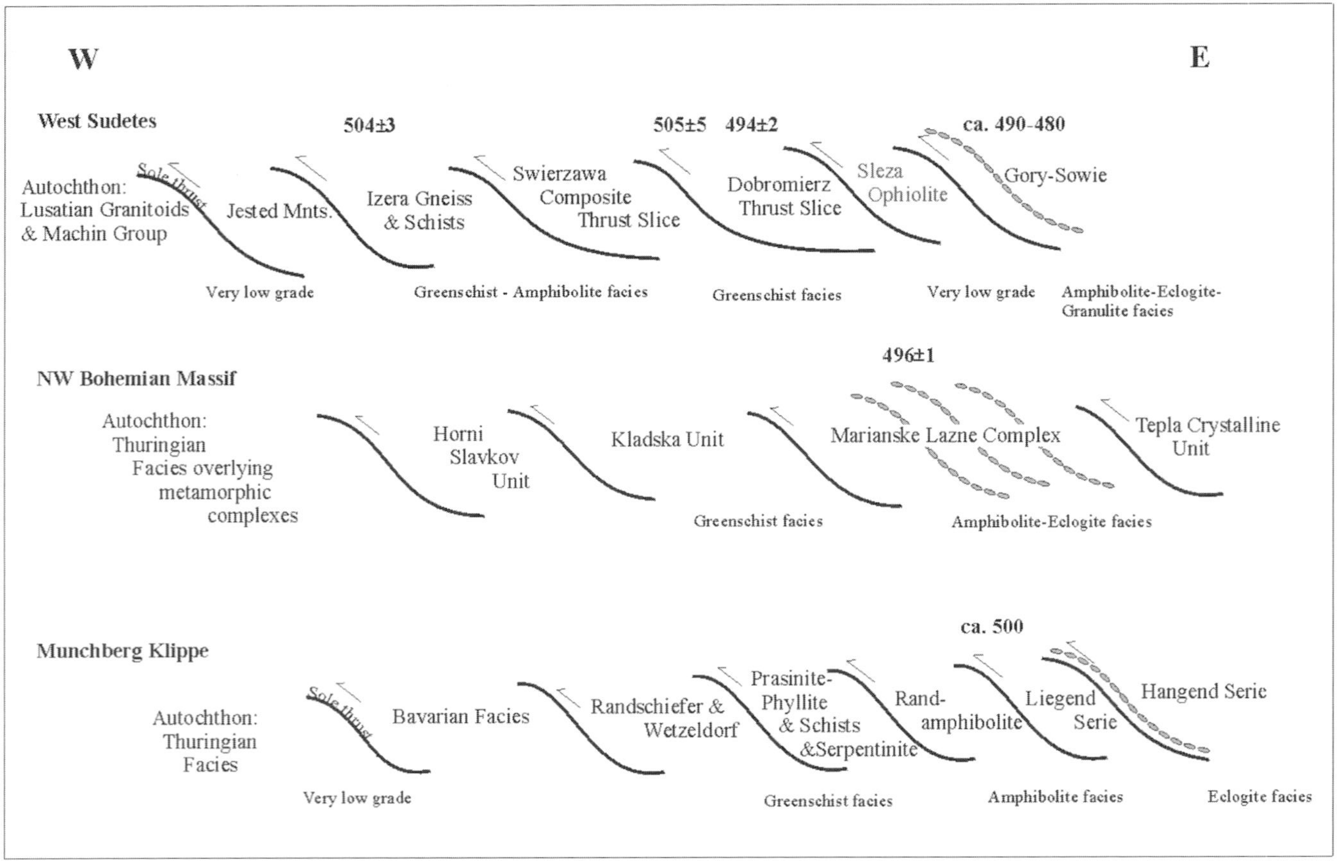

Fig. 10. Schematic comparison of tectonic imbrication sequence, metamorphic grade, and main lithologies of nappe stacks in the North Bohemian Massif. Selected protolith ages of early Palaeozoic magmatism are indicated (see text for relevant citations).

indicate that they were contemporaneous with gabbroic magmatism in the MLC. They are enriched metatholeiitic to meta-alkalic in character and thus are comparable to the Kladská Unit meta-alkaline basalts. The Münchberg klippe features variably depleted metatholeiites of MORB affinity in the Hangend Serie, Liegend Serie and Rand-Amphibolit Serie (Franke *et al.* 1995 and references therein). These are comparable in composition to the metatholeiites and low-Ti metabasalts of the MLC. Protolith ages in the Münchberg Massif (Gebauer & Grüenfelder 1979; Stosch & Lugmair 1990) once again confirm the importance of a significant Cambro-Ordovician magmatic event.

The current stacking order of the Variscan nappe pile sequence, as exemplified by Münchberg but with parallels across the North Bohemian Massif, allows a tentative reconstruction of the southern Saxothuringian and northern Teplá-Barrandian continental margins. Franke (2000) described these elements, from NW to SE, as: (1) (para)-autochthon, (2) shelf sediments resting on Cadomian basement, (3) Cadomian basement, typically affected by Devonian metamorphism, (4) Cadomian basement and early Palaeozoic magmatic rocks (typically alkali in character), (5) early Palaeozoic mafic MORB-like magmatism which has been subjected to *c.* 400 Ma eclogite metamorphism and (6) Cadomian basement of the Teplá-Barrandian Unit (only occurring in the central and western Bohemian Massif).

Conclusions

Protoliths of the MLC were formed during a Cambro-Ordovician mafic dominated rift-related event that eventually produced oceanic crust. Geochemical characterization of the metabasic lithologies indicates that several separate sources were used. Partial melting of both depleted and enriched asthenosphere sources and also depleted subcontinental lithosphere produced several independently fractionating melt batches. Contamination by pelagic sediment was limited. It is proposed that the metamorphosed ultrabasic to basic lithologies represent oceanic lithosphere generated at a spreading centre that interacted with an upwelling mantle plume. This tectonic setting is plausible not only because it provides an explanation for the variety of magmatic sources but is also in keeping with the genesis of contemporaneous mafic dominated bimodal magmatism across the North Bohemian Massif.

This lower Palaeozoic rifting event was not restricted to the North Bohemian Massif, but was widespread across a broad sector of the European Variscides. Contemporaneous magmatism with similar geochemical attributes may be found in basement lithologies of the North Bohemian Massif, Maures Massif, Corsica and Sardinia, Massif Central and the NW Iberian Massif (e.g. Ricci & Sabatini 1978; Dubussion *et al.* 1989; Pin 1990; Pin & Marini 1993; Dallmeyer *et al.* 1995; Briand *et al.* 1995; Santos Zalduegui *et al.* 1996; Crowley *et al.* 2000*b*; Fernández-Suárez *et al.* 2000). This suggests these areas were subjected to, or developed in, a similar lower Palaeozoic tectonic and magmatic event and so once formed part of the same continental (Gondwanan) margin, or at least were mutually proximal terranes immediately prior to this event. This also explains why some separate tectonostratigraphic units within the European Variscides (e.g. Saxothuringian, North Moldanubian and North Teplá-Barrandian) experienced the same magmatic event. A combination of tensional forces and an upwelling mantle plume resulted in the dispersion of the ATA from the northern flanks of Gondwana and its subsequent fragmentation culminating in the development of a network of narrow seaways.

Late Devonian subduction-related eclogite to granulite metamorphism in the MLC records the demise of the Saxothuringian seaway. This convergent event was also a widespread tectonic phenomenon (Maluski & Patočka 1997; Bosse *et al.* 2000; Marheine *et al.* 2000); it may be correlated with other high-grade metamorphic events to the east and also westward across much of Europe. The MLC is unique in that it forms the largest exposure of such rocks in the Variscides and so provides unequivocal proof for the development and destruction of an early Palaeozoic seaway.

EU funding of the Palaeozoic Amalgamation of Central Europe (PACE) research network is gratefully acknowledged (CORDIS TMR contract number: ERBFMRXCT970136). B. Briand and H. Maluski are thanked for their thorough reviews.

References

Beard, B. L., Medaris, L. G., Johnson, C. M., Jelínek, E., Tonika, J. & Riciputi, L. R. 1995. Geochronology and geochemistry of eclogites from the Mariánské-Lázně Complex, Czech Republic: implications for Variscan orogenesis. *Geologische Rundschau*, **84**, 552–567.

Bendl, J. & Patočka F. 1995. The Rb-Sr isotope geochemistry of the metamorphosed bimodal volcanic association of the Rýchory Mts Crystalline Complex, West Sudetes, Bohemian Massif. *Geologia Sudetica*, **29**, 3–18.

BLUNDY T. & HOLLAND J. 1994. Non-ideal interaction in calcic amphiboles and their bearing on amphibole-plagioclase thermometry. *Contributions to Mineralogy and Petrology*, **116**, 433–447.

BORKOWSKA, M, HAMEURT, J. & VIDAL P. 1980. Origin and age of the Izera gneisses and Rumburk granites in the Western Sudetes. *Acta Geologica Polonica*, **30**, 121–149.

BOSBACH, D., STOSCH, H. G. & SEIDEL, E. 1991. Magmatic and metamorphic evolution of metagabbros in the Münchberg Massif, NE Bavaria. *Contributions to Mineralogy and Petrology*, **107**, 112–123.

BOSSE, V., BALLEVRE, M., FERAUD, G. & PEUCAT, J. J. 2000. Petrological and geochronological constraints in the Ile de Groix blueschists (Armorican Massif, France). *In*: Variscan-Appalachian dynamics: the building of the Upper Palaeozoic basement. Basement Tectonics 15, A Coruna, Spain, *Program and Abstracts,* 63–66.

BOWES, D. R. & AFTALION, M. 1991. U-Pb zircon isotope evidence of Early Ordovician and Late Proterozoic units in the Mariánské-Lázně Complex, Central European Hercynides. *Neues Jahrbuch Mineralogisches, Mh.*, **7**, 315–326.

BRIAND, B., BOUCHARDON, J., OUALI, H., PIBOULE, M. & CAPIEZ, P. 1995. Geochemistry of bimodal amphibolitic-felsic gneiss complexes from eastern Massif Central, France. *Geological Magazine*, **132**, 321–337.

CHÁB, J. 1997. Granitic and gabbroic intrusions. *In*: VRÁNA, S. & ŠTĚDRÁ, V. (eds) *Geological model of western Bohemia related to the KTB borehole in Germany*, Czech Geological Survey, Prague, 90–101.

CHÁB, J., ŠRÁMEK, J., POKORNÝ, L., CHLUPÁČOVÁ, M., MANOVÁ, M., VEJNAR, Z., WALDHAUSROVÁ, J. & ŽÁČEK, V. 1997. The Teplá-Barrandian Unit. *In*: VRÁNA, S. & ŠTĚDRÁ, V. (eds) *Geological model of western Bohemia related to the KTB borehole in Germany*, Czech Geological Survey, Prague, 80–82.

CHALOUPSKÝ , J., CHLUPÁČ, I., MAŠEK, J., WALDHAUSROVÁ, J. & CHÁB, J. 1995. Bohemicum, Stratigraphy. *In*: DALLMEYER, D., FRANKE, W. & WEBER, K. (eds) *Pre-Permian geology of central and western Europe*. Springer, Berlin, Heidelberg, New York, 379–391.

CHLUPÁČ , I., HAVLÍČEK, V., KŘÍŽ, J., KUKAL, Z. & ŠTORCH, P. 1992. *In*: *Palaeozoic of the Barrandian (Cambrian-Devonian)*. Czech Geological Survey, Prague, 1–296.

COCKS, L. R. M. 2000. The Early Palaeozoic geography of Europe. *Journal of the Geological Society, London*, **157**, 1–10.

COCKS, L. R. M. & FORTEY, R. A. 1982. Faunal evidence for oceanic separations in the Palaeozoic of Britain. *Journal of the Geological Society, London*, **139**, 465–478.

COCKS, L. R. M., MCKERROW, W. S. & VAN STAAL, C. R. 1997. The margins of Avalonia. *Geological Magazine*, **134**, 627–636.

COX, K. G. 1988. The Karoo Province. *In*: Macdougal, J.D. (ed.) *Continental flood basalts*. Kluwer academic Press, Amsterdam, 239–272.

CROWLEY, Q. G., SESTON, R. AND WINCHESTER, J. A. 1999. Palaeozoic volcanism and nappe structure in the Bohemian Massif: Münchberg and the Polish Sudetes compared. *In*: *Old crust – new problems: geodynamics and utilization 89th annual meeting of the Geologische Vereinigung Terra Nostra*, **1**, 71–72.

CROWLEY, Q. G., FLOYD, P. A. & WINCHESTER, J.A. 2000*a*. Early Palaeozoic magmatism in the European Variscides: evidence for plume related continental break-up. Geological Society, London, GEOSCIENCE 2000 abstracts, 46.

CROWLEY, Q. G., FLOYD, P. A., WINCHESTER, J.A., FRANKE, W. & HOLLAND, J. G. 2000*b*. Early Palaeozoic rift-related magmatism in Variscan Europe: fragmentation of the Armorican Terrane Assemblage. *Terra Nova*, **12**, 171–180.

DALLMEYER, R. D., FRANKE, W. & WEBER, K. 1995. *The Variscan belt of Central and Eastern Europe*. Springer, Berlin, Heidleberg, NewYork.

DALLMEYER, R. D. & URBAN, M. 1998. Variscan vs. Cadomian tectonothermal activity in northwestern sectors of the Teplá-Barrandian zone, Czech Republic: constraints from $^{40}Ar/^{39}Ar$ ages. *Geologische Rundschau*, **87**, 94–106.

DEPAOLO, D. J. 1981. Trace element and isotopic effects of combined wallrock assimilation and fractional crystallisation. *Earth and Planetary Science Letters*, **53**, 189–202.

DUBUISSON, G., MERCIER, J.-C.C., GIRARDEAU, J. AND FRISON, J.-Y. 1989. Evidence for a lost ocean in Variscan terranes of the western Massif Central, France. *Nature*, **337**, 729–732.

FALK, F., FRANKE, W. & KURZE, M. 1995. Saxothuringian Basin: autochton and nonmetamorphic nappe units: stratigraphy. *In*: Dallmeyer, D., Franke, W., Weber, K. (eds) *Pre-Permian geology of central and western Europe*. Springer, Berlin, Heidelberg, New York, 219–234.

FERNÁNDEZ-SUÁREZ, J., GUTIÉRREZ, G. G., JENNER, G. A. & TUBRETT, M. N. 2000. New ideas on the Proterozoic-Early Palaeozoic evolution of NW Iberia: insights from U-Pb detrital zircon ages. *Precambrian Research*, **102**, 185–206.

FLOYD, P. A. & CASTILLO, P. R. 1992. Geochemistry and petrogenesis of Jurrasic ocean crust basalts, OPD led 129, Site 801. *In*: LARSON, R., LANCELOT, Y. *et al.* (eds) *Proceedings of ODP, scientific results*, **46**, 137–150.

FLOYD, P. A. & WINCHESTER, J. A. 1978. Identification and discrimination of altered and metamorphosed volcanic rocks using immobile elements. *Chemical Geology*, **21**, 291–306.

FLOYD, P. A., WINCHESTER, J. A., CIESIELCZUK, J. LEWANDOWSKA, A., SZCZEPANSKI, J. & TURNIAK, K. 1996. Geochemistry of early Palaeozoic amphibolites from the Orlica-Snieznik dome, Bohemian Massif: petrogenesis and palaeotectonic aspects. *Geologische Rundschau*, **85**, 225–238.

FLOYD, P. A., WINCHESTER, J. A., SESTON, R., KRYZA, R. & CROWLEY, Q. G. 2000. Review of geochemical variation in Lower Palaeozoic metabasites from the NE Bohemain Massif: intracratonic rifting and plume-ridge interaction. *In*: FRANKE, W., ALTHERR, R., HAAK, V. & ONCKEN, O. (eds) *Orogenic processes: quantification and modelling in the Variscan Belt of Central Europe*, Geological Society of London Special Publication **179**, 155–174.

FLOYD, P. A., CROWLEY, Q. G., WINCHESTER, J. A., KOZDROJ, W. & CYMERMAN, Z. 2002. Pillow lavas in the Ślęża Ophiolite and their geochemical relationship to other Sudetic mafic components. *In*: WINCHESTER, J. A., PHARAOH, T. C. & VERNIERS, J. (eds) *Palaeozoic Amalgamation of Central Europe*, Geological Society of London Special Publications, **201**, 197–215.

FRANKE, W. 1989. Tectonostratigraphic units in the Variscan belt of Europe. *Geological Society of America*, **230**, 67.

FRANKE, W. 1995. Rhenohercynian foldbelt: autochthon and nonmetamorphic nappe units – stratigraphy. *In*: DALLMEYER, D., FRANKE, W. & WEBER, K. (eds) *Pre-Permian geology of central and western Europe*. Springer, Berlin, Heidelberg, New York, 33–49.

FRANKE, W. 2000. The mid-European segment of the Variscides: tectono-stratigraphic units, terrane boundaries and plate tectonic evolution. *In*: FRANKE, W., ALTHERR, R., HAAK, V. & ONCKEN, O. (eds) *Orogenic processes: quantification and modelling in the Variscan Belt of Central Europe*, Geological Society of London Special Publication **179**, 35–62.

FRANKE, W. & ENGEL, W. 1986. Synorogenic sedimentation in the Variscan Belt of Europe. *Bulletin de la Societé geologique de france*, **8**, 25–33.

FRANKE, W., KREUZER, M., OKRUSCH, M., SCHÜSSLER, U. & SEIDEL, E. 1995. Stratigraphy, structure and igneous activity. *In*: DALLMEYER, D., FRANKE, W. & WEBER, K. (eds) *Pre-Permian geology of central and western Europe*. Springer, Berlin, Heidelberg, New York, 277–294.

FRANZ, G., THOMAS, S. & SMITH, D. C. 1986. High pressure phengite decomposition in the Weissenstein eclogite, Münchberger Gneiss Massif, Germany. *Contributions to Mineralogy and Petrology*, **92**, 71–85.

FURNES, H., KRYZA, R., MUSZYNSKI, A., PIN, C. & GARMANN, L. B. 1994. Geochemical evidence of progressive, rift-related early Palaeozoic volcanism in the western Sudetes. *Journal of the Geological Society, London*, **151**, 91–109.

GASS, I. G. 1980. The Troodos massif: its role in the unravelling of the ophiolite problem and its significance in the understanding of constructive plate margin processes, *Panayiotou*, 23–35.

GEBAUER, D. & GRÜENFELDER, M. 1979. Geological development of the Hercynian Belt of Europe based on age and origin of the high-grade and high-pressure mafic and ultramafic rocks. Abstracts of the 5th International conference on geochronology, cosmochronology and isotope geology, Nikko National Park, Japan, 111–112.

GRAHAM C. M. & POWELL R. 1984. A garnet-hornblende geothermometer: calibration, testing, and application to the Pelona schist, Southern California. *Journal of Metamorphic Geology*, **2**, 13–31.

GROMET, L. P., DYMEK, R. F., HASKIN, L. A. & KOROTEV, R. L. 1984. The 'North American Shale Composite': its compilation, major and trace element characteristics. *Geochimica et Cosmochimica Acta*, **48**, 2469–2482.

HAVLÍČEK , V., VANĚK, J. & FATKA, O. 1994. Perunica microcontinent in the Ordovician (its position in the Mediterranean Province, series division, benthic and pelagic associations). *Sborník Geologických Věd, Geologie*, **46**, 23–56.

HOPSON, C. A., COLEMAN, R. G., GREGÓRY, R. T., PALLISTER, J. S. & BAILEY, E. H. 1981. Geologic section through the Semail ophiolite and associated rocks along a Muscat-Ibra transect, southeastern Oman Mountains. *Journal of Geophysical Research*, **86**, 2527–2544.

JAMES, S. D., PEARCE, J. A. & OLIVER, R. A. 1987. The geochemistry of Lower Proterozoic Willyama Complex volcanics, Broken Hill Block, New South Wales. In: PHARAOH, T. C., BECKINSALE, R. D. & RICKARD, D. (eds) *Geochemistry and Mineralization of Proterozoic Volcanic Suites*. Geological Society, London, Special Publications, **33**, 395–408.

JANARDHAN, A. S., NEWTOWN, R. C. & HENSEN, E. C. 1982. The transformation of amphibolite facies gneiss to charnockite in Southern Karnataka and Northern Tamil Nadu, India. *Contributions to Mineralogy and Petrology*, **79**, 130–149.

JELÍNEK, E. & ŠTĚDRÁ, V. 1997. Petrology and geochemistry of the Mariánské-Lázně Complex. *In*: VRÁNA, S. & ŠTĚDRÁ, V. (eds) *Geological model of western Bohemia related to the KTB borehole in Germany*, Czech Geological Survey, Prague, 63–65.

JELÍNEK, E., ŠTĚDRÁ, V. & CHÁB, J. 1997. The Mariánské-Lázně Complex. *In*: : VRÁNA, S. & ŠTĚDRÁ, V. (eds) *Geological model of western Bohemia related to the KTB borehole in Germany*, Czech Geological Survey, Prague, 66–70.

KACHLÍK, V. 1997*a*. The Kladská Unit. *In*: VRÁNA, S. & ŠTĚDRÁ, V. (eds) *Geological model of western Bohemia related to the KTB borehole in Germany*, Czech Geological Survey, Prague, 70–80.

KACHLÍK, V. 1997*b*. Contact-metamorphic rocks from the mantle of the Lestkov Pluton and their significance for reconstruction of the tectono-metamorphic development of the Teplá-Barrandian area. *In*: *Geoscience Research Reports for 1996*. Czech Geol. Survey, Prague, 81–83.

KACHLÍK, V. & PATOČKA, F. 1998. Lithostratigraphy and tectonomagmatic evolution of the Zelezny Brod Crystalline Unit: some constraints for the palaeotectonic development of the W Sudetes (NE Bohemian Massif). *Geolines*, **6**, 34–35.

KOHN M. J. & SPEAR F. S. 1990. Two new geobarometers for garnet amphibolites, with applications to SE Vermont. *American Mineralogist*, **5**, 89–96.

KOSSMAT, F. 1927. Gliederung des varistischen Gebirgsbaues Abhandlungen des Sächsischen Geologischen Landesamtes, **1**, 1–39.

KREUZER, H., SEIDEL, E., SCHÜSSLER, U., OKRUSCH, M., LENZ, K. L. & RASCHKA, H. 1989. K-Ar geochronology of different tectonic units at the northwestern margin of the Bohemian Massif. *Tectonophysics*, **157**, 149–178.

KUKAL, Z. 1971. Sedimentology of Cambrian deposits

of the Barrandian area (Central Bohemia), *Sborník Geologických Věd, Geologie,* **20**, 53–100.

LINNEMANN, U., GEHMLICH, M., TICHOMIROVA, M., BUSCHMANN, B., NASDALA, L., JONAS, P., LÜTZNER, H. & BOMBACH, K. 2000. From Cadomian subduction to Early Palaeozoic rifting: evolution of the Saxo-Thuringia at the margin of Gondwana in the light of single zircon geochronology and basin development (central European Variscides, Germany). *In*: FRANKE, W., ALTHERR, R., HAAK, V. & ONCKEN, O. (eds) *Orogenic processes: quantification and modelling in the Variscan Belt of Central Europe*, Geological Society of London Special Publications, **179**, 131–154.

MALUSKI, H. & PATOČKA, F. 1997. Geochemistry and $^{40}Ar/^{39}Ar$ geochronology of the mafic metavolcanic rocks from the Rýchory Mountains complex (west Sudetes, Bohemian Massif); palaeotectonic significance. *Geological Magazine*, **134**, 703–716.

MARHEINE, D., PATOČKA, F., KACHLÍK V. MALUSKI, H., ZELAZNIEWICZ, A. 2000. Geochronological constraints on the tectonometamorphic development of the west Sudetes (Bohemian Massif). *In*: *Joint meeting of EUROPROBE (TESZ) and PACE projects, Zakopane / Holy Cross Mountains, Poland*, Polish Academy of Sciences, 62–63.

MARHEINE, D., KACHLÍK, V., MALUSKI, H., PATOČKA, F. & ZELAZNIEWICZ, A. 2002. New Ar-Ar ages in the west Sudetes (Bohemian Massif) – constraints on the Variscan polyphase tectonothermal development. *In*: WINCHESTER, J.A., PHARAOH, T. C. & VERNIERS, J. (eds) *Palaeozoic Amalgamation of Central Europe*, Geological Society of London Special Publication, **201**, 133–156.

MATTE, P. 1986. Tectonic and plate tectonic model for the Variscan belt of Europe. *In*: Variscan-Appalachian dynamics: the building of the Upper Palaeozoic basement. Basement Tectonics 15, A Coruna, Spain, *Program and Abstracts,* 254–257. *Tectonophysics*, **126**, 329–374.

MATTE, P., MALUSKI, H., RAJLICH, P. & FRANKE, W. 1990. Terrane boundaries in the Bohemian Massif: Results of large-scale Variscan shearing. *Tectonophysics*, **177**, 151–170.

MCLENNAN, S. M. 1989. Rare earth elements in sedimentary rocks: influence of provenance and sedimentary processes. *In*: B. R. Lipin and G. A. McKay (eds) *Geochemistry and mineralogy of rare earth elements, Reviews in Mineralogy*, **21**, 169–200.

NAKAMURA, N. 1974. Determination of REE, Ba, Fe, Mg, Na and K in carbonaceous and ordinary chondrites. *Geochimica et Cosmochimica Acta*, **38**, 757–775.

OKRUSCH, M., MATTHES, S., KLEMD, R., O'BRIEN, P. J. & SCHMIDT, K. 1991. Eclogites at the northwestern margin of the Bohemian Massif: a review. *European Journal of Mineralogy*, **3**, 707–730.

PATOČKA, F. 1987. The geochemistry of mafic metavolanics: implications for the origin of the Devonian massive sulfide deposites at Zlaté Hory, Czechoslovakia. *Mineralium Deposita*, **22**, 144–150.

PATOČKA, F. & KACHLÍK, V. 1998. Early Palaeozoic intracontinental rifting in the Bohemian terrane: correlation of volcanic rocks of the central West Sudetes and Barrandian. *Schriften des Staatlichen Museums für Mineralogie und Geologie zu Dresden*, **9**, 175–176.

PATOČKA, F. & SMULIKOWSKI, W. 1997. Petrology and geochemistry of metabasic rocks of the Rychorý Mts-Rudawy Janowickie Mts Complex (West Sudetes, NE Bohemian Massif). *Polskie Towarzystwo Mineralogiczne*, **9**, 146–150.

PATOČKA, F. & VALENTA, J. 1996. Geochemistry of the Late Devonian intermediate to acid metavolcanic rocks from the southern part of the Vrbno Group, the Jeseníky Mts (Moravo-Silseian Belt, Bohemian Massif, Czech Republic): paleotectonic implications. *Geolines*, **4**, 42–54.

PATOČKA, F., VLAŠMINSKÝ, P. & BLECHOVÁ, K. 1993. Geochemistry of Early Paleozoic volcanics of the Barrandian Basin (Bohemian Massif, Czech Republic): implications for paleotectonic reconstructions, *Jahrbuch Geologische*, **136**, 873–896.

PATOČKA, F., DOSTAL, J. & PIN, C. 1997. Early Palaeozoic intracontinental rifting in the central west Sudetes, Bohemian Massif: geochemical and Sr-Nd isotopic study on felsic-mafic metavolcanics of the Rýchory Mts. complex. *Terra Nova* (abstract supplement **1**), 144–145.

PEARCE, J. A. 1996. A user's guide to basalt discrimination diagrams. *In*: WYMAN, D. A. (ed.) *Trace element geochemistry of volcanic rocks: applications for massive sulphide exploration*. Geological Association of Canada, short course notes, **12**, 79–113.

PEARCE, J. A. & CANN, J. R. 1973. Tectonic setting of basaltic volcanic rocks determined using trace element analyses. *Earth and Planetary Science Letters*, **19**, 290–300.

PEARCE, J. A. & NORRY, M. J. 1979. Petrogenetic implications of Ti, Zr, Y and Nb variations in volcanic rocks. *Contributions to Mineralogy and Petrology*, **69**, 33–47.

PHARAOH, T. C. 1999. Palaeozoic terranes and their lithospheric boundaries within the Trans-European Suture Zone (TESZ): a review. *Tectonophysics*, **314**, 17–41.

PIN, C. 1990. Variscan oceans: ages, origins and geodynamic implications inferred from geochemical and radiometric data. *Tectonophysics*, **177**, 215–227.

PIN, C. & MARINI, F. 1993. Early Ordovician continental break-up in Variscan Europe: Nd-Sr isotope and trace element evidence from bimodal igneous associations of Southern Massif Central, France. *Lithos*, **29**, 177–196.

POKORNÝ, L. 1993. *The regional geophysical survey in western Bohemia; a review*. Czech Geological Survey, Prague.

REHKAMPER, M. & HOFMANN, A. W. 1997. Recycled ocean crust and sediment in Indian Ocean MORB. *Earth and Planetary Science Letters*, **147**, 93–106.

RICCI, C. A. & SABATINI, G. 1978. Petrogenetic affinity and geodynamic significance of metabasic rocks from Sardinia, Corsica and Provence. *Neues Jahrbuch Mineralogisches Monatshefte*, **1**, 28–38.

SANTOS ZALDUEGUI, J. F., SCHÄRER, U., GIL IBARGUCHI, J. I. & GIRARDEAU, J. 1996. Origin and evolution of the Paleozoic Cabo Ortegal ultramafic-mafic complex (NW Spain): U-Pb, Rb-Sr and Pb-Pb isotope data. *Chemical Geology*, **129**, 281–304.

SESTON, R., WINCHESTER, J.A., PIASECKI, M. A. J., CROWLEY, Q. G., FLOYD, P. A. 2000. A structural model for the Western-Central Sudetes: a deformed stack of Variscan thrust-bounded slices. *Journal of the Geological Society, London*, **157**, 1155–1167.

SMITH, R E. & SMITH, S. E. 1976. Comments on the use of Ti, Zr, Y, Sr, K, P and Nb in the classification of basaltic magmas. *Earth and Planetary Science Letters*, **32**, 114–120.

SÖLLNER, F., KÖHLER, H. & MÜLLER-SOHNIUS, D. 1981. Rb/Sr-Alterbestimmungen an Gesteinen der Münchberger Gneismasse (MM), NE-Bayern. Teil 1. Gesamtge-steinsdatierungen,. *Neues Jahrbuch Mineralogisches Abhandlungen*, **141**, 90–112.

SPEAR F. S. 1995. *Metamorphic phase equilibria and pressure-temperature-time paths*. Mineralogical Society of America, Washington.

ŠRÁMEK, J. 1994. *Regional gravity maps. Final report of the project 2100*. Czech Geological Survey, Prague.

ŠTĚDRÁ, V. 1996. *Retrograde evolution of the Mariánské Lázně Complex basic rocks*. Final report of the project 205/94/1455 of the Czech Grant Agency, MS Archive CGS, Prague, 129.

ŠTĚDRÁ, V. 1997. Geophysics of the Mariánské-Lázně Complex. *In*: VRÁNA, S. & ŠTĚDRÁ, V. (eds) *Geological model of western Bohemia related to the KTB borehole in Germany*, Czech Geological Survey, Prague, 61–63.

ŠTĚDRÁ, V. & CHÁB 1997. The Mariánské-Lázně Complex, discussion and conclusions. *In*: VRÁNA, S. & ŠTĚDRÁ, V. (eds) *Geological model of western Bohemia related to the KTB borehole in Germany*, Czech Geological Survey, Prague, 70.

STOSCH, H. G. & LUGMAIR, G. W. 1990. Geochemistry and evolution of MORB-type eclogites from the Münchberg Massif, southern Germany. *Earth Planet Science Letters*, **99**, 230–249.

SUN, S. S. & MCDONOUGH, W. F. 1989. Chemical and isotopic systematics of oceanic basalts: implications for mantle composition and processes. *In*: SAUNDERS, A. D. & NORRY, M. J. (eds) *Magmatism in the ocean basins*. Geological Society, London, Special Publications, **42**, 313–345.

TAIT, J, SCHATZ, M., BACHTASDE, V. & SOFFEL, H. 2000. Palaeomagnetism and Palaeozoic palaeogeography of Gondwana and European terranes. *In*: FRANKE, W., ALTHERR, R., HAAK, V. & ONCKEN, O. (eds) *Orogenic processes: Quantification and modelling in the Variscan Belt of Central Europe*, Geological Society of London Special Publications, **179**, 21–34.

TAYLOR, S. R. & MCLENNAN, S. M. 1981. The composition and evolution of the continental crust: rare element evidence from sedimentary rocks. *Philosophical Transactions of the Royal Society*, **A301**, 381–399.

TAYLOR, S. R. & MCLENNAN, S. M. 1985. *The continental crust: its composition and evolution*. Blackwell, Oxford, pp 312.

TURNER, S. P., PLATT, J. P., GEORGE, R. M. M., KELLEY, S. P., PEARSON, D. G. & NOWELL, G. M. 1999. Magmatism associated with orogenic collapse of the Betic-Alboran domain, SE Spain. *Journal of Petrology*, **40**, 1011–1036.

VON QUADT, A. 1990. U-Pb zircon and Sm-Nd analysis on metabasites from the KTB pilot bore hole. *KTB report*, **90–4**, 545.

WALDHAUSROVÁ, J. 1997. Geochemistry of volvanites (metavolcanites) in the western part of the TBU Precambrian and their geotectonic setting. *In*: S. Vrána and V. ŠTĚDRÁ (eds) *Geological model of western Bohemia related to the KTB borehole in Germany*, Czech Geological Survey, Prague, 85–90.

WINCHESTER, J. A., FLOYD, P. A., CHOCYK, M., HORBOWY, K. & KOZDROJ, W. 1995. Geochemistry and tectonic environment of Ordovician meta-igneous rocks in the Rudawy Janowickie complex, SW Poland. *Journal of the Geological Society, London*, **152**, 105–115.

WINCHESTER, J. A., FLOYD, P. A., AWDANKIEWICZ, M. PIASECKI, M. A. J. AWDANKIEWICZ, H. GUNIA, P. & GLIWICZ, T. 1998. Geochemistry and tectonic significance of metabasite suites in the Góry Sowie Block, SW Poland. *Journal of the Geological Society, London*, **155**, 155–164.

ZIEGLER, P. A. 1990. Geological atlas of western and central Europe. 2nd edition, Shell Internationale Petroleum Maatschappij B.V./Geological Society of London/Elsevier, Amsterdam.

ZULAUF, G., DÖRR, W., FIALA, J. & VEJNAR, Z. 1997. Late Cadomian crustal tilting and Cambrian transtension in the Teplá-Barrandian Unit (Bohemian Massif, Central European Variscides), *Geologische Rundschau*, **86**, 571–584.

Ślęża Ophiolite: geochemical features and relationship to Lower Palaeozoic rift magmatism in the Bohemian Massif

P. A. FLOYD[1], R. KRYZA[2], Q. G. CROWLEY[1], J. A. WINCHESTER[1] & M. ABDEL WAHED[3]

[1]*School of Earth Sciences & Geography, Keele University, Staffordshire, ST5 5BG, UK (e-mail: floyd.pa@virgin.net)*

[2]*Institute of Geological Sciences, University of Wrocław, ul. Cybulskiego 30, 50–205 Wrocław, Poland*

[3]*Tanta University, Faculty of Science, Geology Department, Tanta, Egypt*

Abstract: The Ślęża Ophiolite is one of several thrust-bounded crustal slices dominated by metabasites in the western Sudetes. The apparent field association of serpentinites, gabbros and amphibolitic components led previous workers to consider that this lithological assemblage represented an ophiolite sequence. Fieldwork suggests that the ophiolite is now highly inclined, partly overturned, so that an ophiolitic pseudostratigraphy can be deduced, grading from serpentinites and gabbros in the south to metabasite lavas in the north. The recent discovery of pillow lava structures (at Gozdnica Hill, to the west of Sobótka town) confirms that the volcanic top of the ophiolite lies in the northern section, as might be expected from the ophiolite model.

The gabbros have undergone greenschist facies metamorphism with the random development of low-grade amphibole. The volcanic portion of the sequence comprise metamorphosed dolerites and basalts partly within the contact aureole of the Variscan Strzegom-Sobótka granite. Previous work dated plagiogranites associated with the gabbros at about 400–420 Ma (U–Pb zircon ages).

Geochemical data suggest that the gabbros are distinct and apparently not comagmatic with the volcanic section of sheeted dykes and lavas. The gabbros, in particular, although very depleted in incompatible elements are dissimilar to supra-subduction zone ophiolites, exhibiting instead N-MORB-like light REE depleted patterns. Depletion is both a feature of the cumulate character of many of the gabbros, as well as a source effect (especially the uniformly low Nb content). The metabasalts and metadolerites, on the other hand, are a well-evolved single comagmatic suite with high incompatible element contents, Zr/Y approximately 3–4, and generally flat to light REE-depleted patterns. The geochemical dichotomy of the plutonic and volcanic segments calls into question a simple interpretation of the body as a single-stage coherent stratiform ophiolite.

Chemical comparison with Sudetic metabasites from within the nearby Rudawy-Janowickie and Kacazawa Complexes shows that the Ślęża metabasites have a number of features in common, including the presence of both low-Ti (gabbros) and high-Ti (dykes and lavas) chemical groups. The correlation of the gabbros, dykes and lavas with the low-Ti and high-Ti (Main Series) metatholeiites respectively, seen throughout the Bohemian Massif, as well as the Sudetes, places them within the regional collage of Palaeozoic crustal blocks separated by the Saxothuringian Seaway.

Comparison with Bohemian Massif metabasites also indicates that sediment contamination of the Ślęża Ophiolite sources was not an important process and that an enriched plume source played no part in the generation of the ophiolitic melts. The two Ślęża chemical groups were derived from variably depleted asthenospheric mantle sources. Simple modelling suggests that the volcanic segment could have been derived by 10–15% partial melting of a depleted N-MORB source, whereas the plutonic segment represents around 30% partial melting of a more depleted source. To develop varying degrees of depletion in an oceanic environment, the two sources could be related via incremental partial melting of a shallow MORB-type source.

The central European Variscides record the development of various extensional sedimentary basins, partly floored by ocean crust, that opened and closed between late Neoproterozoic and mid-Carboniferous times (Franke 1989*a*; Franke *et al.* 1995). Major episodes of continental rifting are recognized in the Cambro-Ordovician and Devonian when the Cadomian crust at the leading edge of Gondwana underwent fragmentation, with the

From: Winchester, J. A., Pharaoh, T. C. & Verniers, J. 2002. *Palaeozoic Amalgamation of Central Europe*. Geological Society, London, Special Publications, **201**, 197–215. 0305-8719/02/$15.00

variable development of oceanic crust (Pin 1990). During the subsequent Variscan orogeny the Palaeozoic sequences and basement remnants were metamorphosed and deformed (Franke 1989*b*). In tectonic terms this region of the Variscides (largely represented by the Bohemian Massif and western Sudetes) is now broadly interpreted as a collage of crustal blocks or exotic terranes, separated by ductile shear zones, generated by the collision between Gondwana (and Gondwana-derived fragments) and Baltica during the Palaeozoic (e.g. Franke 1989*a*; Matte *et al.* 1990; Oliver *et al.* 1993; Cymerman & Piasecki 1994; Cymerman *et al.* 1997; Franke & Żelaźniewicz 2000).

During progressive rifting of the Gondwana margin, extensive Lower Palaeozoic magmatic activity took place with the generation of bimodal mafic–felsic volcanic suites (Perekalina 1981; Narębski 1993; Furnes *et al.* 1989, 1994; Crowley *et al.* 2001). The presence of ophiolites also indicates that crustal attenuation was sufficient for the development of small basins or seaways floored by ocean crust (e.g. Finger & Steyrer 1995).

Objectives

This paper is concerned with a Palaeozoic ophiolite and its relationship with the mafic units of rift-related bimodal magmatism in other crustal segments. The Ślęża ultramafic–mafic complex in the Sudetes (Fig.1) was initially recognized as a typical ophiolitic assemblage by Majerowicz (1979) and still represents one of the best preserved and most complete fragments of Palaeozoic oceanic crust within the Variscan belt (Pin *et al.* 1988; Majerowicz & Pin 1994).

However, in spite of intense recent studies, relatively little detailed work has been done on the internal geochemical variation, although several authors noticed considerable diversity within the complex (Majerowicz 1981; Narębski & Majerowicz 1985; Pin *et al.* 1988; Majerowicz 1994; Majerowicz & Pin 1994; Dubińska & Gunia 1997; Majerowicz 2000). Here we present and discuss new geochemical data, obtained independently, and preliminarily interpreted by the Wrocław University research team (Abdel Wahed 1999; Kryza & Abdel Wahed 2000) and the 'PACE' research team (Floyd *et al.* 2000*a*).

Sample database and analytical methods

The data of the Wrocław team comprises thirty new major and trace element analyses of plutonic and subvolcanic/volcanic members of the Ślęża ophiolite, representing: metagabbros, mafic and ultramafic cumulates from Ślęża itself (20 samples), together with metadolerites and metabasalts from the Gozdnica, Wiezyca and Stolna hills (10 samples). The analyses were carried out in the Laboratories of Neutron Activation, Quebec University, and in McGill University, Quebec, Canada. Major elements and Rb, Sr, Ni, V, Zr, Nb, Cu, Zn, Pb, Y and Ga were determined by XRF on pressed powder pellets at McGill University; Cr, Sc, Ba, Co, Th, La, Ce, Nd, Sm, Eu, Tb, Ho, Yb, Lu, As, Cs, Sb, Se, Ta, Hf and U were analysed by INAA in the Laboratories of Neutron Activation, Quebec University (Abdel Wahed 1999).

The PACE team data was produced in the Geochemical Laboratory of the School of Earth Sciences and Geography, Keele University, by standard XRF methods for major oxides (using glass beads) and a range of trace elements (on pressed powder pellets). All samples (8 metagabbros and cumulates; 19 metadolerites and metabasalts) were analysed on an ARL 8420 X-ray fluorescence spectrometer calibrated against both international and internal Keele standards of appropriate composition. Details of methods, accuracy and precision are given in Floyd & Castillo (1992).

Table 1 lists the analytical data, according to lithology, discussed in this paper. Analytical compatibility for the XRF data generated in the two laboratories was achieved by obtaining acceptable and recommended results for the same international standards. Also, gabbros and basalt samples collected from virtually the same locations produced similar results. The combined database also includes earlier published analyses by Pin *et al.* (1988).

Geological setting and age

The Ślęża Ophiolite is exposed in the central part of the Fore-Sudetic Block in SW Poland and is the largest ultramafic–mafic body marginal to the gneisses and migmatites of the Góry Sowie Block (Fig. 1a). From bottom to top of the complex (from south to north, Fig. 1b) it comprises a complete ophiolite pseudostratigraphic sequence (cf. Coleman 1977): (1) serpentinized and tectonized peridotites; (2) pyroxene- and amphibole-rich rocks (ultramafic cumulates); (3) metagabbros (largely mafic cumulates); (4) metadolerites and metabasalts (probably sheeted dykes and lavas of the volcanic sequence); and (5) dark radiolaria-bearing metacherts (sedimentary cover). Apart from the basic members, small bodies of plagiogranites and rodingites are found (Majerowicz 1979, 1994, 2000; Majerowicz & Pin 1994; Dubińska &

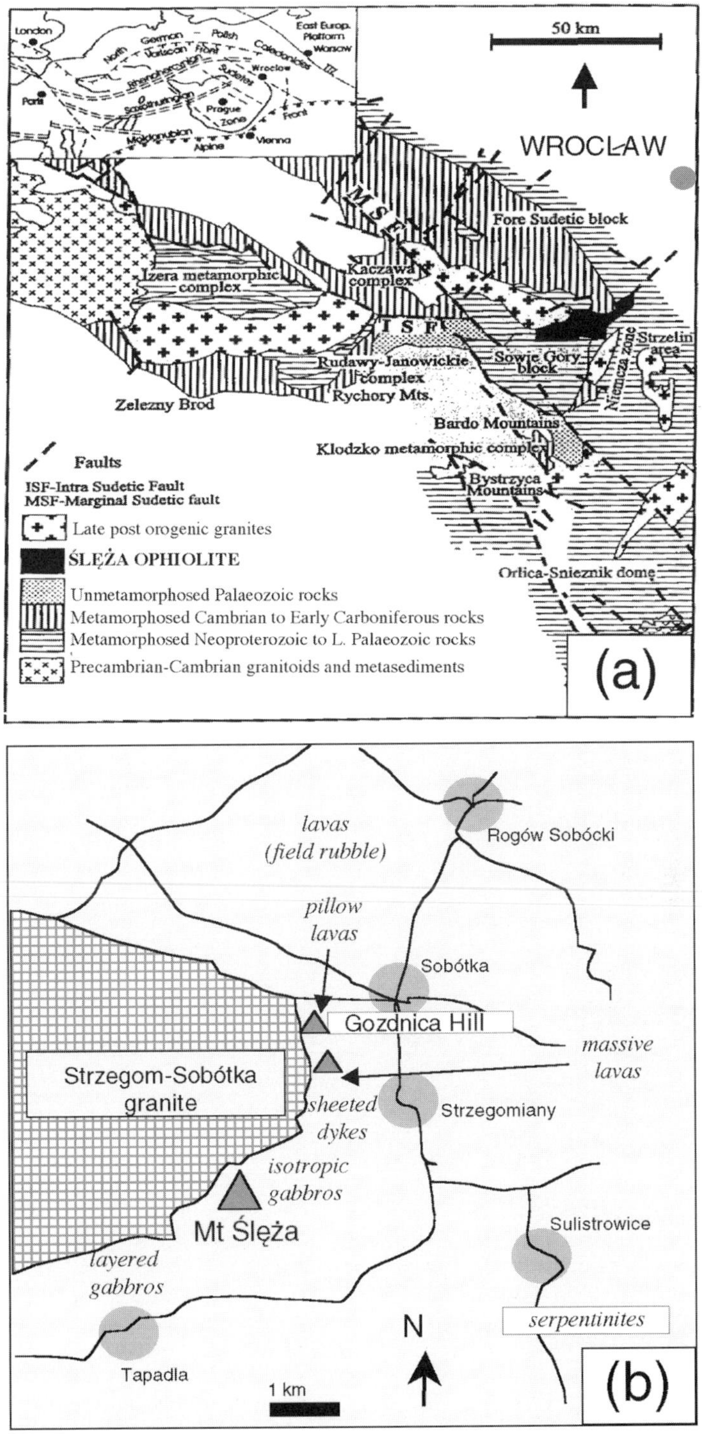

Fig. 1. (a) Location of the Ślęża Ophiolite within the Sudetes; insert shows position of the Sudetes within the Bohemian Massif and relative to the Variscan zones of Europe. (b) General relationship and position of the different magmatic units of the Ślęża Ophiolite.

Table 1. Ślęża Ophiolite geochemical data

Sample	SL99-6	SL99-7	17H	17X	17XV	63C	64H	SL99-1	SL99-2	SL99-3	SL99-5	SL99-8	SL99-9	SL99-11	10A	19C	44A	54	65B	67A	70B	7G
Lithology	ultramaf.	ultramaf.	ultramaf.	ultramaf.	ultramaf.	ultramaf.	ultramaf.	gabbro	gabbro	gabbro	gabbro	gabbro	gabbro	gabbro	gabbro	gabbro	gabbro	gabbro	gabbro	gabbro	gabbro	gabbro
Laboratory	K	K	W	W	W	W	W	K	K	K	K	K	K	K	W	W	W	W	W	W	W	C-F
Major oxides (wt.%)																						
SiO2	38.83	38.58	40.11	44.34	40.68	49.73	50.93	49.88	49.43	50.38	50.75	51.70	50.77	45.25	45.67	49.35	46.08	50.21	49.95	50.20	48.37	50.55
TiO2	0.03	0.01	0.13	0.18	0.10	1.13	0.94	0.34	0.34	0.39	0.40	0.67	0.85	3.44	0.19	0.55	0.10	0.50	0.46	0.39	0.46	0.90
Al2O3	1.26	1.11	22.02	7.59	23.72	3.39	4.00	17.49	16.43	16.18	17.24	15.52	16.07	12.83	19.21	15.48	21.50	15.75	15.34	19.45	16.51	14.75
Fe2O3t	6.48	7.66	5.10	8.87	3.91	14.38	10.20	5.32	6.34	6.13	5.79	8.46	9.38	18.79	8.94	9.24	8.33	8.22	9.29	7.33	8.77	7.21
MnO	0.08	0.09	0.07	0.12	0.06	0.27	0.22	0.11	0.12	0.12	0.12	0.17	0.16	0.21	0.12	0.15	0.11	0.13	0.15	0.13	0.14	0.16
MgO	40.08	39.59	12.48	26.21	10.24	12.28	14.76	9.36	10.65	10.04	8.63	8.30	7.72	6.27	12.21	10.48	10.34	9.74	9.28	6.71	10.63	8.75
CaO	0.80	0.38	15.10	6.02	16.40	17.50	18.30	14.92	13.95	14.13	14.40	12.00	11.56	10.83	9.81	12.90	9.37	12.90	12.70	12.40	12.10	13.30
Na2O	0.09	0.03	0.44	0.15	0.56	0.67	0.57	2.10	1.81	2.11	2.04	2.81	2.90	2.42	2.22	2.25	2.82	2.50	2.73	3.21	2.44	3.20
K2O	0.02	0.02	0.12	0.01	0.17	0.05	0.04	0.05	0.07	0.09	0.14	0.09	0.11	0.10	0.07	0.08	0.05	0.14	0.12	0.07	0.10	
P2O5	0.00	0.00	0.01	0.02	0.01	0.06	0.02	0.01	0.01	0.01	0.01	0.03	0.03	0.06	0.02	0.04	0.01	0.02	0.01	0.02	0.02	
LOI	12.45	12.05	4.99	6.98	4.76	0.75	0.58	0.85	1.01	1.01	0.87	0.71	0.71	0.37	2.87	1.19	2.88	0.81	0.78	0.72	1.76	0.68
Total	100.12	99.51	100.12	99.72	100.27	98.95	99.67	100.43	100.16	100.59	100.39	100.46	100.26	100.57	100.55	100.90	100.86	100.20	100.00	99.99	100.53	99.00
Trace elements (ppm)																						
As			0.54	22.86	1.30	0.41	1.10								0.23	0.24	0.19	0.26	0.23	0.36	0.15	
Ba	5	6	8	9	7	11	7	6	4	14		5	10	18	14	7	15	9	8	12	6	
Cl	7	4						59	36	96	34	41	15	42								
Co			33	45	78	27	41		68						35	55	40	46	44	25	35	
Cr	1910	1916	1109	1636	2702	835	849	804	803	424	279	95	84	124	273	273	369	439	503	651	652	95
Cs			0.34	0.30	0.06	0.16	0.15		0.16						0.09	0.13	0.14	0.11	0.15	0.15	0.83	
Cu	1	1	69	72	158	53	86	54	62	58	57	33	45	202	39	114	107	208	188	33	55	
Ga	1	1	13	7	9	13	14	14	12	13	13	15	15	21	23	10	14	15	13	16	10	
Hf			0.40	0.98	0.04	0.73	0.55		0.29						4.37	3.51	0.32	0.60	0.73	0.15	0.02	
Nb			3.7	2.7	0.0	3.6	4.0	3.0	0.2	3.0	3.0	1.0	1.0		2.6	1.8	3.9	3.3	4.0	3.4	3.7	
Ni	2255	2267	140	168	1479	114	173	136	174	128	96	60	50	82	76	95	146	114	141	113	501	65
Pb			0.00	2.00	0.00	0.00	1.00	6.00	0.50	4.00	4.00	9.00	5.00	8.00	2.00	2.00	0.00	0.00	0.00	0.00	2.00	
Rb	3.00	3.00	3.00	3.00	1.00	4.00	6.00	4.00	1.33	5.00	10.00	5.00	5.00	4.00	3.30	2.00	3.00	3.00	3.00	7.00	6.00	2.00
S	187	153						86	153	30	44	34	40	137								
Sb			0.13	0.90	5.85	0.18	0.16								0.04	0.48	0.20	0.24	0.11	0.29	2.27	
Sc			43	118	14	53	44		44						46	129	43	52	47	31	10	
Se			0.53	0.37	0.18	0.35	0.33								2.54	2.50	0.27	0.33	0.29	0.36	0.10	
Sr			131	10	4	119	88	150	129	135	155	125	138	160	104	13	89	121	88	194	363	129
Ta			0.09	0.08	0.02	0.09	0.04		0.28						0.12	0.08	0.05	0.12	0.07	0.06	0.01	
Th			0.02	0.07	0.03	0.03	0.04		0.09						0.09	0.06	0.04	0.02	0.06	0.03	0.01	
U			0.04	0.06	0.03	0.04	0.06		0.01						0.05	0.05	0.06	0.03	0.06	0.03	0.02	
V	52	41	154	393	88	200	165	131	148	144	164	225	264	1260	385	472	156	187	165	121	54	270
Y	2	2	14	28	3	17	14	7	7	9	8	14	15	21	49	53	14	16	17	11	6	
Zn	33	33	37	31	41	35	47	29	36	42	32	54	60	101	146	59	43	39	35	27	19	
Zr	7	3	13.00	25.00	1.00	19.00	14.00	12	12	13	15	24	32	44	161.00	96.00	11.00	15.00	24.00	7.00	1.00	30
REE (ppm)																						
La			0.03	0.03	0.11	2.40	0.44		0.45						0.30	0.77	0.22	0.57	0.13	0.21	0.18	1.01
Ce			0.30	0.50	0.60	15.60	9.40		1.35						1.10	6.20	0.60	4.20	2.70	1.30	3.00	5.92
Pr									0.26													
Nd			0.65	0.72	0.30	9.72	3.60		1.67						0.76	2.93	0.49	1.54	1.67	0.78	1.87	3.69
Sm			0.18	0.16	0.15	5.04	2.30		0.74						0.29	1.23	0.11	0.98	0.86	0.65	0.92	1.48
Eu			0.13	0.02	0.07	1.63	0.87		0.44						0.33	0.65	0.36	0.52	0.47	0.65	0.52	0.65
Gd									1.18													1.93
Tb			0.05	0.07	0.04	1.64	0.81		0.23						0.06	0.36	0.02	0.33	0.27	0.18	0.28	
Dy									1.47													2.60
Ho			0.12	0.10	0.10	2.20	1.10		0.32						0.16	0.50	0.03	0.40	0.40	0.30	0.40	
Er									0.84													1.46
Tm									0.14													
Yb			0.26	0.26	0.24	5.67	2.84		0.76						0.27	1.26	0.04	1.08	1.05	0.83	1.02	1.40
Lu			0.04	0.04	0.04	0.77	0.38		0.13						0.04	0.18	0.01	0.15	0.15	0.12	0.15	0.19

Table 1. Ślęża Ophiolite geochemical data – *continued*

Sample	18G	IPEG	BRAZ	7A	21	22A	23A	26	46B	58C	62A	SL99-10	SL99-4	SL99-12	SL99-13	SL99-14	SL99-15	SL99-16	II21	IV21	1A	5C
Lithology	gabbro	gabbro	gabbro	gabbro	gabbro	gabbro	gabbro	gabbro	gabbro	gabbro	gabbro	dyke (dol)	dyke (gab)	dolerite	dolerite	dolerite	dolerite	dolerite	dolerite	dolerite	dolerite	dolerite
Laboratory	C-F	C-F	C-F	W	W	W	W	W	W	W	W	K	K	K	K	K	K	K	C-F	C-F	W	W
Major oxides																						
SiO2	49.10	50.00	50.90	49.21	50.76	49.56	49.00	48.44	50.18	50.04	51.59	48.10	49.11	49.40	49.05	48.83	46.85	48.75	46.00	48.40	50.33	49.72
TiO2	0.70	1.30	0.50	0.56	0.58	0.99	0.46	0.52	0.33	0.47	0.41	2.80	0.48	2.80	3.25	3.46	3.47	2.86	2.75	2.50	1.99	2.41
Al2O3	13.70	4.00	15.50	16.82	14.90	15.16	19.36	15.89	20.09	16.25	15.18	11.21	12.93	13.24	13.04	12.65	13.97	13.80	15.15	14.60	14.62	13.60
Fe2O3t	9.21	11.61	5.90	7.33	7.38	12.06	8.12	9.59	6.04	6.77	8.27	21.11	8.90	14.48	15.02	15.67	16.08	14.17	17.91	13.76	12.77	15.01
MnO	0.16	0.26	0.12	0.13	0.14	0.19	0.11	0.15	0.10	0.13	0.14	0.33	0.16	0.21	0.22	0.27	0.22	0.21	0.30	0.28	0.21	0.20
MgO	11.65	15.50	8.80	9.28	9.60	8.16	8.45	9.76	6.47	8.76	9.72	3.03	12.93	6.73	6.43	6.26	5.92	6.76	10.30	8.20	6.52	6.69
CaO	11.80	16.10	15.20	13.70	13.70	10.90	11.40	13.50	13.50	15.00	11.80	8.55	12.70	8.82	9.06	8.62	9.43	9.53	4.30	8.50	10.20	9.39
Na2O	2.20	0.70	2.50	2.09	2.24	3.17	3.10	2.29	2.88	2.23	3.00	3.57	1.56	3.58	3.47	3.42	3.52	3.26	2.40	3.00	3.53	3.91
K2O				0.06	0.07	0.08	0.07	0.08	0.11	0.09	0.08	0.22	0.20	0.17	0.27	0.14	0.13	0.11			0.09	0.13
P2O5				0.03	0.03	0.06	0.04	0.02	0.02	0.02	0.01	1.26	0.01	0.24	0.29	0.28	0.30	0.24			0.21	0.24
LOI	0.88	1.04	0.23	1.26	1.33	0.77	1.15	1.00	1.21	1.11	0.85	0.26	1.32	0.32	0.33	0.31	0.34	0.33	1.59	0.87	0.62	0.27
Total	98.72	99.72	99.25	99.83	100.09	100.05	100.55	100.40	100.40	100.28	100.33	100.44	100.30	99.99	100.43	99.91	100.23	100.02	99.33	99.05	99.97	100.26
Trace elements																						
As				0.10	0.41	0.19	0.18	0.08	0.46	0.19	0.15										0.35	0.20
Ba				9	11	11	12	9	13	18	20	10	24	22	13	17	41	21			4	8
Cl												202	50	43	34	21	22	2				
Co				38	39	27	34	33	38	39	29	52			50						46	40
Cr	400	600	5.90	171	179	208	215	233	234	247	255	45	296	172	175	191	124	208	340	170	10	22
Cs				0.11	0.10	0.23	0.13	0.08	0.13	0.15	1.12	0.16			0.23						0.08	0.13
Cu				139	20	60	22	64	114	127	26	61	43	22	23	23	25	34			48	82
Ga				19	18	14	17	12	18	19	11	32	12	17	25	21	21	18			15	18
Hf				3.99	4.16	0.26	0.52	0.28	4.47	4.22	0.05	1.25			0.69						0.03	0.95
Nb				2.7	2.8	3.6	3.9	3.6	2.9	3.1	3.4	5.9	1.0	2.0	3.9	1.0	2.0	1.0	2.0	2.0	4.5	3.4
Ni	135	130	95	59	62	72	129	93	57	69	334	18	122	56	56	56	46	61	100	80	257	78
Pb				2.00	2.00	0.00	1.00	1.00	1.50	2.00	1.00	0.66	7.00	6.00	0.16	10.00	10.00	7.00			0.00	2.00
Rb	3.00		3.00	4.00	4.00	7.00	4.00	3.00	4.00	3.00	9.00	1.41	10.00	6.00	2.48	4.00	5.00	5.00	12.50	4.00	3.00	4.00
S												2532	21	28	32	25	30	201				
Sb				0.22	0.23	0.37	0.00	0.11	0.19	0.07	2.51										0.23	0.08
Sc				41	44	56	30	42	46	47	8	31			43						3	58
Se				2.14	2.15	0.32	0.71	0.08	1.33	2.33	0.09										0.13	0.51
Sr	90		107	205	184	160	157	122	162	127	410	164	21	128	166	151	103	121	31	95	159	111
Ta				0.05	0.09	0.05	0.03	0.04	0.13	0.07	0.01	0.58			0.36						0.04	0.05
Th				0.09	0.11	0.07	0.03	0.04	0.02	0.13	0.01	0.35			0.13				0.17	0.17	0.02	0.05
U				0.10	0.05	0.05	0.04	0.07	0.06	0.07	0.01	0.12			0.04						0.03	0.08
V	445	540	270	338	356	180	123	149	376	360	40	70	216	436	466	532	503	432	450	455	19	334
Y				44	47	15	13	13	49	49	5	166	10	58	66	65	69	55			5	22
Zn				99	104	22	41	39	33	282	19	116	52	48	52	58	57	51			35	55
Zr	12			150.00	158.00	8.00	20.00	9.00	159.00	171.00	0.00	319	12	183	204	202	211	173	157	166	1.00	28.00
REE (ppm)																						
La	0.71	0.93	1.21	0.42	0.69	0.92	0.60	0.49	0.34	0.51	0.18	16.90			6.80				4.57	4.86	3.68	4.84
Ce	4.60	7.48	7.74	4.80	4.60	7.70	3.50	4.80	3.20	3.90	2.50	56.00			20.50				20.70	21.90	18.90	28.20
Pr												10.70			3.84							
Nd	1.75	4.74	2.54	1.88	2.59	2.96	2.56	2.34	0.98	1.64	1.50	60.10			20.50				15.20	16.45	12.00	15.00
Sm	0.67	2.18	1.40	0.94	1.23	1.60	0.90	1.07	0.66	0.86	0.82	19.07			6.68				5.52	5.84	4.13	5.24
Eu	0.36	0.74	0.62	0.19	0.67	1.24	0.61	0.31	0.58	0.57	0.55	5.22			2.13				1.77	2.03	1.80	2.48
Gd	0.89	3.03	2.07									24.90			8.78				6.37	6.74		
Tb				0.29	0.36	0.49	0.27	0.35	0.23	0.26	0.30	4.19			1.58						1.10	1.43
Dy	1.12	4.11	2.35									25.25			9.78				7.87	8.08		
Ho				0.40	0.50	0.70	0.30	0.50	0.30	0.40	0.30	5.25			2.09						1.40	1.70
Er	0.61	2.30	1.24									13.60			5.57				4.49	4.71		
Tm												2.16			0.92							
Yb	0.59	2.10	1.15	0.99	1.33	1.87	0.91	1.16	0.74	1.03	0.91	11.94			5.17				4.64	4.88	3.86	4.97
Lu	0.09	0.29	0.16	0.14	0.19	0.26	0.13	0.17	0.10	0.14	0.14	1.91			0.81				0.73	0.74	0.54	0.70

Table 1. Ślęża Ophiolite geochemical data – *continued*

Sample	6A	29A	2A	SL99-77	SL99-18	SL99-19	2B	6D	6D1	6E	6F	SL99-20	SL99-21	SL99-22	SL99-23	SL99-24	SL99-25	98-43A	98-43B	98-44A	98-44B
Lithology	dolerite	dolerite	dolerite	basalt	basalt	basalt	basalt	basalt	basalt	basalt	basalt	basalt	basalt	basalt	basalt	basalt	basalt	basalt	basalt	basalt	basalt
Laboratory	W	W	W	K	K	K	W	W	W	W	W	K	K	K	K	K	K	K	K	K	K
Major oxides																					
SiO2	46.63	49.52	49.23	49.40	44.91	48.85	49.76	48.86	48.65	52.64	47.81	48.14	47.69	51.64	48.69	48.16	48.54	50.37	48.63	44.10	47.33
TiO2	2.34	2.76	2.20	3.08	2.39	3.11	2.09	2.95	2.91	2.91	2.36	2.63	2.41	0.66	2.46	2.39	2.97	1.97	1.65	4.25	3.13
Al2O3	14.53	13.37	14.01	13.88	13.02	13.29	14.64	12.45	12.41	13.38	14.28	13.88	14.28	15.14	11.55	11.49	12.87	13.46	14.22	10.61	11.73
Fe2O3t	15.20	16.47	14.98	14.11	13.64	15.08	13.99	18.35	18.39	13.99	15.40	13.42	14.06	8.33	14.33	14.39	15.08	11.80	11.51	19.32	16.60
MnO	0.27	0.24	0.22	0.29	0.26	0.25	0.21	0.33	0.33	0.36	0.39	0.22	0.20	0.17	0.25	0.25	0.24	0.18	0.17	0.24	0.22
MgO	7.98	6.22	6.92	7.52	8.22	6.41	6.65	6.28	6.14	6.55	8.53	6.53	8.27	8.37	8.98	9.83	6.48	8.17	6.90	6.81	6.20
CaO	10.60	8.91	9.79	7.82	7.38	9.61	9.72	9.12	9.25	7.04	8.78	10.79	9.15	11.97	10.12	10.64	9.57	6.48	10.49	9.83	9.29
Na2O	2.84	3.92	3.48	3.71	1.94	3.03	3.74	2.60	2.52	3.89	3.28	3.15	2.97	2.86	2.54	2.47	3.10	3.84	3.21	1.60	2.85
K2O	0.11	0.22	0.10	0.10	0.07	0.12	0.09	0.13	0.14	0.08	0.10	0.13	0.11	0.09	0.16	0.17	0.21	0.08	0.10	0.04	0.05
P2O5	0.32	0.30	0.19	0.26	0.21	0.27	0.21	0.33	0.33	0.22	0.19	0.15	0.17	0.03	0.18	0.17	0.24	0.19	0.17	0.47	0.35
LOI	0.46	0.56	0.40	0.41	7.67	0.29	0.42	0.65	0.53	0.78	0.39	0.41	0.51	0.54	0.58	0.55	0.40	3.54	2.73	2.36	1.81
Total	99.95	101.05	100.21	100.58	99.71	100.31	100.30	100.45	99.99	100.62	100.16	99.45	99.82	99.80	99.84	100.51	99.70	100.08	99.79	99.64	99.58
Trace elements																					
As	0.34	0.17	0.15				0.18	0.20	0.12	0.13	0.23										
Ba	12	6	19	8	10	11	7	9	15	17	7	9	11	23	16	27	201	11	30	22	5
Cl				5	15	25							2		1	1	256	5		93	91
Co	47	45	36				51	24	36	43	29	52	48		50						
Cr	67	69	92	171	367	162	120	136	147	165	165	224	237	257	275	306	150	170	285	28	31
Cs	0.11	0.10	0.15				0.36	0.08	0.27	0.14	0.07	0.04	0.09		0.15						
Cu	118	42	62	24	24	48	62	60	84	81	78	34	45	45	33	37	28	29	44	45	29
Ga	22	22	20	19	20	21	13	18	18	14	16	21	19	18	17	18	21	14	16	19	20
Hf	6.33	6.53	6.03				0.12	5.66	3.48	0.32	0.23	1.31	0.68		0.55						
Nb	2.2	2.6	2.5	1.0	1.0	1.0	4.2	3.6	2.7	3.9	3.9	2.8	2.8	1.0	2.2	1.0	1.0	3.0	4.0	11.0	7.0
Ni	36	49	41	56	100	48	248	24	61	108	80	66	78	81	95	101	57	55	61	47	36
Pb	3.00	3.00	3.00	9.00	7.00	11.00	0.00	2.00	2.70	1.00	0.00	0.29	0.79	8.00	1.73	11.00	13.00				
Rb	3.00	3.00	5.00	6.00	5.00	3.00	4.00	3.00	4.00	4.00	3.00	1.16	1.07	4.00	4.02	6.00	7.00	10.00	7.00	6.00	7.00
S				222	60	903						22	23	562	35	30	311				
Sb	0.08	0.07	0.24				0.12	0.06	0.54	0.04	0.09										
Sc	45	45	50				7	47	39	44	28	43	39		48						
Se	4.87	3.39	2.72				0.12	2.69	1.64	0.21	0.28										
Sr	127	119	146	104	50	160	122	97	181	89	183	227	324	168	146	139	149	143	201	191	146
Ta	0.15	0.08	0.12				0.02	0.14	0.07	0.03	0.03	0.30	0.28		0.21						
Th	0.18	0.21	0.06				0.02	0.24	0.14	0.03	0.03	0.17	0.15		0.13						
U	0.10	0.10	0.04				0.03	0.11	0.05	0.03	0.04	0.14	0.06		0.05						
V	398	411	400	465	455	458	38	413	287	170	131	396	386	381	453	462	487	406	324	604	487
Y	70	70	58	59	54	64	7	56	39	14	12	56	51	50	54	48	57	46	37	103	78
Zn	212	206	61	114	1034	135	45	101	83	41	31	108	109	106	147	126	134	112	94	102	84
Zr	245.00	248.00	216.00	188	160	195	4.00	218.00	125.00	10.00	8.00	173	157	154	144	139	170	136	122	285	229
REE (ppm)																					
La	5.03	6.82	4.89				4.58	7.51	7.48	6.31	4.81	5.30	4.70		4.10						
Ce	23.20	40.30	24.50				22.00	40.40	41.20	33.20	22.80	16.80	14.90		12.80						
Pr												3.16	2.80		2.44						
Nd	14.55	17.82	14.87				13.45	22.51	21.12	17.75	15.63	17.20	15.40		13.50						
Sm	4.98	6.69	4.91				4.63	7.64	7.50	6.10	4.77	5.66	5.09		4.60						
Eu	2.35	2.60	2.16				2.03	3.29	3.23	2.29	1.89	1.82	1.62		1.49						
Gd												7.57	6.77		6.21						
Tb	1.30	1.75	1.30				1.19	1.99	1.98	1.57	1.23	1.37	1.23		1.12						
Dy												8.61	7.59		7.07						
Ho	1.80	2.20	1.60				1.50	2.50	2.50	2.00	1.50	1.85	1.64		1.52						
Er												5.00	4.36		4.19						
Tm												0.82	0.70		0.70						
Yb	4.72	6.11	4.59				4.31	7.08	7.17	5.67	4.74	4.67	3.92		4.10						
Lu	0.66	0.82	0.63				0.60	0.98	0.99	0.79	0.65	0.72	0.59		0.66						

Initial lithology nomenclature used, although all samples are variably metamorphosed
Analytical laboratories: K = Keele; W = Wrocław; C-F = Clemont–Ferrand

Gunia 1997). The long 'missing' pillow lavas of the ophiolite sequence have recently been discovered near the top of the pseudostratigraphic succession (Floyd *et al.* 2000*a*).

The Ślęża Ophiolite has a west–east trending fault contact with the Góry Sowie gneiss to the south, and is intruded by the post-tectonic Strzegom-Sobótka granite (*c.* 280 Ma; Pin *et al.* 1988) to the NW (Fig. 1b). To the north and east its contact with the metamorphic rocks of the Fore-Sudetic Block is masked by Cenozoic cover. In a recent tectonic review of the Bohemian Massif (Winchester *et al.* 2002) the Ślęża Ophiolite represents the structural top of a series of nappes thrust westwards during the Variscan orogeny.

The protolith ages of the Ślęża ophiolite remain controversial. A Sm–Nd whole-rock isochron age of 353±21 Ma, with εNd +8.8±0.1, was obtained by Pin *et al.* (1988) and interpreted as the magmatic age of the mafic rocks and thus, the age of local ocean crust formation. A similar age (351±16 Ma, with εNd +8.5±0.1) was obtained for the nearby Nowa Ruda Ophiolite, south of the Góry Sowie Block (Pin *et al.* 1988). However, U–Pb zircon ages from plagiogranites within the gabbroic segment of the ophiolite produced older ages of 423±20 Ma (Oliver *et al.* 1993), and abraded single zircon ages from a rodingitized plagiogranite (albitite), obtained an upper intercept age of 400±3 Ma (Żelaźniewicz *et al.* 1998). Both ages are interpreted as the age of formation of the plagiogranites within the gabbros. As the Pin *et al.* (1988) data included four Ślęża gabbros on the whole rock isochron, there is clearly a discrepancy in the age of (at least) the plutonic portion of the ophiolite. Although no independent ages exist for the volcanic rocks, the two main segments (volcanic v. plutonic) may be of different ages. As there is now geological information which suggests that the *c.* 350 Ma age may be too young (Franke & Żelaźniewicz 2000), having been reset by the intrusion of the Variscan Strzegom-Sobótka granite, the older *c.* 400 Ma age is preferred for the plutonic segment.

Previous studies

Previous petrological and geochemical studies of the Ślęża ultramafic–mafic complex (e.g. Majerowicz 1994; Majerowicz & Pin 1994) revealed a range of important features typical of an ophiolite sequence. First of all, the lithological assemblage, its petrography and spatial arrangements appeared to correspond to the classical ophiolite pseudostratigraphy (cf. Coleman 1977). Preliminary geochemical studies, including Sm–Nd isotopes (Pin *et al.* 1988), showed that the metabasites were all tholeiitic, derived from a depleted mantle source, and characteristic for an ophiolite formed at a normal mid-ocean ridge.

Detailed petrological studies (e.g. Majerowicz & Pin 1989, 1994) combined with stable isotope characteristics (Jędrysek *et al.* 1989) indicated the probable role of ocean floor metamorphism in the petrogenesis of the ophiolite. Dubińska & Gunia (1997), based on both the geochemistry of rodingites and boninite-type mafic dykes within the serpentinites and analysis of the serpentinization and rodingitization processes, proposed several evolutionary stages, including emplacement in a mid-ocean ridge environment, and subsequent (minor) magmatic and metamorphic processes in a subduction-related fore-arc setting. The involvement of an arc in the formation of the ophiolite is controversial; a suggestion previously rejected on geochemical grounds by Pin *et al.* (1988).

Tectonic models of the Ślęża Ophiolite, confirm its general pseudostratigraphy, with a younging direction from south to north (Fig. 1b). The sequence may have been overturned subsequently, although this interpretation is questionable (Jamrozik 1989; Majerowicz 1979) largely due to poor exposure at critical junctions. According to Mierzejewski & Abdel Wahed (2000), the serpentinites at the base of the ophiolite display evidence of two deformation episodes and tectonic displacements, mainly along low angle, east-dipping, faults: an earlier top to west or WNW thrusting along listric thrust faults, and a later top to east or ESE normal faulting, due to relaxation of thrust blocks. In the metabasites, Abdel Wahed (1999, 2000) distinguished three deformation events: (a) D_1: west-directed tectonic transport inferred from the orientation of magmatic layering which dips to ESE and WSW, and defines a possible fold with SSW dipping axis (equivalent to the earlier deformation event in the serpentinites); (b) D_2: top to NNE shearing along WNW–ESE striking and southward dipping shear planes, and producing a SW-dipping stretching lineation; (this event is thought to correspond to synmetamorphic NNE-directed thrusting along the SE margin of the Bohemian Massif); (c) D_3: brittle deformation represented by slickensides and accompaning striae of various orientation.

Lithological units and petrography

The mafic members of the Ślęża Ophiolite can be broadly subdivided, into a plutonic group (comprising the metagabbros and mafic cumulates) and a volcanic group (composed of metadolerites and metabasalts in the upper (northern) part of the sequence, including strictly subvolcanic units). Locally at Tapadła Pass, within a contact zone about 100 m wide between the serpentinites and metagabbros, are various pyroxene–amphibole-rich ultramafic cumulates (Majerowicz & Pin 1989, 1994). The boundary between the plutonic and volcanic groups is a wide, poorly exposed zone of mafic rocks with highly variable textures. Typically, various fragments of metagabbros and dolerites, with ophitic, subophitic, diabasic, porphyritic and glomeroporphyritic textures, are embedded in fine-grained or even an aphanitic groundmass (Majerowicz 1994; Majerowicz & Pin 1994). Whether these features are indicative of some type of 'transitional boundary' between the two groups of the ophiolite or characteristic of the roof zone of the gabbros with dyke roots, is open to question. Near the assumed base of the volcanic group, however, rather indistinct structures resembling sheeted dykes have been described (Majerowicz & Pin 1989).

Metagabbros and metacumulates

Typically, the metagabbros are massive, medium- to coarse-grained rocks, with ophitic and subophitic textures. However, they display considerable textural and compositional variation, including compositional and grain-size layering, pegmatoid segregations, and localized deformational structures, e.g. faint foliation and usually cm-scale discrete shear zones (Majerowicz & Pin 1994; Majerowicz *et al.* 2000; Abdel Wahed 1999, 2000).

Weakly deformed metagabbros are composed of variably uralitized clinopyroxene and saussuritized plagioclase, together with minor opaques (mostly ilmenite), titanite, epidote and chlorite. The igneous clinopyroxene (diallage augite and diopside) is usually replaced, to varying degrees, by pale green amphibole ranging in composition from Mg-hornblende to actinolite. Relatively fresh and tabular plagioclase has a composition of An_{50-70}, but is often partly replaced by epidote-group minerals and more sodic plagioclase. The metagabbros are, in places, strongly, but not penetratively, foliated and in such varieties the clinopyroxene is nearly completely replaced by secondary amphibole and chlorite, and the plagioclase is strongly saussuritized.

Majerowicz (1979) and in Majerowicz & Pin (1994) described magmatic layering in the gabbro, including centimetre-scale mineral compositional banding, as well as larger-scale 'cryptic layering' expressed as upwards decreasing XMg values in clinopyroxene and An contents in plagioclase, in the gabbro member of the ophiolite. According to Abdel Wahed (1999, 2000), two types of macrorhythmic magmatic layering can be distinguished: (a) modal layering, composed of the alternation of light (plagioclase-rich) and dark (clinopyroxene and Fe–Ti oxide-rich) few centimetre-scale bands; and (b) grain-size layering, with the repeated alternation of medium- and coarse-grained gabbro, as 10–70 cm thick layers. The magmatic layering has two dip directions: ESE and WSW, interpreted by that author as a probable F_1 fold, with a SSW-dipping axis, representing the earliest deformation (D_1) in the gabbro.

The ultramafic cumulates at Tapadla Pass are represented by massive and mostly fine-grained, colour-banded rocks, composed of variable proportions of serpentinized olivine, relics of orthopyroxene and clinopyroxene, tremolite and abundant serpentine minerals, with associated Cr-spinel, iron oxides, chlorite, talc, epidote and carbonates. Their bulk chemical compositions correspond to ultrabasic rocks, and they were interpreted by Majerowicz & Pin (1994) as ultramafic cumulates. Other varieties of ultramafic rocks form dyke-like or irregular bodies within the metagabbros and were described by Abdel Wahed (1999, 2000) as 'metapyroxenites'. They consist predominantly of coarse-grained clinopyroxene (diallage, up to 5 cm in size), and subordinate uralitic amphibole, chlorite, epidote, ilmenite (often as leucoxene) and titanite. Most probably, this rock type represents a clinopyroxene-rich cumulate.

Metadolerites and metabasalts

Mafic rocks of the volcanic member of the ophiolite vary from aphyric and phyric medium-grained metadolerites, to finer grained aphanitic metabasalts (Majerowicz & Pin 1989). As in the metagabbros, the dominant secondary mafic mineral is amphibole and according to Majerowicz & Pin (1994), ranges in composition from Mg-hornblende, through actinolitic hornblende, to actinolite.

The metadolerites are dark greenish-grey in colour and massive with evident well preserved relict igneous textures (diabasic, intergranular and ophitic). They are composed of subidiomorphic and idiomorphic plagioclase laths of variable An contents (from andesine to albite), and green to pale-green amphibole in the form of interstitial grains and acicular aggregates. Larger felted masses of pale green to blue-green amphibole appear to have replaced originally zoned(?) clinopyroxene. Locally, relicts of clinopyroxene are preserved. Many of the metadolerites are plagioclase-phyric with the phenocrysts having undergone variable granular recrystallization, together with the development of brown granules (lepidocrocite after Fe ore?) in their cores. Ilmenite, epidote, chlorite and minor quartz are found as accessories.

A rare metadolerite dyke (sample SL99-10) intruding the gabbro sequence of the ophiolite is composed of strongly pleochroic yellow to blue-green amphibole clots (after pyroxene) and scattered fibres, together with partial recrystallized plagioclase laths. It is unusual in that it contains numerous large apatite prisms and ilmenite grains.

The fine-grained metabasalts are dark greenish-grey rocks composed of phenocrysts of plagioclase (*c*. An_{50-35}) and rarer clinopyroxene set in a microcrystalline groundmass of greenish amphibole, plagioclase, ilmenite, leucoxene, epidote and minor quartz. Very fine-grained to aphanitic varieties of metabasalts are similarly common. Microscopically, they also display well-preserved igneous textures, ranging from granular, aphyric to glomeroporphyritic. The main components include greenish amphibole, with dark green to pale green pleochroism, and subidiomorphic microphenocrysts of plagioclase (*c*. An_{50-30}). Both minerals can also occur as fine-grained aggregates, together with Fe–Ti oxides, chlorite and epidote, in the groundmass. The rim segments of pillow lava metabasalts retain textural evidence for rapid quenching with fork-terminated plagioclase microlites and variolitic fans of originally clinopyroxene, now replaced by amphibole or chlorite. The interiors of pillows are better crystallized and largely composed of small felted masses of amphibole and granular plagioclase.

Chemical alteration effects

All the analysed metabasite samples show varying degrees of low-grade mineralogical alteration, but retain sufficient relict features to discern their original lithology, as detailed above. However, even very mild alteration can be expected to have caused selected element mobility, especially involving the large-ion-lithophile (LIL) elements (e.g. Hart *et al.* 1974; Humphris & Thompson 1978; Thompson 1991). LIL element (e.g. K, Na, Rb, Ba, Sr) abundances are often highly variable, together with most major elements and ratios (e.g. FeO*/MgO), and are unreliable as indicators of petrogenetic relationships or tectonic discrimination. This is particularly true for the volcanic sequence of ophiolites where mineralogical and chemical alteration by submarine hydrothermal processes are well known (e.g. Gass & Smewing 1973; Pearce & Cann 1973; Spooner & Fyfe 1973; Smewing & Potts 1976). In the Ślęża Ophiolite it is likely that this initial submarine alteration has been overprinted subsequently by Variscan regional metamorphism and deformation. However, in either case characteristic magmatic inter-element relationships are often maintained by those elements that are considered relatively immobile during alteration, such as high field strength (HFS) elements and the rare earth elements (REE) (e.g. Pearce & Cann 1973; Smith & Smith 1976; Floyd & Winchester 1978). The interpretation of magmatic relationships and sources is thus dependent on these elements. Under some circumstances, such as the extensive carbonatization of metabasites, even the REE and HFS elements can be mobilized or their abundances diluted (e.g. Hynes 1980), although no carbonate-bearing samples were included in this study.

A further complication in the interpretation of the Ślęża Ophiolite metabasites is the possibility of metasomatism by the adjacent Variscan Strzegom-Sobótka granite. Studies of contact metasomatism by granites have suggested that this process may often be limited to simple hydration (e.g. Pitcher & Sinha 1958), but in Variscan granites from SW England (e.g. Floyd *et al.* 1993), extensive chemical alteration adjacent to the contact may occur, with the addition of LIL elements, F, Cl, B and possibly the light REE (e.g. Floyd & Fuge 1973; Mitropoulos 1982; Stone & Awad 1988). Metabasite samples closest to the assumed position of the Strzegom-Sobótka granite contact were collected about 200 m away and exhibit similar LIL element contents and normalized patterns to those further away, so contact metasomatism is not considered a significant feature in this study.

Geochemical groups and characteristics

Irrespective of the apparent pseudostratigraphy of the ophiolite, the plutonic (metagabbros) and volcanic (metadolerite/metabasalt sheeted dykes and lavas) segments appear to define two separate chemical groups that may not be directly related. The two lithological/structural groups define different trends with variable Ti–V and Zr–Y ratios (Fig. 2). Kryza & Abdel Wahed (2000) also identified other trace element ratios that distinguish the two groups. Normalized rare earth element (REE) patterns (Fig. 3) also differ in terms of overall abundance and pattern form, although variable light REE depletion ($[La–Yb]_N < 1$) is characteristic of both

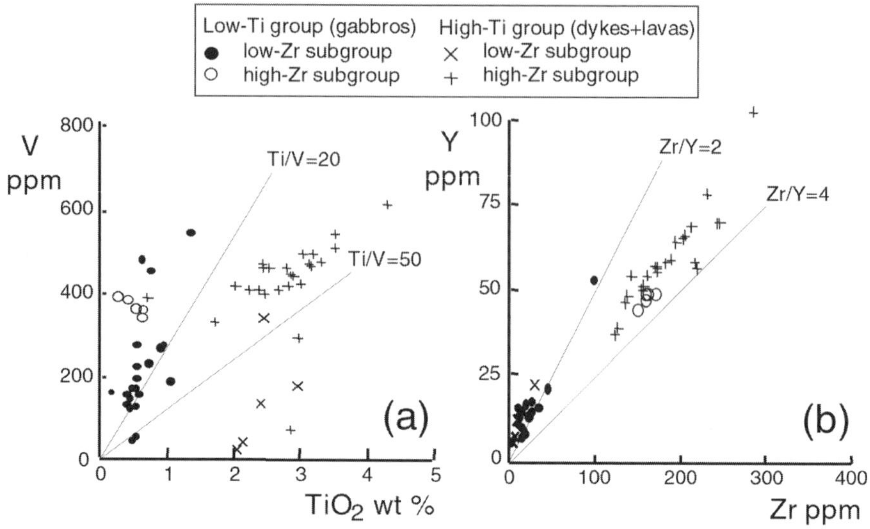

Fig. 2. Chemical distinctions between the volcanic and plutonic segments of the Ślęża Ophiolite. (a) V–TiO_2 diagram and Ti–V ratios (discrimination ratios after Shervais 1982), (b) Y–Zr diagram and Zr–Y ratios.

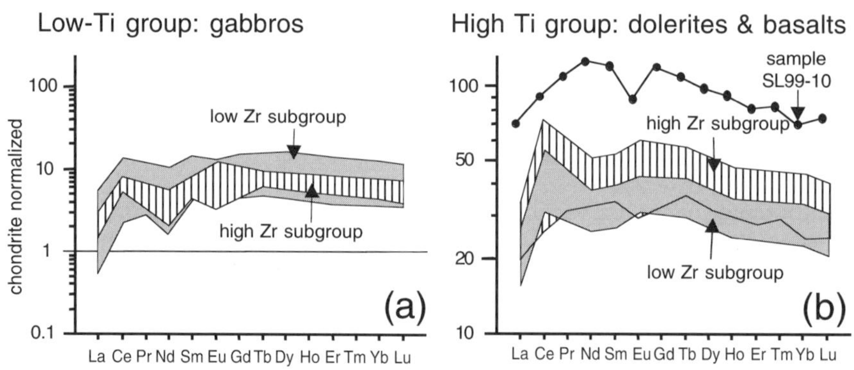

Fig. 3. Normalized REE plots showing the overlap in the range of abundances for the Zr subgroups within the (a) gabbros (low Ti group), and (b) dolerites and basalts (high-Ti group), respectively. Note the markedly enhanced REE abundances for the high-Ti group in general and especially the highly fractionated apatite-bearing sample SL99–10 (diagram (b)).

groups. In terms of the general level of incompatible elements, the metagabbros are strongly depleted relative to the more evolved sheeted dykes and lavas. The highly depleted characteristic is, in part, a reflection of the cumulate nature of many of the metagabbros, although only some have high Ni (>150 ppm) and Cr (>400 ppm) contents that are suggestive of olivine–pyroxene cumulates. On the basis of the chemical dichotomy of the Ślęża Ophiolite segments we identify a low-Ti group (comprising the metagabbros) and a high-Ti group (representing the metadolerites and metabasalts). A further subdivision may be made on Zr content: a number of gabbros with high Zr contents have similar Zr–Y ratios to the lavas and dykes, whereas a few lavas having very low Zr contents plot with the metagabbros (Fig. 2b; Zr–Y plot). This allows four chemical groups to be identified, with both low- and high-Ti groups having a low- and high-Zr subgroup respectively; this fourfold subdivision is maintained in the chemical plots illustrated in this paper.

As previously recognized (Pin *et al.* 1988) the Ślęża Ophiolite has an overall MORB-type chemistry, even allowing for the highly depleted

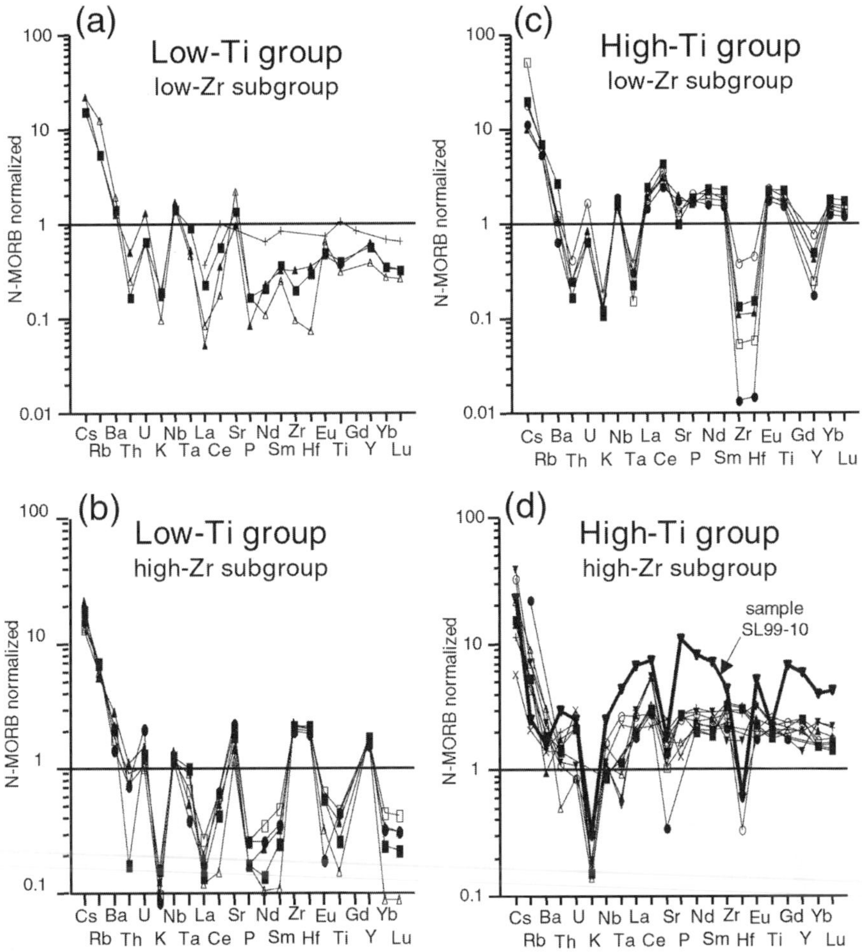

Fig. 4. N-MORB normalized plots for the four chemical subgroups within the Ślęża Ophiolite. Relative to a normalized value of 1, fundamental differences are evident between the plutonic portion (low-Ti group; diagrams (a) and (b)) and the volcanic portion (high-Ti group; diagrams (c) and (d)) of the ophiolite. Normalization factors from Sun & McDonough (1989).

(almost arc-like) nature of many of the metagabbros. The characteristic features of the four chemical groups normalized relative to N-MORB are shown in Figure 4. Apart from the more mobile LIL group of elements, the low-Ti group shows variable, but typically highly depleted patterns (normalized values <1). The exceptions are the presence of (i) positive Sr and Eu anomalies reflecting plagioclase accumulation, and (ii) strong positive Zr, Hf and Y anomalies, probably reflecting cumulate zircon, that defines the high-Zr subgroup. On the other hand, most dykes and lavas of the high-Ti group exhibit enriched or evolved, but parallel, patterns to N-MORB. Negative Sr anomalies represent the effect of plagioclase fractionation, whereas strong negative Zr+Hf+Y anomalies are a dominant feature of the low-Zr subgroup. The most evolved sample of the high-Zr subgroup (SL99–10; Fig. 4) shows strong positive anomalies for the light REE, P and Y, as might be expected from the presence of abundant cumulate apatite. This feature is also supported by the 'm'-shaped REE pattern with a negative Eu anomaly (Fig. 3) for the whole rock that mimics that of apatite.

The chemical relationship between the Ti groups and their respective Zr subgroups is largely a reflection of zircon and apatite extraction and accumulation. The high-Zr subgroup of the low-Ti gabbros is a consequence of local accumulation, whereas the low-Zr subgroup of

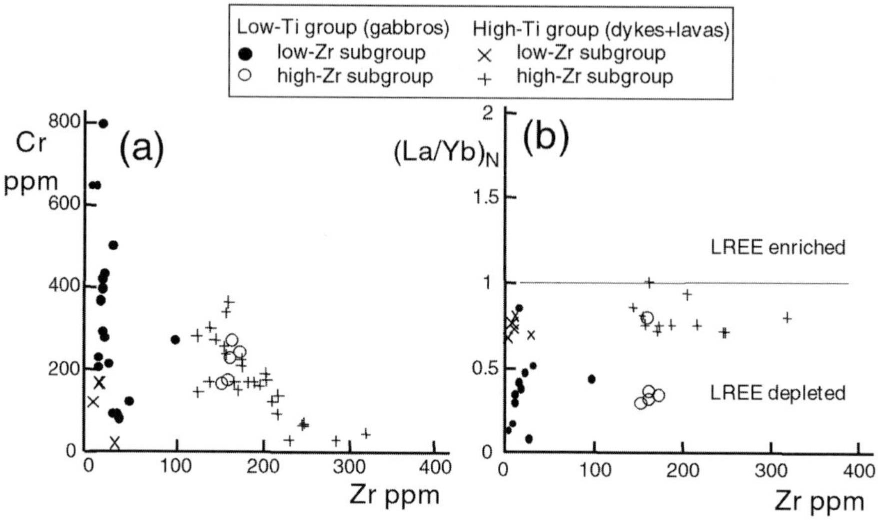

Fig. 5. Chemical differences between the four subgroups in terms of (a) variable fractionation (Cr–Zr plot) and (b) partial melting (La/Yb–Zr plot). All the Ślęża samples are characterized by being light REE depleted, whereas some of the gabbros have Cr contents indicative of pyroxene cumulates.

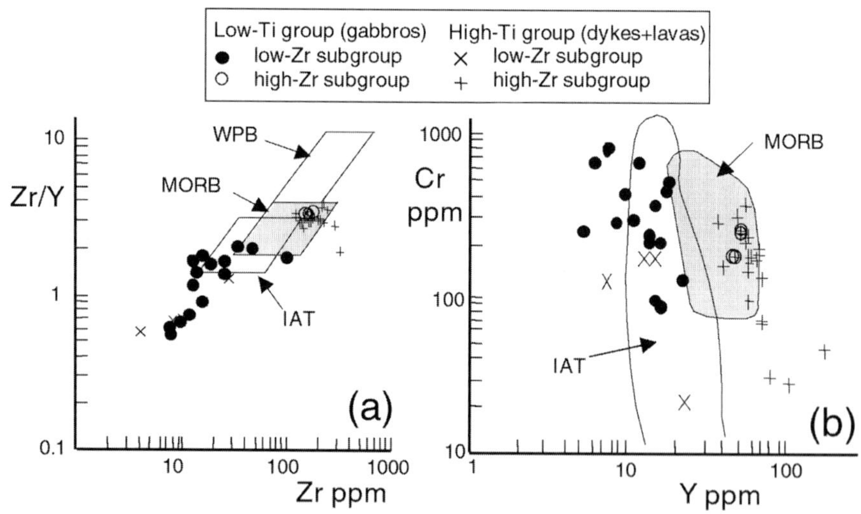

Fig. 6. Chemical discrimination of the eruptive environment of the Ślęża Ophiolite. These plots suggest that the volcanic segment metabasites (high-Ti group) are typical of a spreading ridge (with MORB), whereas the plutonic segment gabbros (low-Ti group) have more depleted 'island arc' (with IAT) affinities. Discrimination fields in (a) after Pearce & Norry (1979) and (b) after Pearce (1980).

the high-Ti metadolerites/metabasalts could represent the composition of residual liquids left after accessory mineral extraction. Although zircon is the most likely host for Zr, its presence is rare in the basalts even though they are very well-fractionated basic compositions; pyroxene could also be another possible host mineral. Figure 5 shows that the two chemical groups define different mafic fractionation paths (Cr–Zr plot) and exhibit different degrees of light-to-heavy REE fractionation (La/Yb–Zr plot). Zircon accumulation during general mafic fractionation would increase the Zr content for the low-Ti group, whereas residual liquids after zircon extraction would have their Zr contents markedly reduced, as shown for the low-Zr metadolerites/metabasalts. Although zircon tends to concentrate the heavy REE relative to the light (Rollinson 1993) this feature would not change the $(La–Yb)_N$ ratio to any marked degree if the proportion of zircon (or apatite) fractionating was relatively low. As seen in Figure 5 the normalized light-to-heavy ratio is approximately the same for the respective Zr subgroups; note also the general overlap of normalized REE patterns for the Zr-subgroups in Figure 3.

The high-Ti group (lavas and dykes) also exhibits two other chemical features which have a bearing on their eruptive environment:

(a) Although they are generally evolved in terms of incompatible element contents, Nb is especially low (often 1–2 ppm) and appears out of context relative to the other abundances. This feature was also noted by Pin *et al.* (1988), and might suggest an arc-related setting (supra-subduction zone) for the Ślęża Ophiolite. However, on the basis of Nd isotope systematics this was ruled out in favour of a normal ridge-spreading environment for Ślęża (Pin *et al.* 1988). Nb contents (1–2 ppm) and La–Nb ratios (1–3) for the lavas and dykes are comparable with average N-MORB (2.5 ppm and *c.* 1 respectively; Sun & McDonough 1989) and suggest derivation from a depleted MORB-type source rather than an arc source. Values for the gabbros, on the other hand, are an order of magnitude lower. In terms of the chemical discrimination of the tectonic environment, the differences between the gabbros, and dykes and lavas are exhibited in Figure 6, with the former apparently displaying arc-like features and the latter typical N-MORB features.

(b) Many of the lavas and dykes are especially rich in Fe (>12wt % total Fe as FeO*) and Ti (>2.5wt % TiO_2) and have some of the characteristics of ferrobasalts from Pacific Ocean rifts (Natland 1980), although none match the highly fractionated Ti-rich ferrobasalts from Iceland discussed by Flower *et al.* (1982) (Fig. 7). The especially Fe-rich nature of some Ślęża lavas may be significant as the propagating tips of MOR-spreading centres also display this characteristic (Clague & Bunch 1976).

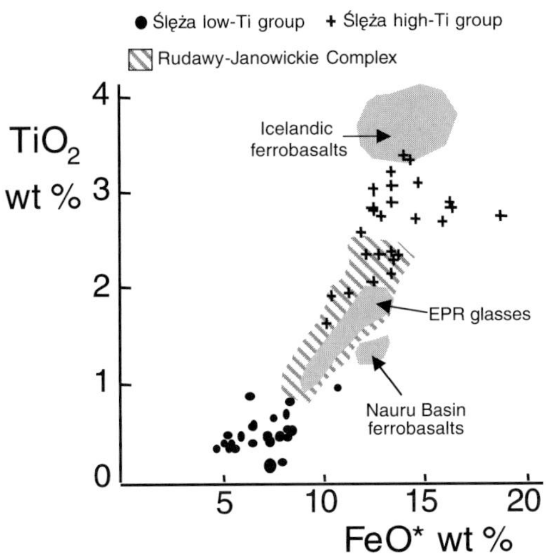

Fig. 7. Comparison of chemically-evolved Ślęża ferrobasalts with metabasites from the adjacent Rudawy–Janowickie Complex (Winchester *et al.* 1995), and oceanic ferrobasalts from the East Pacific Rise (EPR) (Natland 1980), the Nauru Basin (western Pacific) (Floyd 1986), and subaerial Iceland (Flower *et al.* 1982).

Chemical comparison with Bohemian Massif metabasites and ophiolitic sequences

Previous overviews of the chemical features of Bohemian Massif metabasites, especially from the Sudetes (Floyd *et al.* 2000*b*) and the Mariánské Lázně Complex (Crowley *et al.* 2000), have identified two main groups of metatholeiites – a low-Ti group and a high-Ti group – that were probably derived from different mantle sources. As indicated above, the Ślęża ophiolite also exhibts this chemical dichotomy. Within the lavas of the tectonically adjacent Kaczawa and Rudawy–Janowickie crustal blocks both low- and high-Ti groups are a characteristic feature (Fig. 8), and together with Mariánské Lázně Complex, this suggests that these chemical groups are a common feature of Lower Palaeozoic Bohemian mafic volcanism which links the various (now fragmented) blocks together in a single, long-lived, magmatic province. In terms of Ti–V ratios (Fig. 8) the Ślęża lavas and dykes are similar to the Rudawy-Janowickie or Mariánské Lázně high-Ti metabasites, although the former are far more chemically evolved with higher incompatible element abundances. Due to the apparent significant age difference between the Ślęża Ophiolite (*c.* 400 Ma) and the Sudetes magmatism (*c.* 490 Ma) is it unlikely that the Ślęża lavas once represented the evolved top (say) of the Rudawy-Janowickie lava pile. However, the Sudetes magmatic event was long lived and covered the full age range noted above. For example, volcanic activity started at *c.* 505 Ma (Oliver *et al.* 1993; Bendl & Patočka 1995) and persisted throughout the Ordovician and Silurian, and possibly until the

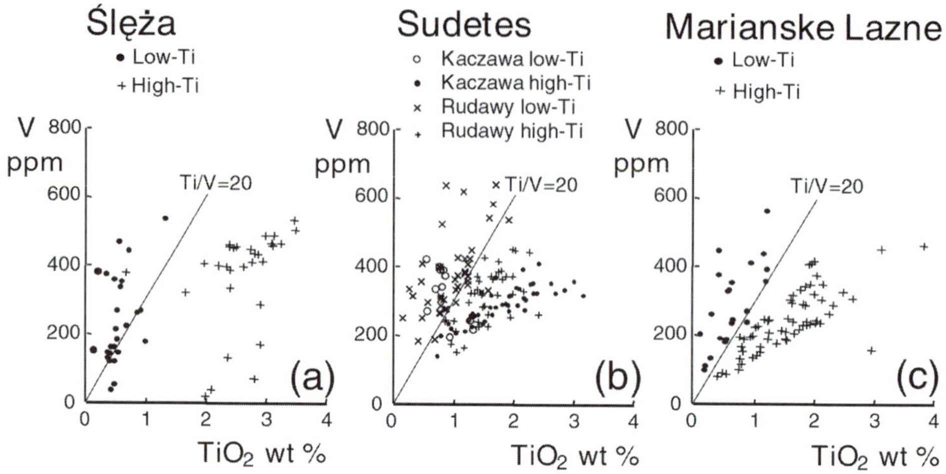

Fig. 8. Comparison of low and high-Ti metabasite groups in the Ślęża ophiolite (plot a) with similar groups in the Sudetes (plot b: data from Furnes *et al.* 1994; Winchester *et al.* 1995) and Mariánské Lázně (plot c: data from Crowley *et al.* 2000).

early Devonian (e.g. Chlupáč 1997; Kachlík & Patočka 1998; Patočka *et al.* 2000).

Apart from the Ślęża Ophiolite (and possibly also Mariánské Lázně Complex; Crowley *et al.* 2000) none of the other Bohemian Massif metabasite sequences show a magmatic pseudostratigraphy or typical ophiolite features. The chemical distinctions between the Ślęża gabbros and the dykes and lavas, and their correspondence to the characteristic chemical groups of the Bohemian Massif, calls into question whether the Ślęża plutonic and volcanic segments can still be considered part of a single comagmatic ophiolite unit. Consideration of incompatible element abundances and ratios of gabbros, dykes and lavas of Neotethyan ophiolites from the eastern Mediterranean, often shows good covariance between all lithological members (e.g. Zr–Y plot, Fig. 9), and not the chemical gap displayed by the Ślęża segments (Fig. 2). It seems likely that the Ślęża plutonic and volcanic segments are not petrogenetically related via any simple fractionation pathway, but like the Bohemian Massif low- and high-Ti metabasite groups were probably derived from different mantle sources (Floyd *et al.* 2000b).

The plutonic and volcanic segments were not necessarily formed in different portions of the ocean crust. There is some field evidence to suggest they may have been closely associated initially. For example, the highly evolved dyke SL99–10 belongs to the high-Ti group and could chemically represent a highly fractionated feeder to the lavas and sheeted dykes. However, as it is structurally low in the magmatic pseudostratigraphy and cuts the low-Ti gabbros, both segments must have been in close proximity at the spreading centre. Similar dykes are not found in the volcanic portion indicating that the two segments may now be separated by a dislocation zone. Furthermore, it suggests that part of the lava sequence may be marginally younger than the gabbros dated at *c.* 400 Ma. This tentative scenario is clearly an oversimplification of a complex spreading centre that would have involved different mantle melting regimes feeding fractionating magma chambers at various times. Subsequent tectonic dislocation would have divorced plutonic and volcanic segments, such that they cannot necessarily be considered comagmatic parent/daughter pairs, even though they are now associated in the field.

Mantle sources

Floyd *et al.* (2000*b*) suggested that the low- and high-Ti metabasites groups of the Bohemian Massif have been derived from a sediment-contaminated lithospheric source and a MORB-like asthenospheric source enriched by a mantle plume, respectively. The chemical correspondence of the plutonic and volcanic segments of the Ślęża Ophiolite with these low- and high-Ti groups suggests a similar derivation of basaltic melts from these two sources, but at a much later date. Figure 10 compares the chemical groups from the Sudetes and exhibits a horizontal trend in the low-Ti metatholeiites due to sediment contamination of a lithospheric source (decreasing Ti/Zr) and a near vertical trend in

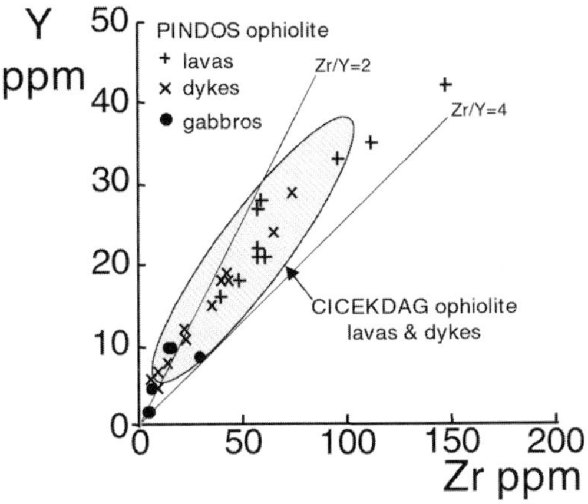

Fig. 9. Zr–Y plot showing the distribution of different lithological units within the Pindos (data from Capedri *et al.* 1980) and Cicekdag (data from Yaliniz *et al.* 2000) ophiolites. Note that when compared with a similar plot in Figure 2, the Ślęża lavas are more evolved with higher Zr and Y abundances, as well as being distinct from the gabbros.

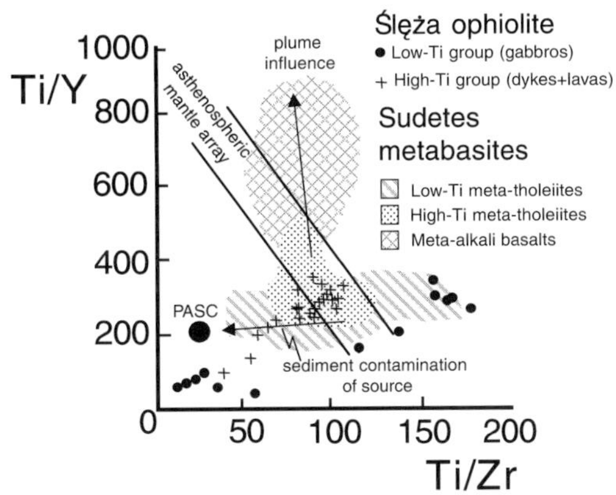

Fig. 10. Comparison of Ślęża Ophiolite groups with chemical subdivisions of Sudetes metabasites, and the relative influence of sediment contamination in the source and enrichment due to a plume. PASC: post-Archaean sediment composite (composition from Taylor & McLennan 1985).

meta-alkali basalts representing the influence of an enriched mantle plume (high Ti/Y) on a depleted asthenospheric MORB-type (low Ti/Y) source. Apart from the apatite and zircon cumulates with very low Ti–Y and Ti–Zr ratios (Fig. 10), the Ślęża high–Ti group clusters around the MORB source in the asthenospheric array, whereas the low-Ti group is derived from a high Ti–Zr (>150) depleted source. This diagram again emphasizes the difference in the degree of depletion of the sources that generated the Ślęża low- and high-Ti melts. Furthermore, any sediment contamination of a lithospheric source was probably very limited and an enriched plume was not involved in the generation of the Ślęża MOR basalts which are all depleted N-type.

Although there are chemical similarities between the Sudetes metabasites and the Ślęża Ophiolite, were the same two distinct sources (lithosphere and depleted asthenosphere) involved in the generation of the Ślęża low- and high-Ti groups? The chemical correspondence is not definitive and other mantle models can be envisaged that link the two Ślęża groups in an ocean ridge setting. For example, the two chemical groups could have been derived from the same asthenospheric source that underwent different degrees of depletion during incremental mantle melting. Melting of depleted asthenosphere under a spreading ridge would produce the high-Ti group of MORB lavas. To allow for their high incompatible element abundances some crustal ponding of melt would be required to produce significant fractionation. The mantle residue left after MORB extraction would be further depleted and would become the highly depleted source of the low-Ti group melts. These processes are tentatively modelled in Figure 11a. The Cr–Y plot shows that by back-extrapolation the fractionation trend of the lavas and dykes they could have been initially generated by about 10–15% partial melting of a depleted N-MORB source. The low-Ti gabbros, on the other hand, could then be generated by about 15–30% partial melting of a mantle residue after 20% MORB had been extracted (curves from Murton 1989). In the companion REE ratio plot (Fig. 11b), the lavas and dykes represent melts from the N-MORB source (values from Sun & McDonough 1989), whereas the gabbros can be derived by *c.* 30% partial melting of a more depleted source. The progressive partial melting curve represents the non-modal melting of a depleted spinel lherzolite source composed of $ol_{50}opx_{25}cpx_{20}sp_5$ in the proportions of $ol_{15}opx_{30}cpx_{50}sp_5$ (cf. Condie & Harrison 1976) with La/Sm = 0.50 and Sm/Yb = 0.81. Extraction of high volume MORB melts and subsequent source depletion would be a feature of spreading centres and thereby generate both low- and high-Ti melts in close proximity, as suggested by the Ślęża ophiolitic components.

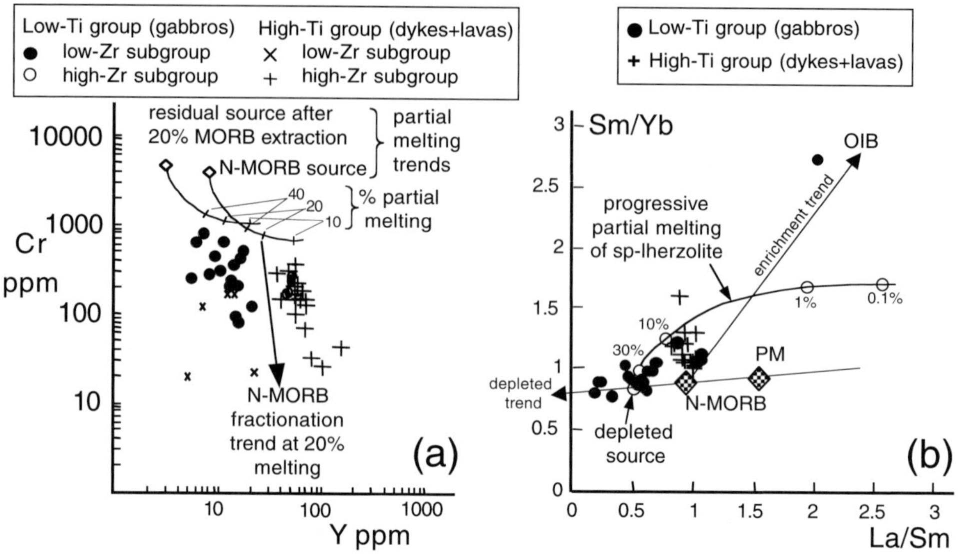

Fig. 11. Preliminary modelling of depleted MORB-like sources for the generation for the low- and high-Ti groups in the Ślęża Ophiolite: see text for details. N-MORB and residue partial melting curves from Murton (1989); initial spinel lherzolite source after Condie & Harrison (1976). OIB (Ocean island basalt), MORB (Mid-ocean ridge basalt) and PM (primitive mantle) values from Sun & McDonough (1989).

Summary and conclusions

1. As demonstrated by several earlier studies, the Ślęża Ophiolite exhibits a typical lithological variation and pseudostratigraphy which grades from serpentinites and gabbros in the south to sheeted dykes and lavas in the north. The recent discovery of pillow lava structures (at Gozdnica Hill, to the west of Sobotka town) confirms that the volcanic top of the ophiolite lies in the northern section. The ophiolite appears to be steeply inclined and possibly overturned. Recent U–Pb zircon ages obtained from plagiogranites associated with the gabbros give ages of *c.* 400–420 Ma, much younger than the main Sudetes mafic–felsic magmatism at *c.* 490 Ma.

2. Geochemical data suggests that the gabbros form a distinct group and are unlikely to be comagmatic with the volcanic section of sheeted dykes and lavas. The gabbros are very depleted in incompatible elements, which, in part, reflects their cumulate nature. The metabasalts and metadolerites, on the other hand, are a well-evolved single comagmatic suite with high incompatible element abundances. The geochemical dichotomy of the plutonic and volcanic segments calls into question a simple interpretation of the body as a single-stage coherent stratiform ophiolite.

3. Both the plutonic and volcanic segments of the ophiolite exhibit N-MORB-like light REE depleted to flat patterns, together with uniformly low Nb contents and La–Nb ratios. A spreading ridge is considered the appropriate eruptive environment, although the presence of many high-Fe lavas suggests a nearby propagating tip fed by chemically evolved melts from a fractionated magma chamber.

4. The Ślęża Ophiolite exhibits two chemical groups that have their analogues within the metabasites of the Bohemian Massif as a whole: low-Ti group (gabbros) and high-Ti group (sheeted dykes and lavas). Both groups can be further subdivided into low- and high-Zr subgroups which reflect zircon and apatite extraction and accumulation.

5. Comparison with Bohemian Massif metabasites indicates that sediment contamination of the Ślęża Ophiolite sources was not important and that an enriched plume source played no part in the generation of ophiolitic melts.

6. The two Ślęża chemical groups were derived from variably depleted asthenospheric mantle sources. Modelling suggests that the volcanic segment could have been derived by 10–15% partial melting of a depleted N-MORB source, whereas the plutonic segment represents around 30% partial melting of a more depleted source. To develop varying degrees of depletion in an oceanic environment, the two sources

could be related via incremental partial melting of a shallow MORB-type source.

Funding for the research collaboration in this project was partly supported by the EU TMR Network project 'Palaeozoic Amalgamation in Central Europe' (PACE – Contract number: ERBFMRXCT97-0136) co-ordinated by J. A. Winchester at Keele University, UK. Fieldwork and sampling were greatly assisted by Z. Cymerman and W. Kozdrój, while field details concerning the gabbros were provided by M. Abdel Wahed.

References

ABDEL WAHED, M. 1999. The Ślęża Ophiolite (SW Poland): petrological and structural evolution. *Unpublished PhD thesis*, Institute of Geological Sciences, Wrocław, 302pp.

ABDEL WAHED, M. 2000. Deformation events and shear localization within the Ślęża metagabbros (Lower Silesia, SW Poland). *In*: MIERZEJEWSKI, M. (ed.) *Tectonics of the* Ślęża *ophiolite and its influence on the distribution of some mineral ores and groundwater*. Field Conference, Uniwersytet Wrocławski, 42–48.

BENDL, J. & PATOČKA, F. 1995. The ^{87}Rb-^{87}Sr isotope geochemistry of the metamorphosed bimodal volcanic association of the Rýchory Mts Crystalline Complex, West Sudetes, Bohemian Massif. *Geologia Sudetica*, **29**, 3–18.

CAPEDRI, S., VENTURELLI, G., BOCCHI, G., DOSTAL, J., GARUTI, G. & ROSSI, A. 1980. The geochemistry and petrogenesis of an ophiolitic sequence from Pindos, Greece. *Contributions to Mineralogy and Petrology*, **74**, 189–200.

CHLUPÁČ, I. 1997. Palaeozoic ichnofossils in phyllites near Zelezny Brod, northern Bohemia. *Journal of the Czech Geological Society*, **42**, 75–94.

CLAGUE, D. A. & BUNCH, T. E. 1976. Formation of ferrobasalts at East Pacific midocean spreading centres. *Journal of Geophysical Research*, **81**, 4247–4256.

COLEMAN, R. G. 1977. *Ophiolites*. Springer, Berlin, 229pp.

CONDIE, K. C. & HARRISON, N. M. 1976. Geochemistry of the Archaean Bulawayan Group, Midlands Greenstone belt, Rhodesia. *Precambrian Research*, **3**, 253–271.

CROWLEY, Q. G., FLOYD, P. A., ŠTĚDRÁ, V., WINCHESTER, J. A. & HOLLAND, J. G. 2000. The Mariánské Lázně Complex: remnants of a subducted early Palaeozoic seaway. *In*: *Variscan-Appalachian dynamics: the building of the Upper Palaeozoic basement*. Basement Tectonic Program and Abstracts, **15**, 86–89, A Coruna, Spain.

CROWLEY, Q. G., FLOYD, P. A., WINCHESTER, J. A., FRANKE, W. & HOLLAND, J. G. 2001. Early Palaeozoic rift-related magmatism in Variscan Europe: fragmentation of the Armorican Terrane Assemblage. *Terra Nova*, **12**, 171–180.

CYMERMAN, Z. & PIASECKI, M. A. J. 1994. The terrane concept in the Sudetes, Bohemian massif. *Geological Quarterly*, **38**, 191–210.

CYMERMAN, Z., PIASECKI, M. A. J. & SESTON, R. 1997. Terranes and terrane boundaries in the Sudetes, northeast Bohemian Massif. *Geological Magazine*, **134**, 717–725.

DUBIŃSKA, E. & GUNIA, P. 1997. The Sudetic ophiolite: current view on its geodynamic model. *Geological Quarterly*, **41**, 1–20.

FINGER, F. & STEYRER, H. P. 1995. A tectonic model for the Eastern Variscides: indicators from the chemical study of amphibolites in the southeastern Bohemian Massif. *Geologica Carpathica*, **46**, 137–150.

FLOWER, M. F. J., PRITCHARD, R. G., BRERM, G., CANN, J. R., DELANEY, J., EMMERMAN, R., GIBSON, I. L., OAKLEY, P. J., ROBINSON, P. T. & SCHMINKE, H. U. 1982. Chemical stratigraphy, Iceland Research Drilling Project, Reydarfjordur, Eastern Iceland. *Journal of Geophysical Research*, **87**, 6489–6510.

FLOYD, P. A. 1986. Petrology and geochemistry of oceanic intraplate sheet-flow basalts, Nauru basin, Deep Sea Drilling Project Leg 89. *In*: MOBERLY, R., SCHLANGER, S. O. *et al.* (eds) *Initial Reports of the DSDP*, **89**, 471–497.

FLOYD, P. A. & CASTILLO, P. R. 1992. Geochemistry and petrogenesis of Jurassic ocean crust basalts, ODP Leg 129, Site 801. *In*: LARSON, R., LAUNCELOT, Y., *et al.* (eds) *Proceedings of ODP, Scientific Results*, **129**, 361–388. College Station, Texas.

FLOYD, P. A. & FUGE, R. 1973. Distribution of F and Cl in some contact and regionally metamorphosed Cornish greenstones. *Proceedings of the Ussher Society*, **2**, 483–488.

FLOYD, P. A. & WINCHESTER, J. A. 1978. Identification and discrimination of altered and metamorphosed volcanic rocks using immobile elements. *Chemical Geology*, **21**, 291–306.

FLOYD, P. A., EXLEY, C. S. & STYLES, M. T. 1993. *Igneous rocks of South-west England*. Geological Conservation Review Series, **5**, 256pp. Chapman & Hall, London.

FLOYD, P. A., CROWLEY, Q. G., WINCHESTER, J. A., KOZDROJ, W. & CYMERMAN, Z. 2000*a*. Pillow lavas in the Ślęża ophiolite and their geochemical relationship to other mafic components in the Sudetes. *PACE meeting abstracts*, Zakopane, Poland, 16–22 September 2000, 10.

FLOYD, P. A., WINCHESTER, J. A., SESTON, R., KRYZA, R. & CROWLEY, Q. G. 2000*b*. Review of geochemical variation in Lower Palaeozoic metabasites from the NE Bohemian Massif: intracratonic rifting and plume-ridge interaction. *In*: FRANKE, W., HAAK, V., ONCKEN, O. & TANNER, D. (eds) *Orogenic processes: Quantification and modelling in the Variscan Belt*, Geological Society of London, Special Publication, **179**, 155–174.

FRANKE, W. 1989*a*. Tectonostratigraphic units in the Variscan belt of Central Europe. *Geological Society of America Special Paper*, **230**, 67–90.

FRANKE, W. 1989*b*. Variscan plate tectonics in Central Europe – current ideas and open questions. *Tectonophysics*, **169**, 221–228.

FRANKE, W. & ŻELAŹNIEWICZ, A. 2000. The eastern termination of the Variscides: terrane correlation and kinematic evolution. *In*: FRANKE, W., ALTHERR, R., HAAK, V. & ONCKEN, O. (eds) *Orogenic processes: Quantification and modelling in the Variscan Belt*, Geological Society of London, Special Publication, **179**, 63–86.

FRANKE, W., DALLMEYER, D. & WEBER, K. 1995. XI: Geodynamic evolution. *In*: DALLMEYER, D., FRANKE, W. & WEBER, K. (eds) *Pre-Permian geology of central and western Europe*. Springer, Berlin, Heidelberg, New York, 579–593.

FURNES, H., KRYZA, R. & MUSZYŃSKI, A. 1989. Geology and geochemistry of Early Palaeozoic volcanics of the Swierzawa Unit, Kaczawa Mts., W. Sudetes, Poland. *Neues Jahrbuch fur Geologie und Palaontologie, Monatshefte*, **3**, 136–154.

FURNES, H., KRYZA, R., MUSZYŃSKI, A., PIN, C. & GARMANN, L. B. 1994. Geochemical evidence for progressive rift-related early Palaeozoic volcanism in the western Sudetes. *Journal of the Geological Society of London*, **151**, 91–109.

GASS, I. G. & SMEWING, J. D. 1973. Intrusion, extrusion and metamorphism at constructive margins: evidence from the Troodos massif, Cyprus. *Nature, London*, **242**, 26–29.

HART, S. R., ERLANK, A. J. & KABLE, E. J. D. 1974. Sea floor basalt alteration: some chemical and Sr isotopic effects. *Contributions to Mineralogy and Petrology*, **44**, 219–230.

HUMPHRIS, S. E. & THOMPSON, G. 1978. Trace element mobility during hydrothermal alteration of oceanic basalts. *Geochimica et Cosmochimica Acta*, **42**, 127–136.

HYNES, A. 1980. Carbonitization and mobility of Ti, Y, and Zr in Ascot Formation metabasalts, S.E. Quebec. *Contributions to Mineralogy and Petrology*, **75**, 79–87.

JAMROZIK, L. 1989. Pozycja geologiczna strefy mineralizacji rudnej Strzegomiany-Kunow w intruzji gabrowej Ślęży (ofiolit Ślęży) *In*: NIEKIEWICZ, J. (ed.) *Ofiolit Ślęży i jego mineralizacja rudna*. Symposium Wrocław-Sobotka, 4–6 Sept., 1989, 154–161.

JĘDRYSEK, M. O., MAJEROWICZ, A., JASIESKA, B. & HAAS, S. 1989. The migration of ocean water into the upper mantle. Evidence from δD and ^{18}O values for minerals of rodingitic rocks of the ophilolitic complex of Ślęża, SW Poland. *Mineralogia Polonica*, **20**, 91–107.

KACHLÍK, V. & PATOČKA, F. 1998. Lithostratigraphy and tectonomagmatic evolution of the Żelezny Brod crystalline unit: some constraints for the palaeotectonic development of the W. Sudetes (NE Bohemian Massif). *Geolines*, **6**, 34–35.

KRYZA, R. & ABDEL WAHED, M. 2000. The Ślęża mafic complex: contrasting geochemistry of plutonic and subvolcanic/volcanic rocks. *Mineralogical Society of Poland, Special Papers*, **17**, 1–3.

MAJEROWICZ, A. 1979. Grupa górska Ślęży a waspó czesne problemy geologiczne ofiolitów. *Field Conference*, Nova Ruda 8–9 Sept. 1979, Uniwersytet Wrocławski, 1–2.

MAJEROWICZ, A. 1981. Rock series of the Ślęża Mt. group in the light of petrologic studies of ophiolite complex. *In*: NARĘBSKI, W. (ed.) *Ophiolites and initialites of northern border of the Bohemian massif. Guidebook of Excursion* **2**, 172–179, Potsdam-Freiberg.

MAJEROWICZ, A. 1994. Textural features and symptoms of ocean-floor metamorphism in the top part of the Ślęża ophiolite (SW Poland). *Archiwum Mineralogiczne, Warszawa*, **50**, 97–139.

MAJEROWICZ, A. 2000. Uwagi o koncentracji ilmentu w ofiolicie Ślęża, na tlebada petrologicznych. *Archiwum Mineralogiczne, Warszawa*, **53**, 12–15.

MAJEROWICZ, A. & PIN, C. 1989. Recent progress in petrologic study of the Ślęża ophiolite complex. *In*: Narębski, W. & MAJEROWICZ, A. (eds) *Lower and Upper Palaeozoic metabasites and ophiolites of the Polish Sudetes*. Guide book of excursions in Poland. Wroclaw University, 34–72.

MAJEROWICZ, A. & PIN, C. 1994. The main petrological problems of the Mt Ślęża ophiolite complex, Sudetes, Poland. *Zentralblatt Geologie und Palaontologi*, **9**, 989–1018.

MAJEROWICZ, A., KRYZA, R., WRÓBLEWSKA, G. 2000. Diallagite pegmatitoids from Mt. Ślęża gabbro. *In*: MIERZEJEWSKI, M. (ed.) *Tectonics of the Ślęża ophiolite and its influence on the distribution of some mineral ores and groundwater*. Field Conference, Uniwersytet Wrocławski, 49–54.

MATTE, P. H., MALUSKI, H., RAJLICJ, P. & FRANKE, W. 1990. Terrane boundaries in the Bohemian massif: result of large-scale Variscan shearing. *Tectonophysics*, **177**, 151–170.

MIERZEJEWSKI, M. & ABDEL WAHED, M. 2000. Overlapping thrust faults within the Gogoeów-Jordanów serpentinite massif (SW Poland): evidence for the obduction of the Ślęża ophiolite complex. *In*: MIERZEJEWSKI, M. (ed.) *Tectonics of the Ślęża ophiolite and its influence on the distribution of some mineral ores and groundwater*. Field Conference, Uniwersytet Wrocławski, 29–41.

MITROPOULOS, P. 1982. REE patterns in the metasedimentary rocks of the Land's End aureole, Cornwall. *Chemical Geology*, **23**, 265–280.

MURTON, B. J. 1989. Tectonic controls on boninite genesis. *In*: SAUNDERS, A. D. & NORRY, M. J. (eds) Magmatism in the ocean basins. *Geological Society of London, Special Publication*, **42**, 347–377.

NARĘBSKI, W. 1993. Lower to Upper Palaeozoic tectonomagmatic evolution of the NE part of the Bohemian Massif. *Zentralblatt für Geologie und Paläontologie*, **1**, 961–972.

NARĘBSKI, W. & MAJEROWICZ, A. 1985. Ophiolites of the Sowie Gory Mts framework and Lower Palaeozoic initialites of the Polish Sudetes. *In*: DOBRETSOV, N. I. (ed.) *Riphean Lower Palaeozoic ophiolites of North Eurasia*, 86–105. Nauka, Novosibirsk.

NATLAND, J. H. 1980. Effect of axial magma chambers beneath spreading centers on the composition of basaltic rocks. *In*: ROSENDAHL, B. R., HEKINIAN, R. *et al. Initial Reports DSDP*, Washington (US Printing Office), **54**, 833–850.

OLIVER, G. J. L., CORFU, F. & KROGH, T. E. 1993. U-Pb ages from Poland: evidence for a Caledonian suture zone between Baltica and Gondwana. *Journal of the Geological Society of London*, **150**, 355–369.

PATOČKA, F., KACHLÍK, V. & FAJST, M. 2000. Mafic-felsic to mafic-ultramafic Early Palaeozoic magmatism of the West Sudetes (NE Bohemian Massif): the South Krkonoše Complex. *Zentralblatt für Geologie Wissenshaft*, **28**, 177–210.

PEARCE, J. A. 1980. Geochemical evidence for the genesis and eruptive setting of lavas from Tethyan ophiolites. *In*: PANAYIOTYOU, A. (ed.) *Proceedings of the International Ophiolite Symposium*, Nicosia, Cyprus, 261–272.

PEARCE, J. A. & CANN, J. R. 1973. Tectonic setting of basaltic volcanic rocks determined using trace element analyses. *Earth and Planetary Science Letters*, **19**, 290–300.

PEARCE, J. A. & NORRY, M. J. 1979. Petrogenetic implications of Ti, Zr, Y, and Nb variations in volcanic rocks. *Contributions to Mineralogy and Petrology*, **69**, 33–47.

PEREKALINA, T. V. 1981. Variscan volcanism in central and western Europe. *Geologie en Mijnbouw*, **60**, 17–31.

PIN, C. 1990. Variscan oceans: ages, origins and geodynamic implications inferred from geochemical and radiometric data. *Tectonophysics*, **177**, 215–227.

PIN, C., MAJEROWICZ, A. & WOJCIECHOWSKA, I. 1988. Upper Palaeozoic oceanic crust in the Polish Sudetes: Nd-Sr isotope and trace element evidence. *Lithos*, **21**, 195–209.

PITCHER, W. S. & SINHA, R. S. 1958. The petrochemistry of the Ardara aureole. *Quarterly Journal of the Geological Society of London*, **113**, 393–408.

ROLLINSON, H. 1993. Using geochemical data: evaluation, presentation, interpretation. Longman Group, UK, 352pp.

SHERVAIS, J. W. 1982. Ti-V plots and the petrogenesis of modern and ophiolitic lavas. *Earth and Planetary Science Letters*, **59**, 101–118.

SMEWING, J. D. & POTTS, P. J. 1976. Rare-earth abundances in basalts and metabasalts from the Troodos Massif, Cyprus. *Contributions to Mineralogy and Petrology*, **57**, 245–258.

SMITH, R. E. & SMITH, S. E. 1976. Comments on the use of Ti, Zr, Y, Sr, K, P and Nb in classification of basaltic magmas. *Earth and Planetary Science Letters*, **32**, 114–120.

SPOONER, E. T. C. & FYFE, W. S. 1973. Sub-seafloor metamorphism, heat and mass transfer. *Contributions to Mineralogy and Petrology*, **42**, 287–304.

STONE, M. & AWAD, N. T. I. 1988. Behaviour of trace alkali and other elements at the Tregonning granite-pelite contacts. *Proceedings of the Ussher Society*, **7**, 47–51.

SUN, S. S. & MCDONOUGH, W. F. 1989. Chemical and isotopic systematics of oceanic basalts: implications for mantle composition and processes. *In*: SAUNDERS, A. D. & NORRY, M. J. (eds) *Magmatism in the ocean basins*. Geological Society, London, Special Publications, **42**, 313–345.

TAYLOR, S. R. & MCLENNAN, S. M. 1985. *The continental crust: its composition and evolution*. Blackwell, Oxford.

THOMPSON, G. 1991. Metamorphic and hydrothermal processes: basalt-seawater interactions. *In*: FLOYD, P. A. (ed.) *Oceanic basalts*. Blackie, Glasgow, 148–173.

WINCHESTER, J. A., FLOYD, P. A., CHOCYK, M., HORBOWY, K. & KOZDROJ, W. 1995. Geochemistry and tectonic environment of Ordovician meta-igneous rocks in the Rudawy Janowickie Complex, SW Poland. *Journal of the Geological Society of London*, **152**, 105–115.

WINCHESTER, J. A. & PACE TMR Network team. 2002. Palaeozoic amalgamation of Central Europe: new results from recent geological and geophysical investigations. *Tectonophysics*, in press.

YALINIZ, K. M., FLOYD, P. A. & GÖNCÜOGLU, M. C. 2000. Geochemistry of volcanic rocks from the Cicekdag Ophiolite, Central Anatolia, Turkey, and their inferred tectonic setting within the northern branch of the Neotethyan Ocean. *In*: BOZKURT, E., WINCHESTER, J. A. & PIPER, J. D. A. (eds) Tectonics and magmatism in Turkey and the surrounding area. *Geological Society of London, Special Publication*, **173**, 203–218.

ŻELAŹNIEWICZ, A., DÖRR, W. & DUBIŃSKA, E. 1998. Lower Devonian oceanic crust from U-Pb zircon evidence and Eo-Variscan event in the Sudetes. *Terra Nostra*, **98**, 174–176.

Coronitic metagabbros of the Mariánské Lázně Complex and Teplá Crystalline Unit: inferences for the tectonometamorphic evolution of the western margin of the Teplá-Barrandian Unit, Bohemian Massif

V. ŠTĚDRÁ[1], V. KACHLÍK [2], & R. KRYZA [3]

[1]*Czech Geological Survey, Klárov 3, 118 21 Praha 1, Czech Republic*
e-mail: stedra@cgu.cz

[2]*Faculty of Science, Charles University, Albertov 6, 128 43 Praha 2, Czech Republic*

[3]*Geological Institute, University of Wroclaw, Poland*

Abstract: Bodies of coronitic metagabbro occur in the SW Mariánské Lázně Complex (MLC) and the adjacent Teplá Crystalline Unit (TCU) on the western margin of the Teplá-Barrandian Unit (TBU), Bohemian Massif. The characteristic structural, geochemical, petrographic, and metamorphic features of five groups of metagabbros and related rocks are presented, compared with other metabasites of the MLC and Zone of Erbendorff-Vohenstrauss (ZEV), and used to constrain the tectonometamorphic evolution of the western part of the TBU. The metagabbros are considered to be a younger intrusive member of the complicated lower crustal tectonic stack of Upper Proterozoic to Early Palaeozoic age which is formed by the Mariánské Lázně Complex and the Teplá Crystalline Unit together. It is proposed that a significant part of the metamorphic evolution of some parts of these units took place before the emplacement of metagabbros and granitoids at around 496–516 Ma. The sequence of metamorphic events is interpreted to have been as follows. Deep burial of primitive MORB type tholeiitic rocks (a) metamorphosed up to eclogite facies, followed by (b) uplift to lower crustal levels so that the partially exhumed rocks were juxtaposed with other lower/middle crustal rocks. Thermal relaxation (c) followed, with an episode of extension recorded in L-tectonites of amphibolite facies. Once this lithologically variegated stack was welded together, it was intruded by the Upper Cambrian–Lower Ordovician granitoids and gabbros (d). This pre-Variscan metamorphic event may be expressed at the supracrustal level by an unconformity between Upper Cambrian and Lower Ordovician rocks in the Barrandian. The final configuration of the units was established during the Variscan collision of the Teplá Barrandian terrane with Saxothuringia (e) in which the rocks of the MLC and TCU were thrust to the NW over the Saxothuringian para-autochthon. The accompanying metamorphic event reached upper amphibolite facies. The thermally relaxed rocks cooled rapidly, and pre-existing thrust planes were re-activated during the final extensional collapse.

The tectonic and metamorphic evolution of individual segments of the Bohemian Massif (Central European Variscides, Fig. 1a) has been the subject of detailed research for the last decade. New geochemical and geochronological data has been obtained from various segments of the Saxothuringian, Moldanubian, and the Moravo-Silesian Zones, whereas geochronological and isotope information from the Teplá-Barrandian Unit and the Mariánské Lázně Complex (Kastl & Tonika 1984) in the western part of the Bohemian Massif is still limited. Results of investigations of the southwestern margin of the Teplá-Barrandian Unit (TBU) have been published by teams from Germany (e.g. KTB Reports 1991–1995, Schüssler *et al.* 1992; Zulauf 1994*a*, 1997) and from the Czech Republic (Žáček 1994; Vrána & Štědrá 1997). This work was based on discoveries resulting from deep drilling (KTB) near Windischeschenbach in Bavaria and from contemporaneous related projects. A new body of structural, geochemical and isotope data on metabasic rocks from the Mariánské Lázně Complex and the Teplá Crystalline Unit has now been collected within the frame of the 'Palaeozoic Amalgamation of Central Europe' (PACE) project. Most of the new geochemical data are discussed in a wide context of mafic metamorphic rocks from the NW margin of the Bohemian Massif in other papers within this volume. This paper focuses on a specific area and a part of its unique rock assemblage.

The TBU at its western and southwestern

From: WINCHESTER, J. A., PHARAOH, T. C. & VERNIERS, J. 2002. *Palaeozoic Amalgamation of Central Europe.* Geological Society, London, Special Publications, **201**, 217–236. 0305-8719/02/$15.00

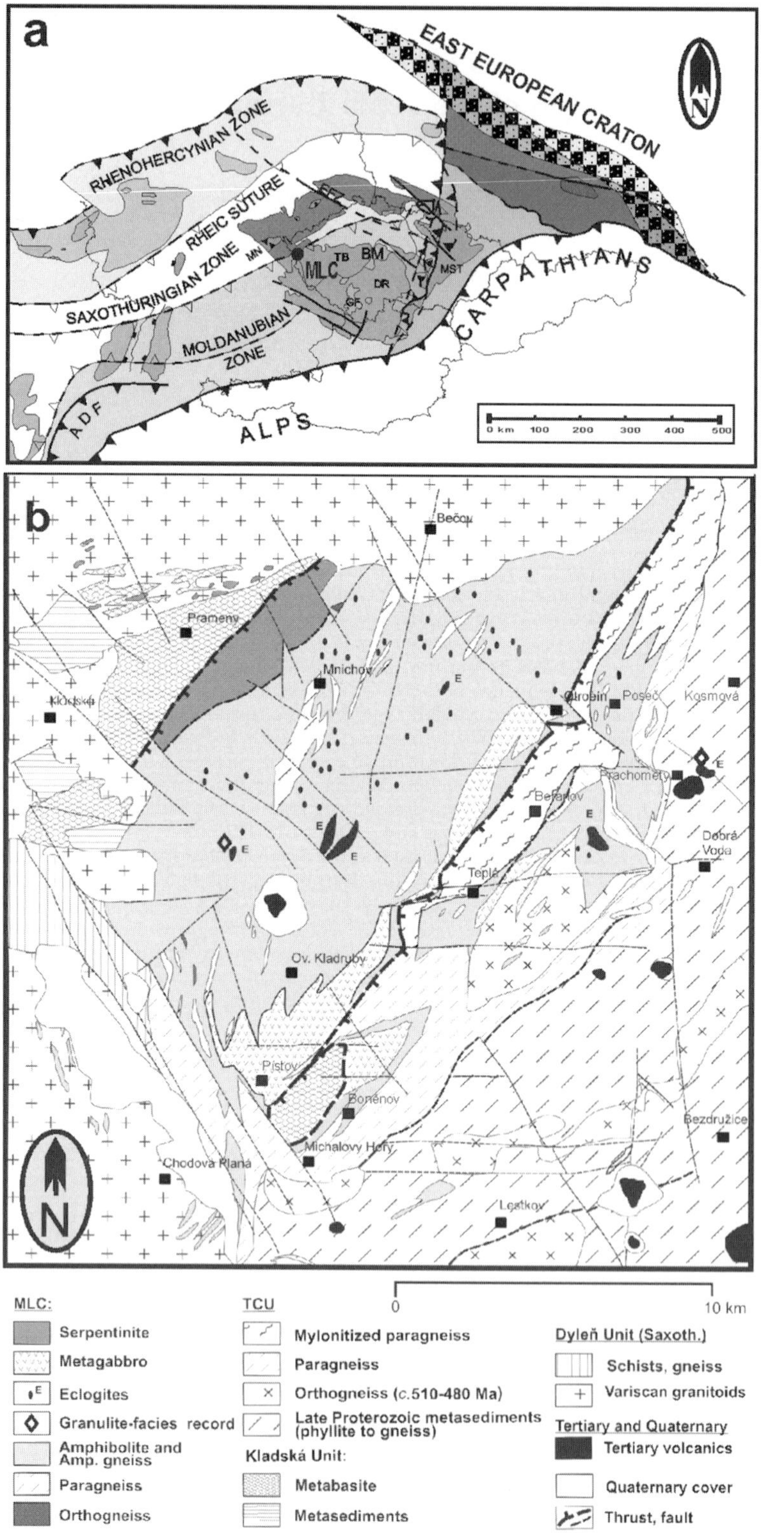
a
EAST EUROPEAN CRATON
RHENOHERCYNIAN ZONE
RHEIC SUTURE
SAXOTHURINGIAN ZONE
MOLDANUBIAN ZONE
CARPATHIANS
ALPS
ADF
MLC
TB
BM
DR
MN
MST
N
0 km 100 200 300 400 500
b
Bečov
Prameny
Mnichov
Kladská
Poseč
Kosmová
Prachomety
Beranov
Dobrá Voda
Teplá
Cv. Kladruby
Pístov
Bonenov
Bezdružice
Michalovy Hory
Chodová Planá
Lestkov
N
0 10 km
MLC:
Serpentinite
Metagabbro
Eclogites
Granulite-facies record
Amphibolite and Amp. gneiss
Paragneiss
Orthogneiss
TCU
Mylonitized paragneiss
Paragneiss
Orthogneiss (c.510-480 Ma)
Late Proterozoic metasediments (phyllite to gneiss)
Kladská Unit:
Metabasite
Metasediments
Dyleň Unit (Saxoth.)
Schists, gneiss
Variscan granitoids
Tertiary and Quaternary
Tertiary volcanics
Quaternary cover
Thrust, fault

margin is composed of Upper Proterozoic crystalline basement overlain by folded unmetamorphosed Cambrian to Middle Devonian sedimentary rocks, in turn covered by the platform Permo-Carboniferous, Mesozoic and Tertiary sediments. Cambro-Ordovician magmatism (Zulauf *et al.* 1997; Dörr *et al.* 1995, 1996, 1998) resulted in the emplacement of granitoids and gabbroic bodies along the SW margin of the TBU. The westernmost edge of the TBU is formed by the Teplá Crystalline Unit (TCU), underlain by the high-grade crystalline, mostly basic and intermediate rocks of the Mariánské Lázně Complex (MLC). The poorly exposed MLC has been interpreted as an exotic meta-ophiolitic unit with its own specific lithological and metamorphic characteristics (Kastl & Tonika 1984; Beard *et al.* 1995). It contains numerous relict high pressure and/or high temperature basic and intermediate rocks derived from several different magma sources (Beard *et al.* 1995) enclosed in upper amphibolite-facies host rocks. It was interpreted as the remains of a tectonically reworked and obliterated suture between the TBU and the Saxothuringian Unit (Jelínek *et al.* 1997).

The adjacent part of the Saxothuringian Zone is composed of three units. These are: the Slavkov Crystalline Unit (Mlčoch 1997) forming a roof pendant above the Karlovy Vary granite pluton, the low-grade lower Palaeozoic metasediments of the Cheb-Dyleň Crystalline Unit (Fiala & Vejnar 1994), and the metasedimentary Kladská Unit with intercalations of mostly alkaline amphibolites (Kachlík 1993). The latter unit is directly juxtaposed against the MLC (Fig. 1b), and ascribed, together with the former unit, to the para-autochthonous segment of a passive continental margin (Fiala & Vejnar 1997).

Scattered bodies of coronitic metagabbros, showing well preserved magmatic fabric and a metamorphic and deformational overprint of variable intensity, are characteristically found within the complex tectonic boundary zone between the medium pressure and temperature Teplá Crystalline Unit and the Mariánské Lázně Complex comprising high pressure and temperature relics. The metagabbros were interpreted to be lower-grade analogues of the relict eclogitized and granulitized rocks described by previous investigators as occurring in the MLC and TCU (Beard *et al.* 1995; Jelínek *et al.* 1997). The earlier research by Czech geologists was concerned with the Cadomian stage of the evolution of the area (Kettner 1917; Zoubek 1948; Žáček & Cháb 1993). In contrast, recent work based on Ar/Ar and Sm–Nd geochronology has favoured a Variscan age for the peak metamorphic overprint in the MLC and TBU (Dallmeyer & Urban 1994; Beard *et al.* 1995). The Mariánské Lázně Complex and the adjacent crystalline parts of the Teplá-Barrandian Unit, together with the Saxothuringian Kladská Unit, are often mentioned in the context of correlations with other units bearing serpentinites, amphibolites, eclogites, orthogneisses and metagabbros. These include the Münchberg Massif, Zone of Erbendorff-Vohenstrauss (ZEV), and crystalline units in the northern Sudetes (e.g. Weber & Vollbrecht 1989; O'Brien *et al.* 1997; Kachlík 1993, 1997; Beard *et al.* 1995). As a contribution to this broad discussion, new information on the setting, petrology and geochemistry of the coronitic metagabbros occurring in the Mariánské Lázně Complex–Teplá Crystalline Unit area, which has never before been described in detail, is presented below.

The garnetiferous metagabbros are treated as a group independent of other basic rocks occurring in the MLC (Štědrá & Kryza 1999). A detailed comparison of chemical composition, metamorphic history and deformational features in these gabbroic rocks, together with an interpretation of their geological setting, is now given. The new data enable the history of emplacement of most of the metabasic rocks to be resolved, and also show similarities and differences in their subsequent metamorphic paths. As a result, knowledge of the tectonometamorphic evolution of this part of the Bohemian Massif can be improved, and some answers regarding the correlation of terranes within the collage of Mid-European Variscides can be provided.

Geological setting of the MLC and TCU

Along its western and southwestern margin, the Teplá-Barrandian basement is formed of polymetamorphosed pelitic metasediments with at

Fig. 1. (**a**) Position of the Mariánské Lázně complex in the Bohemian Massif. Abbreviations: MLC, Mariánské Lázně Complex; MN, Münchberg Nappe; BM, Bohemian Massif; TB, Teplá-Barrandian Unit; GF, Gföhl Unit; DR, Drosendorf Unit; MST, Moravo-Silesian Terrane; EFZ, Elbe Fault Zone; ADF, Alpine Deformation Front. (**b**) Geological map of the Mariánské Lázně Complex and the adjacent part of the Teplá Crystalline Unit. It shows the position of the adjacent Kladská Unit as proposed by Kachlík (1994), the Teplá Crystalline Unit according to Cháb *et al.* (1997), and lithological boundaries mostly based on Tonika (1998) and new mapping.

least two obliquely overlapping systems of regional metamorphic zones. In both events the kyanite and sillimanite zones were reached (Vejnar 1982; Žáček *et al.* 1993; Zulauf 1997). The sequence includes metamorphosed and deformed granitoids and gabbros of Cambro-Ordovician (Dörr *et al.* 1995, 1998; Bowes & Aftalion 1991) or unknown age, Variscan granitoids (e.g. Šmejkal 1964; Zulauf 1994*b*, 1995; Siebel *et al.* 1997), amphibolites interlayered with metasediments and mylonitized rocks, and younger volcanic and sedimentary rocks. The Polom, Lestkov, Telecí Potok, Hanov, and Teplá orthogneiss bodies in the TCU (Cháb 1997) show an intimate spatial association with bodies of garnetiferous metagabbro studied here. Contact aureoles around the granitoid plutons are overprinted by a younger (?Variscan) regional metamorphism (Kachlík 1997; Zulauf *et al.* 1999).

The Mariánské Lázně Complex displays a complicated tectonic imbrication of amphibolites, amphibole gneisses, mylonitized paragneisses and orthogneisses, metagabbros and serpentinized ultrabasic rocks. Minor boudins of eclogites and granulites enclosed in reworked amphibolite-facies rocks are diagnostic features of the unit. The MLC metamorphic rocks of igneous origin formed in contrasting settings, and their magma sources, even within the basic rocks, are diverse. The metamorphic history recorded in eclogites, granulites, metagabbros, and gneisses indicates differences in P–T–t paths of these different rock groups (Štedrá 2001).

Along the NW–SE trending boundary zone between the MLC and TCU, approximately 15 km wide and 25 km long, extending from Planá in the SE to Útvina in the NE, metagabbro bodies are common and associated with orthogneisses hosted by refoliated paragneisses. The supposed tectonic boundary separating the TCU, predominantly made up of metasediments, from the MLC, formed mainly of basic rocks (Žáček & Cháb 1993), does not substantially influence the spatial distribution of the individual gabbro types, with the exception of Mg-rich varieties that may be restricted to the MLC. The geological setting of the wider area, including the MLC and TCU, is given in more detail by Crowley *et al.* (2002).

Lithology of the MLC and TCU rocks

The Mariánské Lázně crystalline complex consists of several main groups of metamorphic rocks (Fig. 1b). The effects of upper and middle amphibolite facies metamorphism, as well as later retrogression, were very heterogeneous. Beginning with the high-grade types, the mafic and felsic rocks forming the Mariánské Lázně Complex are listed below.

Relict eclogites and granulites with well preserved re-equilibrated high pressure and temperature mineral assemblages, and granulitized eclogites showing only the partial imprint of upper amphibolite to granulite facies conditions, form relict boudins and small isolated bodies enclosed in highly deformed amphibolite-facies host rocks. They probably formed from magmatic protoliths of pyroxenite, gabbro, or diorite composition from several sources. They are partially overprinted by amphibolite-facies metamorphism. Several types of relict high-grade rocks can be distinguished (Štědrá 1996, 2001; Jelínek *et al.* 1997). These rocks include omphacite–garnet–quartz–rutile±hornblende± kyanite eclogites, low-alkali eclogites with a pyroxene–actinolite–garnet–quartz assemblage, garnet pyroxenites (clinopyroxene-garnet± amphibole–plagioclase–rutile), epidote-rich eclogites, Al-rich granulites (plagioclase–garnet–omphacite–kyanite–zoisite–edenite–rutile–spinel–zircon–xenotime), and granulitized eclogites (clinopyroxene-plagioclase–garnet-edenite/pargasite-kyanite-orthopyroxene-quartz–titanite–sapphirine–corundum–anorthite –rutile-sillimanite).

These rocks are usually enclosed in voluminous amphibolites and amphibole gneisses, which range from mafic amphibolites consisting of a tschermakite to pargasite–plagioclase–garnet±clinopyroxene–apatite–rutile–titanite–ilmenite assemblage, through intermediate garnetiferous±amphibole gneisses with a substantial proportion of plagioclase, to felsic, quartz-rich and in places kyanite- and garnet-bearing gneisses. The intermediate and felsic types are mostly derived from protoliths of diorite to tonalite composition. The almost total obliteration of early high-grade metamorphism, rarely indicated by relict minerals and textures in some of these rocks, probably affected the low-K felsic rocks occurring in the central part of the unit.

The third group of rocks is distinct in having well-preserved primary magmatic fabrics. Coronitic metagabbros with variable mineral composition (clinopyroxene–plagioclase–ilmenite –apatite±orthopyroxene±olivine±amphibole± biotite±zircon) are mostly coarse- to medium-grained, whereas metadolerites (clinopyroxene–plagioclase–amphibole–ilmenite–apatite) and metaporphyries (Na-plagioclase xenocrysts–plagioclase–clinopyroxene–amphibole–ilmenite) display fine-grained matrices.

Varied quartz–feldspathic orthogneisses and

plagioclase-rich mobilisates with relict igneous fabrics form both concordant layers and lens-shaped bodies enclosed in amphibolites in zones of intense shear deformation. They are also found as discordant irregular injections in partially anatectic amphibolitic rocks. The mineral assemblage comprises Na-plagioclase–quartz–garnet–muscovite–rutile–titanite–apatite–zircon ±amphibole±diopside±muscovite±biotite± epidote±clinozoisite. However, several different generations of felsic rocks appear to make up this group.

Garnetiferous biotite–sillimanite±kyanite paragneisses, micaceous gneisses and garnetiferous mylonitized paragneisses are intercalated with amphibolites and amphibolite gneisses, mostly in the southern part of the MLC and along narrow, intensely sheared zones within basic rocks in the centre of the complex. Felsic gneisses and zircon-rich paragneisses occasionally form boudins surrounded by highly sheared amphibolites, especially in the Teplá and Prameny Potok valleys. Paragneisses predominate in the TCU; they are intensely foliated and sheared and reached the sillimanite and kyanite zones in the NW part adjacent to the MLC. Banded andradite-bearing low-K quartzites containing magnetite were also found in blocks near Boněnov, Kosmová and Prachomety in this NW part of the TCU.

Serpentinized and retrogressed ultramafic rocks of unknown age form the structural base of the MLC unit in its NW part along the border with the Kladská Unit (Kachlík 1993). Rarely, slightly different olivine-rich types with Cr-spinel have also been found in blocks along the opposite southeastern margin of the MLC near Boněnov. Due to the poor preservation of peak metamorphic mineral assemblages, it is difficult to determine the original proportions of rock forming mafic minerals. The chemical composition of samples analysed suggests that some ultrabasic types may have originated as cumulates in gabbroic magma chambers, and some as typical sub-oceanic mantle lherzolites and harzburgites (Fiala 1958; Crowley *et al.* 2002).

Younger igneous intrusions occurring in the area include granite dykes related to late Variscan granites, subvolcanic dykes of Tertiary age. Quaternary sediments overlie deeply weathered bedrock especially in the areas with flat relief.

Sampling and analytical methods

Sampling by the Czech Geological Survey and teams from Charles University preceded the studies made as part of the PACE project. New collections were made by the first author (1994–1996) in the course of structural mapping in the MLC, and investigations during the following three years yielded structural, petrological and mineralogical data to support the findings summarized below. Further samples were collected in 1999 for geochemical and geochronological studies framed by the international PACE project in co-operation with Keele University. The data sets from samples collected by the first two authors before 1999, and during the joint fieldwork have been combined to establish the petrogenetic affinities and original tectonic setting of metagabbroic rocks. The REE were analysed in the laboratory of the Czech Geological Survey by wet and FAAS methods. Major and trace elements were analysed by XRF spectrometry (ARL 8420) at Keele University. In the diagrams, data from 23 samples of metagabbros, amphibolites, granulites and eclogites from Štědrá (2001) and the selected metagabbro data from Crowley *et al.* (2002) are used as the reference sets.

Petrographic description of gabbroic rocks

Metagabbros

Metagabbros of the MLC–TCU area form a huge belt along the SE margin of the MLC, and occur as minor bodies located in the MLC interior and disseminated in the paragneisses and orthogneisses of the TCU. This suite of rocks has a relatively simple mineralogy consisting of (a) plagioclase–clinopyroxene–ilmenite–apatite ±amphibole, or (b) clinopyroxene–orthopyroxene–plagioclase–ilmenite–apatite±olivine in the more mafic types. Both are affected by garnet and amphibole overgrowths (Figs 2 and 3). Variations in primary and secondary mineral composition correspond to differences in major and trace element contents.

An inner belt of scattered bodies of preserved metagabbros, mostly enveloped by deformed, medium to coarse grained amphibolites derived from them, extends from near Pístov towards Otročín in the MLC. The clinopyroxene–orthopyroxene–olivine–plagioclase–ilmenite–apatite–rutile types and clinopyroxene –plagioclase–ilmenite(–rutile-amphibole ±zoisite ±kyanite) type are common in this belt. Mg-rich two-pyroxene and olivine-rich gabbros of Group 1 are typical of this internal part of the MLC, although members representing all other the geochemically distinct groups 2–5 are also present here. Plagioclase from the clinopyroxene-bearing metagabbro near Otročín is progressively zoned from bytownite cores

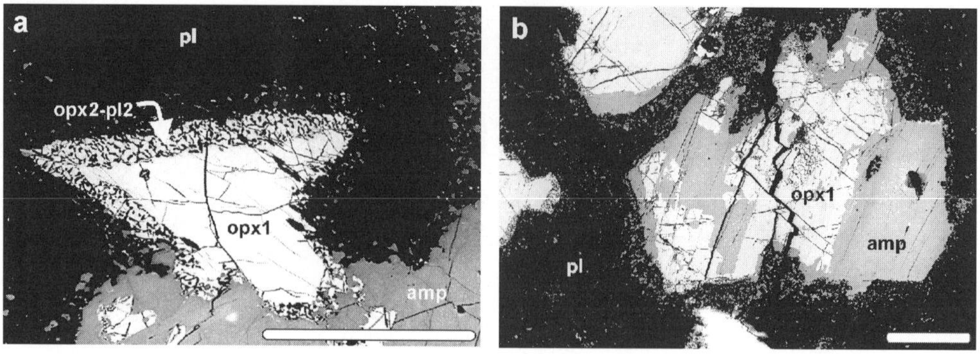

Fig. 2. Back-scattered image of Ti-rich orthopyroxene-bearing metagabbro, showing the symplectitic reaction corona between magmatic orthopyroxene (white) and plagioclase (black) consisting of enstatite similar in composition to the magmatic one and plagioclase (**a**). The phase shown in grey is brown Mg-hornblende replacing enstatite (**a, b**).

(An_{85}) to less calcic labradorite and andesine rims ($An_{55–48}$), plagioclase composition mostly varies between labradorite cores and andesine rims. Recrystallized plagioclase coexisting with metamorphic amphibole at contacts between clinopyroxene and plagioclase has a composition of An_{45}. Most metagabbros from the MLC show the effect of increasing pressure which led to the growth of partial or complete garnet and amphibole coronas around clinopyroxene, orthopyroxene and ilmenite. Rare olivine-bearing Mg-rich types display replacement of olivine by orthopyroxene that preceded the common pressure-dependent garnet growth. Corona garnet is generally rich in almandine (< 55 mol %) and grossular (18–30 mol %) as it often replaces anorthite from its outer contact with mafic phases; spessartine and pyrope components vary according to the source minerals in other basic rocks.

In the zone between Teplá, Kladruby, Beranov, and Poseč, numerous metagabbro bodies are distributed within the metasediments and orthogneisses of the Teplá Crystalline Unit. They contain andesine ($An_{30–50}$) as the main mineral, together with subsidiary clinopyroxene, orthopyroxene, amphibole, biotite and olivine. The main ore minerals present in the metagabbros are ilmenite, pyrite, chalcopyrite, pyrrhotite, and pentlandite (Goliáš 1994), together with small amounts of secondary goethite. All the Ti-rich, Rb-rich, alkali-rich and low-Mg Groups 2–5, in which garnet coronas are variably developed, occur both in the Teplá Unit, and the MLC (cf. Svobodová & Ulrych 1993).

Metadolerites

Medium- to fine-grained metadolerites with ophitic texture were found in blocks and outcrops in the vicinity of the coarse-grained metagabbros. The mineralogy of the metadolerites is mostly similar to that of coarse-grained types, although the fine-grained clinopyroxene–plagioclase–biotite rock from near Teplá is enriched in mafic minerals, ilmenite and other opaque minerals relative to the adjacent coarse-grained metagabbro. The magmatic assemblage is formed by euhedral prismatic plagioclase, and anhedral xenocrysts of pyroxene and brownish hornblende. Garnet forms irregular coronas similar to those observed in the coarse-grained metagabbros. In garnetiferous coronitic metadolerites, plagioclase is compositionally zoned from andesine (An_{35}) cores to oligoclase (An_{15}) rims. Some medium to fine grained basic rocks with ophitic fabrics, found only in blocks, may have originated as marginal facies or chilled margins of larger gabbro intrusions.

Porphyritic metagabbro

Porphyritic gabbro–diorite was found in weathered blocks east of Pístov. It contains subhedral phenocrysts of albite-twinned oligoclase set in a relatively uniform plagioclase–clinopyroxene (amphibole, Fe-epidote)–biotite matrix. The matrix is extensively amphibolized and retrogressed, with secondary growth of amphibole, chlorite, Fe-rich epidote and clinozoisite. Recrystallized narrow mantles of oligoclase phenocrysts and more calcic equant matrix

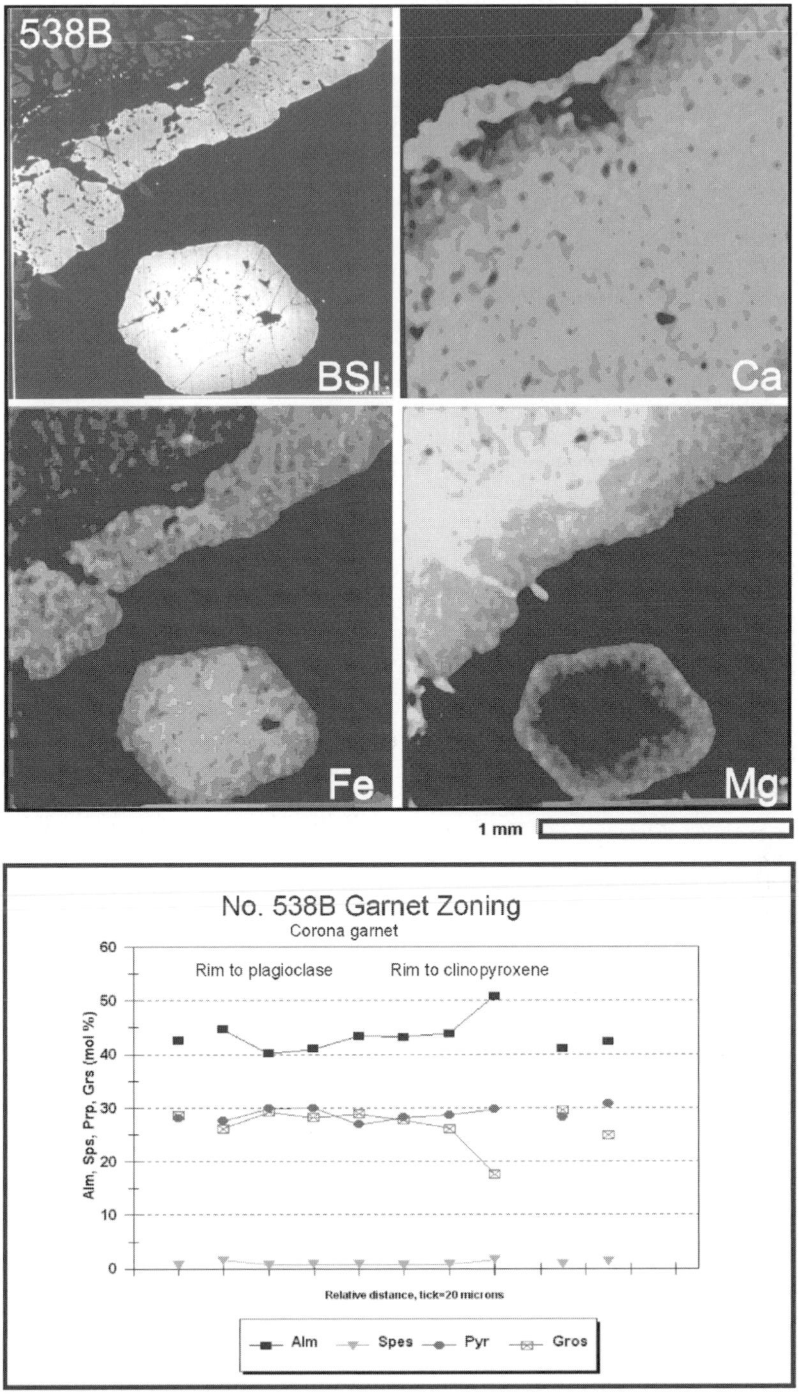

Fig. 3. The back-scattered image (upper left) and qualitative element distribution maps show the contrasting domainal zoning pattern in corona garnet growing between clinopyroxene and plagioclase, and in a euhedral garnet grain growing at the expense of plagioclase, as is indicated by constant Ca distribution in this grain (upper right). Note the inverted zoning of Mg in the both types. The bottom diagram provides quantitative information about the proportion of garnet end-members in the corona type.

Table 1. *Characteristics of the metagabbro groups 1–5 from the MLC–TCU area and their main compositional features*

Metagabbro group	Primary minerals	Relative enrichment	Relative depletion
1	**cpx–opx–pl–ol–ilm–ap**	**Mg,** Nb,	Ti, REE
2	**cpx–pl–ilm–ap**	**Ti,** Eu, Gd, Th	Rb, Zr, Nb, Ce, Ho, Er, Tm, HREE, SiO_2, K_2O
3	**pl–cpx–ilm–ap**	**alkalis,** Ti, Zr, Al, REE, SiO_2, Sr	Mg-Number, Pr
4	**opx–cpx–pl–ilm**	Zr, La, Ce, HREE	**Mg**
5	**cpx–opx–pl–bt (brown hbl)**	**Rb,** Zr, K_2O, Ba, Pb, Nd, Y, Cr, REE	**Mg,** Pr, Er

plagioclase grains alternate with poikilitic metamorphic amphibole, enclosing bleb-like quartz and plagioclase inclusions. Rare aggregates of Mg-chlorite, talc and Fe–Mg amphibole suggest the pre-existence of a primary Mg-rich mineral.

Amphibolites derived from gabbroic protoliths

Late upper amphibolite facies metamorphism overprints many basic rock types, transforming them to amphibolites. In some cases, amphibolized metagabbros can be distinguished from other amphibolized rocks by differences in their accessory minerals, garnet composition and matrix textures. Equilibrated garnet amphibolites, lacking textural evidence of a magmatic protolith or high-grade metamorphism, are rare within the MLC. The garnet amphibolites of the MLC mainly consist of hornblende, plagioclase, garnet, quartz, clinopyroxene, and accessory orthopyroxene, rutile, ilmenite, zircon, monazite, and apatite. Fe-epidote, prehnite, chlorite, calcite, titanite and clinozoisite occur as retrogressive minerals replacing garnet, biotite, zoisite, ilmenite, rutile and anorthite, and in younger hydrothermal veins. The rocks display a wide range of metamorphic fabrics from fine-grained types with a variable degree of preferred mineral orientation, to coarse-grained rocks with large amphibole and garnet poikiloblasts.

Compared to garnets from eclogites (pyrope$_{25-40}$, almandine$_{40-50}$, grossular$_{18-30}$, spessartine <1.5 mol%), amphibolite garnets have a lower pyrope (13–18 %) and grossular (15–25 %) component and higher almandine component (50–62%). Grains analysed show very little variation in composition, i.e. around 4 mol % for individual end-members. The alignment of quartz, amphibole, titanite or rutile inclusions in some garnet porphyroblasts provides evidence of local penetrative refoliation and rotation. Fine-grained amphibolites without large garnets mostly display perfect planar and linear alignment and banded or laminated fabrics. They contain substantially less garnet and more minerals typical of the lower amphibolite facies, including euhedral titanite. Apparently, they formed as a result of intense deformational reworking of the marginal parts of metagabbros, eclogites, and (especially) garnet amphibolites. Garnet was subject to retrograde reactions producing late actinolite, chlorite and plagioclase pseudomorphs and aggregates in sheared and altered parts of the mafic bodies.

Chemical composition of the metagabbros

The metagabbroic rocks (Table 2) occurring in the MLC and TCU are mostly subalkaline. Chemical characteristics of metagabbros and the other basic rocks from the MLC and their sources are provided in more detail by Crowley *et al.* (2002).

In the discrimination diagram FeO/MgO v. TiO_2 for subalkaline rocks (Miyashiro 1975), most metagabbros follow a tholeiitic trend, whereas MLC omphacite–garnet eclogites tend to follow a calc-alkaline trend. Five groups of metagabbros were distinguished based on combined mineralogical and chemical criteria: small variations occur in either primary mineral composition (olivine, orthopyroxene, biotite), or in immobile and mobile components and elements (Mg, Fe, Ti, Sr, Rb, alkalis, and Zr contents, Table 1). Groups 1, 4, and 5 often show a transitional composition, and thus indicate an evolution from Mg-rich to Mg-poor types. Groups 4 and 5 are similar, but may be divided into Rb-poor (Group 4) and Rb-rich types (Group 5, Fig. 4), probably reflecting the effect of metamorphism and fluid circulation. Group 2 metagabbros, enriched in primary biotite, ilmenite and rarely orthopyroxene, show extreme enrichment in Ti, depletion in Zr, Y (Figs 5, 6) and (generally) in all the LREE and

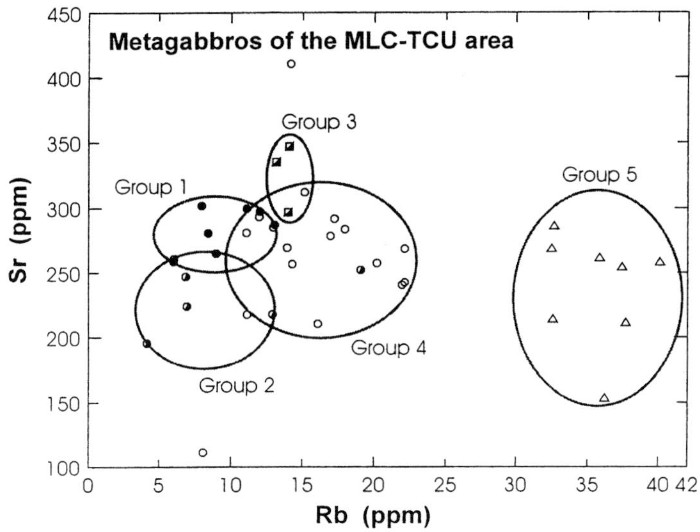

Fig. 4. Rb–Sr plot for whole-rock composition of the MLC/TCU metagabbros, showing the compositional differences between the five groups of metagabbros distinguished in the MLC/TCU area. Although Rb mobility during metamorphism must be taken into account, this grouping is also seen in other compositional plots.

the last members of HREE (Fig. 7). They have markedly constant Fe/Mg and SiO_2. Group 3 (plagioclase-rich metagabbro and diorites) displays high contents of SiO_2, alkalis, Al_2O_3, Zr, and LREE, and has the lowest Mg/Fe ratio of all the groups.

The metagabbro composition, corresponding to the gabbro–diorite series, shows affinities to rocks transitional between enriched mid-ocean ridge and within-plate sources. An exclusively WPB setting is indicated by the HFSE contents in the MLC metagabbros and other basic rocks, depicted in the Y–Zr–Nb ternary diagram (Meschede 1986) in Figure 5. The variability of metagabbros from this area, partially influenced also by heterogeneous metamorphic overprint, is therefore substantially wider than was previously presented (Svobodová 1993; Jelínek *et al.* 1997; Cháb 1997).

Sm–Nd isotope analyses of two metagabbros (ε_{Nd} 6,4 and 6,6; $^{87}Sr/^{86}Sr > 0.7037$, $^{143}Nd/^{144}Nd$ 0.512902–0.512939, Beard *et al.* 1995) indicate that they did not originate at an evolved mid-oceanic rift. Trace- and REE-contents in the metagabbros from the TCU–MLC boundary, in addition to their Sm–Nd isotopic ratios, indicate a within-plate, sub-continental source rather than MORB. Ti-rich and mixed metagabbros are comparable to WPB, whereas other groups show closer geochemical links with enriched and transitional mid-oceanic gabbros. REE patterns are more compatible with the patterns typical of lower crustal rock than those of N-, E-, or T-MORB (Taylor & McLennan 1985; Wilson 1989).

Two samples of fine-grained dolerite were analysed. Dolerite varieties are significantly different from coarse-grained metagabbros in both major and trace element contents, particularly in FeO/MgO, Zr, Ti, SiO_2 and alkalis. The distinctions suggest that finer grained intrusive bodies, and perhaps also fine-grained chilled margins formed as a result of episodes of intrusive activity from differentiated fractions of the mafic melt.

Major element compositions of other sub-alkaline mafic and intermediate rocks from the MLC (cf. Crowley *et al.* 2002) correspond also to those of gabbro and diorite, and are similar to some of the metagabbros in many respects. Variations in major and trace elements and REE have enabled a detailed resolution of individual rocks types (Štědrá 2001). As a result, the garnet amphibolites are now considered to be retrogressed derivatives either of high pressure rocks in the central MLC, or of metagabbros along its boundary with the TCU. Their chemical composition, especially of the well re-equilibrated garnet amphibolites, is similar to the metagabbros. Alkaline basic rocks are not typical of the MLC; but are typical of tectonic segments of the Kladská Unit (Kachlík 1994).

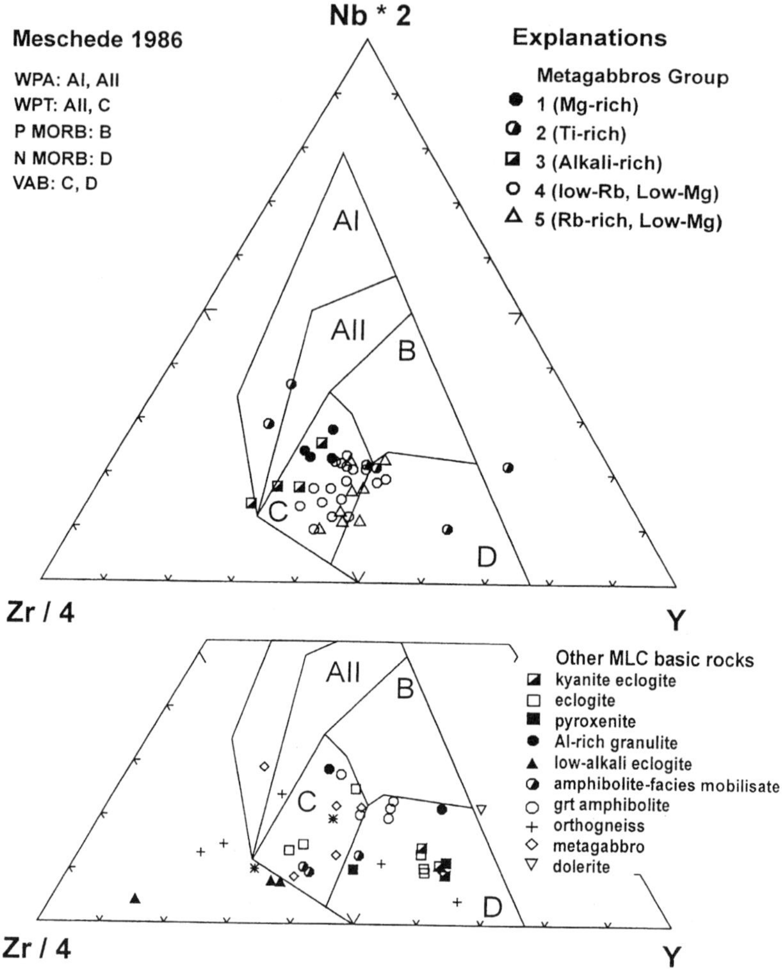

Fig. 5. Trace element plot (Meschede 1986) showing possible within-plate environments generating gabbroic magma. Compositional differences between metagabbros (upper part) and more variable other high pressure and temperature rocks and amphibolites (lower part) from the MLC/TCU area is also shown.

Metamorphic and structural evolution of the metagabbros in the MLC/ TCU area

Mineral composition of metagabbros

Primary magmatic pyroxenes (Table 3) originated in the parent mafic magma. Both orthopyroxene (enstatite) and clinopyroxene (Na-augite, augite, diopside, hedenbergite) are constituents of the metagabbros and gabbronorites from the southern and eastern part of the MLC. These magmatic pyroxenes co-exist predominantly with plagioclase, olivine, ilmenite, hornblende and biotite. Many clinopyroxene grains, affected by metamorphism, contain a dusting of oriented exsolution lamellae of ilmenite. The exsolution of Ti-oxide may indicate the pressure increase preceding amphibolitization (Adam & Green 1994); the ilmenite lamellae have survived even when the pyroxene became hydrated and completely replaced by secondary amphibole (Table 3). Magmatic enstatite in two-pyroxene, biotite-bearing Ti-rich metagabbros is overgrown by a younger orthopyroxene–plagioclase reaction corona (Fig. 2a), or by brown amphibole (Fig. 2b). Olivine in the best preserved Mg-rich metagabbros is surrounded by a peritectic reaction corona of enstatite. In some cases it is completely replaced by granular aggregates of enstatite; as in rocks similar to coronitic

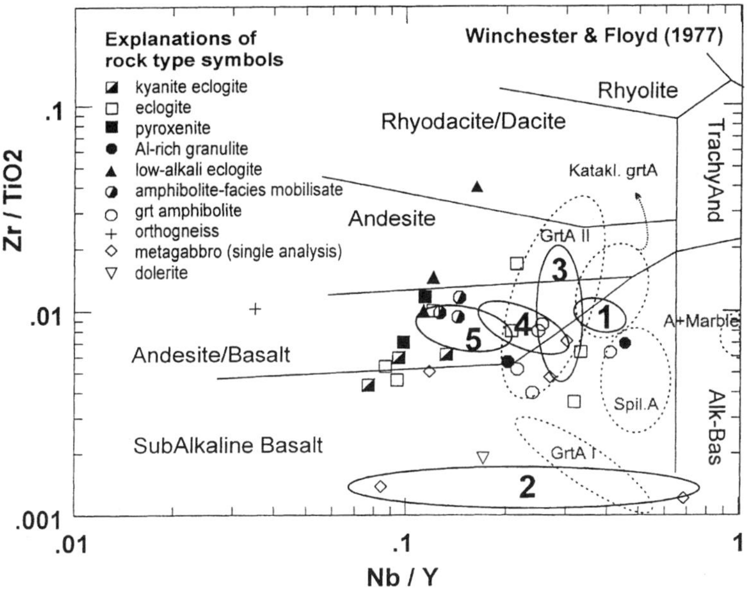

Fig. 6. A Nb–Y v. Zr–TiO_2 discrimination diagram (Winchester & Floyd 1977) displays the trace element composition of five groups of metagabbros (fields marked by bold numbers 1–5). Other, mostly basic, rocks from the MLC are also shown to illustrate their compositional variations. Dashed ellipses show compositions of metabasites from the KTB providing a comparison of KTB amphibolite and metagabbroic rocks with the metagabbros of the MLC–TCU area, and illustrate the absence of eclogite-related rocks in ZEV. Data: MLC, Štědrá (2001); ZEV, Brätz and Okrusch (1996).

gabbronorites (so-called 'hyperites') described by Kryza & Pin (2002) from the Góry Sowie in the NE part of the Bohemian Massif.

Jelínek (1993), Kopecký & Sokol (1993), and Svobodová (1993) stated that the primary mineral assemblage in the metagabbros included magmatic hornblende. Two generations of primary calcic amphiboles were noted by Svobodová & Ulrych (1993); early small euhedral crystals with higher MgO/FeO, and later large poikiloblasts of cummingtonite and pargasitic hornblende with lower MgO/FeO ratios. Present investigations revealed no undoubtedly magmatic amphibole in the coarse-grained gabbros from the MLC. Amphibole phenocrysts always contain relict clinopyroxene in their cores, or show metamorphic overgrowth.

Replacement amphibole–plagioclase aggregates resembling recrystallized and hydrated plagioclase–clinopyroxene symplectites after HP clinopyroxenes from eclogites occur rarely in some metagabbros, where they are associated with recrystallized magmatic plagioclase. These aggregates provide evidence of a small pressure decrease during gabbro exhumation, although the rocks probably never equilibrated under eclogite facies conditions.

Garnet (Table 3) growth resulting from increasing pressure is of two types: asymmetric garnet coronas around mafic minerals in contact with plagioclase, and large rounded symmetrical grains nucleating within plagioclase domains. Each type shows marked contrasts in its compositional zoning patterns, and they differ from one another in consumption of cations depending on the host domain. Corona garnets surrounding mafic minerals in contact with plagioclase vary from those showing only Ca increase, to those in which Mg decreases and Fe is irregularly distributed. Euhedral garnets in the plagioclase matrix show no contrast in the Ca distribution map because the Ca content in garnet and in plagioclase is the same; Mg progressively increases, and Fe decreases (Fig. 3). The homogeneous garnet composition in the recrystallized garnet amphibolites is compatible with the equilibrated state of the amphibolite matrix. It indicates that any heterogeneity of garnet inherited from the coronitic stage was effectively homogenized during amphibolite facies metamorphism.

Metamorphic kyanite and zoisite growing at the expense of Ca-plagioclase in these rocks never achieved the habit and size of poikilitic

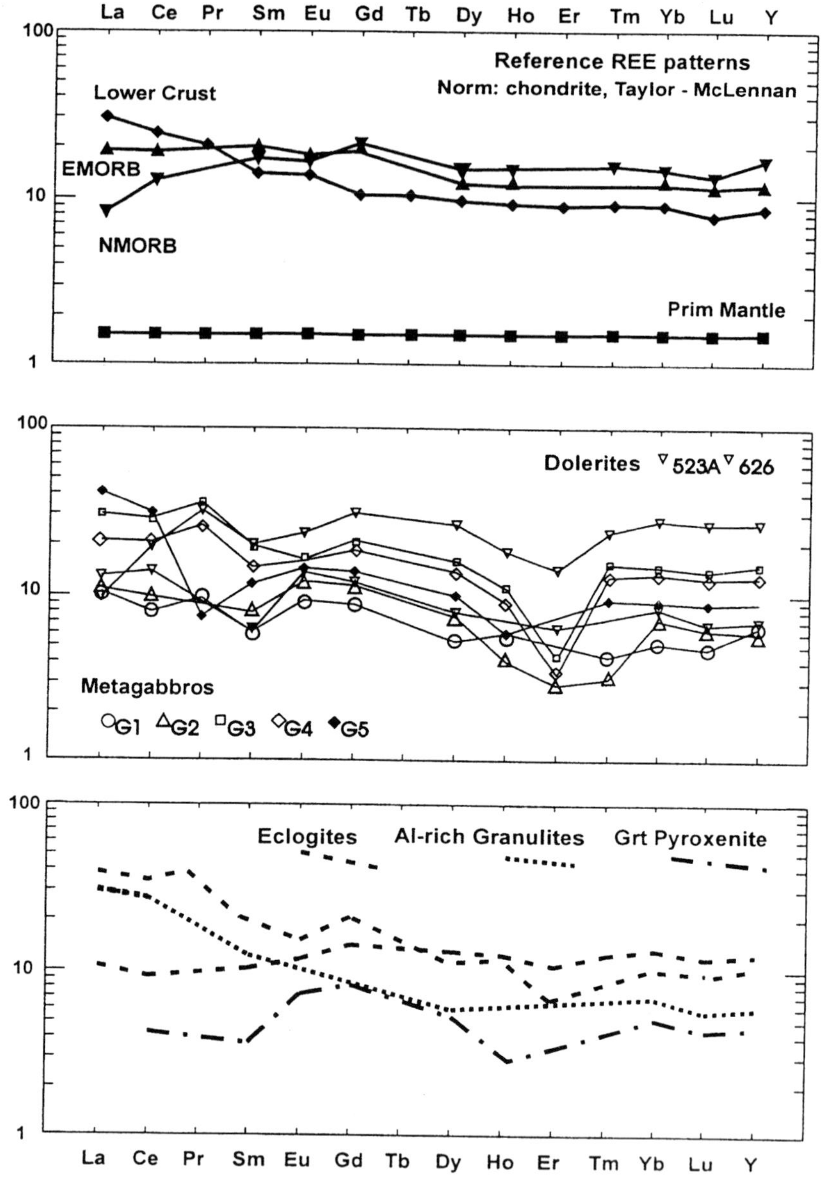

Fig. 7. Rare earth element profiles showing chondrite-normalized composition of metagabbros (middle diagram), eclogites and Al-rich granulites (lower) in comparison to the reference patterns shown in the upper part. A relatively good fit with the lower crustal composition, and limited correlations with MORB-patterns are indicated.

and prismatic kyanite and zoisite observed in eclogites and granulites. They crystallized within the Ca-rich zones of magmatic plagioclase as thin prisms and acicular crystals arranged in regular grids (cf. Zulauf 1997). In some cases, they were replaced by anorthite. These minerals are also preserved as inclusions in the corona garnet which replaces the remaining plagioclase matrix.

Structural features of metagabbro bodies

The structural evolution of the metagabbros is relatively simple, mostly showing thrust- and

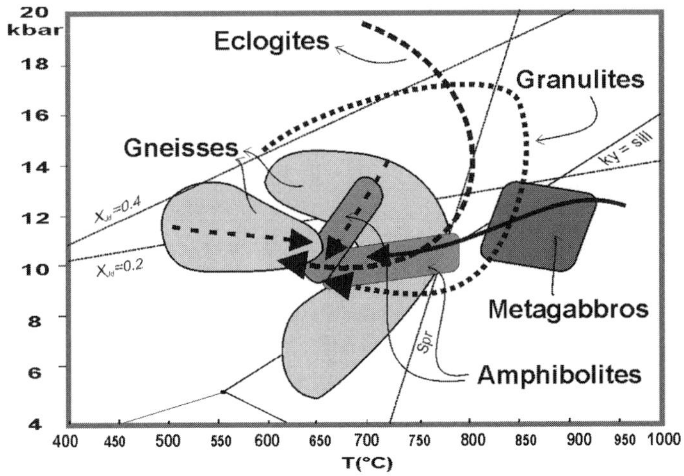

Fig. 8. P–T diagram showing contrasting P–T–t paths of metagabbros, eclogites, granulites, amphibolites and amphibolite gneisses from the Mariánské Lázně Complex, based on detailed thermobarometry of basic rocks after Štědrá (2001). X_{Jd} isopleths are after Gasparik and Lindsley (1980); the 'Spr' line corresponds to the lower limit of stability of sapphirine after Ackermand *et al.* (1975).

fault-related deformation under upper amphibolite and lower-grade metamorphic facies. Metre-scale folding, boudinage, and even the L-tectonites typical of the earlier eclogites, gneisses and amphibolites in the internal parts of the MLC (Štědrá 2001) are mostly absent in the metagabbro-dominated southeastern margin. No analogues of relict foliation and lineation preserved in high pressure rocks have been found in metagabbro bodies or garnet amphibolite bodies probably derived from them.

Metamorphic evolution of the metagabbros

Some metagabbro bodies caused strong contact metamorphism of the surrounding host rocks, for example in the SW part of the Lestkov pluton (Kachlík 1997), which intruded along the major NE–SW shear zone forming the boundary of the biotite/staurolite/kyanite zones. Whereas the granitoids of the Lestkov pluton caused contact metamorphism at hornblende hornfels facies with a peak assemblage consisting of muscovite–biotite–cordierite–garnet, adjacent to metagabbros assemblages record pyroxene hornfels facies conditions. A mineral assemblage with orthopyroxene, unaltered cordierite, biotite and garnet was found near the contact of a minor gabbro intrusion. The field relationships between metagabbros and granitoids and their contact metamorphic aureoles show that intrusion of both rock groups was probably coeval.

Medium pressure metamorphism of the gabbroic rocks reached a maximum at upper amphibolite/lower granulite facies conditions. The gabbros probably intruded at middle crustal levels, where the relict high pressure rocks were already enclosed by amphibolite-facies host rocks. There is multiple evidence of static recrystallization of metagabbros in low-strain domains at around 600–730 °C and pressures between 8–11 kbar corresponding to the upper amphibolite facies. Deformation affected the garnetiferous metagabbros in several stages before, during and after growth of corona garnet (cf. also Zulauf 1997), as seen in highly sheared samples with post-deformational garnet growth, and rotated and cracked asymmetrical garnet porphyroclasts in the other intensely deformed samples.

Results of conventional thermobarometry extensively applied to the MLC basic rocks (Štědrá 2001) indicate differences in early metamorphic histories of eclogites, granulites, gneisses and metagabbros from the area studied (Fig. 8). Two distinct pressure–temperature retrograde fields for garnet amphibolites may reflect small differences in whole-rock composition, different starting mineral equlibria preceding amphibolitization, and, consequently, also different retrograde reactions recorded in amphibolitized eclogites and metagabbros. Subsequent local equilibration of the late amphibole–garnet–plagioclase–quartz assemblage from metagabbros and other rocks types reflects decompression and rapid cooling of the consolidated MLC/TCU block during Variscan exhumation.

Table 2. *Whole-rock chemical composition of metagabbros from the MLC/TCU area*

Sample	S.321	S.324	S.325	S.326	S.327	S.328	S.329	S.330	S.331	S.334	S.335	S.336
location	TCU	TCU	TCU	TCU	MLC	MLC	MLC	MLC	MLC	MLC	TCU	MLC
group no	5	4	5	4	4	4	4	4	4	4	1	1
SiO_2	50.18	48.60	54.03	49.97	48.92	50.54	47.93	47.94	45.39	44.17	47.47	48.42
TiO_2	1.84	1.45	1.09	1.32	1.40	1.54	1.62	1.73	3.49	2.28	1.51	1.06
Al_2O_3	16.99	15.59	16.63	17.15	17.27	17.86	17.55	15.79	13.73	14.10	15.83	17.58
$Fe_2O_3^T$	10.51	10.61	8.17	9.52	10.26	9.62	10.07	11.37	14.35	11.65	10.81	8.69
MnO	0.17	0.17	0.14	0.15	0.16	0.15	0.16	0.17	0.22	0.20	0.16	0.14
MgO	6.29	8.44	6.42	7.83	8.70	6.26	7.97	8.42	5.63	5.92	11.69	9.40
CaO	8.42	11.75	7.53	10.03	10.10	9.94	9.42	9.35	12.02	14.94	8.46	10.48
Na_2O	3.65	1.81	3.26	2.79	2.83	3.31	3.04	3.68	3.20	3.04	2.98	2.77
K_2O	1.05	0.19	1.07	0.44	0.40	0.73	0.64	0.77	0.71	0.62	0.50	0.42
P_2O_5	0.26	0.17	0.15	0.13	0.16	0.19	0.21	0.17	0.48	0.39	0.23	0.12
LOI	1.35	0.99	1.65	1.09	0.19	0.30	1.50	0.69	0.65	2.99	0.67	0.23
TOTAL	100.70	99.76	100.14	100.41	100.38	100.45	100.11	100.09	99.87	100.30	100.30	99.31
Ba	144	94	244	101	149	232	164	316	204	103	83	138
Cl	50	0	50	17	23	43	37	19	71	53	18	50
Cr	92	348	214	263	231	232	134	329	193	210	303	298
Cu	29	53	19	31	45	38	33	65	43	78	50	38
Ga	17	16	16	15	19	19	17	17	24	19	15	14
Nb	4	6	4	4	5	6	6	3	20	16	3	6
Ni	45	130	41	58	140	39	102	102	85	131	266	147
Pb	6	7	8	8	3	5	6	7	6	8	2	2
Rb	40	8	32	11	12	17	17	17	14	15	13	11
S	152	34	115	460	314	333	135	425	476	107	490	292
Sr	256	110	280	279	294	279	288	297	362	281	286	297
Th	3	2	3	4	1	1	3	3	3	3	2	1
V	229	272	231	196	172	186	181	262	372	223	160	137
Zn	106	87	81	77	76	78	83	81	121	105	88	66
Zr	171	92	128	120	113	159	128	114	251	192	162	93
La	11.43	2.69	12.54	5.20	5.95	11.65	6.64	<0,2	11.84	12.13	5.81	4.62
Ce	26.59	11.01	31.04	16.04	17.37	32.52	21.77	14.44	44.29	33.93	19.28	14.98
Pr	4.38	2.23	4.00	2.34	2.59	4.68	3.47	3.07	7.05	5.17	3.11	2.33
Nd	18.14	9.76	18.55	11.20	12.24	20.58	15.29	11.08	28.52	21.69	13.78	9.16
Sm	4.41	2.29	3.99	2.45	2.65	3.97	3.18	2.19	5.49	4.30	2.94	1.67
Eu	1.52	1.06	1.28	1.09	1.10	1.42	1.31	1.13	2.16	1.76	1.19	0.860
Gd	6.00	4.88	5.68	3.93	4.08	6.15	5.15	4.72	8.49	6.86	4.63	3.27
Tb	<0.70	<0.70	0.70	<0.70	<0.70	<0.70	<0.70	<0.70	<0.70	<0.70	<0.70	<0.70
Dy	5.93	4.92	5.13	3.69	3.84	5.59	4.70	4.17	7.26	5.79	4.20	2.77
Ho	0.96	0.74	0.81	0.47	0.60	0.78	0.54	0.49	1.00	1.00	0.75	0.44
Er	0.73	0.67	0.57	<0.40	<0.40	0.41	<0.40	<0.40	1.15	0.55	<0.40	<0.40
Tm	0.51	0.51	0.52	0.38	0.36	0.52	0.41	0.35	0.51	0.51	0.36	0.27
Yb	3.52	3.29	3.19	2.31	2.21	3.29	2.76	2.49	4.08	3.35	2.52	1.66
Lu	0.51	0.48	0.46	0.32	0.31	0.47	0.40	0.35	0.58	0.48	0.36	0.24
Y	32.37	28.23	28.57	19.27	19.82	30.08	25.59	22.37	39.19	32.18	22.87	14.51

Sample	S.340	S.341	S.342	S.343	S.344	S.345	438	442	456	597B	608A	626
location	TCU	TCU	TCU	TCU	TCU	MLC	MLC	MLC	MLC	MLC	MLC	MLC
group no	1	4	4	3	5	1	2	2	4	5	2	4
SiO_2	51.95	48.74	49.43	50.04	51.36	47.86	48.45	48.69	50.45	51.57	46.92	49.91
TiO_2	0.93	2.79	1.83	1.40	2.00	1.80	2.08	4.54	1.33	1.25	2.91	2.59
Al_2O_3	16.77	15.33	17.58	21.48	15.53	16.15	15.64	13.96	16.41	17.97	13.44	12.73
$Fe_2O_3^T$	8.93	13.74	10.56	7.70	11.77	11.44	12.56	11.91	10.91	8.41	16.78	16.84
MnO	0.18	0.21	0.17	0.13	0.19	0.19	0.19	0.21	0.17	0.13	0.26	0.26
MgO	8.42	6.69	5.87	3.84	6.32	8.10	7.73	7.91	6.73	5.86	7.23	5.67
CaO	9.45	8.59	9.24	10.10	8.75	9.67	9.04	10.52	8.22	8.69	9.82	9.12
Na_2O	2.80	3.05	3.42	3.43	3.15	2.98	2.88	2.44	3.88	3.19	1.61	2.43
K_2O	0.24	0.79	0.62	0.49	1.21	0.44	0.74	0.25	0.60	1.14	0.72	0.12
P_2O_5	0.03	0.24	0.27	0.17	0.26	0.23	0.21	0.05	0.17	0.14	0.24	0.21
LOI	0.59	0.45	0.54	0.51	0.08	1.01	0.89	0.29	0.99	1.55	0.42	0.54
TOTAL	100.28	100.62	99.53	99.29	100.62	99.86	100.39	100.76	99.87	99.91	100.35	100.43

Table 2. *Continued*

Sample location group no	S.340 TCU 1	S.341 TCU 4	S.342 TCU 4	S.343 TCU 3	S.344 TCU 5	S.345 MLC 1	438 MLC 2	442 MLC 2	456 MLC 4	597B MLC 5	608A MLC 2	626 MLC 4
Ba	86	230	160	143	379	171	153	77	153	388	269	29
Cl	10	50	44	55	60	26	44	6	78	42	18	0
Cr	164	99	184	55	178	142	138	224	216	179	194	80
Cu	26	40	32	21	40	40	37	43	30	28	58	56
Ga	17	21	21	20	20	17	18	15	19	17	18	19
Nb	0	11	9	8	8	5	7	4	3	5	4	0
Ni	63	43	44	20	44	69	59	53	63	62	60	35
Pb	0	4	7	3	14	4	11	4	11	8	13	9
Rb	6	22	15	13	38	11	19	7	14	32	28	5
S	399	857	170	332	62	154	12	479	78	19	172	597
Sr	258	241	309	331	213	215	251	225	406	211	88	187
Th	0	0	4	0	4	1	5	2	3	6	3	2
V	244	265	198	143	245	206	244	588	216	168	505	479
Zn	66	112	102	63	104	91	96	65	139	72	129	146
Zr	28	164	139	109	152	161	99	55	130	145	133	145
La	1.26	14.94	11.81	7.64	16.58	9.36	4.98	<0.2	7.43	12.99	<0.2	3.55
Ce	4.59	41.25	29.60	18.86	41.48	25.31	19.54	3.09	28.89	32.97	6.75	18.32
Pr	<0.8	6.57	4.52	2.97	6.22	3.90	3.22	<0.8	2.93	4.54	2.61	4.30
Nd	3.48	27.27	19.30	12.67	26.12	17.04	14.42	2.29	16.24	19.32	6.05	16.24
Sm	1.13	4.14	3.82	2.62	5.22	4.01	3.24	<0.30	4.42	4.21	2.81	4.54
Eu	0.70	2.16	1.53	1.22	1.58	1.44	1.26	0.38	1.50	1.17	1.22	2.00
Gd	1.55	8.47	5.87	4.11	7.73	6.20	5.26	1.52	5.50	6.04	7.22	9.31
Tb	<0.70	<0.70	<0.70	<0.70	<0.70	<0.70	<0.70	<0.70	<0.70	<0.70	<0.70	<0.70
Dy	1.41	7.72	5.17	3.71	7.17	5.79	4.70	1.14	4.22	5.73	8.71	9.89
Ho	0.91	1.01	0.78	0.49	1.11	0.95	0.76	0.22	0.69	0.79	1.53	1.52
Er	<0.40	1.87	0.81	<0.40	1.79	0.99	0.41	0.87	0.57	1.05	2.75	3.48
Tm	<0.20	0.44	0.47	0.32	0.68	0.51	0.44	<0.20	0.41	0.43	0.94	0.83
Yb	1.09	4.76	3.15	2.21	4.22	3.52	2.87	1.01	2.78	3.40	6.05	6.78
Lu	0.15	0.69	0.46	0.33	0.60	0.52	0.39	0.12	0.39	0.49	0.93	0.99
Y	7.47	42.99	28.39	19.59	39.32	33.02	26.02	5.83	23.62	31.68	53.24	59.14

Major elements (wt %) and trace elements (ppm) were analysed by Q. G.Crowley, Keele University; REE by CGS laboratory, chief analyst V. Sixta

Summary and conclusions

The main characteristics of the basic rocks from the Mariánské Lázně Complex and the Teplá Crystalline Unit indicate that formation of these rocks cannot be related exclusively to either rift-related or oceanic environments. There are several different sources of the rocks forming the MLC and TCU, which correspond to their poly-metamorphic and multiphase tectonic evolution. The evolutionary scheme proposed below partially explains the pre-Ordovician stage of the geological history of the area, and provides an alternative to the approaches favouring a single Cambro-Ordovician magmatic event (Crowley *et al.* 2002) followed by the Variscan high pressure metamorphic overprint (Dallmeyer & Urban 1994; Beard *et al.* 1995) for the MLC and the adjacent units related to the TBU.

The metagabbros are geochemically similar to some relict high pressure and temperature basic rocks and amphibolites in the MLC and TCU (cf. Crowley *et al.* 2002). Largely immobile trace elements provide evidence of compositional heterogeneity in the metagabbros, which mostly correspond to rocks derived from within-plate environments. By contrast the varied group of the MLC eclogites and granulites show E-MORB, N-MORB, WPB, and uncertain geochemical characteristics. The metagabbros with compositions close to N-MORB and E-MORB contain slightly reduced Ni, Cr, and Co, and elevated LREE and LIL elements when compared to the standard values (Taylor & McLennan 1985). The differences may, however, partly result from alteration by metamorphic fluids. The REE distributions in metagabbros from the five groups distinguished display slightly variable patterns. REE compositions indicate compatibility with standard E-MORB REE patterns or average lower continental crust, thus also favouring a possible WPB setting of these rocks.

The petrological features and field structural relationships of the metagabbros and other basic rocks from the Mariánské Lázně Complex and the Teplá Crystalline Unit, supported by the

Table 3. *Selected microprobe analyses of the main rock-forming minerals from the representative metagabbro and metadolerite samples*

PYROXENES from metagabbros Group 1 (538), 2 (442, 540B), 4 (540C, 541), 5 (609A) and metadolerite (473C)

Analysis	442/06	442/07	538B/22	538B/27	538B/28	538B/30	540B/13	540B/18	540B/29	540C/18	540C/38	540C/42	541A/01	541a/04	609A/05	609A/09
SiO_2	52.72	55.10	54.63	52.33	53.38	54.77	52.75	52.69	52.53	53.53	52.18	53.18	52.98	52.54	52.52	53.49
TiO_2	0.00	0.00	0.00	0.55	0.23	0.00	0.06	0.39	0.21	0.03	0.05	0.34	0.28	0.34	0.36	0.22
Al_2O_3	0.93	3.64	1.18	3.75	2.02	0.42	1.44	0.71	3.01	2.20	0.43	2.44	1.65	3.71	4.70	2.04
Cr_2O_3	0.00	0.00	0.00	0.00	0.00	0.00	0.07	0.03	0.45	0.09	0.00	0.52	0.00	0.00	0.04	0.08
FeO	23.62	9.98	18.84	6.59	5.88	17.46	23.49	22.10	7.16	7.67	26.21	8.72	6.64	8.60	14.54	7.97
MnO	0.60	0.25	0.25	0.00	0.00	0.18	0.52	0.43	0.15	0.26	0.40	0.08	0.28	0.00	0.36	0.10
MgO	19.93	17.64	24.58	13.57	13.88	25.50	21.64	22.17	13.64	13.30	18.88	12.81	13.90	14.12	14.31	13.74
CaO	1.70	11.99	0.28	20.85	23.93	0.13	0.41	1.10	21.72	21.65	0.40	20.47	23.17	20.08	11.64	21.74
Na_2O	0.00	0.42	0.00	1.30	0.89	0.21	0.00	0.00	1.00	1.09	0.64	1.20	0.56	0.75	0.88	0.76
Total	99.49	99.02	99.75	98.95	100.21	98.67	100.38	99.62	99.88	99.83	99.18	99.74	99.47	100.13	99.33	100.13
Wo	1.46	21.39	0.00	38.79	45.67	0.00	0.00	0.14	40.28	41.86	0.78	39.46	44.39	36.59	19.38	41.75
En	56.77	51.26	67.20	36.79	37.36	70.52	58.21	61.40	36.89	36.57	53.14	35.24	38.10	38.54	40.83	37.61
Fs	37.74	16.28	29.44	10.03	8.88	27.25	36.75	34.34	10.87	11.84	41.38	13.46	10.21	13.17	23.28	12.24
Jd	0.00	3.76	0.00	6.27	4.97	1.50	0.00	0.00	5.11	7.35	4.38	5.63	2.49	3.52	4.95	4.30

GARNETS

Analysis	442/01	442/03	473c/02	473c/06	473c/09	473c/11	473c/13	538B/01	538B/06	538B/07	538B/33	540C/21	540C/23	540C/28	540C/29	540C/31
SiO_2	38.52	38.17	39.56	39.31	39.44	39.61	39.35	39.10	38.89	39.18	39.35	38.69	38.80	38.64	38.73	38.65
TiO_2	0.00	0.25	0.00	0.00	0.00	0.00	0.00	0.00	0.00	0.00	0.19	0.00	0.07	0.11	0.14	0.16
Al_2O_3	20.84	20.69	22.00	21.73	21.82	22.09	21.79	21.57	21.25	21.57	21.44	21.75	22.07	21.70	21.53	21.39
Cr_2O_3	0.00	0.00	0.00	0.00	0.00	0.00	0.00	0.00	0.00	0.00	0.00	0.00	0.00	0.00	0.08	0.00
FeO	24.29	24.81	21.23	23.69	20.96	21.14	21.46	20.28	20.60	20.76	20.18	22.38	22.14	23.22	22.98	24.83
MnO	1.44	2.15	0.39	0.52	0.37	0.32	0.29	0.36	0.41	0.38	0.63	0.56	1.05	0.89	0.71	1.10
MgO	4.73	4.97	7.02	7.03	7.93	7.60	7.69	7.49	7.15	7.62	8.24	5.99	5.24	5.66	5.87	5.33
CaO	9.84	8.14	10.01	8.28	9.37	9.48	9.20	10.65	10.71	10.38	9.43	9.80	10.47	9.43	10.09	8.81
Total	99.65	99.18	100.22	100.55	99.89	100.25	99.79	99.44	99.01	99.87	99.45	99.17	99.84	99.65	100.13	100.28
Alm	55.26	53.50	45.22	50.03	44.15	44.82	45.41	42.56	43.39	43.24	42.33	48.46	48.15	50.10	48.73	53.06
Spes	3.31	4.70	0.85	1.12	0.80	0.69	0.63	0.77	0.87	0.80	1.33	1.23	2.31	1.95	1.52	2.38
Pyr	19.20	19.09	26.63	26.46	29.77	28.73	29.01	28.03	26.85	28.27	30.82	23.12	20.31	21.78	22.20	20.30
Gros	28.68	21.99	27.30	22.39	25.29	25.75	24.95	28.63	28.89	27.69	24.99	27.20	29.02	25.86	26.92	23.79

Table 3. *Continued*

AMPHIBOLES

Analysis	442/2	442/8	442/9	473c/05	473c/07	538B/10	538B/16	538B/34	540B/20	540B/21	541A/06	541A/14	541A/26	609A/04	609A/12	609A/14
SiO_2	46.21	48.05	42.86	45.25	43.98	41.64	54.97	41.55	46.46	43.97	54.85	47.97	51.47	49.92	48.15	50.02
TiO_2	0.99	0.73	1.68	1.04	1.22	0.31	0.00	0.00	1.70	2.56	0.24	0.86	0.34	0.57	0.71	0.55
Cr_2O_3	0.31	0.00	0.00	0.00	0.00	0.00	0.00	0.00	0.00	0.06	0.00	0.00	0.00	0.08	0.00	0.00
Al_2O_3	10.37	8.56	13.15	11.76	12.52	15.68	1.23	16.50	9.16	10.83	2.46	9.84	6.02	6.05	8.47	6.75
CaO	12.13	12.35	12.08	11.95	11.77	11.64	5.47	11.70	11.55	11.78	13.19	12.74	12.83	11.28	11.27	10.77
FeO	12.63	11.41	12.14	11.51	11.38	10.62	16.78	9.43	13.60	12.99	9.64	11.47	10.13	14.44	14.98	14.35
MgO	12.69	13.98	11.83	13.67	12.93	12.90	17.95	13.41	12.92	12.55	17.00	13.44	15.44	13.02	11.75	13.09
MnO	0.00	0.00	0.00	0.00	0.00	0.00	0.52	0.00	0.43	0.07	0.00	0.00	0.00	0.24	0.25	0.39
K_2O	0.48	0.38	0.92	0.55	0.70	0.34	0.00	0.22	0.47	0.66	0.00	0.12	0.00	0.11	0.27	0.18
Na_2O	1.35	1.05	1.62	1.76	1.99	2.62	0.30	2.39	1.72	2.01	0.39	1.13	0.78	1.24	1.11	1.06
TOTAL	97.16	96.51	96.27	97.48	96.48	95.75	97.22	95.19	98.02	97.47	97.75	97.57	97.01	96.95	96.96	97.16
Si^{4+}	6.75	6.99	6.36	6.51	6.45	6.13	7.92	6.09	6.75	6.46	7.76	6.94	7.37	7.28	7.05	7.27
Ti^{4+}	0.11	0.08	0.19	0.11	0.13	0.03	0.00	0.00	0.19	0.28	0.03	0.09	0.04	0.06	0.08	0.06
Al^{3+}	1.79	1.47	2.30	1.99	2.16	2.72	0.21	2.85	1.57	1.88	0.41	1.68	1.02	1.04	1.46	1.16
Ca^{2+}	1.90	1.93	1.92	1.84	1.85	1.84	0.84	1.84	1.80	1.86	2.00	1.97	1.97	1.76	1.77	1.68
Mg^{1+}	2.76	3.03	2.62	2.93	2.83	2.83	3.86	2.93	2.80	2.75	3.58	2.90	3.30	2.83	2.57	2.84
Mn^{2+}	0.00	0.00	0.00	0.00	0.00	0.01	0.06	0.01	0.05	0.01	0.00	0.00	0.00	0.03	0.03	0.05
K^{1+}	0.09	0.07	0.17	0.10	0.13	0.06	0.00	0.04	0.09	0.12	0.00	0.02	0.00	0.02	0.05	0.03
Na^{1+}	0.38	0.30	0.47	0.49	0.56	0.75	0.08	0.68	0.48	0.57	0.11	0.32	0.22	0.35	0.31	0.30
Fe^{3+}	0.20	0.23	0.22	0.63	0.32	0.43	0.02	0.53	0.38	0.26	0.10	0.00	0.13	0.35	0.33	0.27
Fe^{2+}	1.34	1.16	1.29	0.75	1.07	0.88	2.01	0.62	1.28	1.33	1.04	1.39	1.08	1.41	1.51	1.47
S	15.33	15.26	15.52	15.36	15.52	15.68	15.00	15.59	15.38	15.52	15.02	15.30	15.12	15.13	15.16	15.12
Fe_{tot}	1.54	1.39	1.51	1.38	1.40	1.31	2.02	1.16	1.65	1.60	1.14	1.39	1.21	1.76	1.83	1.74
Fe^{3+}/Fe_{Tot}	0.13	0.17	0.14	0.46	0.23	0.33	0.01	0.46	0.23	0.16	0.09	0.00	0.11	0.20	0.18	0.16
Fe/(Fe+Mg)	0.36	0.31	0.37	0.32	0.33	0.32	0.34	0.28	0.37	0.37	0.24	0.32	0.27	0.38	0.42	0.38

PLAGIOCLASES

Analysis	442/10	473c/04	473c/04a	473c/08	538B/24	538B/29	540B/22	540B/25	540B/26	541a/13	541a/15	541a/20	541a/25	541a/28	541a/29	609A/07
SiO_2	54.31	62.34	63.49	64.37	57.23	54.00	56.39	46.59	54.25	58.86	56.37	54.35	53.11	47.46	48.38	54.60
Al_2O_3	28.26	23.36	22.72	22.07	25.69	27.96	27.19	33.92	28.51	25.67	27.33	28.50	29.17	33.05	32.24	28.53
CaO	11.69	5.51	4.52	3.90	9.08	11.85	8.98	16.90	10.79	8.31	10.27	11.81	12.95	17.36	16.58	10.57
Na_2O	4.74	7.88	8.68	8.97	6.07	4.65	6.04	1.69	5.13	6.69	5.73	4.77	4.37	1.91	2.44	5.49
K_2O	0.15	0.32	0.33	0.32	0.00	0.00	0.08	0.01	0.06	0.00	0.00	0.00	0.00	0.00	0.00	0.02
Total	99.15	99.40	99.73	99.63	98.07	98.47	98.69	99.10	98.75	99.53	99.70	99.43	99.59	99.77	99.64	99.21
#An	57.18	27.36	21.94	19.03	45.25	58.45	44.88	84.62	53.54	40.70	49.77	57.79	62.06	83.40	78.98	51.50
#Ab	41.94	70.76	76.17	79.10	54.75	41.55	54.63	15.30	46.09	59.30	50.23	42.21	37.94	16.60	21.02	48.41
#Or	0.87	1.88	1.89	1.87	0.00	0.00	0.49	0.08	0.37	0.00	0.00	0.00	0.00	0.00	0.00	0.09

geochemical data, favour the emplacement of basic and felsic igneous rocks into already established late Proterozoic or early Cambrian lower to middle crust with the MLC and TCU already juxtaposed.

An Early Ordovician emplacement age of metagabbros is indicated by a U-Pb zircon age from the Výškovice metagabbro (Bowes & Aftalion 1991), and by U-Pb zircon ages of the corresponding metagabbros and garnet amphibolites from the KTB drill hole in the Zone of Erbendorff–Vohenstrauss in Germany. An age of 490 ± 5.5 Ma for these basic rocks was obtained by von Quadt (1997), and similar ages were also presented by Söllner & Miller (1994), Grauert *et al.* (1994), and Hölzl & Kohler (1994). Although a single protolith age of both metagabbros and eclogites from the MLC/TCU area cannot be excluded due to the lack of geochronological data, the chemical, metamorphic, and structural features of metagabbros and high pressure rocks reveal their contrasting history (Fig. 8). Moreover, a pre-Variscan Ordovician high pressure and temperature metamorphic and anatectic event was reported by O'Brien *et al.* (1997) for the mafic rocks from the Zone of Erbendorff–Vohenstrauss. A significant time interval between burial (presumably mid-Ordovician) and Devonian decompression of garnet-bearing paragneisses from the KTB borehole is documented by Reinhardt (1997).

It is proposed that the oldest basement rocks of the western part of the TBU are relics of tectonically active lower crust. An age around 540 Ma has already been indicated for the Cadomian stage of TBU history (Zulauf *et al.* 1999). This age is also considered to apply to the early evolution of the high-grade metamorphic rocks in the MLC/TCU area, which were intruded by metagabbros in the Early Ordovician. Several possible original tectonic settings can be discussed, the two examples being (1) a deep reworked part of the continental accretionary wedge of Gondwana, and (2) the subducted segment of the active continental margin with incorporated back-arc members and metasedimentary and magmatic oceanic rocks. A model of subduction and exhumation of part of the back-arc basin presented by Chemenda *et al.* (2001) may be used to explain the latter example of burial and exhumation of the pre-Ordovician rock sequence forming part of the MLC/TCU area at present.

Early Ordovician mafic magmatism postdated the first part of the tectonic evolution outlined. After the emplacement of felsic and mafic granitoids, the rock sequence experienced Variscan stacking, uplift and thrusting, during which it achieved its present form and latest tectonic features. At this stage, the metagabbros, together with other basic and intermediate rocks, experienced a widespread amphibolite facies overprint. Consequently, in contrast to the previously published interpretation of the MLC evolution, the main high pressure and temperature event recorded in the MLC and TCU was pre-Variscan (possibly Cadomian) and not Variscan (cf. Beard *et al.* 1995). A medium-pressure peak of the early Devonian metamorphic event is also supported by the K–Ar cooling ages on hornblendes and micas from the Teplá-Domazlice Zone (Kreuzer *et al.* 1992). During late Variscan post-orogenic extension and related thrusting and faulting, the high-grade parts of the MLC and TCU were juxtaposed against the upper crustal, low-grade metasediments exposed across most of the Teplá-Barrandian Unit.

The work benefited from the EU-funded TMR project PACE (No. ERBFMRXCT970136), and the Czech Grant Agency project No. 205/94/1455. The authors are grateful to Q. G. Crowley, P. A. Floyd, and V. Sixta for the chemical analyses provided, to J. Frýda and I. Vavřín for assistance with microprobe work at the CGS, Prague, and to C. Halls and J. A. Winchester for their help with the improvement of English style throughout the text.

References

ACKERMAND, D., SEIFERT, F. & SCHREYER, W. 1975. Instability of sapphirine in high pressures. *Contribution to Mineralogy and Petrology*, **50**, 79–92.

ADAM, J. & GREEN, T. H. 1994. The effect of pressure and temperature on the partitioning of Ti, Sr and REE between amphibole, clinopyroxene and basanitic melts. *Chemical Geology*, **117**, 219–233.

BEARD, B. L., MEDARIS, L., JOHNSON, C. M., JELÍNEK, E. & TONIKA, J. 1995. Geochronology and geochemistry of eclogites from the Mariánské Lázně Complex, Czech Republic: Implications for Variscan Orogenesis. *Geologische Rundschau*, **84**, 552–567.

BOWES, D. R. & AFTALION, M. 1991. U-Pb zircon isotope evidence of Early Ordovician and Late Proterozoic units in the Mariánské Lázně Complex, Central European Hercynides. *Neues Jahrbuch für Mineralogie*, **7**, 315–326.

BRÄTZ, H. & OKRUSCH, M. 1996. Die Metabasite der KTB-Hauptbohrung: Petrographische und geochemische Untersuchungen am Kernmaterial aus dem Teufenbereich zwischen 6242 and 8080 m. *Geologica Bavarica*, **101**, 53–112.

CHÁB, J. 1997. The Teplá Barrandian Unit: The granitic and gabbroic intrusions. *In*: VRÁNA, S. & ŠTĚDRÁ, V. (eds) Geological model of western Bohemia related to the KTB borehole in Germany. *Journal of Geological Sciences, Geology*, **47**, 56–57.

CHÁB, J., ŠRÁMEK, J., POKORNÝ, L., CHLUPÁČOVÁ, M., MANOVÁ. M., VEJNAR, Z., WALDHAUSEROVÁ, J. & ŽÁČEK, V. 1997: The Teplá-Barrandian Unit. *In*: VRÁNA, S. & ŠTĚDRÁ, V. (eds) Geological model of western Bohemia related to the KTB borehole in Germany. *Journal of Geological Sciences, Geology,* **47**, 80–104.

CHEMENDA, A., HURPIN, D. & STEPHAN, J. F. 2001. Physical modelling of arc-continental collision preceded by oceanic subduction with active back-arc basin: constraints on the mechanism of deep burial and exhumation of HP/LT rocks. *EUG XI Abstracts*, 430, Strasbourg.

CROWLEY, Q. G., FLOYD, P. A., ŠTĚDRÁ, V., WINCHESTER, J. A., KACHLÍK, V. & HOLLAND, J. G. 2002. The Mariánské-Lázně Complex, NW Bohemian Massif: development and destruction of an early Palaeozoic seaway. *In*: WINCHESTER, J. A., PHARAOH, T. C. & VERNIERS, J. (eds) *Geological Society, London, Special Publications*, **201**, 177–195.

DALLMEYER, R. D. & URBAN, M. 1994. Variscan vs. Cadomian tectonothermal evolution within the Teplá-Barrandian zone, Bohemian massif, Czech Republic: $^{40}Ar/^{39}Ar$ mineral and whole-rock slate/phyllite ages. *Journal of the Czech Geological Society*, **39**, 21–22.

DÖRR, W., FIALA, J., PHILIPPE, S., VEJNAR, Z. & ZULAUF, G. 1995. Cadomian vs. Variscan imprints in the Teplá-Barrandian unit. Part A: U-Pb ages of the pre-Variscan granitoids, in 4. Kolloquium *'Orogene Prozesses'*, 29.–30.3.1995, Jena.

DÖRR, W., FIALA, J., PHILIPPE, S., VEJNAR, Z. & ZULAUF, G. 1996. Evidence for a pervasive Cambrian magmatism in the Teplá-Barrandian: Continental break-up?, *Terra Nostra*, **96/2**, 39–43.

DÖRR, W., FIALA, J., VEJNAR, Z. & ZULAUF, G. 1998. U-Pb zircon ages and structural development of metagranitoids of the Teplá crystalline complex – evidence for pervasive Cambrian plutonism within the Bohemian Massif (Czech Republic), *Geologische Rundschau*, **87**, 135–149.

FIALA, J. 1958. Hlavní typy hornin v širším okolí Pramenů v Císařském lese. *Geologické Práce*, **50,** 1–55, Bratislava.

FIALA, J. & VEJNAR, Z. 1994. Metamorphic mineral isograds and metamorphic events in the Saxothuringicum-Moldanubicum transition zone. *Terra Nostra,* **94**, 35–36.

FIALA, J. & VEJNAR, Z. 1997. The Teplá-Barrandian Unit: The Cheb-Dyleň Crystalline unit, relationship to the Moldanubian unit. *In*: VRÁNA, S. & ŠTĚDRÁ, V. (eds) Geological model of western Bohemia related to the KTB borehole in Germany. *Journal of Geological Sciences, Geology,* **47**, 56–57.

GASPARIK, T. & LINDSLEY, D. H. 1980. Phase equilibria at high pressure of pyroxenescontaining monovalent and trivalent ions. *In*: PREWIT, C. T. (ed.) *Pyroxenes.* Mineralogical Society of America Reviews in Mineralogy, **7**, Washington, 309–339.

GOLIÁŠ, V. 1994. *Rudní minerály mariánskolázeňského metabazitového komplexu.* Diploma thesis, MS, Charles University, Prague.

GRAUERT, B., MAUMANN, A. & KLEEMAN, U. 1994. Altersbestimmung einer HAT-mylonitzone im Grenzbereich ZEV/Moldanubikum, KTB Report, **90–4**, 551, Niedersächsischen Landesamt für Bodenforschung, Hannover, Germany.

HÖLZL, S., KOHLER, H. 1994. Zirkondatierungen an Gesteinen der KTB, KTB Report, **94–2**, B35, Niedersächsischen Landesamt für Bodenforschung, Hannover, Germany.

JELÍNEK, E. 1993. *Geochemie a geochronologie eklogitů mariánskolázeňského komplexu.* Final report of the project 'Geological model of western Bohemia in relation to deep borehole KTB in FRG' (in Czech).

JELÍNEK, E., ŠTĚDRÁ, V. & CHÁB, J. 1997. The Mariánské Lázně Complex. *In*: VRÁNA S. & ŠTĚDRÁ, V. (eds) Geological model of western Bohemia related to the KTB borehole in Germany. *Journal of Geological Sciences, Geology*, **47**, 61–70.

KACHLÍK, V. 1993. The evidence for Late Variscan nappe thrusting of the Mariánské Lázně Complex over the Saxothuringian terrane (W. Bohemia). *Journal of Czech Geological Society*, **38,** 46–58.

KACHLÍK, V. 1994. The Kladská Unit – petrological and structural evidence for Variscan thrusting of the MLC over the Saxothuringian Terrane (W. Bohemia). *KTB-Report*, **94-3**, 19–32.

KACHLÍK, V. 1997. Contact metamorphic rocks from the mantle of the Lestkov Pluton and their significance for reconstruction of the tectonometamorphic development of the Teplá-Barrandian Unit. in Czech, *Geoscience Research Reports for 1996*, CGS Prague.

KASTL, E. & TONIKA, J. 1984. The Mariánské Lázně metaophiolite complex (west Bohemia). *Krystalinikum*, **17**, 59 – 76.

KETTNER, R. 1917. Versuch einer Stratigraphischen Einteilung des Bohmischen Algonkiums. *Geologische Rundschau*, **8**, 169–188.

KOPECKÝ, L. & SOKOL A. 1993. *Geochemie hlavních horninových typů mariánsko-lázeňského komplexu.* Partial final report of the project 'Geological model of western Bohemia in relation to deep borehole KTB in FRG' (in Czech), CGS Prague.

KREUZER, H., VEJNAR, Z., SCHÜSSLER, U., OKRUSCH, M. & SEIDEL, E. 1992. K-Ar dating in the Teplá-Domazlice zone at the western margin of the Bohemian Massif. *In*: Z. KUKAL (ed.) *Proceedings of the 1st International Conference on the Bohemian Massif*, CGS Prague. 168–175.

KRYZA R. & PIN C. 2002. Mafic rocks in a deep-crustal segment of the Variscides (the Gory Sowie, SW Poland): evidence for crustal contamination in an extensional setting. *International Journal of Earth Sciences*, in press.

MESCHEDE, S. 1986. A method of discriminating between different types of mid-ocean ridge basalts and continental tholeiites with the Nb-Zr-Y diagram. *Chemical Geology*, **56**, 207–218.

MLČOCH, B. 1997. The Horní Slavkov Crystalline Unit. *In*: VRÁNA, S. & ŠTĚDRÁ, V. (eds) Geological model of western Bohemia related to the KTB borehole in Germany. *Journal of Geological Sciences, Geology*, **47**, 57–58.

MIYASHIRO, A. 1975. Classification, characteristics and origin of ophiolites. *Journal of Geology,* **83**, 249–281.

O'BRIEN, J., DUYSTER, J., GRAUERT, B., SCHREYER, W., STÖCKHERT, B. & WEBER, K. 1997. Crustal evolution of the KTB site: From oldest relics to the late Hercynian granites. *In*: The KTB Deep Drill Hole, *Journal of Geophysical Research*, **102**, No. B8, AGU, Washington. 18203–18220.

QUADT, A. VON 1997. U-Pb zircon and Sr-Nd-Pb whole-rock investigations from the continental deep drilling KTB. *Supplement Geologische Rundschau*, **86,** 258–271.

REINHARDT, J. 1997. Thermobarometry and P-T path of garnet-aluminosilicate-bearing gneisses from the KTB drill core (Continental Deep Drilling Project, Germany). *Geologische Rundschau,* **86**, S167–S183.

SCHÜSSLER, U., VEJNAR, Z., OKRUSCH, M., ROSE, S. & SEIDEL, E. 1992. Geochemistry of metabasites and gabbroic rocks from the Teplá-Domazlice Zone. *In*: Z. KUKAL (ed.). *Proceedings of the 1st International Conference on the Bohemian Massif*, 168–175, CGS Prague.

SIEBEL, W., TRZEBSKI, R., STETTNER, G., HECHT, L., CASTEN, U., HÖHNDORF, A. & MÜLLER, H. 1997. Granitoid magmatism of the NW Bohemian massif revealed: gravity data, composition, age relation and phase concept. *Geologische Rundschau*, **86**, supplement S45–S63.

SÖLLNER, F. & MILLER, H. 1994. U-Pb systematics on zircon from chlorite gneiss of metavolcanic layer V4 (7260–7800 m) from the KTB-Hauptbohrung. *KTB Report*, **94–2**, B31, Niedersächsischen Landesamt für Bodenforschung, Hannover, Germany.

ŠMEJKAL, V. 1964. Absolute age of some magmatic and metamorphic rocks from the Bohemian massif according to K-Ar method. *Sborník geologických věd*, **G4**, 121–136.

ŠTĚDRÁ, V. 1996. Retrograde evolution of the Mariánské Lázně Complex basic rocks. – *Final report of the project 205/94/1455 of the Czech Grant Agency*, MS Archive CGS Prague, 129 pages.

ŠTĚDRÁ, V. 2001. *Tectonometamorphic evolution of the Mariánské Lázně Complex, Western Bohemia, based on the study of metabasic rocks.* PhD thesis, 136 pages, Charles University, Prague.

ŠTĚDRÁ, V. & KRYZA, R. 1999. Gabbros in accretionary prisms of the NW Bohemian Massif: A comparative petrogenetic study of the MLC and Gory Sowie (1). Abstract, PACE 4th meeting, Copenhagen.

SVOBODOVÁ, J. 1993. *Gabbroic rocks of the Mariánské* Lázně *Metabasite Complex*. MS Diploma Thesis, in Czech, Charles University, Prague.

SVOBODOVÁ, J. & ULRYCH, J. 1993. Geochemie gabroidních hornin mariánsko-lázeňského metabazitového komplexu a charakteristika jejich hlavních horninotvorných minerálů. *Final report, Project Geological Model of western Bohemia in relation to the KTB deep drilling in Germany.* CGS Prague, 93.

TAYLOR, S. R. & MCLENNAN, S. M. M. 1985. *The continental crust: Its composition and evolution*. 312 pp, Blackwell, Oxford.

TONIKA, J. 1998. Geological map on 1:50 000 scale, Mariánské Lázně sheet. CGS Prague.

VEJNAR, Z. 1982. The regional metamorphism of psammitic-pelitic rocks in the Domazlicearea. English summary. *Journal of Geological Sciences, Geology,* **37**, 9–70.

VRÁNA, S. & ŠTĚDRÁ V. (eds) 1997. Geological model of western Bohemia related to the KTB borehole in Germany. *Journal of Geological Sciences, Geology*, **47,** CGS Prague.

WEBER, K. & VOLLBRECHT, A. 1989. Crustal structure at the KTB drill site, Oberpfalz. *In*: EMMERMAN, R. & WOHLENBERG, J. (eds) The German continental deep drilling programme *(KTB)*, 5–35, Springer Verlag, Berlin.

WILSON, M. 1989. *Igneous petrogenesis*. Harper Collins Academic, London, 466 pp.

WINCHESTER, J. A. & FLOYD, P. A. 1977: Geochemical discrimination of different magma series and their differentiation products using immobile elements. *Chemical Geology*, **20**, 325–343.

ŽÁČEK, V. 1994. Garnets and metamorphic evolution of the Teplá crystalline complex, Western Bohemia. *Zentralblatt für Geologie und Palaeontologie,* **1**, 847–856.

ŽÁČEK, V. & CHÁB, J. 1993. Garnet compositional zoning and P-T condition of metamorphism, Teplá Crystalline complex, W. Bohemia. Geological model of the western Bohemia in relation to the deep borehole KTB in FRG. – Abstracts 1992, ČGÚ Praha, 35–36.

ŽÁČEK, V., SLABÝ, J. & CHÁB, J. 1993. *The metamorphic evolution of the Teplá Crystalline Unit. – Final report of the project 'Geological Model of western Bohemia in relation to the KTB deep drilling in Germany',* CGS Prague (in Czech).

ŻELAŹNIEWICZ, A. 1999. Structural relationships within the Münchberg Gneiss Complex. *Terra Nostra,* **99/1**, 213.

ZOUBEK, V. 1948. Notes on stratigraphy of crystalline complexes of the Bohemian Massif I., II. (in Czech). *Sborník Státního geologického Ústavu ČSR*, **13**, 468–498.

ZULAUF, G. 1994*a*. The tectonometamorphic evolution of the western part of the Teplá-Barrandian (Bohemian Massif, Czech Republic). *Journal of the Czech Geological Society*, **39**, 125–126.

ZULAUF, G. 1994*b*. Ductile normal faulting along the West Bohemian Shear Zone (Moldanubian/Teplá-Barrandian boundary: evidence for late Variscan extensional collapse in the Variscan Internides. *Geologische Rundschau,* **83**, 276–292.

ZULAUF, G. 1995. Cadomian and Variscan tectonothermal events in the SW part of the Teplá-Barrandian Unit (Bohemian Massif, Czech Republic). *Zentralblatt für Geologie und Palaeontologie,* **2**, 1515–1528.

ZULAUF, G. 1997. *Von der Anchizone bis zur Eclogitfazies.* – habilitation work, Giessen.

ZULAUF, G., DÖRR, W., FIALA, J. & VEJNAR, Z. 1997. Late Cadomian crustal tilting and Cambrian transtension in the Teplá Barrandian Unit (Bohemian Massif, Central European Variscides). *Geologische Rundschau*, **86**, 571–584.

ZULAUF, G., SCHITTER, F., RIEGLER, G., FIALA, F. & VEJNAR, Z. 1999. Age constraints on the Cadomian evolution of the Teplá Barrandian Unit (Bohemian Massif) through electron microprobe dating of metamorphic monazite. *Zeitschrift der Deutschen Geologischen Gesellschaft*, **150**, 627–639.

Collage tectonics in the northeasternmost part of the Variscan Belt: the Sudetes, Bohemian Massif

P. ALEKSANDROWSKI & S. MAZUR

Uniwersytet Wrocławski, Instytut Nauk Geologicznych, Cybulskiego 30, 50–205 Wrocław, Poland (e-mail: palex@ing.uni.wroc.pl)

Abstract: A synthesis of published and new data is used to interpret the Sudetic segment of the Variscan belt as having formed by the accretion of four major and two or three minor terranes. From west to east the major terranes are (1) Lusatia-Izera Terrane, exposing Armorican continental basement reworked by Ordovician plutonism and Late Devonian–Carboniferous collision, showing Saxothuringian affinities; (2) composite Góry Sowie-Kłodzko Terrane characterized by multistage evolution (Silurian subduction, mid- to late Devonian collision, exhumation and extension, Carboniferous deformational overprint), with analogues elsewhere in the Bohemian Massif, Massif Central and Armorica; (3) Moldanubian (Gföhl) Terrane comprising the Orlica-Śnieżnik and Kamieniec massifs, affected by Early Carboniferous high-grade metamorphism and exhumation and (4) Brunovistulian Terrane in the East Sudetes, set up on Avalonian crust and affected by Devonian to late Carboniferous sedimentation, magmatism and tectonism. The main terranes are separated by two smaller ones squeezed along their boundaries: (1) Moravian Terrane, between the Moldanubian and Brunovistulian, deformed during Early Carboniferous collision, and (2) SE Karkonosze Terrane of affinities to the Saxothuringian oceanic realm, sandwiched between the Lusatia-Izera and Góry Sowie-Kłodzko (together with Teplá-Barrandian) terranes, subjected to high pressure-metamorphism and tectonized during Late Devonian–Early Carboniferous convergence. The Kaczawa Terrane in the NW, of oceanic accretionary prism features, metamorphosed and deformed during latest Devonian–Early Carboniferous times, may either be a distinct unit unrelated to closure of the Saxothuringian Ocean or represent a continuation of the SE Karkonosze Terrane.

The Variscan Belt of central Europe, together with the adjoining fragment of the Trans-European Suture Zone defines a composite tectonic collage finally assembled during the Carboniferous. It comprises units derived from Avalonia, from the Armorican Terrane Assemblage and others of uncertain (northern Gondwana?, Baltica?) provenance (e.g. Pharaoh 1999; Belka *et al.* 2000; Franke 2000; Tait *et al.* 2000, Aleksandrowski *et al.* 2000; Winchester & PACE 2002). One of its key fragments, of as yet poorly constrained geological structure and evolution, is the northeasternmost segment of the Variscides: the Sudetes area in southwestern Poland and northern Bohemia. Geologically, the area is located on the NE margin of the Bohemian Massif; it shows a complex structure consisting of a mosaic of geologically distinct, fault-bounded pre-Permian units affected by contractional, strike-slip and extensional tectonics and characterized by abrupt changes in the dominant structural trends. The apparently independent geological evolution of most geological units in the Sudetes, combined with the occurrence of ophiolitic bodies along some of their boundaries, or of igneous rocks with MORB-type geochemical signature, as well as of high pressure (HP) metamorphic rocks (blueschists and eclogites) strongly suggests that the area comprises a number of distinct tectonostratigraphic terranes, separated by tectonic sutures and major faults/shear zones. A possible terrane arrangement in this region was a matter of a lively discussion during the last decade (Matte *et al.* 1990; Aleksandrowski 1990, 1995; Oliver *et al.* 1993; Cymerman & Piasecki 1994; Franke *et al.* 1995*a*; Cymerman *et al.* 1997; Franke & Żelaźniewicz 2000). We reassess the evidence published to date and combine it with our new data from several Sudetic units. An updated terrane model of the Sudetes is proposed, based on the assumption that most of the main tectonic units known from elsewhere in the Variscan Belt continue into this area. It rejects some and modifies other previous terrane concepts. In conclusion, a possible setting for accretion of the Sudetic terranes is given. We identify and discuss many unsolved problems of Sudetic/Variscan geology, the answers to which may significantly improve knowledge of the tectonic structure and evolution of the Variscan Belt of central Europe.

From: WINCHESTER, J. A., PHARAOH, T. C. & VERNIERS, J. 2002. *Palaeozoic Amalgamation of Central Europe.* Geological Society, London, Special Publications, **201**, 237–277. 0305-8719/02/$15.00

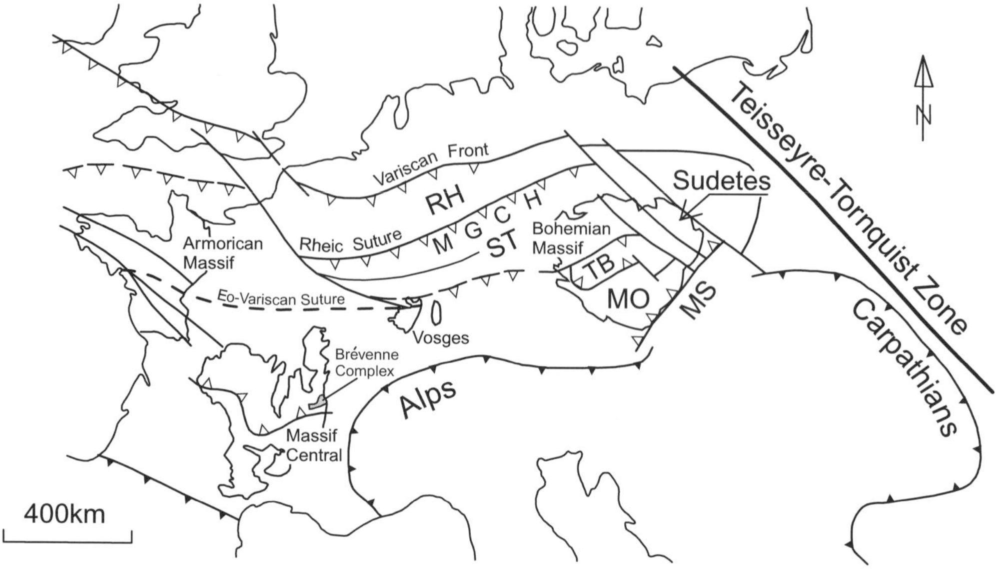

Fig. 1. Tectonic setting of Sudetes in the Variscan Belt. MGCH, Mid-German Crystalline High; MO, Moldanubian; MS, Moravo-Silesian; RH, Rhenohercynian; ST, Saxothuringian; TB, Tepla-Barrandian.

Tectonic setting

The Sudetes, a mountainous area on the northern margin of the Bohemian Massif, together with the southern part of the Silesian-Lusatian Plain (in the geological sense collectively termed here the Sudetic region or, simply, the Sudetes), represents the northeasternmost outcropping segment of the Central European Variscan Belt (Figs 1 and 2) and exposes strongly deformed and metamorphosed complexes of the Variscan internides. To the NE, across the Middle Odra Fault Zone, these complexes adjoin Carboniferous unmetamorphosed flysch and molasse succession of the Variscan externides, buried below the thick Permo-Mesozoic sequence of the German–Polish Basin and underlain by a poorly documented Devonian and low-grade metamorphosed Lower Palaeozoic basement (Grad *et al.* 2002). The Variscan externides and their basement merge into the Trans-European Suture Zone, a broad and complex zone of Palaeozoic terrane accretion separating the Phanerozoic lithosphere of central and eastern Europe from the Precambrian East European Platform (Pharaoh 1999).

The Sudetic region extends between the WNW–ESE trending Middle Odra Fault Zone in the NE and the parallel Elbe Fault Zone in the SW (Fig. 2). To the SE, it approaches the Miocene Carpathian Front and to the NW it merges with the Lusatian Massif. The area is usually divided into the West Sudetic domain or Lugicum (Suess 1926), with prevailing NW–SE structural trends and a smaller, East Sudetic domain (partly equivalent to Suess's Silesicum) with predominantly NNE–SSW structures. The West Sudetic, Lugian domain reveals complex geology, consisting of a variety of distinct, fault-bounded Palaeozoic tectonic units with usually unclear mutual geometrical and genetic interrelationships. It is bisected by a prominent WNW–ESE to NW–SE fault zone, the Intra-Sudetic Fault, parallel to the Middle Odra and Elbe Faults. In this paper, we concur with some workers (e.g. Teisseyre *et al.* 1957) and distinguish the eastern part of the Lugian domain as the Central Sudetes, to account for its distinct geological features. The East Sudetes, located further to the SE and east, show a generally regular, zonal geological structure with dominant trends roughly at right angles to those prevailing in the West and Central Sudetes.

The NW–SE trending Sudetic Boundary Fault divides the Sudetes into the Sudetic mountains in the south and the Sudetic foreland in the north; this topographically based distinction is the result of Late Tertiary (Alpine) uplift of the southern side of the fault. The depositional, magmatic and deformational evolution of the main structural units of the Sudetes is summarized diagrammatically in Figure 3 and the distribution of these units is shown in Figure 2.

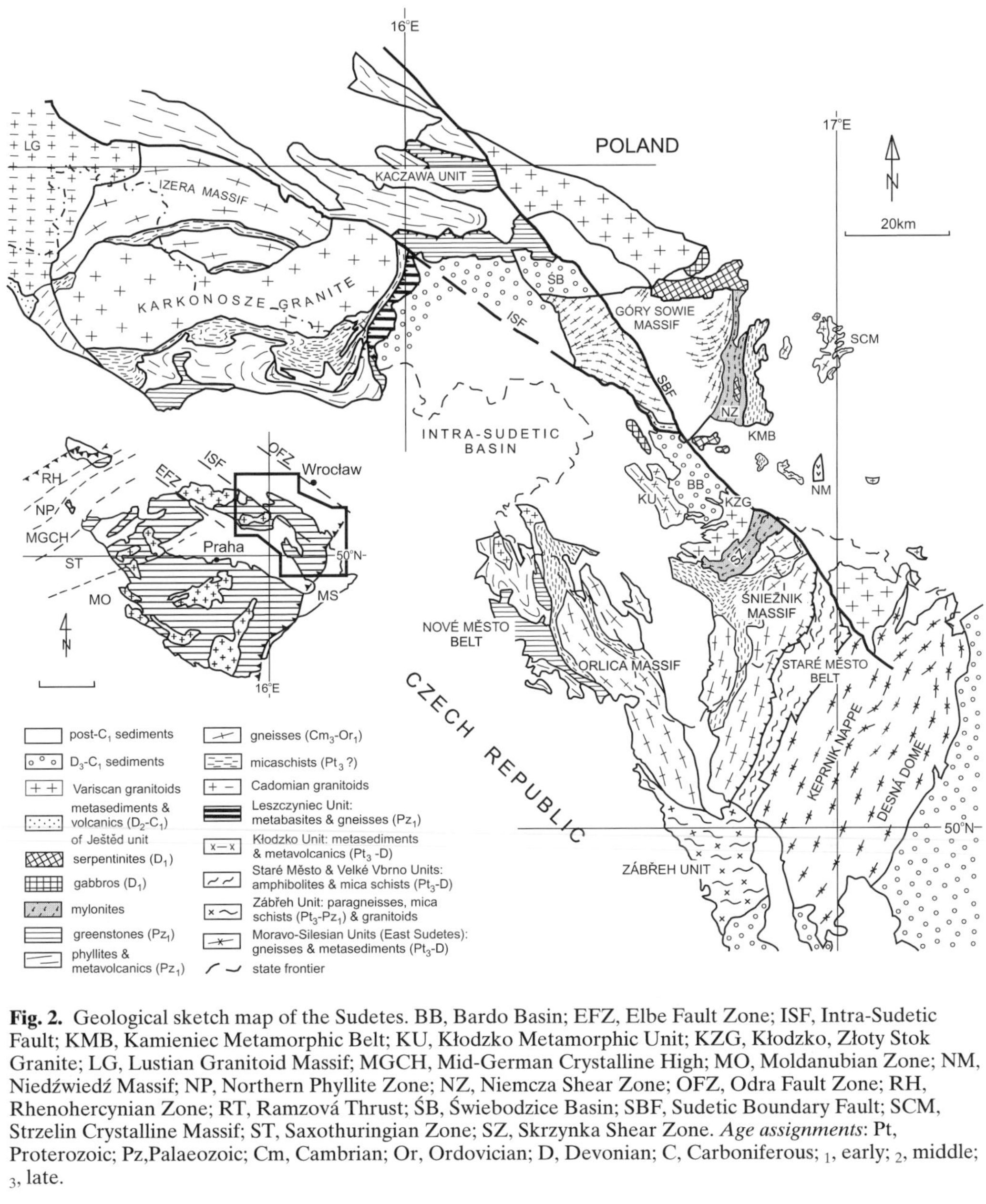

Fig. 2. Geological sketch map of the Sudetes. BB, Bardo Basin; EFZ, Elbe Fault Zone; ISF, Intra-Sudetic Fault; KMB, Kamieniec Metamorphic Belt; KU, Kłodzko Metamorphic Unit; KZG, Kłodzko, Złoty Stok Granite; LG, Lustian Granitoid Massif; MGCH, Mid-German Crystalline High; MO, Moldanubian Zone; NM, Niedźwiedź Massif; NP, Northern Phyllite Zone; NZ, Niemcza Shear Zone; OFZ, Odra Fault Zone; RH, Rhenohercynian Zone; RT, Ramzová Thrust; ŚB, Świebodzice Basin; SBF, Sudetic Boundary Fault; SCM, Strzelin Crystalline Massif; ST, Saxothuringian Zone; SZ, Skrzynka Shear Zone. *Age assignments*: Pt, Proterozoic; Pz,Palaeozoic; Cm, Cambrian; Or, Ordovician; D, Devonian; C, Carboniferous; $_1$, early; $_2$, middle; $_3$, late.

A controversy has persisted since the 1920s as to whether the West and Central Sudetes are an extension of the Saxothuringian Zone of the German Variscides (Kossmat 1927) or pertain to a distinct crustal domain, referred to as Lugicum (Suess 1926; Stille 1951). The East Sudetes have been included in the Moravo-Silesian Zone of the Variscides (Suess 1912, 1926), which is often believed either to continue into the Rhenohercynian Zone in Germany (e.g. Engel & Franke 1983; Franke 1989) or to represent a fragment of a separate, SE branch of the Variscan Belt (e.g. Dvořák & Paproth 1969; Matte 1986, 1991). The Moravo-Silesian Zone is underlain by crystalline Cadomian basement of the Brunovistulian Block (Dudek 1980; Schulmann & Gayer 2000; Friedl *et al.* 2000).

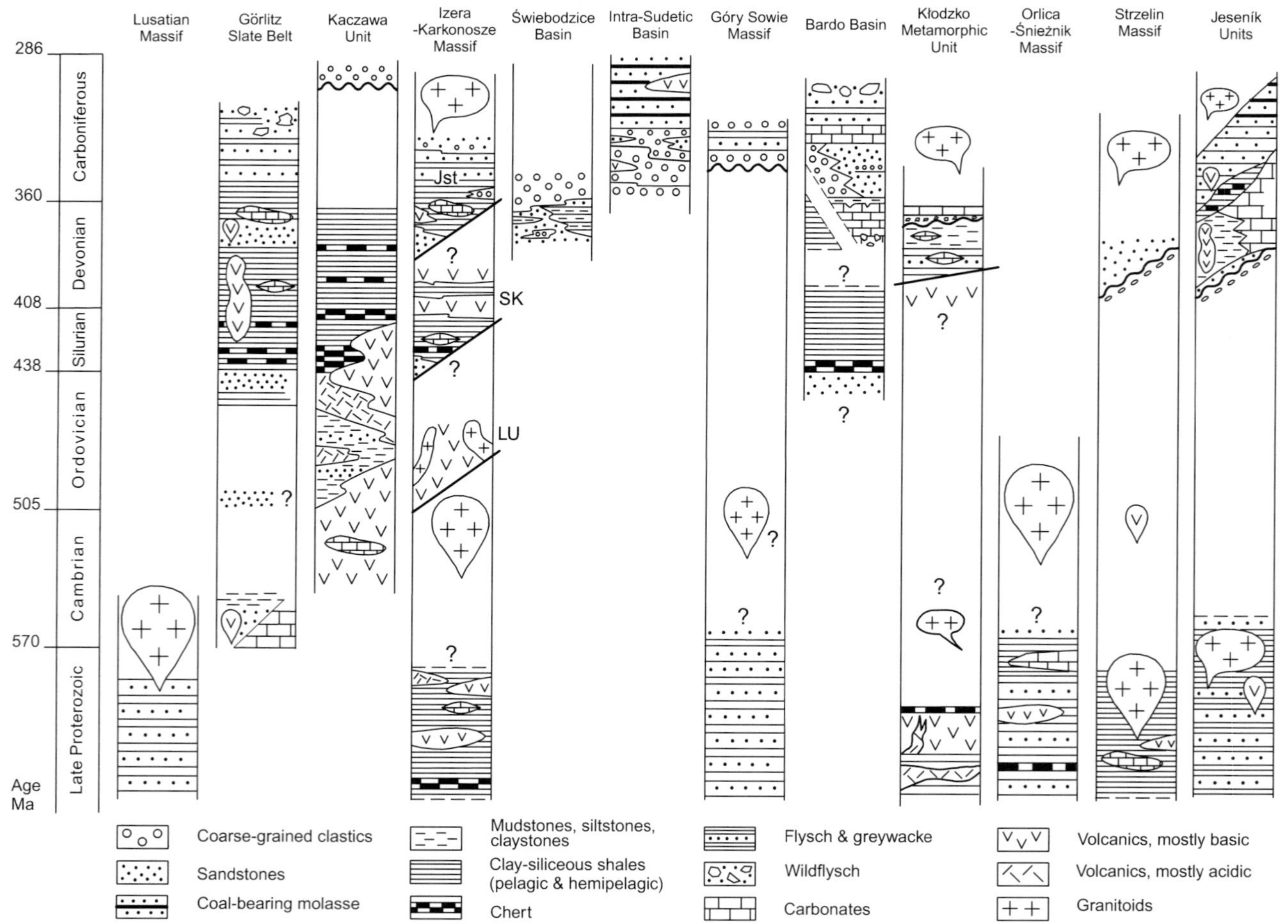

Fig. 3. Simplified stratigraphic columns of the main structural units of the Sudetes (modified from Aleksandrowski *et al.* 2000). Metamorphic rocks are represented by their sedimentary and igneous protoliths. LU, Leszczyniec Unit; SK, South Karkonosze Unit; Jst, Ještěd Unit.

Geology of the Sudetes: an overview

Below, we review the basic geological features of particular Sudetic structural units from NW to SE. Units are more comprehensively described where their specific geological aspects are significant for possible palaeotectonic reconstructions and understanding the sequence of terrane accretion, or where up-to date descriptions in the existing literature are lacking.

West Sudetes

The West Sudetes extend from the Lusatian Massif in the NW to the eastern rims of the Karkonosze-Izera Massif and the Kaczawa Unit (Figs 1 and 2).

Lusatian Massif and Görlitz Slate Belt. The Lusatian Massif comprises Cadomian, Neoproterozoic to Early Cambrian granitoids within non- to low-grade metamorphic Neoproterozoic turbiditic greywackes (e.g. Kröner *et al.*1994; Linnemann *et al.*1998). The Görlitz Slate Belt (e.g. Urbanek *et al.*1995), sometimes inappropriately referred to as the 'Görlitz Syncline', adjoins the Lusatian Massif in the NE and contains fragments of dismembered Lower Cambrian to Lower Carboniferous sequences partly embedded in Lower Carboniferous flysch (cf. Linnemann & Buschmann 1995). Its tectonic fabric seems to be analogous to that of the Kaczawa Unit.

Kaczawa Unit. The Kaczawa Unit comprises several folded thrust-sheets composed of numerous thrust slices and significant mélange bodies. The thrust sheets and mélanges involve various fragments of a sedimentary–volcanic succession of Late Cambrian/Ordovician through Late Devonian (Famennian) age (Baranowski *et al.* 1987, 1990). The Kaczawa succession is made up of low-grade metamorphosed (in blueschist overprinted by greenschist facies) siliciclastics, volcaniclastics, carbonates, basic and acid volcanics, pelagic clay and siliceous shales and flysch, and is interpreted as an end Devonian/Early Carboniferous accretionary prism complex (Baranowski *et al.* 1987, 1990; Kryza & Muszyński 1992; Collins *et al.* 2000; Seston *et. al.* 2000). The lower part of the Kaczawa succession, comprising Cambrian?–Ordovician shallow marine sedimentary rocks and bimodal volcanics of within-plate geochemical signature, is interpreted to have accumulated in an initial rift underlain by continental crust (Baranowski *et al.* 1990; Kryza & Muszyński 1992; Kryza 1993; Furnes *et al.* 1994; Seston *et al.* 2000). The upper part of the succession is dominated by a thick monotonous sequence of tholeiitic E-MORB to N-MORB pillowed basalts accompanied by black graptolitic shales in the Silurian and by siliceous and clayey slates and cherts in the Devonian parts of the profile. It was probably emplaced in an oceanic rift environment (Furnes *et al.* 1994).

The high pressure metamorphism is thought to have taken place by about 360 Ma, by analogy with rocks dated in the east Karkonosze suture by Maluski & Patočka (1997). The Kaczawa thrust sheets were emplaced during latest Devonian(?)-earliest Carboniferous(?) times in a top-to-NW to WNW shearing regime synchronous with greenschist facies metamorphism. Subsequently (at least in the southeastern part of the unit) they were affected by extensional collapse of top-to-ESE kinematics and folded and refolded 2–3 times in progressively more brittle and cool conditions (Kryza *et al.*1998; Seston *et al.* 2000), probably contemporaneously with recurrent strike-slip displacements on the adjacent Intra-Sudetic Fault (Aleksandrowski 1995; Aleksandrowski *et al.*1997).

The Karkonosze-Izera Massif. The Karkonosze-Izera Massif includes the Karkonosze Granite Pluton, dated at 329 ± 17 Ma (Rb–Sr whole rock isochron; Duthou *et al.* 1991) and its metamorphic envelope (Fig. 4). The envelope comprises four different structural units. From base to top these are: (1) the Izera-Kowary, (2) Ještěd, (3) South Karkonosze and (4) Leszczyniec, the two latter units containing a tectonic suture (for a comprehensive review see Mazur & Aleksandrowski 2001*a*).

The Izera-Kowary Unit is composed mainly of the Upper Cambrian/Lower Ordovician (Borkowska *et al.* 1980; Oliver *et al.* 1993; Korytowski *et al.* 1993) Izera (Rumburk) Granite, in most part transformed by a subsequent, Late Devonian to Early Carboniferous deformation into the Izera/Kowary Granite Gneiss and Gneiss (equivalent to the Krkonoše Gneiss on the Czech side of the massif, dated at around 500 Ma by Oliver *et al.* 1993 and Kröner *et al.* 2001). Its other significant component is mica schist representing remains of its Neoproterozoic(?) envelope (Velká Úpa Group on the Czech side and Czarnów Formation on the Polish side; Chaloupský 1965; Teisseyre 1973). These rocks underwent medium pressure metamorphism under upper greenschist – lower amphibolite facies conditions (Żaba 1984; Oberc-Dziedzic 1987; Kryza & Mazur 1995). To the east and south, the Izera-Kowary rocks plunge below

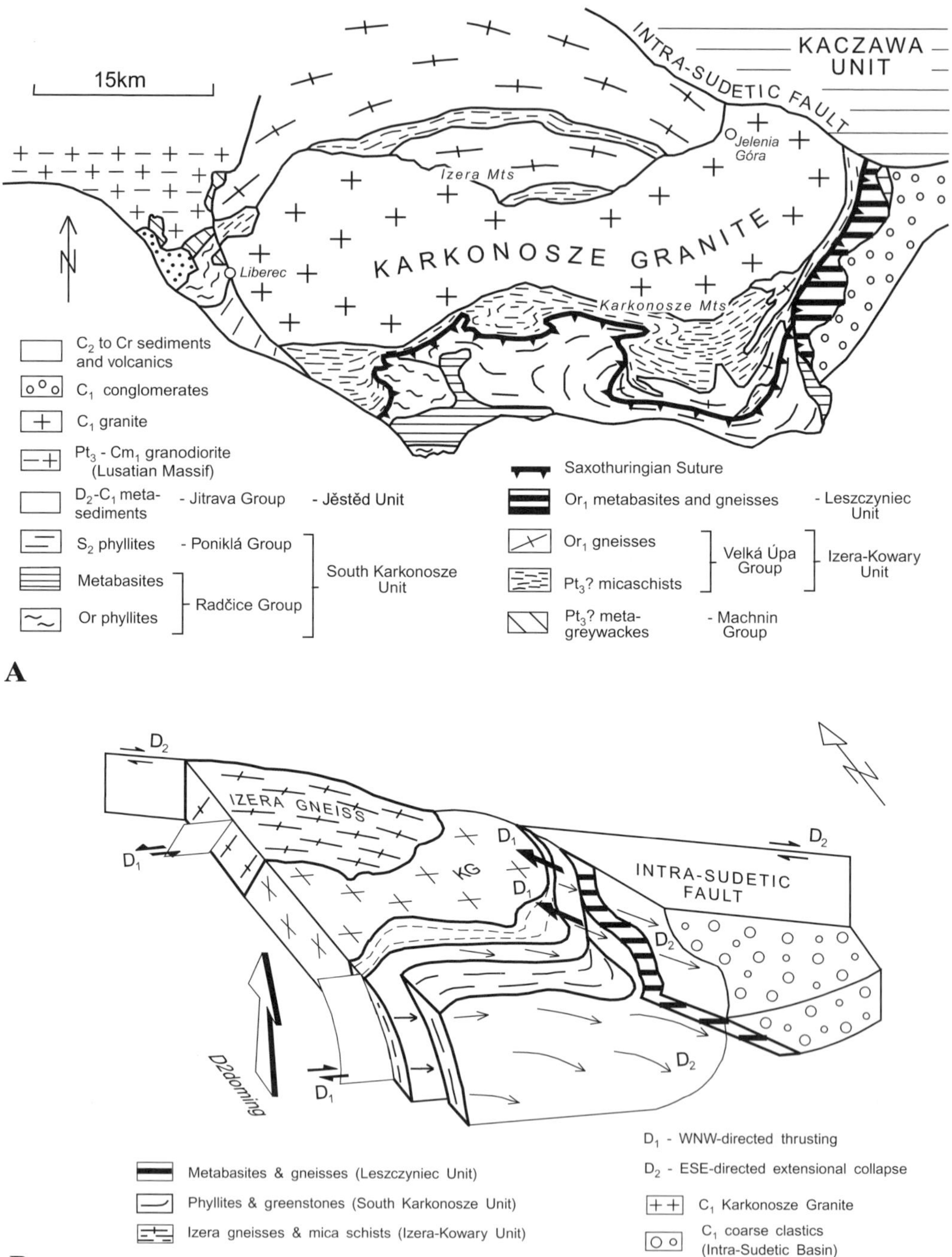

Fig. 4. Geology of the Karkonosze-Izera Massif: (**a**) sketch map, (**b**) block diagram showing deformation partitioning and sense of tectonic transport during deformation events D_1 and D_2 (modified from Aleksandrowski *et al.* 1997). *Age assignments* as in Figure 2.

metamorphic complexes of the South Karkonosze and the Leszczyniec units affected by HP metamorphism. Towards the west, the Izera-Kowary granite and gneiss reveal intrusive contacts with Cadomian granitoids of the Lusatian Massif (Ebert 1943; Domečka 1970).

The Ještěd Unit, a small fault-bounded block exposed at the SW edge of the Karkonosze-Izera Massif, comprises Middle to Upper Devonian shallow marine to hemipelagic sediments with minor volcanics (Jítrava Group of Chaloupský 1989), subjected to very weak, low temperature–medium pressure metamorphism, passing upwards into greywackes and conglomerates of probably Tournaisian to (?)Early Viséan Culm facies rocks (Chlupáč & Hladil 1992; Chlupáč 1993).

The South Karkonosze Unit comprises several hundred metres of metamorphosed Ordovician to (?)Devonian sedimentary rocks (Chlupáč 1993, 1997) accompanied by bimodal, mostly basic volcanics (Figs 3 and 4). Described as the Poniklá and Radčice groups on the Czech side (Chaloupský 1989) and as the Niedamirów Formation on the Polish side (Kryza & Mazur 1995), the sequence contains highly differentiated marine sedimentary rocks and is characterized by close proximity of various facies assemblages typical of a neritic through hemipelagic to pelagic environment. The metavolcanic rocks of the Radčice Group, including the Železný Brod Complex and the Rýchory Mountains metavolcanic succession, and those of the Poniklá Group, represent a differentiated magmatic suite, ranging from predominantly felsic rocks (dated at 501 +/–8 Ma using Rb–Sr whole rock method; Bendl & Patočka 1995) of within-plate geochemical signature to basic lavas and pyroclastics of P-type MORB affinities (Bendl *et al.* 1997; Patočka & Smulikowski 1998, 2000). The basic metavolcanics of the Radčice Group in the Železný Brod and Rýchory Mountains preserve a record of early blueschist facies metamorphism (Wieser 1978; Cháb & Vrána 1979; Guiraud & Burg 1984; Kryza & Mazur 1995; Smulikowski 1995; Patočka *et al.* 1996) dated at about 360 Ma, followed by a greenschist overprint of about 340 Ma ($^{40}Ar/^{39}Ar$ method; Maluski & Patočka 1997). The South Karkonosze Unit probably represents a tectonically dismembered, once vertically continuous and laterally diversified sedimentary succession of an extensive marine basin (Mazur & Aleksandrowski 2001*a*).

The Leszczyniec Unit is represented by a differentiated suite of mafic and felsic rocks of volcanic and plutonic origin (Teisseyre 1973; Kryza *et al.* 1995; Mazur & Kryza 1995*b*), dated at *c.* 500 Ma (U–Pb zircon method; Oliver *et al.* 1993) and showing N-MORB affinities (Kryza *et al.* 1995; Winchester *et al.* 1995). The Leszczyniec Unit tapers out to the south and does not continue into the Rýchory Mountains (Mazur 1995; Mazur & Aleksandrowski 2001*a*).

The concept of a nappe structure for the Karkonosze-Izera Massif was introduced by Kodym & Svoboda (1948) and partly accepted by Oberc (1961, 1972). A new nappe model for the massif was proposed in several recent papers (Mazur 1995; Mazur & Kryza 1996; Seston *et al.* 2000; Mazur & Aleksandrowski 2001*a*). The nappe structure is inferred from contrasting metamorphic paths (Kryza & Mazur 1995) and the record of different tectonic settings (Mazur & Aleksandrowski 2001*a*) shown by the four constituent units of the Karkonosze-Izera Massif. The Izera-Kowary Unit (together with the adjacent Lusatian Massif) represents pre-Variscan continental crust of Saxothuringian affinity. A small fragment of its original sedimentary cover, probably parautochthonous (or representing the lowermost nappe?) is the Ještěd Unit, tectonically sandwiched between the overlying South Karkonosze Unit and the underlying Izera-Kowary rocks. A structurally higher position is occupied by the South Karkonosze Unit, probably composed of lower order thrust sheets or slices with mutually similar structural and metamorphic histories. The uppermost position in the pile is held by the Leszczyniec Unit, characterized by the NNE–SSW structural grain, entirely different from the WNW–ESE structural trend in the remaining part of the Karkonosze-Izera Massif (Mazur & Aleksandrowski 2001*a*).

The South Karkonosze and Leszczyniec nappes, comprising blueschist facies rocks and MORB-type magmatic complexes, are tectonically emplaced on top of the parautochthonous Izera-Kowary unit of continental basement features. The nappe pile was formed at the turn of Late Devonian/Early Carboniferous time due to northwestward thrusting and in general, shows metamorphic inversion (Mazur & Aleksandrowski 2001*a*). It was subsequently modified by Early Carboniferous, southeasterly-directed extensional collapse and, eventually, intruded by the Karkonosze Granite (Mazur 1995; Mazur & Aleksandrowski 2001*a*).

Central Sudetes

The Central Sudetes comprise structural units that occur east of the Karkonosze-Izera Massif and the Kaczawa Unit and west of the Velké

Vrbno and Branná units of the East Sudetes as well as west of the Strzelin Crystalline Massif in the eastern part of the Fore-Sudetic Block.

Góry Sowie Massif. The Góry Sowie Massif is made up of amphibolite facies paragneisses thought to have been derived from uppermost Proterozoic protoliths (Gunia 1985; but see Moczydłowska 1995, who questions the fossil determinations of Gunia). They are accompanied by migmatites and orthogneisses and contain small ultramafic and granulitic bodies (Kryza 1981; Żelaźniewicz 1990) recording pressure and temperature conditions typical of mantle and lower crustal levels. Granulite formation at pressures of 18–20 kbar and temperatures above 900 °C (Kryza *et al.* 1996) was dated, using the U–Pb zircon method, at about 400 Ma (O'Brien *et al.* 1997), corresponding closely with a Sm–Nd age of the associated, mantle-derived ultramafics (Brueckner *et al.* 1996). This high grade metamorphic event was followed by relatively rapid uplift of the massif, associated with contemporaneous decompression under amphibolite facies conditions. The time interval for this medium pressure–high temperature metamorphism is estimated at around 385–370 Ma on the basis of U–Pb monazite, xenotime and zircon ages and Rb–Sr data (van Breemen *et al.* 1988; Bröcker *et al.* 1998; Timmermann *et al.* 2000). The final exhumation of the Góry Sowie Massif is recorded by gneissic pebbles in uppermost Devonian–lowermost Carboniferous conglomerates of the Świebodzice Basin (Porębski 1981, 1990).

Świebodzice Basin. The Upper Devonian (upper Frasnian–Famennian) to lowermost Carboniferous Świebodzice Basin ('Depression') succession, up to 4000 m thick, occurs in a small, rhomboidal, fault-bounded block, consisting mainly of polymict conglomerates. In the lower part, they are interbedded with thick fossiliferous mudstones containing sandstone turbidites and rare limestone lenses. These deposits reflect mostly gravity-flow sedimentation within a slope-type fan-delta complex which invaded a rapidly subsiding basin, bounded to the SW and south by an active fault system, of probable strike-slip displacement component (Porębski 1981, 1990).

Bardo Basin. The Bardo Basin succession comprises unmetamorphosed Upper Devonian limestone and Lower Carboniferous flysch strata, capped by wildflysch deposits (Wajsprych 1978, 1986). The wildflysch contains large olistoliths of Lower Palaeozoic and Devonian deep marine sediments (Haydukiewicz 1990). The Bardo Basin succession was folded at the turn of Early/Late Carboniferous into east–west trending folds and intruded by the Kłodzko-Złoty Stok Granitoid Pluton. Late Carboniferous refolding produced NE–SW to north–south trending folds superimposed on the older east–west structures (Oberc 1972).

Central Sudetic Ophiolite. The Central Sudetic Ophiolite comprises several mafic/ultramafic bodies that crop out along the northern (Ślęża Ophiolite), eastern (Szklary and Braszowice ultramafic bodies) and southwestern (Nowa Ruda Gabbro-Diabase Massif) rims of the Góry Sowie Massif. The Devonian exhumation of the Central Sudetic Ophiolite is constrained by the occurrence of a pre-Upper Devonian erosional surface cutting the Nowa Ruda Massif. This situation refers directly only to the Nowa Ruda Ophiolite, but may concern also the other circum-Góry Sowie ophiolites, as all these ophiolite bodies show similar geochemical characteristics (Pin *et al.* 1988; Gunia 1997) and are probably related.

For several years an inferred Devonian age of obduction and exhumation of the Ślęża Ophiolite contradicted the first isotopic dating of its protolith. Whole-rock Sm–Nd analysis of six samples of the ophiolitic gabbro indicated an age of 353 +/– 21 Ma, and ten samples of gabbro from the Nowa Ruda Gabbro-Diabase Massif, on the SW side of the Góry Sowie Massif, yielded an age of 351 ± 16 Ma (Pin *et al.* 1988). However, Oliver *et al.* (1993) determined a Silurian igneous age (420 + 20/–2 Ma) for the Ślęża ophiolite on the basis of U–Pb zircon measurements on a gabbro sample. A recent U–Pb age of 400 + 4/–3 Ma (Żelaźniewicz *et al.* 1998) was obtained on abraded zircons from rodingitized plagiogranite of the Ślęża Massif. The latter date seems to be the most reliable approximation of the age of ophiolite igneous crystallization, whereas the Sm–Nd ages may reflect subsequent overprints of tectonic emplacement and cooling.

Kłodzko Metamorphic Unit. The Kłodzko Metamorphic Unit (Kłodzko Massif) comprises six tectonic subunits (Figs 5 and 6), representing (from base to top):

(1) the Mały Bożków Subunit comprising Middle Devonian (Givetian) progradational shelf sequence (Hladil *et al.* 1999);
(2) a mélange body of unknown age, defining the Łączna Subunit (Mazur & Kryza 1999);
(3) the Bierkowice Subunit, composed of Palaeozoic(?) mafic volcanics with intraplate basalt geochemical signature (Kryza *et al.* 2000);

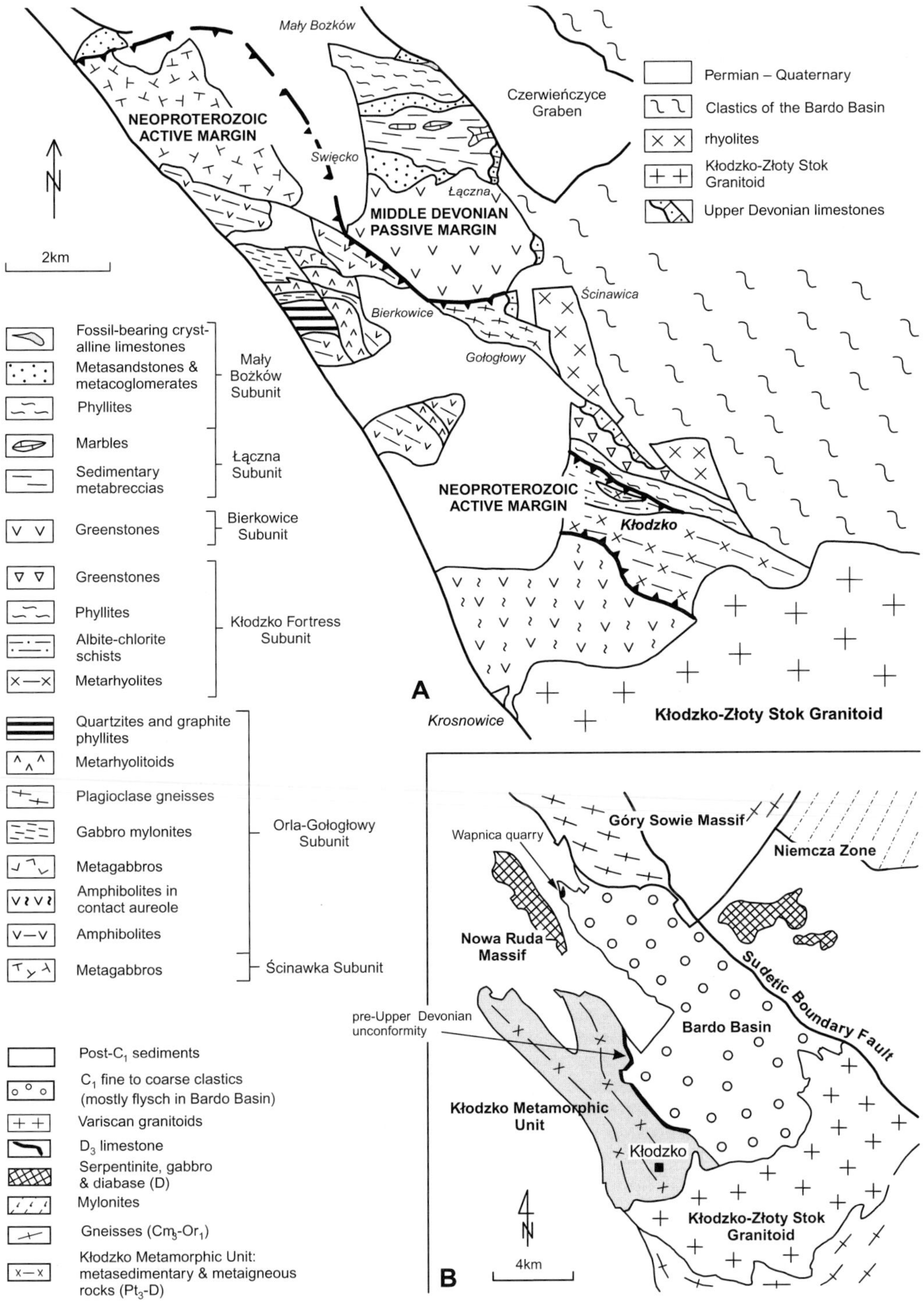

Fig. 5. Sketch geological map showing palaeotectonic interpretation of the Kłodzko Metamorphic Unit (after Mazur, unpublished data). *Inset*: geological setting within the Góry Sowie–Kłodzko domain of Central Sudetes.

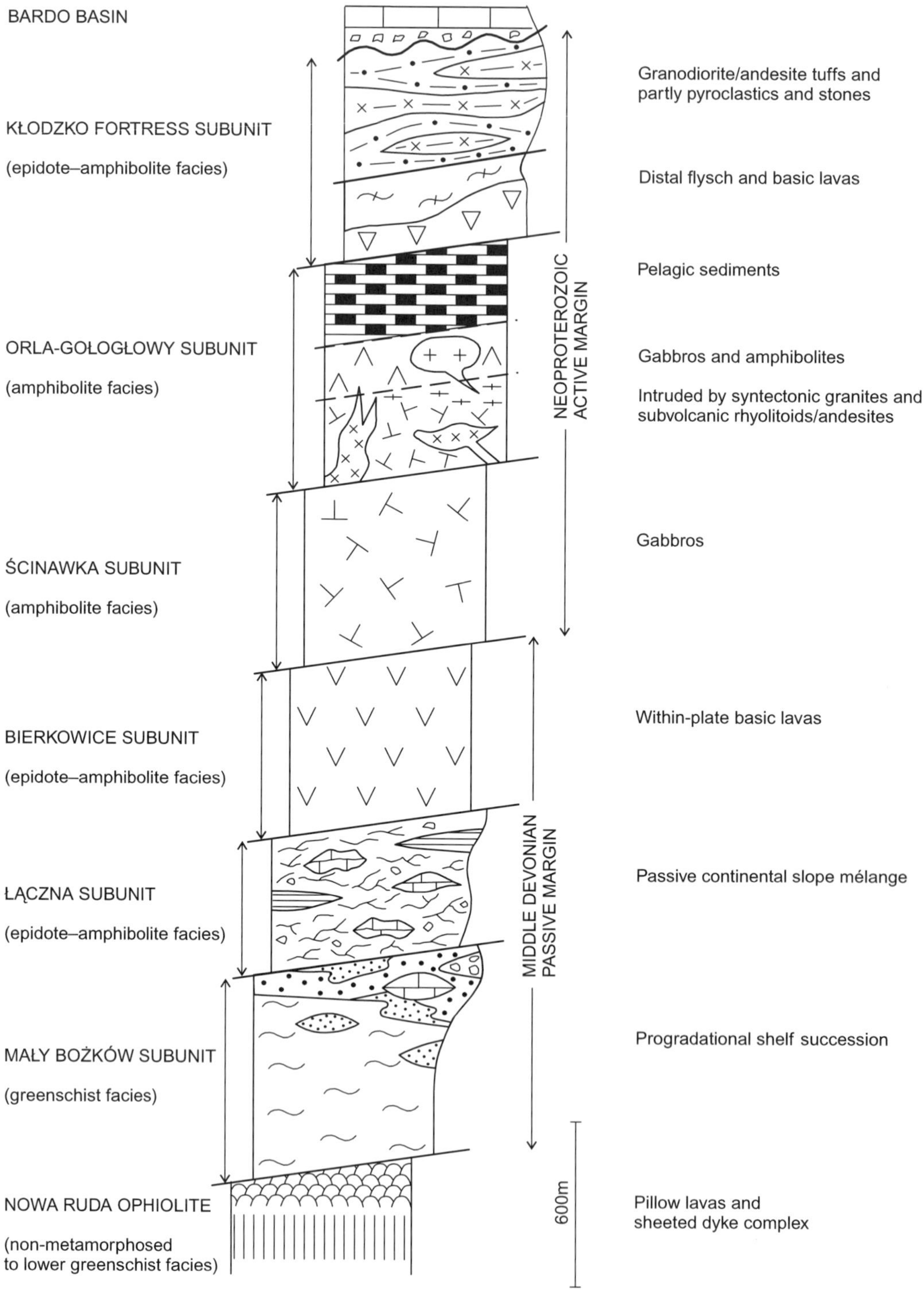
BARDO BASIN
KŁODZKO FORTRESS SUBUNIT
(epidote–amphibolite facies)
ORLA-GOŁOGŁOWY SUBUNIT
(amphibolite facies)
ŚCINAWKA SUBUNIT
(amphibolite facies)
BIERKOWICE SUBUNIT
(epidote–amphibolite facies)
ŁĄCZNA SUBUNIT
(epidote–amphibolite facies)
MAŁY BOŻKÓW SUBUNIT
(greenschist facies)
NOWA RUDA OPHIOLITE
(non-metamorphosed
to lower greenschist facies)
NEOPROTEROZOIC
ACTIVE MARGIN
MIDDLE DEVONIAN
PASSIVE MARGIN
600m
Granodiorite/andesite tuffs and
partly pyroclastics and stones
Distal flysch and basic lavas
Pelagic sediments
Gabbros and amphibolites
Intruded by syntectonic granites and
subvolcanic rhyolitoids/andesites
Gabbros
Within-plate basic lavas
Passive continental slope mélange
Progradational shelf succession
Pillow lavas and
sheeted dyke complex

(4) MORB-type gabbro of the Ścinawka Subunit (Kryza *et al.* 2000), showing affinities to
(5) the Orla-Gołogłowy Subunit, composed of gabbro and MORB-type mafic volcanics (Kryza *et al.* 2000), intruded by granitoids and accompanied by deep marine sediments; some subvolcanic felsic rocks being preliminarily dated as Neoproterozoic (U-Pb method on zircons; K. Turniak 2001, pers. com.), and
(6) distal flysch with basaltic lavas, accompanied by pyroclastic sandstones and dacitic/andesitic tuffs, the latter of Neoproterozoic age (K. Turniak 2001, pers. com.), composing the Kłodzko Fortress Subunit.

Thus, the lower three subunits of the Kłodzko Metamorphic Unit comprise rocks which are, at least in part, of Middle Devonian age, whereas the remaining upper three subunits include Neoproterozoic plutonic and volcanic rocks. The tectonic contacts between the subunits were inferred mostly from their contrasting metamorphic paths (Kryza *et al.* 2000), different lithostratigraphic contents and palaeoenvironmental affinities. The metamorphic grade increases up-profile, from greenschist to amphibolite facies, except for the highest subunit that shows epidote–amphibolite facies metamorphism. All the subunits of the Kłodzko Metamorphic Unit record the same sequence of deformation, in part shared with the Bardo Basin succession: (1) WNW-directed ductile thrusting; (2) folding into east–west-trending folds; and (3) dextral strike-slip shearing along WNW–ESE direction, associated with exhumation in a transpressive regime. These three deformation events were recorded only in the metamorphic rocks of the Kłodzko Unit. The following two events also affected the sedimentary succession of the Bardo Basin: (4) WNW–ESE-directed sinistral strike-slip synchronous with the emplacement of the Kłodzko-Złoty Stok Granitoid intrusion; (5) intense folding of the Bardo Basin and the adjacent part of the Kłodzko Metamorphic Unit due to north–south compression. The first deformation event, the top-to-WNW thrusting, resulted in nappe formation in the Kłodzko Metamorphic Unit. The entire nappe pile rests on top of the essentially unmetamorphosed Nowa Ruda Ophiolite and is unconformably covered by the younger sedimentary sequence of the Bardo Basin.

Intra-Sudetic Basin. The Intra-Sudetic Basin is a syn- to post-orogenic, relatively large (60 × 25 km) intramontane basin, initiated during Tournaisian time. The basin remained active throughout the Carboniferous and Permian periods, accumulating molasse up to 11 000 m (Nemec *et al.* 1982; Dziedzic & Teisseyre 1990). The Early Carboniferous sedimentation reflected rapid orogenic uplift of the basin's surroundings, accompanied by extensional collapse and exhumation of the freshly deformed Sudetic crystalline complexes.

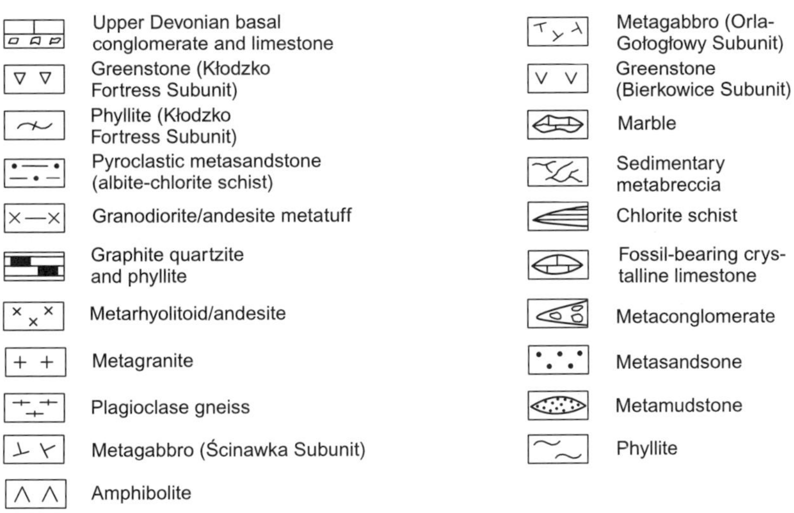

Fig. 6. Lithotectonic log of the Kłodzko Metamorphic Unit (after Mazur, unpublished data).

Orlica-Śnieżnik Massif. The Orlica-Śnieżnik Massif is composed mostly of amphibolite-grade orthogneisses and subordinate staurolite-grade variegated schists of presumably Neoproterozoic protoliths with eclogite and granulite lenses (e.g. Don *et al.* 1990). Emplacement of the magmatic precursor to all textural varieties of orthogneisses was dated at approximately 500 Ma, using various U–Pb and Pb/Pb methods (Oliver *et al.* 1993; Turniak *et al.* 2000; Kröner *et al.* 2001). The Carboniferous tectonothermal phenomena included medium- to high-grade metamorphism and intense synmetamorphic deformation, accompanied by exhumation of high-grade rocks. High pressure and temperature granulite facies metamorphism is dated at around 360–369 Ma, using the U–Pb method on zircons (Klemd & Bröcker 1999), and Sm–Nd whole rock ages for eclogites range from 350 to 330 Ma (Brueckner *et al.* 1991). The age of a later high temperature–medium or low pressure phase of metamorphism, which resulted in partial migmatization of Orlica-Śnieżnik rock complexes has been determined by SHRIMP dating of metamorphic rims on zircons at about 342 Ma (Turniak *et al.* 2000). Ar/Ar cooling ages for the Orlica-Śnieżnik Massif are in the range of 340–330 Ma (Steltenpohl *et al.* 1993; Maluski *et al.* 1995; Marheine *et al.* 2000). The structure of the Orlica-Śnieżnik Massif is poorly understood. The common view is that the massif represents a gneissic dome, in which the gneisses crop out in antiforms, whereas 'mantling' schists are preserved in synforms. There is a little evidence, however, that the Orlica-Śnieżnik Massif is composed of a number of folded thrust sheets, as are the East Sudetic units adjacent to the SE. This question awaits further research.

Nové Město Belt. The Nové Město Belt is located in the western Góry Orlickie (Orlica Mts) and adjoins the Orlica-Śnieżnik Massif to the NE (Fig. 2). It is composed mostly of phyllites, greenstones and amphibolites, whose ages remain unknown, but which are traditionally viewed as Late Proterozoic by comparison to the Teplá-Barrandian domain of the Bohemian Massif (Chaloupský *et al.* 1995). The amphibolites crop out in an approximately continuous belt, 1.5 to 5 km wide, that follows the contact with the Orlica-Śnieżnik Massif (Fig. 7). According to the geochemical results of Opletal *et al.* (1990) and Floyd *et al.* (1996), they show MORB-type affinities. On the eastern side, the contact is mostly accompanied by a mica schist belt up to 2 km wide. Two late-tectonic granitoid intrusions, the Olešnice and Kudowa plutons, are emplaced in the contact zone. A third late-tectonic intrusion, the Nový Hrádek Granodiorite, is entirely surrounded by phyllites in plan view. Small granitoid bodies are also common in the contact zone further to the south. Metamorphic grade increases in the Nové Město Belt from greenschist facies in the west, to amphibolite facies along the boundary with the Orlica-Śnieżnik Massif. Metamorphic isograds are roughly parallel to the boundary with the Orlica-Śnieżnik Massif and the contacts with the granitoid intrusions.

An important structural discontinuity along the contact between the Nové Město Belt and the Orlica-Śnieżnik Massif is suggested by the structural study by Fajst (1976), who documented divergent structural trends on both sides of the contact. The Orlica-Śnieżnik Massif is characterized by approximately north–south structural grain whereas the Nové Město Belt shows NW–SE structural trends. The two structural patterns are locally superimposed within the contact zone. Across the entire NW part of the Orlica-Śnieżnik Massif and the Nové Město Belt, they are uniformly overprinted by younger east–west trending folds.

The fabric of the Nové Město Belt bears a record of D_1 top-to-ESE ductile thrusting. The D_2 kinematics are NNW–SSE-directed dextral strike-slip at the contact of the Nové Město Belt with the Orlica-Śnieżnik Massif mica schists and gneisses (the Uhřinov Shear Zone), changing gradually into top-to-NNE shearing in the core of the Massif. The foliation in the whole area is refolded by east–west trending south-vergent F_3 folds.

A primary contact between the Nové Město Belt and the Orlica-Śnieżnik Massif was probably represented by a thrust characterized by top-to-ESE kinematics (Mazur & Aleksandrowski 2001*b*), which juxtaposed phyllites and amphibolites of the Nové Město Belt, differing in their metamorphic grade. The original contact was subsequently folded and reactivated as a major dextral shear zone. Dextral displacements resulted in the subsequent juxtaposition of the Nové Město and the Orlica-Śnieżnik units. A similar deformation sequence including top-to-ESE thrusting and succeeding dextral shear was described by Přikryl *et al.* (1996) further to the east in the Orlica-Śnieżnik Massif. The Nové Město Belt partly continues to the SE into the Zábřeh metamorphic belt (Opletal *et al.* 1980).

Niemcza and Skrzynka Shear Zones. The Niemcza Shear Zone, extending along the eastern edge of the Góry Sowie Massif (see Fig. 8), was interpreted by Scheumann (1937) to contain mylonitized gneisses. Based on detailed

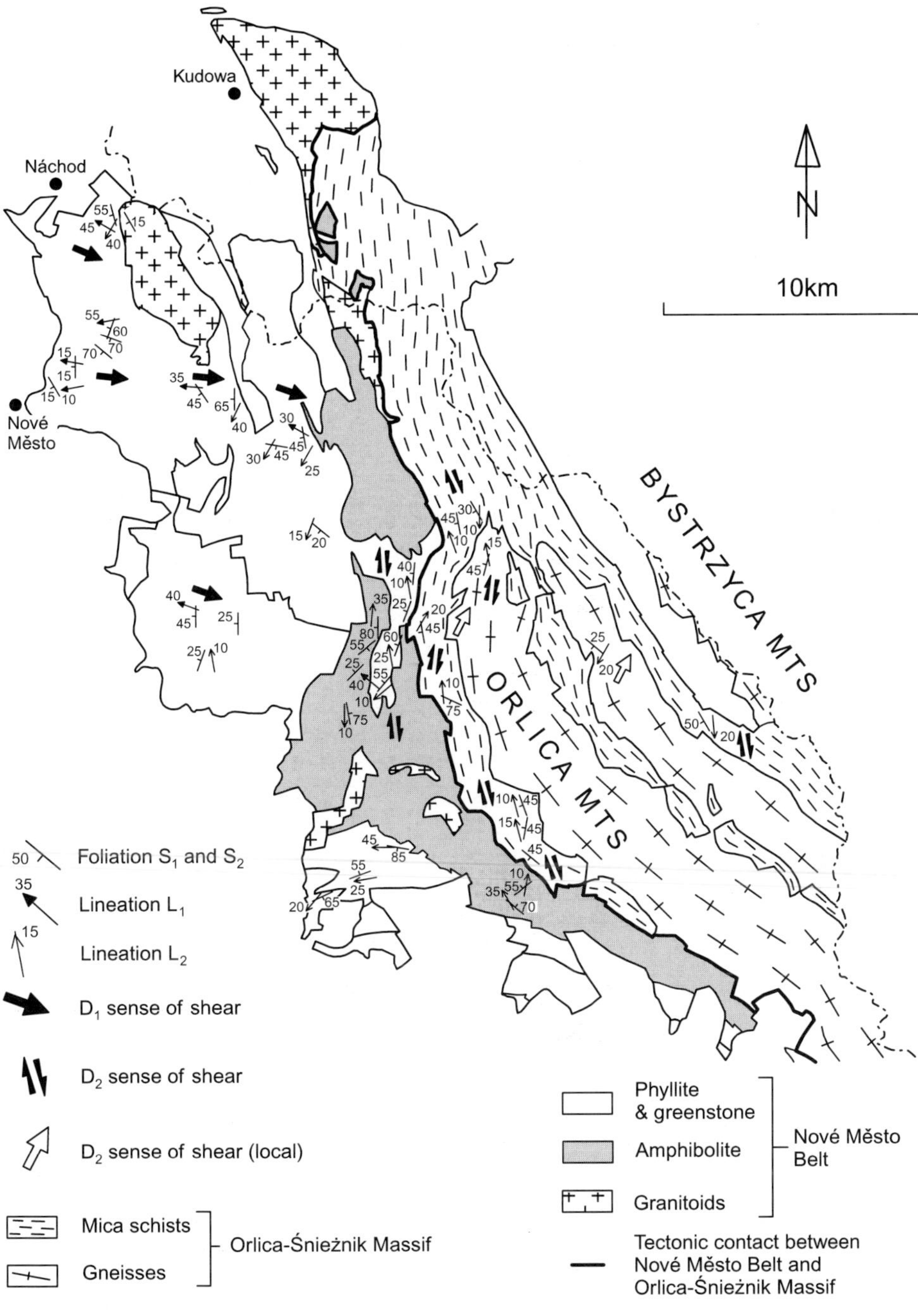

Fig. 7. Schematic geological and tectonic map of the Nové Město Belt (after Aleksandrowski & Mazur, unpublished data).

study, Mazur & Puziewicz (1995) showed that the Niemcza Zone represented a 5 km wide, left-lateral strike-slip ductile shear belt, separating the Góry Sowie Massif from the Kamieniec Metamorphic Belt. The Niemcza mylonites were derived from the Góry Sowie Gneiss (Scheumann 1937; Mazur & Puziewicz 1995; but see e.g. Franke & Żelaźniewicz 2000, for an

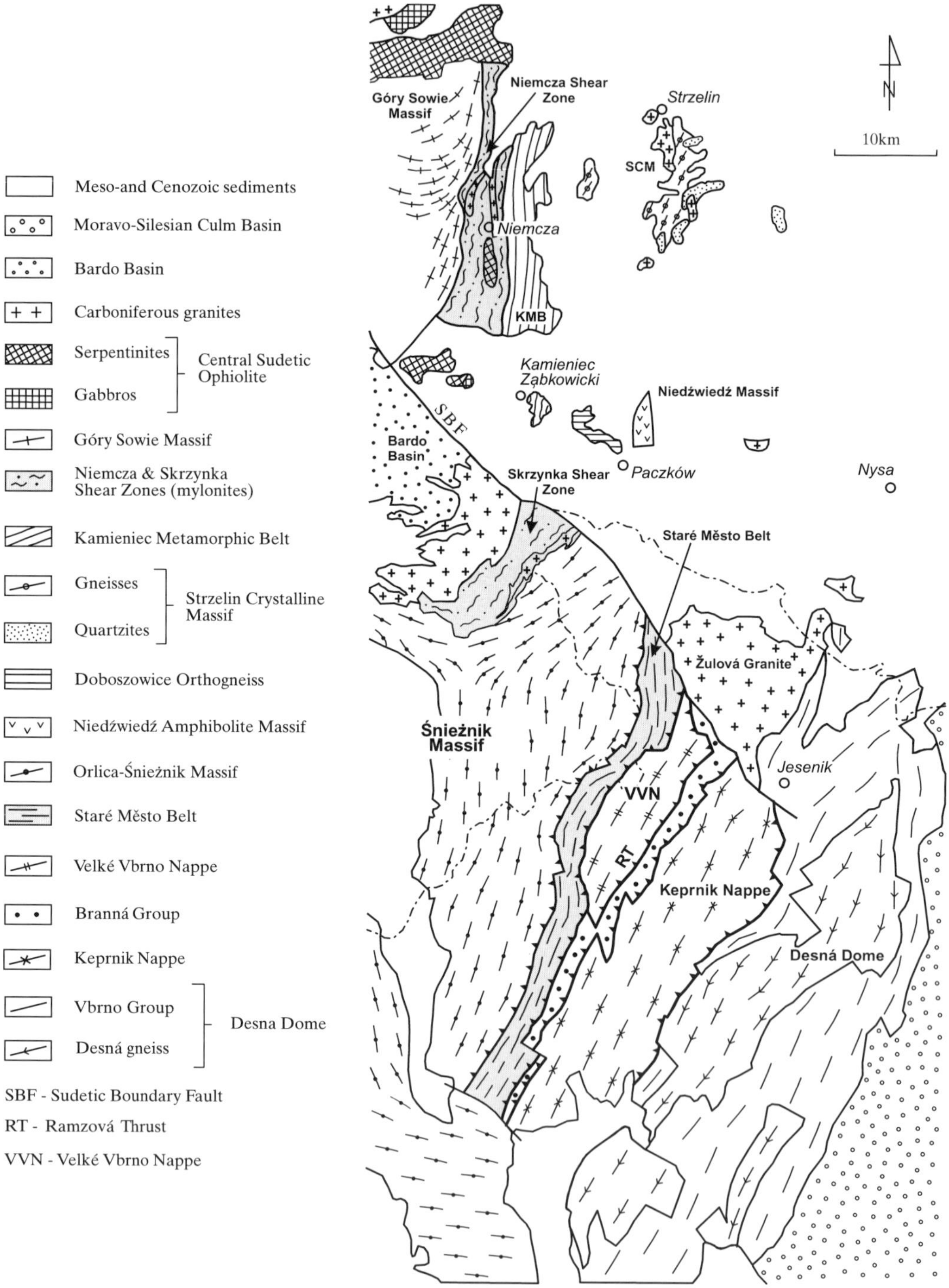

Fig. 8. Geological sketch map of the East Sudetes.

alternative interpretation of the Niemcza Zone rocks as metagreywackes, following Bederke 1929) and include minor lenses of non-mylonitic gneisses, amphibolites and quartzo-graphitic schists. The mylonites occur as high- and low-temperature varieties, formed in amphibolite and greenschist facies conditions respectively (Mazur & Puziewicz 1995). The Niemcza Shear Zone contains numerous small bodies of undeformed to little-deformed granitoids and syenites/diorites. One late kinematic granodiorite was dated, using the U–Pb method on zircons, at about 340 Ma (Oliver *et al.* 1993; Kröner & Hegner 1998). Earlier fabric is completely obliterated in the mylonites, so that no superposition of structures can be recorded. However, in the southern part of the adjoining Kamieniec Metamorphic Belt, older structures showing a top-to-the-NE sense of shear on a shallow westerly dipping foliation are locally preserved and overprinted by a common top-to-the-SW or sinistral fabric. Hence, by analogy, the left-lateral motion in the Niemcza Zone (or top-to-the-south motion in places where the foliation is shallow-dipping) seems to be younger and superimposed on the earlier fabric of regional extent, related to the Early Carboniferous top-to-the NE thrusting and dextral shearing on the SE edge of the Bohemian Massif (Rajlich, 1987; Schulmann *et al.* 1991; Fritz & Neubauer, 1993).

The Skrzynka (or Złoty Stok-Skrzynka) Shear Zone, trending NNE–SSW to NE–SW (Fig. 8), is approximately 12 km long and 4 km wide and constitutes the boundary between the sedimentary rocks of the Bardo Basin and the Kłodzko-Złoty Stok Granitoid Massif to the NW and the Śnieżnik Massif to the SE (e.g. Don *et al.* 1990). The Skrzynka Shear Zone exposes various blastomylonites, mylonites, cataclasites, gneisses and schists. The regional amphibolite-facies metamorphism is thermally overprinted along the contact with the Kłodzko-Złoty Stok Granitoid Massif. The penetrative, steeply dipping foliation within the Skrzynka Zone parallels its boundaries and contains a subhorizontal to shallow SW plunging mineral stretching lineation. Cymerman (1992, 1996) showed that the main deformation in the Skrzynka Zone was a left-lateral, ductile strike-slip non-coaxial shear. The latest stages of the shearing seem to have postdated the emplacement of both the Kłodzko-Złoty Stok pluton and another small granitoid body near to the SE boundary of the Skrzynka Zone (the Jawornik Granitoid), since the margins of the two plutons underwent mylonitization.

Kamieniec Metamorphic Belt. The Kamieniec Metamorphic Belt crops out as a north–south elongated schist belt east of the Góry Sowie Massif and of the Niemcza Shear Zone (Fig. 8). It comprises mica schists containing intercalations of quartzo-feldspathic schists and marbles and subordinate lenses of quartzo-graphitic and amphibolitic schists and eclogites. The Kamieniec Belt consists of two tectonic units with different metamorphic grades. The first unit comprises mica schists containing kyanite, garnet, staurolite and late andalusite porphyroblasts, whereas the second one is composed of mica schists with albite porphyroblasts (Mazur & Józefiak 1999).

Staré Město Belt. The Staré Město Belt forms a tectonic boundary separating the Central from the East Sudetes (or the Lugian domain from the Moravo-Silesian domain). It is a stack of thrust sheets adjacent in the SE to the Orlica-Śnieżnik Massif (Fig. 8). Occupying a NNE–SSW elongated outcrop zone about 40 km long and 4–5 km wide, the Staré Město Belt comprises high-grade metasediments, banded felsic amphibolite rocks with lenses of spinel peridotites at their base, sheared gabbros and a syntectonic tonalite intrusion (Parry *et al.* 1997; Schulmann & Gayer 2000). The peridotite bodies are considered to be tectonic slices extracted from the mantle by thick-skinned thrusting (Parry *et al.* 1997; Schulmann & Gayer 2000). U–Pb zircon dating of the banded amphibolites and metagabbros yielded Cambrian ages (510–500 Ma, Kröner *et al.* 2000). Analogous ages are widespread, as well, in the Orlica-Śnieżnik Massif and in many igneous rock suites of the Central and West Sudetes (Oliver *et al.* 1993; Turniak *et al.* 2000; Kröner *et al.* 2001). East of the Staré Město Belt, however, Neoproterozoic protolith ages (from 684–546 Ma) of the basement orthogneisses and of the paragneisses have been recorded (van Breemen *et al.* 1982; Kröner *et al.* 2000). Thus the Staré Město Belt is the easternmost unit of the Central Sudetes; east of it units of different affinities occur. The Staré Město rocks have been interpreted as products of a Cambro-Ordovician initial rift (Parry *et al.* 1997; Štípská *et al.* 2001; Kröner *et al.* 2000), based on their igneous association. This conclusion is corroborated by crustal contamination of MORB-type amphibolites, suggesting the emplacement of their protoliths in an ensialic rift zone (Floyd *et al.* 1996). At the same time, medium pressure–high temperature granulite facies metamorphism, synchronous with the intrusion of Cambro-Ordovician magmatic suite and associated with pressures of 7–10 kbar and temperatures of 800–850 °C (Štípská *et al.* 2001), must have been accompanied by an exceptionally high heat flow typical of thinned continental crust (Parry *et al.* 1997; Štípská *et al.* 2001). The age of the crustal thinning and the associated extensional tectonics has been determined on metamorphic zircons from high-grade migmatitic metapelite at 507 ± 7 Ma (Kröner *et al.* 2000). The Variscan deformation and metamorphism in the Staré Město Belt, as well as the thrusting of the belt over the East Sudetes, was roughly contemporaneous with the intrusion of a syntectonic tonalite dyke dated at 339 ± 7 Ma, using the Pb/Pb method on zircons (Parry *et al.* 1997).

The Staré Město Belt shows features typical of a suture zone. The juxtaposition of dissimilar crustal domains of the East and Central Sudetes along the Staré Město thrust belt must have been triggered by exhumation of plate fragments, that, earlier, had been subducted westwards, below the active margin of the West Sudetes during Early Carboniferous times (Schulmann & Gayer 2000). The possible continuation of the Staré Město Belt northward, across the Sudetic Boundary Fault, into the Fore-Sudetic area, remains problematic. The scarcity of outcrops emerging from beneath a thick Cenozoic cover prevents any definite correlations, but the Staré Město Belt probably continues into the Niedźwiedź Amphibolite Massif.

Niedźwiedź Amphibolite Massif. Despite the rare outcrops, the Niedźwiedź Massif is relatively well known from borehole data (Cymerman & Jerzmański 1987, Jerzmański 1992). It comprises a 1.5 km-thick succession of MORB-type amphibolites and metagabbros (Awdankiewicz 2001) that were subjected to high-grade metamorphism, locally leading to partial melting of the metabasites (Puziewicz & Koepke 2001). The Niedźwiedź Massif is apparently overthrust from the west by a medium- to high-grade association of paragneisses, amphibolites and hornblende gneisses (Mazur & Józefiak 1999).

East Sudetes

The East Sudetes form part of a collision-related belt of deformation and metamorphism, nearly 50 km wide and 300 km long, that occupies the eastern margin of the Bohemian Massif, from Lower Austria, through Moravia, to Silesia. The belt is composed of nappe piles that crop out from below the upper plate of the collision zone (represented in the south by Moldanubian and in the north by Central Sudetic rock complexes) in three tectonic half-windows: of Thaya (Czech: Dyje), of Svratka (German: Schwarzawa) and of the East Sudetes. The nappes exposed in the two former windows were distinguished as Moravian units, whereas those cropping out in the East Sudetes were termed Silesian units (Suess 1912, 1926; Dudek 1980). Recently, the eastern Bohemian collisional belt, together with an extensive Devonian–Carboniferous sedimentary basin to the east, are usually described under the general name of the Moravo-Silesian Zone, Belt or domain (e.g. Dallmeyer *et al.* 1995; Franke & Żelaźniewicz 2000) or as the Moravian Terrane (Matte *et al.* 1990; Matte 1991). Significant differences between the structure of the collision zone in Moravia and that in Silesia, have been reported, however, in a number of recent papers (e.g. Schulmann *et al.* 1991; Schulmann & Gayer 2000).

In its Sudetic segment, the collision zone is represented by the East Sudetic pile of nappes (Fig. 8), overridden from the west by the Central Sudetic Orlica-Śnieżnik Massif and the Staré Město Belt. From top to bottom, or in plan view from the west to the east, the East Sudetic nappe pile comprises the highest Velké Vbrno Nappe ('upper allochthon' of Schulmann & Gayer 2000) and the Keprník Nappe ('lower allochthon'), separated from each other by the detached Branná Unit. The Keprník Nappe rests in turn on parautochthonous gneisses of the Desná Dome (Schulmann & Gayer 2000) covered by the probably allochthonous Devonian metasedimentary Vrbno Group.

Velké Vrbno Nappe. Directly east of the Staré Město Belt, there occurs the Velké Vbrno Nappe, composed of orthogneisses and metasediments (banded amphibolite rocks, kyanite–staurolite mica schists, graphite schists and quartzites, dolomitic marbles and biotite paragneisses) metamorphosed under upper-amphibolite facies conditions. The gneisses were dated at about 574 Ma (Kröner *et al.* 2000). The presence of relict eclogites (Žáček 1996; Schulmann & Gayer 2000) points to an older stage of high pressure–low temperature metamorphism. A protolith of the supracrustal series must have been a variegated, volcanosedimentary sequence of unknown age, traditionally considered to be Early Palaeozoic (e.g. Květoň 1951).

Branná Unit and Keprník Nappe. The Velké Vbrno Nappe overrides a narrow belt of highly sheared Devonian metasediments named the

Branná Unit, metamorphosed under greenschist facies conditions. It mostly comprises shallow water polymict metaconglomerates, quartzites, crystalline limestones, sericite and sericite–graphite phyllites, porphyroids and calcsilicate schists. The Branná Unit is overthrust on the crystalline Keprník Nappe, made up of a large granitic orthogneiss body (U–Pb zircon-dated at around 546 Ma; van Breemen *et al.* 1982) with subordinate staurolite-bearing metapelites, calcsilicate rocks and quartzites (Schulmann & Gayer 2000).

Desná Dome. The easternmost tectonic unit of the Silesian domain, known as the Desná Dome, contains the relatively monotonous metasedimentary Devonian succession of the Vrbno Group (Fig. 8), overlying Desná gneisses and mylonites with U–Pb zircon ages of 570–650 Ma (Kröner and colleagues data reported by Schulmann & Gayer 2000). The Vrbno Group comprises a succession of predominantly deep-water siliciclastic, siliceous and calcareous slates, up to 3000 m thick associated with abundant metavolcanics (Svoboda *et al.* 1966), metamorphosed under greenschist facies conditions and markedly different from the shallow-water Devonian of the Branná Unit and from platform Devonian rocks of the Brno Massif and the basement underlying the Upper Carboniferous Silesian Coal Basin and the Miocene Carpathian nappes. The Vrbno Group was presumably overthrust on the Brunovistulian gneissic basement now exposed within the Desná Dome. The metavolcanic rocks of the Vrbno Group are interpreted as originally deposited in an extensional basin related to crustal thinning. The supra-subduction geochemical signature of some metavolcanic rocks may indicate arc and back-arc (the latter initiated as intracontinental rift) tectonic settings (Patočka & Valenta 1996). Palaeontological dating of crystalline limestones suggests that much of the Vrbno volcano-sedimentary succession was formed between end Givetian and latest Frasnian times (Hladil 1986).

Strzelin Crystalline Massif. Towards the north, the East Sudetic nappe pile plunges below a cover of Cenozoic sediments. The crystalline rocks emerge at the surface only within the Strzelin Crystalline Massif, 40 km south of Wrocław. The massif mainly consists of gneisses accompanied by minor mica schists and amphibolites with a within-plate geochemical signature (Szczepański & Oberc-Dziedzic 1998). These rocks are tectonically interleaved with quartzites similar to those known from the Desna Dome and Branná Unit, which are therefore also assumed to be of Early to Middle Devonian age. Varied gneisses are derived from different protoliths, dated at 504 ± 3 Ma (Oliver *et al.* 1993) and at about 570 to 600 Ma (Oberc-Dziedzic *et al.* 2001). The metamorphic rocks of the Strzelin Massif are intruded by abundant granitoids ranging in age from around 350 to 330 Ma (dated by the Rb–Sr whole rock method, Oberc-Dziedzic *et al.* 1996).

Interpretation and discussion

The structural units of the Sudetes described above represent various palaeotectonic environments, geological evolutionary paths and various, often unclear, affinities. Here, we present and, where necessary, discuss the interpretation of particular units, as adopted in the recent literature, or proposed by us.

West Sudetes

Lusatian and Karkonosze-Izera massifs and Görlitz Slate Belt: the Saxothuringian passive margin. The Lusatian Massif represents the Neoproterozoic parautochthonous continental basement of the epicontinental part of the Saxothuringian Basin (Linnemann *et al.* 1998, 2000; Franke & Żelaźniewicz 2000; Mazur & Aleksandrowski 2001*a*). Similarly, the adjacent gneisses and mica schists of the Izera-Kowary Unit in the Karkonosze-Izera Massif, are interpreted as the pre-Variscan (Early Palaeozoic) continental crust partly underlying the Saxothuringian Basin (Franke & Żelaźniewicz 2000; Mazur & Aleksandrowski 2001*a*), subsequently involved in Variscan deformation near to a collision zone extending along the Saxothuringian Suture (see below). This continental crust must have once underlain the passive margin of the Saxothuringian epicontinental area (Mazur & Aleksandrowski 2001*a*). The Izera gneisses are Early Ordovician intrusions into an older basement of Cadomian Lusatian granitoids (Domečka 1970) dated at between *c.* 590–545 Ma (Kröner *et al.* 1994). Structurally, therefore, the NW part of the Karkonosze-Izera Massif constitutes one common element together with the Lusatian Massif. The intensity of deformation and metamorphism gradually increases eastwards, i.e. towards the suture zone, starting from practically undeformed Cadomian granitoids of Lusatia (Ebert 1943).

The sedimentary succession of the NE passive margin of the epicontinental part of Saxothuringia is represented by the rocks of the

Ještěd Unit (Karkonosze-Izera Massif; Figs 2 and 4; Mazur & Aleksandrowski 2001*a*) and of the thrust-folded and strike-slip sheared Görlitz Slate Belt (cf. Hirschmann 1966; Brause & Hirschmann 1969; Urbanek *et al.* 1995), together with its eastward continuation up to the vicinities of Gryfów Śląski (the Lubań subunit, previously considered as the western part of the Kaczawa Unit; see discussion below).

Saxothuringian Suture in the southern and eastern Karkonosze Mountains. The suture of the SE Karkonosze Mountains is probably a fragment of the Saxothuringian Suture (equivalent names: the Münchberg-Teplá Suture (Matte *et al.* 1990); or Teplá/Saxothuringian (Mazur & Aleksandrowski 2001*a*)), defining a boundary between the Saxothuringian domain to the NW and the Teplá-Barrandian domain to the SE (e.g. Matte *et al.* 1990; Matte 1991, 1998; Franke *et al.* 1995*a*). The Saxothuringian Suture is known from exposures near to Mariánské Lázně and in the Münchberg nappe pile in the western part of the Bohemian Massif. In that area the suture is defined by eclogites derived from *c.* 500 Ma MOR-type magmatic rocks and accreted continental slope sediments of the Saxothuringian Bavarian facies (Beard *et al.* 1995; Franke *et al.* 1998). The suture continues to the NE, below the cover of Mesozoic–Cenozoic rocks of the Cheb (Eger) Graben, up to the upper Elbe Valley, where it is presumably displaced to the SE on the dextral Elbe Fault Zone together with the entire Teplá-Barrandian domain (cf. Rajlich 1987; Pin *et al.* 1988; Matte *et al.* 1990; Aleksandrowski 1990, 1995), to crop out in the southern and eastern Karkonosze Mountains. Further to the NE, the Saxothuringian Suture is cut by the dextral strike-slip Intra-Sudetic Fault Zone (see Aleksandrowski 1990, 1995; Aleksandrowski *et al.* 1997) so that no direct continuation of it can be identified.

An age equivalent of the Mariánské Lázně Complex in the eastern Karkonosze-Izera Massif is the Leszczyniec Unit (Figs 2 and 4), metamorphosed under relatively high pressure epidote–amphibolite facies (Kryza & Mazur 1995) and showing *c.* 500 Ma protolith age (Oliver *et al.* 1993). It is interpreted as an obducted fragment of the Saxothuringian sea floor (Mazur & Aleksandrowski 2001*a*). The underlying South Karkonosze Unit comprises a volcanosedimentary basin infill metamorphosed under blueschist facies conditions. The rocks bear record of changing tectonic environments from initial rifting during the Ordovician to a mature oceanic basin during the Silurian (Patočka & Smulikowski 1998; 2000). Both the Leszczyniec and South Karkonosze units are derived from a hypothetical root zone of the SE Karkonosze nappe pile, currently buried below the Carboniferous and Permian deposits of the Intra-Sudetic and Fore-Karkonosze basins. These units can be considered together as a separate ‘oceanic’ South Karkonosze Terrane composed of fragments of the floor and sedimentary succession of the Saxothuringian Sea (Mazur & Aleksandrowski 2001*a*).

Kaczawa Unit: a rift to oceanic succession involved in an accretionary prism. The Kaczawa Unit exposes a volcanic–sedimentary succession that records a transition from an initial rift in the Ordovician, to a mature ocean in Silurian times (Furnes *et al.* 1994). As an accretionary prism (Baranowski *et al.* 1990) with a significant proportion of rocks of oceanic affinities, the Kaczawa Unit is, in a way, analogous to a tectonic suture: it separates distinct continental domains of different provenance (the Lusatian Massif together with the Görlitz Slate Belt in the west and the Góry Sowie Massif in the east). A suture zone proper, related to the Kaczawa thrust stack, might be expected to occur within a hypothetical root zone that must be concealed somewhere to the east, probably below the Świebodzice Basin and/or the Góry Sowie Massif.

No direct relationship between the Görlitz and Kaczawa successions. Traditionally, the Kaczawa Unit has been considered to continue westward into the Görlitz Slate Belt (e.g. Jaeger 1964; Hirschmann 1966; Brause & Hirschmann 1969; Urbanek *et al.* 1995). This view, however, though widely held due to geographical proximity of both units, does not seem to be correct, as indicated by important differences in stratigraphy, age of deformation and variation of metamorphic grade between rock complexes of the two units.

The Görlitz Palaeozoic succession is unmetamorphosed or very slightly metamorphosed, in contrast to that of the Kaczawa Unit, which, over its entire extensive outcrop area shows greenschist facies metamorphism, obliterating relicts of an earlier high pressure–low temperature event. The Görlitz succession begins with Lower Cambrian carbonates, sandstones and scarce volcanics capped by trilobite-bearing shales (Freyer 1977; Urbanek *et al.* 1995), whereas rocks of such age are not known in the Kaczawa Mountains. Although abundant in the Kaczawa Unit, no Upper Cambrian(?)–Lower Ordovician volcanics are known from the

Görlitz Slate Belt; the scarce Ordovician rocks there are siliciclastics. The Silurian section in the Görlitz succession, comparable to that of the Thuringian facies of the Saxothuringian Zone (Hirschmann 1966), is composed of siliceous and 'alum shales', intercalated at the top with tuffs, quartzites, greywacke and rare limestone (Urbanek *et al.* 1995): the latter intercalations are not present in Silurian rocks in the Kaczawa Unit, which instead contains significant amounts of mafic MORB-type volcanics. The Devonian section in the Görlitz succession is more terrigenous than that in the Kaczawa Mountains, and is represented by a monotonous series of alternating quartzites, pelites and greywackes with rare limestones and basic volcanic rocks (Urbanek *et al.* 1995), whereas the latter are not yet described from the Devonian of the Kaczawa Unit.

The Lower Carboniferous of the Görlitz Slate Belt is composed of normal flysch, accompanied towards the top by chert, limestone and significant conglomerate intercalations (Urbanek *et al.* 1995) and, also of wildflysch deposits containg fragments of Ordovician through Lower Carboniferous rocks as olistoliths (Thomas 1990). Linnemann & Buschmann (1995) and Linnemann *et al.* (1998) even interpreted the entire Görlitz Slate Belt as a Lower Carboniferous flysch embedding various sized fragments of dismembered Lower Cambrian to Lower Carboniferous sequences. The south-verging tight folds and steep schuppen structure formation in the Görlitz Slate Belt is dated at middle Viséan and its final stage considered as late Viséan (Hirschmann 1966; Brause & Hirschmann 1969).

Lower Carboniferous rocks have not yet been recognized unequivocally in the central and eastern part of the Kaczawa Unit. Early Carboniferous ages, are, nevertheless, suspected for some of the mélange bodies in that region, considered to be of tectonic origin (Baranowski *et al.* 1987, 1990; Collins *et al.* 2000). At the same time, however, greenschist facies Kaczawa rock fragments are abundantly present as pebbles in Late Tournaisian(?)–Early Viséan conglomerates of the northern rim of the Intra-Sudetic Basin (A. K. Teisseyre 1968, 1971, 1975) and, less abundantly, in the Świebodzice Basin (H. Teisseyre 1968). Therefore, the metamorphism, deformation and exhumation of at least part of the Kaczawa Unit must have taken place by Early Viséan times.

At the same time, conodont-proven Lower Carboniferous deposits, reaching up to the lower part of the upper Viséan, have been documented by Chorowska (1978) from the far-western part of the Kaczawa Unit, near Gryfów Śląski and Lubań. Palaeozoic rocks crop out there along an elongated narrow belt adjacent to the Intra-Sudetic Fault to the south. Further west, this belt merges into the Görlitz Slate Belt. According to Milewicz *et al.* (1989, see also Cymerman in press), this belt of Palaeozoic rocks is tectonically divided NE of Gryfów Śląski, into two segments with differing metamorphic grades. The eastern segment shows greenschist facies metamorphism, typical of the Kaczawa rocks and is included in the Pilchowice Subunit, extending east to near Jelenia Góra. The western segment, however, distinguished as the Lubań Subunit (Fig. 9), shows only a very weak metamorphic imprint. The Viséan limestone in Rząsiny, occurring as fragments within a chaotic deposit (olistostrome; Chorowska 1978), comes from the Lubań Subunit. The timing of tectonic deformation inferred from this finding does not match the relationships known from the Kaczawa Unit further east; instead it corresponds well with those determined for the Görlitz Slate Belt (see above). We therefore suggest, that the Lubań Subunit probably represents the easternmost part of the Görlitz Slate Belt, extended and sheared along the northern wall of the Intra-Sudetic Fault Zone. Thus, the discovery of upper Viséan sediments there should not directly affect conclusions about the timing of deformation and metamorphism within the Kaczawa Unit.

Central Sudetes

Based on contrasting geological histories, the Central Sudetes can be subdivided into two domains. The northwestern domain, includes the Góry Sowie Massif and the Kłodzko Metamorphic Unit, together with the surrounding Central Sudetic ophiolitic bodies and the Bardo and Świebodzice sedimentary basins and, probably much of the NE part of the basement to the Intra-Sudetic Basin. The southeastern domain corresponds to the Orlica-Śnieżnik Massif and the Staré Město Belt in the Sudetic Mountains and the Kamieniec Metamorphic Belt in the Fore-Sudetic Block. The two domains are separated by the major strike-slip Skrzynka and Niemcza shear zones.

Kłodzko Metamorphic Unit: contrasting palaeotectonic elements juxtaposed in a nappe pile. The Kłodzko Metamorphic Unit juxtaposes rocks that have formed at various times in two contrasting tectonic environments: a passive continental margin of (at least partly) Middle Devonian age and a Neoproterozoic(?) active

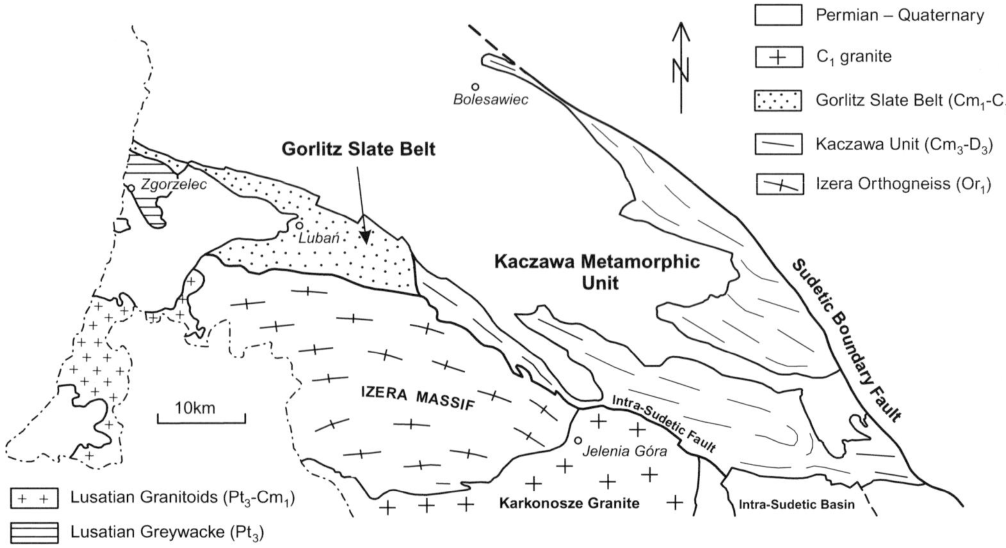

Fig. 9. Schematic map showing the suggested extent of the Kaczawa Unit and the Görlitz Slate Belt in the West Sudetes. *Age assignments* as in Figure 2.

continental margin (Figs 5 and 6). The structurally lower NE part of the Kłodzko Unit is, thus, represented by passive margin Middle Devonian sediments and volcanics (the Mały Bożków, Łączna and Bierkowice subunits). The upper SW part comprises rocks yielding Neoproterozoic ages and representing fragments of a supracrustal magmatic arc sequence interlayered with distal flysch sediments (the Kłodzko Fortress Subunit) and relics of a plutonic–volcanic complex associated with deep marine sedimentary rocks, possibly of back-arc origin (the Orla-Gołogłowy and Ścinawka subunits). Further to the NW, the Nowa Ruda Ophiolite is overridden by the Kłodzko Unit nappe pile. The stacking of nappes must have taken place in Middle to Late Devonian times, i.e. approximately synchronously with the obduction of the Nowa Ruda Ophiolite. Subsequently, the entire nappe pile was buried below the clastic deposits of the Bardo Basin. Sedimentation continued there from Upper Devonian to the beginning of Late Carboniferous times, i.e. until the folding of the sedimentary sequence and emplacement of the Kłodzko-Złoty Stok Granitoid Pluton.

Góry Sowie Massif: Cadomian continental crust subducted to mantle depths in Late Silurian/ Early Devonian and exhumed in Late Devonian times. The thermobarometric data and radiometric age constraints on relict granulitic and eclogitic rocks (Kryza *et al.* 1996; O'Brien *et al.* 1997; Brueckner *et al.*1991, 1996) accompanying widespread paragneisses and migmatites, indicate that the Góry Sowie granulites represent 'type I granulites' of Pin & Vielzeuf (1983), probably derived from continental crust subducted in the Variscan Belt to mantle depths at around 430–400 Ma (Vielzeuf & Pin 1989). Subsequently, the granulites experienced a distinct decompressive event (4–10 kbar), under continuing high temperatures (600°–700 °C), probably coinciding with the peak of metamorphism and anatexis in the surrounding gneisses (Kryza *et al.* 1996). The cooling ages of about 385–370 Ma (van Breemen *et al.* 1988; Bröcker *et al.* 1998) represent the end of metamorphism in the Góry Sowie Massif prior to the Late Devonian. Gneissic clasts in the Late Devonian strata of the Świebodzice Basin show that Góry Sowie gneisses were exhumed during the Famennian (Porębski 1981, 1990).

Position of Kłodzko Metamorphic Unit and of Central Sudetic Ophiolites relative to Góry Sowie Massif. The spatial interrelationships between the Góry Sowie Massif on one side and the Kłodzko Metamorphic Unit and the Central Sudetic Ophiolite on the other side are not clear, since their contacts are not exposed. From Viséan times, the Góry Sowie Massif abundantly supplied the marginal parts of the Bardo Basin with gneissic pebbles which accumulated in a system of coarse clastic fans that developed along the SW edge of the Góry Sowie Massif (Wajsprych 1978, 1986; Haydukiewicz 1990).

The mapped relationships show that the Bardo Basin is in part floored by the Kłodzko Metamorphic and the Nowa Ruda Ophiolite rocks. However the only borehole believed to have penetrated through the Bardo Basin fill, was stopped after drilling several tens of metres in typical Góry Sowie gneiss (Chorowska *et al.* 1986). Therefore, it cannot be excluded that all three: the Kłodzko Metamorphic Massif, the Nowa Ruda Ophiolite and the Góry Sowie gneisses constituted a pre-Carboniferous basement on which the Bardo Basin succession was laid down.

Based on gravimetric data, the Central Sudetic ophiolites have been thought to extend below the Góry Sowie Massif (Znosko 1981). However, on recent gravimetric maps (e.g. Królikowski & Petecki 1995) as well as on the gravimetric and magnetic data referred to by Znosko (1981), the maximum values of positive Bouguer anomalies or positive magnetic anomalies roughly coincide with the ophiolite outcrops at the northern, eastern and southern margins of the Góry Sowie Massif. The massif itself coincides with high unidirectional negative gradients in gravimetric and magnetic fields, with no positive anomalies below the gneisses. Therefore, a more reasonable conclusion is that the Góry Sowie Massif is not underlain by dense and highly magnetic ophiolitic rocks. Moreover, the Góry Sowie Massif has been uplifted since the beginning of the Carboniferous, bounded by high-angle faults. Thus, any ophiolites that underlie the Góry Sowie Massif would have been buried more deeply in the areas adjacent to the massif than below the massif itself. However, the ophiolite may have been thrust over the gneisses. The spatial relationship of the Góry Sowie gneisses and the Sudetic ophiolites remains unsolved and requires further study.

Nové Město Belt: the sheared NE margin of Teplá-Barrandian domain? Rocks of the Teplá-Barrandian domain are believed by some workers to continue northeastwards into the basement of the North Bohemian Cretaceous Basin (Malkovský 1979; see also Chaloupský 1989; Pin *et al.* 1988; and Matte *et al.* 1990). This implies that the Nové Město Belt may be the northeasternmost fragment of the Teplá-Barrandian (Chaloupský 1989) emerging from below the deposits of the North Bohemian Basin. The NE edge of the Nové Město Belt, defined by the dextral Uhřinov Shear Zone (Mazur & Aleksandrowski 2001*b*), is an important tectonic boundary that separates low-grade phyllites, greenstones and amphibolites, from geologically different amphibolite-grade gneisses and mica schists of the Orlica-Śnieżnik Massif. The Nové Město rocks show similarities in lithology and metamorphic grade to the Teplá Complex in the interior of the Bohemian Massif (Chaloupský *et al.* 1995), whereas the Orlica-Śnieżnik rocks have affinities to the Gföhl Unit of the Moldanubian Zone of the Bohemian Massif (discussed in some detail below). The structural characteristics of this tectonic contact between the Nové Město Belt, and the Orlica-Śnieżnik Massif are roughly similar to those of the contact between the Teplá-Barrandian and Moldanubian domains exposed in the Železné Hory Mountains. A primary thrust contact in the latter area was extensively modified by a normal top-to-NW ductile shear zone (Pitra *et al.* 1994). The orientation of the stretching lineation at Železné Hory is analogous to that in the Orlica Mountains and suggestive of a strike-slip dextral shearing component. The prevailing strike-slip and down-dip sense of displacement in the Orlica and Železné Hory mountains, respectively, is determined from the present-day attitude of foliation. This has been strongly reorientated by late domal uplift, which is particularly distinct in the footwall of the Nové Město Belt. The lithological and geochemical characteristics of the amphibolite belt along the NE flank of the Nové Město Belt may indicate a tectonic suture (Opletal *et al.* 1990; Floyd *et al.* 1996).

Góry Sowie–Kłodzko domain of Central Sudetes: an area of pre-Upper Devonian subduction, collision and exhumation

The Kłodzko Metamorphic Massif, together with the Góry Sowie Massif, the Świebodzice and Bardo basins, the Central Sudetic Ophiolite massifs as well as the basement of the NE part of the Intra-Sudetic Basin, represent the large Central Sudetic domain that contains a record of pre-Late Devonian Eo-Variscan deformation and exhumation. This area is collectively termed here the Góry Sowie–Kłodzko domain (and below is distinguished as a terrane). The effects of the pre-Late Devonian tectonism were subsequently overprinted there by Early Carboniferous deformation, so widespread in the Sudetes and the entire Bohemian Massif. One of the most typical features of the Góry Sowie–Kłodzko domain is the widespread pre-upper Devonian unconformity.

Pre-Upper Devonian unconformity. The pre-Upper Devonian erosional surface that planes the top of the Nowa Ruda Ophiolite Massif and of the Kłodzko Metamorphic Unit is characteristic of the NW part of Central Sudetes. The crystalline metamorphic or igneous basement is overlain by unmetamorphosed upper Devonian basal conglomerates and limestones. This relationship was considered by Bederke (1924, 1929) as a record of the Caledonian Orogeny in the Sudetes. In modified form, Bederke's conclusions can be found in many recent publications (e.g. Don 1984, 1990; Oliver *et al.* 1993; Johnston *et al.* 1994; Kröner & Hegner 1998; Kröner *et al.* 2000). However, many lines of evidence suggest that the main tectonic and thermal events occurred in the Sudetes during end-Devonian and Early Carboniferous times (e.g. Aleksandrowski 1994; Żelaźniewicz & Franke 1994; Aleksandrowski *et al.* 2000). These include: (1) continuous Palaeozoic basinal sedimentation of the Kaczawa succession, lasting until the end of the Devonian; (2) Carboniferous metamorphic ages in the Karkonosze-Izera and Orlica-Śnieżnik complexes; and (3) widespread Carboniferous granitoid plutonism.

The type localities of the pre-Upper Devonian unconformity in the Sudetes are outcrops exposing the contact of Upper Devonian strata of the Bardo Basin with the underlying metamorphics of the Kłodzko Massif and mafic plutonic rocks of the Nowa Ruda Ophiolite Massif. The crystalline basement rocks of the Kłodzko Metamorphic Unit and of the Nowa Ruda Massif were described by Bederke (1924) to be in sedimentary contact with unmetamorphosed Upper Devonian and Tournaisian thin basal conglomerates, calcareous breccias and limestones up to 60 m thick. The most complete profile of the Upper Devonian succession is exposed at Mt Wapnica in Dzikowiec, adjacent to the Nowa Ruda Massif (Fig. 5), a part of the circum-Góry Sowie, Central Sudetic Ophiolite belt. At its base is a calcareous sedimentary breccia comprising large boulders of gabbro (e.g. Gürich 1900; Mazur 1987), which are believed to have been derived from the crystalline basement of the Nowa Ruda Massif (Gürich 1900, 1902; Dathe 1900). The sedimentary contact of the basal breccia with the gabbroic basement was documented by Gürich (1900) in the Dzikowiec (Ebersdorf) quarry, which is no longer exposed. The age of the breccia was provisionally estimated as upper Frasnian, based on problematic macrofauna (Gürich 1902; Bederke 1929). The size and quantity of the gabbro pebbles decreases rapidly upwards as the sedimentary breccia grades into the Basal Limestone and, then, into the partly nodular Main Limestone (Mazur 1987). The age of the latter is palaeontologically well constrained by conodonts as Famennian (Freyer 1968; Chorowska 1974). The Main Limestone passes upwards into the upper Famennian *Clymenia* Limestone (Weyer 1965; Freyer 1968; Chorowska 1974). The uppermost part of the calcareous succession forms the Lower Tournaisian limestone representing the ammonoid zone *Gattendorfia crassa* (Schindewolf 1937; Weyer 1965), consistent with the conodont fauna (Freyer 1968; Chorowska 1974; Dzik 1997). In most places, the limestones are discordantly overlain by a thick sequence of Culm sandstones (Dathe 1900; Bederke 1924) considered to be latest Tournaisian in age (Głuszak & Tomaś 1993). This contact probably represents a local erosional discordance (Mazur 1987) rather than a significant thrust plane (the Kłodzko Thrust of Bederke 1924). The base of the sandstones is usually marked by a thin layer of black shales, which are dated by conodonts as Tournaisian (Haydukiewicz 1981). The fragmentary outcrops of Upper Devonian strata at the boundary of the Kłodzko Metamorphic Unit and the Bardo Basin are comparable to those exposed at Mt Wapnica (Bederke 1924; Gunia 1977). The specific feature of the Mt Wapnica profile is the unconformable contact of Upper Devonian rocks with gabbro of the Nowa Ruda Massif: in all other localities the Upper Devonian is in contact with the metamorphic basement of the Kłodzko Unit. Bederke (1924) described eight localities of Devonian sedimentary rocks along the western edge of the Bardo Basin. However, except in the Mt Wapnica section, the angular unconformity has not been exposed since the 1950s. The unconformity was recently excavated in two of Bederke's localities around Kłodzko (Kryza *et al.* 1999). At each site, the metamorphic rocks were found to be in unconformable contact with the overlying basal sedimentary breccias and conglomerates with no evidence of tectonic disturbance. Near the Gołogłowy quarry the Upper Devonian rocks overlying the unconformity, show an upward transition into the Bardo Basin succession (Haydukiewicz 1981).

The angular unconformity cutting the Kłodzko Metamorphic Unit and the Nowa Ruda Ophiolite Massif must have formed during a relatively narrow time interval of about 10 Ma between the early Givetian and late Frasnian. This timing is constrained by the late Frasnian/Famennian age of the limestones directly overlying the basal conglomerates (Gürich 1902; Bederke 1929; Gunia 1977) and by the early Givetian age of a coralline fauna from the greenschist facies crystalline limestones of the Kłodzko Metamorphic

Unit (underlying the unconformity) at Maly Bożków (Hladil *et al.* 1999), previously interpreted as Late Silurian (Gunia & Wojciechowska 1971). The existence of this unconformity implies that at the turn of Middle and Late Devonian times, freshly deformed and metamorphosed rocks were exposed and onlapped by deposits of the Bardo sequence, which were folded during latest Viséan/Namurian times (Oberc 1972). Apart from the Kłodzko/Nowa Ruda/Bardo units, the pre-Upper Devonian unconformity most probably defines the floor of the Świebodzice Basin, although no details are known about its basement.

Multistage evolution of the Góry Sowie–Kłodzko domain. The geological data from the Góry Sowie–Kłodzko domain of the Central Sudetes indicate multistage evolution defined by three stages of convergence and two stages of extension. The first, Late Silurian/Early Devonian convergence was recorded in the high pressure–high temperature metamorphism of the Góry Sowie Massif. The resulting granulites, about 400 Ma old (O'Brien *et al.* 1997) are evidence for a continental subduction, which was presumably due to late Silurian/Early Devonian collision. However, the tectonically isolated position of the Góry Sowie Massif within younger metamorphic complexes does not allow a reconstruction of Silurian/Early Devonian accretion in the Central Sudetes.

The Late Silurian/Early Devonian convergence was followed by a relatively long period of extension, resulting in the opening of an oceanic basin (whose floor supplied the protolith for the Central Sudetic ophiolites). The 400 Ma age of these rocks (Żelaźniewicz *et al.* 1998), seems to be the most probable from among the several determinations (Pin *et al.* 1988; Oliver *et al.* 1993), and shows that extension took place in earliest Devonian times, contemporaneously with the uplift and exhumation of the Góry Sowie Massif. A wide range of cooling ages for the Góry Sowie gneisses (van Breemen *et al.* 1988; Bröcker *et al.* 1998; Timmermann *et al.* 2000) reflects their gradual exhumation during the Early and Middle Devonian. The extension lasted until the late Middle Devonian, as documented by the early Givetian age of the Mały Bożków succession in the Kłodzko Metamorphic Unit (Hladil *et al.* 1999).

A successive compressional stage took place between the early Givetian and Famennian. It resulted in obduction of the Central Sudetic ophiolites and in deformation and metamorphism of the Kłodzko Unit, which was finally thrust upon the Nowa Ruda Ophiolite.

The termination of the Middle/Late Devonian stage of accretion in the Central Sudetes is constrained by the late Frasnian or earliest Famennian age of the pre-upper Devonian unconformity. The unconformity defines the beginning of the successive extensional stage, leading to the formation of the Bardo Basin. Its subsidence and the accompanying sedimentation lasted until the turn of the Early/Late Carboniferous, when inversion took place and folding affected the basin fill, probably synchronously with the intrusion of the Kłodzko- Złoty Stok Granitoid Pluton. The latter compressional stage did not result in any metamorphic or thick-skinned tectonic effects in the Bardo Basin or in the other units of the Góry Sowie–Kłodzko domain; however, it had a significant influence on the adjacent Orlica-Śnieżnik Massif (see below).

Transregional correlations. A record of pre-Late Devonian deformation and exhumation, typical of the Góry Sowie–Kłodzko domain, is present also in few other areas of the Bohemian Massif, such as the basement of the North Bohemian Basin and the Münchberg Massif. In the North Bohemian Cretaceous Basin, whose basement is often interpreted to lie within the Teplá-Barrandian domain (Malkovsky 1979; Chaloupsky *et al.* 1995), boreholes east of Hradec Kralove reveal unmetamorphosed Upper Devonian to Lower Carboniferous succession resting transgressively on low-grade schists of uncertain, Lower Palaeozoic(?) age (Chlupač & Zikmundová 1976; Čech *et al.* 1989). At the same time, within the allochthonous units of the Münchberg Massif, there occur approximately 395 Ma eclogites (Stosch & Lugmair 1990), previously dated at *c.* 425–410 Ma (Gebauer & Grünenfelder 1979). They underwent cooling and exhumation by Famennian (*c.* 365 Ma) times, i.e. before the final stacking of the Münchberg nappe pile. This exhumation is believed to have taken place in a relatively narrow zone between the colliding Teplá-Barrandian domain to the SE and the Saxothuringian passive margin, resulting in the sedimentation of Givetian clastics on top of the Barrandian sedimentary succession and of Famennian greywackes at the southern flank of the Fichtelgebirge (Franke *et al.*1995*b*; Franke *et al.*1998; Franke 2000; Franke & Stein 2000).

However, the Central Sudetic Góry Sowie–Kłodzko domain seem to be the only area in the Bohemian Massif known to have recorded effects of Palaeozoic tectonic evolution in which early collision was followed by later rifting and opening of a basin underlain by oceanic crust. A record of polycyclic tectonic evolution similar to that from the Góry Sowie–Kłodzko domain of the Central Sudetes is known in areas of the Variscan Belt outside the Bohemian Massif. Close analogues occur in France, where a belt of Upper Silurian to Early Devonian high-grade metamorphic rocks extends from the southern part of the Armorican massif to the NE Massif Central and the Vosges (e.g. Faure *et al.* 1997; Shelley & Bossière 2000 and references therein). It defines the Massif Central suture (Matte 1991, 1998), also called Eo-Variscan Suture (e.g. Faure *et al.* 1997), which is interpreted to have originated due to Late Silurian continental subduction, which was also responsible for the high pressure metamorphism (e.g. Pin & Vielzeuf 1988; Pin 1990). This was followed by crustal extension from Mid-Devonian time, leading to the formation of Brévenne Rift in the NE part of the Massif Central and similar smaller structures in the Armorican Massif and the southern Vosges. This 'mid-Variscan' stage of the orogenic evolution was terminated by Late Variscan compression ('the Hercynian Event'; e.g. Faure *et al.* 1997). These geological relationships in the 'Moldanubian Zone' of France led Ziegler (1986), Eisbacher *et al.* (1989) and Pin (1990) to assume a succession of two orogenic cycles during formation of that part of the Variscan Belt.

High pressure metamorphic rocks of the French Variscides known as the 'Leptyno-Amphibolite Complex' (e.g. Pin & Vielzeuf 1983; Ledru *et al.* 1989) yield ages ranging from 440 to 400 Ma (Pin & Peucat 1986; Paquette *et al.* 1995). During exhumation, the Silurian high pressure metamorphosed rocks underwent almost isothermal decompression, leading to partial melting and migmatization (Santallier *et al.* 1994), an event dated by various methods at 385–380 Ma (Pin & Peucat 1986; Costa & Maluski 1988; Boutin *et al.* 1995). The migmatization ages point indirectly to rapid uplift and correspond well to stratigraphic data. The latter suggest the Eo-Variscan complexes were already exposed at the surface in Middle Devonian times. They are unconformably covered by Givetian unmetamorphosed sandstones and limestones that occur in a zone extending from the Armorican Massif through the northern Massif Central to the Vosges (cf. Faure *et al.* 1997).

The development of the Devonian Brévenne-Violay-Beaujolais rift in the NE part of the Massif Central (Sider & Ohnenstetter 1986; Sider *et al.* 1986; Ohnenstetter & Sider 1988; Leloix *et al.* 1999) is evidence for Devonian extension of the Eo-Variscan complexes. The Devonian rocks exposed in the rift comprise mafic and acidic volcanic rocks, gabbro, diabase, trondhjemite, serpentinized ultramafic rocks, siltstones, cherts greywackes and sandstones with gabbro clasts (cf. Leloix *et al.* 1999). They form several tectonic units and reveal variable metamorphic grade, from greenchist to amphibolite facies. The tectonic units are thrust to the NW, over gneissic basement (Affoux Gneiss). The entire nappe pile shows tectonic inversion of metamorphic grade, with metamorphism increasing from greenschist to amphibolite facies from the base to top of the profile and from the NW to the SE in plan view.

The deformed and metamorphosed Devonian rocks are covered by Late Tournaisian–Early Viséan volcanosedimentary formations, which are in part unmetamorphosed and undeformed, but locally underwent deformation and weak greenschist facies metamorphism due to post-thrusting dextral wrenching. They are, in turn, unconformably overlain by entirely undeformed and unmetamorphosed mid- and upper Viséan formations.

The evolution of the the Brévenne-Violay-Beaujolais Complex shows a number of similarities with the Central Sudetes and, in particular, with the Kłodzko Metamorphic Unit together with the Nowa Ruda Ophiolite. In this comparison the position of the Góry Sowie Massif can be hypothetically compared to that of the pre-Early Devonian Affoux Gneiss in the Brévenne rift basement. The Devonian rocks of the Brévenne rift sedimentary infill record extension that followed the Eo-Variscan convergence, although the age of deformation is a little different in both areas. In the Kłodzko Metamorphic Unit these processes took place during the Frasnian, whereas in the Brévenne-Violay-Beaujolais Complex they occurred at the Famennian/Tournaisian boundary (Pin & Paquette 1998). A common feature of both areas is early Carboniferous subsidence with the development of sedimentary basins above freshly deformed and metamorphosed complexes. However, the NE Massif Central recorded an intra-Viséan tectonic event, whereas in the Central Sudetes the folding of the Bardo Basin infill took place at the turn of Early and Late Carboniferous times.

East Sudetes: fragment of a Variscan collisional belt along the eastern margin of the Bohemian Massif

The East Sudetes lie within a Variscan collision zone at the eastern margin of the Bohemian Massif, that separates the Brunovistulian (or Brunosilesian) domain in the east and the units comprised in the Moldanubian and West and Central Sudetic (Lugicum) domains in the west (Matte *et al.* 1990; Schulmann *et al.* 1991; Fritz & Neubauer 1993; Schulmann & Gayer 2000). The Early Carboniferous collision followed westward subduction of the Brunovistulian passive margin, below the Moldanubian and Central Sudetic domains. The collision resulted in large-scale eastward overthrusting of the latter domains onto nappe complexes of the underthrust lower plate (e.g. Suess 1912; Matte *et al.* 1990; Schulmann & Gayer 2000). The Moravian, more southerly segment of this collision zone has been considered in the literature as different and distinct from that of the East Sudetes. However, a question arises if, indeed, there are no close analogies between the contents, structure and development of both segments of the Moravo-Silesian Zone.

Southern Moravian equivalent? In Moravia, the collision zone comprises three main elements thrust towards the east. The uppermost position in the thrust stack is occupied by the Moldanubian 'upper plate', exposed in the west, which overrides the Moravian Nappes and the parautochthonous Brunovistulian 'lower plate' extending far to the east. The Moravian supracrustal thrust sheets appear exotic with respect to the Brunovistulian basement and its sedimentary cover. They contain variegated metasediments, in early papers named the Inner and Outer Phyllites, separated from each other by the Biteš Orthogneiss (Suess 1912, 1926). The Inner Phyllites are greenchist facies metapelites with intercalations of crystalline limestones, quartzites, graphitic schists, metagreywackes and metabasites. The Outer Phyllite comprises amphibolite facies biotite- and two-mica paragneisses with abundant mica schists, crystalline limestones, amphibolites, quartzites, graphite schists and calc-silicate rocks. The age of this sequence is not exactly known; it is usually referred to as Neoproterozoic or Early Palaeozoic.

The adjacent sedimentary cover of the Brunovistulian domain, exposed within the Brno Massif and in the central part of the Svratka Window, overthrust by the Moravian Nappes, is represented by Middle Devonian to Tournaisian shallow water limestones resting directly on basal clastics and the crystalline basement (Dvorak 1995). The 586 Ma (Friedl *et al.* 1998) age of the Biteš Gneiss is contemporary with that of the Neoproterozoic granitoids from the Brunovistulian basement dated at 560–590 Ma (van Breemen *et al.* 1982; Friedl *et al.* 1998). However, as the geochemical characteristics of the Biteš Gneiss do not conform exactly to those of the Brno Massif granitoids (Hanžl 1994), the Moravian units may be a distinct, narrow domain, squeezed in between the Moldanubian and Brunovistulian elements (the upper and lower plates) and not detached and overthrust fragments of the subducted Brunovistulian margin.

Possible affinities between the East Sudetes and Moravia. The structure of the metamorphic ('Silesian') domain in the East Sudetes, seems to differ significantly from that in Moravia. Schulmann & Gayer (2000) contend that in the boundary zone of our Central Sudetes and the East Sudetes, the Central Sudetic 'upper plate' (Lugicum) is juxtaposed across the Staré Město Suture with a nappe complex derived directly from the subducted Brunovistulian passive margin. No additional tectonic elements between the Central Sudetes and the Brunovistulian-derived units are distinguished that could be correlated with the Moravian Nappes further south. Hence all of the metamorphic nappe units of the East Sudetes (Silesian units) are considered to be derived from the Brunovistulian domain.

Independently of its uncertain age, the lithological content of the supracrustal series in the high-grade metamorphic Velké Vrbno Nappe, occurring directly east of the Staré Město Suture, differs from that of the Devonian sequences exposed further east. Therefore, we follow earlier interpretations (e.g. Zapletal 1932, 1950; Mísař *et al.* 1983; Mísař & Urban 1995), in suggesting that the Velké Vbrno Nappe, as a likely northward continuation and/or structural equivalent of the Moravian Nappes, may represent a distinct terrane located between the Central Sudetes and the Brunovistulian domain.

Further east, the Branná succession has been considered as an original sedimentary cover of the Keprník Nappe (e.g. Schulmann & Gayer 2000), despite being completely detached from its basement. However, the different facies in the presumed Devonian metasediments in the Branná Group and those of the basinal Devonian Vrbno Group in the Desná Dome, and the contrastingly low grade of metamorphism in the Branná rocks compared to the amphibolite-grade metamorphism in both the adjacent Velké Vrbno and Keprník units, suggest that the Branná succession belongs to the 'exotic' Moravian domain between the Brunovistulian and Central Sudetic/Moldanubian domains. In such a model, the Velké Vrbno Nappe will be the upper member and the Branná Unit the lower member of a northern extension of the Moravian nappe pile, showing inverted metamorphic gradient. The Keprník Nappe, adjacent to the east, will thus represent the uppermost member in the separate stack of nappes that involve the reactivated Brunovistulian basement. Hence, the basal thrust of the Branná Unit will mask a cryptic suture between the Moravian and Brunovistulian domains.

The basin in which the Vrbno Group sediments were deposited can hardly be correlated with the Staré Město Suture zone. The latter comprises exclusively Cambrian–Ordovician mafic igneous rocks, as well as older paragneisses of unknown age and synorogenic Carboniferous tonalites. Also, clastic rocks at the base of the Vrbno Group indicate that it must have been deposited in an open basin in Early/Middle Devonian time. Therefore the Vrbno Group may represent sediments of a marine basin which separated the Brunovistulian and Moravian domains. By contrast, the Staré Město Belt corresponds to a separate oceanic basin delimiting the Central Sudetic and Moravian units.

Terranes in the Sudetes

Previous terrane models

Models suggesting several terrane configurations proposed since the early 1990s for the NE Bohemian Massif, can be differentiated into two groups. Some of the models are based on the assumption that the tectonics of the Sudetic area were shaped mainly by the Variscan tectonism during Devonian–Carboniferous times and that the main tectonostratigraphic units of the central European Variscan Belt, mostly as defined by Kossmat (1927), continue into the Sudetes. The other models assume a significant or even predominant role of Caledonian orogenic events during Ordovician through Early–Middle Devonian times and introduce exotic terranes into the Sudetes, that do not show affinities to the main Variscan tectonostratigraphic units.

The idea of a direct continuation of the main Variscan units (though not necessarily understood as mutually significantly diplaced tectonostratigraphic terranes) into the Sudetes has been presented systematically by Franke and coauthors (e.g. Behr *et al.* 1984; Franke 1989; Franke *et al.* 1993, 1995a; Franke & Żelaźniewicz 2000). They consider the West and Central Sudetes to be an easterly continuation of the Saxothuringian Zone, and the East Sudetes an extension of the Rhenohercynian Zone of the Variscides.

The terrane model of Matte *et al.* (1990, repeated in Matte 1991 and 1998 and mostly accepted by Pharaoh 1999), has also suggested a direct continuation into the Sudetes, along a roughly SW–NE trend, of the main Variscan terranes/tectonostratigraphic zones from the western and central Bohemian Massif, as well as from Germany and France. The Saxothuringian Zone of the German Variscides has, thus, been extended into the Lusatian Massif, most of the Karkonosze-Izera Massif, Kaczawa Unit and the tectonic substrate of the Góry Sowie Massif, including the Niemcza Shear Zone. The southern and eastern margins of the Karkonosze-Izera Massif, the eastern Kaczawa Unit and the Góry Sowie Massif, together with the surrounding ophiolites, have been incorporated into the Münchberg-Teplá Terrane. The basement of the Intra-Sudetic Basin, as well as the Nové Město and Kłodzko metamorphic units have been ascribed to the Barrandian Terrane of the central Bohemian Massif. The Gföhl Terrane of the Moldanubian domain of the SE Bohemian Massif has been extrapolated into the Orlica-Śnieżnik Massif of the Sudetes, whereas the East Sudetes (Jeseník metamorphic area and the Carboniferous basin to the east) have been included in the Moravian Terrane. Matte *et al.* (1990) considered that terrane assembly occurred during the Variscan orogeny at 390 to 300 Ma.

Another terrane configuration is implicit in the strike-slip tectonic model of the Sudetes by Aleksandrowski (1990, 1995). The eastern extension of the Saxothuringian domain, located in the part of the West and Central Sudetes south of the Intra-Sudetic Fault was assumed to continue into the Orlica-Śnieżnik Massif. The Kaczawa Unit and the Góry Sowie Massif may have been displaced on the latter fault from the Northern Phyllite Zone at the southern rim of the Rhenohercynian domain in Germany and from the area of the Mid-German Crystalline High, respectively. A combination of the effects of Devonian–Carboniferous strike-slip tectonics with the ideas of Matte *et al.* (1990) about the general NE–SW structural grain of the Sudetic collage was suggested by Aleksandrowski (1998) and Aleksandrowski *et al.* (2000, fig. 1).

Among hypotheses ascribing the formation of the Sudetes to the Caledonian orogeny, early concepts (Don 1984, 1990), though not conceived in terms of tectonostratigraphic terranes or mobilistic tectonic solutions, assumed nevertheless, that the portion of the West Sudetes located north of the Intra-Sudetic Fault had undergone the 'Hercynian' orogenic cycle, whereas that located south of the fault had showed an entirely different geological evolution, being consolidated by the Caledonian orogeny.

Ideas that, unlike in the Variscan belt proper, the Caledonian orogeny played a predominant role in the Sudetes during Ordovician through Early-Mid Devonian times (Oliver *et al.* 1993; Johnson *et al.* 1994), led these authors to place a Gondwana-Baltica, Caledonian Tornquist Suture along the Intra-Sudetic Fault zone and to recognize a number of (exotic?) pre-Variscan terranes in the West Sudetes, apparently unrelated to the well-established major Variscan tectonostratigraphic zones. The terranes included a central, 'Sudetic Batholith Terrane', comprising all the *c.* 500 Ma metagranites, surrounded and, partly, cross-cut by the adjacent Rudawy Janowickie, Kłodzko, Góry Sowie and Kaczawa terranes. A partly similar proposal (Cymerman & Piasecki 1994; Cymerman *et al.* 1997), incorporating also some elements from the concept of Matte *et al.* (1990), explained the evolution of the Sudetes in terms of superposition of successive orogenic cycles. Cymerman *et al.* (1997) distinguished a late Proterozoic, Lusatian Terrane from a Caledonian, Saxothuringian one, comprising the Karkonosze-Izera Massif and the western Kaczawa Unit. The Saxothuringian Terrane was believed to be separated by a Caledonian suture along the Rudawy Janowickie and the 'Kaczawa Line' from the mostly Silurian Central Sudetic Terrane, including the eastern Kaczawa Unit and the basement to the Intra-Sudetic and Świebodzice basins, the Kłodzko Metamorphic Unit and the circum-Góry Sowie ophiolites together with the Niemcza Shear Zone and the tectonic substrate to the Góry Sowie. The Central Sudetic Terrane, tectonically overlain by a high-grade Góry Sowie Terrane, was interpreted as having accreted in Early Carboniferous times to the Moldanubian Terrane, represented in the Sudetes by the Orlica-Śnieżnik Massif.

A distinct view on the Variscan tectonics of central Europe, including the Sudetes, has been held by Krohe (1996). He argued that the original Armorica- and Avalonia-derived terranes in this area were assembled during Late Ordovician to Early Devonian times and were later intensely sheared, displaced and reshuffled along numerous strike-slip faults by Late Devonian–Early Carboniferous intracontinental tectonism. As a result, the recent puzzle of small fault-bounded crustal blocks comprising the central European Variscides define an assemblage of splinters and slivers broken off from original larger terranes, whose fragmentation precludes reconstruction of any consistent primary larger units or zones.

A comprehensive critical assessment of the evidence invoked in support of the Caledonian orogeny in the Sudetes and of the terrane arrangement suggested by Oliver and collaborators (Oliver *et al.* 1993; Johnston *et al.* 1994) has recently been produced by Aleksandrowski *et al.* (2000; see also Aleksandrowski 1994 and Żelaźniewicz & Franke 1994). In the light of available data, most of this evidence and of the conclusions based on it (in particular, the location of the Tornquist Suture in the Sudetes) cannot be sustained. Similar reservations apply to the model of Cymerman & Piasecki (1994) and Cymerman *et al.* (1997). Their terrane distribution also overlooked the significance of the Intra-Sudetic Fault, whereas a number of the inferred terrane boundaries do not seem to follow any real geological lines (e.g. their southern and NW boundaries of the Central Sudetic Terrane and the boundary between their Lusatian and Saxothuringian terranes). The regional kinematic model of of Cymerman & Piasecki (1994) and Cymerman *et al.* (1997), assuming, for example significant southward tectonic transport of the Góry Sowie allochthonous massif does not fit the regional tectonic context. On the other hand, these authors rightly related the Orlica-Śnieżnik Massif to the Gföhl Unit of the Moldanubian Zone (after Matte *et al.* 1990) and assembled several units into their Central Sudetic Terrane.

Among the terrane models published to date, that of Matte *et al.* (1990) most closely fits the available geological data. Our criticisms concern a few particular, relatively detailed solutions adopted in it. The Münchberg-Teplá Terrane of these authors has already been discussed critically (Mazur & Aleksandrowski 2001*a*). This terrane does not appear to continue, as assumed by Matte *et al.* (1990), north of the Intra-Sudetic Fault into the eastern Kaczawa Unit, the Góry Sowie Massif and, in particular, the Central Sudetic Ophiolites, as the latter show a different age from that of the mafic rocks in the Münchberg-Teplá (or South Karkonosze; Mazur & Aleksandrowski 2001*a*) Terrane. The Niemcza Shear Zone, included by Matte *et al.* (1990) in the Saxothuringian Terrane, in our opinion does not seem to fit there. Matte *et al.* (1990) did not extend their model onto the NE Fore-Sudetic Block.

Terrane boundaries: tectonic sutures and strike-slip shear zones

Rock associations typical of tectonic suture zones occur in the Sudetes, generally, within at least three areas/zones. The first one extends along the southern and eastern margins of the Karkonosze-Izera Massif. It is typified by Cambrian/Ordovician protolith ages of MORB-type basalt complexes and by blueschist facies metamorphism in late Devonian time. It resulted from the closure of the Saxothuringian oceanic basin. The second suture zone comprises the Central Sudetic ophiolitic bodies, which show Late Silurian(?) to Mid-Devonian protolith ages and have experienced pre-Famennian deformation and metamorphism. To which Variscan suture this ophiolite belt corresponds is, as yet, unclear. The third suture zone is the Staré Město Belt with Cambrian/Ordovician ages of igneous protoliths, metamorphosed and deformed during Carboniferous collision. This suture zone can be either correlated with the Rheic (Rheno-Hercynian) Suture, bounding the Rheno-Hercynian domain of the Variscides (i.e. Finger *et al.* 1998), considered a part of a suture related to the SE branch of the Variscan orogen (cf. Matte 1986, 1991; Matte *et al.* 1990) or can constitute a separate, local suture between the Central Sudetic domain and the Brunovistulian Block. One candidate for a suture zone is the boundary between the Nové Město Belt and the Orlica-Śnieżnik Massif. A hypothetical and cryptic suture, may be concealed below the Świebodzice Basin and/or the Góry Sowie Gneiss Massif and related to the Kaczawa thrust stack. Another candidate for a cryptic suture is the basal thrust of the Branná Unit in the East Sudetes. The above sutures almost always follow thrust zones/belts (sometimes, as is the case with the East Karkonosze or the NE boundary of the Nové Město Belt, the original thrusts were later reactivated in normal or strike-slip regime), corresponding to terrane boundaries. The remaining terrane boundaries in the Sudetes are main strike-slip ductile shear or fault zones, notably the Intra-Sudetic Fault zone and the Niemcza and Skrzynka shear zones (Aleksandrowski *et al.* 1997) or a combination of thrust with strike-slip tectonics, as in the case of the Kłodzko Metamorphic Unit.

Terranes

From the preceding overview of the particular structural units and discussion of various aspects of their geology, the Sudetes may be divided into distinct tectonostratigraphic terranes of different provenance and evolution. From NW to SE, we propose the following tectonostratigraphic terranes to be distinguished on the NE margin of the Bohemian Massif (Fig. 10).

(1) The Lusatia-Izera Terrane comprises the Lusatian Massif and the Kowary-Izera Unit of the Karkonosze-Izera Massif, which both correspond to crystalline basement of the passive margin of Saxothuringian epicontinental area. The Görlitz Slate Belt and the Ještěd Unit are made up of Variscan-deformed sedimentary successions deposited on this passive margin.

(2) The South-East Karkonosze Terrane is thrust over the Lusatia-Izera Terrane and comprises the South Karkonosze and Leszczyniec units of the Karkonosze-Izera Massif. These units contain the (meta)sedimentary–volcanic succession of the Saxothuringian oceanic basin and the adjacent oceanic basin floor, with the latter overthrust upon the basin infill during Late Devonian–Early Carboniferous times. This terrane contains a fragment of the Saxothuringian Suture.

(3) The Kaczawa Terrane is separated from the Lusatia-Izera Terrane by the strike-slip Intra-Sudetic Fault zone and from the Góry Sowie–Kłodzko Terrane by an inferred thrust contact (Seston *et al.* 2000). It contains a Palaeozoic rift-to-oceanic volcanosedimentary succession involved in a Variscan accretionary prism, which formed during the latest Devonian-earliest Carboniferous(?). The affinities of the Kaczawa succession are still unclear (Saxothuringian or Rheic oceanic basin infill?).

(4) The composite Góry Sowie–Kłodzko Terrane includes the Góry Sowie Massif and the NE part of the Kłodzko Metamorphic Unit, the Świebodzice and Bardo basins, the Central Sudetic ophiolites, the Niemcza Shear Zone, and NE part of the basement to the Intra-Sudetic Basin. These units represent a variety of palaeotectonic environments, including subducted continental crust (Góry Sowie), Middle Devonian passive margin (Kłodzko Metamorphic Unit), oceanic crust (Central Sudetic ophiolites) and synorogenic basins (Świebodzice and Bardo). The principal common feature for the units of this terrane is Eo-Variscan, pre-Upper Devonian deformation, metamorphism and exhumation, which took place significantly earlier than in other Sudetic terranes. This terrane shows some affinities with parts of the NE Massif Central and Armorican Massif in France as well as with the Teplá-Barrandian Terrane that extends well beyond the Sudetes into the central Bohemian Massif.

(5) The Teplá-Barrandian Terrane, in its Sudetic segment, is interpreted here to

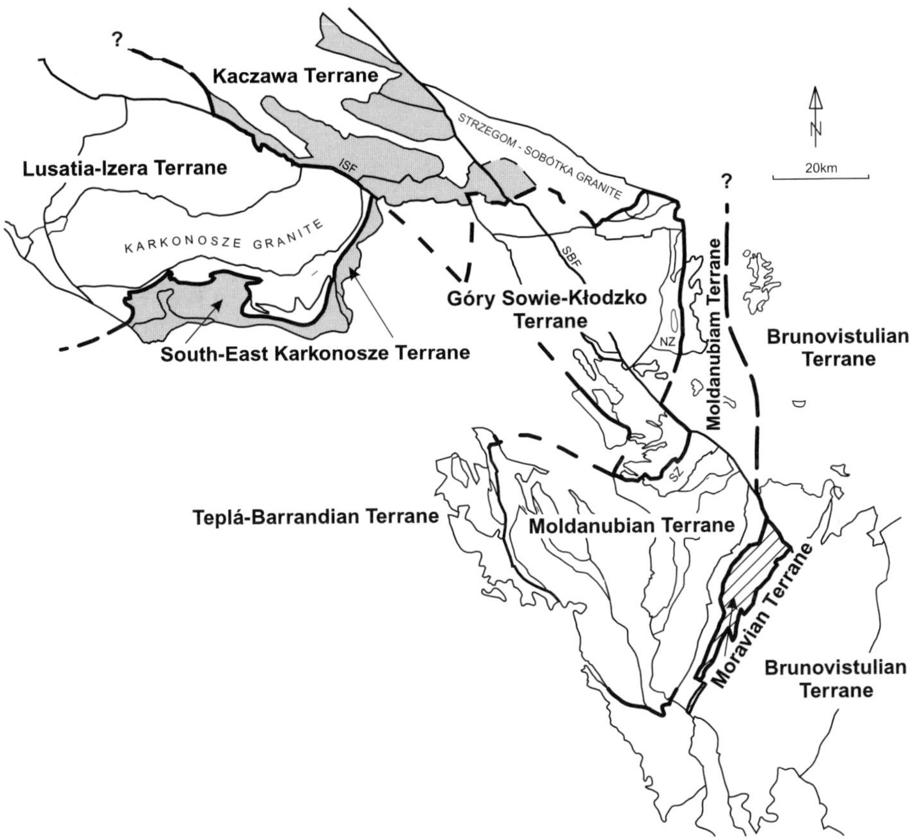

Fig. 10. Tectonostratigraphic terranes in the Sudetes.

incorporate the Nové Město Belt, Zábřeh Unit and the SW part of the basement of the Intra-Sudetic Basin (cf. Mazur & Aleksandrowski 2001*b*), which crops out in the SW part of the Kłodzko Metamorphic Unit. It cannot be excluded that the Góry Sowie-Kłodzko Terrane is part of the Teplá-Barrandian.

(6) The Moldanubian (Gföhl) Terrane in the Sudetes comprises the Orlica-Śnieżnik Massif, the Kamieniec Metamorphic Belt and the Staré Město Belt. Its western boundary with the Góry Sowie-Kłodzko Terrane bounds the sinistral Niemcza and Skrzynka shear zones; with the Teplá-Barrandian Terrane it is a primary thrust, later converted into a dextral ductile fault, and with the Moravian Terrane it is a thrust belt. The rock complexes of the terrane display prominent Early Carboniferous collision-related medium- to high-grade (high pressure-high temperature and high temperature-medium to low pressure) metamorphism and deformation, analogous to that of the Gföhl Terrane of the Moldanubian domain in the Bohemian Massif further south.

(7) The Moravian Terrane is inferred to occur between the Moldanubian and Brunovistulian terranes, by analogy with the situation further south in Moravia. This narrow terrane is represented by the Velké Vrbno and Branná thrust units of the East Sudetes.

(8) The Brunovistulian Terrane comprises the Keprník Nappe and Desná Dome and the successive, easterly located units of the East Sudetes, including the Culm basin and the coal-bearing molasse basin at the East-Sudetic foreland, both basins established on Brunovistulian basement of Avalonian affinities (Friedl *et al.* 2000; Finger *et al.* 2000). Within the Fore-Sudetic Block, the Brunovistulian Terrane includes also the Strzelin Crystalline Massif of similar characteristics.

The proposed terrane distribution is in many respects similar to those earlier suggested by

Matte *et al.* (1990), Cymerman & Piasecki (1994) and Cymerman *et al.* (1997). However, certain solutions incorporate new field and laboratory data and use different criteria to distinguish the tectonostratigraphic terranes.

Discussion

'Oceanic' Kaczawa Terrane and the Central Sudetic ophiolites: Saxothuringian or Rheic realm? Similar geological characteristics of the Kaczawa and South Karkonosze Terranes suggest that they may represent a common 'oceanic' terrane derived from the Saxothuringian ocean. On the other hand, the correlation between them is hampered by the still fairly incomplete knowledge of the stratigraphy of both areas and by the fact that they are separated by a major strike-slip shear zone of the Intra-Sudetic Fault. Large-scale dextral displacement up to about 300 km during Late Devonian-Carboniferous times was proposed for this fault zone (Aleksandrowski 1990, 1995). The displacement was believed to have moved the Kaczawa Unit and Góry Sowie Massif from their original locations thought to be the Northern Phyllite Zone and Mid-German Crystalline High, respectively, of the German Variscides (Figs 1 and 2) and to have emplaced them in their present position, adjacent to units of different provenance SW of the Intra-Sudetic Fault (Aleksandrowski 1990, 1995; Aleksandrowski *et al.* 1997). Indeed, there is a general resemblance between the south Karkonosze unit and that of the Kaczawa terrane. Likewise, correlation of the Kaczawa and Góry Sowie with the Northern Phyllite and Mid-German Crystalline High, still cannot be rejected, and if correct, would imply a possible linkage between the Central Sudetic ophiolites and similar bodies of the Lizard-Rhenish suture (see discussion below).

Góry Sowie–Kłodzko Terrane: deformed, exhumed and subsequently rifted prior to the main collision. The Central Sudetes within the Góry Sowie–Kłodzko Terrane are typified by certain features that make this area different from all other segments of the Bohemian Massif. Unlike the Teplá-Barrandian domain or Münchberg Massif, the Central Sudetes contain Late Silurian/Early Devonian high pressure granulites in the Góry Sowie Massif, comparable to the type I granulites of Pin & Vielzeuf (1988) considered as typical of Eo-Variscan complexes. The Central Sudetes also record late Silurian(?)-Devonian rifting (up to the oceanic stage), documented by approximately 400 Ma ages of the circum-Góry Sowie ophiolites (Żelaźniewicz *et al.* 1998). Similarly, no pre-Famennian angular unconformity on top of the obducted ophiolites and mid-Devonian metasediments is known from other places in the Bohemian Massif.

Is the Góry Sowie–Kłodzko Terrane in the Central Sudetes an exotic element in the Bohemian Massif, as tacitly assumed by Cymerman & Piasecki (1994) and Cymerman *et al.* (1997)? In our opinion, a better solution is to attempt to correlate this area with appropriate Variscan segments outside the Sudetes. On the basis of numerous mutual similarities, the Góry Sowie–Kłodzko Terrane can be compared to the basement of the North Bohemian Basin within the Teplá-Barrandian Terrane, to the Münchberg Massif and, in particular, to the 'Moldanubian Zone' in the NE part of the French Massif Central and southern Armorican Massif. If such a parallel were correct, the Central Sudetic Ophiolite could be expected to represent an obducted relict of closure of a local basin underlain by oceanic crust, probably of back-arc origin. The formation of this basin would reflect Early Devonian extension and rifting, otherwise relatively widespread in the Variscan belt. The main difficulty, however, in trying to correlate the Central Sudetes with the NE Massif Central is the lack of a direct spatial link between the two areas.

An alternative interpretation for the Central Sudetic Ophiolite, based on its protolith age, is to locate it in the Rheno-Hercynian (Rheic) suture (cf. Finger *et al.* 1998). The recent dating of the Lizard Ophiolite (Clark *et al.* 1998; 397 ± 2 Ma, U–Pb zircon) yielded almost identical results to those obtained from the Central Sudetic Ślęża Ophiolite. The Rheic suture may continue into the Sudetes either due to oroclinal bending of the Variscan belt (e.g. Engel & Franke 1983; Franke 1989) or as the result of dextral displacement on large-scale strike-slip faults (Aleksandrowski 1990, 1995).

Irrespective of the assumed interpretation, the Early Devonian high-grade metamorphism in the Góry Sowie must have been the effect of continental crust subduction due to the collision of Avalonia and Armorica from Silurian to Devonian times. However, as no similar ages of high-grade metamorphism have so far been reported from the Rheno-Hercynian or Mid-German Crystalline zones, it is not possible to correlate them with the Sudetes.

Niemcza and Skrzynka strike-slip shear zones: an intra-Central Sudetic terrane boundary. The Central Sudetes are cut by the Niemcza and Skrzynka late-orogenic sinistral shear zones

(Mazur & Puziewicz 1995; Cymerman 1996; Aleksandrowski *et al.* 1997). In separating the Góry Sowie–Kłodzko Terrane to the west and north, from the Orlica-Śnieżnik Massif and the Kamieniec Metamorphic Belt, i.e. the Gföhl Terrane to the east and south, they separate the NW domain with features similar to those of the NE Massif Central and Teplá-Barrandian from the SE domain which is analogous in many respects to the Moldanubian zone in the Moravia and Austria. The Niemcza and Skrzynka shear zones can be considered together as a sinistral shear belt that splits the Central Sudetes into terranes of dissimilar deformation, metamorphism and exhumation ages. The sinistral displacements along the Niemcza and Skrzynka shear zones presumably accommodated much strain and protected the area west of them from intense tectonism caused by the collision of the Moldanubian and Moravian terranes to the east.

In contrast to other Central Sudetic units, which had already undergone tectonism and exhumation prior to the Late Devonian, the Orlica-Śnieżnik Massif mainly records the effects of the Carboniferous convergent stage. This may have resulted from possible obliteration of any intra-Devonian tectonic events by high-grade metamorphism that affected the Orlica-Śnieżnik rocks during Late Devonian–Early Carboniferous times. Possible intra-Devonian exhumation of some subunits of the Orlica-Śnieżnik Massif is consistent with the largely neglected discovery by Kasza (1964) of unmetamorphosed conglomerates of unknown age wedged between thrust slices involving the Śnieżnik crystalline rocks.

Affinities of the Orlica-Śnieżnik Massif and Kamieniec Metamorphic Belt with the Moldanubian domain of the SE Bohemian Massif. The analogies between the Orlica-Śnieżnik Massif and the Gföhl Nappe in the Czech and Austrian parts of the Moldanubian Zone were described by Matte *et al.* (1990). Further evidence that the two areas are equivalent is provided by new zircon SHRIMP dating of gneisses from the Orlica-Śnieżnik and from the Gföhl units (Friedl *et al.* 2000; Turniak *et al.* 2000). The gneissic protoliths in both areas show nearly identical ages of around 500 Ma, and the high temperature and moderate to low pressure metamorphism, resulting in migmatization, has been determined, again in both areas, at 340 Ma. The rocks of the Kamieniec Metamorphic Belt on the Fore-Sudetic Block also show numerous similarities to Gföhl rocks exposed further south. The higher-grade Kamieniec metamorphic complex is in many respects analogous to the mica schist zone at the sole of the Moldanubian nappes.

The prolongation of the collision zone from the East Sudetes into the Fore-Sudetic Block. Both terrane boundaries from the East Sudetes (i.e. that separating the Moldanubian and Moravian, as well as that between the Moravian and Brunovistulian terranes) presumably continue to the north, over the Fore-Sudetic Block. The Moldanubian/Moravian boundary follows the eastern limit of the Niedźwiedź Massif, whereas the Moravian/Brunovistulian one is presumably buried below Cenozoic deposits further to the east. The largest outcrop zone of crystalline basement in the eastern part of the Fore-Sudetic Block is the Strzelin Crystalline Massif located to the east of that of Niedźwiedź. The Strzelin Massif rocks show an overall similarity to the basement of the Desná Dome, in containing both Devonian quartzites of imprecise age and Neoproterozoic orthogneisses dated at 600–570 Ma (Oberc-Dziedzic *et al.* 2001). Inherited zircons from these gneisses yielded ages of around 1230 and 1880 Ma (Oberc-Dziedzic *et al.* 2001), pointing to their possible Avalonian affinities.

Possible provenance of the Brunovistulian Terrane. Currently, a lively discussion continues about the problem whether the Brunovistulian microcontinent represents an easternmost Avalonian fragment (Moczydłowska 1997; Friedl *et al.* 2000, Finger *et al.* 2000), or a distinct Gondwana-derived terrane, which docked to Baltica during Early Cambrian times (Belka *et al.* 2000; Valverde-Vaquero *et al.* 2000). The two alternative solutions are based on different interpretation of palaeontological, geochronological and palaeomagnetic data. Both of these solutions imply the existence of an extensive ocean between the Moldanubian and Brunovistulian domains during Early Palaeozoic and Devonian times. If such an extensive oceanic domain existed between the Moldanubian and Brunovistulian terranes during the Early Palaeozoic, which of the two basins that later closed along the eastern edge of the Bohemian Massif could have been this ocean? Most probably, it was not the basin that separated the Moravian and Brunovistulian terranes, since this one opened only as late as the Devonian. Therefore, the remnant of the postulated ocean should be the Staré Město Suture.

Rotation of the Brunovistulian Terrane? The terrane model proposed here does not take into account the clockwise rotation by 90° to 135° of the Brunovistulian basement with respect to the German segments of the Variscan belt, postulated by Krs *et al.* (1994) and Tait *et al.* (1996) to have occurred since the Late Devonian. This rotation was used by Tait *et al.* (1996) to support the oroclinal model of the eastern termination of the belt. However, the Devonian limestones used for magnetic measurements have undergone significant tectonic strain associated with ductile thrusting and intense folding (Kettner 1949; Kalvoda & Melichar 1999), and we believe this has not been sufficiently accounted for in both the interpretations assuming significant rotation. Hence, the problem of possible rotation requires further study.

Conclusion: the accretionary setting of the Sudetic terranes

The Sudetic segment of the Variscan Orogenic Belt formed by accretion of the following terranes: the Moldanubian (Gföhl), Góry Sowie–Kłodzko and NE Teplá-Barrandian, making up the Central Sudetes and the Lusatia-Izera and Brunovistulian terranes located in the West and East Sudetes, respectively. Between these large units are smaller terranes compressed along their boundaries, namely the Moravian Terrane between the Moldanubian and Brunovistulian, and the SE Karkonosze and Kaczawa terranes between the Lusatia-Izera and Teplá on one side and the Góry Sowie–Kłodzko on the other side. The Moravian Terrane in the East Sudetes is inferred to be an extension of the likewise named terrane known from the tectonic windows in southern Moravia, whereas the SE Karkonosze and Kaczawa terranes probably represent the Saxothuringian oceanic realm.

Amalgamation of the Sudetic segment of the Variscan Belt continued from Late Silurian to Early Carboniferous times. It was not a continuous process of terrane convergence, but rather a sequence of several distinct orogenic events separated by stages of extension and rifting. In the Late Silurian an amalgamation of the composite Góry Sowie–Kłodzko Terrane began (as inferred by analogy with the situation in the Massif Central and Armorica) during the Eo-Variscan orogenic cycle. The Góry Sowie Massif represents a fragment of continental crust that had been deeply subducted. By analogy with the geological relationships in the Massif Central and Brittany, there may have been NW-directed (?) subduction of a small oceanic domain together with a fragment of a Gondwana-derived continental crust below the southern, active margin of Armorica (Faure *et al.* 1997).

Following Eo-Variscan compression, Devonian extension took place in various segments of the Variscan Belt. In the Sudetes it resulted in opening of two basins: the Central Sudetic oceanic basin floored by oceanic crust, now preserved as the Central Sudetic Ophiolite, and the Vrbno Group basin, separating the Moravian and Brunovistulian terranes. In Middle to Late Devonian times, the Central Sudetic oceanic basin was closed and its floor partly obducted as the Central Sudetic Ophiolite. At the same time the Kłodzko Metamorphic Unit underwent deformation and metamorphism. Soon afterwards, during the Late Devonian, a new sedimentary cycle commenced in the NW Central Sudetes, recorded by deposition in the Bardo Basin. Simultaneously, in the western part of the Sudetes a SE-directed subduction of the Saxothuringian Ocean continued below the active margin of the already amalgamated Teplá-Barrandian and Góry Sowie-Kłodzko terranes. In the east, the subduction of the oceanic basin separating the Moldanubian and Moravian terranes, probably commenced in Late Devonian times.

At the turn of Devonian/Carboniferous times a collision started between the Saxothuringian epicontinental domain, represented in the Sudetes by the Lusatia-Izera Terrane, and the already amalgamated Teplá-Barrandian and Góry Sowie–Kłodzko terranes, which brought about the Saxothuringian Suture zone. In the West Sudetes this also led to northwestward thrusting in the Karkonosze-Izera Massif, including the overthrusting of the SE Karkonosze Terrane. The accretion of the Sudetic area was completed by early/mid Carboniferous times, when collision between the Moldanubian Terrane (at that time already amalgamated with the West and Central Sudetes) and the Moravian and Brunovistulian terranes occurred. The Staré Město Suture zone developed at that time, together with the nappe structure of the East Sudetes.

Financial support from Wrocław University grants 2022/W/ING/01 and 1017/S/ING/01 and the EU TMR Network 'Palaeozoic Amalgamation of Central Europe', is gratefully acknowledged. Many thanks are due to John Winchester, Christian Pin, and Jan Zalasiewicz for their helpful comments and corrections made on the text.

References

ALEKSANDROWSKI, P. 1990. Early Carboniferous strike-slip displacements at the northeast periphery of the Variscan Belt in Central Europe. *In*: *International Conference on Paleozoic Orogens in Central Europe* (IGCP Program 233: Terranes in the Circum-Atlantic Paleozoic Orogens), *Abstracts*, 7–10. Göttingen-Giessen.

ALEKSANDROWSKI, P. 1994. Discussion on U-Pb ages from SW Poland: evidence for a Caledonian suture zone between Baltica and Gondwana. *Journal of the Geological Society, London*, **151**, 1049–1050.

ALEKSANDROWSKI, P. 1995. The significance of major strike-slip displacements in the development of Variscan structure of the Sudetes, SW Poland. *Przegląd Geologiczny,* **43**, 745–754. [in Polish; English abstract].

ALEKSANDROWSKI, P. 1998. Pre-Variscan and Variscan terranes in SW Poland. *In:* U. LINNEMANN *et al.* (eds) *Schriften des Staatlichen Museums für Mineralogie und Geologie zu Dresden,* **9,** 93–95 (Abstracts Intern. Geological Conference: *Pre-Variscan Terrane Analysis of Gondwanan Europe*. Dresden, Germany).

ALEKSANDROWSKI, P., KRYZA, R., MAZUR, S. & ŻABA, J. 1997. Kinematic data on major Variscan strike-slip faults and shear zones in the Polish Sudetes, northeast Bohemian Massif. *Geological Magazine*, **133**, 727–739.

ALEKSANDROWSKI, P., KRYZA, R., MAZUR, S., PIN, C. & ZALASIEWICZ, J. A. 2000. The Polish Sudetes: Caledonian or Variscan ? *Transactions of the Royal Society, Edinburgh*, **90**, 127–146.

AWDANKIEWICZ, H. 2001. Palaeozoic mafic magmatism in the Sudetes region: further constraints from the geochemistry of the Niedźwiedź Massif metabasites (Fore-Sudetic Block), SW Poland. *EUG XI, Strasbourg, Journal of Conference Abstracts* **6**, 580.

BARANOWSKI, Z., HAYDUKIEWICZ, A., KRYZA, R., LORENC, S., MUSZYŃSKI, A. & URBANEK, Z. 1987. The development of the eastern part of the Góry Kaczawskie structure. *In:* BARANOWSKI, Z.,GROCHOLSKI, A., MALINOWSKI, J., OBERC, J. & PORĘBSKI, S. (eds) *The Field Guide of 58th Meeting of the Polish Geological Society, Wałbrzych*. Instytut Geologiczny & Uniwersytet Wrocławski, Krakow, 61–114 [in Polish].

BARANOWSKI, Z., HAYDUKIEWICZ, A., KRYZA, R., LORENC, S., MUSZYŃSKI, A., SOLECKI, A. & URBANEK, Z. 1990. Outline of the geology of the Góry Kaczawskie (Sudetes, Poland). *Neues Jahrbuch für Geologie und Paläontologie, Abhandlungen*, **179**, 223–257.

BEARD, B. L., MEDARIS JR, L. G., JOHNSON, C. M., JELINEK, E., TONIKA, J. & RICIPUTI, L. R. 1995. Geochronology and geochemistry of eclogites from the Mariánské Lázně complex, Czech Republic: implications for Variscan orogenesis. *Geologische Rundschau*, **84**, 552–567.

BEDERKE, E. 1924. Das Devon in Schlesien und das Alter der Sudetenfaltung. *Fortschritte der Geologie und Paläeontologie*, **7**, 1–50.

BEDERKE, E. 1929. Die varistische Tektonik der mittleren Sudeten. *Fortschritte der Geologie und Paläeontologie*, **23**, 429–524.

BEHR, H. J., ENGEL, W., FRANKE, W., GIESE, P. & WEBER, K. 1984. The Variscan belt in central Europe: main structures, geodynamic implications, open questions. *Tectonophysics* **109**, 15–40.

BELKA, Z., AHRENDT, H., FRANKE, W. & WEMMER, K. 2000. The Baltica-Gondwana suture in central Europe: evidence from K-Ar ages of detrital muscovites and biogeographical data. *In:* FRANKE, W., HAAK, V., ONCKEN, O. & TANNER, D. (eds) *Orogenic processes: Quantification and modelling in the Variscan Belt.* Geological Society, London, Special Publications, **179**, 87–102.

BENDL, J. & PATOČKA, F. 1995. The Rb-Sr isotope geochemistry of the metamorphosed bimodal volcanic association of the Rýchory Mts crystalline complex, West Sudetes, Bohemian Massif. *Geologia Sudetica*, **29**, 3–18.

BENDL, J., PATOČKA, F. & PIVEC, E. 1997. The ^{87}Rb-^{86}Sr isotope geochemistry of the blueschists and greenschists metavolcanics of the Rýchory Mts. crystalline complex, West Sudetes, Bohemian Massif. *Geologia Sudetica*, **30**, 3–20.

BORKOWSKA, M., HAMEURT, J. & VIDAL, P. 1980. Origin and age of Izera gneisses and Rumburk granites in the Western Sudetes. *Acta Geologica Polonica*, **30**, 121–46.

BOUTIN, R., MONTIGNY, R. & THUIZAT, R. 1995. Chronologie K-Ar et ^{40}Ar/^{39}Ar du métamorphisme et du magmatisme des Vosges. Comparaison avec les massifs varisques avoisinants. *Géologie de la France*, **1**, 3–25.

BRAUSE, H. & HIRSCHMANN, G. (eds) 1969. *Alt- und Vorpaläozoikum des Görlitzer Schiefergebirges und der westlichsten Westsudeten*. Exkursionsführer, Deutsche Gesellschaft für geologische Wissenschaften, Berlin, 1–115.

BRÖCKER, M., ŻELAŹNIEWICZ, A. & ENDERS, M. 1998. Rb-Sr and U-Pb geochronology of migmatitic layers from the Góry Sowie (West Sudetes, Poland): the importance of Mid-Late Devonian metamorphism. *Journal of the Geological Society, London*, **155**, 1025–36.

BRUECKNER, H. K., MEDARIS, L. G. JR. & BAKUN-CZUBAROW, N. 1991. Nd and Sr age of isotope patterns from Variscan eclogites of the eastern Bohemian Massif. *Neues Jahrbuch für Mineralogie, Abhandlungen*, **163**, 169–193.

BRUECKNER, H. K., BLUSZTAJN, J. & BAKUN-CZUBAROW, N. 1996. Trace element and Sm-Nd ‘age’ zoning in garnets from peridotites of the Caledonian and Variscan mountains and tectonic implications. *Journal of Metamorphic Geology*, **14**, 61–73.

ČECH, S., HAVLÍČEK, V. & ZIKMUNDOVÁ, J. 1989. The Upper Devonian and Lower Carboniferous in north-eastern Bohemia (based on boreholes in the Hradec Králové area). *Věstník Ustředního Ústavu geologického*, **64**, 65–76.

CHÁB, J. & VRÁNA, S. 1979. Crossite-actinolite amphiboles of the Krkonoše-Jizera crystalline complex and their geological significance. *Věstník Ustředního Ústavu geologického*, **54**, 143–150.

CHALOUPSKÝ, J. 1965. Metamorphic development of the Krkonoše crystalline complex. *Krystalinikum*, **3**, 31–54.

CHALOUPSKÝ, J. (ed.) 1989. *Geology of the Krkonoše and Jizerské hory Mts*. Ustřední Ústav geologický, Praha. 288 p. [in Czech, English summary].

CHALOUPSKÝ, J., CHLUPÁČ, I., MAŠEK, J., WALDHAUSROVÁ, J. & CHÁB, J. 1995. VII. Moldanubian Region, B. Teplá-Barrandian Zone, 1.Stratigraphy. *In*: DALLMEYER, R. D., FRANKE, W. & WEBER, K. (eds) *Pre-Permian geology of Central and Eastern Europe,* Springer, Berlin, 379–391.

CHLUPÁČ, I. 1993. Stratigraphic evaluation of some metamorphic units in the N part of the Bohemian Massif. *Neues Jahrbuch für Geologie und Paläontologie, Abhandlungen*, **188,** 363–388.

CHLUPÁČ, I. 1997. Palaeozoic ichnofossils in phyllites near Železný Brod, northern Bohemia. *Journal of the Czech Geological Society*, **42**, 75–94.

CHLUPÁČ, I. & HLADIL, J. 1992. New Devonian occurrences in the Ještěd Mts, North Bohemia. *Časopis pro Mineralogii a Geologii*, **37**, 185–191.

CHLUPÁČ, I. & SVOBODA, J. 1963. Geological relationships in the Konice-Mladeč Devonian on the Drahany Upland. *Sbornik Ustředního ústavu geologického*, **28**, 347–386 [in Czech].

CHLUPÁČ, I. & ZIKMUNDOVÁ, J. 1976. The Devonian and Lower Carboniferous in the Nepašice borehole in East Bohemia. *Věstník Ustředního Ústavu geologického*, **51**, 269–278.

CHOROWSKA, M. 1974. *The Upper Devonian of vicinities of Kłodzko in the light of conodont study*. PhD thesis, Polish Geological Institute [in Polish].

CHOROWSKA, M. 1978. Viséan limestones in the metamorphic complex of the Kaczawa Mts (Sudetes). *Annales de la Societé géologique de Pologne*, **48**, 245–261 [in Polish, English summary].

CHOROWSKA, M., MILEWICZ, J. & RADLICZ, K. 1986. Preliminary results of borehole Zdanów IG-1. *Przegląd Geologiczny*, **10**, 596–597 [in Polish].

CLARK, A. H., SCOTT, D. J., SANDEMAN, H. A. & BROMLEY, A. V. 1998. Siegenian generation of the Lizard ophiolite: U-Pb zircon age data for plagiogranite, Porthkerris, Cornwall. *Journal of Geological Society, London,* **155**, 595–598.

COLLINS, A. S., KRYZA, R. & ZALASIEWICZ, J. 2000. Microfabric fingerprints of Late Devonian – Early Carboniferous subduction in the Polish Variscides, the Kaczawa Complex., Sudetes. *Journal of Geological Society, London,* **157**, 283–288.

COSTA, S. & MALUSKI, H. 1988. Datations par la méthode $^{40}Ar/^{39}Ar$ de matériel magmatique et métamorphique paléozoïque provenant du forage de Couy-Sancerre (Cher, France). Programme GPF. *Comptes Rendus de l'Académie des Sciences, Paris*, **306**, 351–356.

CYMERMAN, Z. 1992. Rotational ductile deformations in the Śnieżnik Metamorphic Complex. *Geological Quarterly*, **36**, 393–420.

CYMERMAN, Z. 1996. The Złoty Stok – Trzebieszowice regional shear zone: the boundary of terranes in the Góry Złote Mts (Sudetes). *Geological Quarterly*, **40**, 89–118.

CYMERMAN, Z. Structural analysis and Variscan tectonic development of the Kaczawa metamorphic unit. *Prace Państwowego Instytutu Geologicznego* [in Polish, English summary] in press.

CYMERMAN, Z. & JERZMAŃSKI, J. 1987. The metamorphic complex in the eastern part of the Fore-Sudetic block at Niedźwiedź, near to Ziębice. *Kwartalnik Geologiczny*, **31**, 239–262 [in Polish].

CYMERMAN, Z. & PIASECKI, M. A. J. 1994. The terrane concept in the Sudetes, Bohemian Massif. *Geological Quarterly*, **38**, 191–210.

CYMERMAN, Z. & PIASECKI, M. A. J. & SESTON, R. 1997. Terranes and terrane boundaries in the Sudetes, northeast Bohemian Massif. *Geological Magazine*, **134**, 717–725.

DALLMEYER, R. D., FRANKE, W. & WEBER, K. 1995. (eds) *Pre-Permian Geology of Central and Eastern Europe,* Springer, Berlin.

DATHE, E. 1900. Die Lagerungsverahältnisse des Oberdevon und Kulm am Kalkberge bei Ebersdorf in Schlesien. *Jahrbuch der Königlich Preussischen Geologischen Landesanstalt und Bergakademie*, **21**, 214–237.

DOMEČKA, K. 1970. Pre-Variscan granitoids of the West Sudetes. *Sborník Geologických Věd*, **18**, 161–189 [in Czech, English abstract].

DON, J. 1984. The Caledonides and Variscides in the Western Sudetes. *Przegląd Geologiczny*, **8–9**, 459–68 [in Polish].

DON, J. 1990. The differences in Paleozoic facies-structural evolution of the West Sudetes. *Neues Jahrbuch für Geologie und Paläontologie, Abhandlungen*, **179**, 307–328.

DON, J., DUMICZ, M., WOJCIECHOWSKA, I. & ŻELAŹNIEWICZ, A. 1990. Lithology and tectonics of the Orlica–Śnieżnik Dome, Sudetes: recent state of knowledge. *Neues Jahrbuch für Geologie und Paläontologie, Abhandlungen*, **179**, 159–188.

DUDEK, A. 1980. The crystalline basement block of the Outer Carpathians in Moravia: Bruno-Vistulicum. *Rozpravý České Akademie Věd*, **90**, 1–85.

DUTHOU, J.-L, COUTURIE, J.-P., MIERZEJEWSKI, M. P. & PIN, C. 1991. Age determination of the Karkonosze granite using isochrone Rb-Sr whole rock method. *Przegląd Geologiczny*, **2,** 75–79 [in Polish, English abstract].

DVOŘÁK, J. 1995. VIII. Moravo-Silesian Zone. B.1. Stratigraphy. *In*: DALLMEYER, R. D., FRANKE, W. & WEBER, K. (eds) *Pre-Permian geology of Central and Eastern Europe,* Springer, Berlin, 477–489.

DVOŘÁK, J. & PAPROTH, E. 1969. Über die Position und Tektogenese des Rhenoherzynikums und des Sudeticums in de mitteleuropäischen Varisciden. *Neues Jahrbuch für Geologie und Paläontologie, Monatshefte*, **18**, 65–88.

DZIEDZIC, K. & TEISSEYRE, A. K. 1990. The Hercynian molasse and younger deposits in the Intra-Sudetic Depression, SW Poland. *Neues Jahrbuch für Geologie und Paläontologie, Abhandlungen,* **179**, 285–305.

DZIK, J. 1997. Emergence and succession of Carboniferous conodont and ammonoid communities in the Polish part of the Variscan sea. *Acta Palaeontologica Polonica*, **42**, 57–170.

EBERT, H. 1943. *Das granitische Grundgebirge der öslichen Lausitz*. Preisschriften der Fürstliche Jablonowski Gesellschaft, **58**. Leipzig. pp 119.

EISBACHER, G. H., LÜSCHEN, E. & WICKERT, F. 1989.

Crustal-scale thrusting and extension in the Hercynian Schwarzwald and Vosges, Central Europe. *Tectonics*, **8**, 1–21.

ENGEL, W. & FRANKE, W. 1983. Flysch sedimentation: its relation to tectonism in the European Variscides. *In*: MARTIN, H. & EDER, F. W. (eds) *Intracontinental fold belts*. Springer, Berlin, 289–321.

FAJST, M. 1976. A new unconformity in the Precambrian of the Bohemian Massif. *Časopis pro mineralogii a petrologii*, **21**, 257–275 [in Czech].

FAURE, M., LELOIX, CH. & ROIG, J.-Y. 1997. L'évolution polycyclique de la chaîne hercynienne. *Bulletin de la Societé géologique de France*, **168**, 695–705.

FINGER, F., VON QUADT, A. & STEYRER, H. P. 1998. The ophiolite chain along the western Moravo-Silesian plate margin – A trace of the Rheic suture? *Acta Universitatis Carolinae*, **42**, 244–245.

FINGER, F., HANŽL, P., PIN, C., VON QUADT, A., PIN. C. & STEYRER, H. P. 2000. The Brunovistulian: Avalonian Precambrian sequence at the eastern end of the Central European Variscides?. *In*: FRANKE, W., HAAK, V., ONCKEN, O. & TANNER, D. (eds) *Orogenic processes: Quantification and modelling in the Variscan Belt*. Geological Society, London, Special Publications, **179**, 103–112.

FLOYD, P. A., WINCHESTER, J. A., CIESIELCZUK, J., LEWANDOWSKA, A., SZCZEPAŃSKI, J. & TURNIAK, K. 1996. Geochemistry of early Palaeozoic amphibolites from the Orlica-Śnieżnik dome, Bohemian massif: petrogenesis and palaeotectonic aspects. *Geologische Rundschau*, **85**, 225–238.

FRANKE, W. 1989. Tectonostratigraphic units in the Variscan belt of central Europe. *Geological Society of America Special Paper*, **230**, 67–90.

FRANKE, W. 2000. The mid-European segment of the Variscides: tectonostratigraphic units, terrane boundaries and plate tectonic evolution. *In:* FRANKE, W., HAAK, V., ONCKEN, O. & TANNER, D. (eds) *Orogenic processes: Quantification and modelling in the Variscan Belt*. Geological Society, London, Special Publications, **179**, 35–61.

FRANKE, W. & STEIN, E. 2000. Exhumation of high-grade rocks in the Saxothuringian Belt: geological constraints and geodynamic concepts. *In:* FRANKE, W., HAAK, V., ONCKEN, O. & TANNER, D. (eds) *Quantification and modelling in the Variscan Belt*. Geological Society, London, Special Publications, **179**, 337–354.

FRANKE, W. & ŻELAŹNIEWICZ, A. 2000. The eastern termination of the Variscides: terrane correlation and kinematic evolution. *In:* FRANKE, W., HAAK, V., ONCKEN, O. & TANNER, D. (eds) *Quantification and modelling in the Variscan Belt*. Geological Society, London, Special Publications, **179**, 63–86.

FRANKE, W., ŻELAŹNIEWICZ, A., PORĘBSKI, S. J. & WAJSPRYCH, B. 1993. Saxothuringian zone in Germany and Poland: differences and common features. *Geologische Rundschau*, **82**, 583–99.

FRANKE, W., DALLMEYER, R. D. & WEBER, K. 1995*a*. XI. Geodynamic evolution. *In* DALLMEYER, R. D., FRANKE, W. & WEBER, K. (eds) *Pre-Permian geology of Central and Eastern Europe,* Springer, Berlin, 579–593.

FRANKE, W., KREUZER, H., OKRUSCH, M., SCHÜSSLER, U. & SEIDEL, E. 1995*b*. V. Saxothuringian Basin. C. Exotic Metamorphic Nappes. 1. Stratigraphy, Structure and Igneous Activity. *In* DALLMEYER, R. D., FRANKE, W. & WEBER, K. (eds) *Pre-Permian geology of Central and Eastern Europe,* Springer, Berlin, 277–294.

FRANKE, W., DÖRR, W., SCHÄFER, J., ZULAUF, G. 1998. Exhumation of HP rocks in the Saxothuringian belt: sedimentary and tectonic record. *Acta Universitatis Carolinae*, **42**, 249–250.

FREYER, G. 1968. Konodontenfunde aus dem Oberdevon und Unterkarbon von Dzikowiec Kłodzki (Ebersdorf) und Gołogłowy (Hollenau) in Dolny Śląsk (Niederschlesien). *Jahrbuch für Geologie (Berlin)*, **17**, 60–67.

FREYER, G. 1977. Das Unterkambrium von Görlitz, seine geologische Erforschung und Fossilführung – ein Überblick. *Abhandlungen und Berichte Naturkundemuseums, Görlitz*, **51**, 1–14.

FRIEDL, G., MCNAUGHTON, N. J., FLETCHER, I. R. & FINGER, F. 1998. New SHRIMP-zircon ages for orthogneisses from the southeastern part of the Bohemian Massif (Lower Austria). *Acta Universitatis Carolinae*, **42,** 251–252.

FRIEDL, G., FINGER, F., MCNAUGHTON, N. J. & FLETCHER, I. R. 2000. Deducing the ancestry of terranes: SHRIMP evidence for South America-derived Gondwana fragments in central Europe. *Geology*, **28**, 1035–1038.

FRITZ, H. & NEUBAUER, F. 1993. Kinematics of crustal stacking and dispersion in the south-eastern Bohemian Massif. *Geologische Rundschau*, **82**, 556–565.

FURNES, H., KRYZA, R., MUSZYŃSKI, A., PIN, C. & GARMANN, L. B. 1994. Geochemical evidence for progressive, rift-related early Paleozoic volcanism in the western Sudetes. *Journal of the Geological Society, London*, **151**, 91–109.

GEBAUER, D. & GRÜNENFELDER, M. 1979. U-Pb zircon and Rb-Sr mineral dating of eclogites and their country rocks. Example: Münchberg Gneiss Massif, norhteast Bavaria. *Earth and Planetary Science Letters*, **42**, 35–44.

GŁUSZAK, A. & TOMAŚ, A. 1993. Age of the Nowa Wieś Formation (Bardzkie Mts, Middle Sudetes, SW Poland). *Annales Societatis Geologorum Poloniae*, **62**, 293–308.

GRAD, M., GUTERCH, A. & MAZUR, S. 2002. Seismic refraction evidence for crustal structure in the central part of the Trans-European Suture Zone in Poland. *In:* WINCHESTER, J. A., PHARAOH, T. C. & VERNIERS, J.(eds) *Palaeozoic Amalgamation of Central Europe*. Geological Society, London, Special Publication, **201**, 295–309.

GUIRAUD, M. & BURG, J.-P. 1984. Mineralogical and petrological study of a blueschist metatuff from the Železný Brod crystalline complex, Czechoslovakia. *Neues Jahrbuch für Mineralogie, Abhandlungen*, **149**, 1–12.

GUNIA, P. 1997. Petrology of ultrabasic rocks of the Góry Sowie block. *Acta Universitatis Wratislaviensis, Prace Geologiczno-Mineralogiczne*, **65**, 1–78 [in Polish].

GUNIA, T. 1977. The Late Devonian fauna of the Kłodzko area, Middle Sudetes. *Acta Universitatis Wratisalviensis, Prace Geologiczno-Mineralogiczne*, **6**, 16–39 [in Polish, English abstract].

GUNIA, T. 1985. Geological position of the Sowie Góry block and its influence on the palaeogeography of the Palaeozoic complexes of Central Sudetes. *Geologia Sudetica*, **20**, 83–119 [in Polish, English summary].

GUNIA, T. & WOJCIECHOWSKA, I. 1971. On the age of limestones and phyllites from Mały Bożków, Central Sudetes. *Geologia Sudetica*, **5**, 137–164.

GÜRICH, G. 1900. Über Gabbro im Liegendes des Oberdevonischen Kalkes von Ebersdorf bei Neurode in der Grafschaft Glatz. *Zeitschrift der Deutschen Geologischen Gesselschaft*, **52**, 161–164.

GÜRICH, G. 1902. Zur Diskussion über das Profil von Ebersdorf. *Zeitschrift der Deutschen Geologischen Gesselschaft*, **54**, 57–65.

HANŽL, P. 1994. The correlation between the Nectava gneisses and granodiorites of the northern part of the Brno massif. *Věstník Českého Geologického Ústavu*, **69**, 73–80.

HAYDUKIEWICZ, J. 1981. The pelagic Tournaisian succession in the SW part of the Góry Bardzkie Mts. *Geologia Sudetica*, **16**, 219–225 [in Polish, English summary].

HAYDUKIEWICZ, J. 1990. Stratigraphy of Paleozoic rocks of the Góry Bardzkie and some remarks on their sedimentation. *Neues Jahrbuch für Geologie und Paläontologie, Abhandlungen*, **179**, 275–84.

HIRSCHMANN, G. 1966. Assyntische und variszische Baueinheiten im Grundgebirge der Oberlausitz (unter spezieller Berücksichtung der geologie des östlichen Görlizter Schiefergebirges). *Freiberger Forschungshefte*, **C 212**, 1–146.

HLADIL, J. 1986. Preliminary report on microfacies and age of carbonate intercalations in the Janovice-Ruda borehole (the Vrbno Group, Nizky Jesenik Mts.). *Zpravy o geologických vyzkumech v roce 1984*, 76–77 [in Czech].

HLADIL, J., MAZUR, S., GALLE, A. & EBERT, J. 1999. Revised age of the Mały Bożków limestone in the Kłodzko metamorphic unit (Early Givetian, late Middle Devonian): implications for the geology of the Sudetes. *Neues Jahrbuch für Geologie und Paläontologie., Abhandlungen*, **211**, 329–353.

JAEGER H., 1964. Monograptus hercynicus in den Westsudeten und das Alter der Westsudeten-Hauptfaltung. *Geologie*, **13**, 249–273.

JERZMAŃSKI J., 1992. Niedźwiedź IG 1 and Niedźwiedź IG 2. *Profile głębokich otworów wiertniczych Państwowego Instytutu Geologicznego*, **75**, 1–109 [In Polish; English abstract].

JOHNSTON, J. D., TAIT, J. A., OLIVER, G. J. H. & MURPHY, J. C. 1994. Evidence for a Caledonian orogeny in Poland. *Transactions of The Royal Society of Edinburgh, Earth Sciences*, **85**, 131–142.

KALVODA J. & MELICHAR R. 1999. Palaeozoic sediments of the Drahany Upland. *Geolines*, **8**, 88–89.

KALVODA, J., BÁBEK, O. & BRZOBOHATÝ, R. 1998. *Historical Geology*. Palacký University, Olomouc [in Czech].

KASZA, L. 1964. Geology of the upper basin of Biała Lądecka stream. *Geologia Sudetica*, **1**, [in Polish, English summary].

KETTNER, R. 1949. Geological structure of the northern part of the Moravian karst and adjacent areas. *Rozpravy II třídy České Akademie Věd a Umění* **59**(11), 1–29. [in Czech].

KLEMD, R. & BRÖCKER, M. 1999. Fluid influence on mineral reactions in ultrahigh-pressure granulites: a case study in the Śnieżnik Mts (West Sudetes, Poland). *Contributions to Mineralogy and Petrology*, **136**, 358–373.

KODYM, O. & SVOBODA, J. 1948. Caledonian nappe structure of the Krkonoše and Jizerské hory. *Sborník Státního Geologického Ústavu*, **15**, 109–160 [in Czech].

KORYTOWSKI, A., DÖRR, W. & ŻELAŹNIEWICZ, A. 1993. U-Pb dating of (meta)granitoids in the NW Sudetes (Poland) and their bearing on tectonostratigraphic correlation. *Terra Nova*, **5**, 331.

KOSSMAT, F. 1927. Gliederung des varistischen Gebirgsbaues. *Abhandlungen Sächsischen Geologischen Landesamts*, **1**, 1–39.

KROHE, A. 1996. Variscan tectonics of central Europe: Postaccretionary intraplate deformation of weak continental lithosphere. *Tectonics* **15**, 1364–1388.

KRÓLIKOWSKI, C. & PETECKI, C. 1995. *Gravimetric map of Poland, 1: 1000 000*. Państwowy Instytut Geologiczny, Warszawa.

KRÖNER, A. & HEGNER, E. 1998. Geochemistry, single zircon ages and Sm-Nd systematics of granitoid rocks from the Góry Sowie (Owl Mts, Polish West Sudetes): evidence for early Palaeozoic arc-related plutonism. *Journal of the Geological Society, London*, **155**, 711–724.

KRÖNER, A., HEGNER, E., HAMMER, J., HAASE, G., BIELICKI, K.-H., KRAUSS, M. & EIDAM, J. 1994. Geochronology and Nd-Sr systematics of Lusatian granitoids: significance for the evolution of the Variscan orogen in east-central Europe. *Geologische Rundschau*, **83**, 357–76.

KRÖNER, A., ŠTÍPSKÁ, P., SCHULMANN, K. & JAECKEL, P. 2000. Chronological constraints on the pre-Variscan evolution of the northeastern margin of the Bohemian Massif, Czech Republic. *In:* FRANKE, W., HAAK, V., ONCKEN, O. & TANNER, D. (eds) *Orogenic processes: Quantification and modelling in the Variscan Belt*. Geological Society, London, Special Publications, **179**, 175–197.

KRÖNER, A., JAECKEL, P., HEGNER, E. & OPLETAL, M. 2001. Single zircon ages and whole-rock Nd isotopic systematics of early Palaeozoic granitoid gneisses from the Czech and Polish Sudetes (Jizerské hory, Krkonoše and Orlice-Sněžnik Complex). *International Journal of Earth Sciences (Geologische Rundschau)*, **90**, 304–324.

KRS, M., HLADIL, J., KRSOVÁ, M. & PRUNER, P. 1994. Palaeomagnetic evidence for Variscan palaeotectonic rotation of Moravian Devonian rocks. *Geologický výzkum na Moravé a Slezku* **1994**, 53–57. [in Czech].

KRYZA, R. 1981. Migmatization in gneisses of the northern part of the Sowie Góry, Sudetes. *Geologia Sudetica*, **14**, 7–100 [in Polish, English summary].

KRYZA, R. 1993. Basic metavolcanic rocks of the central Kaczawa Mts (Sudetes): a petrological study. *Acta Universitatis Wratislaviensis, Prace Geologiczno-Mineralogiczne*, **39**, 1–139. [in Polish; English summary].

KRYZA, R. & MAZUR, S. 1995. Contrasting metamorphic paths in the eastern margin of the Karkonosze-Izera Block, SW Poland. *Neues Jahrbuch für Mineralogie, Abhandlungen*, **169**, 157–192.

KRYZA, R. & MUSZYŃSKI, A. 1992. Pre-Variscan volcanic-sedimentary succession of the central southern Góry Kaczawskie, SW Poland: outline geology. *Annales Societatis Geologorum Poloniae*, **62**, 117–140.

KRYZA, R., MAZUR, S. & PIN, C. 1995. Leszczyniec meta-igneous complex in the eastern part of the Karkonosze-Izera Block, Western Sudetes: trace element and Nd isotope study. *Neues Jahrbuch für Mineralogie, Abhandlungen*, **170**, 59–74.

KRYZA, R., PIN, C. & VIELZEUF, D. 1996. High pressure granulites from the Sudetes (SW Poland): evidence of crustal subduction and collisional thickening in the Variscan Belt. *Journal of Metamorphic Geology*, **14**, 531–546.

KRYZA, R., ALEKSANDROWSKI, P. & MAZUR, S. 1998 The structural and metamorphic evolution of the Kaczawa Complex (West Sudetes): a tentative model. *EUROPROBE TESZ project and PACE TMR network meetings. Programme and Abstracts:* 25. Czech Geological Survey, Prague 4–6.10.1998.

KRYZA, R., MAZUR, S. & ALEKSANDROWSKI, P. 1999. Pre-Late Devonian unconformity in the Kłodzko area excavated: a record of Eo-Variscan metamorphism and exhumation in the Sudetes. *Geologia Sudetica*, **32**, 127–137.

KRYZA, R., MAZUR, S. & PIN, C. 2000. Geochemistry of Palaeozoic meta-igneous rocks of the Kłodzko Metamorphic Unit (Central Sudetes): evidence of complex emplacement setting or tectonic mosaic? *Polish Mineralogical Society Special Papers*, **17**, 201–203.

KVĚTOŇ, P. 1951. Report on geological mapping of some occurrences of Inner Phyllites, the Moravian Group. *Věstník Ustředního Ústavu geologického*, **26**, 59–61 [in Czech].

LEDRU, P., LARDEAUX, J.-M., SANTALLIER, D., AUTRAN, A., QUENARDEL, J.-M., FLOC'H, J-P., LEROUGE, G., MAILLET, N., MARCHAND, J. & PLOQUIN, A. 1989. Où sont les nappes dans le Massif central français? *Bulletin de la Societé géologique de France*, **8V**, 605–618.

LELOIX, C., FAURE, M. & FEYBESSE, J.-L. 1999. Hercynian polyphase tectonics in the northeast French Massif Central: the closure of the Brévenne Devonian-Dinantian rift. *International Journal of Earth Sciences*, **88**, 409–421.

LINNEMANN, U. & BUSCHMANN, B. 1995. Die cadomische Diskordanz im Saxothuringikum (oberkambrisch-tremadocische overlap-Sequenzen). *Zeitschrift für Geologischen Wissenschaften*, **23,** 707–727.

LINNEMANN, U., GEHMLICH, M., TICHOMIROVA, M. & BUSCHMANN, B. 1998. The Peri-Gondwanan basement of the Saxothuringian Composite Terrane. *In*: LINNEMANN, U., HEUSE, T., FATKA, O., KRAFT, P., BROCKE, R., & ERDTMANN, B.-D. (eds) *Pre-Variscan terrane analysis of 'Gondwanan Europe'*, International Geological Conference, Excursion Guides & Abstracts, *Schriften Des Staatlichen Museums für Mineralogie und Geologie zu Dresden*, **9**, 7–13.

LINNEMANN, U., GEHMLICH, M., TICHOMIROVA, M., BUSCHMANN, B., NASDALA, L., JONAS, P., LUTZNER, H. & BOMBACH, K. 2000. From Cadomian subduction to Early Palaeozoic rifting: the evolution of Saxo-Thuringia at the margin of Gondwana in the light of single zircon geochronology and basin development (Central European Variscides, Germany). *In:* FRANKE, W., HAAK, V., ONCKEN, O. & TANNER, D. (eds) *Orogenic processes: Quantification and modelling in the Variscan Belt.* Geological Society, London, Special Publications, **179**, 131–153.

MALKOVSKÝ, M. 1979. *Tectogenesis of the Bohemian Massif platform cover.* Knihovna Ústředniho Ústavu geologického, **53.** Praha. 176 pp. [in Czech].

MALUSKI, H. & PATOČKA, F. 1997. Geochemistry and ^{40}Ar-^{39}Ar geochronology of the mafic metavolcanic rocks from the Rýchory Mountains complex (west Sudetes, Bohemian Massif): paleotectonic significance. *Geological Magazine*, **134**, 703–716.

MALUSKI, H., RAJLICH, P. & SOUČEK, J. 1995. Pre-Variscan, Variscan and early Alpine thermo-tectonic history of the north-eastern Bohemian Massif: $^{40}Ar/^{39}Ar$ study. *Geologische Rundschau*, **84**, 345–58.

MARHEINE, D., PATOČKA, F., KACHLIK V., MALUSKI, H. & ŻELAŹNIEWICZ, A. 2000. Geochronological constraints on the tectonometamorphic development of the West Sudetes (Bohemian Massif). *Joint Meeting of Europrobe (TESZ) and PACE Projects, Zakopane, Poland, September 16–23, 2000, Abstracts Volume,* 62–63.

MATTE, P. 1986. Tectonic and plate tectonics model for the Variscan Belt of Europe. *Tectonophysics*, **177**, 329-74.

MATTE, P. 1991. Accretionary history and crustal evolution of the Variscan Belt in Western Europe. *Tectonophysics*, **196**, 309–337.

MATTE, P. 1998. Continental subduction and exhumation of HP rocks in Paleozoic orogenic belts: Uralides and Variscides. *Geologiska Föreningens i Stockholm Förhandlingar*, **120**, 209–222.

MATTE, P., MALUSKI, H., RAJLICH, P. & FRANKE, W. 1990. Terrane boundaries in the Bohemian Massif: results of large-scale Variscan shearing. *Tectonophysics*, **177**, 151–170.

MAZUR, S. 1987. The structural geology of Upper Devonian – Lower Carboniferous succession in the Dzikowiecquarry, Góry Bardzkie Mts. *Przegląd Geologiczny*, **6**, 327–332 [in Polish].

MAZUR, S. 1995. Structural and metamorphic evolution of the country rocks at the eastern contact of the Karkonosze granite in the southern Rudawy Janowickie Mts and Lasocki Ridge. *Geologia Sudetica*, **29**, 31–98 [in Polish, English summary].

MAZUR, S. & ALEKSANDROWSKI, P. 2001*a*. The Tepla(?)/Saxothuringian suture in the Karkonosze-Izera massif, Western Sudetes, Central European Variscides. *International Journal of Earth Sciences*, **90**, 341–360.

MAZUR, S. & ALEKSANDROWSKI, P. 2001*b*. A Variscan suture zone along the sheared SW boundary of the the Orlica-Śnieżnik massif (the Sudetes, NE Bohemian Massif)? *EUG XI, Strasbourg, Journal of Conference Abstracts*, **6** (2001), 632.

MAZUR, S. & JÓZEFIAK, D. 1999. Structural record of Variscan thrusting and subsequent extensional collapse in the mica schists from vicinities of Kamieniec Ząbkowicki, Sudetic foreland, SW Poland. *Annales Societatis Geologorum Poloniae*, **69**, 1–26.

MAZUR, S. & KRYZA, R. 1996. Superimposed compressional and extensional tectonics in the Karkonosze-Izera Block, NE Bohemian Massif. *In*: ONCKEN, O. & JANSSEN, C. (eds) *Europe and Other Regions*. Basement Tectonics, **11**, Kluwer, Dordrecht, 51–66.

MAZUR, S. & KRYZA, R. 1999. Preliminary report on the metamorphosed melange in the Kłodzko metamorphic complex (West Sudetes, SW Poland). *Mineralogical Society of Poland, Special Papers*, **14,** 102–104.

MAZUR, S. & PUZIEWICZ, J. 1995. Mylonites of the Niemcza Zone. *Annales Societatis Geologorum Poloniae*, **64**, 23–52 [in Polish, English summary].

MILEWICZ, J., SZAŁAMACHA, J. & SZAŁAMACHA, M. 1989. *Geological Map of Poland 1: 200 000, Explanations.* Sheet Jelenia Góra. Wydawnictwa Geologiczne, Warszawa.

MÍSAŘ, Z. & URBAN, M. 1995. VIII. Moravo-Silesian Zone. A. Introduction. *In* DALLMEYER, R.D., FRANKE, W. & WEBER, K. (eds) *Pre-Permian geology of Central and Eastern Europe,* Springer, Berlin, 469–473.

MÍSAŘ, Z., DUDEK, A., HAVLENA, V. & WEISS, J. 1983. *Geology of CSSR, I. Bohemian Massif.* Státní Pedagogické Nakladatelství, Praha. 333 pp. [in Czech].

MOCZYDŁOWSKA, M. 1995. Neoproterozoic and Cambrian successions deposited on the East European Platform and Cadomian basement area in Poland. *Studia Geophisica et Geodetica*, **39**, 276–85.

MOCZYDŁOWSKA, M. 1997. Proterozoic and Cambrian successions in Upper Silesia: an Avalonian terrane in southern Poland.*Geological Magazine*, **134**, 679–89.

NEMEC, W., PORĘBSKI, S. J. & TEISSEYRE, A. K. 1982. Exploratory notes to the lithotectonic molasse profile of the Intra-Sudetic Basin, Polish part (Sudety Mts, Carboniferous–Permian). *Veröffentlichungen Zentralinstituts für Physik der Erde, Akademie der Wissenschaften der DDR*, **66**, 267–278.

OBERC, J. 1961. An outline of the geology of the Karkonosze-Izera Block. *Zeszyty Naukowe Uniwersytetu Wrocławskiego*, **IV**, seria **B, 8**, 139–170.

OBERC, J. 1972. Geological structure of Poland 4, Tectonics 2, *Sudetes*. *In:* POŻARYSKI, W. (ed.) *Geological structure of Poland 4, Tectonics 2,* Wydawnictwa Geologiczne, Warszawa, 307 p. [in Polish].

OBERC-DZIEDZIC, T. 1987. The development of gneisses and granites in the eastern part of the Izera crystalline unit in the light of textural investigations. *Acta Universitatis Wratislaviensis*, **997**, *Prace Geologiczno-Mineralogiczne*, **13**, 1–181 [in Polish, English summary].

OBERC-DZIEDZIC, T., PIN, C., DUTHOU, J.-L. & COUTURIE, J.-P. 1996. Age and origin of the Strzelin granitoids (Fore-Sudetic Block, Poland): $^{87}Rb/^{86}Sr$ data. *Neues Jahrbuch für Mineralogie, Abhandlungen*, **171**, 187–198.

OBERC-DZIEDZIC, T., KLIMAS, K., KRYZA, R. & FANNING, M. 2001. SHRIMP zircon geochronology of the Neoproterozoic Strzelin gneiss: evidence for the Moravo-Silesian zone affinity of the Strzelin masssif, Fore-Sudetic block, SW Poland. *Geolines*, **13**, 96–97.

O'BRIEN, P. J., KRÖNER, A., JAECKEL, P., HEGNER, E., ŻELAŹNIEWICZ, A. & KRYZA, R. 1997. Petrological and isotopic studies on Palaeozoic high-pressure granulites, Góry Sowie Mts, Polish Sudetes. *Journal of Petrology*, **38**, 433–456.

OHNENESTETTER, M. & SIDER, H. 1988. Contraintes géochimiques apportées par le magmatisme sur le développement du bassin marginal ensialique du Beaujolais au Dévonien. *Bulletin de la Societé géologique de France*, **(8)IV** (3), 499–510.

OLIVER, G. J. H., CORFU, F. & KROGH, T. E. 1993. U-Pb ages from SW Poland: evidence for a Caledonian suture zone between Baltica and Gondwana. *Journal of the Geological Society, London*, **150**, 355–369.

OPLETAL, M., DOMEČKA, K. ČECH, S., ČUTA, M., FAJST, M., HOLUB, V., KAČURA, K., LÍBALOVÂ, J., POŠMOURNÝ, K., SEKYRA, J.,, STŘÍDA, M., ŠALANSKÝ, K., ŠULCEK, Z., TÁSLER, R. & VALEČKA, J. 1980. *Geology of the Orlica Mountains*. Ústřední Ústav Geologický, Praha. 202 p. [in Czech; English summary].

OPLETAL, M., JELINEK, E., PEČINA, V., POŠMOURNÝ, K. & POUBOVÁ, E. 1990. Metavolcanites of the SE part of the Lugicum, their geochemistry and geotectonic interpretation. *Sborník Geologických Věd, Geologie*, **45**, 37–64.

PAQUETTE, J. L., MONCHOUX, P. & COUTURIER, M. 1995. Geochemical and isotopic study of a norite-eclogite transition in the European Variscan belt: implications for U-Pb zircon systematics in metabasic rocks. *Geochimica et Cosmochimica Acta*, **59**, 1611–1622.

PARRY, M., ŠTÍPSKÁ, P., SCHULMANN, K., HROUDA, F., JEŽEK, J. & KRÖNER, A. 1997. Tonalite sill emplacement at an oblique plate boundary: northeastern margin of the Bohemian Massif. *Tectonophysics*, **280**, 61–81.

PATOČKA, F. & SMULIKOWSKI, W. 1998. The geochemical correlation of the East Krkonoše Complex metabasites (West Sudetes, Czech Republic and Poland): relation to the Cambrian-Ordovician northern Gondwana breakup: a review. *Acta Universitatis Carolinae*, **42**, 485–488.

PATOČKA, F. & SMULIKOWSKI, W. 2000. Early Palaeozoic

intracontinental rifting and incipient oceanic spreading in the Czech/Polish East Krkonoše/Karkonosze Complex, West Sudetes (NE Bohemian Massif) *Geologia Sudetica*, **33**, 1–15.

PATOČKA, F. & VALENTA, J. 1996. Geochemistry of the late Devonian intermediate to acid metavolcanic rocks from the southern part of the Vrbno Group, the Jeseníky Mts (Moravo-Silesian Belt, Bohemian Massif, Czech Republic): palaeotectonic implications. *Geolines*, **4**, 42–54.

PATOČKA, F., PIVEC, E. & OLIVERIOVÁ, D. 1996. Mineralogy and petrology of mafic blueschists from the Rýchory Mts crystalline complex (Western Sudetes, Bohemian Massif). *Neues Jahrbuch für Mineralogie, Abhandlungen*, **170**, 313–320.

PHARAOH, T. C. 1999. Palaeozoic terranes and their lithospheric boundaries within the Trans-European SutureZone (TESZ): a review. *Tectonophysics,* **314**, 17–41.

PIN, C. 1990. Variscan oceans: Ages, origins and geodynamic implications inferred from geochemical and radiometric data. *Tectonophysics*, **177**, 215–227.

PIN, C. & PAQUETTE, J-L. 1998. A mantle-derived bimodal suite in the Hercynian Belt: Nd isotope and trace element evidence for a subduction-related origin of the Late Devonian Brévenne metavolcanics, Massif Central (France). *Contributions to Mineralogy and Petrology*, **129**, 222–238.

PIN, C. & PEUCAT, J. J. 1986. Ages des épisodes de métamorphisme paléozoïques dans le Massif Central et le Massif armoricain. *Bulletin de la Societé géologique de France*, **8 II**, 461–469.

PIN, C. & VIELZEUF, D. 1983. Granulites and related rocks in Variscan median Europe: a dualistic interpretation. *Tectonophysics*, **93**, 47–74.

PIN, C. & VIELZEUF, D. 1988. Les granulites de haute pression d'Europe moyenne témoins d'une subduction éo-hercynienne. Implications sur l'origine des groupes leptyno-amphiboliques. *Bulletin de la Societé géologique de France*, **8 IV**, 13–20.

PIN, C., MAJEROWICZ, A. & WOJCIECHOWSKA, I. 1988. Upper Palaeozoic oceanic crust in the Polish Sudetes: Nd-Sr isotope and trace element evidence. *Lithos*, **21**, 195–209.

PITRA, P., BURG, J.-P., SCHULMANN, K. & LEDRU, P. 1994. Late orogenic extension in the Bohemian Massif: petrostructural evidence in the Hlinsko region. *Geodinamica Acta*, **7**, 15–30.

PORĘBSKI, S. J. 1981. Świebodzice succession, Upper Devonian-lowermost Carboniferous, Western Sudetes: a prograding mass-flow dominated fan-delta complex. *Geologia Sudetica*, **16**, 101–92 [in Polish, English summary].

PORĘBSKI, S. J. 1990. Onset of coarse clastic sedimentation in the Variscan realm of the Sudetes (SW Poland): an example from the Upper Devonian-Lower Carboniferous Świebodzice succession. *Neues Jahrbuch für Geologie und Paläontologie, Abhandlungen*, **179**, 259–74.

PŘIKRYL, R., SCHULMANN, K. & MELKA, R. 1996. Perpendicular fabrics in the Orlické Hory orthogneisses (western part of the Orlice-Snežnik Dome, Bohemian Massif) due to high temperature E-W deformational event and late lower temperature N-S overprint. *Journal of the Czech Geological Society*, **41**, 156–166.

PUZIEWICZ, J. & KOEPKE, J. 2001. Partial melting of garnet-hornblende granofels and the crystallisation of igneous epidote in the Niedźwiedź Amphibolite Massif (Fore-Sudetic Block, SW Poland). *Neues Jahrbuch für Mineralogie, Monatshefte*, **12**, 529–547.

RAJLICH, P. 1987. Variszische duktile Tektonik im Böhmischen Massiv. *Geologische Rundschau*, **76**, 755–786.

SANTALLIER D., LARDEAUX, J.-M., MARCHAND, J. & MARIGNAC, C. 1994. Metamorphism. *In:* KEPPIE, J. D. (ed.) *Pre-Mesozoic geology of France and related areas*. Springer Verlag, Berlin, 324–334.

SCHEUMANN, K. H. 1937. Zur Frage nach dem Vorkommen von Kulm in der Nimptscher Kristallinzone. *Zeitschrift für Kristallographie, Mineralogie und Petrographie, Abteilung B: Mineralogische und Petrographishe Mitteilungen*, **49**, 216–40.

SCHINDEWOLF, O. H. 1937. Zur Stratigraphie und Paläontologie der Wocklumer Schichten (Oberdevon). *Abhandlungen Preussichen Geologischen Landesanstalt, Neue Folge*, **179**, 1–132.

SCHULMANN, K. & GAYER, R. 2000. A model for a continental accretionary wedge developed by oblique collision: the NE Bohemian Massif. *Journal of Geological Society, London,* **157**, 401–416.

SCHULMANN, K., LEDRU, P., AUTRAN, A. MELKA, R., LARDEAUX, J.M., URBAN, M. & LOBKOWICZ, M. 1991. Evolution of nappes in the eastern margin of the Bohemian Massif: A kinematic interpretation. *Geologische Rundschau*, **80**, 73–92.

SESTON, R., WINCHESTER, J. A., PIASECKI, M. A. J., CROWLEY, Q. G. & FLOYD, P. A. 2000. A structural model for the western-central Sudetes: a deformed stack of Variscan thrust sheets. *Journal of Geological Society, London,* **157**, 1155–1167.

SHELLEY, D. & BOSSIÈRE, G. 2000. A new model for the Hercynian Orogen of Gondwanan France and Iberia. *Journal of Structural Geology*, **22**, 757–776.

SIDER, H. & OHNENESTETTER, M. 1986. Field and petrological evidence for the development of an ensialic marginal basin related to the Hercynian orogeny in the Massif Central, France. *Geologische Rundschau*, **75**, 421–443.

SIDER, H., LEISTEL, J-M. & GAGNY, C. 1986. Réflections sur l'évolution géodynamique de la chaîne hercynienne au Paléozoïque dans le Nord-Est du Massif central français. *Bulletin de la Societé géologique de France*, **(8)II**, 637–644.

SMULIKOWSKI, W. 1995. Evidence for glaucophane-schist facies metamorphism in the East Karkonosze complex, West Sudetes, Poland. *Geologische Rundschau*, **94**, 720–37.

STELTENPOHL, M. G., CYMERMAN, Z., KROGH, E. J. & KUNK, M. J. 1993. Exhumation of eclogitized continental basement during Variscan lithospheric delamination and gravitational collapse, Sudety Mountains, Poland. *Geology*, **21**, 1111–14.

STILLE, H. 1951. Das mitteleuropäische varistische

Grundgebirge im Bilde des gesamteuropäischen. *Beihefte zum Geologischen Jahrbuch*, **2**, 1–138.

ŠTÍPSKÁ, P., SCHULMANN, K. & KRÖNER, A. 2001. Tectonometamorphic evolution of the Velké Vrbno unit – evidence of subduction of the western margin of the Silesian domain. *Geolines*, **10**, 71–72.

STOSCH, H. G. & LUGMAIR, G. W. 1990. Geochemistry and evolution of the MORB-type eclogites from the Münchberg Massif, southern Germany. *Earth and Planetary Science Letters*, **99**, 230–249.

SUESS, F. E. 1912. Die moravischen Fenster und ihre Beziehung zum Grundgebirge des Hohes Gesenkes. *Denkschriften der österreichischen Kaiserlichen Akademie der Wissenschaften, mathematisch–naturwissenschaftliche Klasse*, **88**, 541–631.

SUESS, F. E. 1926. *Intrusionstektonik und Wandertektonik im variszischen Grundgebirge*. Borntraeger, Berlin. 268 p.

SVOBODA, J. (ed.) 1966. *Regional Geology of Czechoslovakia. I. Bohemian Massif.* Publishing House of the Czechoslovak Academy of Sciences, Praha, 668 p.

SZCZEPAŃSKI, J. & OBERC-DZIEDZIC, T. 1998. Geochemistry of amphibolites from the Strzelin crystalline massif, Fore-Sudetic Block, SW Poland. *Neues Jahrbuch für Mineralogie, Abhandlungen*, **173**, 23–40.

TAIT, J. A., BACHTADSE, V. & SOFFEL, H. 1996. Eastern Variscan fold belt: Paleomagnetic evidence for oroclinal bending. *Geology* **24**, 871–874.

TAIT, J. A., SCHÄTZ, M., BACHTADSE, V. & SOFFEL, H. 2000. Palaeomagnetism and Palaeozoic palaeogeography of Gondwana and European terranes. *In*: FRANKE, W., HAAK, V., ONCKEN, O. & TANNER, D. (eds) *Orogenic processes: Quantification and modelling in the Variscan Belt.* Geological Society, London, Special Publications, **179**, 21–34.

TEISSEYRE, A. K. 1968. The Lower Carboniferous of the Intrasudetic Basin; a study in sedimentary petrology and basin analysis. *Geologia Sudetica*, **4**, 221–298.

TEISSEYRE, A. K. 1971. Sedimentology of the Culm of Ciechanowice and palaeogeography of the lowest Culm of the Intrasudetic basin. *Geologia Sudetica,* **5**, 237–280.

TEISSEYRE, A. K. 1975. Sedimentology and palaeogeography of the Culm alluvial fans in the western Intrasudetic Basin. (Central Sudetes, SW Poland). *Geologia Sudetica*, **9**, 1–135.

TEISSEYRE, H. 1968. Stratigraphy and tectonics of the Świebodzice depression. *Biuletyn Instytutu Geologicznego*, **222**, 71–106.

TEISSEYRE, H., SMULIKOWSKI, K. & OBERC, J. 1957. *Regional Geology of Poland.* Vol. III. *The Sudetes.* Pt.1. *Pre-Tertiary complexes*. Polskie Towarzystwo Geologiczne – Państwowe Wydawnictwo Naukowe, Kraków. 300 p. [in Polish].

TEISSEYRE, J. H. 1973. Metamorphic rocks of the Rudawy Janowickie and Lasocki Range. *Geologia Sudetica*, **8**, 7–120 [in Polish, English summary].

THOMAS, U. 1990. Unterkarbone Wildflysch-Ablagerungen im Südteil der DDR. *Zeitschrift für Angewandte Geologie*, **36**, 182–184.

TIMMERMANN, H., PARRISH, R. R., NOBLE, S. R. & KRYZA, R. 2000. New U-Pb monazite and zircon data from the Sudetes Mountains in SW Poland: evidence for a single-cycle Variscan orogeny. *Journal of the Geological Society, London*, **157**, 265–268.

TURNIAK, K., MAZUR, S. & WYSOCZANSKI, R. 2000. SHRIMP zircon geochronology and geochemistry of the Orlica-Śnieżnik gneisses (Variscan belt of Central Europe) and their tectonic implications. *Geodinamica Acta*, **13**, 1–20.

URBANEK, Z., ŻELAŹNIEWICZ, A., KEMNITZ, H., HERMSDORF, N. & LINNEMANN, U. 1995. VI. Western Sudetes (Lugicum). B. Stratigraphy. *In*: Dallmeyer, R. D., Franke, W. & Weber, K. (eds) *Pre-Permian geology of Central and Eastern Europe*. Springer, Berlin, 315–27.

VALVERDE-VAQUERO, P., DÖRR, W., BELKA, Z., FRANKE, W., WISZNIEWSKA, J. & SCHASTOK, J. 2000. U-Pb single-grain dating of detrital zircon in the Cambrian of central Poland: implications for Gondwana versus Baltica provenance studies. *Earth and Planetary Science Letters,* **184**, 225–240.

VAN BREEMEN, O., AFTALION, M., BOWES, D. R., DUDEK, A., MÍSAŘ, Z., POVONDRA, P. & VRÁNA, S. 1982. Geochronological studies of the Bohemian Massif, Czechoslovakia, and their significance in the evolution of Central Europe. *Transactions of the Royal Society, Edinburgh, Earth Sciences,* **73,** 89–108.

VAN BREEMEN, O., BOWES, D. R., AFTALION, M. & ŻELAŹNIEWICZ, A. 1988. Devonian tectonothermal activity in the Sowie Góry gneissic block, Sudetes, SW Poland: evidence from Rb-Sr and U-Pb isotopic studies. *Annales Societatis Geologorum Poloniae,* **58**, 3–19.

VIELZEUF, D. & PIN, C. 1989. Geodynamic implications of granulitic rocks in the Hercynian belt. *In*: DALY, J. S., CLIFF, R. A. & YARDLEY, B. W. D. (eds) *Evolution of metamorphic belts*. Geological Society, London, Special Publications, **43**, 343–348.

WAJSPRYCH, B. 1978. Allochthonous Paleozoic rocks in the Viséan of the Bardzkie Mts, Sudetes. *Annales de la Societé géologique de Pologne,* **58**, 99–127 [in Polish, English summary].

WAJSPRYCH, B. 1986. Sedimentary record of tectonic activity on a Devonian-Carboniferous continental margin, Sudetes. *In*: TEISSEYRE A. K. (ed.) *IAS 7th Regional Meeting, Excursion Guidebook,* 141–64. Ossolineum, Kraków & Wrocław.

WEYER, D. 1965. Zur Ammonoideen-Fauna der Gattendorfia-Stufe von Dzikowiec (Ebersdorf) in Dolny Śląsk (Niederschlesien), Polen. *Berichte der Geologischen Gesellschaft der DDR*, **10**, 443–464.

WIESER, T. 1978. Glaucophane schists and associated rocks of Kopina Mt. (Lasocki range, Sudetes). *Mineralogia Polonica*, **9**, 35–56.

WINCHESTER, J. A., FLOYD, P. A., CHOCYK, M., HORBOWY, K. & KOZDRÓJ, W. 1995. Geochemistry and tectonic environment of Ordovician metaigneous rocks in the Rudawy Janowickie

Complex, SW Poland. *Journal of the Geological Society, London*, **152**, 105–115.

WINCHESTER, J. A. & PACE TMR NETWORK TEAM 2002. Palaeozoic amalgamation of Central Europe: new results from recent geological and geophysical investigations. *Tectonophysics*, in press.

ŻABA, J. 1984. Genesis and metamorphic evolution of gneisses and granitoids of the Izera Stóg massif, Western Sudetes. *Geologia Sudetica*, **19**, 89–195 [in Polish, English summary].

ŽÁČEK, V. 1996. Retrograded eclogite from the Staré Město Belt, NE margin of the Bohemian Massif. *Journal of the Czech Geological Society*, **41**, 167–175.

ZAPLETAL, K. 1932. *Geology and petrography of Moravia and Silesia with emphasis on economic deposits*. Vlastivedecke publikace moravskó-slezské, **1**, Brno. 284 p. [in Czech].

ZAPLETAL, K. 1950. *Review of the geology of Silesia*. Přirodovědeckí Sborník Ostravského kraje, **11** (appendix), 1–40 [in Czech].

ŻELAŹNIEWICZ, A. 1990. Deformation and metamorphism in the Góry Sowie gneiss complex, Sudetes, SW Poland. *Neues Jahrbuch für Geologie und Paläontologie, Abhandlungen*, **179**, 129–157.

ŻELAŹNIEWICZ, A. & FRANKE, W. 1994. Discussion on U-Pb ages from SW Poland: evidence for a Caledonian suture zone between Baltica and Gondwana. *Journal of the Geological Society, London*, **151**, 1050–1052.

ŻELAŹNIEWICZ, A., DÖRR, W. & DUBIŃSKA, E. 1998. Lower Devonian oceanic crust from U-Pb zircon evidence and Eo-Variscan event in the Sudetes. *In* FRANKE, W. & HERZBERG, S. (eds) *SPP 'Orogene Processe'* 7. Kolloquium, Giessen. *Terra Nostra*, **98/2**, 174–176.

ZIEGLER, P. A. 1986. Geodynamic model for the Paleozoic crustal consolidation of Western and Central Europe. *Tectonophysics*, **126**, 303–328.

ZNOSKO, J. 1981. The problem of oceanic crust and of ophiolites in the Sudetes. *Bulletin de l'Académie Polonaise des Sciences, série Sciences de la Terre*, **29**, 185–197.

Structure and evolution of the Bohemian Arc

W. FRANKE[1] & A. ŻELAŹNIEWICZ[2]

[1]*Institut für Geowissenschaften der Universität, D-35390 Gießen, Germany e-mail: wolfgang.franke@geolo.uni-giessen.de*

[2]*Instytut Nauk Geologicznych PAN, Podwale 75, 50–449 Wrocław, Poland*

Abstract: Tectonic zones and palaeogeographic units (terranes) in the German segment of the Variscides correlate with equivalents in the Sudetes at the NE margin of the Bohemian Massif. This correlation defines an arcuate structure with an opening angle of about 90°. The structure is truncated to the SE by a crustal scale, NE-trending fault zone with dextral transpression, the Moldanubian Thrust (MT). The arc cannot have been formed by northeastward indentation of the Bohemian Massif, since there is no evidence of a fault zone on the NW flank of the notional indenter, and little evidence for northeastward tectonic transport. Kinematic and age constraints on the main fault zones instead suggest that the structural array was formed by a complex sequence of events. Northwestward displacement along the margin of the East European Platform (EEP) with clockwise rotation was followed by large southwestward movements along the Moldanubian Thrust, and renewed northwestward displacement along the SW margin of the East European Platform.

The Variscan Belt of Europe represents an orogenic collage composed of Avalonia, a complex Armorican Terrane Assemblage, and a suspect 'Moldanubia' Terrane, which possibly formed part of Gondwana mainland (see the latest summary in Franke 2000). Closure of narrow oceans or seaways between these terranes has produced separate orogenic belts, which largely correspond to the Rheno-Hercynian, the Saxothuringian and Moldanubian Zones defined by Kossmat (1927). Westwards, some of these tectonic belts can be traced through France into Iberia (e.g. Matte 1986, 1991; Franke 1989). Toward the east, the Variscan Belt abuts against the SW margin of the East European Platform (Fig. 1).

Recent studies have established firm correlations between the German segment of the Variscan Belt and the West Sudetes (eastern margin of the Bohemian Massif in the Czech Republic and Poland): it appears that the West Sudetes contains equivalents of the Saxothuringian Belt, and outliers of an intervening median massif, the Teplá-Barrandian Unit (Franke *et al.* 1993; Franke & Żelaźniewicz 2000; Floyd *et al.* 2000; Crowley *et al.* 2000; Aleksandrowski *et al.* 1997). The resulting structural pattern is a much disrupted arcuate structure, which forms the eastern termination of the Variscides.

In this paper, we summarize the tectonic and palaeogeographic subdivision of the Bohemian Massif, add some new structural observations, and attempt a synthesis of the tectonic evolution of the 'Bohemian Arc'. For the correlation of events at different crustal levels, we refer to the time tables of Menning *et al.* (2000) and McKerrow & van Staal (2000).

Geological framework

General setting

The structural complexity of the Bohemian Massif and adjacent areas has led some authors to propose a small-scale mosaic of microplates (e.g. Oliver *et al.* 1993; Cymerman 2000). However, detailed evaluation of new findings and comparisons with the Variscan basement in Germany are compatible with a simple model, in which the evolution of the north and east parts of the Bohemian Massif is attributed to one Variscan orogenic cycle. It commenced with Cambro-Ordovician rifting in Cadomian crust, which created a narrow ocean between the Saxothuringian and the Bohemian Terranes of the Armorican Terrane Assemblage (Franke *et al.* 1993). The alternative concept of an early Ordovician magmatic arc taken as evidence of a 'Caledonian' orogenic event (e.g. Oliver *et al.* 1993; Kröner & Hegner 1998) has been discussed and dismissed by numerous authors (e.g. Aleksandrowski 1994; Aleksandrowski *et al.* 2000; Franke & Żelaźniewicz 2000; Żelaźniewicz & Franke 1994; Crowley *et al.* 2000; Timmermann *et al.* 2000).

Variscan convergence was performed by Devonian subduction and early Carboniferous

From: WINCHESTER, J. A., PHARAOH, T. C. & VERNIERS, J. 2002. *Palaeozoic Amalgamation of Central Europe.* Geological Society, London, Special Publications, **201**, 279–293. 0305-8719/02/$15.00

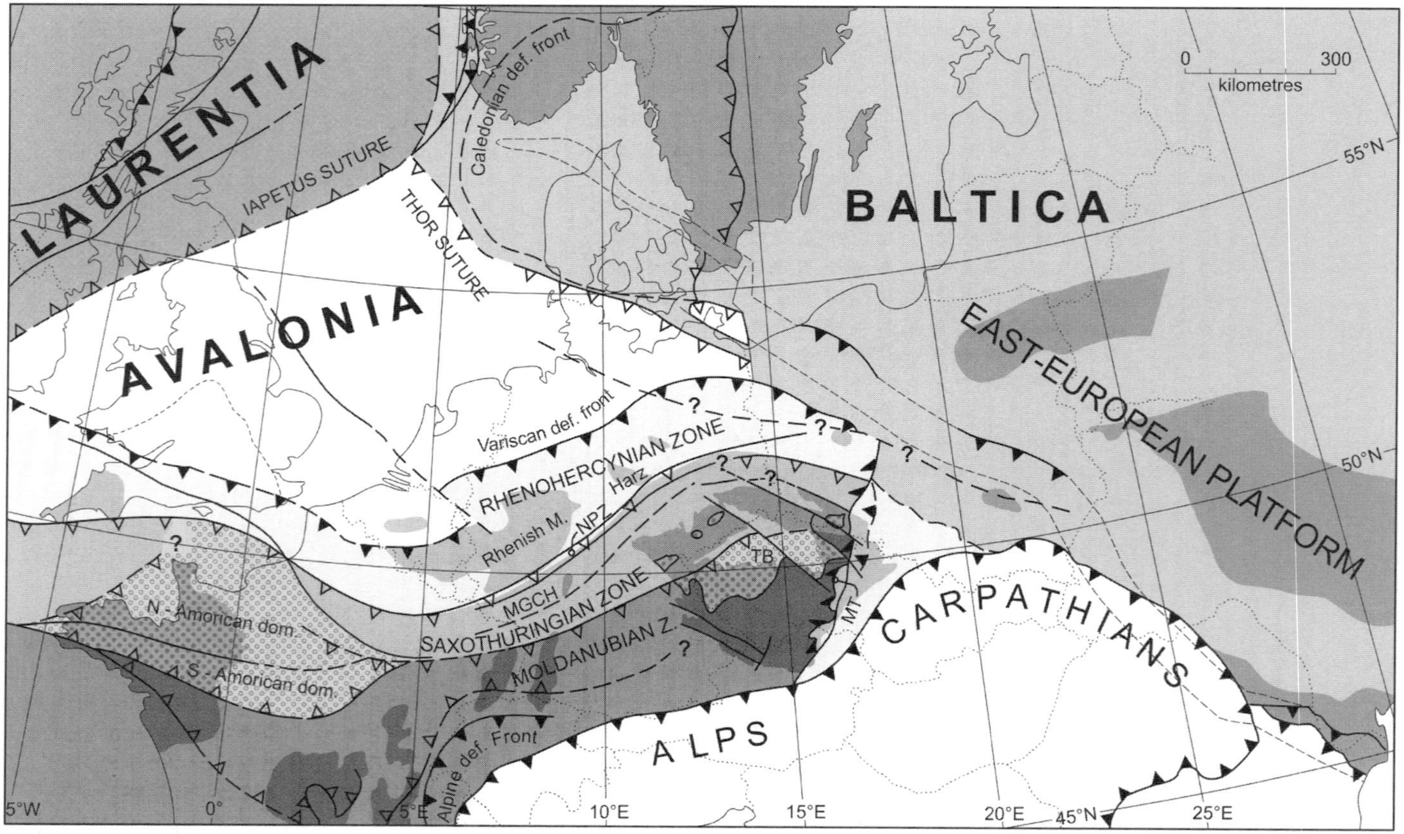

Fig.1. Structural map of north-central Europe for Late Carboniferous/Permian time (adapted from the map compiled by J. A.Winchester adapted from the TESZ working group map). Excluding Laurentia and Baltica, shades are: White, Avalonia; light grey, tectonic units derived from northern parts of the Armorican Terrane Assemblage; stipple, Bohemian and Armorican terranes (SE and western parts of the Armorican Terrane Assemblage); dark grey, ?north Gondwana. Basement outcrops in darker shades. Abbreviations: MGCH, Mid-German Crystalline High; MT, Moldanubian Thrust; NPZ, Northern Phyllite Zone; TB, Teplá-Barrandian.

collision. Tectonic structures reveal a bilateral symmetry: Subduction/underthrusting was grossly directed to the SE on the northern flank of the Variscides (Rheno-Hercynian and Saxo-thuringian Belts), and to the NW on the southern flank (Moldanubian Belt). The Teplá-Barrandian Unit in the heart of the Bohemian Massif acted as a median massif. Shear criteria in metamorphic rocks on both sides of the Teplá-Barrandian block are often sub-parallel with the main tectonic boundaries and reveal a regime of dextral transpression (Franke 2000).

On a larger scale, the eastern parts of the collisional belt (West Sudetes, east of the Elbe Fault Zone) define an arcuate structure with an opening angle of about 80°. Part of this effect is caused by stepwise dextral displacement along the NW-trending Elbe and Intra-Sudetic Fault Zones, and probably also on the Odra and Dolsk Fault zones (EFZ, ISFZ, OFZ, DFZ, Fig. 2). These faults parallel the SW margin of the East European Platform (EEP, Fig. 1) and controlled the formation of large pull-apart basins (Intra-Sudetic and Świebodzice Basins).

Correlation of stretching lineations and shear sense in the Saxothuringian and the Sudetes suggests some block rotation. In the Erzgebirge, the main ductile lineation indicates transport to the west (e.g. Konopasek *et al.* 2001). In the Izera region north of the Karkonosze granite and further east in the Kaczawa Mountains, main ductile transport is to the NW (Aleksandrowski *et al.* 1997; Seston *et al.* 2000). Northwestward ductile transport has also been reported from the Orlica Mountains (Cymerman 2000). This array would indicate rotation through about 65°, an angle similar to that defined by the boundaries of the tectonic units. However, it is uncertain whether all these lineations have the same tectonic significance, and if they are of the same age.

In areas to the NE of the Intra-Sudetic Fault Zone (IFZ), recent palaeomagnetic studies have been used to argue against rotation of these areas in time after the Silurian. However, a palaeopole from the Silurian/Early Devonian Góry Sowie ophiolite may also represent late Carboniferous remagnetization (Jeleńska *et al.* 1995). A supposedly Silurian mafic sill from the southwestern Holy Cross Mountains (Nawrocki 2000) lacks direct evidence of its age.

Unequivocal evidence for rotation is available for the Moravo-Silesian Belt to the SE of the Moldanubian Thrust (MT), which has been rotated clockwise through at least 90° (with respect to the Rheno-Hercynian Belt in Germany) since the Devonian. This is documented by palaeomagnetic studies (Krs *et al.* 1995; Tait *et al.* 1994). The palaeogeographic zonation within the Moravo-Silesian trends toward the NE (Hladil *et al.* 1999), thus including almost a hairpin bend with the palaeogeographic zonation of the Rheno-Hercynian Belt in Germany.

The Elbe and Intra-Sudetic Fault Zones are truncated by a crustal-scale fault zone, the 'Moldanubian Thrust' (MT) first recognized by F. E.Suess (1912), which evolved in a regime of dextral transpression (Schulmann *et al.* 1994; Schulmann & Gayer 2000; Misař & Urban 1995). This thrust cuts off equivalents of the Mid-German Crystalline High, the Saxothuringian Belt and the Moldanubian Belt against which it juxtaposes, at a right angle, the Moravo-Silesian Belt (Figs 1–3). Displacement along the Moldanubian Thrust must be larger than the width of the tectonic zones it truncates, i.e. must exceed 400 km.

The geological base map (Fig. 2)

The geological map of the Bohemian Massif of Franke & Żelaźniewicz (2000) needs modification in the western part of the Orlica-Śnieżnik Unit and the basement of the Intra-Bohemian Cretaceous basin. Mazur & Aleksandrowski (2001) have drawn attention to a fault zone separating the high-grade Orlica gneisses to the NE from the greenschist-grade Nové Město Unit to the SW (Olešnice-Uhřinov Fault). Along this fault zone, ductile dextral displacement has been overprinted by normal faulting. To the south of a minor east–west-displacement, this fault probably continues between the Zábřeh complex (of unknown palaeogeographic affinity) and the Śnieżnik gneisses. Further SE, it is probably continued in the Bušin fault, which shows southwestward normal faulting under greenschist conditions (Franke & Żelaźniewicz, unpubl. observations). We interpret the earlier, dextral movements along the fault as accommodating pull-apart at the southwestern margin of the Intra-Sudetic basin formed along the Intra-Sudetic Fault.

To the south of the Nové Město Unit, a one kilometre-wide, subvertical, ductile shear zone thrown down to the north (Rychnov Fault Zone) has been identified. This fault splays off from the Orlica Mountains. towards the WNW and changes its orientation to a southwestward trend east of Rychnov. From there it can be correlated, across the Cretaceous cover, with the Hlinsko normal fault, that is with the southern boundary fault of the Teplá-Barrandian block against the Moldanubian. These relationships support the assignments of rocks to the north of the

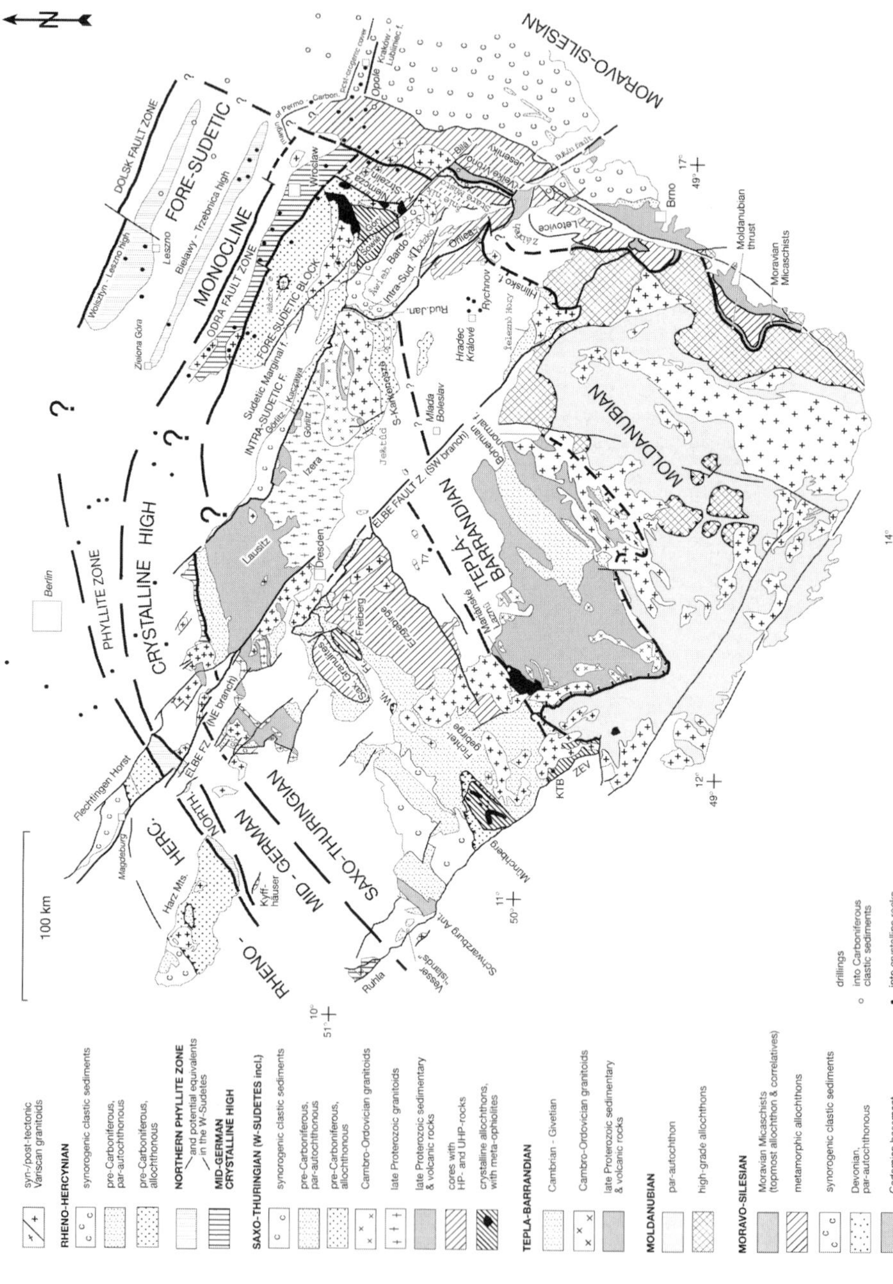

Fig. 2. Simplified geological map of the Bohemian Massif (after Franke & Żelaźniewicz 2000).

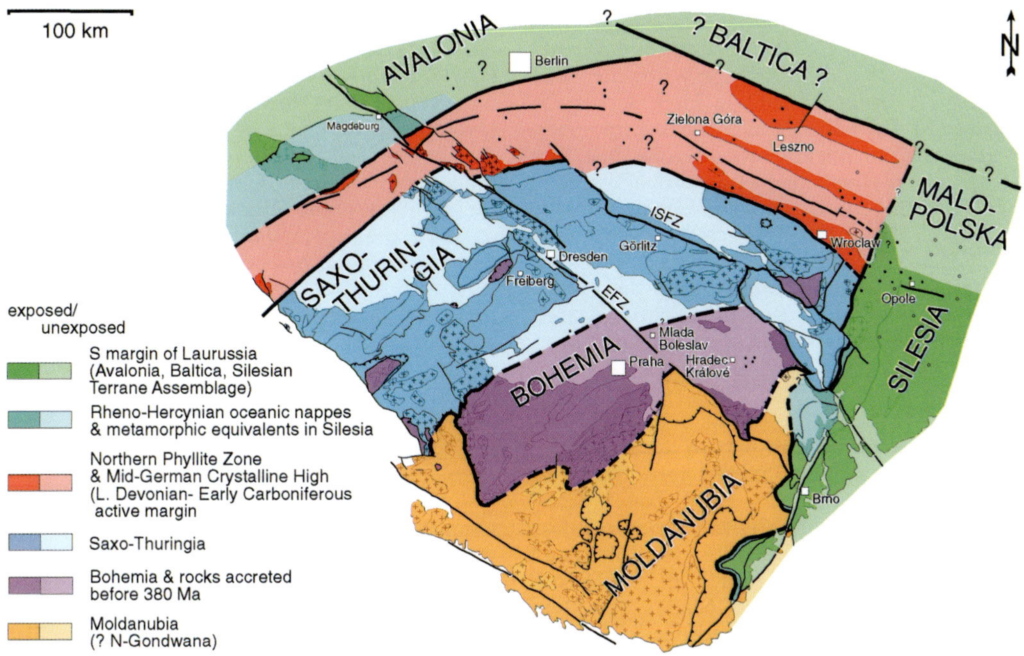

Fig. 3. Terrane map of the Bohemian Massif and adjacent areas (after Franke & Żelaźniewicz 2000). Geological boundaries as in Figure 2.

Rychnov Fault to the Teplá-Barrandian Unit. The curvature in the fault east of Rychnov parallels curved magnetic anomalies on the magnetic map of the Czech Republic. This pattern, together with bends in the exposed Hlinsko Fault, point to late refolding during SW/NE-directed compression.

Palaeogeographic assignment of tectonic units

The palaeogeographic interpretation of the Bohemian Massif is shown in Figure 3, and discussed below, in order from north to south. Unless cited otherwise, details are discussed in Franke (2000) and Franke & Żelaźniewicz (2000).

Avalonia

The periphery of the Bohemian arcuate structure is occupied by palaeogeographic units which were situated, at least from the late Lower Devonian onwards, at the southern margin of Laurussia. Such a position is well documented for the Devonian sediments of the Rheno-Hercynian Belt, which were deposited on the southern margin of the Avalonian segment of Laurussia. Correlatives on the SE margin of the Bohemian Massif are discussed in the end of this section.

Northern Phyllite Zone (NPZ) and Mid-German Crystalline High (MGCH)

The Northern Phyllite Zone and the Mid-German Crystalline High (Fig. 2) are tectonic assemblages of metamorphic rocks, which evolved from the southern margin of Avalonia, a Silurian/Devonian magmatic arc and from a northern member of the Armorican Terrane Assemblage (Franke 2000). The pre-Carboniferous palaeogeography of these units is poorly constrained, because protolith ages are much too scarce. However, the Northern Phyllite Zone and the Mid-German Crystalline High formed a northwestward growing active margin, formed in a regime of grossly southeastward subduction (e.g. Oncken 1997). It is characterized by Early Carboniferous pressure-dominated metamorphism in the Northern Phyllite Zone and part of the Mid-German Crystalline High (Massonne 1995; Okrusch 1995) and Devonian to early Carboniferous subduction-related magmatism

(Altherr *et al.* 1999). Although it lacks resolution for the Early Palaeozoic, this snapshot at a late stage of evolution of the Northern Phyllite Zone and Mid-German Crystalline High serves the purpose of our paper, which focuses upon tectonic displacements in Carboniferous time. The Mid-German Crystalline High probably correlates with the concealed basement high of the Odra Fault Zone. The greenschist grade rocks of the Bielawy-Trzebnica and Wolsztyn-Leszno Highs invite comparison with the Northern Phyllite Zone, but might also represent a suspect terrane of the Armorican Terrane Assemblage.

Saxothuringia

Units within the western part of the Bohemian Massif may easily be traced into the West Sudetes (east of the Elbe Fault Zone). Saxothuringia is represented by the par-autochthon of the Saxothuringian Belt W of the Elbe Fault Zone, by the Lausitz-Izera Unit, the South Krkonoše Unit (see also Mazur & Aleksandrowski 2001) and the Wądroże Window in the Fore-Sudetic Block. These foreland units have been accreted during the Late Viséan (e.g. Marheine *et al.* 2002) to the overriding Bohemia Terrane.

Eclogites and granulites formed around 340 Ma occur in the Erzgebirge and Granulitgebirge (Saxonian Granulites) of the Saxothuringian foreland and in the Orlica-Śnieżnik dome (see compilations in DEKORP & Orogenic Processes Working Groups 1999; Franke & Stein 2000 and Marheine *et al.* 2002). These rocks probably record subduction of Saxothuringian continental crust and subsequent hydraulic expulsion into the foreland (Franke & Stein 2000; Henk 2000; Krawczyk *et al.* 2000).

The southeasternmost part of Saxothuringia is probably represented by the narrow Staré Město Belt to the SE of the Orlica-Śnieżnik Unit. It records early Palaeozoic rifting (Kröner *et al.* 2000), which characterizes the Saxothuringian tectonic belt (and especially the foreland, i.e. the Saxothuringian Terrane).

Bohemia and rocks accreted to its northwestern margin

Bohemia (mainly exposed in the Teplá-Barrandian block) is characterized by Late Proterozoic sedimentary and volcanic rocks, deformed and metamorphosed during the Cadomian orogeny, and unconformably overlain by Cambrian and Ordovician sediments (Chlupáč 1993; Pouba & Skoček 2000; Zulauf 1997).

East of the Elbe Fault Zone, the concealed basement rocks known from drillings NE of Hradec Králové (Čech *et al.* 1989) and the isolated exposures of the Zvičin hills west of Králův Dvůr are also assigned to Bohemia. We propose the the same interpretation for the greenschist grade rocks to the north of the Rychnov fault, on the SW flank of the Orlické hory (see above).

Along the northwestern margin of Bohemia and in equivalent parts of the Sudetes, there are subduction-related, high pressure metamorphic rocks which evolved from oceanic or thinned continental crust of the Saxothuringian narrow ocean. Such rocks are known from the Saxothuringian Münchberg Klippe (395 Ma; review in Franke *et al.* 1995) and the Góry Sowie (402 Ma; O'Brien *et al.* 1997; Brueckner *et al.* 1996).

Another group of tectonic units spread out along the Saxothuringian/Bohemian suture zone underwent metamorphism and deformation during the Late Devonian and earliest Carboniferous (*c.* 380–350 Ma). Such rocks occur in the lower parts of the Saxothuringian Klippen in Germany, in the NW part of the Teplá-Barrandian block, in the East Krkonoše Unit and in the Góry Sowie (see compilations in Franke *et al.* 1995; Franke 2000; Franke & Żelaźniewicz 2000; Marheine *et al.* 2002). A similar age of deformation and metamorphism is likely for blueschists in the South Krkonoše Unit and the higher grade, upper parts of the Kaczawa tectonic pile to the NE of the Intra-Sudetic Fault, but remains to be proven. The Devonian metamorphic ages occur in units both structurally below as well as above the metamorphosed mafic and ultramafic rocks, which are taken to trace the Saxothuringian Suture (eclogites and serpentinites within the Münchberg thrust stack and in the Marianské Lazne meta-ophiolite at the NW margin of the Teplá-Barrandian block). Where the meta-ophiolites are missing, it is difficult to attribute the metamorphic rocks either to the Saxothuringian or the Bohemian Terrane, so that Franke & Żelaźniewicz (2000, fig. 6) have referred to all rocks with Devonian metamorphism as 'suture zone Saxothuringia/Bohemia'.

In the revised terrane map of the Bohemian Massif (Fig. 3), a more differentiated approach is proposed. The best-preserved example of Saxothuringian subduction/collision, the Münchberg Klippe, shows downward younging and decreasing grade of metamorphism within the thrust stack, which reflects northwestward accretion of foreland units (review in Franke *et al.* 1995). Rocks of amphibolite and higher metamorphic grades with mineral ages (zircon, hornblende, white mica) of more than 380 Ma occur in rocks close to the meta-ophiolites. In Fig. 3, only these higher grade and older metamorphic rocks have been assigned to the category "Bohemian Terrane and rocks accreted to it before 380 Ma". This applies to the Góry Sowie and to higher grade thrust sheets in the Saxothuringian Klippen (the lower grade, lower thrust sheets are too small to figure separately on the map).

These lower grade thrust sheets with younger metamorphic ages are taken to indicate derivation from the foreland of the Saxothuringian tectonic belt (Saxothuringia Terrane). This applies to the lower parts of the Saxothuringian Klippen in Germany, to the East Karkonosze Unit and its equivalents in the Kaczawa Mountains and probably also to the Kłodzko Unit. The lower parts of the Rudawy Janowickie and Kaczawa Units, which underlie meta-ophiolites (Seston *et al.* 2000) are likewise assigned to the foreland (Saxothuringia).

Silesian Terrane Assemblage

The Moravo-Silesian Belt contains Devonian and Carboniferous sequences that were deposited on a complex array of older rocks, in which it is possible to distinguish three terranes separated by NW- to WNW-trending faults (Silesian, Małopolska and Łysogóry Terranes). These terranes are summarized, in this paper, as the Silesian Terrane Assemblage (STA). The palaeogeographic affinities of the Silesian Terrane Assemblage and neighbouring parts of Baltica have been assessed by studies of zircons (Compston *et al.* 1995; Finger *et al.* 2000; Oberc-Dziedzic *et al.* 2001;Valverde-Vaquero *et al.* 2000; Żelaźniewicz *et al.* 2001), of detrital white micas (Belka *et al.* 2000, 2002) and by biogeographic data (Belka *et al.* 2002). It appears that the members of the Silesian Terrane Assemblage were derived from northwestern parts of Gondwana, became detached during the Cambrian and arrived at the Baltic margin at different times between the Cambrian and Silurian. In Devonian time, the Silesian Terrane Assemblage was firmly anchored to the southern margin of Laurussia.

The record of the Moravo-Silesian Belt has much in common with that of the Rheno-Hercynian Belt: Devonian rifting and mafic volcanism, Middle and Late Devonian reef carbonates, Early Carboniferous synorogenic clastic sediments, an Early Carboniferous external carbonate platform, and Late Carboniferous coal-bearing molasse Germany (e.g. Engel & Franke 1983; Havlena 1966; Schulmann & Gayer 2000). The NW part of the Jeseníký segment contains Devonian and early Carboniferous deeper water sediments, whereas the SE part represents a shelf (Dvořák 1995; Hladil *et al.* 1999). Deepening toward the internal (now northwestern) part of the Bohemian Arc conforms with the palaeogeographic pattern of the Rhenohercynian Belt in Germany.

The northwesternmost and structurally highest unit of the Jeseniky segment, the Velke Vrbno thrust sheet, contains eclogite relicts (Schulmann & Gayer 2000). This unit is probably continued, to the SW of the Bušin Fault, in the Zábřeh and Letovice complexes, and by the thin band of the 'Moldanubian Micaschists', which form the top of the Moravo-Silesian tectonic pile in the Svratka and Thaya Windows, immediately below the Moldanubian Thrust (e.g. Misař & Urban 1995; Schulmann *et al.* 1995). The high pressure rocks contained in these units probably represent deeper parts of the Rheno-Hercynian/Moravo-Silesian basin, which were subducted, then exhumed and overthrust on the Moravo-Silesian foreland. The southwestward narrowing of these rocks is due to tectonic excision by the Moldanubian Thrust.

Large scale structure

Taken altogether, the palaeogeographic zonation of the Bohemian Massif delineates an arc structure, which is slightly disrupted by the Elbe and Intra-Sudetic Fault Zones (see Fig. 2). Although the German and the Sudetic parts of the Saxothuringian Belt define an opening angle of about 90°, correlation of the Moravo-Silesian rocks with the Rheno-Hercynian rift in Germany implies almost a hairpin bend (Fig. 3). It appears, that about half of this curvature is primary.

Finger & Steyrer (1995) have proposed that a tight bend also exists within the Moldanubian Unit, where orthogneisses within the Variegated Unit are taken as equivalents of the Moravian Biteš Gneisses. However, these orthogneisses show different ages. We therefore prefer the view of Tollmann (1982), in which the Moldanubian and Moravo-Silesian are interpreted as different palaeogeographic units, and the Moldanubian Thrust is interpreted to truncate an older, intra-Moldanubian thrust stack.

Timing of post-collisional fault zones

The activity of the Intra-Sudetic Fault (ISF) is constrained by the infill of the Intra-Sudetic Basin (ISB), which represents a pull-apart structure controlled by the Intra-Sudetic Fault

(Aleksandrowski *et al.* 1997; Franke & Żelaźniewicz 2000). Miospores from the deepest part of the sequence are not older than the TS spore zone (Turnau *et al.* in prep.: approximately Viséan 2b to 3a). U–Pb zircon SHRIMP dating of a tuff in Australia, which ranges within this interval, has yielded an age of 332.3 ±2.2 Ma (see the correlation chart in Menning *et al.* 2000). A marine ingression higher up in the Intra-Sudetic Basin occurred during the *Goniatites crenistria* zone of the goniatite zonation, which corresponds to an isotopic age of about 326 Ma (Trapp, cited in Menning *et al.* 2000). The overlying intramontane clastic sequence with coal seams in the Upper Carboniferous extends into the Permian, although the main subsidence occurred in Late Viséan time (Dziedzic & Teisseyre 1990).

Direct isotopic evidence from the Intra-Sudetic Fault is available in Marheine *et al.* (2002). Mylonites from the fault zone yielded Ar/Ar white mica ages of 333 ± 3 and 324 ± 3 Ma. In addition, several white mica ages between 336 and 332 Ma have been obtained from the eastern part of the South Krkonoše and from the East Krkonoše Complex (Marheine *et al.* 2002). These areas show late-tectonic extension with movement down to the SE. These extensional faults and the cooling ages in their footwall most probably represent shoulder uplift to the NW of the subsiding Intra-Sudetic Basin. The isotopic ages fit well with the biostratigraphic record and confirm tectonic activity along the Intra-Sudetic Fault between about 333 and 325 Ma (Late Viséan).

The Elbe Fault Zone (EFZ) displaces the Teplá-Barrandian block (Bohemia) towards the SE. A marker is provided by correlation of the important normal fault at the southeastern boundary of the Teplá-Barrandian (Zulauf 1997) with the Hlinsko normal fault (Fig. 2), which implies an offset of about 70 km. Further to the NW, in the Mid-German Crystalline High (MGCH), displacement is much smaller. This apparent discrepancy is probably explained by the fact that the present-day tectonic structures of the Mid-German Crystalline High and areas to the NW largely post-date the main activity of the Elbe Fault Zone. A similar consideration might apply to the Intra-Sudetic Fault, but the area NE of the Lausitz-Izera block is not exposed. To the SE, both the Elbe and the Intra-Sudetic Fault are clearly truncated by the Moldanubian Thrust.

The age of the Moldanubian Thrust is best constrained by the infill of the Moravo-Silesian foreland basin, which received detritus from the Moldanubian rocks in the hanging wall. High grade metamorphism and rapid uplift in the Moldanubian nappes occurred between about 340 and 335 Ma (see the compilation in Franke 2000). Heavy minerals and pebbles derived from the Moldanubian granulite nappes do not occur before the latest Viséan (Myšlejovice Fm.; Hartley & Otava 2001). These youngest flysch sediments of the Moravo-Silesian foreland basin have yielded single grain Ar/Ar detrital mica ages of 340–331 Ma (Schneider *et al.* 1999). The isotopic age of the Viséan/Namurian boundary is approximately 325 Ma, fittingly younger than the age of the detrital minerals in the latest Viséan sediments. These relationships indicate that the present-day juxtaposition of blocks along the Moldanubian Thrust was largely accomplished at about 325 Ma. In the Desná and Keprník domes of the Moravo-Silesian Belt, which underlie the Moldanubian Thrust, Maluski *et al.* (1995) obtained Ar/Ar ages on white mica and biotite ranging between 310 and 280 Ma. Such late events might well have been accomodated within the present-day tectonic assemblage.

Taking into account the error bars on the isotopic data, it is not possible to differentiate between the ages of activity of the Intra-Sudetic Fault and the Moldanubian Thrust. However, the map pattern clearly reveals that the main displacement along the Moldanubian Thrust postdates that of the Intra-Sudetic and Elbe Faults.

Discussion

Statement of problem

The structural observations discussed above reveal three chapters of tectonic evolution:

- late Devonian to early Carboniferous ductile dextral transpression along the Variscan sutures
- block rotation and transverse dextral shearing in the Sudetes, leading to an arc-like structure with an opening angle of about 90° . Formation of intra-montane pull-apart basins along the NW-trending fault zones
- truncation of the arc by dextral transpression along the NE-trending Moldanubian Thrust.

The resulting arc-like-structure is contained, today, in an embayment defined by the Avalonian crust of the Rheno-Hercynian Belt to the NW, the East European Platform to the NE, and the Silesian Terrane Assemblage to the SE. Any plate-tectonic model must explain the arc-like arrangement of terranes and its truncation by the Moldanubian Thrust.

Model A: Northeastward displacement of parts of the Bohemian Massif

Although dextral displacement along the Moldanubian Thrust amounts to at least 400 km, there is no indication of a continuation of the Moldanubian Thrust into the East European Platform. Thus the present position of rocks contained in the core of the arc may have been brought about by northeastward, dextral transpression of central parts of the Bohemian Massif against a stable Silesian Terrane Assemblage to the SE, which itself was firmly anchored against the East European Platform. In this model, the core of the arc would either have been accomodated into a pre-existing embayment on the SW margin of Laurussia, or else represent an indenter moving northeastwards over the margin of the East European Platform. In both these cases, the Bohemian block would have been bounded by a dextral transpressive fault to the SE (the Moldanubian Thrust) and some sinistral equivalent on its NW flank.

The following points may be raised against this view:

- There is no evidence of any sinistral counterpart to the Moldanubian Thrust in the NW part or to the NW of the Bohemian Massif. Some sinistral movements in the Northern Phyllite Zone have been dated by K–Ar on white mica at 323 ± 4 and 308 ± 4 Ma (Klügel 1997). They therefore post-date the main activity of the Moldanubian Thrust.
- Northeastward movement of the Bohemian Massif towards a relatively stable East European Platform could only be explained by tectonic escape of the Massif as an orogenic wedge, which is not backed up by the configuration of main fault zones in central Europe (Fig.1).
- Within the exposed Bohemian Massif, there are no indications of NE-directed ductile tectonic transport. Stretching lineations in the Sudetes are generally oriented NW–SE, with a frequent sense of shear top to the NW.
- Geophysical and borehole evidence indicate that the SW margin of the East European Platform does not extend beyond the Elbe Line (Fig. 1, see also Krawczyk *et al.* 1999). Therefore, the overlap of Variscan deformation on Baltic basement does not exceed about 100 km, whereas dextral movements along the Moldanubian Thrust are in excess of 400 km.

One kinematic alternative – dextral, southwestward displacement of the Silesian Terrane Assemblage along the Moldanubian Thrust relative to a stable Bohemian Massif – is ruled out by the right angle at which the thrust abuts against the East European Platform: in this configuration, any southwestward movement of the Silesian Terrane Assemblage would have produced an extensional corridor between it and the East European Platform, for which there is no evidence.

Model B: Alternating displacements along NW- and NE-trending fault zones

Problems inherent with the above concepts may be overcome if large-scale tectonic shearing along NE-trending fault zones occurred in a position to the south or at the southern margin of the East European Platform. The important dextral displacements along the Variscan sutures between about 380 and 340 Ma definitely required such a position. This also applies to the dextral displacement of at least 400 km along the Moldanubian Thrust after 325 Ma. Both these episodes of dextral, orogen-parallel shearing cannot have been performed within the present-day tectonic configuration.

Matte (2001) has even suggested that the Moldanubian Thrust continues toward the SW through the Variscan basement of the Alps into southern France. In his paper, segments of a Variscan suture zone contained in the Massif de Maures (west of Nice) and in Sardinia are interpreted as continuations of the suture zone between Bohemia and Moldanubia. Such a large offset of about 1000 km would definitely imply that the Moldanubian Thrust was formed to the south of the Baltic part of Laurussia, where dextral shearing along NE-trending faults was unimpeded.

We therefore evaluate a model, in which the terranes now contained in the Variscides of central Europe were originally situated to the SE of Baltica (in present-day co-ordinates; Fig. 4a). The large-scale fault zones are attributed to the westward displacement of Gondwana with respect to Laurussia (Arthaud & Matte 1977; Tapponier 1977). This displacement occurred during the late stage of the Variscan collision in an environment of dextral transpression. Transpression was partitioned into two sets of dextral shear zones: one orogen-parallel (grossly

SW/NE) and another one in NW/SE, which transsects the orogenic trend. This latter fault system was mechanically guided by the NW-trending margin of the East European Platform, but may additionally have functioned as Riedel shears associated with the main, orogen-parallel faults. Furthermore, we propose that the kinematic evolution implied an alternation of orogen-parallel and transverse shearing.

When Gondwana impinged upon the Variscan terrane assemblage, Variscan terranes were displaced towards the NW along the Elbe and the Intra-Sudetic Fault Zones and possibly also along concealed faults within the Trans-European Suture Zone (TESZ; Fig. 4b). An early stage of the Bohemian Arc was formed by a combination of stepwise dextral displacements and clockwise rotation of blocks set between these faults.

During continued westward displacement of Gondwana, the NW-trending shear zones were truncated by a new, NE-trending zone of dextral transpression, the Moldanubian Thrust (Fig. 4c). The Silesian Terrane Assemblage was displaced towards the SW, and, thereby, rotated clockwise.

Later still, the Moldanubian Thrust was displaced by renewed dextral translation along the southwestern, Baltic margin of Laurussia (Fig. 4d). Shearing may have occurred either along the Tornquist-Teisseyre Fault Zone, along the Holy Cross Dislocation and possible northwestern continuation (Dolsk Fault Zone, Fig. 1). It was during this late stage, that the Variscan deformation fronts on the NW and the SE of the Bohemian Massif acquired their present-day positions. They were therefore not affected by the northwestward displacement. Some later change in the direction of compression then 'drove' the Bohemian block to the NE, with limited overthrusting on the Polish Caledonides and the margin of Baltica.

Testing the late stage of model B

The SE margin of Baltica trends from the Danube Delta across the Crimean Peninsula to the north of the Greater Caucasus (e.g. Görür *et al.* 1997), about 1000 km to the SE of the present-day position of the Moldanubian Thrust. Hence, model B requires northwestward displacements of the Moldanubian Thrust after its main period of activity (after about 325 Ma). Basic support for such large-scale northwestward displacements of the Silesian Terrane Assemblage along the Baltica Margin comes from palaeomagnetic findings Lewandowski 1993, 1995; Mizerski 1995). The following observations also permit a quantitative estimate of the notional displacement:

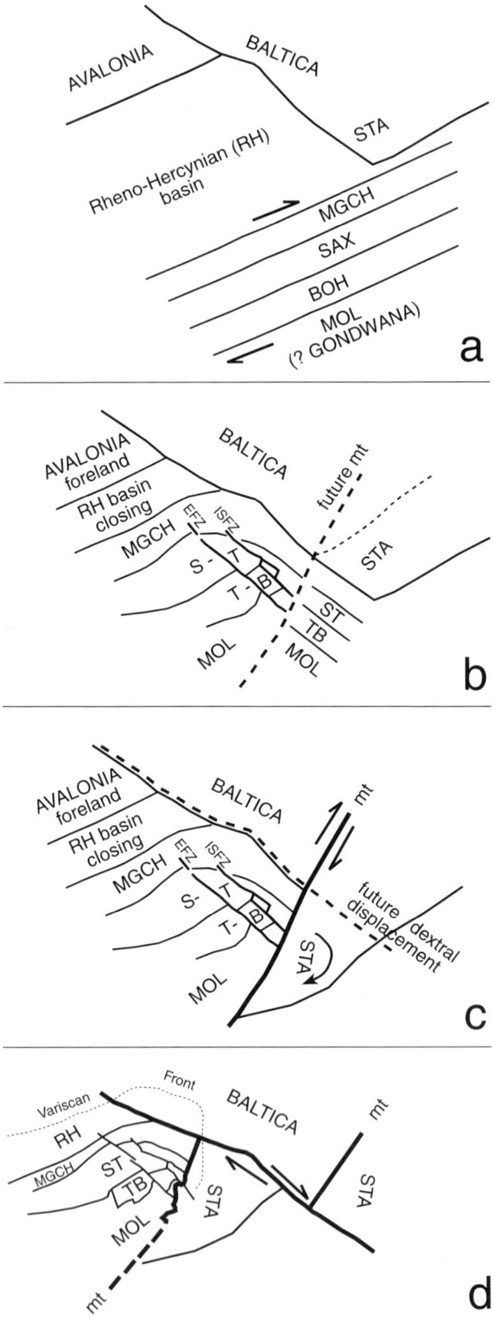

Fig. 4. Plate-kinematic sketch sequence illustrating the formation of the Bohemian Arc. (a) 380–340 Ma; (b) *c.* 335 Ma; (c) ≥ 325 Ma; (d) *c.* 305 Ma. Mt, Moldanubian thrust; MGCH, Mid-German Crystalline High; MOL, Moldanubia Terrane resp. Moldanubian Zone (in d); RH, Rheno-Hercynian Belt; ST, Saxothuringian Terrane resp. Saxothuringian Belt (in d); STA, Silesian Terrane Assemblage; TB, Teplá-Barrandian Block.

- An important part of shortening within the Rheno-Hercynian belt occurred after about 325 Ma, and amounted to a minimum of 175 km (Oncken *et al.* 2000).
- Southeastward backthrusting of the Mid-German Crystalline High over the Saxothuringian Belt contributes a conservative estimate of about 100 km for the Late Carboniferous tectonic activity (Schäfer *et al.* 2000).
- Within the Bohemian Massif, NW and NE-trending faults active during the late Carboniferous and Permian form a conjugate set which has accommodated north–south directed shortening of some some tens of kilometres (Handler *et al.* 1991; Urban & Synek 1995).

Together, these increments amount to a minimum of about 300 km.

Further displacements have been proven, but are difficult to quantify:

- Some southeastward ductile thrusting has been documented for the SE part of the Moldanubian Belt in Austria, and dated by Ar/Ar on biotite at 323 ± 7 Ma (Matte *et al.* 1985).
- Recent palaeogeographic and tectonic studies in the Rheno-Hercynian Belt have revealed continental allochthonous units between the oceanic Giessen/Harz Nappe and the par-autochthon, which Oncken *et al.* (2000) did not consider in their conservative estimate of shortening (Franke 2000). This may add another 100 km to the amount of thrusting.
- Late Carboniferous/Permian dextral movements also occurred along various fault zones within the Trans-European Suture Zone (e.g. Thybo 1997).
- Late Cretaceous inversion tectonics in the northern foreland of the Alps (e.g. Ziegler 1987) likewise required dextral displacements along the TESZ.
- Late Carboniferous basin inversion by NW/SE-directed shortening has also been documented for northern parts of the British Isles (Chadwick *et al.* 1993).

Lastly, the Moldanubian Thrust might have been displaced to the NW by reactivation of the British segment of the Iapetus suture, which might be responsible for the preservation of higher structural levels in Scandinavia as compared with Scotland.

Taken altogether, the southern segment of the Moldanubian Thrust has been displaced to the NW, more recently than about 325 Ma, for several hundreds of kilometres. Since the quantitative constraints yield only conservative estimates, the observed displacements are at least in the same order of magnitude as the 1000 km required by model B.

Conclusions

Terrane correlation and evaluation of fault activities of Variscan Units to the SW of the East European Platform permit discrimination of two basic models:

(A) Large-scale, northeastward movement of the Bohemian Massif as an indenter, limited to the SE by the Moldanubian Thrust. Although this model cannot be entirely ruled out, it has fundamental deficiencies: any large sinistral shear zone on the NW flank of the indenter remains to be identified, and there is no evidence of large-scale northeastward tectonic transport.

(B) Tectonic units now contained in the central European segment of the Variscides were originally situated to the south of Baltica, where they underwent orogen-parallel dextral shearing in Devonian and earliest Carboniferous time. Later, these units were dextrally displaced and rotated clockwise by northwestward movement along the SW margin of the East-European Platform (TESZ and parallel faults). This incipient arc structure was then truncated in the south by long-distance southwestward transport of units south of the Moldanubian Thrust. Renewed displacement toward the NW across a distance of about 1000 km brought the thrust into its present position. This model conforms to the sequence of events, and the demonstrable displacement of the Moldanubian toward the NW amounts at least to some hundreds of kilometres.

Other palaeogeographic tests of our model would require discussion of a still wider geological framework. If the Variscan terrane assemblage, in Late Devonian/Early Carboniferous time, was situated in a position south of the Baltic segment of Laurussia, 'along-strike' continuations of these terranes should exist along the southern margin of Laurussia, e.g. in Turkey and areas further east. Variscan basement outcrops in Turkey have been compared with possible counterparts in central Europe (Görür *et al.* 1997). Similarly, the tectonic concept for the Moldanubian Thrust proposed by Matte (2001) needs further confirmation: in this model equivalents of the Bohemian and Saxothuringian Terranes should be present to the NE of the Variscan suture linking the Massif de Maures (south France) with Sardinia, i.e. within the basement of the Alps. It also remains uncertain, whether the late Viséan and younger transpressional movements at the eastern and southern margins of the Bohemian Massif were caused by the impingement of a continental block to the south (Gondwana), or else by oblique northwestward subduction of oceanic crust. All these issues would require extensive discussion of the Balkan Peninsula and Turkey, as well as of the Variscan basement in the Alps, and therefore go beyond the frame of this paper.

The authors gratefully acknowledge travel funding by the International Bureau of the German Ministry of Education and Science (BMBF) and the Polish Academy of Sciences, which has enabled scientific exchange between the authors since 1989. Results and views presented in this paper also profited from discussions in a priority programme of Deutsche Forschungsgemeinschaft ('Orogenic Processes') and the EU-funded TMR programme PACE. We especially wish to acknowledge helpful comments by E. Stein (who also helped with the preparation of the diagrams) and J. A. Winchester, as well as constructive reviews by B. Leveridge and V. Kachlík.

References

ALEKSANDROWSKI, P. 1994. Discussion on U-Pb ages from SW Poland: evidence for a Caledonian suture zone between Baltica and Gondwana. *Journal of the Geological Society, London*, **151**, 1049–1050.

ALEKSANDROWSKI, P., KRYZA, R. & ŻABA, J. 1997. Kinematic data on major Variscan strike slip faults and shear zones in the Polish Sudetes, northeast Bohemian Massif. *Geological Magazine*, **134**, 727–739.

ALEKSANDROWSKI, P., KRYZA, R., MAZUR, S., PIN, C. & ZALASIEWICZ, J. 2000. The Polish Sudetes: Caledonian or Variscan? *Transactions of the Royal Society of Edinburgh: Earth Sciences*, **90**, 127–146.

ALTHERR, R., HENES-KLAIBER, U., HEGNER, E., SATIR, M. & LANGER, C. 1999. Plutonism in the Variscan Odenwald (Germany): from subduction to collision. *International Journal of Earth Sciences*, **88**, 422–443.

ARTHAUD, F. & MATTE, P. 1977. Late Paleozoic strike-slip faulting in Southern Europe and Northern Africa: result of a right lateral shear zone between the Appalachians and the Urals. *Geological Society of America Bulletin*, **88**, 1305–1320.

BELKA, Z., AHRENDT, H., FRANKE, W. & WEMMER, K. 2000. The Baltica-Gondwana suture in central Europe: evidence from K-Ar ages of detrital muscovites and biogeographical data. *In*: FRANKE, W., HAAK, V., ONCKEN, O. & TANNER, D. (eds) *Orogenic processes: Quantification and modelling in the Variscan Belt*. Geological Society, London, Special Publications, **179**, 87–102.

BELKA, Z., VALVERDE-VAQUERO, P., AHRENDT, H., WEMMER, K., FRANKE, W. & SCHÄFER, J. 2002. Accretion of first Gondwana-derived terranes at the margin of Baltica. . *In*: WINCHESTER, J. A., PHARAOH, T. C. & VERNIERS, J. (eds) *Palaeozoic amalgamation of Central Europe*. Geological Society, London, Special Publications, **201**, 19–36.

BRUECKNER, H. K., BLUSZTAJN, J. & BAKUN-CZUBAROW, N. 1996. Trace element and Sm-Nd 'age' zoning in garnets from peridotites of the Caledonian and Variscan Mountains and tectonic implications. *Journal of Metamorphic Geology*, **14**, 61–73.

ČECH, S., HAVLIČEK, V. & ZIKMUNDOVA, J. 1989. The Upper Devonian and Lower Carboniferous in north-eastern Bohemia (based on boreholes in the Hradec Králové area). *Věstník Ústředniho Ústavu Geologickeho*, **64/2**, 65–76.

CHADWICK, R. A., EVANS, D. J. & HOLLIDAY, D. W. 1993. The Maryport fault: the post-Caledonian tectonic history of southern Britain in microcosm. *Journal of the Geological Society of London*, **150**, 247–250.

CHLUPÁČ, I. 1993. *Geology of the Barrandian: A Field Trip Guide*. Senckenbergische Naturforschende Gesellschaft & Czech Geological Survey. Verlag Waldemar Kramer, Frankfurt am Main, 163 p.

COMPSTON, W., SAMBRIDGE, M. S., REINFRANK, R. F., MOCZYDLOWSKA, M., VIDAL, G. & CLAESSON, S. 1995. Numerical ages of volcanic rocks and the earliest faunal zone within the Late Precambrian of east Poland. *Journal of the Geological Society, London*, **152**, 599–611.

CROWLEY, Q. G., FLOYD, P. A., WINCHESTER, J. A., FRANKE, W. & HOLLAND, J. G. 2000. Early Palaeozoic rift-related magmatism in Variscan Europe: fragmentation of the Armorican Terrane Assemblage. *Terra Nova*, **12**, 171–180.

CYMERMAN, Z. 2000. Palaeozoic orogeneses in the Sudetes: a geodynamic model. *Geological Quarterly*, **44**, 59–80.

DEKORP & OROGENIC PROCESSES WORKING GROUPS (DÜRBAUM H.-J., F. W., STEIN E., KRAWCZYK, C. M., LÜSCHEN E., OETTINGER G., ONCKEN O., SCHUSTER K., HAAK V., ENDERLE U., PRODEHL C., BERGER H., KRENTZ O., POLOM U., PROCHNOW U., THOMAS M., THOMAS R., KRONER U.) 1999. Structure of the Saxonian Granulites: Geological and geophysical constraints on the exhumation of high-pressure/high-temperature rocks in the mid-European Variscan belt. *Tectonics*, **18**, 756–773.

DVOŘÁK, J. 1995. Moravo-Silesian Zone: autochthon: stratigraphy. *In*: DALLMEYER, D., FRANKE, W. & WEBER, K. (eds) *Pre-Permian geology of Central and Western Europe*. Springer, Berlin, Heidelberg, New York, 475–489.

DZIEDZIC, K. & TEISSEYRE, A. K. 1990. The Hercynian molasse and younger deposits in the Intra-Sudetic Depression, SW Poland. *Neues Jahrbuch für Geologie und Paläontologie, Abhandlungen*, **179**, 285–305.

ENGEL, W. & FRANKE, W. 1983. Flysch sedimentation: Its relation to tectonism in the European Variscides. *In*: MARTIN, H. & EDER, F. W. (eds) *Intracontinental fold belts. Case studies in the Variscan Belt of Europe and the Damara Belt in Namibia*. Springer, Berlin-Heidelberg-New York, 289–321.

FINGER, F. & STEYRER, H. P. 1995. A tectonic model for the Eastern Variscides: Indications from a chemical study of amphibolites in the South-Eastern Bohemian Massif. *Geologica Carpathica*, **46**, 137–150.

FINGER, F., HANZL, P., PIN, C., VON QUADT, A. & STEYRER, H. P. 2000. The Brunovistulian: Avalonian Precambrian sequence at the eastern end of the Central European Variscides? *In*: FRANKE, W., HAAK, V., ONCKEN, O. & TANNER, D. (eds) *Orogenic processes: Quantification and modelling in the Variscan Belt*. Geolocical Society, London, Special Publications, **179**, 103–112.

FLOYD, P. A., WINCHESTER, J., SESTON, R., KRYZA, R. & CROWLEY, Q. G. 2000. Review of geochemical variation in Lower Palaeozoic metabasites from the NE Bohemian Massif: intracratonic rifting and plume-ridge interaction. *In*: FRANKE, W., HAAK, V., ONCKEN, O. & TANNER, D. (eds) *Orogenic processes: Quantification and modelling in the Variscan Belt*. Geological Society, London, Special Publications, **179**, 155–174.

FRANKE, W. 1989. Tectonostratigraphic units in the Variscan belt of central Europe. *Geological Society of America, Special Paper*, **230**, 67–90.

FRANKE, W. 2000. The mid-European segment of the Variscides: tectonostratigraphic units, terrane boundaries and plate tectonic evolution. *In*: FRANKE, W., HAAK, V., ONCKEN, O. & TANNER, D. (eds) *Orogenic processes: Quantification and modelling in the Variscan Belt*. Geological Society, London, Special Publications, **179**, 35–62.

FRANKE, W. & STEIN, E. 2000. Exhumation of high-grade rocks in the Saxothuringian Belt: geological constraints and geodynamic concepts. *In*: FRANKE, W., HAAK, V., ONCKEN, O. & TANNER, D. (eds) *Orogenic processes: Quantification and modelling in the Variscan Belt*. Geological Society, London, Special Publications, **179**, 337–354.

FRANKE, W. & ŻELAŹNIEWICZ, A. 2000. The eastern termination of the Variscides: terrane correlation and kinematic evolution. *In*: FRANKE, W., HAAK, V., ONCKEN, O. & TANNER, D. (eds) *Orogenic processes: Quantification and modelling in the Variscan Belt*. Geological Society, London, Special Publications, **179**, 63–86.

FRANKE, W., ŻELAŹNIEWICZ, A., PORĘBSKI, S. J. & WAJSPRYCH, B. 1993. Saxothuringian zone in Germany and Poland: differences and common features. *Geologische Rundschau*, **82**, 583–599.

FRANKE, W., KREUZER, H., OKRUSCH, M., SCHÜSSLER, U. & SEIDEL, E. 1995. Saxothuringian Basin: Exotic metamorphic nappes: stratigraphy, structure, and igneous activity. *In*: DALLMEYER, D., FRANKE, W. & WEBER, K. (eds) *Pre-Permian geology of Central and Western Europe*. Springer, Berlin, Heidelberg, New York, 275–294.

GÖRÜR, N., MONOD, O., SENGÖR, A. M. C., TÜYSÜZ, O., YIGITABAS, E., SKANIC, M. & AKKÖK, R. 1997. Place of the 'Palaeozoic of Istanbul' (Turkey) in the frame of the Variscan Belt. *Terra Nova, Abstract supplement no. 1*, **9**, 101.

HANDLER, R., BRANDMAYR, M. & WALLBRECHER, E. 1991. The Rodl shear zone in the southern Bohemian Massif. *Zentralblatt für Geologie und Paläontologie*, **1991/1**, 69–86.

HARTLEY, A. J. & OTAVA, J. 2001. Sediment provenance and dispersal in a deep marine foreland basin: the Lower Carboniferous Culm Basin, Czech Republic. *Journal of Geological Society, London*, **158**, 137–150.

HAVLENA, V. 1966. The Upper Silesian Basin. *In*: ZOUBEK, V. (ed.) *Regional geology of Czechoslovakia. part I: The Bohemian Massif*. Geological Survey of Czechoslovakia, Publishing House of the Czechoslovak Academy of Sciences, Prague, 414–433.

HENK, A. 2000. Foreland-directed lower-crustal flow and its implications for the exhumation of high-pressure/high-temperature rocks. *In*: FRANKE, W., HAAK, V., ONCKEN, O. & TANNER, D. (eds) *Orogenic processes: Quantification and modelling in the Variscan Belt*. Geological Society, London, Special Publications, **179**, 355–368.

HLADIL, J., MELICHAR, R., OTAVA, A., KRS, M., MAN, O., PRUNER, P., CEJCHAN, P. & OREL, P. 1999. The Devonian in the Easternmost Variscides, Moravia: a Holistic Analysis Directed Towards Comprehension of the Original Context. *Abhandlungen der Geologischen Bundesanstalt*, **54**, 27–47.

JELEŃSKA, M., KĄDZIAŁKO-HOFMOKL, M., EDEL, J. B., JAMROZIK, L. PETERSEN, N. & SOFFEL, H. 1995. Palaeomagnetic investigations of the Palaeozoic circum-Sowie Góry Mountains ophiolitic belt in the Sudetes, Poland. *Geophysical Journal International*, **122**, 658–674.

KLÜGEL, T. 1997. Geometrie und Kinematik einer variszischen Plattengrenze. Der Südrand des Rhenoherzynikums im Taunus. *Geologische Abhandlungen Hessen*, **101**, 215 p.

KONOPÁSEK, J., SCHULMANN, K. & LEXA, O. 2001. Structural evolution of the central part of the Krusne hory (Erzgebirge) Mountains in the Czech Republic – evidence for changing stress regime during Variscan compression. *Journal of Structural Geology*, **23**, 1373–1392.

KOSSMAT, F. 1927. Gliederung des varistischen Gebirgsbaues. *Abhandlungen des Sächsischen Geologischen Landesamtes*, **1**, 1–39.

KRAWCZYK, C. M., STILLER, M. & DEKORP Basin Research Group, 1999. Reflection seismic constraints on Paleozoic crustal structure and Moho beneath the NE German Basin. *Tectonophysics*, **314**, 241–253.

KRAWCZYK, C. M., STEIN, E., CHOI, S., OETTINGER, G., SCHUSTER, K., GÖTZE, H.-J., HAAK, V., ONCKEN, O., PRODEHL, C. & SCHULZE, A. 2000. Geophysical constraints on exhumation mechanisms of high-pressure rocks: the Saxothuringian case between the Franconian Line and Elbe Zone. *In*:

FRANKE, W., HAAK, V., ONCKEN, O. & TANNER, D. (eds) *Orogenic processes: Quantification and modelling in the Variscan Belt*. Geological Society, London, Special Publications, **179**, 303–322.

KRÖNER, A. & HEGNER, E. 1998. Geochemistry, single zircon ages and Sm-Nd systematics of granitoid rocks from the Góry Sowie (Owl Mts), Polish west Sudetes: evidence for early Palaeozoic arc-related plutonism. *Journal of the Geological Society, London*, **155**, 711–724.

KRÖNER, A., STIPSKÁ, P., SCHULMANN, K. & JAECKEL, P. 2000. Chronological constraints on the pre-Variscan evolution of the northeastern margin of the Bohemian Massif, Czech Republic. *In*: FRANKE, W., HAAK, V., ONCKEN, O. & TANNER, D. (eds) *Orogenic processes: Quantification and modelling in the Variscan Belt*. Geological Society, London, Special Publications, **179**, 175–198.

KRS, M., KRSOVA, M. & PRUNER, P. 1995. Paleomagnetism and paleogeography of Variscan formations of the Bohemian Massif: a comparison with other regions in Europe. *Studia geophysica and geodaetica*, **39**, 309- 319.

LEWANDOWSKI, M. 1993. Paleomagnetism of the Paleozoic rocks of the Holy Cross Mts (Central Poland) and the origin of the Variscan orogen. *Publications of the Instiute of Geophysics, Polish Academy of Sciences*, **A23**, 3–85.

LEWANDOWSKI, M. 1995. Paleomagnetic constraints for Variscan mobilism of the Upper Silesian and Malopolska Massifs, southern Poland – reply. *Geological Quarterly*, **39**, 282–292.

MALUSKI, H., RAJLICH, P. & SOUČEK, J. 1995. Pre-Variscan, Variscan and early Alpine thermotectonic history of the north-eastern Bohemian Massif: $^{40}Ar/^{39}Ar$ study. *Geologische Rundschau*, **84**, 345–358.

MARHEINE, D., KACHLÍK, V., MALUSKI, M., PATOČKA, F. & ŻELAŹNIEWICZ, A. 2002. The Ar-Ar ages from the West Sudetes (NE Bohemian Massif): constraints on the Variscan polyphase tectonothermal development. *In*: WINCHESTER, J. A., PHARAOH, T. C. & VERNIERS, J. (eds) Palaeozoic amalgamation of Central Europe. *Geological Society, London, Special Publications*, **201**, 133–156.

MASSONNE, H.-J. 1995. Rhenohercynian foldbelt: metamorphic units (Northern Phyllite Zone) – metamorphic evolution. *In*: DALLMEYER, R. D., FRANKE, W. & WEBER, K. (eds) *Pre-Permian geology of Central and Eastern Europe*. Springer, Berlin, Heidelberg, New York, 132–137.

MATTE, P. 1986. Tectonic and plate tectonic model for the Variscan belt of Europe. *Tectonophysics*, **126**, 329–374.

MATTE, P. 1991. Accretionary history and crustal evolution of the Variscan Belt in Western Europe. *Tectonophysics*, **196**, 309–337.

MATTE, P. 2001. The Variscan collage and orogeny (480–290 Ma) and the tectonic definition of the Armorica microplate: a review. *Terra Nova*, **13**, 122–128.

MATTE, P., MALUSKI, H. & ECHTLER, H. 1985. Cisaillements ductiles varisques vers l'Est – Sud-Est dans les nappes du Waldviertel (Sud-Est du Massif de Bohème, Autriche). Données microtectoniques et radiométriques 39Ar/40Ar. *Comptes Rendues Académie des Sciences, Paris*, **301**, 721–726.

MAZUR, S. & ALEKSANDROWSKI, P. 2001. A Variscan Suture Zone along Sheared SW Boundary of the Orlica-Śnieżnik Massif (The Sudetes, NW Bohemian Massif)? *In*: *European Union of Geosciences XI*. Abstract Volume. Strasbourg, Cambridge Publications, 632.

MCKERROW, W. S. & VAN STAAL, C. R. 2000. The Palaeozoic time scale reviewed. *In*: FRANKE, W., HAAK, V., ONCKEN, O. & TANNER, D. (eds) *Orogenic processes: Quantification and modelling in the Variscan Belt*. Geological Society, London, Special Publications, **179**, 5–8.

MENNING, M., WEYER, D., DROZDZEWSKI, G., VAN AMEROM, H. W. J. & WENDT, I. 2000. A Carboniferous Time Scale 2000: Discussion and Use of Geological Parameters as Time Indicators from Central and Western Europe. *Geologisches Jahrbuch Hannover*, **A156**, 3–44.

MISAR, Z., and URBAN, M. 1995. Moravo-Silesian Zone – Introduction. *In*: DALLMEYER, D., FRANKE, W., and WEBER, K. (eds) *Pre-Permian Geology of Central and Western Europe*. Springer, Berlin, Heidelberg, New York, 469–473.

MIZERSKI, W. 1995. Geotectonic evolution of the Holy Cross Mts in central Europe. *Biuletyn Państwowego Instytutu Geologicznego*, **372**, 5–47.

NAWROCKI, J. 2000. Late Silurian paleomagnetic pole from the Holy Cross Mountains: constraints for the post-Caledonian tectonic activity of the Trans-European Suture Zone. *Earth and Planetary Science Letters*, **179**, 325–334.

OBERC-DZIEDZIC, T., KLIMAS, K., KRYZA, R. & FANNING, M. 2001. SHRIMP zircon geochronology of the Neoproterozoic Strzelín gneiss: evidence for the Moravo-Silesian Zone affinity of the Strzelín Massif, Fore-Sudetic Block, SW Poland. *Geolines*, **13**, 96–97.

O'BRIEN, P. J., KRÖNER, A., JAECKEL, P., HEGNER, E., ŻELAŹNIEWICZ, A. & KRYZA, R. 1997. Petrological and isotopic studies on Paleozoic high-pressure granulites, Góry Sowie Mts, Polish Sudetes. *Journal of Petrology*, **38**, 433–456.

OKRUSCH, M. 1995. Mid-German Crystalline High: metamorphic evolution. *In*: DALLMEYER, D., FRANKE, W. & WEBER, K. (eds) *Pre-Permian geology of Central and Western Europe*. Springer, Berlin, Heidelberg, New York, 201–213.

OLIVER, G. J. H., CORFU, F. & KROGH, T. E. 1993. U-Pb ages from SW Poland: evidence for a Caledonian suture zone between Baltica and Gondwana. *Journal of the Geological Society of London*, **150**, 355–369.

ONCKEN, O. 1997. Transformation of a magmatic arc and an orogenic root during oblique collision and its consequences for the evolution of the European Variscides (Mid-German Crystalline Rise). *Geologische Rundschau*, **86**, 2–20.

ONCKEN, O., PLESCH, A., WEBER, J., RICKEN, W. & SCHRADER, S. 2000. Passive margin detachment during arc-continent collision (Central European Variscides). *In*: FRANKE, W., HAAK, V., ONCKEN,

O. & TANNER, D. (eds) *Orogenic processes: Quantification and modelling in the Variscan Belt.* Geological Society, London, Special Publications, **179**, 199–216.

POUBA, Z. & SKOČEK, V. (eds) 2000. Neoproterozoic of the Barrandian (Czech Republic). *Bulletin of the Czech Geological Survey*, **75/3**, 360 p.

SCHÄFER, F., ONCKEN, O., KEMNITZ, H. & ROMER, R. 2000. Upper-plate deformation during collisional orogeny: a case study from the German Variscides (Saxothuringian Zone). *In*: FRANKE, W., HAAK, V., ONCKEN, O. & TANNER, D. (eds) *Orogenic processes: Quantification and modelling in the Variscan Belt.* Geological Society, London, Special Publications, **179**, 281–302.

SCHNEIDER, D., HANDLER, R., NEUBAUER, F., KALVODA, J. & TOMEK, C. 1999. New $^{39}Ar/^{40Ar}$ single- und multi-grain ages of detrital white mica from the Moravo-Silesian zone (Czech Republic). *Terra Nostra*, **99**, 238–239.

SCHULMANN, K. & GAYER, R. 2000. A model for a continental accretionary wedge developed by oblique collision: the NE Bohemian Massif. *Journal of the Geological Society, London*, **157**, 401–416.

SCHULMANN, K., MELKA, R., LOBKOWICZ, M. Z., LEDRU, P., LARDEAUX, J.-M. & AUTRAN, A. 1994. Contrasting styles of deformation during progressive nappe stacking at the southeastern margin of the Bohemian Massif (Thaya Dome). *Journal of Structural Geology*, **16**, 355–370.

SCHULMANN, K., LOBKOWICZ, M., MELKA, R. & FRITZ, H. 1995. Moravo-Silasian Zone: allochthonous units: structure. *In*: DALLMEYER, D., FRANKE, W. & WEBER, K. (eds) *Pre-Permian geology of Central and Western Europe*. Springer, Berlin, Heidelberg, New York, 530–540.

SESTON, R., WINCHESTER, J. A., PIASECKI, M. A. J., CROWLEY, Q. G. & FLOYD, P. A. 2000. A structural model for the Western-Central Sudetes: a deformed stack of Variscan thrust sheets. *Journal of the Geological Society, London*, **157**, 1155–1168.

SUESS, F. E. 1912. Die moravischen Fenster und ihre Beziehung zum Grundgebirge des Hohen Gesenke. *Denkschriften Akdemie der Wissenschaften Wien, mathematisch-naturwissenschaftliche Klasse*, **88**, 541–631.

TAIT, J. A., BACHTADSE, V. and SOFFEL, H. 1994. Eastern Variscan fold belt: Paleomagnetic evidence for oroclinal bending. *Geology*, **24**, 871–874.

TAPPONNIER, P. 1977. Évolution tectonique du système alpin en Méditerranée: poinconnement et écrasement rigide-plastique. *Bulletin de la Société géologique de France*, **7**, 437–460.

THYBO, H. 1997. Geophysical characteristics of the Tornquist Fan area, northwest Trans-European Suture Zone: indication of Late Carboniferous to Early Permian dextral transtension. *Geological Magazine*, **134**, 597–606.

TIMMERMANN, H., PARRISH, R. R., NOBLE, S. R. & KRYZA, R. 2000. New U-Pb monazite and zircon data from the Sudetes Mountains in SW Poland: evidence for a single-cycle Variscan orogeny. *Journal of the Geological Society of London*, **157**, 265–268.

TOLLMANN, A. 1982. Großräumiger variszischer Deckenbau im Moldanubikum und neue Gedanken zum Variszikum Europas. *Geotektonische Forschungen*, **64**, 1–91.

URBAN, M. & SYNEK, J. 1995. Moldanubian Zone: structure. *In*: DALLMEYER, D., FRANKE, W. & WEBER, K. (eds) *Pre-Permian geology of Central and Western Europe*. Springer, Berlin, Heidelberg, New York, 429–443.

VALVERDE-VAQUERO, P., DÖRR, W., BELKA, Z., FRANKE, W., WISZNIEWSKA, J. & SCHASTOK, J. 2000. U-Pb single-grain dating of detrital zircon in the Cambrian of central Poland: implications for Gondwana versus Baltica provenance studies. *Earth and Planetary Science Letters*, **184**, 225–240.

ŻELAŹNIEWICZ, A. & FRANKE, W. 1994. Discussion on U-Pb ages from SW Poland: evidence for a Caledonian suture zone between Baltica and Gondwana. *Journal of the Geological Society, London*, **151**, 1050–1052.

ŻELAŹNIEWICZ, A., GRAD, M. & GUTERCH, A. 2001. Late Proterozoic through Early Palaeozoic Crustal Additions at the Teisseyre-Tornquist Margin of Baltica: Geological and Geophysical data from Poland. *ESF EUROPROBE Meeting Ankara, Abstracts, 96–97.*

ZIEGLER, P. 1987. Late Cretaceous and Cenozoic intraplate compressional deformations in the Alpine foreland – a geodynamic model. *Tectonophysics*, **137**, 389–420.

ZULAUF, G. 1997. Von der Anchizone bis zur Eklogitfazies: Angekippte Krustenprofile als Folge der cadomischen und variscischen Orogenese im Teplá-Barrandium (Böhmische Masse). *Geotektonische Forschungen*, **89**, 1–302.

Seismic refraction evidence for crustal structure in the central part of the Trans-European Suture Zone in Poland

M. GRAD[1], A. GUTERCH[2] & S. MAZUR[3]

[1]*Institute of Geophysics, University of Warsaw, Pasteura 7, 02-093 Warsaw, Poland*
email: mgrad@mimuw.edu.pl

[2]*Institute of Geophysics, Polish Academy of Science, Ks. Janusza 64, 01-452 Warsaw, Poland*

[3]*Institute of Geological Sciences, University of Wrocław, Maxa Borna 9, 50-204 Wrocław, Poland*

Abstract: The results of seismic investigations obtained for the Trans-European Suture Zone (TESZ) show the presence of relatively low velocity rocks ($V_p < 6.1$ kms^{-1}), of sedimentary, metamorphic or volcanic origin, down to a depth of 20 km; high velocity (V_p = 6.8–7.3 kms^{-1}) lower crust, the Moho at a depth of approximately 30–33 km; and a high-velocity ($V_p > 8.3$ kms^{-1}) uppermost mantle. The transition of the crustal structure is seen across a 200 km wide zone. The three-layered crystalline crust of Baltica changes over this distance into the two-layered crust of Palaeozoic (Variscan) Europe, due to the disappearance of the lowest layer ($V_p \sim 7.1$ kms^{-1}) and tapering off of the Baltican/cratonic wedge. The seismic profiles suggest that the lower crust ($V_p \sim 7.1$ kms^{-1}) in the transition zone represents the attenuated Baltica margin underthrust towards the SW beneath the Avalonian accretionary wedge. The latter corresponds to the low-velocity upper crust ($V_p <$ 6.1 kms^{-1}) characteristic of the German–Polish Caledonides. Consequently, the high-velocity reflective lower crust of Baltica affinity extends approximately 200 km to the SW of the Teisseyre-Tornquist Zone within the basement of the Palaeozoic Platform. The Avalonian upper/middle crust is confined in the SW against the WNW–ESE trending Dolsk Fault. To the SW of the Odra Fault, a typical Variscan crust is detected which shows two-layer structure and relatively low P-wave velocities. The WNW–ESE Odra Fault, approximately parallel to the Dolsk Fault, splits the Variscan domain into the Variscan externides buried beneath the Palaeozoic Platform in the NE and the Variscan internides of the Sudetes in the SW. We interpret both the Odra and Dolsk Faults as dextral strike-slip features that cross cut the NE termination of the Variscan Orogen parallel to the Teisseyre-Tornquist Zone. In a relatively small area, they juxtapose three crustal domains representing, successively, the Variscan internides, externides and the Variscan foreland.

Geology of the area

The structure and evolution of the contact between Precambrian Europe in the NE and younger, Phanerozoic terranes in the SW is still one of the most interesting geological and tectonic problems of Europe. The Trans-European Suture Zone (TESZ) represents the most prominent lithospheric boundary in Europe north of the Alpine–Carpathian orogenic front. It is a broad and complex zone of terrane accretion separating ancient lithosphere of the Baltic Shield and East European Craton from the younger lithosphere of western and southern Europe (Pharaoh 1999). This 2000-km long boundary zone between the two crustal domains is a deep-seated structure, also clearly defined in the whole lithosphere (Zielhuis & Nolet 1994; Schweitzer 1995). It transects the European continent from the North Sea in the NW, to the Black Sea in the SE (Fig. 1). Over most of its length the TESZ is concealed beneath younger basins filled with Permo-Mesozoic and Cenozoic sedimentary sequences. However, available borehole data, combined with the results of numerous geophysical experiments indicate a polyphase structural evolution of the TESZ, comprising several phases ascribed to successive Caledonian, Variscan and Alpine orogenic events (Dadlez *et al.* 1994, 1995; Berthelsen 1998; Pharaoh 1999). They resulted in a structurally complex zone where Palaeozoic terranes, including Eastern Avalonia, Małopolska and Bruno-Silesia, were accreted to the SW margin of the Baltic Shield and the East European Craton (EEC) and then transformed by several deformation stages, into the basement of the European Palaeozoic Platform (Berthelsen 1992, 1998). From a plate tectonic perspective, the TESZ can, in a general sense, be correlated

From: WINCHESTER, J. A., PHARAOH, T. C. & VERNIERS, J. 2002. *Palaeozoic Amalgamation of Central Europe.* Geological Society, London, Special Publications, **201**, 295–309. 0305-8719/02/$15.00

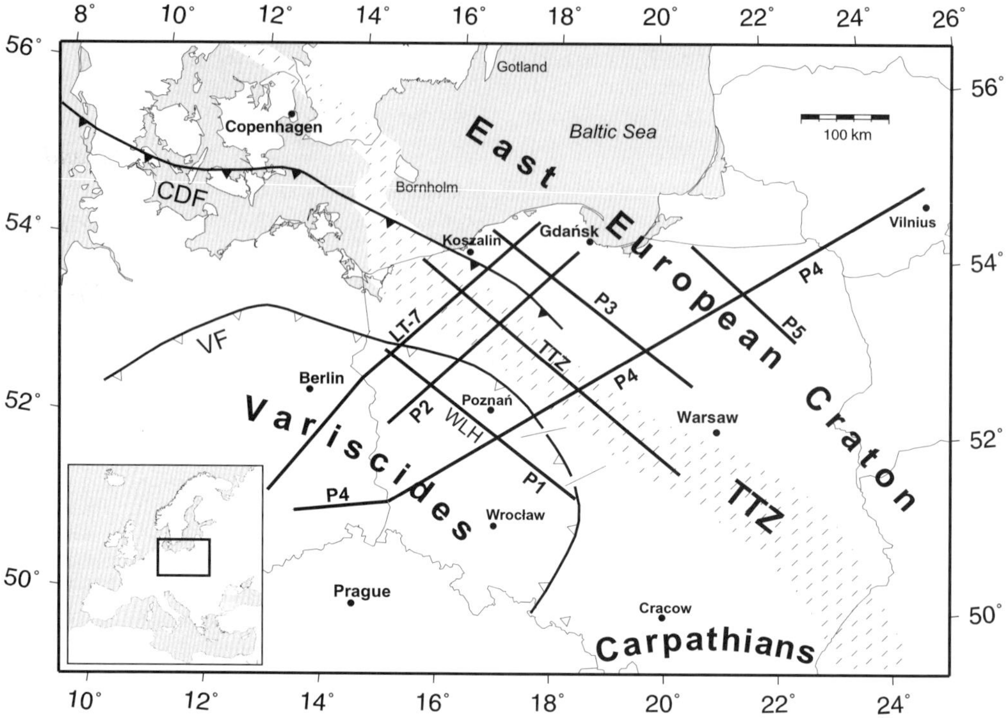

Fig. 1. Location of seismic refraction and wide-angle reflection profiles in Poland on the background of main tectonic units of Trans-European Suture Zone (TESZ) in Central Europe. CDF, Caledonian Deformation Front; TTZ, Teisseyre–Tornquist Zone; W-H, Wolsztyn High; VF, Variscan Front. LT-7, TTZ, P1–P5, deep seismic sounding refraction profiles.

across the Atlantic Ocean, with the Appalachian Orogen, where its equivalent is probably represented by the boundary zone between the Avalon terrane to the SE and the North American Craton to the NW (Hatcher 1989; Keller & Hatcher 1999). Indeed, the Appalachian-Ouachita Orogen was shaped by a succession of Palaeozoic tectonic events comparable to that recorded in the TESZ (Keller & Hatcher 1999).

The TESZ represents the SW boundary of the Baltica palaeocontinent, which comprises the concealed East European Craton and exposed Baltic and Ukrainian Shields. Palaeoproterozoic crystalline basement of the East European Craton, adjacent to the Polish segment of the TESZ, is mostly composed of granulites, migmatites, anorthosites and granite-gneisses (Ryka 1982). These rocks form NE–SW trending belts, sharply truncated by the TESZ (Bogdanova *et al.* 1996). The platform sedimentary cover at the SW margin of the EEC comprises three successions dated as Late Vendian–Early Palaeozoic, Devonian–Carboniferous and Permo-Mesozoic respectively. They are separated by regional unconformities and overlain by a thin Cenozoic cover. Their development in a pericratonic setting was closely related to Caledonian, Variscan and Alpine periods of tectonic activity along the TESZ. The Lower Palaeozoic sequence is characterized by the greatest thickness and lateral extent along the SW margin of the EEC. Its lower portion, represented by the Late Vendian–Middle Ordovician succession of thin shelf sediments, was formed in a passive continental margin setting at the SW edge of Baltica (Poprawa *et al.* 1999). As a consequence of its subsequent collision with Avalonia, the shallow marginal basin was succeeded by a rapidly subsiding late Silurian foredeep, developed on the SW Baltica margin, in front of the Caledonian orogen (Poprawa *et al.* 1999). The total thickness of Silurian fine-grained sediments filling this foreland basin locally exceeds 4500 m.

The Palaeozoic Platform of SW Poland extends between the Variscan Sudetes in the SW, the Carpathian thrust front in the south and the East European Craton margin in the NE. The SW boundary of the EEC is defined by the Teisseyre–Tornquist Zone (TTZ) which is several tens of kilometres wide (Fig.1). In present denotation, the Teisseyre–Tornquist Zone corresponds to the NE part of the TESZ, directly following the East Europeam Craton margin. The crystalline basement of the Palaeozoic Platform presumably comprises several terranes accreted to the SW margin of Baltica during Caledonian and Variscan orogenic events (Pożaryski 1990; Franke 1990) or, alternatively, can be partly derived from Baltica (Berthelsen 1992, 1998). Although part of the Palaeozoic Platform of Poland is situated to the SW of the Variscan deformation front (Fig. 1), it remains unclear if the Variscan tectonics show thick- or thin-skinned characteristics in that area (Jensen *et al.* 1999).

Several drillings have penetrated the Early Palaeozoic basement of the Platform in Western Pomerania. It is mainly represented by monotonous dark shales of Ordovician to Silurian age (Bednarczyk 1974; Tomczyk 1968; Teller & Korejwo 1968). These rocks were intensely deformed and, in places, slightly metamorphosed in late Silurian times (Dadlez 1974, 1978), presumably due to the collision of Baltica and Avalonia. They may be roughly equivalent to the deformed Ordovician strata known from the island of Rügen and interpreted as a fragment of a Caledonian thrust-and-fold belt (Berthelsen 1992; Giese *et al.* 1997). Both in Rügen and Pomerania Lower Palaeozoic sediments may represent more proximal parts of the same Caledonian foreland basin whose distal part onlaps the SW margin of the EEC (Poprawa *et al.* 1999). The Lower Palaeozoic of Western Pomerania is unconformably overlain by an essentially undeformed Devonian to Carboniferous platform succession up to a few thousand metres thick (e.g. Żelichowski 1987; Matyja 1993). The latter conceals the position of the Caledonian deformation front in NW Poland which, according to incomplete borehole data, approximately coincides with the eastern margin of the TESZ (e.g. Pożaryski 1990; Dadlez *et al.* 1994).

To the SW of the Variscan deformation front, the basement of the Palaeozoic Platform is documented only from a small area of the Wolsztyn High. Low grade phyllites and quartzites were found there in some deep boreholes (Oberc 1972). Their age, not precisely estimated, is generally considered as Early Palaeozoic (Krawczyńska-Grocholska & Grocholski 1976). The Lower Palaeozoic rocks must be overlain unconformably by Upper Palaeozoic strata which lack evidence of metamorphism. The borehole documentation of Devonian rocks at the base of the Upper Palaeozoic sequence is incomplete; the total thickness and extent of Devonian strata remaining unknown. The overlying Carboniferous succession is much better explored and comprises folded but unmetamorphosed flysch sediments (Żelichowski 1964; Grocholski 1975; Wierzchowska-Kicułowa 1984). The flysch series accumulated in a Variscan foreland basin (Grocholski 1975; Pożaryski *et al.* 1992), reaching a maximum thickness of approximately 2000 m (Wierzchowska-Kicułowa 1984). Its northern and northeastern limit defines the location of the Variscan deformation front under the Permo-Mesozoic cover (Jubitz *et al.* 1986). The upper stratigraphic extent of the flysch succession is controversially interpreted as Lower Westphalian (Wierzchowska-Kicułowa 1984; Pożaryski *et al.* 1992) or Lower Namurian (Karnkowski & Rdzanek 1982; Karnkowski 1999). The flysch succession is overlain by a few hundred metres of Upper Carboniferous molasse deposits.

Folded syn- to post-orogenic Carboniferous sediments are unconformably overlain by the thick Permo-Mesozoic cover of the German-Polish Basin. Onset of sedimentation in the earliest Permian, as well as further progressive subsidence during the Mesozoic were apparently connected with the development of a continental rift basin along the SW margin of the EEC (Jowett & Jarvis 1984; Ziegler 1990; Kutek 1997). The thickness of the Permo-Mesozoic strata increases gradually from about 2000 m near to the margin of the Sudetes, to more than 10 000 m in the Mid-Polish Trough (e.g. Dadlez *et al.* 1995). The latter represents a NW–SE elongated basin, being the axial zone of the rift structure, which parallels the SW edge of the East European Craton and approximately coincides with the Teisseyre–Tornquist Zone which underlies the trough (Dadlez 1989; Ziegler 1990; Kutek 1997).

Crustal and lower lithospheric structure of the TESZ

The Trans-European Suture Zone area in northern Poland has been investigated over the last decade by a number of seismic refraction and wide-angle reflection profiles: LT-7, TTZ and P1-P5 profiles made within the POLONAISE'97 experiment (Guterch *et al.* 1994, 1999; Grad *et al.* 1999; Jensen *et al.* 1999; Środa and POLONAISE Working Group 1999; Wilde-Piórko *et al.*

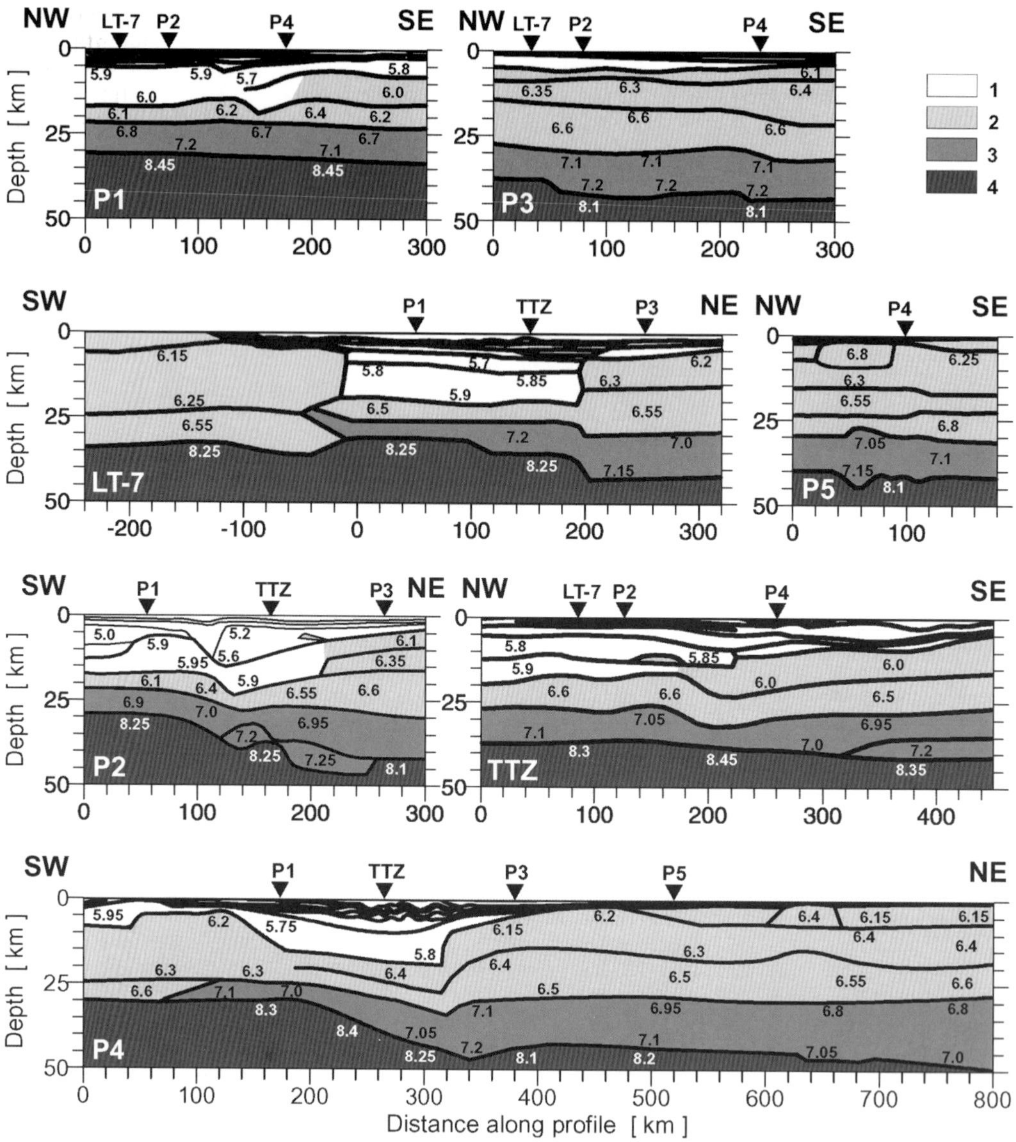

Fig. 2. Collection of velocity models beneath LT-7, TTZ and P1–P5 POLONAISE'97 profiles in the TESZ area. Compiled from Guterch *et al.* 1994; Grad *et al.* 1999; Jensen *et al.* 1999; Środa *et al.* 1999; Krysiński *et al.* 2000; Czuba *et al.* 2001; Janik *et al.* 2002. 1, sediments and metamorphosed sediments with $V_p < 6.0$ kms^{-1}; 2, crystalline crust with $V_p > 6.0$ kms^{-1}; 3, high velocity lower crust with $V_p = 6.8$–7.3 kms^{-1}; 4, mantle with $V_p > 8.0$ kms^{-1}. Black arrows show intersection with other profiles; 'zero' of P4 profile corresponds to Polish–German border. V_p velocities in kms^{-1}.

Fig. 3. Results of 3-D P-wave tomographic inversion for POLONAISE '97 data. Upper rectangle shows a portion of the model with P1–P5 profiles location; distance X and Y in km (for bottom and left edge) and geographical coordinates (for the right and upper edge). Thick pink dotted line shows SW edge of the EEC; red dotted line shows present extent of the Polish Rotliegend Basin (Karnkowski 1999); dark pink area shows Kętrzyn anorthosite massif within EEC (Czuba *et al.* 2001). Next two rectangles show horizontal slices of the velocity distribution at depths Z = 10 and 40 km, respectively. In the bottom vertical slice of the 3-D velocity model is shown (along line Y = 130 km shown as blue line on the map).

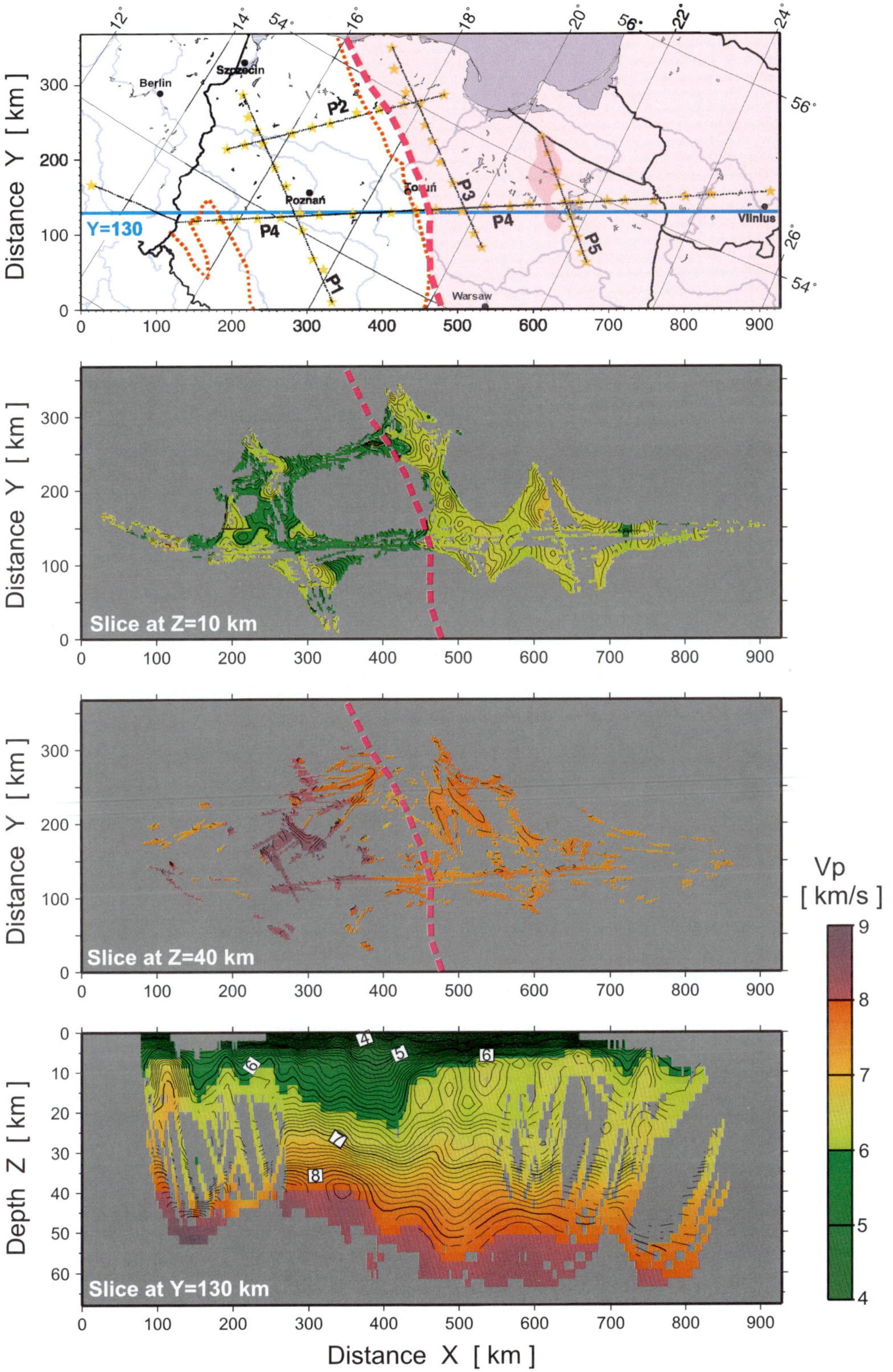
Distance Y [km]
Berlin
Szczecin
Poznań
Toruń
Warsaw
Vilnius
Y=130
P1
P2
P3
P4
P5
Slice at Z=10 km
Slice at Z=40 km
Slice at Y=130 km
Depth Z [km]
Distance X [km]
Vp
[km/s]

1999; Krysiński *et al.* 2000; Czuba *et al.* 2001; Jensen 2001; Janik *et al.* 2002). The collection of all discussed profiles is shown in Fig. 2. Seismic data provided by the LT-7 line in NW Poland show that the crustal thickness in the TESZ is intermediate between that of the East European Craton to the east (about 42 km) and that (approximately 30 km) in the area to the SW of the Variscan Front (Guterch *et al.* 1994; Guterch & Grad 1996). This initial finding provided the framework for the succeeding seismic survey undertaken under the POLONAISE'97 project. The main results of this experiment are summarized below.

The crustal structure of the East European Craton (EEC) is represented by profiles P3, P5 and northeastern parts of profiles P4 and LT-7. All models of the crust for this area are characterized by nearly horizontal uniform seismic structure. The crystalline crust consists of three parts: upper, middle and lower with P-wave velocities of 6.1–6.4, 6.5–6.7 and 7.0–7.2 km s^{-1}, respectively (Figs 2 and 5). The crystalline basement lies at the depth 0.5–5 km and plunges strongly in a SW direction, almost perpendicular to the edge of the craton. In the northwestern part of the profile P5, a body with high seismic velocities of about 6.6 km s^{-1} was found at depth range 2–10 km. It coincides with the rapakivi-like and anorthosite Mazurian complex, well known from borehole data. The depth of the Moho boundary ranges from 39–45 km in northeastern Poland, reaching 50 km beneath Lithuania. The sub-Moho P-wave velocity is 8.05–8.1 kms^{-1}.

The crustal structure of the Palaeozoic Platform beneath the Polish Basin is represented by profile P1, the southwestern parts of profiles P2, P4 and LT-7, and by profile TTZ. The latter runs NW–SE through the Polish Basin, parallel to its elongation. The similarly oriented profile P1 is situated immediately to the SW of the Variscan Front. The remaining profiles P4, P2 and LT-7 are approximately perpendicular to the P1 and TTZ lines and to the margin of the EEC. In general, the P-wave velocities of the upper crust in the Palaeozoic Platform, between the EEC edge and the Wolsztyn High, are low (<6.1–6.2 km s^{-1}) down to 20 km of depth with low reflectivity (Guterch *et al.* 1992). Low velocities are interpreted to be indicative for volcanosedimentary rocks subjected to only low-grade metamorphism. In contrast, the lower crust has a P-wave velocity of 6.5 – 6.8 – 7.3 km s^{-1}, a high velocity gradient, and strong, ringing reflectivity. This layer has a distinctly laminar seismic structure and is interpreted as being associated with the Moho – a flat and horizontal first-order discontinuity (Jensen *et al.* 1999; Guterch *et al.* 1992). The velocity of the sub-Moho uppermost mantle is high (> 8.2 – 8.3 km s^{-1}). The high velocity lower crust was previously found beneath a number of refraction profiles at the Fore-Sudetic Monocline (Guterch *et al.* 1986, 1992). The SW limit of this kind of lower crust coincides in SW Poland with the Odra Fault. Typical two-layer Variscan-type crust was observed SW of the Odra Fault beneath international profile VII and at approximately –40 km beneath profile LT-7 (for location LT-7 profile see Fig. 1; Guterch *et al.* 1986, 1994).

The 3-D velocity model of the crust, obtained using a tomographic inversion method, shows substantial horizontal variations of the structure across the study area (Fig. 3). The results of the modelling reflect the complex structure of the Earth's crust in the TESZ region and provide insights into the physical characteristics of tectonic units juxtaposed in that area. The horizontal slices of the 3-D velocity distribution presented in Figure 3 clearly show the differences between the areas of the Palaeozoic Platform to the SW, the Teisseyre–Tornquist Zone, and the EEC to the NE (Środa *et al.* 2002). At a depth of 10 km, velocities of less than 6.0 km s^{-1} typical of sedimentary rocks are observed in the Polish Basin and in parts of the Palaeozoic Platform, while in the EEC area P-wave velocities at the same depth exceed 6 kms^{-1}. The latter value is characteristic of crystalline rocks of the upper crust. In the slice cutting the model at 40 km depth, the velocity distribution is distinctly different between the Palaeozoic Platform, which is characterized by velocities of 8 km s^{-1} or greater, and the EEC with velocities of about 7 km s^{-1}. Velocities in the EEC area are consistent with the presence of lower crustal rocks, while velocities significantly higher than 8.0 km s^{-1} in the Palaeozoic Platform area indicate the upper mantle. This variability is also conspicuous on the vertical slices in the Y-plane, especially for Y = 130 km, cutting the model near the location of the P4 profile. From SW to NE, the slice crosses the Palaeozoic Platform with a relatively thin crust (about 35 km), the Teisseyre-Tornquist Zone comprising a deep asymmetrical Polish Basin underlain by a thin crystalline layer, and finally, the East European Craton with rare sedimentary cover and crust about 45 km thick.

The lower lithosphere under the TESZ is characterized by the presence of a series of seismic reflectors in depths ranging from the Moho to about 90 km (Guterch *et al.* 1994; Grad *et al.* 2002). A seismic reflector generally occurs

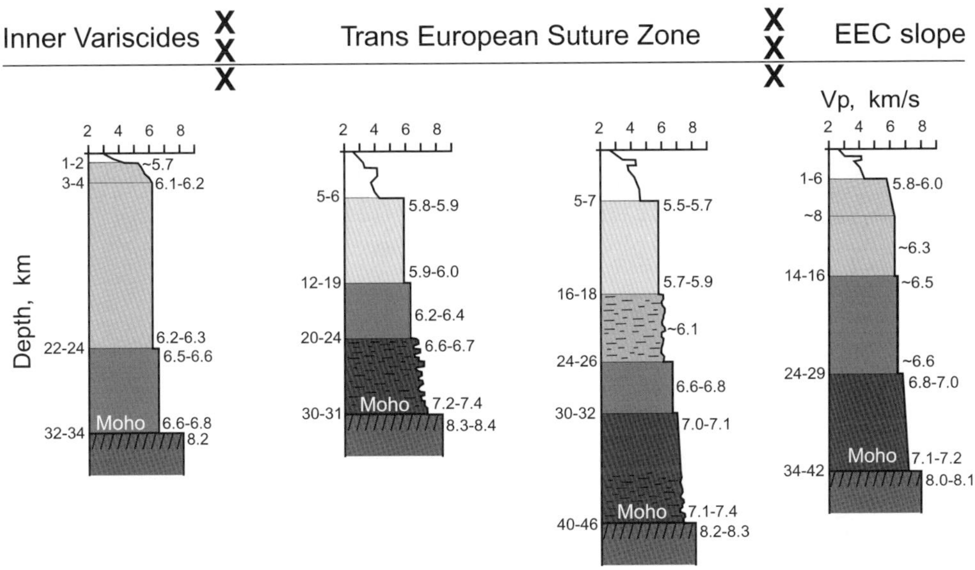

Fig. 4. Simplified one-dimensional model representations for the crust of Variscides, TESZ and EEC slope.

about 10 km below the Moho and the reflectivity of the uppermost mantle is stronger beneath the Palaeozoic Platform and TESZ than beneath the East European Craton (Grad *et al.* 2002). Since this stratification disappears laterally on both sides of the TESZ towards the East European Platform and Palaeozoic Europe, it seems directly linked to the lithospheric mobility at the transition zone. Consequently, the low velocity upper mantle layer may represent an upwelling of relatively lighter material into the mobile zone at the mantle-rooted craton margin, due to thermal softening of the lithosphere during the Late Palaeozoic/Mesozoic.

A tomographic analysis of the shear wave velocity structure of the mantle under Europe shows that, in Poland, the TESZ is a deep-seated structure separating regions with high S-wave velocities beneath the Precambrian Platform from low velocity regions under the Palaeozoic Platform (Zielhuis & Nolet 1994). Another confirmation of the TESZ influence on the propagation of regional seismic waves was obtained from observations of several hundred earthquakes and explosions located in Europe. To explain the blockage of the seismic energy propagation of regional seismic events across the TESZ, the structural anomaly between eastern and western Europe must reach down to a depth of at least 200 km (Schweitzer 1995).

Seismic structure of the TESZ and its geological interpretation

Contrasting seismic structure of the crust accompanied by a large Moho step indicates essential differences between the East European Craton and the basement of the Palaeozoic Platform juxtaposed along the TESZ. Their contact corresponds to a distinct vertical discontinuity penetrating the upper and middle crust along the boundary between the Teisseyre–Tornquist Zone and the SW margin of the EEC (Figs 2 and 3). Therefore, at higher crustal levels, the TTZ seems to represent an important geological boundary between Baltica and terranes accreted to its SW margin during the Palaeozoic. However, in the lower crust this boundary is not evident and the EEC bottom layer appears to continue further to the SW of the TTZ (Fig. 2).

The seismic structure of the basement underlying Central Poland west of the EEC margin displays several common features. It comprises three main layers, including a thick upper crust and a reflective lower crust characterized by high P-wave velocities (Fig. 4). The seismic structure of this crustal domain is very different from those of the EEC to the NE and of the Variscan orogen to the SW. Its remarkable characteristics have been already determined in earlier seismic surveys (Guterch *et al.* 1984, 1986, 1994, 1999). The SW boundary of the three-layer basement of Central Poland is reflected on

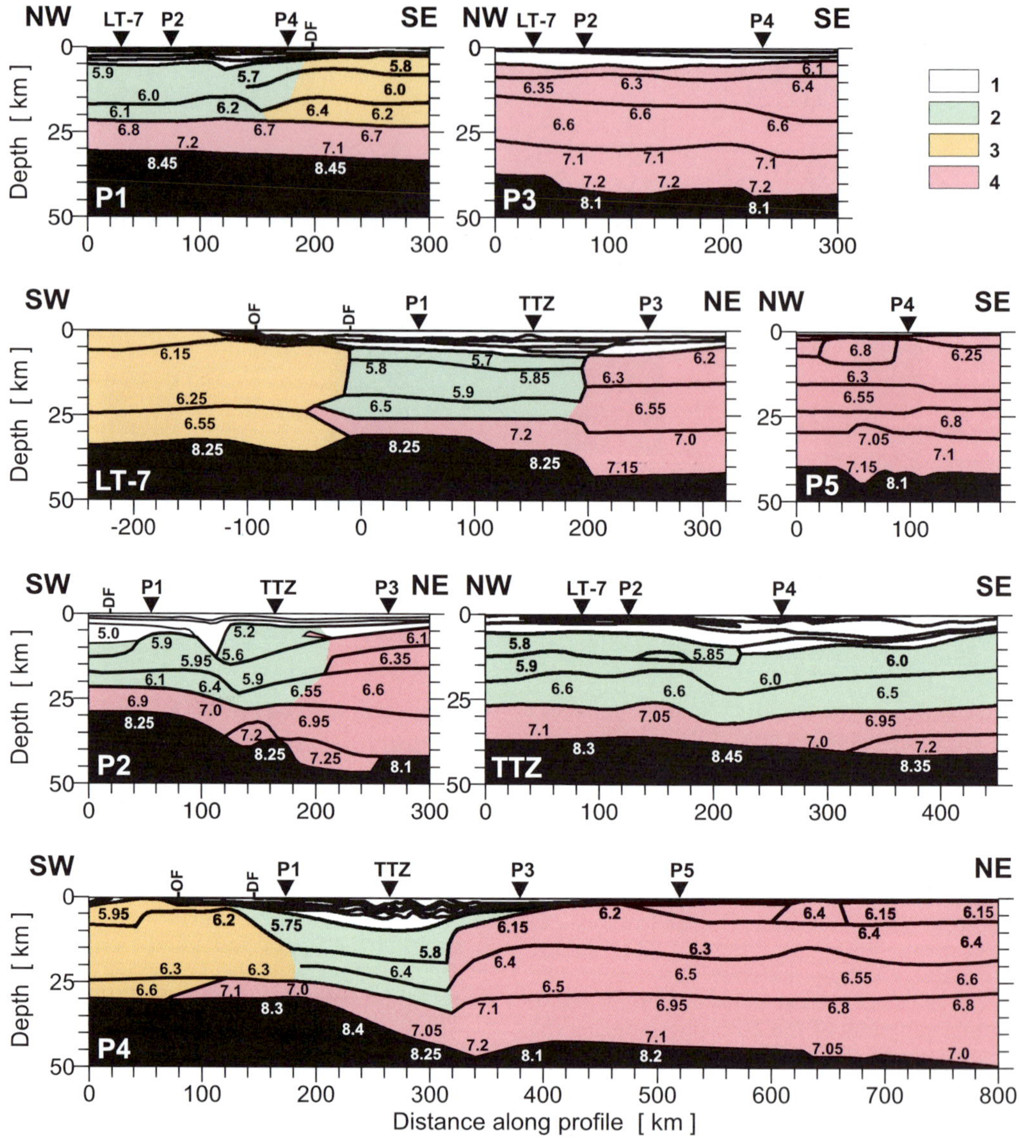

Fig. 5. Classification of crustal types and their extent in TESZ area. 1, Palaeozoic (?) and younger sediments; 2, Avalonia; 3, Variscides; 4, Baltica. DF, Dolsk Fault; OF, Odra Fault. For more explanations see Figure 2.

seismic profiles as a relatively sharp vertical discontinuity (Fig. 2). It seems to coincide with the WNW–ESE trending Dolsk Fault documented beneath the Permo-Mesozoic cover by borehole data (Znosko 1979; Wierzchowska-Kicułowa 1984). This fault defines the NE boundary of the Wolsztyn High and separates it from the deeply buried basement further to the NE (Fig. 1).

The seismic structure of the crust between the TTZ and the Dolsk Fault generally resembles that known from northern Germany and southern Denmark (Aichroth *et al.* 1992; Rabbel *et al.* 1995; Abramovitz *et al.* 1998). It is typical of the basement domain that extends between the CDF in the NE and the Elbe Lineament (EL) to the SW. Like the basement of Central Poland, that domain shows low P-wave seismic velocities down to a depth of about 20 km and relatively high velocities below. Furthermore, its lower crust is characterized by high vertical velocity gradient and strong, ringing reflectivity (Aichroth *et al.* 1992; Rabbel *et al.* 1995;

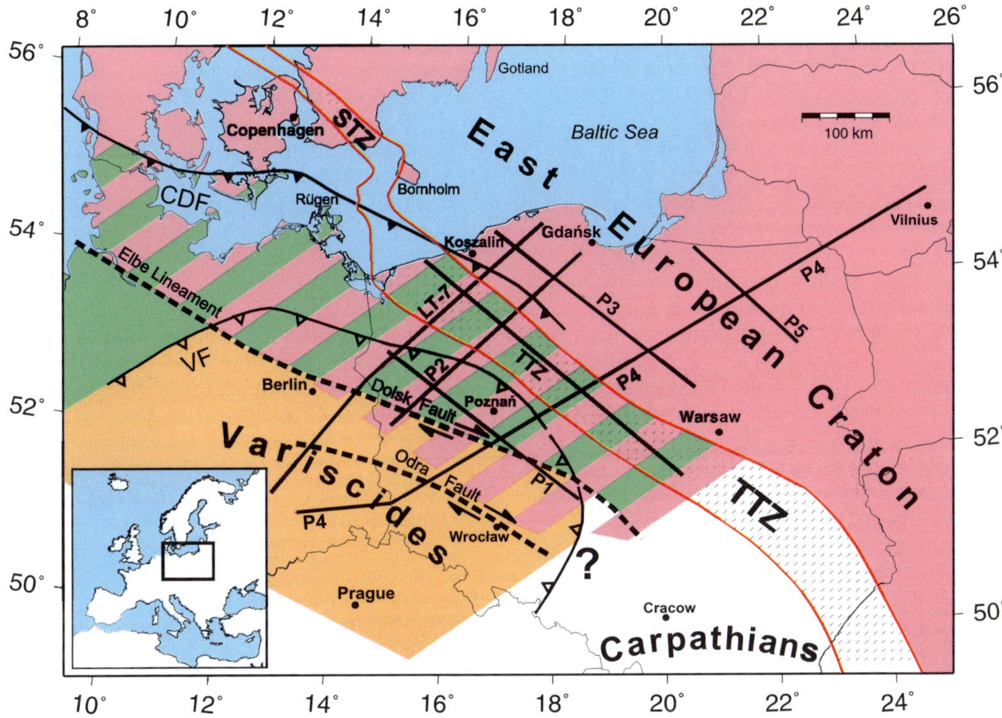

Fig. 6. The extent of the different types of the crust in TESZ area. Precambrian crust of the East European Craton (pink), Avalonian crust, not affected by the Variscan deformation and characterized by relatively low seismic velocities (Vp < 6.0 kms^{-1}) down to about 20 km depth (green) and two-layer Variscan crust (orange) including Avalonian crust influenced by the Variscan thick-skinned tectonics between the Dolsk and Odra Faults and the Armorica crust in the Variscan internides to the SW of the Odra Fault. Mixed green (or orange) and pink belts mean Avalonian (or Variscan) crust underlined by high velocity lower crust (Vp = 6.8–7.3 kms^{-1}) of Baltica.

Abramovitz *et al.* 1998). These similarities have been already indicated by Jensen *et al.* (1999) with reference to the profile P1.

Possible affinities of the crustal block to the NE of the Elbe Lineament have produced an extensive discussion. Its peculiar geophysical features are different from those typical of eastern Avalonia to the SW of the Elbe Lineament. The Elbe Lineament defines the NE boundary of the high gravity values (Bachmann & Grosse 1989) and upper crustal conductivity anomaly (ERCEUGT Group, 1992) characteristic of the Lüneburg Plain. At the same time, P-wave velocities in the lower crust are obviously higher to the NE of the Elbe Lineament (Aichroth *et al.* 1992; Rabbel *et al.* 1995) than to the SW of it. Furthermore, the lineament separates domains with different basement tectonic features (Betz *et al.* 1987). Because of that the Elbe Lineament has recently been recognized as an important crustal boundary (Rabbel *et al.* 1995), possibly even as the NE boundary of Avalonia (Tanner & Meissner 1996). It was supposed to separate the Avalonian crust from that of the East Elbe Suspect Terrane (Franke 1995).

An alternative explanation for the peculiar geophysical features demonstrated by the basement blocks underlying northern Germany and southern Denmark was proposed by Berthelsen (1998) and Abramovitz *et al.* (1998). They interpret the thick upper crust of low seismic velocity as being of Avalonian origin whereas the lower crust of high velocity is thought to represent attenuated Baltica margin. The latter was, presumably, underthrust beneath Avalonia during its Early Palaeozoic collision with Baltica. The attenuated Baltica lower crust is expected to extend far to the SW behind the Caledonian Deformation Front (CDF) and the Teisseyre–Tornquist Zone (Berthelsen 1998; Pharaoh 1999). This concept seems to be confirmed by seismic profiles west of Denmark showing mid-crustal reflections constantly inclined to the SW at an angle of 10–12° (MONA LISA Working Group 1997*a,b*). The inclined reflectors are interpreted as the Baltica/Avalonia contact defined by Berthelsen (1998) as the Thor Suture. Further east, similar reflectors gently dipping to the SW beneath the NE German Basin were documented by DEKORP-BASIN Research Group (Krawczyk *et al.* 1999). Accordingly, the high-velocity lower crust of possible Baltica origin is shown to continue southwestwards as far as the Elbe Lineament and the basin depocentre below the NE German mainland (Thybo 1990; Aichroth *et al.* 1992; Rabbel *et al.* 1995; Krawczyk *et al.* 1999). The geophysical evidence for an Avalonia accretionary wedge overthrusting the Baltica margin is also supported by the occurrence of Baltica-derived sediments in the Loissin Borehole south of the CDF (Dallmeyer *et al.* 1999).

Results of the POLONAISE'97 experiment seem to support the southwestward continuation of the attenuated Baltica margin within the basement of the Palaeozoic Platform in the central part of the TESZ in Poland. Seismic profiles LT-7 and P4, approximately orthogonal to the Caledonian Deformation Front and Teisseyre–Tornquist Zone, show the lower crust similar to that of the EEC, extending far to the SW and forming the bottom part of the Palaeozoic Platform basement (Fig. 4). The high-velocity reflective lower crust of Baltica affinity tapers off approximately 200 km to the SW of the TTZ, south of the Dolsk Fault (Figs 5 and 6). Seismic data do not show, however, whether this fault extends down into the lower crust and if it directly terminates the EEC margin (Figs 2 and 4).

The middle and upper crust in the central part of the TESZ in Poland between the TTZ and the Dolsk Fault is up to 20 km thick and shows relatively low P-wave velocities (Fig. 4). The latter are generally distinctive of low-grade metamorphosed volcanosedimentary successions. Similar characteristics of the upper crust are documented further NW beneath northern Germany and southern Denmark (Thybo 1990; Aichroth *et al.* 1992; Abramovitz *et al.* 1998) and interpreted as indicative for the eastern Avalonia accretionary wedge (Abramovitz *et al.* 1998; Berthelsen 1998). Hence, the seismic data provided by the POLONAISE'97 experiment allow correlation between the crustal structure in northern Germany and central Poland (Jensen *et al.* 1999). Both areas seem to share a common basement representing the easternmost section of Avalonia underthrust by the attenuated Baltica margin. The eastward extension of the Avalonian upper crust in Poland is ultimately terminated against the TTZ. Results of seismic experiments do not support the concept of Berthelsen (1998) regarding the Polish Caledonides as deformed foredeep and shelf sequences, originally deposited over the Baltica margin. Therefore, it is rather unlikely that the Thor Suture in central–western Poland runs along the SW rim of underthrusted Baltica crust, far to the SW of the Caledonian Deformation Front (Berthelsen 1998).

The seismic structure of the crust between the Teisseyre–Tornquist Zone and the Dolsk Fault is very different from that characteristic of the Variscan belt (cf. Aichroth *et al.* 1992). The Variscan crust displays a two-layer structure with low P-wave velocities continuing through the whole crust down to Moho (Fig. 5). Since the Variscan Front (VF) is situated to the north of the Dolsk Fault (Jubitz *et al.* 1986), it overlaps the area underlain by Avalonian crust (Fig. 5). The Variscan deformation did not, however, affect the seismic structure of the basement to the NE of the Dolsk Fault. Consequently, it suggests a thin-skinned style of Variscan deformation in the area between the Dolsk Fault and the Variscan Front. In contrast, the basement of the Palaeozoic Platform to the SW of the Dolsk Fault reveals a seismic structure comparable with other Variscan areas of Central Europe. Seismic profiles P4, LT-7 and P1 indicate that the crust occurring between the Dolsk Fault in the NE and the Odra Fault in the SE shows features similar to those found in the crust known from the eastern section of the Rheno-Hercynian Zone (cf. Dadlez 1997). Therefore, a thick-skinned type of the Variscan structure can be inferred for the crustal domain enclosed between the Dolsk and Odra Faults. Its tectonic fabric may have been analogous to that detected in the eastern Rhenish Massif (Franke *et al.* 1990) but the lack of reflection seismic data from SW Poland prevent any further conclusions.

Geological model of the basement structure in the central part of the TESZ

The TESZ structure in Central Poland was mostly shaped by the Early Palaeozoic collision between Avalonia and Baltica. The Avalonia accretionary wedge is overthrust onto the Baltica margin which, in consequence, is deeply concealed beneath the Palaeozoic Platform about 150–200 km to the SW of the Teisseyre–Tornquist Zone. The SW boundary of the high-velocity Baltica lower crust does not coincide with the Dolsk Fault (Fig. 6). Nevertheless, this fault represents an important discontinuity in the upper and middle crust, visible on the profiles P1, P4 and LT-7 (Figs 2 and 5). It separates the middle and upper crust with velocities typical of the Avalonian basement in the NE from the thick fairly uniform crust of the Variscan type in the SW (see Fig. 4). The segment of the Platform between the TTZ and the Dolsk Fault probably represents the easternmost section of the Avalonia terrane. The basement of the platform drilled in western Pomerania consists of intensely deformed very low-grade to unmetamorphosed Lower Palaeozoic sediments probably representing the Caledonian thrust-fold belt (Dadlez 1974, 1978). These rocks, generally comparable with those of the Rügen Caledonides, are unconformably overlain by thick Upper Palaeozoic platform cover representing the Variscan foreland developed on the crustal domain to the NE of the Dolsk Fault. Nevertheless, relatively thin Carboniferous flysch and molasse strata overlap the southern part of the platform to the north of the Dolsk Fault. Their extent seems to outline the location of the Variscan Front in Central Poland. Despite the deformation recorded in the Carboniferous succession, the Variscan event did not affect the seismic structure of its basement extensively. Therefore, we suggest that the flysch and molasse succession might be overthrust onto its foreland along a relatively shallow décollement. Alternatively, it may represent a marginal part of the syn- to post-orogenic Variscan basin extending over the platform succession. Despite the interpretation, the position of the Variscan Front does not coincide with the Dolsk Fault, which is the NE boundary of the upper and middle crust showing a seismic structure typical of the Variscan externides. Thus, in terms of deep crustal structure, theVariscan Front is not the boundary between the Variscides and the Avalonia foreland.

The part of the Palaeozoic Platform with characteristics like that of the Variscan foreland in North Germany is bounded on the SW by the NW–SE trending Dolsk Fault. Further SW, the crust shows seismic structure distinctive of the Variscan belt in Central Europe. Furthermore, several drillings located between the Dolsk and Odra Faults documented a 2000 m thick succession of Carboniferous non-metamorphosed flysch and molasse sediments underlying the Permo-Mesozoic cover. Consequently, the area of SW Poland between the Odra and Dolsk Faults represents a part of the Variscan externides. Indeed, any evidence of late Palaeozoic metamorphism or intensive plutonism is generally missing in that area. Moreover, the Upper Palaeozoic unmetamorphosed succession is directly underlain by folded and epizonally metamorphosed Lower Palaeozoic series, documented by boreholes in the area of the Wolsztyn High (Krawczyńska-Grocholska & Grocholski 1976). The latter probably represents an elevated fragment of the low-grade Caledonian basement. Outcrops of such basement are virtually unknown within the Variscan internides. In contrast, the Rheno-Hercynian Zone extending north of the Rheic Suture can involve detached fragments of the Avalonian basement. In fact, Lower Palaeozoic massifs in the Ardennes represent pieces of the pre-Variscan basement developed on Avalonian crust and incorporated into the Rheno-Hercynian externides. However, the seismic structure of the crust to the SW of the Dolsk Fault resembles rather the eastern section of the Rheno-Hercynian Zone where pre-Variscan basement is practically unknown except for scarce and equivocal borehole documentation (Krebs 1978; Teichmüller 1978). If the analogies to the Rheno-Hercynian Zone are true, the deep basement between the Dolsk and Odra Faults comprises Avalonian crust affected by thick-skinned Variscan deformation. On seismic profiles the Dolsk fault appears as a sharp vertical discontinuity and can hardly be interpreted as the Variscan frontal thrust. Probably it represents a strike-slip fault which juxtaposes the Variscan foreland against the Variscan externides and it cuts off and terminates the zone of thick-skinned thrusts at the front of the orogen. The overall geometry of the Variscan belt (cf. Arthaud & Matte 1977) suggests a dextral displacement along the Dolsk Fault, which may continue towards the NW into the Elbe Lineament of northern Germany (Fig. 6).

The SW part of the Palaeozoic Platform, interpreted above as a fragment of the Variscan exeternides, represents a relatively narrow domain. In the SW, it is bounded by the WNW–ESE trending Odra Fault, which is the boundary with the Variscan internides of the Sudetes. The unmetamorphosed Carboniferous flysch succession occurring to the NE below the Permo-Mesozoic cover is juxtaposed across the Odra Fault against the medium-grade metamorphic complexes intruded by numerous Carboniferous granitoids (Oberc-Dziedzic *et al.* 1999). Since the Odra Fault appears on seismic profiles as an important vertical discontinuity (Fig. 6), it probably represents a strike-slip structure as well. Like the Dolsk Fault, the general geometry of the Variscan orogen in Central Europe implies a dextral kinematic history for the Odra Fault

(Aleksandrowski 1990, 1995). The occurrence of a thick Permo-Mesozoic succession on the NNE side of this fault indicates its subsequent post-Variscan activity as a dip-slip feature.

The Palaeozoic Platform in the central part of the TESZ in Poland represents the NE termination of the Variscan orogen crosscut by dextral strike-slip faults parallel to the Teisseyre–Tornquist Zone. The results of deep seismic sounding provided by the POLONAISE'97 experiment combined with geological evidence allow identification of two large-scale tectonic boundaries. These are the Odra and Dolsk Faults which, on a relatively small area, juxtapose three crustal domains representing successively the Variscan internides, externides and the Variscan foreland (Fig. 6). Thus late Variscan strike-slip tectonics have modified the primary structure of the TESZ, originally shaped by the early Palaeozoic collision of Avalonia and Baltica and subsequently by the late Palaeozoic overthrust of the Variscan Front. The timing of the strike-slip displacements is not well constrained, but significant difference in the thickness of the Lower Carboniferous/Namurian sequence on the opposite sides of the Dolsk Fault (Żelichowski 1980) suggests post-Namurian activity.

Conclusions

The Teisseyre–Tornquist Zone (TTZ), corresponds to a distinct vertical discontinuity zone penetrating the upper and middle crust along the SW margin of the East European Craton (EEC). However, in the lower crust this boundary is not evident and the EEC bottom layer appears to continue further to the SW of the TTZ. The high-velocity reflective lower crust of Baltican affinity tapers off approximately 200 km to the SW of the EEC edge.

The seismic structure of the basement underlying Central Poland between the EEC margin and the Dolsk Fault comprises three main layers, including a thick upper crust and a reflective lower crust characterized by high P-wave velocities. The seismic structure of this transitional domain is considerably different from those of the EEC to the NE, and the Variscan orogen to the SW. It is typical of the basement that extends between the Caledonian Deformation Front and the Elbe Lineament.

The middle and upper crust between the TTZ and the Dolsk Fault is up to 20 km thick and shows relatively low P-wave velocities, distinctive of low-grade metamorphosed volcano-sedimentary successions. Its characteristics are considered indicative of the accretionary wedge of eastern Avalonia. Since the Avalonia accretionary wedge is overthrust onto the SW Baltica margin, it is deeply concealed beneath the Palaeozoic Platform about 150–200 km to the SW of the TTZ.

The Dolsk Fault appears on seismic profiles as a sharp vertical discontinuity and can hardly be interpreted as the Variscan frontal thrust. Probably it represents a strike-slip fault which juxtaposes the Variscan foreland against the Variscan externides and it cuts off and terminates the zone of thick-skinned thrusts at the front of the orogen. The overall geometry of the Variscan belt suggests a dextral kinematic history for the Dolsk Fault.

The basement to the SW of the Odra Fault reveals a seismic structure closely comparable with other Variscan areas of Central Europe. It displays a two-layer structure with low P-wave velocities continuing through the whole crust down to the Moho (Fig. 5). The WNW–ESE Odra Fault, approximately parallel to the Dolsk Fault, splits the Variscan domain into the Variscan externides buried beneath the Palaeozoic Platform in the NE and the Variscan internides of the Sudetes in the SW. Since the Odra Fault appears on seismic profiles as an important vertical discontinuity, it probably represents a strike-slip structure too.

The Palaeozoic Platform in the central part of the TESZ in Poland represents the NE termination of the Variscan orogen cross cut by dextral strike-slip faults parallel to the TTZ. These are the Odra and Dolsk Faults which in a relatively small area juxtapose three crustal domains representing successively the Variscan internides, externides and the Variscan foreland.

The authors are grateful to two referees, Dr A. Żelaźniewicz and Dr A. Lassen, for their constructive criticism and helpful comments, and Dr P. Aleksandrowski for his review of the manuscript. Part of this work was also funded under the EU TMR Network Palaeozoic Amalgamation of Central Europe.

References

ABRAMOVITZ, T., THYBO, H. & MONA LISA WORKING GROUP 1998. Seismic structure across the Caledonian Deformation Front along MONA LISA profile 1 in the southeastern North Sea. *Tectonophysics*, **288**, 153–176.

AICHROTH, B., PRODEHL, C. & THYBO, H. 1992. Crustal structure along the Central Segment of the EGT from seismic refraction studies. *Tectonophysics*, **207**, 43–64.

ALEKSANDROWSKI, P. 1990. Early Carboniferous strike-slip displacements at the northeast periphery of the Variscan Belt in Central Europe. *In*:

International Conference on Paleozoic Orogens in Central Europe (IGCP Program 233: Terranes in the Circum-Atlantic Paleozoic Orogens), Göttingen-Giessen, *Abstracts*, 7–10.

ALEKSANDROWSKI, P. 1995. The significance of major strike-slip displacements in the development of Variscan structure of the Sudetes (SW Poland). *Przegląd Geologiczny,* **43**, 745–754 (In Polish, English abstract).

ARTHAUD, F. & MATTE, P. 1977. Late Paleozoic strike-slip faulting in southern Europe and northern Africa: Result of a right-lateral shear zone between the Appalachians and the Urals. *Geological Society of America Bulletin,* **88**, 1305–1320.

BACHMANN, G. H. & GROSSE, S. 1989. Struktur und Entstehung des Norddeutschen Beckens: Geologische und geophysikalische Interpretation einer verbesserten Schwerekarte. *Niedersächsische Akademie Geowissenschaften Veröffentlichungen*, **2**, 23–47.

BETZ, D., FÜHRER, G., GREINER, G. & PLEIN, E. 1987. Evolution of the Lower Saxony Basin. *Tectonophysics*, **137**, 127–170.

BEDNARCZYK, W. 1974. The Ordovician in the Koszalin-Chojnice region (Western Pomerania). *Acta Geologica Polonica*, **24**, 581–600.

BERTHELSEN, A. 1992. From Precambrian to Variscan Europe. *In*: BLUNDELL, D. J., FREEMAN, R., MUELLER, S. (eds) *A continent revealed – the European geotraverse*. Cambridge University Press, 153–164.

BERTHELSEN, A. 1998. The Tornquist Zone northwest of the Carpathians: an intraplate-pseudosuture. *Geologiske Föreningen Förhandlinger*, **120**, 223–230.

BOGDANOVA, S., PASKEVICH, I, K., GORBATSCHEV, R. & ORYLUK, M. I. 1996. Riphean rifting and major Palaeoproterozoic crustal boundaries in the basement of the East European Craton: geology and geophysics. *Tectonophysics*, **268**, 1–21.

CZUBA W., GRAD M., LUOSTO U., MOTUZA G., NASEDKIN V. & POLONAISE P5 WORKING GROUP 2001. Crustal structure of the East European Craton along POLONAISE'97 P5 profile. *Acta Geophysica Polonica*, **49**, 145–168.

DADLEZ, R. 1974. Tectonic position of Western Pomerania (North-Western Poland) prior to the Upper Permian. *Biuletyn Instytut Geologicznego*, **274**, 49–87.

DADLEZ, R. 1978. Sub-Permian rock complexes in the Koszalin-Chojnice Zone. *Kwartalnik Geologiczny*, **22**, 269–301 (in Polish, English summary).

DADLEZ, R. 1989. Epicontinental Permian and Mesozoic basins in Poland. *Kwartalnik Geologiczny*, **33**, 175–198.

DADLEZ, R. 1997. Seismic profile LT-7 (northwest Poland): geological implications. *Geological Magazine*, **134**, 653–659.

DADLEZ, R., KOWALCZEWSKI, Z. & ZNOSKO, J. 1994. Some key problems of the pre-Permian tectonics of Poland. *Geological Quarterly*, **38**, 169–189.

DADLEZ, R., NARKIEWICZ, M., STEPHENSON, R.A., VISSER, M.T.M. & VAN WEES, J.-D. 1995. Tectonic evolution of the Mid-Polish Trough: modelling implications and significance for central European geology. *Tectonophysics*, **252**, 179–195.

DALLMEYER R.D., GIESE U., GLASMACHER U. & PICKEL W. 1999. First ^{40}Ar-^{39}Ar age constraints for the Caledonian evolution of the Trans-European Suture Zone in NE Germany. *Journal of the Geological Society, London*, **156**, 279–290.

ERCEUGT GROUP 1992. An electrical resistivity crustal section from the Alps to the Baltic Sea (central segment of the EGT). *Tectonophysics*, **207**, 123–139.

FRANKE, D. 1990. Der präpermische Untergrund der Mitteleuropäischen Senke Fakten und Hypothesen. *Niedersächsische Akademie der Geowissenschaften Veröffentlichungen*, **4**, 19–75.

FRANKE, D. 1995. The Caledonian terranes along the south-western border of the East European Platform – evidence, speculation and open questions. The Trans-European Suture Zone: EUROPROBE in Libice 1993, GEE, D. G., BECKHOLMEN, M. (eds) *Studia Geophysica et Geodaetica*, **39**, 241–256.

FRANKE, W., BORTFELD, R. K., BRIX, M., DROZDZEWSKI, G., DÜRBAUM, H. J., GIESE, P., JANOTH, N., JÖDICKE, H., REICHERT, C., SCHERP, A., SCHMOLL, J., THOMAS, R., THÜNKER, K., WEBER, K., WIESNER, M. G. & WONG, H. K. 1990. Crustal structure of the Rhenish Massif: results of deep seismic reflection lines DEKORP 2North and DEKORP 2-North-Q. *Geologische Rundschau*, **79**, 523–566.

GEE, D. G. & ZEYEN, H. 1996. EUROPROBE 1996 – Lithosphere Dynamics – Origin and evolution of Continents. EUROPROBE Secretariat, Uppsala.

GIESE, U., KATZUNG, G., WALTER, R. & WEBER, J. 1997. The caledonian deformation of the Brabant Massif and the Early Palaeozoic in northeast Germany: compared (Brabant Massif and Rügen Palaeozoic compared). *Geological Magazine*, **134**, 637–652.

GRAD, M., JANIK, T., YLINIEMI, J., GUTERCH, A., LUOSTO, U., TIIRA, T., KOMMINAHO, K., ŚRODA, P., HÖING, K., MAKRIS, J. & LUND, C.-E. 1999. Crustal structure of the Mid-Polish Trough beneath the Teisseyre-Tormquist Zone seismic profile. *Tectonophysics,* **314**, 145–160.

GRAD, M., KELLER, G.R., THYBO, H., GUTERCH, A & POLONAISE WORKING GROUP 2002. Lower lithospheric structure beneath the Trans-European Suture Zone from POLONAISE'97 seismic profiles. *Tectonophysics*, in press.

GROCHOLSKI, W. 1975. Variscides of southern Wielkopolska. *Przegląd Geologiczny*, **4**, 171–174 (in Polish, English summary).

GUTERCH, A. & GRAD, M. 1996. Seismic structure of the Earth's crust between Precambrian and Variscan Europe in Poland. *Publications of the Institute of Geophyics, Polish Academy of Science*, **M-18 (273)**, 67–73.

GUTERCH, A., GRAD, M., MATERZOK, R., PAJCHEL, J., PERCHUĆ, E. & TOPORKIEWICZ, S. 1984. Deep structure of the Earth's crust in the contact zone of the Palaeozoic and Precambrian Platform and the Carpathian Mts in Poland. *Acta Geophysica Polonica*, **32**, 25–41.

GUTERCH, A., GRAD, M., MATERZOK, R. & PERCHUĆ, E. 1986. Deep structure of the Earth's crust in the contact zone of the Palaeozoic and Precambrian Platforms in Poland (Tornquist-Teisseyre Zone). *Tectonophysics*, **128**, 251–279.

GUTERCH, A., GRAD, M., MATERZOK, R., PERCHUĆ, E., JANIK, T., GACZYŃSKI, E., DOAN, T.T., BIAŁEK, T., GADOMSKI, D., MŁYNARSKI, S. & TOPRKIEWICZ, S. 1992. Laminated structure of the lower crust in the fore-Sudetic region in Poland, derived from seismic data. *Physics of Earth and Planets International*, **69**, 217–223.

GUTERCH, A., GRAD, M., JANIK, T., MATERZOK, R., LUOSTO, U., YLINIEMI J., LÜCK E., SCHULZE A. & FÖRSTE K. 1994. Crustal structure of the transitional zone between Precambrian and Variscan Europe from new seismic data along LT-7 profile (NW Poland and eastern Germany). *Comptes Rendus Academie des Sciences, Paris*, **319**, serie II, 1489–1496.

GUTERCH, A., GRAD, M., THYBO, H., KELLER, G. R. & THE POLONAISE WORKING GROUP 1999. POLONAISE'97 – an international seismic experiment between Precambrian and Variscan Europe in Poland. *Tectonophysics*, **314**, 101–121.

HATCHER, R. D. 1989. Tectonic synthesis of the U.S. Appalachians. *In*: HATCHER, R. D., THOMAS, W. D. & VIELE, G. W. (eds) *The Appalachian-Quachita orogen in the United States – The Geology of North America* vol. F-2. Geological Society of America, Boulder, CO, 511–535.

JANIK, T., YLINIEMI, J., GRAD, M., THYBO, H., TIIRA, T. & POLONAISE P2 WORKING GROUP 2002. Crustal structure across the TESZ along POLONAISE'97 seismic profile P2 in NW Poland. *Tectonophysics*, in press.

JENSEN, S. L. 2001. POLONAISE'97 Seismic Structure of the Crust and uppermost Mantle in the Trans European Suture Zone. Ph. D. thesis, Faculty of Science, Geological Institute, University of Copenhagen, pp.133+110.

JENSEN, S. L., JANIK, T., THYBO, H. & POLONAISE PROFILE P1 WORKING GROUP 1999. Seismic structure of the Palaeozoic Platform along POLONAISE'97 profile P1 in northwestern Poland. *Tectonophysics*, **314**, 123–143.

JOWETT, E. C. & JARVIS, G. T. 1984. Formation of foreland rifts. *Sedimentary Geology*, **40**, 51–72.

JUBITZ, K. B., ZNOSKO, J. & FRANKE, D. (eds) 1986. Tectonic Map, International Geological Correlation Programme, Project No 86: South-West Border of the East European Platform. *Zentrale Geologisches Institut, Berlin.*

KARNKOWSKI, P. H. 1999. Origin and evolution of the Polish Rotligend Basin. *Polish Geological Institute Special Papers*, **3**, 1–93.

KARNKOWSKI, P. H. & RDZANEK, K. 1982. Remarks on the substratum of Permian in Wielkopolska. *Kwartalnik Geologiczny*, **26**, 327–340 (in Polish).

KELLER, G. R. & HATCHER, R. D. 1999. A summary of crustal structure along the Appalachian Quachita orogenic belt in North America: a comparison with the TESZ. *Tectonophysics*, **314**, 43–68.

KRAWCZYK, C. M., STILLER, M. & DEKORP-BASIN RESEARCH GROUP 1999. Reflection seismic constraints on Palaeozoic crustal structure and Moho beneath the NE German Basin. *Tectonophysics*, **314**, 241–253.

KRAWCZYŃSKA-GROCHOLSKA, H. & GROCHOLSKI, W. 1976. Remarks on Carboniferous from the north-western surroundings of the Fore-Sudetic Block. *Kwartalnik Geologiczny*, **20**, 53–62 (in Polish).

KREBS, W. 1978. Die Kaledoniden im nördlichen Mitteleuropa. *Zeitschrift der Deutschen Geologischen Gesellschaft,* **A 27**, 489–498.

KRYSIŃSKI, L., GRAD, M. & POLONAISE'97 WORKING GROUP 2000. POLONAISE'97 – seismic and gravimetric modelling of the crustal structure in the Polish Basin. *Physics and Chemistry of the Earth (A),* **25**, 355–363.

KUTEK, J. 1997. The Polish Permo-Mesozoic Rift Basin. IGCP Project No. 369: Comparative evolution of Peri-Tethyan Rift Basins, *Abstract Book, 4th Annual Meeting and Fieldtrip, 29 August–3 September, Barcelona, Spain.*

MATYJA, H. 1993. Upper Devonian of Western Pomerania. *Acta Geologica Polonica*, **43**, 8–94.

MONA LISA WORKING GROUP 1997*a*. Deep seismic investigations of the lithosphere in the southeastern North Sea. *Tectonophysics*, **269**, 1–19.

MONA LISA WORKING GROUP 1997*b*. Closure of the Tornquist Sea: Constraints from MONA LISA deep seismic reflection data. *Geology*, **25**, 1071–1074.

OBERC, J. 1972. Geology of Poland IV. Tectonics 2. Sudetes and adjacent areas. Wydawnictwa Geologiczne, Warszawa, 1–307 (in Polish).

OBERC-DZIEDZIC, T., ŻELAŹNIEWICZ, A. & CWOJDZIŃSKI, S. 1999. Granitoids of the Odra Fault Zone: late- to post-orogenic Variscan intrusions in the Saxothuringian Zone, SW Poland. *Geologia Sudetica*, **32**, 55–71.

PHARAOH, T. C. 1999. Palaeozoic terranes and their lithospheric boundaries within the Trans-European Suture Zone (TESZ): a review. *Tectonophysics*, **314**, 17–41.

POPRAWA, P., ŠLIAUPA, S., STEPHENSON, R. & LAZAUSKIENE, J. 1999. Late Vendian-Early Palaeozoic tectonic evolution of the Baltic Basin: regional implications from subsidence analysis. *Tectonophysics*, **314**, 219–239.

POŻARYSKI, W. 1990. The Middle European Caledonides – wrenching orogen composed of terranes. *Przegląd Geologiczny*, **38**, 1–9 (in Polish, English summary).

POŻARYSKI, W., GROCHOLSKI, A., TOMCZYK, H., KARNKOWSKI, P. & MORYC, W. 1992. The tectonic map of Poland in the Variscan epoch. *Przegląd Geologiczny*, **40**, 643–651.

RABBEL, W., FÖRSTE, K., SCHULZE, A., BITTNER, R., RÖHL, J. & REICHERT, J. C. 1995. A high-velocity layer in the lower crust of the North German Basin. *Terra Nova*, **7**, 327–337.

RYKA, W. 1982. Precambrian evolution of the Polish part of the East European Platform. *Geological Quarterly*, **26**, 257–272.

SCHWEITZER, J. 1995. Blockage of regional seismic

waves by the Teisseyre-Tornquist zone. *Geophysical Journal International*, **123**, 260–276.

ŚRODA, P. & POLONAISE P3 WORKING GROUP 1999. P- and S-wave velocity model of the southwestern margin of the Precambrian East European Craton; POLONAISE'97, profile P3. *Tectonophysics,* **314**, 175–192.

ŚRODA, P., CZUBA, W., GRAD, M., GUTERCH, A., GACZYŃSKI, E., POLONAISE WORKING GROUP 2002. Three-dimensional seismic modelling of crustal structure in the TESZ region based on POLONAISE'97 data. *Tectonophysics*, in press.

TANNER, B. & MEISSNER, R. 1996. Caledonian deformation upon southwest Baltica and its tectonic implications: alternatives and consequences. *Tectonics*, **15**, 803–812.

TEICHMÜLLER, M. 1978. Nachweis von Graptolithen-Periderm in geschieferten Gesteinen mit Hilfe kohlenpetrographischer Methoden. *Neues Jahrbuch für Geologie und Paläontologie Monatshefte*, 430–447.

TELLER, L. & KOREJWO, K. 1968. Early Paleozoic deposits in the deep substratum of North-Western Poland. *Acta Geologica Polonica*, **18**, 613–619.

THYBO, H. 1990. A seismic model along the EGT profile – from the North German Basin into the Baltic Shield. *In*: FREEMAN, R. & MUELLER, S. (eds) *Proceedings of the 5th Study Centre on the European Geotraverse Project*. ESF, Strasbourg, 99–108.

TOMCZYK, H. 1968. Silurian stratigraphy in the Baltic region of Poland based on the borehole data. *Kwartalnik Geologiczny*, **12**, 15–36 (in Polish, English summary).

WIERZCHOWSKA-KICUŁOWA, K. 1984. Geology of the Pre-Permian series of the Fore-Sudetic Monocline. *Geologia Sudetica*, **19**, 120–142 (in Polish, English summary).

WILDE-PIÓRKO, M., GRAD, M. & POLONAISE WORKING GROUP 1999. Regional and teleseismic events recorded across the TESZ during POLONAISE'97. *Tectonophysics*, **314**, 161–174.

ZIEGLER, P. A. 1990. *Geological Atlas of Western and Central Europe*. 2nd edn., Shell Internationale Maatschappij B.V./Geological Society of London/ Elsevir, Amsterdam. 239pp.

ZIELHUIS, A. & NOLET, G. 1994. Deep seismic expression of an ancient plate boundary in Europe. *Science* **265**, 79–81.

ZNOSKO, J. 1979. Teisseyre-Tornquist tectonic zone: some interpretative implications of recent geological and geophysical investigations. *Acta Geologica Polonica*, **29**, 365–382.

ŻELICHOWSKI, A. M. 1964. Carboniferous strata in the basement of the Fore-Sudetic Monocline. *Przegląd Geologiczny*, **12**, 224–227 (in Polish).

ŻELICHOWSKI, A. M. 1980. Synoptic profile of Carboniferous in the substratum of the Fore-Sudetic Monocline. *Kwartalnik Geologiczny*, **24**, 942–943.

ŻELICHOWSKI, A. M. 1987. Development of the Carboniferous of the SW margin of East European Platform in Poland. *Przegląd Geologiczny*, **35**, 230–237 (English summary).

Basement structure in the southern North Sea, offshore Denmark, based on seismic interpretation

M. SCHECK[1], H. THYBO[2], A. LASSEN[2], T. ABRAMOVITZ[2] & M. LAIGLE[3]
[1]*GeoForschungsZentrum Potsdam, Telegrafenberg C423, D-14473 Potsdam, Germany*
e-mail: leni@gfz-potsdam.de
[2]*Geological Institute, University of Copenhagen, Øster Voldgade 10, DK-1350 Copenhagen K, Denmark*
[3]*Institut de Physique du Globe de Paris, Boite 89, 4, Place Jussieu, 75252 PARIS Cedex 05, France*

Abstract: Seismic reflection data from the Danish North Sea are interpreted to map the structure of the Palaeozoic basement in the area of the MONA LISA deep seismic lines. Based on a characteristic near-basement reflection, the upper crystalline crust of Baltica of offshore Denmark is traced to the south into the southern Horn-Graben, and to the west to the eastern shoulder of the Central Graben. A two-way-traveltime map of the near-basement horizon and several interpreted seismic sections reveal that three main tectonic events influenced the topography of the basement: (1) a compressional event which could be Caledonian in age; (2) a Palaeozoic extensional event postdating the compressional deformation and expressed in a system of WSW–ESE to W–E striking Palaeozoic half-grabens; and (3) the Permo-Triassic rifting that led to the formation of NNW–SSE to NNE–SSW trending Mesozoic faults of the Horn Graben and the Central Graben which are oriented sub-perpendicular to the Palaeozoic system. Compressive deformation is localized in a narrow zone around and south of the hitherto interpreted Caledonian Deformation Front and foreland deformation on Baltica is suggested as its origin. The seismic image of the Palaeozoic half-grabens indicates that the East North Sea High is an inverted Palaeozoic rift which subsequently was cut by younger Late Palaeozoic to Mesozoic rifts of the Horn and Central Grabens. The timing of the first extensional phase remains speculative, but it predates the Rotliegend unconformity. Some of the older Palaeozoic normal faults may have been reactivated as transfer zones between the different graben segments during the Permo-Mesozoic extension.

The pre-Permian history of the southern North Sea is still a matter of debate since the corresponding rocks are deeply buried below several kilometres of Permian to Cenozoic sediments. Several tectonic events are assumed to have left their imprints. The Caledonian Suture is supposed to be located offshore from Denmark as indicated in Figure 1. This suture is the result of the triple plate collision between Laurentia, Baltica and the Gondwana-derived microcontinent Eastern Avalonia in Late Ordovician–Early Silurian times. Although the geometry of the Caledonian Suture is well defined between Laurentia and Baltica in the north, and in the western Baltic Sea there are still uncertainties concerning the geometry and the exact location of the suture between Avalonia and Baltica (MONA LISA Working Group 1997). The existence of a Caledonian orogen between Avalonia and Baltica has been shown by different basement ages found in well cores sampled across the Caledonian Suture (Frost *et al.* 1981; Nielsen & Japsen 1990). NE of the suture, Precambrian basement ages are 880–825 Ma, whereas to the SW Caledonian ages of 450–415 Ma have been measured in metamorphic basement rocks. In the study area, the Precambrian crystalline basement has been dated in the Ibenholt-1 well (Fig. 1) as 844–781 Ma old (Phillips Petroleum Company 1987 cited in Abramovitz & Thybo 1999) which thus is located on Baltica. However, in this well, the Palaeozoic sequence is missing and the Lower Permian Rotliegend volcanics are lying above crystalline basement. The nature of the suture between Baltica and Avalonia may either be related to orthogonal closure of the Tornquist ocean or to oblique transform-fault related docking of the two plates (MONA LISA Working Group 1997). After the Caledonian orogeny, the area might have been affected by extension due to postorogenic collapse, and subsequently by compression and extension induced by the Variscan orogeny. Moreover, the Palaeozoic sequence and its basement were deformed during the Permian and

From: WINCHESTER, J. A., PHARAOH, T. C. & VERNIERS, J. 2002. *Palaeozoic Amalgamation of Central Europe*. Geological Society, London, Special Publications, **201**, 311–326. 0305-8719/02/$15.00

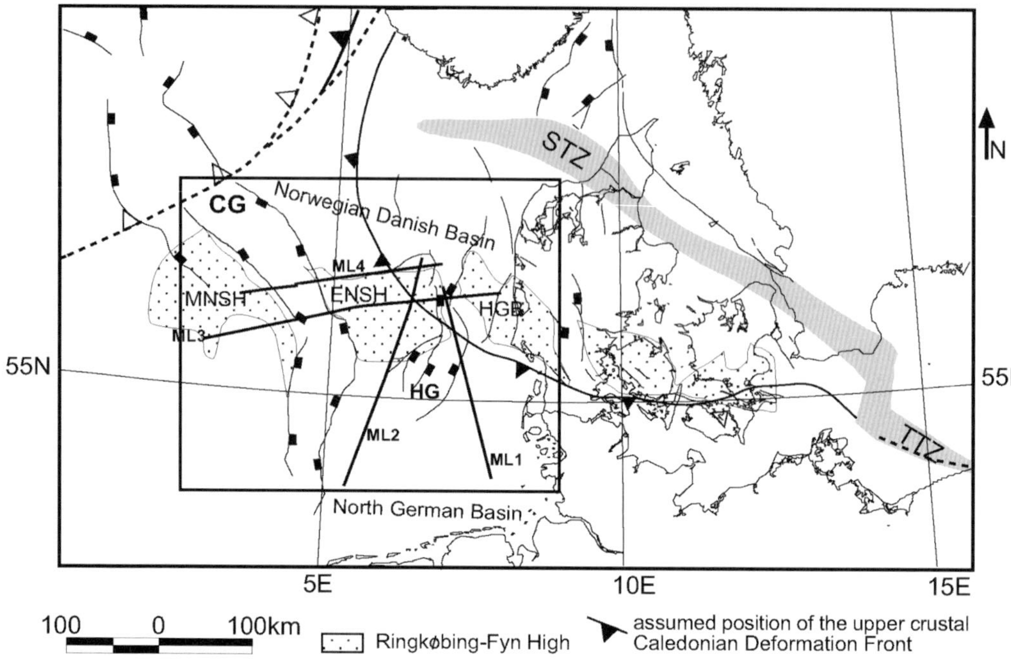

Fig. 1. Regional structural framework of the study area. The upper crustal Caledonian Suture is supposed to be located offshore to the west of Denmark, but Palaeozoic structures are deeply buried below Permian to Cenozoic sediments of the Danish–Norwegian and the North German Basins as well as of the Permo-Triassic rifts of the Horn Graben and Central Graben. MNSH, Mid North Sea High; ENSH, East North Sea High; HGB, Holmsland-Grindsted-Block; CG, Central Graben; HG, Horn Graben; STZ, Sorgenfrei–Tornquist Zone; TTZ, Teisseyre–Tornquist Zone; ML 1 to 4, MONA LISA deep seismic lines 1 to 4.

Mesozoic tectonic phases leading to the evolution of the Norwegian–Danish Basin and of the North German Basin (Southern and Northern Permian Basins after Ziegler 1990) and of the Graben Systems of the Central and Horn Graben. Tectonic inversion took place during the Late Cretaceous and Palaeocene and was followed by regional subsidence. Inversion was strongest along the Sorgenfrei Tornquist Zone, but uplift of the RFH during compression has been described as well (Gemmer *et al.* 2002; Clausen & Huuse 1999). The subsequent Late Cenozoic subsidence was most intense in the central part of the North Sea Basin, while the margins of the entire North Sea Basin were contemporaneously affected by uplift and erosion (Hansen *et al.* 2000; Japsen 1998).

As a first step to unravel the possible relicts of the various tectonic processes, we investigated the structural setting of the Palaeozoic sequence and the Precambrian basement. A characteristic double reflection is interpreted to mark the near-basement reflection of the Lower Palaeozoic sediments on Baltica (Michelsen & Nielsen 1993; Lassen *et al.* 2001). We have constructed a two-way-traveltime (TWT) map of this horizon from interpretation of seismic normal-incidence reflection data recorded to 7 seconds TWT of two industrial seismic surveys: RTD-81 and NP85N (Fig. 2) and the reflection seismic data of the deep seismic MONA LISA project. Interpretations were tied to the available wells in the study area (Nielsen & Japsen 1991). We show some seismic lines (bold lines in Fig. 2) imaging the typical structural appearance of the basement with main emphasis on the pre-Permian part of the sequence and compare them with the pre-stack-depth-migrated data of MONA LISA. The structural characteristics of the basement have implications for the tectonic history of the area.

Basement structure

The presence of Palaeozoic sediments underlying Permian strata has been described below the East North Sea–Ringkobing Fyn High as well as below the Permian Basins and Mesozoic grabens (Cartwright 1990; Vejbaek 1990; Clausen & Korstgård 1993; MONA LISA

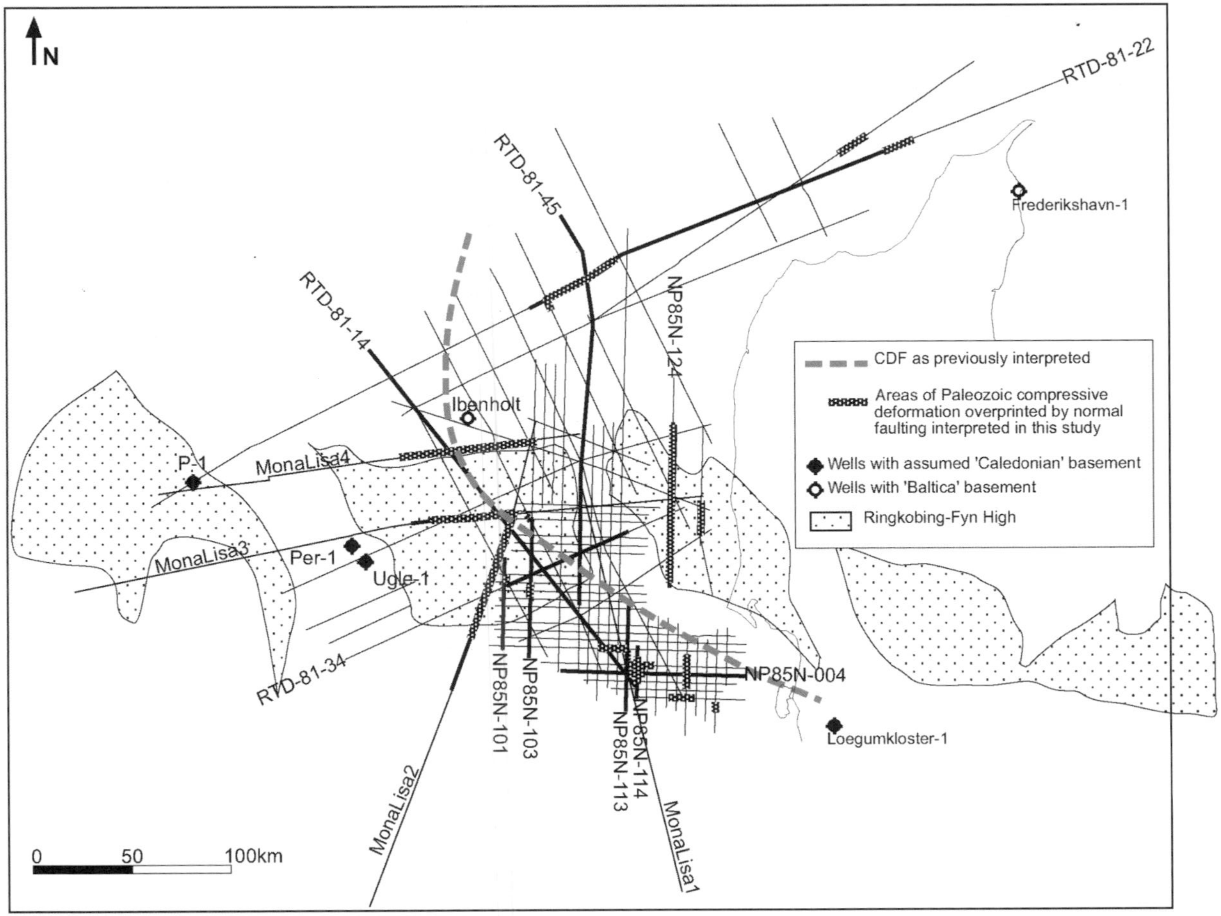

Fig. 2. Overview of the distribution of the seismic normal-incidence reflection data used for interpretation in this study including two industrial seismic surveys: RTD-81 and NP85N and the deep seismic MONA LISA project. The seismic lines shown in Figures 3 to 6 are indicated by bold lines. The stippled signature shows, where we found indications for Palaeozoic compressive deformation. Although the structures in the southern part of the study area can be correlated, the structures found along line RTD-81–22 may be part of another compressive zone located further north.

Working Group 1997; Zhou & Thybo 1997; Abramovitz *et al.* 1998; Nielsen *et al.* 1998; Abramovitz & Thybo 1999). The Palaeozoic sequence comprises an interval of reflections between the Lower Permian Rotliegend unconformity and the crystalline basement. In areas where the Upper Permian Zechstein salt is present, the base Zechstein is addressed as the upper boundary of the investigated interval. The Rotliegend unconformity truncates dipping reflections in the lower Palaeozoic sequence, previously termed 'pre-rift' deposits with respect to the main rifting phase of the Horn Graben (Abramovitz & Thybo 1999).

The main structural characteristic of the pre-Permian Palaeozoic sediments is block-faulting which also affects the near-basement reflection. This near-basement reflection is characterized by a strong continuous double reflection of a low frequency. The reflectivity pattern of the near-basement reflection in the study area resembles that of the Hardeberga sandstone (Michelsen & Nielsen 1993) equivalent to the marine Balka sandstone and a Lower Cambrian sandstone found onshore Denmark (Poulsen 1969). Therefore, we correlate it with an almost identical double reflection, observed to the east of Denmark and in the Baltic Sea, which has been drilled and identified as an interface between Cambrian quartzite and shales beneath the dissected remains of the Silurian foredeep (Lassen *et al.* 2001). Such basement reflections are known from Baltica crust north and east of the Caledonian deformation front. Deep wells in the study area reaching the crystalline basement are located in areas where the Palaeozoic sequence is missing or incomplete. As well data in the study area penetrating the basement reflection is absent, we follow Abramovitz and Thybo (2000) and interpret the near-basement reflection as an offshore equivalent of the top of lower Cambrian sandstone. Its presence is considered as a characteristic attribute of Baltica and could outline the extent of Baltica's upper crust to the south and west.

Examples of our interpretations of the Palaeozoic sediments are imaged in Figure 3 along two regional, roughly NNW–SSE trending seismic lines, from the Norwegian–Danish Basin across the East North Sea High and into the Southern Horn Graben. The Palaeozoic sequence is truncated by a regional unconformity corresponding to the Rotliegend in areas where the Zechstein salt is absent and to the base Zechstein in the Norwegian–Danish Basin. The near-basement reflection appears as a down-faulted horizon which is offset by Mesozoic and Palaeozoic normal faults. Although the Palaeozoic faults terminate upwards at the Rotliegend unconformity, the Mesozoic faults offset the largest part of the Palaeozoic and Mesozoic sequence. In the area of the Norwegian–Danish Basin, the near-basement reflection occurs at about 4–5 seconds TWT along line RTD81–45 (Fig. 3a), but the resolution is increasingly poor to the north due to the masking effect of the Zechstein salt in the Norwegian–Danish Basin. The resolution is good around the Horn Graben and the near-basement reflection is visible as a block-faulted horizon within small half-grabens dipping predominantly to the north. At the southeastern shoulder of the Horn Graben, the near basement reflection is encountered at a very shallow level (1.5–2 seconds TWT) and is predominantly south-dipping within small Palaeozoic half grabens. The Palaeozoic interval is up to 1.7 seconds TWT thick below the Horn Graben and less than 0.1 second TWT on the southwestern shoulder.

In line RTD81–14 (Fig. 3b), the near-basement reflection is encountered at 3–4 seconds TWT below the Norwegian–Danish Basin and appears as a sub-horizontal signal interrupted by normal faults. On the East North Sea High, the near-basement reflection is shallow at 2–3 seconds TWT dipping to the SE in the southeastern half of the high and to the NW in the northwestern part of the high. The Palaeozoic interval is thickest below the Norwegian–Danish Basin and much thinner on the East North Sea High. In the Palaeozoic interval, the East North Sea High appears as a large-scale horst compared to the Norwegian–Danish Basin and the Horn Graben with the central part of the high being the shallowest. The vertical offset along the main border faults of the Horn Graben attains up to 2 seconds TWT. In some cases the Mesozoic faults developed on top of older Palaeozoic faults, but the density of Mesozoic faults is much smaller than that of the Palaeozoic ones, and the vertical offsets are generally larger on the Mesozoic than on the Palaeozoic faults. This indicates a different tectonic origin for the two fault systems.

Figure 4 shows five interpreted, N–S-trending seismic lines from the southern part of the study area. Again, the Rotliegend unconformity truncates block-faulted Palaeozoic reflections. In line NP85N-103 (Fig. 4a), the near basement reflection is dipping to the south. It occurs at 1.5 to 3 seconds TWT on the southern slope of the East North Sea High and shows offsets of up to 1 second TWT along north-dipping Palaeozoic normal faults in small half-grabens. The Mesozoic border fault of the Horn Graben offsets the

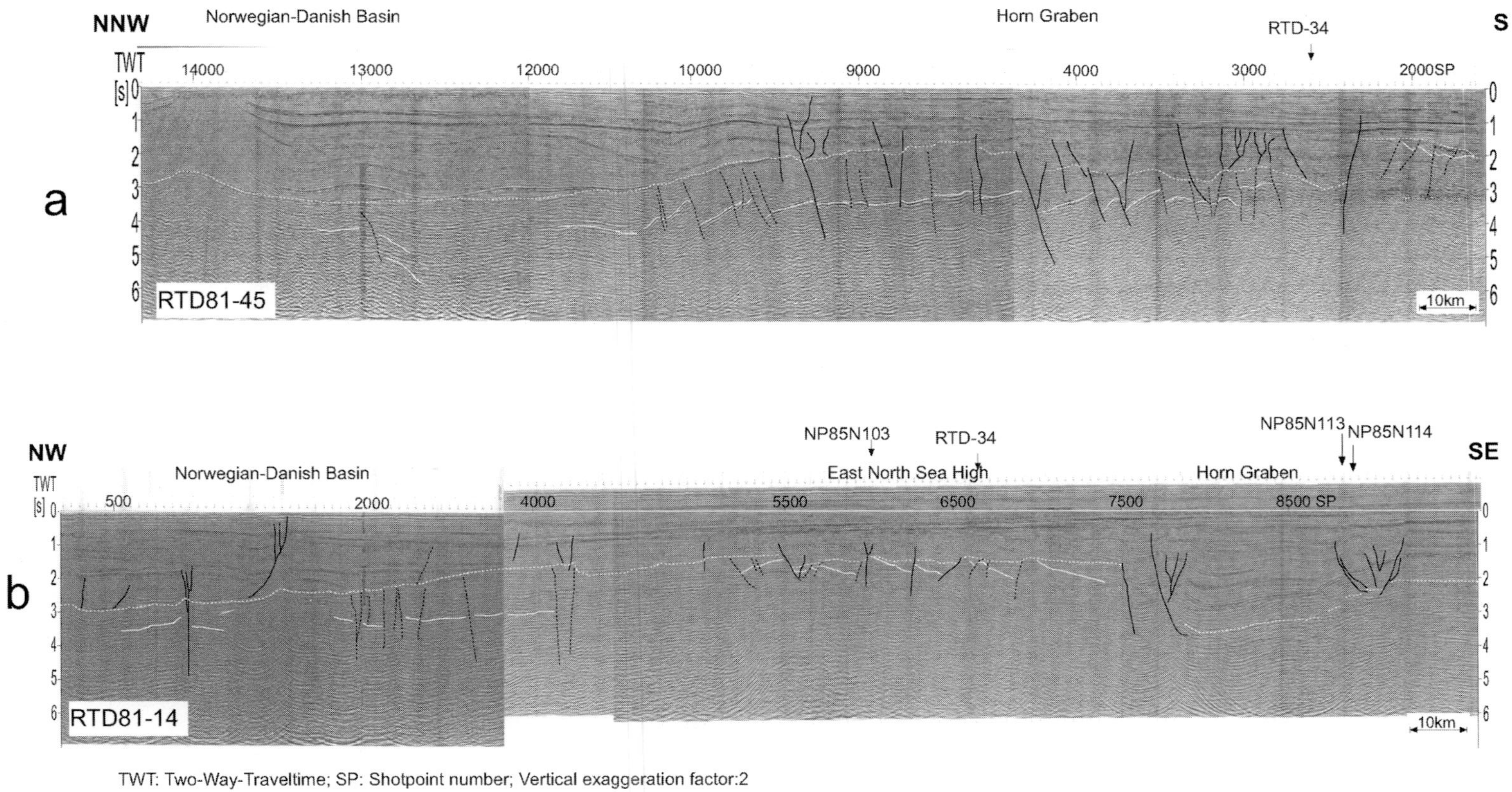

Fig. 3. Interpretations of the Palaeozoic sediments and the block-faulted near-basement reflection along (a) line RTD81–45 and (b) line RTD81–14 from the Norwegian–Danish Basin to the southern Horn Graben. The near-basement reflection (white line) is offset by Mesozoic faults (solid black lines) which offset the largest part of the Palaeozoic and Mesozoic sequence and by Palaeozoic normal faults (dashed black lines) which terminate upwards at the Rotliegend unconformity (dashed white line). The Mesozoic faults show larger offsets and are less densly spaced than the Palaeozoic faults. Below the Norwegian–Danish Basin the near-basement reflection is deeper than on the East North Sea High or below the Horn Graben, and is at a very shallow level below the southwestern shoulder of the Horn Graben. The Palaeozoic interval is thickest below the Norwegian–Danish Basin and below the Horn Graben and thinnest below the graben shoulders and on the East North Sea High. Note that the Paleozoic faults are dipping towards the central part of the high where they are increasingly steep.

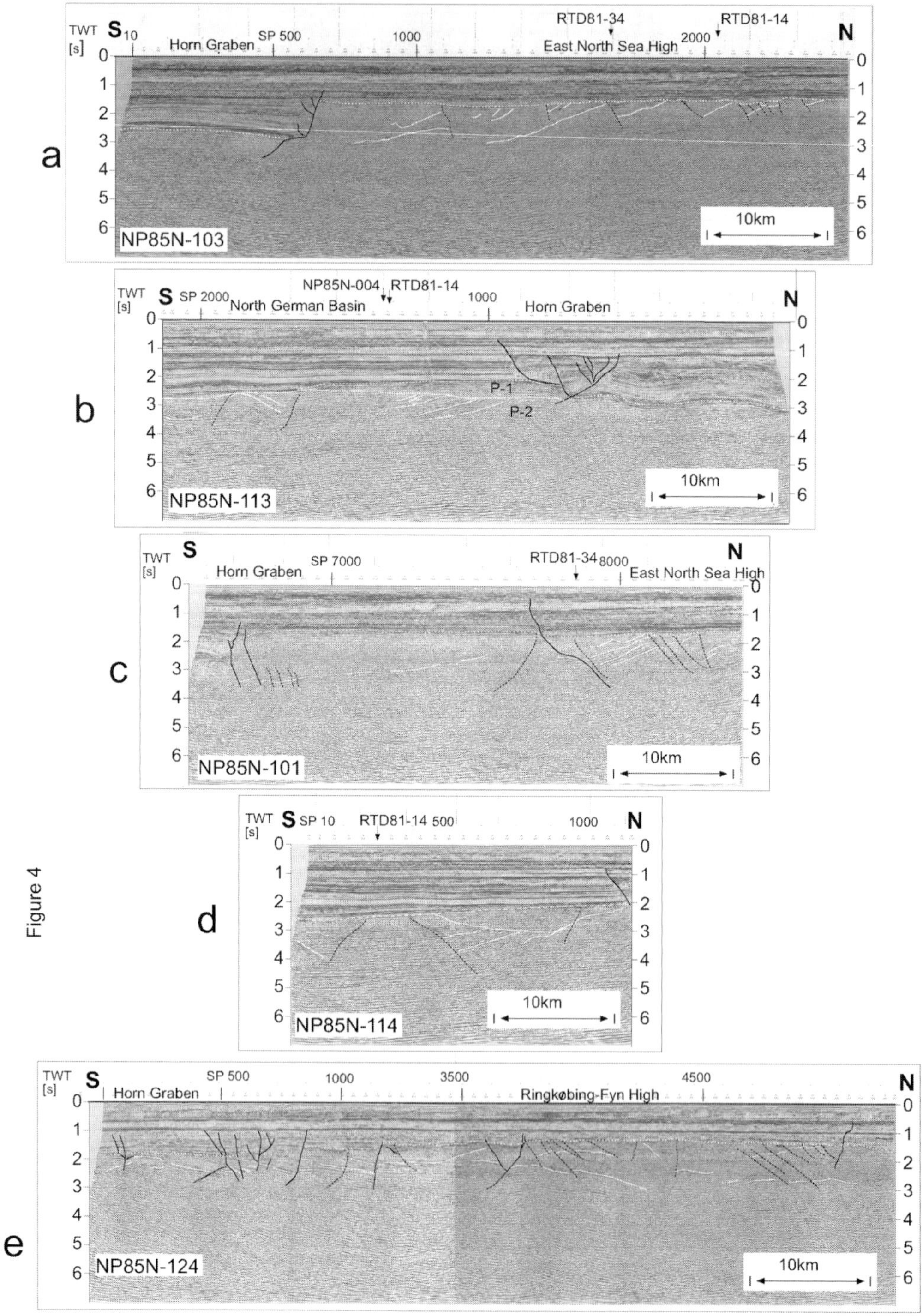
a
TWT [s]
S
N
Horn Graben
SP 500
1000
RTD81-34
East North Sea High
2000
RTD81-14
10km
NP85N-103
b
NP85N-004
RTD81-14
SP 2000
North German Basin
1000
Horn Graben
P-1
P-2
10km
NP85N-113
c
Horn Graben
SP 7000
RTD81-34
8000
East North Sea High
10km
NP85N-101
d
SP 10
RTD81-14
500
1000
10km
NP85N-114
e
Horn Graben
SP 500
1000
3500
Ringkøbing-Fyn High
4500
10km
NP85N-124

Figure 4

Rotliegend unconformity with 1.5 seconds TWT and the near-basement reflection cannot be identified below the Horn Graben where multiples are strong.

Line NP85N-113 (Fig. 4b) images the transition from the southern Horn Graben into the North German Basin. The Palaeozoic sequence lies between 2.5 and 3.5 seconds TWT. In the central part of the section, an intra-Palaeozoic unconformity separates a transparent Upper Palaeozoic layer from a lower Palaeozoic layer containing dipping reflections. The reflections have a similar signature as the near-basement reflection, but appear in a bundle. Abramovitz and Thybo (1999) interpreted the south dipping Palaeozoic reflections as Caledonian thrusts. Yet, at the southern end of line NP85N-113, another set of north-dipping Palaeozoic reflections occurs, however, with only a few individual reflections. The true nature of these reflections remains unclear, they could represent either relicts of Caledonian thrusts and folds or sedimentary surfaces within the Palaeozoic half-grabens. The latter package of reflections is bordered by south-dipping normal faults, indicating an extensional event overprinting the initial structure. Between these normal faults, the reflections are continuous and parallel, suggesting that they represent sedimentary surfaces rather than thrust planes. Similar features have been observed in neighbouring profiles. The areas, where these features have been found are indicated in Figure 2. Other examples are line NP85N-101 (Fig. 4c) and line NP85N-114 (Fig. 4d) displaying similar bundles of south-dipping reflections which could represent compressively deformed Palaeozoic layers overprinted by normal faulting.

Line NP85N-124 (Fig. 4e) is crossing the western part of the Holmsland-Grindsted Block (eastern part of the Ringkøbing-Fyn High). Here, the structural setting of the Palaeozoic sequence is more complicated than in the lines shown in Figure 3. Again, there are several reflections within the Palaeozoic succession between 2 and 3 seconds TWT. If these reflections correspond stratigraphically to one single horizon, their vertical repetition can be explained only by thrusting. However, normal offsets are also indicated. A possible explanation for the observed seismic structures could be originally folded and thrusted horizons overprinted by two generations of normal faults (Palaeozoic and Mesozoic). Nevertheless, these reflections may represent sedimentary surfaces above a north-dipping basement. In the northern part of the section, a series of very densely spaced Palaeozoic north-dipping faults are present which show reverse offsets of the lowermost near-basement reflection.

The two roughly east–west trending seismic lines in Figure 5 cross the southern part of the study area. Line NP85N-004 (Fig. 5a) displays the typical appearance of the near-basement reflection in east–west oriented seismic sections. Over the eastern half of the section, it is subhorizontal between 3 and 4 second TWT and appears as long wavelength folds towards the west. The continuity of the reflection is interrupted by several east-dipping Palaeozoic normal faults which terminate at the Rotliegend unconformity. To the west, the near-basement reflection disappears below the Horn Graben, probably because of strong multiples and/or signal attenuation in the thick sediments.

In line RTD81–34 (Fig. 5b), the near-basement reflection is present on the East North Sea High in small half-grabens between 2 and 3 seconds TWT. The faults between the half-grabens dip to the east along the western slope of the high and to the west along the eastern slope. The near-basement horizon itself is

Fig. 4. Interpreted, north-south-trending seismic lines from the southern part of the study area: (a) In line NP85N-103, the near basement reflection (solid white line) dips to the south at 1.5 to 3 seconds TWT on the southern slope of the East North Sea High. North-dipping Palaeozoic normal faults attain up to 1 second TWT offset. The Mesozoic border fault of the Horn Graben offsets the Rotliegend unconformity with 1.5 seconds TWT. Strong multiples hamper the interpretation of the near-basement reflection below the Horn Graben. (b) In line NP85N-113, the Palaeozoic sequence is at 2.5 to 3.5 seconds TWT across the transition from the southern Horn Graben into the North German Basin and consists of two parts separated by an intra-Palaeozoic unconformity into 2 layers P-1 and P-2. The lower Palaeozoic layer contains both, bundles of north- and south-dipping reflections of similar signature as the near-basement reflection, which could either be remnants of Caledonian thrusts or Palaeozoic bedding planes. The bundled reflections are confined by south-dipping normal faults. Similar bundles of south-dipping reflections offset along Palaeozoic normal faults are observed in profiles (c) line NP85N-101 and (d) line NP85N-114. The south-dipping reflections at 3 seconds near SP 1300 on line NP85N-113 correlate to similar south-dipping structures on MONA LISA line 1.(e) In line NP85N-124 we observe several reflections within the Palaeozoic succession between 2 and 3 seconds TWT from the Holmsland-Grindsted-Block to the westernmost parts of the Horn Graben, which either correspond to one single horizon, and their vertical repetition is due to Caledonian thrusting or represent stratigraphic surfaces above a north-dipping basement.

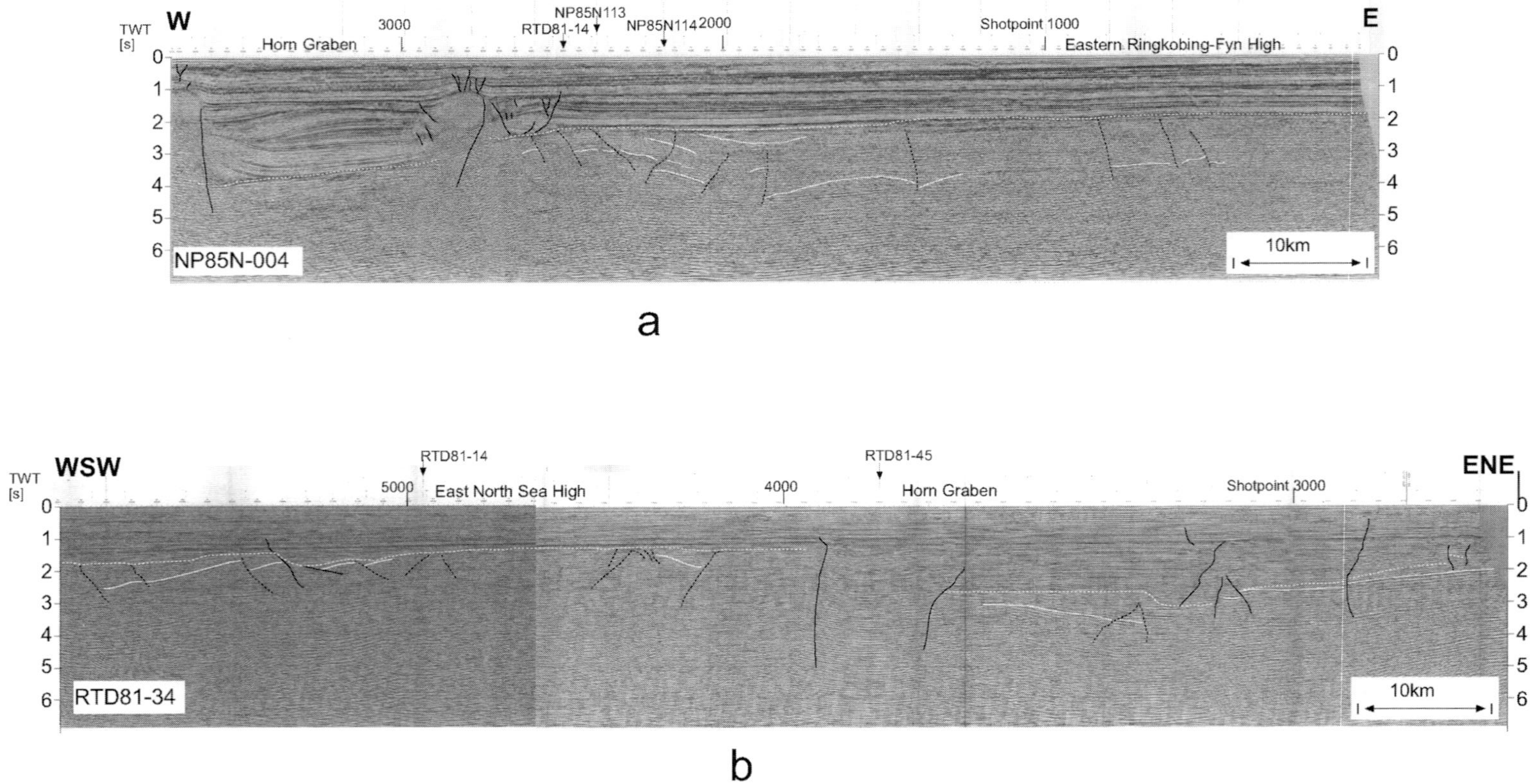

Fig. 5. East–west trending seismic lines from the southern part of the study area. (a) In line NP85N-004, the near-basement reflection is sub-horizontal between 3 and 4 second TWT but appears weakly folded towards the west (between shotpoints 1500 to 2500). Several east-dipping Palaeozoic normal faults transect the long-wavelength folds. (b) Line RTD81–34 shows small Palaeozoic half-grabens between 2 and 3 seconds TWT on the East North Sea High. The half-graben-bounding faults dip towards the centre of the East North Sea High, and the near-basement horizon dips symmetrically away from the high, similar to line RTD81–14 in Figure 3b.

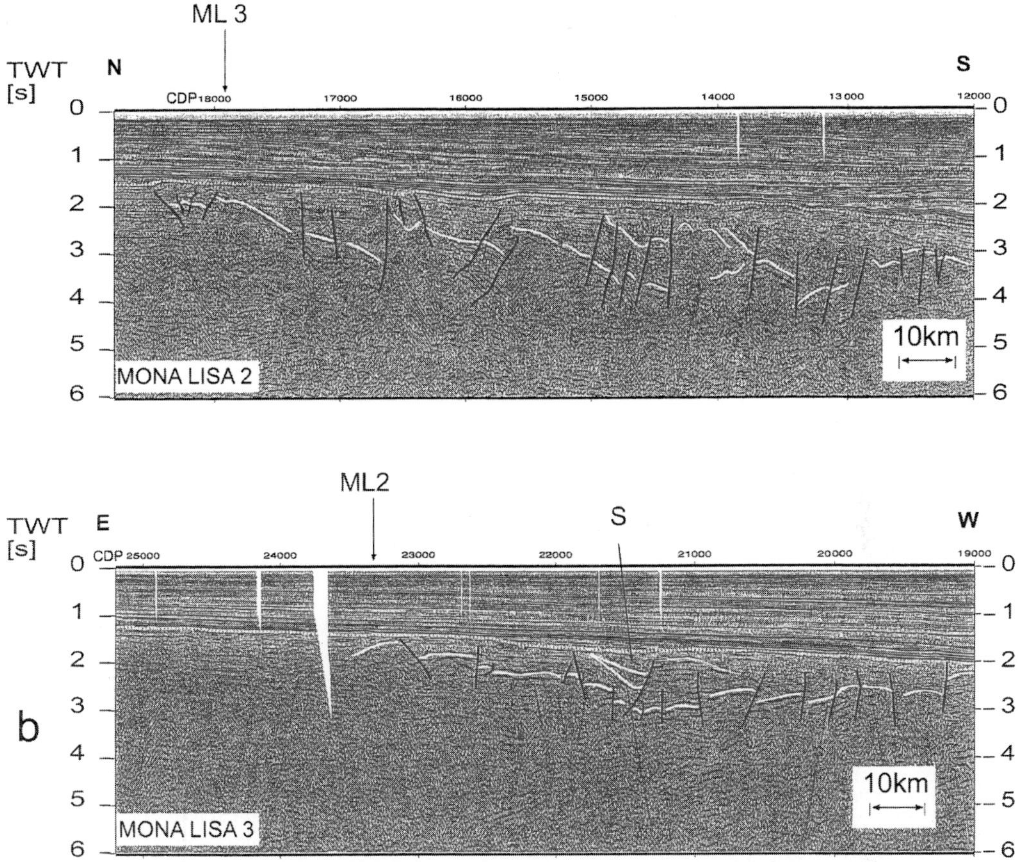

Fig. 6. Pre-stack depth migrated parts of MONA LISA lines 2 and 3 (Laigle 1999) crossing the central part of the East North Sea High converted to time. (a) In line MONA LISA 2, the near-basement reflection consists of south dipping elements between 2 and 4 seconds TWT, which are located at a shallower level towards the centre of the East North Sea High and appear thrust. Additionally, steep, north-dipping Palaeozoic normal faults offset the 'thrust' elements. (b) In line MONA LISA 3, doubling of the near-basement reflection is evident at CDP 21000. As a general trend, the near-basement reflection is sub-horizontal to weakly folded in east–west oriented seismic profiles and Palaeozoic sub-vertical normal faults offset the original structure. Stratigraphic thickening is evident in the upper part of the Palaeozoic (index 'S' between CDP 21500 and 22000) indicating that deposition was syn-tectonic with the formation of the half-grabens.

west-dipping on the western slope of the High and east-dipping along the eastern slope. This symmetry of normal faults dipping towards the centre of the high is similar to line RTD81–14 in a NW–SE-direction (Fig. 3b).

The pre-stack depth migrated parts of MONA LISA lines 2, and 3 (Laigle *et al.* 2000) converted to time are shown in Figure 6. In line 2 (Fig. 6a), the near-basement reflection appears as a succession of south dipping elements between 2 and 4 seconds TWT. In parts (at CDP 16000 and CDP 14000) the reflection appears doubled, a possible consequence of Caledonian thrusting. However, there is also indication for north-dipping normal faults in the Palaeozoic sequence, offsetting older thrusts, which, as in line RTD81-14, are located at shallow levels towards the centre of the East North Sea High. Similar doubling of the near-basement reflection is also obvious in line MONA LISA 3 (Fig. 6b), where a shallow element is visible at CDP 21000 at about 2 seconds TWT, below which a deeper element is present at about 3 seconds TWT. Otherwise, the near-basement reflection appears sub-horizontal or weakly folded in line MONA LISA 3. As in line 2, Palaeozoic normal faults overprint the initial structure. Both lines MONA LISA 2 and MONA LISA 3 cross the

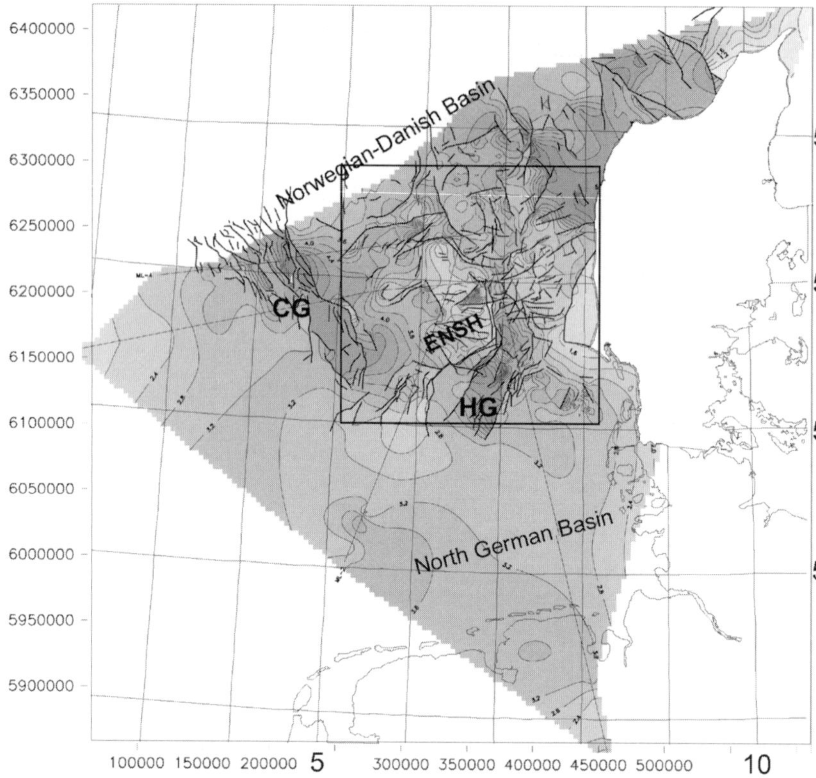

Fig. 7. Two-way-traveltime- map of the near-basement reflection in the Danish North Sea contoured from the interpreted seismic data considering the distribution of faults (recompiled after Vejbaek 1997; Thybo 1997; this study). Dark greyscales indicate deep-lying basement; light colours match shallower depths. Two sets of major faults have been mapped: WSW–ESE to east–west striking Palaeozoic faults are disrupted and offset by NNE–SSW to NNW–SSE striking Mesozoic faults. The rectangle indicates the more detailed close-up of the TWT-map shown in Figure 8. HGB, Holmsland-Grindsted-Block; HG, Horn Graben, ENSH, East North Sea High; CG, Central Graben.

central part of East North Sea High and the angle of the normal faults is steeper (in places sub-vertical) than in the sections across the slopes of the high. In line MONA LISA 3 (Fig. 6b), there is stratigraphic thickening in the upper part of the Palaeozoic sequence, in the graben between CDP 22000 and CDP 21500. This feature was observed in several lines across the study area and shows that the upper part of the Palaeozoic was deposited syntectonically during the formation of the half-grabens.

Although the quality of the seismic data varies across the lines, it is possible to identify the near-basement reflection in most of the sections. The upper part of the double pulse reflection has been picked consistently, in cases of reflection doubling the lower element. The picked TWT have been contoured using automatic interpolation with the programme Earth Vision (release 5.1, Dynamic Graphics Ltd. 2001) with a minimum tension gridding technique (Fig. 7). Additional input to the interpolation process is the distribution of faults, which have been recompiled after the results of this study and published maps (Vejback 1997; Thybo 1997). The faults were considered as discontinuities during interpolation.

The resulting TWT map of the near-basement reflection in the Danish North Sea is shown in Figure 7. From east to west major structures are the Holmsland-Grindsted-Block in the eastern part of the Ringkøbing-Fyn High, the Horn Graben, the East North Sea High and the Central Graben. Two sets of major faults form the structure of the topography of the near-basement horizon: WSW–ENE to east–west striking Palaeozoic faults and NNE–SSW to NNW–SSE striking Mesozoic faults which cross and offset

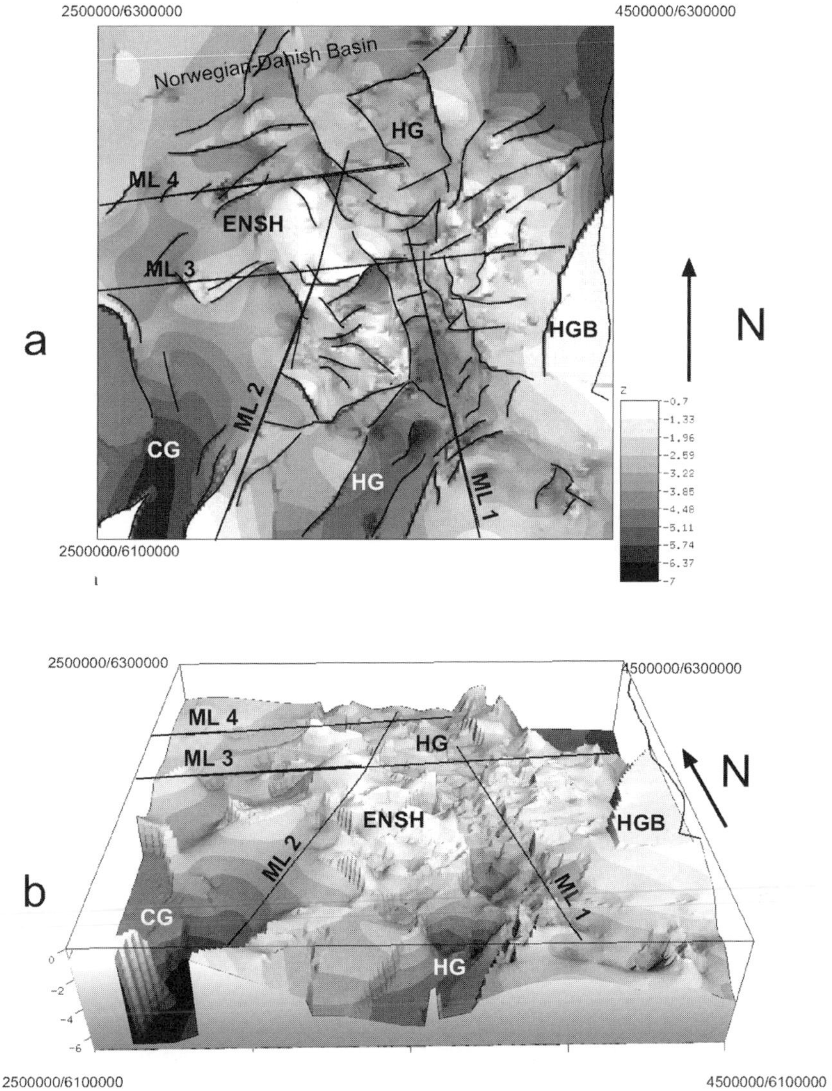

Fig. 8. Shaded relief view of the near-basement horizon in the map segment of highest data density. (a) A view lit from above shows the main faults and images how the main faults of the Horn Graben crosscut and offset the older border faults of the East North Sea High (ENSH) and of the Holmsland-Grindsted-Block (HGB). (b) A 3-D view from south shows interfering fault systems of different ages within the Horn Graben area and on the East North Sea High.

the older ones. The basement topography is clearly demonstrated on a shaded relief view of the near-basement horizon shown in Figure 8, where the segment of the map with the highest data density has been imaged. Figure 8a shows the main faults of the Horn Graben in vertical view and how they crosscut and offset the older border faults of the East North Sea High and of the Holmsland-Grindsted-Block. A three dimensional view from the south (Fig. 8b) shows the complex pattern of intersecting faults within the Horn Graben, in a dense network. It also images the change in strike along the Horn Graben with a NNW–SSE striking northern segment and a NNE–SSW striking southern branch. The East North Sea High is also structured by interfering fault systems of different ages.

Discussion

Analysis of seismic data has revealed the presence of a thick sequence of Palaeozoic sediments below the Permian in the Danish North Sea distributed in WSW–ENE to WNW–ESE striking small (2–15 km wide) half-grabens which are limited downwards by the near-basement reflection. If we consider the near-basement reflection as a characteristic attribute of Baltica, the study area is located entirely on Baltica upper crust. This is consistent with deep seismic data, revealing a 3-layered crust similar to the Baltic crust as far south as the Elbe Line (Thybo 1990; Abramovitz & Thybo 2000), a concept also considered by Tanner and Meissner (1996).

The near-basement reflection in the Danish North Sea area has been affected by at least two significant Palaeozoic tectonic events and by Permian–Mesozoic extension.

Probable Palaeozoic compressive structures are identified in all MONA LISA lines and in some of the commercial data. They are supposed to represent remnants of the Caledonian orogeny. Abramovitz and Thybo (1999) and Thybo (2001) analysed pre-Zechstein structures in the Horn Graben area using parts of the RTD81 and NP85N surveys to map the Rotliegend reflection and present several interpretations of the top basement reflection. They differentiate between two different sets of south-dipping reflections according to their seismic attributes: (1) the near-basement reflector and (2) remnants of Caledonian thrusts (with stronger amplitude). We find that the south dipping reflections in the southern part of the study area may represent old thrusts or, alternatively, bedding planes in rotated blocks of half-grabens. If we assume that they represent Caledonian thrusts, the area affected is a narrow WNW–ESE oriented zone of less than 20 km width. This deformed area is separated from the main area where compressive structures have been mapped (central part of MONA LISA survey) by a zone, where no compressional features are present in the sedimentary sequence. A similar type of structural expression is known from foreland deformation in collisional settings. For example, the Alpine collision affected the Swiss Molasse Basin in a similar way with large areas dominated by normal faulting between areas which show strong compressive deformation (Ziegler 1990). Considering our basic assumption, that the upper crystalline crust in the study area is part of Baltica, it is reasonable to adopt such a tectonic scenario to explain the observations. We therefore modify the suggestion of Abramovitz and Thybo (1999) for the sedimentary sequence that the Caledonian Deformation Front is a 80 km wide deformation belt now preserved between the Rotliegend and basement reflection and in intra-basement half-grabens. However, the Caledonian Deformation Front might be wider at depth as indicated by deep seismic (Abramovitz *et al.* 1998) and potential field interpretation (Zhou & Thybo 1997).

The second Palaeozoic tectonic event which affected the Palaeozoic sediments is crustal extension. This is expressed in numerous WSW–ENE to west–east striking half-grabens which offset the compressive structures as well as the near-basement reflection. The Lower Palaeozoic sediments are generally parallel to the near-basement reflection and attain a thickness of up to 4 km (2 seconds TWT with average P-wave velocity of 4 kms^{-1}). Locally, the Palaeozoic sequence is separated by an unconformity between an upper transparent part and a lower part consisting of tilted blocks. This is consistent with velocity models for line MONA LISA 1 (Abramovitz *et al.* 1998) and line MONA LISA 2 (Abramovitz & Thybo 2000) which show evidence for an upper Palaeozoic layer with P-wave-velocities of 3.9–4.7 kms^{-1} and a lower Palaeozoic layer with P-wave velocities of 4.7–5.6 kms^{-1}. According to these models, pre-Permian deposits in down-faulted half-grabens are 3.5–4.1 km thick below the Horn Graben and their average thickness is increasing southwards beneath the graben (Abramovitz & Thybo 2000). A similar velocity structure was found in line MONA LISA 2 where a 2.5 km thick Palaeozoic sequence with velocities of 4.9–5.6 kms^{-1} has been observed above the basement on the East North Sea High, in accordance with magnetic data (Zhou & Thybo 1997).

Our interpretations show that the total thickness of preserved Palaeozoic sediments varies throughout the study area within half-grabens, depending on the proximity to the East North Sea High and to the shoulders of the Mesozoic Grabens. In the Holmsland-Grindsted Block and below the Horn Graben area the Palaeozoic interval is up to 4 km thick (2 seconds TWT with an average velocity of 4kms^{-1}), although it is markedly thinned on the central part of the East North Sea High and on the shoulders of the Horn Graben and Central Graben.

The regional base Rotliegend unconformity is erosional and truncates the Palaeozoic reflections discordantly. Consequently the area was peneplained prior to the Rotliegend. This is consistent with the results of Zhou and Thybo (1997) who concluded from the presence of Rotliegend volcanic rocks on the East North Sea

High that the Northern and Southern Permian Basins were connected during Rotliegend times. Maximum uplift and erosion of the Paleozoic sequence is observed on the East North Sea High, where Palaeozoic rocks and in places crystalline basement is directly overlain by Mesozoic sediments (Nielsen & Japsen 1991; Abramovitz & Thybo 1999). The differentiation between the pre-Rotliegend erosional event and younger Mesozoic phases of erosion indicated by the present stratigraphic succession is beyond the topic of this paper.

The Palaeozoic faults controlling the small half-grabens dip towards the centre of the East North Sea High and get increasingly steeper towards the central part of the high. The same pattern is seen in the entire Danish Basin north of the Ringkøbing-Fyn-High. This pattern is commonly observed in the central parts of rifts. Paradoxically, in this central part of the East North Sea High, the Palaeozoic succession is both, (1) in the most shallow position due to post-depositional uplift, and (2) in parts completely eroded. This implies, that the East North Sea High is an inverted Palaeozoic rift located between the younger Mesozoic rifts of the Horn Graben and the Central Graben.

Nielsen *et al.* (1998) observe a correlation between negative free-air anomalies and Palaeozoic thickness maxima. In general, they observe that the thickness of the lower Palaeozoic deposits on the Ringkøbing-Fyn High decreases toward the east with the thickest sequence (up to 4 km) observed near the eastern margin of the Central Graben. Palaeozoic sediments are very thin or absent on the shoulders of the Central Graben and of the Horn Graben, which they interpret as a consequence of rift shoulder uplift and erosion during the Triassic (Nielsen *et al.* 1998). These conclusions are supported by this study. They also conclude from the structural setting (block-faulted half-grabens) that the region has been subject to crustal extension, which they relate to a late Carboniferous–Early Permian stretching episode. The Late Carboniferous–Early Permian stress field was dominated by east–west extension (Ziegler 1990). This makes it unlikely that WSW–ENE to west–east striking half-grabens would be initiated. On the other hand, pre-existing WSW–ENE to west–east striking faults would be prone to reactivation as strike-slip faults under appropriate stress in east–west extension. This setting is consistent with Late Carboniferous to Early Permian wrenching in North Central Europe (Arthaud & Matte 1977; Ziegler 1990) and along the Tornquist Fan (Thybo 1997).

As a second possibility, Abramovitz and Thybo (1999) proposed that the Palaeozoic intrabasement half-grabens might have been part of a Caledonian foreland basin on the Baltica plate and suggested that the Horn Graben developed in the original Caledonian foredeep. Although an inherited crustal weakness might act as a controlling factor for localized deformation during the formation of the Horn Graben, we find only sparse evidence of fault reactivation in the seismic data. This could be due to the fact that the strike direction of most Palaeozoic faults is perpendicular to the main Permo-Mesozoic faults of the Horn Graben. Nevertheless, some Palaeozoic faults may well have influenced the initiation of the Horn Graben.

Zhou and Thybo (1997) also interpreted the Palaeozoic deposits as originating from a foreland basin of the Caledonides as already suggested by Berthelsen (1992). They suggest, that the foreland basin was a wide flexural basin, which extended to 100–200 km north of the Caledonian Deformation Front. Zeck *et al.* (1988) document that minimum burial of the basement in Scania and on Bornholm is 5 km. Hence, sedimentation in a deep foreland basin is indicated. North of the Ringkøbing-Fyn High, a 5 km thick Palaeozoic sequence is indicated along the deep seismic refraction lines 1 and 2 of the EUGENO-S experiment (Thybo 1990, 2001; Thybo & Schönharting 1991) of which the main part is probably Lower to Middle Palaeozoic. However, some of the seismic data show, that the extensional phase postdates the compressive phase. This indicates that the foreland was first affected by compression and subsequently by extension. A possible reason for this configuration could be northward migration of the flexural bulge of a down-going Baltic plate during continued subduction which may have caused extension in areas that had been compressively deformed previously.

A third cause for the Palaeozoic graben-formation could be the post-orogenic extensional collapse of the Caledonides. Bearing in mind that some seismic lines image syntectonic deposits in the upper part of the graben-fill, these uppermost sediments should be pre-Devonian in age if Caledonian collapse led to graben subsidence. This is contradicted by the few wells reaching pre-Permian sediments (Nielsen & Japsen 1991) which are all of Carboniferous and thus Late Variscan age.

Alternatively, the Palaeozoic block-faulting might not be related to Caledonian collision at all and a younger, possibly Variscan extensional phase should be considered. Thus, a fourth possible candidate for the graben formation is Early Carboniferous extension prior to the Variscan closure of the Rheic ocean (Glennie & Underhill 1998). This explanation is backed by the observation of half-grabens similar to those in the area of study, in other parts of the North Sea, e.g. on the Mid North Sea High, where NW–SE trending normal faults indicate regional crustal extension during Lower Carboniferous times (Besly 1998). From our point of view, only additional well data penetrating the fill of the half-grabens can provide evidence on the exact timing and tectonic context of their activity.

The third important tectonic event that influenced the basement structure in the Danish North Sea, is the Permo-Triassic rifting which led to the formation of the Horn Graben and the Central Graben (Clausen & Korstgård 1993). This may have initiated the general subsidence of the Norwegian–Danish Basin. The westerly trending Palaeozoic fault systems are offset by northerly striking Mesozoic faults. The strike directions of the mapped faults indicate, that the original Palaeozoic extension was north–south to NNE–SSW directed and thus almost perpendicular to the Permo-Triassic phase for which NNW–SSE to NNE–SSW striking faults indicate an east–west directed extension (Ziegler 1990). The older Palaeozoic faults could have been reactivated during the Permo-Triassic phase as divergent strike-slip faults, thus providing transfer zones between the main north–south oriented normal faults (Thybo 1997).

Traces of Late Cretaceous–Early Tertiary inversion have not been found to have affected the basement, but detailed studies of the Mesozoic–Cenozoic sequence would be necessary to rule out partial reverse reactivation of Early Mesozoic normal faults during the Alpine compressional phases.

Conclusions

The near basement reflection is traced offshore from Denmark to the south into the southern Horn Graben and westwards to the eastern shoulder of the Central Graben. This characteristic reflection indicates that Baltica upper crust extends further south and west than had been assumed previously. The present distribution of Palaeozoic sediments shows uniform thickness throughout the study area within the different half-grabens, except for the East North Sea High and the rift shoulders of the Horn Graben and the Central Graben, where large parts of the Palaeozoic sequence have been eroded at pre-Permian and Mesozoic times.

The near-basement reflection in the Danish North Sea area exhibits the imprints of at least two significant Palaeozoic tectonic events and of a Permian–Mesozoic tectonic event. The observed Palaeozoic structures show compressive deformation of sediments of possible 'pre-Caledonian' age overprinted by a younger Palaeozoic extensional event. Compressive deformation appears to have been very localized, as it is restricted to a narrow zone in the south of the study area which is about 20 km wide, and to the area where the MONA LISA lines intersect. In the southeastern part, the south to SW-dipping reflections may be interpreted as thrust planes or, alternatively, as bedding planes within rotated half-grabens. In the central part of the study area, compressive deformation is indicated along the northern segment of MONA LISA line 2, with thrust planes dipping southward and along the eastern segment of MONA LISA line 3. To explain the Palaeozoic compressive structures we propose that Baltican upper crust was involved in foreland deformation.

The dominant structural feature in the Palaeozoic sequence is block-faulting along WSW–ESE to west–east striking normal faults indicating a north–south to NNE–SSW directed extension. This caused the formation of numerous Palaeozoic half-grabens. The expression of compressive deformation corresponds to far-field foreland deformation but the extensional phase postdates the compressive phase. Therefore, a foreland setting related to Caledonian collision can be ruled out for the extensional phase, which, however, may be related to post-orogenic collapse of the Caledonides. The structure of the Palaeozoic half-grabens on the East North Sea High indicates that the high is an inverted Palaeozoic rift located between the younger Permo-Mesozoic rifts of the Horn Graben and the Central Graben. The timing of this Palaeozoic extensional phase remains speculative. We can conclude that it predates the Rotliegend unconformity and affected the deposition of the uppermost layers in the half-grabens (the upper part of the graben-fill represent syn-tectonic deposits), the age of which could be Carboniferous.

The main Permo-Mesozoic event affecting the basement of the Danish North Sea is +/– east–west directed extension, which led to the formation of the NNW-SSE to NNE-SSW striking Horn Graben and Central Graben. These fault systems offset all older structures. Some of the Palaeozoic normal faults may have been reactivated as transfer zones between the graben segments. We found no obvious indications for Late Cretaceous–Early Tertiary inversion of the basement, but further detailed studies of the Mesozoic–Cenozoic sequence are required to analyse this topic.

The study has been possible due to financial support from the EU-TMR network PACE. We gratefully acknowledge extensive discussions with our colleagues L. Nielsen and A. Ross. We further thank G. Fernandez-Viejo, for allowing us to use the results from the pre-stack depth migrated part of line MONA LISA 4 in the TWT-map generation (Fernandez-Viejo *et al.* 2000). The Geological Survey of Denmark and Greenland (GEUS) provided the industrial seismic data. We are very grateful to Rolf Meissner and Asger Berthelsen who both very constructively reviewed the paper and helped to improve it.

References

ABRAMOVITZ, T. & THYBO, H. 1999. Pre-Zechstein Structures around the MONA LISA deep seismic lines in the southeastern North Sea. *Bulletin of the Geological Society of Denmark*, **45**, 99–116.

ABRAMOVITZ, T. & THYBO, H. 2000. Seismic images of Caledonian, lithosphere-scale collision structures in the southeastern North Sea along Mona Lisa Profile 2. *Tectonophysics*, **317**, 27–54.

ABRAMOVITZ, T., THYBO, H. & MONA LISA Working Group, 1998. Seismic structure of the crust and upper mantle across the Caledonian Deformation Front along MONA LISA profile 1 in the southeastern North Sea. *Tectonophysics*, **288**, 153–176.

ARTHAUD, F. & MATTE, P. 1977. Late Paleozoic strike-slip faulting in southern Europe and Northern Africa: results of right lateral shear zone between the Appalachians and the Urals. *Geological Society of America Bulletin*, **88**, 1305–1320.

BERTHELSEN, A. 1992. Mobile Europe. *In*: BLUNDELL, D., FREEMAN, R. & MUELLER, S (eds) *A continent revealed: The european geotraverse*. Cambridge University Press, Cambridge, 11–32.

BESLY, B. M. 1998. Carboniferous. *In*: GLENNIE, K. W. *Petroleum geology of the North Sea. Basic concepts and recent advances. Fourth Edition*. Blackwell Science Ltd. London, UK, 104–136.

CARTWRIGHT, J. 1990. The structural evolution of the Ringkøbing-Fyn High. *In*: BLUNDELL, D. J. & GIBBS, A. D. (eds) *Tectonic evolution of the North Sea rifts*. Publication 181 of the International Lithosphere Programme, Oxford University Press, 2000–216.

CLAUSEN, O. R. & HUUSE, M. 1999. Topography of the Top Chalk surface on- and offshore Denmark. *Marine and Petroleum Geology*, **16**, 677–691.

CLAUSEN, O. R. & KORSTGÅRD, J. A. 1993. Faults and faulting in the Horn Graben Area, Danish North Sea. *First Break*, **11**, No 4, 127–143.

DYNAMIC GRAPHICS Inc., 2001. Earth Vision, Release 5.1., Alameda, California.

FERNANDEZ-VIEJO, G., THYBO, H. & LAIGLE, M. 2000. Reprocessing of MONA LISA deep seismic reflection data: The Palaeozoic rift in the North Sea. *In*: THYBO, H., ABRAMOVITZ, T., NIELSEN, L., ROSS, A., FERNANDEZ-VIEJO, G., HUSEBYE, E., BERG, B., HESTHOLM, S., KALAND, N., SNARTEMO, D. & FEDERENKO, Y. V. (eds) *The Millenial 9th International symposium on deep seismic profiling of the continents and their margins*, Ulvik, Norway, p. 45.

FROST, R. T. C., FITCH, F. J., & MILLER, J. A. 1981. The age and nature of the crystalline basement of the North Sea Basin. *In*: ILLING, L. V. & HOBSON, G. D. (eds) *Petroleum geology of the continental shelf of North West Europe*. Heyden and Son Ltd. For Institue of Petroleum London, 43–57.

GEMMER, L., NIELSEN, S. B., HUUSE, M. & LYKKE-ANDERSEN, H. 2002. Post-mid Cretaceous eastern North Sea evolution inferred from 3D thermo-mechanical modelling. *Tectonophysics*, **350**, 315–342.

GLENNIE, K. W. & UNDERHILL, J. R. 1998. Origin Development and Evolution of Strucutral Styles. *In*: GLENNIE, K.W. (ed.) *Petroleum Geology of the North Sea. Basic concepts and recent advances. Fourth Edition*. Blackwell Science Ltd. London, UK, 42–84.

HANSEN, D. L., NIELSEN, S. B., & LYKKE-ANDERSEN, H., 2000. The post-Triassic evolution of the Sorgenfrei-Tornquist-Zone – results from thermo-mechanical modelling. *Tectonophysics*, **328**, 245–267.

JAPSEN, P. 1998. Regional velocity-depth anomalies, North Sea Chalk: A record of overpressure and Neogene uplift and erosion. *American Association of Petroleum Geologists, Bulletin*, **82**, 2031–2074.

LAIGLE, M., THYBO, H. & BAYER, U. 2000. New seismic images of Caledonian collision structures beneath the North Sea Basement High (Denmark). *In*: THYBO, H., ABRAMOVITZ, T., NIELSEN, L., ROSS, A., FERNANDEZ-VIEJO, G., HUSEBYE, E., BERG, B., HESTHOLM, S., KALAND, N., SNARTEMO, D. & FEDERENKO, Y. V. (eds) *The Millenial 9th International Symposium on Deep Seismic Profiling of the Continents and their Margins*, Ulvik, Norway, p. 39.

LASSEN, A., THYBO, H. & BERTHELSEN, A. 2001. Reflection seismic evidence for Caledonian deformed sediments above Sveconorwegian basement in the southwestern Baltic Sea. *Tectonics*, **20**, 268–276.

MICHELSEN, O. & NIELSEN, L. H. 1993. Structural development of the Fennoscandian Border Zone, offshore Denmark. *Marine and Petroleum Geology*, **10**, 124–134.

MONA LISA Working Group, 1997. Closure of the Tornquist Sea: Constraints from MONA LISA deep seismic reflection data. *Geology*, **25**, 1071–1074.

NIELSEN, L. H. & JAPSEN, P. 1991. Deep wells in Denmark 1935–1990, *Danish Geological Survey*, **A31**, 179 pp.

NIELSEN, L. H., KLINKBY, L. & BALLING, N. 1998. Seismic evidence for deep Paleozoic sedimentary units in the Ringkøbing-Fyn High offshore Denmark. *Bulletin of the Geological Society of Denmark*, **45**, 1–10.

POULSEN, C. 1969. The Lower Cambrian from Slagelse no. 1, Western Zealand. *Danmarks Undersogelse, II Series*, **93**, 5–29.

TANNER, B. & MEISSNER, R. 1996. Caledonian deformation upon southwest Baltica and its tectonic implications: Alternatives and consequences. *Tectonics*, **15**, No. 4, 803–812.

THYBO, H. 1990. A seismic velocity model along the EGT profile – from the North German Basin into the Baltic shield, *In*: Freeman, R., Giese, P. & Mueller, S. (eds) *The european geotraverse: integrative studies*, European Science Foundation, Strasbourg, 99–108.

THYBO, H. 1997. Geophysical characteristics of the Tornquist Fan area, northwest TESZ: Indications of Late Carboniferous to Early Permian dextral transtension. *Geological Magazine*, **134**, 597–606.

THYBO, H. 2001. Crustal structure along the EGT profile across the Tornquist Fan interpreted from seismic, gravity and magnetic data. *Tectonophysics*, **334**, 155–190.

THYBO, H. & SCHÖNHARTING, G. 1991. Geophysical evidence for Early Permian igneous activity in a transtension environement, Denmark. *Tectonophysics*, **189**, 193–208.

VEJBAECK, O. V. 1990. The Horn graben and its relationship to the Oslo Graben and the Danish basin. *Tectonophysics*, **178**, 29–49.

VEJBAECK, O. V. 1997. Dybe strukturer i danske sedimentære bassiner. *Geologisk Tidsskrift*, **4**, 31 pp.

ZECK, H. P., ANDRIESSON, P. A. M., HANSEN, K., JENSEN, P. K., RASMUSSEN, B. L. 1988. Palaeozoic palaeo-cover of the southern part of the Fennoscandian Shield; fission track constraints. *Tectonophysics*, **149**, 61–66.

ZIEGLER, P. A. 1990. *Geological Atlas of Western and Central Europe*, 2nd edn. Shell International Petroleum Mij., B.v., distributed by Geological Society Publishing House Bath, 239 pp.

ZHOU, S. & THYBO, H. 1997. Pre-Zechstein geology of the southeast North Sea, offshore Denmark – A geophysical perspective. *First Break*, **15**, 387–395.

A compressional wedge model for the Lower Palaeozoic Anglo-Brabant Belt (Belgium) based on potential field data

MANUEL SINTUBIN[1] & MICHEL EVERAERTS[2]

[1]*Structural Geology & Tectonics Group, Katholieke Universiteit Leuven, Redingenstraat 16, B-3000 Leuven, Belgium e-mail: manuel.sintubin@geo.kuleuven.ac.be*

[2]*Observatoire Royal de Belgique, 3 Avenue Circulaire, B-1180 Uccle, Belgium*

Abstract: Structural field observations only allow kinematic inferences to be made for the southern extremity of the predominantly concealed Lower Palaeozoic Anglo-Brabant Belt. On the other hand potential field data (aeromagnetic and Bouguer anomaly maps), enable the field of vision to be enlarged. They not only corroborate the kinematics derived from structural field observations; they also reveal possible key players in the late Silurian to early Devonian deformation event causing the observed orogenic architecture.

This integrated approach has led to a compressional wedge model for the Brabant Massif. It is also proposed that crystalline basement blocks, most probably of Precambrian age, controlled the kinematics. Finally, it is suggested that the development of rift or pull-apart basins in a transtensional intracontinental setting during the Cambrian may have been crucial in the subsequent deformation history of the Anglo-Brabant Belt.

In predominantly concealed orogenic belts it is hard to obtain enough structurally relevant information to provide a consistent image of the kinematics of the deformation events, let alone to propose an integrated tectonic model. This is true for the Anglo-Brabant Belt, which runs from northern England into Belgium (Pharaoh *et al.* 1993, 1995; Winchester *et al.* in press). Contrary to its counterpart, cropping out in Wales (Woodcock *et al.* 1988) and the Lake District (Soper *et al.* 1987), the Anglo-Brabant Belt is completely concealed except for its southeastern extremity in the Brabant Massif in Belgium. The only tools available to overcome the limitations inherent in the poor degree of exposure are potential field data, which enable the large-scale and three-dimensional structural architecture of the concealed basement to be envisaged.

In the southeastern extremity of the Brabant Massif a first integration of structural field observations, made in the limited outcrop area, and potential field data, lead to a number of kinematic constraints (Sintubin 1999). To what extent is the inferred kinematic model representative for the entire Brabant Massif, or even for the entire Anglo-Brabant Belt? In this paper, we attempt to extend the kinematic model for the southeastern part of the Brabant Massif (Sintubin 1999) to the entire Brabant Massif through an advanced and detailed analysis of the potential field data (Everaerts 2000). The resulting compressional wedge model is put into the more regional perspective of the tectonic history of the Eastern Avalonia microcontinent. Throughout the paper we present alternative interpretations, as caution should always be paid in interpreting potential field data, and we note that interpreting potential field data without taking the geology into account remains a questionable enterprise.

Tectonic framework

The Brabant Massif forms the southeastern extremity of the Anglo-Brabant Belt (Pharaoh *et al.* 1993, 1995; Winchester *et al.* in press), the predominantly concealed eastern branch of the slate belt moulded around the Neoproterozoic Midlands Microcraton (Fig. 1). Other parts of this slate belt crop out in the Lake District (Soper *et al.* 1987), Wales (Woodcock *et al.* 1988) and Southern Ireland.

This low-grade metamorphosed slate belt reflects a late Silurian–early Devonian deformation event (Soper & Hutton 1984). Because the accretion of Eastern Avalonia to Baltica and Laurentia is assumed to have occurred in late Ordovician to early Silurian time (Cocks *et al.* 1997; Vecoli & Samuelsson 2001; Verniers *et al.* 2002), the former event has been ascribed to a post-accretionary intracontinental accommodation, probably related to the closure of the Rheic Ocean along the southern margin of Eastern Avalonia (Pickering & Smith 1995; Cocks *et al.* 1997; Rey *et al.* 1997).

As shown on existing lithostratigraphical subcrop maps (Legrand 1968; De Vos *et al.* 1993*b*), the Brabant Massif is characterized by

From: WINCHESTER, J. A., PHARAOH, T. C. & VERNIERS, J. 2002. *Palaeozoic Amalgamation of Central Europe*. Geological Society, London, Special Publications, **201**, 327–343. 0305-8719/02/$15.00

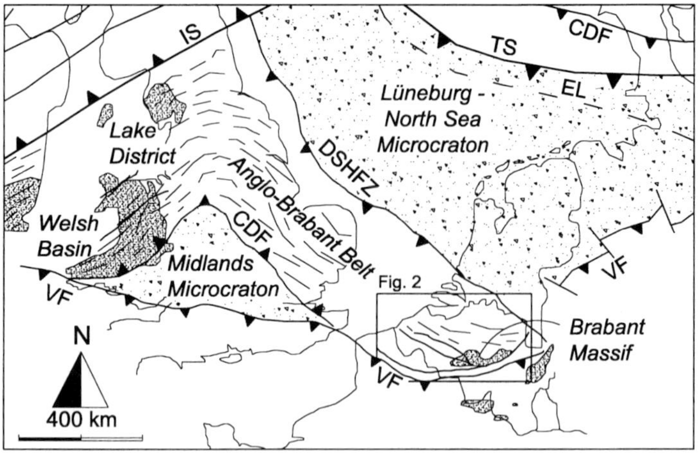

Fig. 1. Tectonic setting of the Brabant Massif at the southeastern extremity of the Anglo-Brabant Belt (after Blundell *et al.* 1992; Pharaoh *et al.* 1995; Pharaoh 1999; Sintubin 1999, Winchester *et al.* in press). IS, Iapetus Suture; TS, Thor Suture; EL, Elbe Lineament; CDF, Caledonian deformation front; VF, Variscan thrust front; DSHFZ, Dowsing-South Hewett fault zone. Hatched areas are the outcrop areas of the British-Belgian Caledonides.

an overall NW–SE structural grain (Fig. 2). To the east this structural grain curves into a more ENE–WSW direction. The massif has a symmetrical disposition with a Cambrian core flanked on both sides by Ordovician–Silurian strata. In the south the Ordovician–Silurian strata form part of a deformed foreland basin (Van Grootel *et al.* 1997; Verniers *et al.* 2002). It is assumed that the northern subcrop area of Ordovician–Silurian strata reflects a deformed foreland basin. To the NW both foreland basins converge to form the Anglian Basin (Pharaoh *et al.* 1995). This symmetrical disposition has been interpreted as an anticlinal culmination (Van Grootel *et al.* 1997; Mansy *et al.* 1999). An alternative model, primarily based on an integration of structural field observations and an advanced analysis of the potential field data, will be presented in this paper.

Kinematic constraints

The very limited outcrop area along the southern margin of the Brabant massif only enables determination of a, possibly biased, kinematic model for just a small part of the massif (Sintubin 1999). Detailed descriptions of the kinematic constraints can be found in Sintubin (1997*a*, 1999), Sintubin *et al.* (1998), Debacker (1999, 2001), Debacker *et al.* (1999, 2001). These kinematic constraints only concern a unique deformation event, in which cleavage and fold development is largely coeval, and which is currently considered to have occurred during late Silurian – early Devonian times (Verniers & Van Grootel 1991; Van Grootel *et al.* 1997; Debacker 2001; Verniers *et al.* 2002).

Three main lithostructural domains can be recognized in the exposed massif (Sintubin 1997*a*, 1999). To the north lies a strongly deformed and exhumed 'steep belt', composed of predominantly Cambrian low-grade metamorphosed metasedimentary series. This 'steep belt' has a pervasive, subvertical cleavage with transposed strata and an arcuate cleavage trajectory. In the field, the 'Asquempont fault zone' (Legrand 1968) has been commonly considered the main reverse fault that delimits the 'steep belt' to the SW. Following a detailed structural survey, Debacker (2001) demonstrated that the fault zone has a completely different origin and can no longer be considered as the superficial trace of the main tectonic break delimiting the 'steep belt' to the SW (see also Verniers *et al.* 2002). To the south of the 'steep belt' is a deformed Ordovician–Silurian foreland basin (Van Grootel et al. 1997). The changing nature of the cleavage-fold transection along the arcuate belt in the foreland basin domain infers the impingement of a 'rigid' body (Soper *et al.* 1987). In the absence of any transection in the central part of the southern arcuate belt, an N10E-orientated bulk shortening direction is assumed. SW of the 'steep belt' is an intermediate domain marked by fold-thrust kinematics (Debacker 1999).

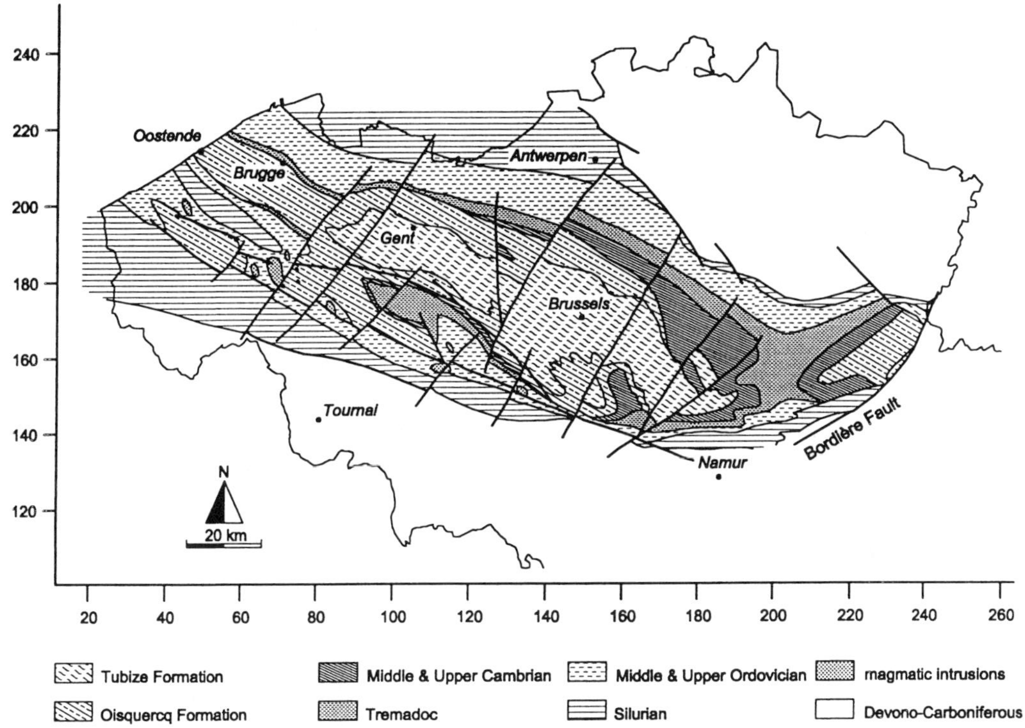

Fig. 2. Lithostratigraphical subcrop map of the Brabant Massif (after Legrand 1968; De Vos *et al.* 1993*b*). Reference system is the Belgian Lambert (in kilometres).

The proposed indenter has been identified with the 'steep belt'. Moreover, during impingement an eastward escape within the 'steep belt' can be assumed, causing its internal arcuate cleavage trajectory. The impingement of the 'steep belt' in the southern domain and the eastward escape within the 'steep belt', causing the arcuate structural grain in both domains, suggests the presence of a rigid buffer SW of the 'steep belt', most probably underlying the intermediate domain. The presence and nature of this rigid buffer cannot be derived from field data, but has been determined based on potential field data (Sintubin 1999).

To the south and SW several observations (see Sintubin 1999) indicate a gradual decrease of deformation up to undeformed Silurian strata conformably overlain by lower Devonian strata (see also Verniers *et al.* 2002), inferring the presence of a deformation front. Moreover, this deformation front, coincides with the transition of anchizonal metamorphism to diagenesis (Van Grootel *et al.* 1997). As in other parts of the belt (Woodcock 1991), this deformation front probably coincides with the edge of a Neoproterozoic cratonic domain (Midlands Microcraton, see Fig. 1), but this awaits confirmation using potential field data.

Acquisition and treatment of potential field data

The Bouguer anomaly map of the study area was prepared using the gravity database of the Royal Observatory of Belgium which contains more than 250 000 data points covering the entire territory of Belgium. On the study area the coverage is 1/km^2 except for the northern part with a coverage of 1/16 km^2. The gravity values onshore were calculated using a density of 2.1 g cm^{-3}. This value corresponds to the density of the Meso-Cenozoic cover on top of the Brabant Massif. A higher density would create artefacts due to an overcorrection of the topography in some areas. Offshore values consist of free air anomalies. All gravity data are measured with respect to the gravity datum of Uccle 1976 and theoretical gravity was computed using the 1980 geodetic reference system formula

(International Association of Geodesy). All data are expressed in terms of *x* and *y* in a planar projection specific for Belgium, called the Belgian Lambert, converted from latitude–longitude. A gridding procedure was performed using the Kriging method (Surfer Golden Software). The grid interval is 1 km. A search radius of 30 km with a linear weighting coefficient is used to avoid biasing in poorly covered areas.

A high-resolution aeromagnetic survey was commissioned in 1994 by the geological surveys of Belgium and Luxembourg and carried out by GEOTERREX. Acquisition with a caesium-vapour magnetometer was digital with a recording interval of 0.1 s, giving a linear resolution of 7 m. Flight lines were 0.5 to 1 km apart and oriented north–south, and tie lines were 10 km apart. Height of flight was 120 m above ground level. This made acquisition above Brussels impossible for safety reasons. After the diurnal correction a constant value of 47950 was subtracted and the international geomagnetic reference field 1985, updated to 1994/10, was also subtracted. The onshore and offshore data were combined and levelled. A 1 km grid was calculated in the same projection as the gravity map. A reduction to the pole was calculated using an inclination of 66° and a declination of 2°W.

Transformation of potential field data in some of their components has now become a very popular tool to obtain more information from the data. All these transformations have been treated in a Fourier domain. The data are first converted from space domain to Fourier domain and mathematical transformations are applied. Afterwards they are transferred again towards the space domain. More information on the filtering techniques can be found in Blakeley (1995).

Analysis of potential field data

Recently, attempts to link structural field observations with the aeromagnetic and Bouguer anomaly maps were made by Sintubin (1997*b*) and Sintubin *et al.* (1998), enabling a kinematic model for the southeastern part of the Brabant Massif to be proposed (Sintubin 1999). Based exclusively on potential field data, Mansy *et al.* (1999) derived a different model for the kinematics in the southeastern part of the Brabant Massif. However, we demonstrate that the potential field data corroborate the kinematic model of Sintubin (1999), which is largely based on structural field observations. A detailed analysis of the potential field data also permits definition of additional tectonic elements and the extension of the tectonic model to the entire, predominantly concealed, Brabant Massif.

Aeromagnetic anomaly map reduced to the pole

The most prominent feature on the total field aeromagnetic anomaly map (Chacksfield *et al.* 1993; Mansy *et al.* 1999) is the pronounced NW–SE-trending positive anomaly in the central part of the Brabant area (Sintubin 1997*b*) (Fig. 3a). This anomaly largely coincides with the Cambrian 'steep belt'. The highest relief and strongest density of aeromagnetic lineaments is situated along its southwestern boundary, which itself is bounded by a pronounced gradient. The latter has commonly been linked to the superficial trace of the Asquempont fault zone (Legrand 1968). Taking into account the findings of Debacker (2001) this assumption is no longer valid (see also Verniers *et al.* 2002). Therefore, to make a clear distinction, the pronounced gradient is named the Asquempont Lineament.

The southern extremity of this boundary anomaly shows a lobate and step-like morphology, possibly caused by a dextral displacement along NW–SE-trending features (Sintubin 1997*b*, 1999; Sintubin *et al.* 1998). The southern apex of this boundary anomaly also coincides with the area of no transection, and thus of orthogonal convergence, in the southern domain (Sintubin 1997*b*, 1999). This entire southwestern boundary anomaly has therefore been interpreted as a dextral transpressional shear zone (Sintubin *et al.* 1998; Sintubin 1999). The coincidence of the southern apex of this shear zone with the area of orthogonal convergence in the southern domain indicates that the Cambrian 'steep belt' may have been the 'rigid' indenter.

South of Brussels, within the Cambrian 'steep belt' the aeromagnetic lineaments show an arcuate pattern, remarkably parallel to the internal cleavage trajectory (Sintubin 1999). This particular pattern is explained by a lateral, eastward, escape (Sintubin 1997*b*, 1999).

In a field survey (De Vos *et al.* 1993*a*) the lower Cambrian Tubize formation has been identified as the most magnetic formation. Therefore the aeromagnetic relief has always been identified with the subcrop of the Tubize formation. However, the magnetic signature of the Tubize formation depends on its lithological composition. In general, only slaty horizons show pronounced magnetic susceptibility; that of coarser-grained horizons is similar to other lithologies in the Brabant Massif. Furthermore,

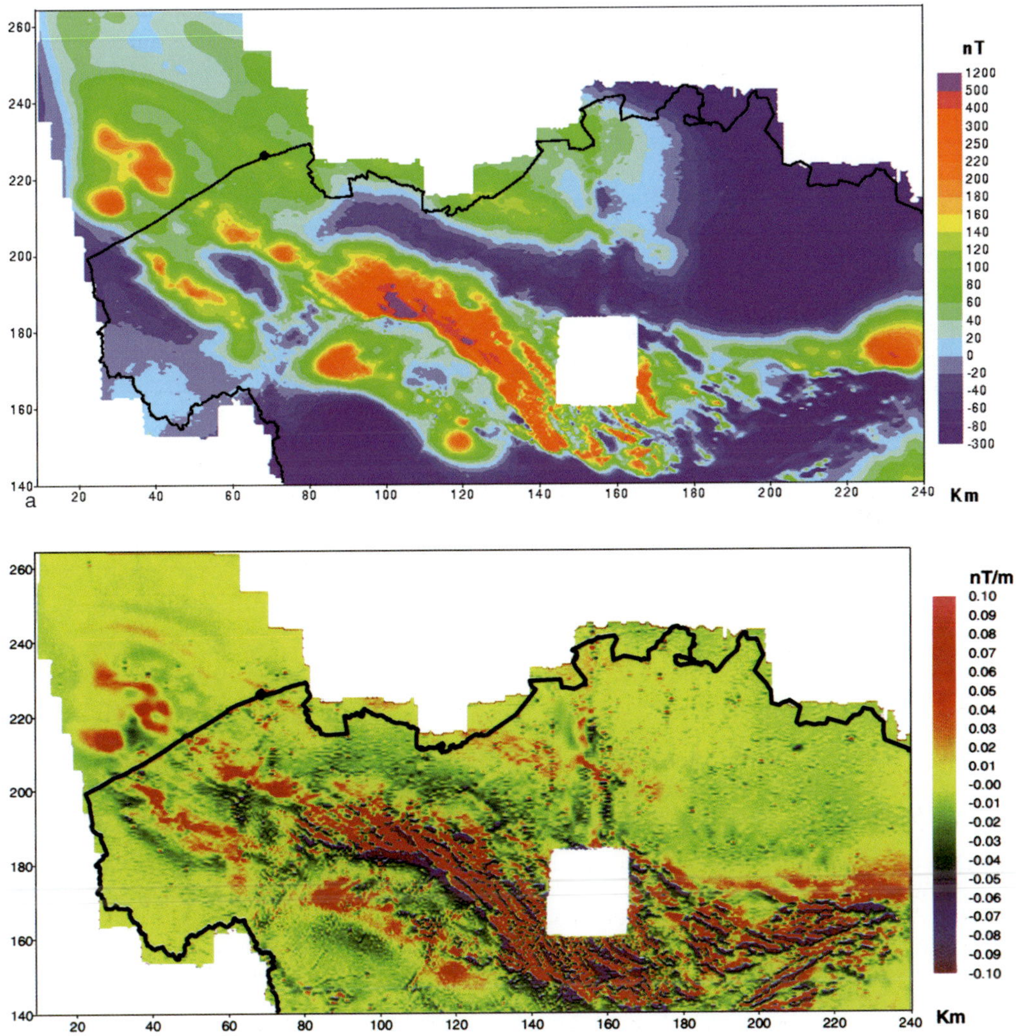

Fig. 3. (a) aeromagnetic anomaly map reduced to the pole (in nT); (b) vertical gradient map of the aeromagnetic field (in nT/m). Reference system is the Belgian Lambert (in kilometres).

other lithostratigraphical horizons not observed in outcrop or subcrop, possibly Precambrian in age (Lee *et al.* 1990) may contribute to the magnetic signal.

To the south in the Ardennes, long-wavelength anomalies are assumed to be related to the presence of a cratonic basement (Busby *et al.* 1993; Chacksfield *et al.* 1993). This cratonic basement is considered as the southwestern continuation of the Midlands Microcraton (Sintubin 1999). Underneath the central part of the Brabant Massif such a cratonic basement is probably absent.

Vertical gradient of the aeromagnetic anomaly map reduced to the pole

On the aeromagnetic vertical gradient map (Fig. 3b) the internal structure of the central part of the Brabant Massif becomes even more obvious. Within single anomalies a series of linear features can be identified. These individual aeromagnetic lineaments have been interpreted to represent the cores of anticlinal structures (e.g. De Vos *et al.* 1993*b*; Mansy *et al.* 1999). However, taking into account the lithological dependence of the magnetic susceptibility and

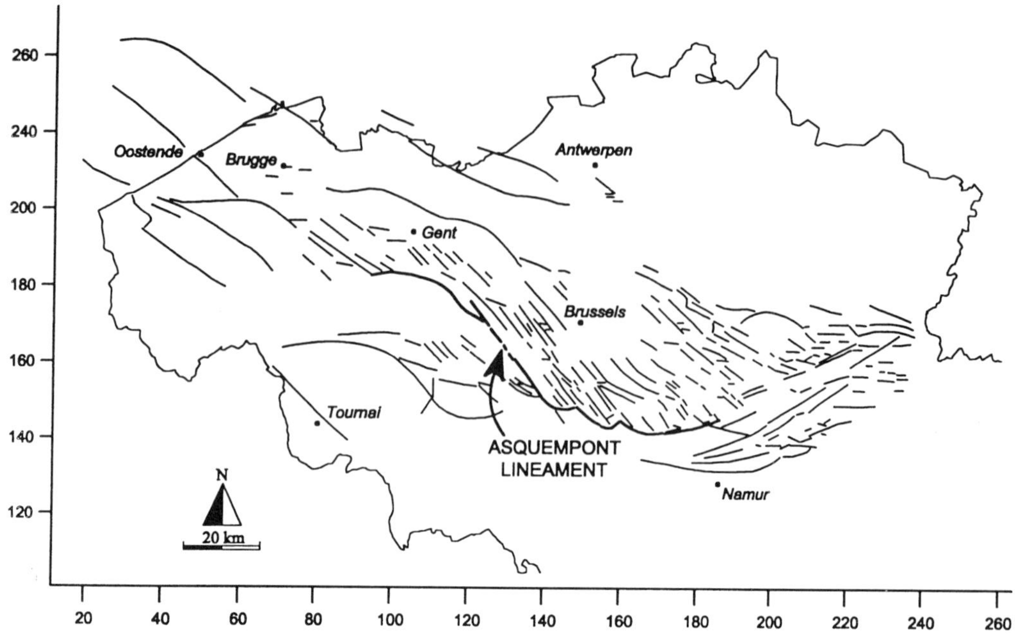

Fig. 4. Line drawing of the aeromagnetic lineaments as apparent on the aeromagnetic anomaly maps. The Asquempont Lineament is defined as the major gradient at the southwestern limit of the aeromagnetic high. Reference system is the Belgian Lambert (in kilometres).

the lithological characteristics of the Tubize formation, there may be alternative explanations for these individual lineaments (Everaerts *et al.* 1998):

(1) a purely sedimentary explanation reflecting the nature of the Tubize formation, consisting of an alternation of fine-grained silty and clayey sequences, rich in metamorphic magnetite, and coarse-grained sequences, poor in magnetite;
(2) a purely geometric explanation, reflecting limbs (instead of cores) of large-scale folds;
(3) a kinematic explanation, reflecting high-strain zones, characterized by a transposition of the strata and a pervasive cleavage development.

The aeromagnetic lineaments are up to 20 km long, follow the arcuate cleavage trajectory and are probably first-order features. We therefore interpret them as reflecting high-strain zones, characterized by strong transposition and a pervasive cleavage development. The development of a pervasive cleavage may be responsible for a passive concentration of magnetite in these high-strain zones, thus generating prominent magnetic lineaments.

Residual versus regional components of the aeromagnetic field

Total field aeromagnetic anomalies may be the result of an overlap of two or more geological features of variable dimensions and/or depths. A spectral factorization enables the distinction of so-called 'regional' and 'residual' components. The separation of potential field data into its components has always been a difficult task (Gupta & Ramani 1980). In recent years, the application of digital filters to achieve this residual-regional separation has become increasingly popular amongst geophysicists. Although in many cases it is purely subjective (Nettelton 1976) the 'regional-residual' definition is still used. Therefore it is important to define objective separation criteria. In spectral factorization, different wavelength populations can be distinguished from the power spectrum of the aeromagnetic map thereby enabling the definition of objective separation criteria. These limits are subsequently used to design filters to separate these components or to enhance or suppress specific wavelengths associated with particular geological features (Everaerts *et al.* 1998).

If applied to the central part of the Brabant

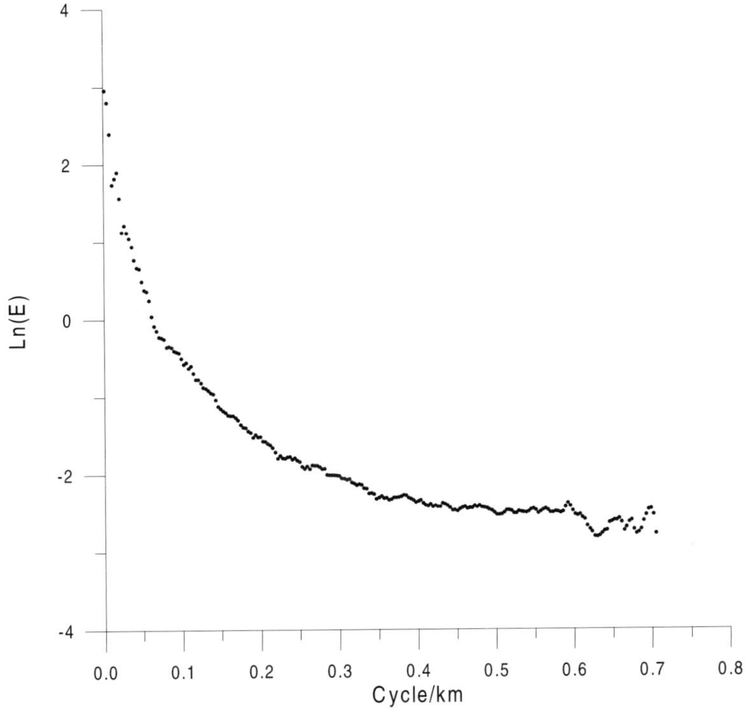

Fig. 5. Frequency spectrum of the aeromagnetic field of the Brabant Massif.

Massif, the power spectrum shows no clear break (Fig. 5) revealing the vertical architecture of the central part of the Brabant Massif. Despite this, two wavelength populations can be identified. In between, a limit, separating the regional and residual components of the aeromagnetic field, is applied arbitrarily. This limit, at 0.2 cycles/km, corresponds to 5 km. Maps are thus generated of the 'residual' high-frequency part of the spectrum (wavelengths less than 5km), reflecting small-scale and/or relatively superficial features, and of the 'regional' low-frequency part of the spectrum (wavelengths more than 5km), reflecting large-scale and/or relatively deep features (Fig. 6).

Both in the case of the 'regional' and 'residual' anomaly maps of the central part of the Brabant Massif, a perfect superposition of the anomalies occurs at different crustal levels. If these components are interpreted as reflecting features at different depths, this perfect superposition infers a predominantly vertical structural architecture of the central part of the Brabant Massif. Major subhorizontal features, such as large-scale nappe structures can be excluded in this part of the massif (Anthoine & Anthoine 1943; Mansy *et al.* 1999).

Bouguer anomaly map

An obvious feature on the Bouguer anomaly map in the Brabant area is an elongated 'chain' of negative anomalies (Fig. 7a). Several observations with respect to these anomalies are important in the overall tectonic model for the Brabant Massif (Sintubin 1999):

(1) the two main negative anomalies together form an elongated, NW–SE-trending feature; these two anomalies seem dextrally displaced with respect to one another along a NW–SE-trending feature (Fig. 8);

(2) the same displacement is apparent for the less pronounced and smaller negative anomalies, situated more to the east; particularly in the latter case the linear features, along which this dextral displacement seems to have taken place, coincide with aeromagnetic lineaments (Figs 4 & 8);

(3) the entire aeromagnetic boundary anomaly forming the southwestern edge of the Cambrian 'steep belt', and interpreted as a dextral transpressional shear zone, coincides with the northeastern limit of the main negative Bouguer anomaly;

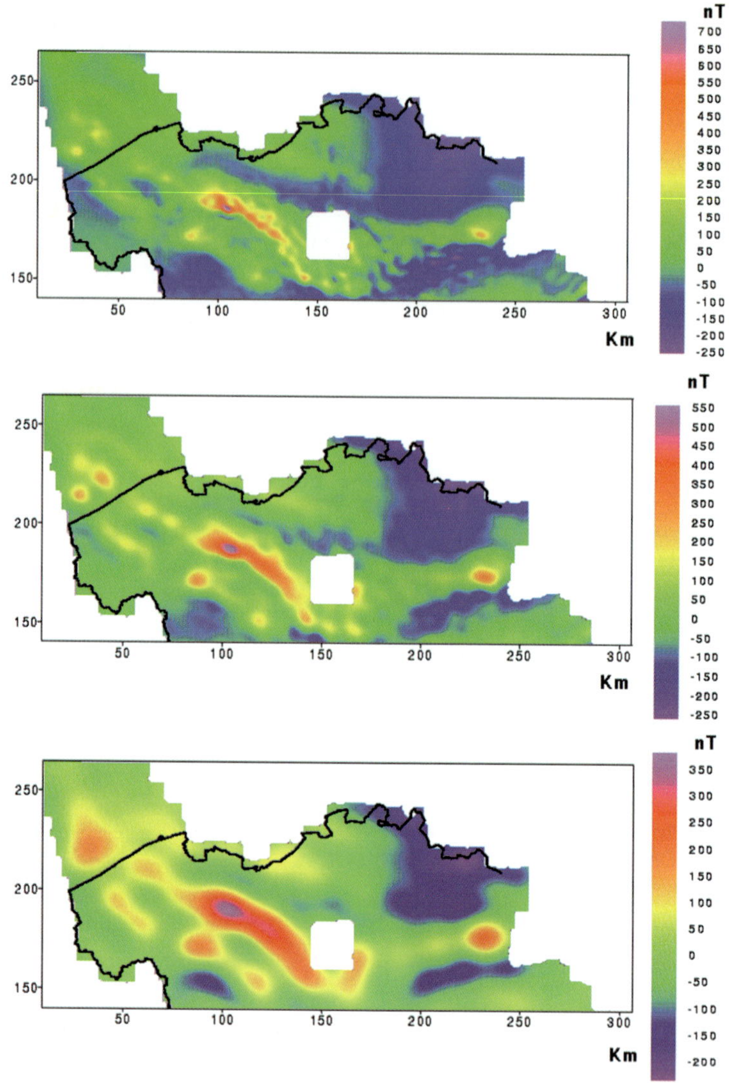

Fig. 6. Regional component of the aeromagnetic field. The maps represent different low-pass filters: 5km for the upper one, 10 km for the middle one, 20 km for the lower one. The aim is to see if the anomaly changes position when gradually neglecting a larger part of the superficial signal (5, 10 & 20 km) and only taking into account the deeper signal. Reference system is the Belgian Lambert (in kilometres).

(4) structural field observations in the area on top of the negative Bouguer anomaly indicate the presence of thin-skinned thrusting (Debacker 1999), defining the intermediate domain in between the central 'steep belt' and the deformed foreland basin.

All these features indicate a clear genetic relationship between the emplacement of the main aeromagnetic boundary anomaly and the negative Bouguer anomaly. Both also seem to show a similar kinematic image (dextral displacement). The geological feature causing the negative Bouguer anomaly not only controlled the emplacement of the geological feature causing the aeromagnetic anomaly but also took part in the deformation event, which eventually resulted in the particular disposition on both aeromagnetic and Bouguer anomaly maps.

Everaerts *et al.* (1996) modelled the main

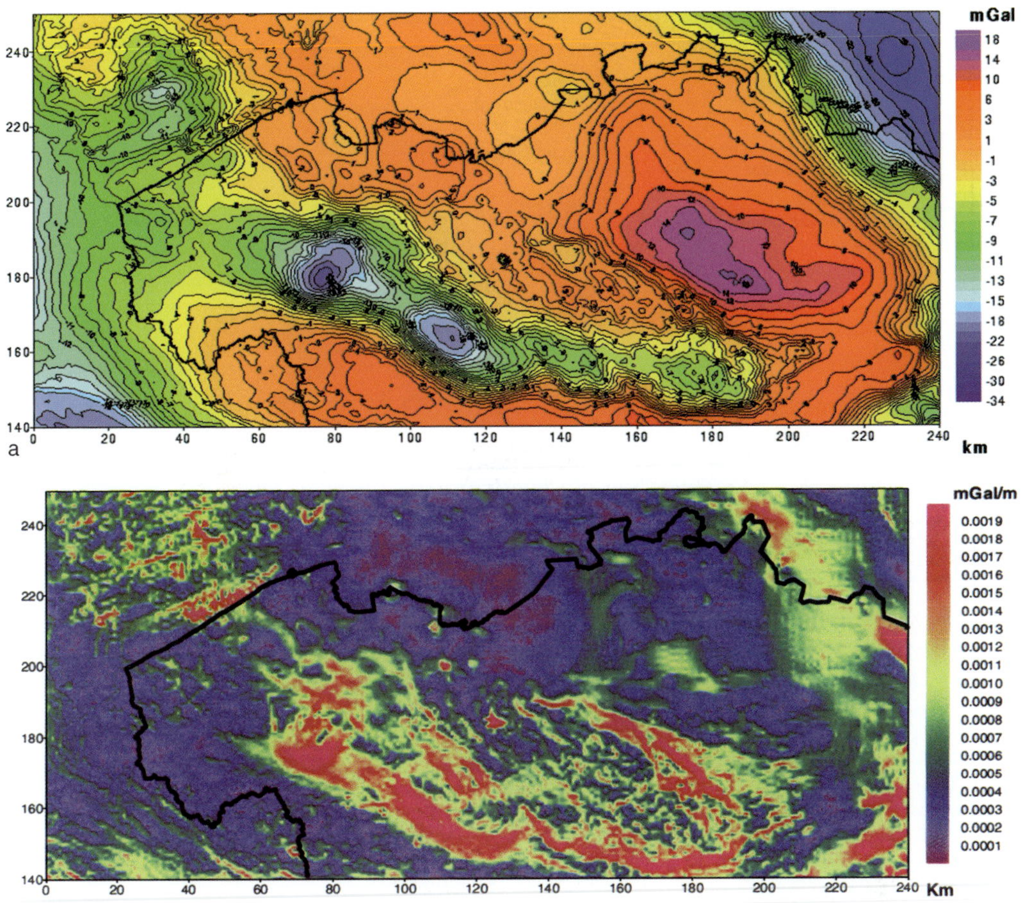

Fig. 7. (a) Bouguer anomaly map (in mGal); (b) the horizontal gradient map of the Bouguer anomaly (in mGal/m). Reference system is the Belgian Lambert (in kilometres).

negative Bouguer anomaly as a steep-sided, probably fault-bounded, body composed of low-density (2.63 g cm^{-3}) material at a minimum depth of 2.5 km. A granitic composition has been proposed. Currently this negative Bouguer anomaly is interpreted to reflect the presence at depth of an elongated, arcuate granitic batholith, considered the source of the Ordovician magmatism in the southwestern part of the Brabant Massif (De Vos 1997; Mansy *et al.* 1999) and the cause of oroclinal bending of the southern domain (Mansy *et al.* 1999). However, the Ordovician magmatic rocks present in outcrop have a higher density than the inferred density of the granitic body, indicating that the magmatic rocks in outcrop probably do not originate from differentiation in the postulated granitic batholith (see also Verniers *et al.* 2002). Furthermore, the Ordovician magmatic rocks have an intermediate (andesitic to dacitic) composition (see André 1991; Verniers *et al.* 2002). If the negative Bouguer anomaly represents a granitic batholith, two magmatic events have to be considered. It is difficult to include a granitic batholith of Ordovician age, thus other alternatives should be considered. Variscan granites (Rabae & Kearey 1997) cannot be considered because of the intimate kinematic relationship with the late Silurian–early Devonian deformation event. There may have been synkinematic emplacement of these granitic bodies in the transpressional shear zone. Alternatively, these bodies may be Precambrian crustal blocks of granitic composition. This alternative is favoured because of geochemical characteristics of the Cambrian–Ordovician terrigeneous material (Vander Auwera & André 1985; André 1991). Finally, a 'rift-like' basin fill within a Precambrian basement cannot be excluded (De Meyer 1984).

To the SW and to the south (see fig. 7 in

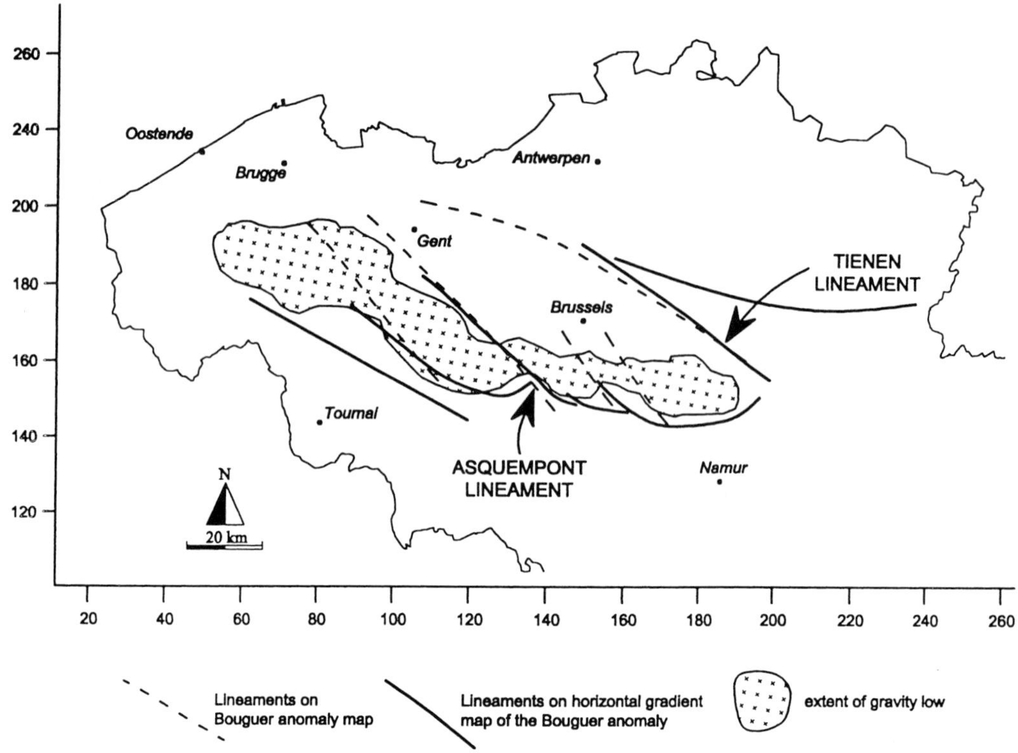

Fig. 8. Line drawing of the main features on the Bouguer anomaly map and the horizontal gradient map of the Bouguer anomaly. The Tienen Lineament is defined based on the horizontal gradient map. Reference system is the Belgian Lambert (in kilometres).

Verniers *et al.* 2002) the large, pronounced gravity lows underneath the Ardennes and the northern part of the Paris Basin, analogous to the Midlands area, are considered indicative for a cratonic basement at depth (Verniers *et al.* 2002). This cratonic basement is considered as the southwestern continuation of the Midlands Microcraton (Sintubin 1999).

Horizontal gradient of the Bouguer anomaly

The horizontal gradient map of the Bouguer anomaly (Fig. 7b) again emphasizes some important tectonic features (Fig. 8). This map shows that the entire Cambrian 'steep belt' is probably fault-bounded, not only along its southwestern side, with the emplacement of the dextral transpressional shear zone, but also along its northeastern side. The lineament coincides with the trace of the so-called 'Tienen fault' of Mortelmans (1955). Analogous to the Asquempont Lineament, a steeply dipping fault is supposed instead of a weakly north-dipping thrust fault as postulated by Mortelmans (1955). Contrary to the Asquempont Lineament, the Tienen Lineament seems rectilinear.

This leads to a symmetrical disposition of the Cambrian 'steep belt', bounded on both sides by steeply dipping fault zones. The Cambrian 'steep belt' may be a sort of pop-up structure. Although the Asquempont Lineament fits in a dextral transpressional regime the Tienen Lineament may be considered as reflecting a purely reverse displacement.

No conclusion can yet be drawn on the link between the two fault systems. A NE–SW trending linear feature can be recognized, striking parallel to the Bordière fault (Legrand 1968). Currently the possibility of a sinistral transfer fault between both fault systems is considered (see Matte 1986). The extent of this transfer fault towards the NE and the SW remains an open question.

Synthesis of potential field data

Several domains can be defined from the potential field data within the Brabant Massif

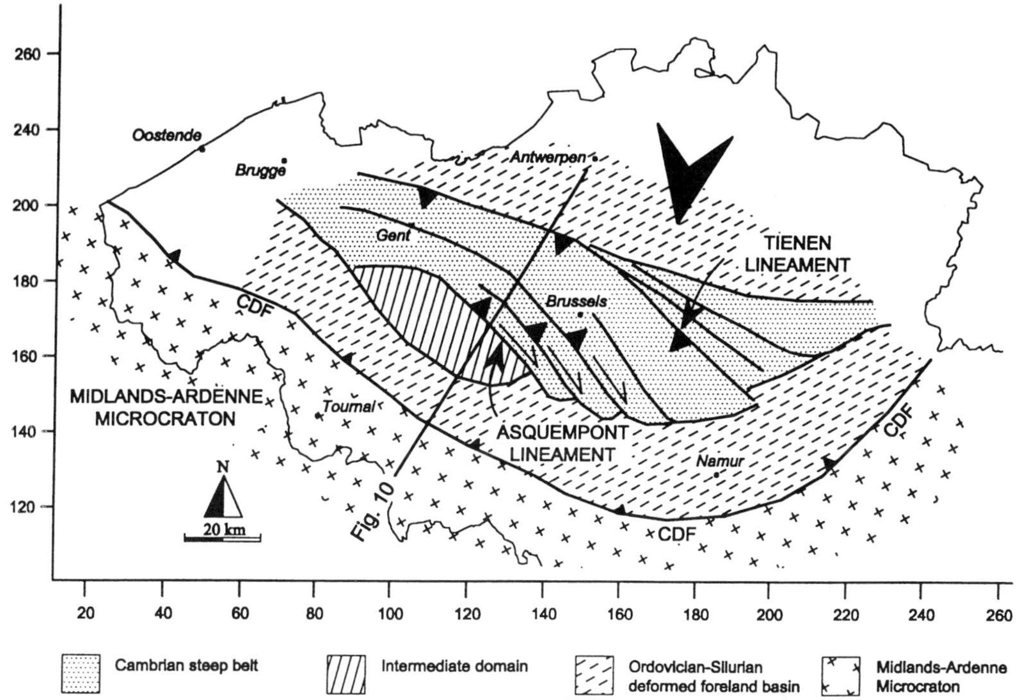

Fig. 9. Conceptual tectonic map of the Brabant Massif with the indication of the different lithostructural domains. The intermediate domain is situated on top of the gravity low, interpreted as a granitic basement block. The arrow indicates the overall shortening direction. The trace of the Caledonian deformation front (CDF) is taken from Sintubin (1999). Reference system is the Belgian Lambert (in kilometres).

(Figs 9 & 10). These domains are separated by lineaments, apparent on both aeromagnetic and Bouguer anomaly maps, and even better on the vertical gradient map of the aeromagnetic field and the horizontal gradient map of the Bouguer anomaly field.

The most prominent feature is the central domain. It forms a magnetic high and is characterized by the presence of distinct aeromagnetic lineaments. The analysis of the residual and regional component of the aeromagnetic field corroborates the inferred vertical architecture of the central 'steep belt'. The presence of cratonic basement at depth is questionable. On the subcrop map this domain coincides with the area of subcrop for the epizonal Tubize metasediments. The central 'steep belt' is delimited on both sides by major lineaments, the Asquempont Lineament in the SW and the Tienen Lineament in the NE.

To the east of the Tienen Lineament, this central domain occupies a broad, triangular area. The aeromagnetic lineaments present show a more ENE–WSW trend (Fig. 4). This is in agreement with the structural grain as apparent on the subcrop map (Fig. 2). This changing trend is interpreted to be caused by escape tectonics (Sintubin 1997*b*, 1999). To the SW of the Asquempont Lineament a domain can be defined which largely coincides with the gravity low, and with the intermediate domain, as identified in the outcrop area. This low-density body is a granitic basement block, possibly Precambrian in age. It acted as a rigid block largely controlling the development and evolution of the central 'steep belt'.

To the south, the long-wavelength aeromagnetic anomaly in the Ardennes area is interpreted as indicating the presence at depth of a cratonic basement. To the North, the presence of a cratonic basement is presumed. Contrary to its southern counterpart no gravity low is observed in this area, most probably indicating a different composition of the basement (denser material). On the subcrop map this area is composed of Cambro-Ordovician metasediments. This domain is still situated on top of the magnetic high. It is therefore interpreted as belonging to the central 'steep belt'. To the north, the Ordovician–Silurian metasediments, belonging to the northern deformed foreland basin subcrop.

The presence of a Precambrian basement?

The oldest strata at outcrop or subcrop in the Brabant Massif belong to the Tubize Formation and are of lowermost Cambrian age (Verniers *et al.* 2001, 2002). The stratigraphical position of the 'barren' Jodoigne Formation, possibly latest Precambrian in age, remains a matter of debate. Also in the Ardenne inliers the oldest material, belonging to the Deville Group, is of lowermost Cambrian age (Vanguestaine 1992; Verniers *et al.* 2001, 2002). It is therefore legitimate to question the presence of a Precambrian basement. Based on the potential field data, the following inferences with respect to the presence of a Precambrian basement can be made:

(1) the long-wavelength anomalies underneath the Ardennes are interpreted as indicative of the presence of a cratonic basement (Busby *et al.* 1993; Chacksfield *et al.* 1993);
(2) the pronounced gravity lows underneath the Ardennes and underneath the northern part of the Paris Basin are also assumed to be indicative for the presence of a cratonic basement (Verniers *et al.* 2002);
(3) the low-density body, responsible for the negative Bouguer anomaly in the south-western part of the Brabant Massif, may be interpreted as a granitic crustal block, of probably Precambrian age.

The following field observations may also contribute to the localization of a Precambrian cratonic basement in subcrop:

(1) the metamorphic contour line, separating the diagenetic zone in the SW of the Brabant Massif, probably coincides with the Caledonian deformation front (Woodcock 1991); to the SW of this deformation front no Caledonian deformation has been observed in the Lower Palaeozoic strata (Verniers & Van Grootel 1991; Van Grootel *et al.* 1997); the presence of persistent shelf sediments in the Silurian is also considered indicative of the presence of a cratonic basement (Verniers *et al.* 2002);
(2) as in the Welsh Basin (Soper *et al.* 1987) the arcuate structural grain of the deformed southern Ordovician–Silurian foreland basin may be considered to be parallel to the deformation front, situated to the south underneath the Variscan Ardenne Allochthon; Sintubin (1999) has defined the Brabant re-entrant as similar to the Midlands Promotory (see Fig. 11); the sinuous limit of this 'rigid' cratonic basement eventually determines the local deformation circumstances.

Based on the geochemical characteristics of the Lower Palaeozoic (Cambrian and Ordovician) sediments (Vander Auwera & André 1985; André 1991) a heterogeneous source, composed of crystalline and basic magmatic rocks, is inferred in the proximity of the deposition centre. The geochemistry reveals both a juvenile, late Proterozoic (<0.9 Ga), source rock and an older, early Proterozoic (>1.9 Ga), source of felsic crystalline rocks. The late Proterozoic source rocks favour of the proximity of a Neoproterozoic crystalline basement, such as the Midlands Microcraton, in which all rocks have an age younger than 1.2 Ga. The widespread presence of granitic fragments in the Lower Cambrian Tubize metasediments (Vander Auwera & André 1985) seem to infer the presence of large volumes of granite within the basement, which acted as source area for the Lower Cambrian sedimentation. Current analysis of Silurian turbiditic sequences indicate a southern source of sediment supply (Verniers & Van Grootel 1991), which may also indicate the presence of a cratonic high towards the south of the Ordovician–Silurian foreland basin (see also Verniers *et al.* 2002).

A Precambrian crystalline basement, underlying the Variscan Ardenne Allochthon possibly in continuity with the Midlands Microcraton (Sintubin 1999) but absent underneath the central 'steep belt' of the Brabant Massif, has some important implications:

(1) the central 'steep belt' of the Brabant Massif formed a compressional wedge, situated in between two crustal basement blocks: the Midlands–Ardennes Microcraton to the SW and the Lüneburg–North Sea Microcraton to the NE; the entire sedimentary, Lower Palaeozoic, cover sequence has been detached from its basement, resulting in significant tectonic thickening;
(2) the basal Variscan décollement, currently considered to be situated within the Lower Palaeozoic rocks (Adams & Vandenberghe 1999; Oncken *et al.* 1999, 2000) may be situated at the interface between a Lower Palaeozoic cover sequence and a Precambrian crystalline basement;
(3) the Ardenne Inliers, composed of Cambrian–Ordovician rocks, are situated on top of the cratonic basement; this

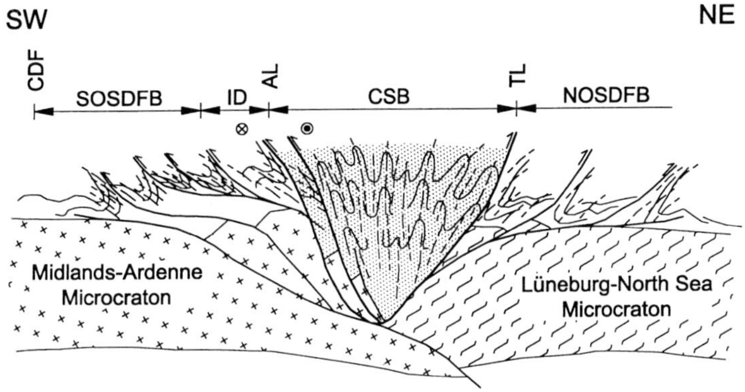

Fig. 10. Conceptual crustal-scale cross-section, indicating the compressional wedge model of the Brabant Massif. The crustal thickness is estimated at 40 to 50 km (cf. Bouckaert *et al.* 1988). SOSDFB, Southern Ordovician–Silurian deformed foreland basin; ID, Intermediate domain; CSB, Cambrian 'steep belt'; NOSDFB, Northern Ordovician–Silurian deformed foreland basin; AL, Asquempont Lineament; TL, Tienen Lineament; CDF, Caledonian deformation front.

disposition may have some important tectonic implications with respect to the ongoing discussion whether or not these Ardennes Inliers have suffered an Ordovician deformation event (e.g. Hugon & Le Corre 1979; Delvaux de Fenffe & Laduron 1984, 1991; Le Gall 1992; Hugon 1983; Piessens & Sintubin 1997; Belanger 1998; Verniers *et al.* 2002):

(a) if the Ardenne Inliers originated from the Lower Palaeozoic cover sequence of the cratonic basement, these Lower Palaeozoic rocks did not suffer any Ordovician deformation; this is very similar to the Cambrian–Ordovician, diagenetic to lower anchizonal, sediments on top of the Midlands Microcraton that lack a tectonic fabric (Lee *et al.* 1990); this would imply that all deformation, observed in the Ardenne Inliers, is Variscan (Hugon & Le Corre 1979; Hugon 1983; Le Gall 1992).

(b) the Lower Palaeozoic material, cropping out in the Ardenne Inliers, is transported to the north due to Variscan thrusting; this may be remnant material from an early Palaeozoic basin, situated south of the cratonic basement underlying the Ardennes and taken up into a north-verging tectonic event, classically considered middle Ordovician in age ('Ardenne Phase' Michot 1980); thus most of the deformation observed in the Ardenne Inliers, is Ordovician in age (Delvaux de Fenffe & Laduron 1991; Piessens & Sintubin 1997; Belanger 1998) and the significance of the Venn and Eifel ramps (Oncken *et al.* 1999, 2000) as southern extremity of the Precambrian cratonic basement has to be considered.

Compressional wedge model

In the first integrated model for the southeastern part of the Brabant Massif (Sintubin 1999) impingement and escape kinematics were inferred in a tectonic regime dominated by an N10E-orientated bulk shortening. The potential field data supported the inferences made from structural field observations. They also helped to determine the key players in the deformation event and, eventually, to propose an overall tectonic model for the Brabant Massif (Figs 9 & 10). Three key elements are derived from the potential field data:

(1) a fault-bounded, crustal, probably granitic, basement block controlled the deformation and was also involved in the deformation; new modelling is currently being performed to refine the model of Everaerts *et al.* (1996), taking into account the results of the current work;

(2) a high-strain zone existed along the northeastern edge of this crustal block, which developed into a dextral transpressional shear zone, due to the partitioning of the N10E-orientated bulk shortening within the NW–SE-trending high-strain zone; this

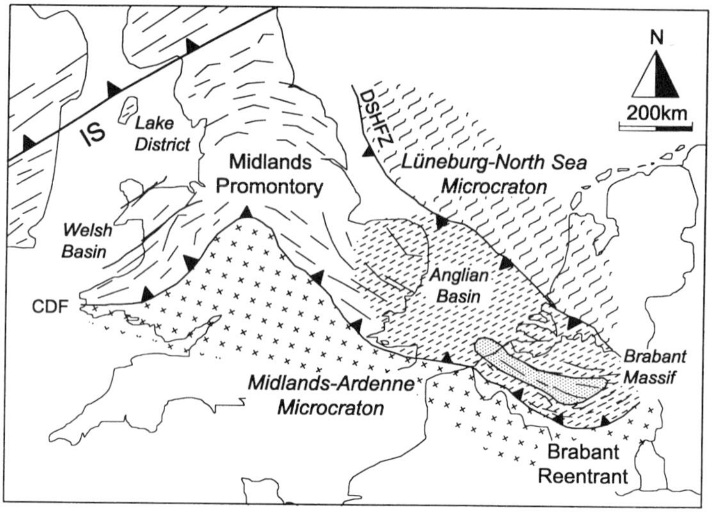

Fig. 11. Reconstruction of the Caledonian tectonic setting of the Brabant Massif as part of the Anglo-Brabant Belt (after Blundell *et al.* 1992; Pharaoh *et al.* 1995; Pharaoh 1999; Sintubin 1999; Winchester *et al.* in press). The Cambrian 'steep belt' may be interpreted as an inverted Cambrian rift basin. IS, Iapetus Suture; DSHFZ, Dowsing-South Hewett fault zone; CDF, Caledonian deformation front. Hatching is identical to Figure 10.

dextral transpressional shear zone eventually generated a 'rigid' indenter, causing arcuate cleavage and fold trajectories during deformation of the, southern, Ordovician–Silurian foreland basin, and also causing the variable transection angle along the cleavage arc;

(3) the Cambrian 'steep belt', probably bordered both sides by steeply dipping fault zones and is not underlain by a cratonic basement; its symmetrical disposition suggests that the overall south-vergence, observed in the field, is merely a coincidence; a vergence divide probably has to be assumed along the axis of the Cambrian 'steep belt'; the Cambrian 'steep belt' is thus interpreted as showing a compressional wedge geometry (Keppie 1993) (Fig. 10), contrary to the classic image of an anticlinal culmination (see e.g. De Vos *et al.* 1993*b*; Van Grootel *et al.* 1997; Mansy *et al.* 1999).

Discussion

On a larger scale the aeromagnetic map shows that the extent of the aeromagnetic high identified as the Cambrian 'steep belt' is limited to the southeastern part of the Anglo-Brabant Belt (Everaerts 2000) (Fig. 11). To the NW, in the Anglian part, this pronounced basement high is missing (Pharaoh *et al.* 1995). The assumption that the Cambrian 'steep belt' is only present in the southeastern part of the Anglo-Brabant Belt, forming a 400 km long and 50 km wide belt, requires explanations about its emplacement and development. Answers may be found in the stratigraphical framework of the Lower Palaeozoic in Belgium (see Verniers *et al.* 2002). The sedimentary history prior to the Tremadoc (approximately 490 Ma) is characterized by a synrift sedimentation in an extensional setting along the Avalonian–Cadomian arc at the northern active plate margin of Gondwana. Sedimentation may have occurred in an isolated rift or pull-apart basin on a Precambrian craton. The southeastern limit of the Brabant Massif, as well as a pronounced aeromagnetic lineament in the North Sea (N2 in Lee *et al.* 1993), may be interpreted as transfer boundary faults bounding this rift basin. When the setting changed to a compressional setting from the early Silurian onwards (Verniers *et al.* 2002), the Cambrian, synrift sedimentary sequence was 'squeezed' out between cratonic crustal blocks, causing a significant shortening, a decoupling of the metasediments from their basement, vertical exhumation, and transposition of the strata, eventually generating the 'steep belt'. Where these rift basins did not develop (e.g. Anglian part of the Anglo-Brabant Belt), shortening probably caused thin-skinned fold-and-thrust tectonics. We assume that the Brabantian 'steep belt' may not have been the only one. It is fair to assume that the Ardenne Inliers, with significant Cambrian synrift sedimentation (see Verniers *et al.* 2002), also represent one or more inverted rift basins. Inversion in these basins occurred much earlier (during the Ordovician), immediately following the synrift phase (Verniers *et al.* 2002).

This tectonic setting is very similar to that suggested by Murphy *et al.* (1999) for Western Avalonia. At the end of the Precambrian a transition occurred from an arc to a transtensional intracontinental environment. Older pull-apart basins within the arc, caused by oblique subduction, closed, new pull-apart basins formed. This transition was caused by the subduction of the Iapetus Ocean spreading ridge and occurred diachronously along the Avalonian–Cadomian active plate margin. It is therefore fair to assume that this transition, creating intracratonic rift basins, also occurred in Eastern Avalonia, but somewhat later than in Western Avalonia. The subduction of the spreading ridge eventually caused the rifting of the peri-Gondwanan microcontinents, such as Avalonia and Armorica, etc. (Crowley *et al.* 2000).

Conclusions

The very limited degree of relevant rock exposure means that potential field data form the key to a tectonic model for the Brabant Massif. Although kinematics are inferred from structural field observations, the potential field data both corroborate these field observations and also reveal other key features, enabling an integrated tectonic model to be developed. It is proposed that the Brabant Massif shows a symmetrical compressional wedge geometry. The fault-bounded Cambrian 'steep belt' forms the backbone of this compressional wedge. The kinematics are controlled by the presence of rigid crystalline basement blocks and the inferred presence of a cratonic basement, both probably Precambrian in age.

It is also suggested that the development of isolated rift or pull-apart basins along the Gondwanan active plate margin during the Cambrian may have played a crucial role in the subsequent deformation history and to a large extent determined the architecture of the southeastern extremity of the Anglo-Brabant Belt.

The current work amply proves that the interpretation of potential field data can only be done when supported by geological evidence and that all alternatives should always be considered. Otherwise such an interpretation becomes a very risky and even questionable enterprise.

This research was backed by the F.W.O. projects N° G.0084.97 and N° G.0094.01. Manuel Sintubin is Research Associate of the Bijzonder Onderzoeksfonds K.U.Leuven. Met steun van de Universitaire Stiching van België voor de illustraties.

References

ADAMS, R. & VANDENBERGHE, N. 1999. The Meuse valley section across the Condroz-Ardennes (Belgium) based on a predeformational sediment wedge. *In*: SINTUBIN, M., VANDYCKE, S. & CAMELBEECK, T. (eds) *Palaeozoic to Recent tectonics in the NW European Variscan Front Zone*. Tectonophysics, **309**, 179–195.

ANDRÉ, L. 1991. The concealed crystalline basement in Belgium and the 'Brabantia' microplate concept: constraints form the Caledonian magmatic and sedimentary rocks. *Annales de la Société géologique de Belgique*, **114**, 117–139.

ANTHOINE, R. & ANTHOINE, P. 1943. Les assises de Mousty et de Villers-la-Ville du bassin supérieur de la Dyle. *Annales de la Société géologique de Belgique*, **66**, 53–170.

BELANGER, I. 1998. Effets d'une déformation hétérogène sur un ensemble mécaniquement anisotrope: le cas du Massif de Rocroi (Ardenne franco-belge). Unpublished Ph. D. thesis, Université Catholique de Louvain.

BLAKELY, R. J. 1995. *Potential theory in gravity and magnetic application*. University Press, Cambridge.

BLUNDELL, D. J., FREEMAN, R. & MUELLER, S. 1992. *A continent revealed. The european geotraverse*. Cambridge University Press, Cambridge, 275.

BOUCKAERT, J., FOCK, W. & VANDENBERGHE, N. 1988. First results of the Belgian Geotraverse 1986 (BELCORP). *Annales de la Société géologique de Belgique*, **111**, 279–290.

BUSBY, J. P., KIMBELL, G. S. & PHARAOH, T. C. 1993. Integrated geophysical/geological modelling of the Caledonian and Precambrian basement of southern Britain. *Geological Magazine*, **130**, 593–604.

CHACKSFIELD, B. C., DE VOS, W., D'HOOGE, L., DUSAR, M., LEE, M. K., POITEVIN, C., ROYLES, C. P. & VERNIERS, J. 1993. A new look at Belgian aeromagnetic and gravity data through image-based display and integrated modelling techniques. *Geological Magazine*, **130**, 583–591.

COCKS, L. R. M., MCKERROW, W. S. & VAN STAAL, C. R. 1997. The margins of Avalonia. *Geological Magazine*, **134**, 627–636.

CROWLEY, Q. G., FLOYD, P. A., WINCHESTER, J. A., FRANKE, W. & HOLLAND, J. G. 2000. Early Palaeozoic rift-related magmatism in Variscan Europe: fragmentation of the Armorican Terrane Assemblage. *Terra Nova*, **12**, 171–180.

DE MEYER, F. 1984. Two structural models for the western flank of the Brabant Massif. *Geophysical Prospecting*, **32**, 37–50.

DE VOS, W. 1997. Influence of the granitic batholith of Flanders on Acadian and later deformation (Brabant Massif, Belgium). *Aardkundige Mededelingen*, **8**, 49–52.

De Vos, W., Poot, B., Hus, J. & El Khayati, M. 1993*a*. Geophysical characterization of lithologies from the Brabant Massif as a contribution to gravimetric and magnetic modelling. *Bulletin de la Société belge de Géologie*, **101**, 173–180.

De Vos, W., Verniers, J., Herbosch, A. & Vanguestaine, M. 1993*b*. A new geological map of the Brabant Massif, Belgium. *Geological Magazine*, **130**, 605–611.

Debacker, T. N. 1999. Folds trending at various angles to the transport direction in the Marcq area, Brabant Massif, Belgium. *Geologica Belgica*, **2**, 159–172.

Debacker, T. N. 2001. Palaeozoic deformation of the Brabant Massif within eastern Avalonia: how, when and why? Unpublished Ph. D. thesis, Universiteit Gent.

Debacker, T. N., Sintubin, M. & Verniers, J. 1999. Cleavage/fold relationships in the Silurian metapelites, southeastern Anglo-Brabant fold belt (Ronquières, Belgium). *Geologie en Mijnbouw*, **78**, 47–56.

Debacker, T. N., Sintubin, M. & Verniers, J. 2001. Large-scale slumping from structural and sedimentary features in the Lower Palaeozoic Anglo-Brabant fold belt, Belgium. *Journal of the Geological Society, London*, **158**, 341–352.

Delvaux de Fenffe, D. & Laduron, D. 1984. Analyse structural au bord sud du Massif de Rocroi (Ardennes françaises). *Bulletin de la Société belge de Géologie*, **93**, 11–26.

Delvaux de Fenffe, D. & Laduron, D. 1991. Caledonian and Variscan structures in the Rocroi-Ardenne Lower Paleozoic basement (Belgium and adjacent countries). *Annales de la Société géologique de Belgique*, **114**, 141–162.

Everaerts, M. 2000. L'interpretation structurale de la Manche au Rhin: apport du filtrage des champs de potentiel. Unpublished Ph. D. thesis, Université Catholique de Louvain.

Everaerts, M., Poitevin, C., De Vos, W. & Sterpin, M. 1996. Integrated geophysical/geological modelling of the western Brabant Massif and structural implications. *Bulletin de la Société belge de Géologie*, **105**, 41–59.

Everaerts, M., Sintubin, M., Lemonne, E. & Herbosch, A. 1998. Correlations of field and aeromagnetic anomaly data in a predominantly concealed basement: examples from the Lower Paleozoic Brabant Massif (Belgium). *Acta Universitatis Carolinae – Geologica*, **42**, 236–237.

Gupta, V. K. & Ramani, N. 1980. Some aspects of regional-residual separation of gravity anomalies in precambrian terrain. *Geophysics*, **45**, 1412–1426.

Hugon, H. 1983. Structures et déformation du Massif de Rocroi (Ardennes). *Bulletin de la Société géologique et minéralogique de Bretagne*, **15**, 109–143.

Hugon, H. & Le Corre, C. 1979. Mise en évidence d'une déformation hercynienne en régime cisaillant progressif dans le Massif de Rocroi (Ardenne). *Comptes Rendus de l'Académie des Sciences, Paris*, **289**, 615–618.

Keppie, J. D. 1993. Synthesis of Palaeozoic deformational events and terrane accretion in the Canadian Appalachians. *Geologische Rundschau*, **82**, 381–431.

Le Gall, B. 1992. The deep structure of the Ardennes Variscan thrust belt from structural and ECORS seismic data. *Journal of Structural Geology*, **14**, 531–546.

Lee, M. K., Pharaoh, T. C. & Soper, N. J. 1990. Structural trends in central Britain from images of gravity and aeromagnetic fields. *Journal of the Geological Society, London*, **147**, 241–258.

Lee, M. K., Pharaoh, T. C., Williamson, J. P., Green, C. A. & De Vos, W. 1993. Evidence on the deep structure of the Anglo-Brabant Massif from gravity and magnetic data. *Geological Magazine*, **130**, 575–582.

Legrand, R. 1968. Le massif de Brabant. *Mémoires pour servir à l'explication des Cartes géologiques et minières de la Belgique*, **9**, 1–148.

Mansy, J.-L., Everaerts, M. & De Vos, W. 1999. Structural analysis of the adjacent Acadian and Variscan fold belts in Belgium and northern France from geophysical and geological evidence. *In*: Sintubin, M., Vandycke, S. & Camelbeeck, T. (eds) Palaeozoic to Recent tectonics in the NW European Variscan Front Zone. *Tectonophysics*, **309**, 99–116.

Matte, P. 1986. La chaîne varisque parmi les chaînes paléozoiques péri atlantiques, modèle d'évolution et position des grands blocs continentaux au Permo-Carbonifère. *Bulletin de la Société géologique de France* **(8)2**, 9–24.

Michot, P. 1980. Le segment tectogénique calédonien belge. *Mémoire de l'Académie royal de Belgique, classe des Sciences. Collection in 8°, 2ème série*, **43**, 1–61.

Mortelmans, G. 1955. Considérations sur la structure tectonique et la stratigraphie du Massif du Brabant. *Bulletin de la Société belge de Géologie*, **64**, 179–218.

Murphy, J. B., Keppie, J. D., Dostal, J. & Nance, R. D. 1999. Neoproterozoic-early Paleozoic evolution of Avalonia. *In*: Ramos, V. A. & Keppie, J. D. (eds) *Laurentia-Gondwana connections before Pangea*. Special Paper **336**. Geological Society of America, Boulder, Colorado, 253–266.

Nettelton, L. L. 1976. *Gravity and magnetic in oil prospecting*. McGraw-Hill Book Co. Inc., New York.

Oncken, O., von Winterfeld, C. & Dittmar, U. 1999. Accretion of a rifted passive margin: The Late Paleozoic Rhenohercynian fold and thrust belt (Middle European Variscides). *Tectonics*, **18**, 75–91.

Oncken, O., Plesch, A., Weber, J., Ricken, W. & Schrader, S. 2000. Passive margin detachment during arc-continent collision (Central European Variscides). *In*: Franke, W., Haak, V., Oncken, O. & Tanner, D. (eds) *Orogenic processes: Quantification and modelling in the Variscan Belt*. Geological Society, London, Special Publications, **179**, 199–216.

Pharaoh, T. C. 1999. Palaeozoic terranes and their

lithospheric boundaries within the Trans-European Suture Zone (TESZ): a review. *Tectonophysics*, **314**, 17–41.

PHARAOH, T. C., MOLYNEUX, S. G., MERRIMAN, R. J., LEE, M. K. & VERNIERS, J. 1993. The Caledonides of the Anglo-Brabant Massif reviewed. *Geological Magazine*, **130**, 561–562.

PHARAOH, T. C., ENGLAND, R. & LEE, M. K. 1995. The concealed Caledonide basement of eastern England an the southern North Sea – a review. *Studia geophysica et geodaetica*, **39**, 330–346.

PICKERING, K. T. & SMITH, A. G. 1995. Arcs and backarc basins in the Early Paleozoic Iapetus Ocean. *The Island Arc*, **4**, 1–67.

PIESSENS, K. & SINTUBIN, M. 1997. Partitioning of Variscan strain in the southern part of the Caledonian Stavelot-Venn Inlier in the Ardenne Allochthon (Belgium). *Aardkundige Mededelingen*, **8**, 135–138.

RABAE, A. M. & KEAREY, P. 1997. Granites of the western Anglo-Brabant Massif. *Annales de la Société géologique de Belgique*, **120**, 217–226.

REY, P., BURG, J.-P. & CASEY, M. 1997. The Scandinavian Caledonides and their relationship to the Variscan belt. *In*: BURG, J.-P. & FORD, M. (eds) *Orogeny through time*. Geological Society, London, Special Publications, **121**, 179–200.

SINTUBIN, M. 1997*a*. Cleavage-fold relationships in the Lower Paleozoic Brabant Massif (Belgium). *Aardkundige Mededelingen*, **8**, 161–164.

SINTUBIN, M. 1997*b*. Structural implications of the aeromagnetic lineament geometry in the Lower Paleozoic Brabant Massif (Belgium). *Aardkundige Mededelingen*, **8**, 165–168.

SINTUBIN, M. 1999. Arcuate fold and cleavage patterns in the southeastern part of the Anglo-Brabant Fold Belt (Belgium): tectonic implications. *In*: Sintubin, M., Vandycke, S. & Camelbeeck, T. (eds) Palaeozoic to Recent tectonics in the NW European Variscan Front Zone. *Tectonophysics*, **309**, 81–97.

SINTUBIN, M., BRODKOM, F. & LADURON, D. 1998. Cleavage-fold relationship in the Lower Cambrian Tubize Group, southeast Anglo-Brabant Fold Belt. *Geological Magazine*, **135**, 217–226.

SOPER, N. J. & HUTTON, D. H. W. 1984. Late Caledonian sinistral displacements in Britain: Implications for a three-plate collision model. *Tectonics*, **3**, 781–794.

SOPER, N. J., WEBB, B. C. & WOODCOCK, N. H. 1987. Late Caledonian (Acadian) transpression in north-west England: timing, geometry and geotectonic significance. *Proceedings of the Yorkshire Geological Society*, **46**, 175–192.

VAN GROOTEL, G., VERNIERS, J., GEERKENS, B., LADURON, D., VERHAEREN, M., HERTOGEN, J. & DE VOS, W. 1997. Timing of magmatism, foreland basin development, metamorphism and inversion in the Anglo-Brabant fold belt. *Geological Magazine*, **134**, 607–616.

VANDER AUWERA, J. &, ANDRÉ, L. 1985. Sur le milieu de dépot, l'origine des matériaux et le facies métamorphique de l'assise de Tubize (Massif du Brabant, Belgique). *Bulletin de la Société belge de Géologie*, **94**, 171–184.

VANGUESTAINE, M. 1992. Biostratigraphie par Acritarches du Cambro-Ordovicien de Belgique et des régions limitrophes: synthèse et perspectives d'avenir. *Annales de la Société géologique de Belgique*, **115**, 1–18.

VECOLI, M. & SAMUELSSON, J. 2001. Reworked acritarchs as provenance indicators in the Lower Palaeozoic of Denmark. *Comptes Rendus de l'Académie des Sciences, Paris*, **332**, 465–471.

VERNIERS, J. & VAN GROOTEL, G. 1991. Review of the Silurian in the Brabant Massif, Belgium. *Annales de la Société géologique de Belgique*, **114**, 163–193.

VERNIERS, J., HERBOSCH, A., VANGUESTAINE, M., GEUKENS, F., DELCAMBRE, B., PINGOT, J.-L., BELANGER, I., HENNEBERT, M., DEBACKER, T. N., SINTUBIN, M. & DE VOS, W. Cambrian–Ordovician–Silurian lithostratigraphic units (Belgium). *Geologica Belgica*, **4**, 5–38.

VERNIERS, J., PHARAOH, T. C., ANDRÉ, L., DEBACKER, T. N., DE VOS, W., EVERAERTS, M., HERBOSCH, A., SAMUELSSON, J., SINTUBIN, M. & VECOLI, M. 2002. The Cambrian to mid Devonian basin development and deformation history of eastern Avalonia, east of the Midlands Microcraton: new data and a review. *In*: WINCHESTER, J. A., PHARAOH, T. C. & VERNIERS, J. (eds) *Palaeozoic Amalgamation of Central Europe*. Geological Society, London, Special Publications, **201**, 47–94.

WINCHESTER, J. A. & PACE TMR Network Team 2002. Palaeozoic Amalgamation of Central Europe: new results from recent geological and geophysical investigations. *Tectonophysics*, in press.

WOODCOCK, N. H. 1991. The Welsh, Anglian and Belgian Caledonides compared. *Annales de la Société géologique de Belgique*, **114**, 5–17.

WOODCOCK, N. H., AWAN, M. A., JOHNSON, T. E., MACKIE, A. H. & SMITH, R. D. A. 1988. Acadian tectonics of Wales during Avalonia/Laurentia convergence. *Tectonics*, **7**, 483–495.

Index

Page numbers in *italics* refer to Figures and page numbers in **bold** refer to Tables.